Lecture Notes in Computer Science 11548

Commenced Publication in 1973
Founding and Former Series Editors:
Gerhard Goos, Juris Hartmanis, and Jan van Leeuwen

More information about this series at http://www.springer.com/series/7407

Michael Khachay · Yury Kochetov ·
Panos Pardalos (Eds.)

Mathematical Optimization Theory and Operations Research

18th International Conference, MOTOR 2019
Ekaterinburg, Russia, July 8–12, 2019
Proceedings

Editors
Michael Khachay
Krasovsky Institute of Mathematics
and Mechanics
Ekaterinburg, Russia

Yury Kochetov
Sobolev Institute of Mathematics
Novosibirsk, Russia

Panos Pardalos
University of Florida
Gainesville, FL, USA

ISSN 0302-9743 ISSN 1611-3349 (electronic)
Lecture Notes in Computer Science
ISBN 978-3-030-22628-2 ISBN 978-3-030-22629-9 (eBook)
https://doi.org/10.1007/978-3-030-22629-9

LNCS Sublibrary: SL1 – Theoretical Computer Science and General Issues

This Springer imprint is published by the registered company Springer Nature Switzerland AG
The registered company address is: Gewerbestrasse 11, 6330 Cham, Switzerland

Preface

This volume contains the refereed proceedings of the 18th international conference on Mathematical Optimization Theory and Operations Research (MOTOR 2019)[1] held during July 8–12, 2019, near Ekaterinburg, Russia.

The conference brings together a wide research community in the fields of mathematical programming and global optimization, discrete optimization, complexity theory and combinatorial algorithms, optimal control and games, and their applications in relevant practical problems of operations research, mathematical economy, and data analysis.

MOTOR 2019 was a successor of the following well-known international and Russian conference series, which were organized in Ural, Siberia, and the Far East:

– Baikal International Triennial School Seminar on Methods of Optimization and Their Applications, BITSS MOPT, was established in 1969 by academic N.N.Moiseev; the 17th event[2] in this series was held in 2017, in Buryatia.
– All-Russian Conference on Mathematical Programming and Applications, MPA, was established in 1972 by academic I.I.Eremin; the 15th conference[3] in this series was held in 2015, near Ekaterinburg.
– International Conference on Discrete Optimization and Operations Research, DOOR, was organized nine times from 1996, and the last event[4] was held in 2016 in Vladivostok.
– International Conference on Optimization Problems and Their Applications, OPTA, has been organized regularly in Omsk since 1997, the seventh event[5] in this series was held in 2018.

Starting from different origins, today these conference series grew very close to each other, having much in common in their research topics, scientific community, and organizers. Therefore, this year the common Program Committee (PC) decided to organize a joint conference inheriting the long history of all the events and to name it the 18th International Conference on Mathematical Optimization Theory and Operations Research (MOTOR).

As per tradition, the main conference scope includes but is not limited to mathematical programming, bi-level and global optimization, integer programming and combinatorial optimization, approximation algorithms with theoretical guarantees and approximation schemes, heuristics and meta-heuristics, optimal control and game theory, optimization problems in function approximation, optimization in machine

---

1 http://motor2019.uran.ru.
2 http://isem.irk.ru/conferences/mopt2017/en/index.html.
3 http://mpa.imm.uran.ru/96/en.
4 http://www.math.nsc.ru/conference/door/2016/.
5 http://opta18.oscsbras.ru/en/.

learning and data analysis, and valuable practical applications in operations research and economics.

In response to the call for papers, MOTOR 2019 received 232 submissions. Out of 170 full papers considered for reviewing (62 abstracts and short communications were excluded because of formal reasons) only 48 papers were selected by the PC for publication. Thus, the acceptances rate for this volume is about 28%. Each submission was reviewed by at least three PC members or invited reviewers, experts in their fields, in order to supply detailed and helpful comments. In addition, the PC recommended to include 50 papers in the supplementary volume after their presentation and discussion during the conference and subsequent revision with respect to the reviewers' comments.

The conference featured ten invited lectures:

- Prof. Olga Battaïa (ISAE-Supaero, Toulouse France), "Decision Under Ignorance: A Comparison of Existing Criteria in a Context of Linear Programming"
- Prof. Oleg Burdakov (Linköping University, Sweden), "Node Partitioning and Cycles Creation Problem"
- Prof. Christoph Dürr (Sorbonne Université, France), "Bijective Analysis of Online Algorithms"
- Prof. Alexander Grigoriev (Maastricht University, The Netherlands), "A Survey on Possible and Impossible Attempts to Solve the Treewidth Problem via ILPs"
- Prof. Mikhail Kovalyov (United Institute of Informatics Problems NASB, Belarus) "No-Idle Scheduling of Unit-Time Jobs with Release Dates and Deadlines on Parallel Machines"
- Prof. Vadim Levit (Ariel University, Israel) "Critical and Maximum Independent Sets Revisited"
- Prof. Bertrand M. T. Lin (National Chiao Tung University, Hsinchu Taiwan), "An Overview of the Relocation Problem"
- Prof. Natalia Shakhlevich (University of Leeds, UK), "On a New Approach for Optimization Under Uncertainty"
- Prof. Angelo Sifaleras (University of Macedonia, Greece), "Exterior Point Simplex-Type Algorithms for Linear and Network Optimization Problems"
- Prof. Vitaly Strusevich (University of Greenwich, UK), "Design of Fully-Polynomial Approximation Schemes for Non-linear Boolean Programming Problems"

The following seven tutorials were given by outstanding scientists:

- Prof. Tatjana Davidović (Mathematical Institute of the Serbian Academy of Sciences and Arts, Serbia), "Distributed Memory-Based Parallelization of Metaheuristic Methods"
- Prof. Stephan Dempe (TU Bergakademie Freiberg, Germany), "Bilevel optimization: The Model and its Transformations"
- Prof. Oleg Khamisov (Melentiev Energy Systems Institute SB RAS, Russia), "The Fundamental Role of Concave Programming in Continuous Global Optimization"
- Prof. Alexander Kononov (Sobolev Institute of Mathematics, Russia), "Primal-dual Method and Online Problems"

- Prof. Nenad Mladenovic (Mathematical Institute SANU, Serbia), "Solving Nonlinear System of Equations as an Optimization Problem"
- Prof. Evgeni A. Nurminski (Far Eastern Federal University, Russia), "Projection Problems and Problems with Projection"
- Prof. Alexander Strekalovsky (Matrosov Institute for System Dynamics and Control Theory SB RAS, Russia), "Modern Methods of Non-convex Optimization"

We would like to thank all the authors for their submissions, as well as all members of the PC and external reviewers for their efforts in providing exhaustive reviews. We thank our sponsors, the Russian Foundation for Basic Research, Higher School of Economics (Campus Nizhny Novgorod), Ural Federal University, and Novosibirsk State University. In addition, we are grateful to Alfed Hofmann, Aliaksandr Birukou, Anna Kramer, and their colleagues from Springer LNCS and CCIS editorial board for their kind and helpful support.

July 2019

Michael Khachay
Yury Kochetov
Panos Pardalos

Organization

Program Committee Chairs

Michael Khachay	Krasovsky institute of Mathematics and Mechanics, Russia
Yury Kochetov	Sobolev Institute of Mathematics, Russia
Panos M. Pardalos	University of Florida, USA

Program Committee

Alexander Afanasiev	IITP RAS, Russia
Edilkhan Amirgaliev	Suleyman Demirel University, Kazakhstan
Anatoly Antipin	Dorodnicyn Computing Centre FRC CSC RAS, Russia
Adil Bagirov	Federation University Australia, Australia
Evripidis Bampis	Sorbonne Université, France
Olga Battaïa	ISAE-Supaero, Toulouse, France
Vitaly I. Berdyshev	Krasovsky Institute of Mathematics and Mechanics, Russia
Vladimir Beresnev	Sobolev Institute of Mathematics, Russia
René van Bevern	Novosibirsk State University, Russia
Givi Bolotashvili	Georgian Technical University, Georgia
Oleg Burdakov	Linköping University, Sweden
Sergiy Butenko	Texas A&M University, USA
Igor Bykadorov	Sobolev Institute of Mathematics, Russia
Alexander G. Chentsov	Krasovsky Institute of Mathematics and Mechanics, Russia
Tatjana Davidović	Mathematical Institute SANU, Serbia
Vladimir Deineko	Warwick University, UK
Stephan Dempe	Freiberg University, Germany
Alexander Dolgui	IMT Atlantique, France
Anton Eremeev	Dostoevsky Omsk State University, Russia
Adil Erzin	Novosibirsk State University, Russia
Yuri G. Evtushenko	Dorodnicyn Computing Centre FRC CSC RAS, Russia
Fedor Fomin	University of Bergen, Norway
Edward Gimadi	Sobolev Institute of Mathematics, Russia
Alexander Gornov	Matrosov Institute of System Dynamics and Control Theory, Russia
Alexander Grigoriev	Maastricht University, The Netherlands
Mikhail Gusev	Krasovsky Institute of Mathematics and Mechanics, Russia
Milojica Jacimović	University of Montenegro, Montenegro
Vyacheslav Kalashnikov	ITESM, Campus Monterrey, Mexico

Valeriy Kalyagin	Higher School of Economics, Russia
Alexander Kazakov	Matrosov Institute of System Dynamics and Control Theory, Russia
Alexander Kel'manov	Sobolev Institute of Mathematics, Russia
Oleg Khamisov	Melentiev Energy Systems Institute, Russia
Andrey Kibzun	Moscow Aviation Institute, Russia
Donghyun (David) Kim	Kennesaw State University, USA
Igor Konnov	Kazan Federal University, Russia
Alexander Kononov	Sobolev Institute of Mathematics, Russia
Vladimir Kotov	Belarusian State University, Belarus
Ilias Kotsireas	University of Waterloo, Canada
Mikhail Y. Kovalyov	United Institute of Informatics Problems NASB, Belarus
Alexander Lazarev	Trapeznikov Institute of Control Sciences, Russia
Vadim Levit	Ariel University, Israel
Bertrand M. T. Lin	National Chiao Tung University, Hsinchu, Taiwan
Nikolay Lukoyanov	Krasovsky Institute of Mathematics and Mechanics, Russia
Vladimir Mazalov	Institute of Applied Mathematical Research of KRC RAS, Russia
Nenad Mladenović	Mathematical Institute, Serbian Academy of Sciences and Arts, Serbia
Yury Nikulin	University of Turku, Finland
Evgeni Nurminski	Far Eastern Federal University, Russia
Leon Petrosyan	Saint Petersburg State University, Russia
Alexander Petunin	Ural Federal University, Russia
Boris T. Polyak	Trapeznikov Institute of Control Sciences, Russia
Leonid Popov	Krasovsky Institute of Mathematics and Mechanics, Russia
Mikhail Posypkin	Dorodnicyn Computing Centre FRC CSC RAS, Russia
Oleg Prokopyev	University of Pittsburgh, USA
Artem Pyatkin	Sobolev Institute of Mathematics, Russia
Soumyendu Raha	Indian Institute of Science, India
Konstantin V. Rudakov	Dorodnicyn Computing Centre FRC CSC RAS, Russia
Kristian Sabo	University of Osijek, Croatia
Leonidas Sakalauskas	University of Vilnius, Lithuania
Eugene Semenkin	Siberian State Aerospace University, Russia
Yaroslav Sergeyev	University of Calabria, Italy
Natalia Shakhlevich	University of Leeds, UK
Alexander Shananin	Moscow Institute of Physics and Technology, Russia
Angelo Sifaleras	University of Macedonia, Greece
Vladimir Skarin	Krasovsky Institute of Mathematics and Mechanics, Russia
Alexander Strekalovsky	Matrosov Institute for System Dynamics and Control Theory, Russia
Vitaly Strusevich	University of Greenwich, UK

Tatiana Tchemisova	University of Aveiro, Portugal
Viktor Ukhobotov	Chelyabinsk State University, Russia
Vladimir N. Ushakov	Krasovsky Institute of Mathematics and Mechanics, Russia
Vladimir V. Vasin	Krasovsky Institute of Mathematics and Mechanics, Russia

Additional Reviewers

Alexander Adelshin
Vera Afreixo
Alexander L. Ageev
Natalia Aizenberg
Elena Akimova
Ekaterina Alekseeva
Ricardo Almeida
Zhazira Amirgaliyeva
Boris Ananyev
Aram V. Arutyunov
Sergey Astrakov
Pasqualy Avella
Yuri Averboukh
Konstantin Avrachenkov
Artem Baklanov
Nuno Bastos
Arnab Basu
Pavel Borisovsky
Endre Boros
Edmund Burke
Valentina Cacchiani
Gruia Calinescu
S. P. Chakrabarty
Pavel Chebotarev
Ilya Chernykh
A. Yu. Chirkov
Dimitrije D. Čvokić
Alexey Danilin
Ivan Davydov
Maxim Demenkov
Vitali Demidenko
Yury Dor
Vladimir Fedorov
Alexander Filatov
Tatiana Filippova
Matteo Fischetti
Sergey Foss
Stefania Funari
Anindita Ganguly
Alexander Gasnikov
Mikhail Gomojunov
Danijel Grahovac
Tatiana Gruzdeva
Mikhail Gudyma
Jae Jin Hwang
Alexei Ignatov
Victor Il'ev
Evgeny Ivanko
Sergey Ivanov
Viktor Izhutkin
Igor' Izmest'ev
Slobodan Jelić
Uwe Kahler
Yuri Kan
Igor Kandoba
Daniel Karapetyan
Margarita Karaseva
Lev Kazakovtsev
Farid Khan
Vladimir Khandeev
Dmitry Khlopin
Elena Khoroshilova
Anatoly Kleimenov
Xenia Klimentova
Konstantin Kobylkin
Sergey Kokovin
Polina Kononova
Elena Kostousova
Julia Kovalenko
Igor Kozin
Denis Krotov
Konstantin Kudryavtsev
Ivana Kuzmanović Ivičić
Sergey Lavlinskii
Pavel Lebedev
Seokjun Lee
Anna Lempert
Tatyana Levanova
Snježana Majstorović
Vittorio Maniezzo
Natalia Martins
Igor Masich
Domagoj Matijević
Oxana Matviychuk
Andrey Melnikov
Elena Musatova
Andrey Naumov
Katherine Neznakhina
Nataliia Obrosova
Timm Oertel
Yuri Ogorodnikov
Andrei Orlov
Idowu Ademola Osinuga
Anna Panasenko
Ilya Panfilov
Artem Panin
Sophie Parragh
Valerii Patsko
Jiming Peng
Alexander Petunin
Aleksandr Plakhov
Roman Plotnikov
Alexander Plyasunov
Nick Pogodaev
Dmitry Pokrovsky

Bernhard Primas
Franz Rendl
Anna Rettieva
Aleksandr Rogozin
Evgeny Rudoy
Pavel Ruzankin
Ivan Ryzhikov
Yaroslav Salii
Marina Sandomirskaia
Alexander Semenov
Daehee Seo
Dmitry Serkov
Vladimir Servakh
Alexander Sesekin
Alexander Shapiro
Jhilakshi Sharma
Vladimir Shenmaier
Alexander Sidorov
Denis Sidorov
Konstantin Siemenikhin
Cristiana Silva
Ruslan Simanchev
Gaurav Singh
Evgeny Skvortsov
Andrei Sleptchenko
Gueorgui Smirnov
Olga Sokolova
Evgenii Sopov
Stepan Sorokin
Vladimir A. Srochko
Vasile Staicu
Predrag Stanimirovic
Vladimir Stanovov
Maxim Staritsyn
Dmitry Stashkov
Fedor Stonyakin
Alena Stupina
Alexander Tarasyev
Galina Timofeeva
Paul Tochilin
Zoran Tomljanović
Ya-Chih Tsai
Olga Tsekhan
Oxana Tsidulko
Yury Tsoy
Inna Urazova
Dragan Urošević
Igor Vasilyev
Stefan Voß
Gyung Soo Woo
Sergei Yakovlev
Elena Yanovskaya
Alexander Yurin
Gennady Zabudsky
Vyacheslav Zalyubovsky
Lidia Zaozerskaya
Vannel Zeufack
Nikolai Zolotykh
Alexander Zyryanov

Industry Section Chairs

Damir Gainanov	Ural Federal University, Russia
Alexander Kurochkin	Sobolev Institute of Mathematics, Russia

Organizing Committee

Alexey Borbunov	Ural Federal University, Russia
Michael Khachay	Krasovsky Institute of Mathematics and Mechanics, Russia
Konstantin Kobylkin	Krasovsky Institute of Mathematics and Mechanics, Russia
Nina Kochetova	Sobolev Institute of Mathematics, Russia
Polina Kononova	Sobolev Institute of Mathematics, Russia
Galina F. Kornilova	Krasovsky Institute of Mathematics and Mechanics, Russia
Maria A. Kostina	Krasovsky Institute of Mathematics and Mechanics, Russia
Timur Medvedev	Higher School of Economics, Russia
Katherine Neznakhina	Krasovsky Institute of Mathematics and Mechanics, Russia
Yuri Ogorodnikov	Krasovsky Institute of Mathematics and Mechanics, Russia
Maxim Pasynkov	Krasovsky Institute of Mathematics and Mechanics, Russia
Maria Poberiy	Krasovsky Institute of Mathematics and Mechanics, Russia

Organizers

Krasovsky Institute of Mathematics and Mechanics, Russia
Sobolev Institute of Mathematics, Russia
Melentiev Energy Systems Institute SB RAS, Russia

Sponsors

Russian Foundation for Basic Research, grant no. 19-07-20007
Higher School of Economics (Campus Nizhny Novgorod)
Ural Federal University
Novosibirsk State University

Abstracts of Invited Talks

Decision Under Ignorance: A Comparison of Existing Criteria in a Context of Linear Programming

Olga Battaïa

ISAE-Supaero, Toulouse, France
Olga.Battaia@isae.fr

Abstract. Decision or optimization problems often arise in an uncertain context. Depending on available information, several approaches have been proposed to model this uncertainty. In this talk, we focus on the case of low knowledge on possible states, namely decision under ignorance. In this case the decision-maker is able to give the set of possible values of optimization problem parameters but she/he is not able to differentiate them. We compare a set of criteria that can be used in this case on the example of a linear programming problem and discuss some possible applications.

Keywords: Decision making · Uncertainty · Linear programming

Node Partitioning and Cycles Creation Problem

Oleg Burdakov

Linköping University, Sweden
oleg.burdakov@liu.se

Abstract. We present a new class of network optimization problems, which extend the classical NP-hard travelling salesman problem. It is formulated as follows. Given a graph with a certain time associated with each node and each arc, a feasible partition of the nodes in subsets is such that, for each subset, there exists a Hamiltonian cycle whose travelling time is below the time associated with each node in the tour. It is required to find a feasible partitioning which minimizes the number of such cycles. Problems of this kind are typical in numerous applications, where services are repeatedly provided for a set of customers. For each customer, there is a critical time within which a service must be repeated. Given the travelling time between the customers, the set of customers is partitioned so that each subset is served by one agent in a cyclic manner without violating any individual critical time requirement. The number of agents is minimized. As an example, we consider a problem, in which a fleet of unmanned aerial vehicles is used for area patrolling.

We introduce an mixed integer programming formulation of the node partitioning and cycles creation problem, and also heuristic algorithms for solving this problem. Results of numerical experiments are presented. (Joint work with: Kai Hoppmann, Thorsten Koch and Gioni Mexi (ZIB, Berlin, Germany)).

Keywords: TSP · Subtours · Integer programming

Bijective Analysis of Online Algorithms

Christoph Dürr

Sorbonne Université, France
christoph.durr@lip6.fr

Abstract. In the online computing framework the instance arrives in form a request sequence, every request must be served immediately, through a decision, which generates some cost. Think at the paging problem for memory caches. The goal in this research area is to identify the best strategy, also called online algorithm. Classically this is done through the competitive analysis, i.e. the performance of an online algorithm is compared with the optimal offline solution. The goal is to find an algorithm which minimizes this ratio over the worst case instance. You would say that algorithm A is better than algorithm B if it has a smaller ratio. However there are situations where two algorithms have the same ratio, still in practice one is better than the other. So people came up with a different technique to compare online algorithms directly with each other, rather than through the optimal offline solution. The bijective analysis is one of them. I would do a survey on this technique, and talk about a related personal work: Best-of-two-worlds analysis of online search, with Spyros Angelopoulos and Shendan Jin.

Keywords: Online algorithms · Bijective analysis

A Survey on Possible and Impossible Attempts to Solve the Treewidth Problem via ILPs

Alexander Grigoriev

Maastricht University, Netherlands
a.grigoriev@maastrichtuniversity.nl

Abstract. We survey a number of integer programming formulations for the pathwidth and for the treewidth problems. The attempts to find good formulations for the problems span the period of 15 years, yet without any true success. Nevertheless, some formulations provide potentially useful frameworks for attacking these notorious problems. Some others are just curious and interesting fruits of mathematical imagination.

Keywords: Treewidth and pathwidth problems · Integer programming

No-Idle Scheduling of Unit-Time Jobs with Release Dates and Deadlines on Parallel Machines

Mikhail Kovalyov

United Institute of Informatics Problems NASB, Belarus
kovalyov_my@yahoo.co.uk

Abstract. While the problem of scheduling unit-time jobs with release dates and deadlines on parallel machines is polynomially solvable via a reduction to the assignment problem, the no-idle requirement destroys this reduction and makes the problem challenging. In the presentation, a number of properties of this problem are reported, and heuristic and optimal algorithms based on these properties are described.

Keywords: Scheduling · Optimal algorithms · Heuristics

Critical and Maximum Independent Sets Revisited

Vadim Levit

Ariel University, Israel
levitv@ariel.ac.il

Abstract. A set of vertices of a graph is independent if no two its vertices are adjacent. A set is critical if the difference between its size and the size of its neighborhood is maximum. Critical independent sets define an important area of research due to their close relationships with the well-known NP-hard problem of finding a maximum independent set. Actually, every critical independent set is contained in a maximum independent set, while a maximum critical independent set can be found in polynomial time. If S is an independent set such that there is a matching from its neighborhood into S, then it is a crown. It is known that every critical independent set forms a crown. A graph is König-Egerváry if every maximum independent set is a crown. Crowns are also accepted as important tools for fixed parameter tractable problems. For instance, the size of the vertex cover can be substantially reduced by deleting both the vertices of a crown and its neighborhood. In this presentation, we discuss various connections between unions and intersections of maximum (critical) independent sets of graphs, which lead to deeper understanding of crown structures, in general, and König-Egerváry graphs, in particular.

Keywords: Maximum independent set · Critical independent set · Parameterized complexity

An Overview of the Relocation Problem

Bertrand M. T. Lin

National Chiao Tung University, Hsinchu, Taiwan
bmtlin@mail.nctu.edu.tw

Abstract. The relocation problem is formulated from a municipal redevelopment project in east Boston. In its abstract form, the relocation problem incorporates a generalized resource constraint in which the amount of the resource returned by a completed activity is not necessarily the same as that the activity has acquired for commencing the processing. We will first introduce the connection of the relocation problem to flow shop scheduling. Several traditional scheduling models with the generalized resource constraints have been proposed investigated. We will review existing results, suggest new models and present several open questions.

Keywords: Relocation problem · Flow shop scheduling

On a New Approach for Optimization Under Uncertainty

Natalia Shakhlevich

University of Leeds, UK
N.Shakhlevich@leeds.ac.uk

Abstract. Research on decision making under uncertainty has a long history of study. Still theoretical findings have strong limitations: stochastic programming requires probability distributions for uncertain parameters which are often hard to specify; robust optimisation essentially relies on worst-case scenarios which can be over-pessimistic and far from realistic scenarios; stability analysis explores optimal solutions which can be hard to find even for well predicted scenarios. As an alternative approach, we propose a new system model based on the concept of resiliency. Resilient solutions are not required to be optimal, but they should keep quality guarantees for the widest range of uncertain problem parameters. The talk illustrates key steps of resiliency analysis considering examples of 0/1 combinatorial optimisation problems.

Keywords: Decision making · Stochastic programming · Resiliency

Exterior Point Simplex-Type Algorithms for Linear and Network Optimization Problems

Angelo Sifaleras

University of Macedonia, Greece
sifalera@uom.gr

Abstract. Two decades of research led to the development of a number of efficient algorithms that can be classified as exterior point simplex-type. This type of algorithms can cross over the infeasible region of the primal (dual) problem and find an optimal solution reducing the number of iterations needed. Thus, such approaches aim to find an efficient way to get to an optimal basis via a series of infeasible ones. In this lecture, we present the developments in exterior point simplex-type algorithms for linear and network optimization problems, over the recent years. We also present other approaches that, in a similar way, do not preserve primal or dual feasibility at each iteration such as the monotonic build-up Simplex algorithms and the criss-cross methods, and also discuss some open research problems.

Keywords: Exterior point algorithms · Simplex-type algorithms · Criss-cross methods

Design of Fully-Polynomial Approximation Schemes for Non-linear Boolean Programming Problems

Vitaly Strusevich

University of Greenwich, UK
V.Strusevich@gre.ac.uk

Abstract. The talk is aimed at describing various techniques used for designing fully-polynomial approximation schemes (FPTAS) for problems of minimizing and maximizing non-linear non-separable functions of Boolean variables, either with no additional constraints or with linear knapsack constraints. Most of the reported results are on optimizing a special quadratic function known as the half-product, which has numerous scheduling applications. Besides, problems with a more general objective and nested linear constraints are considered and a design of an FPTAS based on the *K*-approximation calculus is discussed.

Keywords: FPTAS · *K*-approximation calculus · Half-product

Abstracts of Tutorials

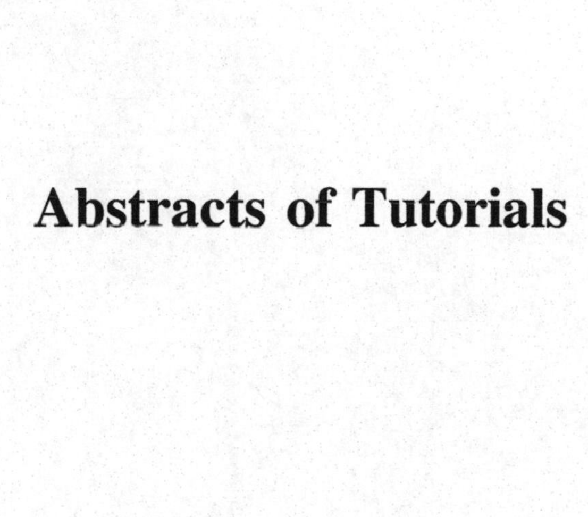

Distributed Memory Based Parallelization of Metaheuristic Methods

Tatjana Davidović

Mathematical Institute of the Serbian Academy of Sciences and Arts, Serbia
tanjad@mi.sanu.ac.rs

Abstract. Metaheuristics represent powerful tools for addressing hard combinatorial optimization problems. However, real life instances usually cannot be treated efficiently by the means of computing times. Moreover, a major issue in metaheuristic design and calibration is to provide high performance solutions for a variety of problems. Parallel metaheuristics aim to address both issues. The main goal of parallelization is to speed up the computations by dividing the total amount of work between several processors. Parallelization of stochastic algorithms, such as metaheuristics may involve several additional goals. Besides speeding up the search (i.e., reducing the search time), it could be possible to: improve the quality of the obtained solutions (by enabling searching through different parts of the solution space); improve the robustness of the search (in terms of solving different optimization problems and different instances of a given problem in an effective manner; robustness may also be measured in terms of the sensitivity of the metaheuristic to its parameters); and solve large-scale problems (i.e., solve very large instances that cannot be even stored in the memory of a sequential machine). A combination of gains may also be obtained: parallel execution can enable an efficient search through different regions of the solution space, yielding an improvement of the quality of the final solution within a smaller amount of execution time. The objective of this talk is to present a state-of-the-art survey of the main ideas and strategies related to the parallelization of metaheuristic methods. Various paradigms related to the development of parallel metaheuristics are explained. Among them, communications, synchronization, and control aspects are identified as the most relevant. Implementation issues are also discussed, pointing out the characteristics of shared and distributed memory multiprocessors as target architectures. All these topics are illustrated by the examples from recent literature related to the parallelization of various meta-heuristic methods, with the focus on distributed memory parallelization of Variable Neighborhood Search (VNS) and Bee Colony Optimization (BCO) using Message Passing Interface (MPI) communication protocol.

Keywords: Parallel metaheuristics · Distributed memory · MPI

Bilevel Optimization: The Model and its Transformations

Stephan Dempe

TU Bergakademie Freiberg, Germany
dempe@math.tu-freiberg.de

Abstract. Bilevel (or hierarchical) optimization problems aim to minimize one function subject to (a subset of) the graph of the solution set mapping of a second, parameter dependent optimization problem. The parameter is the decision variable of the socalled leader, the optimization problem describing the constraints is the problem of the follower. These problems have a large number of applications in science, engineering, economics. To investigate and solve them, they need to be transformed into a single-level optimization problem. For that different approaches can be used.

(1) If the followers problem is regular and convex, it can be replaced using the Karush-Kuhn-Tucker conditions. The result is a so-called Mathematical Program with Equilibrium Constraints. In these nonconvex optimization problems, the Mangasarian-Fromovitz constraint qualification is violated at every feasible point. Solution algorithms converge (under suitable assumptions) to stationary points which are, in general, not related to stationary points of the bilevel optimization problem. To overcome this unpleasant situation, a certain regularization approach can be used. Another approach uses the transformation to a mixed integer (nonlinear) optimization problem.
(2) If the optimal value function of the followers problem is used, a nonconvex, nonsmooth optimization problem arises. Again, the (now nonsmooth) Mangasarian-Fromovitz constraint qualification is violated at every feasible point. If the optimal value function is convex or concave, its approximation is helpful to describe a solution algorithm. Optimality conditions can be derived using partial calmness or a certain penalization approach.
(3) The problem can be reformulated as a generalized Nash equilibrium problem. Topic of the lecture is the introduction of the model together with some surprising properties and a short overview over promising accesses to investigate and solve it.

Keywords: Bilevel optimization · KKT theorem · Constraint qualification

The Fundamental Role of Concave Programming in Continuous Global Optimization

Oleg Khamisov

Melentiev Energy Systems Institute SB RAS, Russia
globopt@mail.ru

Abstract. A comprehensive description of connections between concave programming and other branches of global optimization like Lipschitz optimization, d.c. optimization etc. is given. It is shown that in general solution of almost every global optimization problem can reduced to solution of a sequence of concave programming problems. Modern concave optimization technology including cuts, branch and bounds, branch and cuts and so on as well as the corresponding extensions to different global optimization problems are presented. A part of the talk is devoted to the connection between concave and mixed 0-1 linear programming.

Keywords: Global optimization · Concave programming · Mixed linear programming

Primal-Dual Method and Online Problems

Alexander Kononov

Sobolev Institute of Mathematics, Russia
alvenko@math.nsc.ru

Abstract. The primal-dual method is a powerful tool in the design of approximate algorithms for combinatorial optimization problems. In our tutorial we discuss how this method can be extended to develop online algorithms. The tutorial is based on the survey by N. Buchbinder and J. Naor and the web-presentation by N. Bansal.

Keywords: Combinatorial optimization · Primal-dual method · Online algorithms

Solving Nonlinear System of Equations as an Optimization Problem

Nenad Mladenovic

Mathematical Institute SANU, Serbia
nenadmladenovic12@gmail.com

Abstract. The Nonlinear System of Equations (NSE) problem is usually transformed into an equivalent optimization problem, with an objective function that allows us to find all the zeros. Instead of the usual sum-of-squares objective function, the new objective function is presented as the sum of absolute values. Theoretical investigation confirms that the new objective function provides more accurate solutions regardless of the optimization method used. In addition, we achieve increased precision at the expense of reduced smoothness. In this paper, we propose the continuous variable neighbor-hood search method for finding all the solutions to a NSEs. Computational analysis of standard test instances shows that the proposed method is more precise and much faster than two recently developed methods. Similar conclusions are drawn by comparing the proposed method with many other methods in the literature.

Keywords: System of nonlinear equations · Continuous optimization · Variable neighborhood search · Direct search methods

---

Joint work with Jun Pei, Zorica Drazic, Milan Drazic, and Panos M. Pardalos.

Projection Problems and Problems with Projection

Evgeni A. Nurminski

Far Eastern Federal University, Russia
nurmi@dvo.ru

Abstract. This lecture reviews the state of the art for probably the most common computational operation in applied mathematics—projection, which can be also considered as the problem of finding the least norm element (LNE) in a given subset of a linear vector space. The special attention in the lecture will be given to Euclidean or orthogonal projection, but we plan to discuss another norms as well. Projection is computationally intensive operation even for relatively simple sets like canonical simplexes and special algorithms are a way more efficient than off-the-shelf quadratic programming methods especially for large-scale problems. Large-scale projection problems can be decomposed in different sequential or parallel manner as extension of celebrated Kaczmarz sequential projection procedure and block-row action methods. We discuss also the problem of numerical instability of projection operation which is quite common in such applications as new optimization algorithms, linear programming, machine learning and automatic classification.

Keywords: Projection procedures · Large-scale problems · Decomposition

Modern Methods of Nonconvex Optimization

Alexander Strekalovsky

Matrosov Institute for System Dynamics and Control Theory SB RAS, Russia
strekal@icc.ru

Abstract. We address the nonconvex optimization problem with the cost function and equality and inequality constraints given by d.c. functions. The linear space of d.c. functions possesses a number of very attractive properties. For example, every continuous function can be approximated at any desirable accuracy by a d.c. function and any twice differentiable function belongs to the DC space. In addition, any lower semicontinuous (l.s.c.) function can be approximated at any precision by a sequence of continuous functions. Furthermore, provided that for the optimization problem under study we proposed the new Global Optimality Conditions (GOCs), which have been published in the English and Russian languages. The natural question arises: is it possible to construct a computational scheme based on the GOCs (otherwise, what are they for?) that would allow us not only to generate critical points (like the KKT-vectors) but to escape any local pitfall, which makes it possible to reach a global solution to the problem in question? First of all, we recall that with the help of thc Theory of Exact Penalization, the original d.c. problem was reduced to a problem without constraints. Moreover, it can be readily seen that this penalized problem is a d.c. problem as well. Furthermore, special Local Search Methods (LSMs) were developed and substantiated in view of their convergence features. In addition, the GOCs were generalized for the minimizing sequences in the penalized problem. A special theoretical method was proposed and its convergence properties were studied. We developed a Global Search Scheme (GSS) based on all theoretical results presented above, and, moreover, we were lucky to prove that the sequence produced by the GSS turned out to be minimizing in the original d.c. optimization problem. Finally, we developed a Global Search Method (GSM), combining the special LSM and the GSS proposed. The convergence of the GSM is also investigated under some natural assumptions. The first results of numerical testing of the approach will be demonstrated.

Keywords: Global optimization · d.c. functions

Contents

Bilevel Optimization

Integer Programming

Combinatorial Optimization

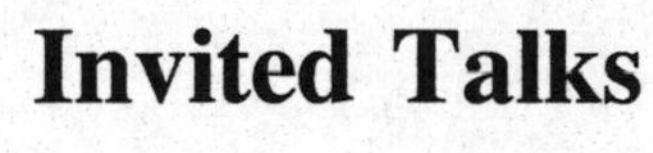

Invited Talks

Critical and Maximum Independent Sets Revisited

Vadim E. Levit[1(✉)] and Eugen Mandrescu[2]

[1] Ariel University, 40700 Ariel, Israel
levitv@ariel.ac.il

[2] Holon Institute of Technology, 5810201 Holon, Israel
eugen_m@hit.ac.il

Abstract. Let G be a simple graph with vertex set $V(G)$.

A set $S \subseteq V(G)$ is *independent* if no two vertices from S are adjacent, and by $\mathrm{Ind}(G)$ we mean the family of all independent sets of G.

The number $d(X) = |X| - |N(X)|$ is the *difference* of $X \subseteq V(G)$, and a set $A \in \mathrm{Ind}(G)$ is *critical* if $d(A) = \max\{d(I) : I \in \mathrm{Ind}(G)\}$ [34].

Let us recall the following definitions:

- $\mathrm{core}(G) = \bigcap\{S : S$ *is a maximum independent set*$\}$ [16],
- $\mathrm{corona}(G) = \bigcup\{S : S$ *is a maximum independent set*$\}$ [5],
- $\ker(G) = \bigcap\{S : S$ *is a critical independent set*$\}$ [18],
- $\mathrm{nucleus}(G) = \bigcap\{S : S$ *is a maximum critical independent set*$\}$ [12]
- $\mathrm{diadem}(G) = \bigcup\{S : S$ *is a (maximum) critical independent set*$\}$ [24].

In this paper we focus on interconnections between ker, core, corona, nucleus, and diadem.

Keywords: Independent set · Critical set · Ker · Core · Corona · Diadem · Matching

1 Introduction

Throughout this paper $G = (V, E)$ is a finite, undirected, loopless graph without multiple edges, with vertex set $V = V(G)$ of cardinality $|V(G)| = n(G)$, and edge set $E = E(G)$ of size $|E(G)| = m(G)$. If $X \subseteq V(G)$, then $G[X]$ is the subgraph of G induced by X. By $G - W$ we mean either the subgraph $G[V(G) - W]$, if $W \subseteq V(G)$, or the subgraph obtained by deleting the edge set W, for $W \subseteq E(G)$. In either case, we use $G - w$, whenever $W = \{w\}$. If $A, B \subseteq V(G)$, then (A, B) stands for the set $\{ab : a \in A, b \in B, ab \in E(G)\}$.

The *neighborhood* $N(v)$ of a vertex $v \in V(G)$ is the set $\{w : w \in V(G)$ *and* $vw \in E(G)\}$, while the *closed neighborhood* $N[v]$ of $v \in V(G)$ is the set $N(v) \cup \{v\}$; in order to avoid ambiguity, we use also $N_G(v)$ instead of $N(v)$. A vertex v is *isolated* if $N(v) = \emptyset$. Let us define $\mathrm{isol}(G)$ as the set of all isolated vertices.

M. Khachay et al. (Eds.): MOTOR 2019, LNCS 11548, pp. 3–18, 2019.
https://doi.org/10.1007/978-3-030-22629-9_1

The *neighborhood* $N(A)$ of $A \subseteq V(G)$ is $\{v \in V(G) : N(v) \cap A \neq \emptyset\}$, and $N[A] = N(A) \cup A$. We may also use $N_G(A)$ and $N_G[A]$, when referring to neighborhoods in a graph G.

A set $S \subseteq V(G)$ is *independent* if no two vertices from S are adjacent, and by $\mathrm{Ind}(G)$ we mean the family of all the independent sets of G. An independent set of maximum size is a *maximum independent set* of G, and the *independence number* $\alpha(G)$ of G is $\max\{|S| : S \in \mathrm{Ind}(G)\}$. Let $\Omega(G)$ denote the family of all maximum independent sets, and let

- $\mathrm{core}(G) = \bigcap\{S : S \in \Omega(G)\}$ [16],
- $\mathrm{corona}(G) = \bigcup\{S : S \in \Omega(G)\}$ [5].

Clearly, $N(\mathrm{core}(G)) \subseteq V(G) - \mathrm{corona}(G)$, and there exist graphs with $N(\mathrm{core}(G)) \neq V(G) - \mathrm{corona}(G)$. The problem of whether $\mathrm{core}(G) \neq \emptyset$ is **NP**-hard [5].

A *matching* is a set M of pairwise non-incident edges of G. If $A \subseteq V(G)$, then $M(A)$ is the set of all the vertices matched by M with vertices belonging to A. A matching of maximum cardinality, denoted $\mu(G)$, is a *maximum matching*.

Recall from [34] the following definitions for a graph G:

- $d(X) = |X| - |N(X)|$ is the *difference* of $X \subseteq V(G)$;
- $d(G) = \max\{d(X) : X \subseteq V(G)\}$ is the *critical difference*;
- $id(G) = \max\{d(I) : I \in \mathrm{Ind}(G)\}$ is the *critical independence difference*;
- if $A \in \mathrm{Ind}(G)$ and $d(A) = id(G)$, then A is a *critical independent set*.

Clearly, $d(G) \geq id(G)$. It was shown in [34] that $d(G) = id(G)$ holds for every graph G. All pendant vertices not belonging to K_2 components are included in every inclusion maximal critical independent set.

For example, let $X = \{v_1, v_2, v_3, v_4\}$ and $I = \{v_1, v_2, v_3, v_6, v_7\}$ in the graph G of Fig. 1. Note that X is a critical set, since $N(X) = \{v_3, v_4, v_5\}$ and $d(X) = 1 = d(G)$, while I is a critical independent set, because $d(I) = 1 = id(G)$. Other critical sets are $\{v_1, v_2\}$, $\{v_1, v_2, v_3\}$, $\{v_1, v_2, v_3, v_4, v_6, v_7\}$.

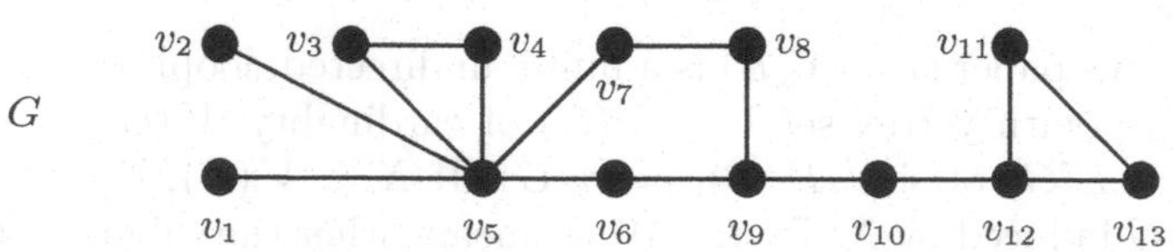

Fig. 1. $\mathrm{core}(G) = \{v_1, v_2, v_6, v_{10}\}$ is a critical set.

It is known that finding a maximum independent set is an **NP**-hard problem [10]. Zhang proved that a critical independent set can be found in polynomial time [34]. A simpler algorithm, reducing the critical independent set problem to computing a maximum independent set in a bipartite graph is given in [1].

Theorem 1. *[6] Each critical independent set can be enlarged to a maximum independent set.*

Theorem 1 leads to an efficient way of approximating $\alpha(G)$ [33]. Moreover, it has been shown that a critical independent set of maximum cardinality can be computed in polynomial time [14]. Recently, a parallel algorithm computing the critical independence number was developed [8].

Recall that if $\alpha(G) + \mu(G) = n(G)$, then G is a *König-Egerváry graph* [9,32]. As a well-known example, each bipartite graph is a König-Egerváry graph as well. Various properties of König-Egerváry graphs can be found in [2–4,11–20,23,26,31].

Theorem 2. *[17] If G is a König-Egerváry graph, M is a maximum matching of G, and $S \in \Omega(G)$, then:*

(i) *M matches $V(G) - S$ into S, and $N(\text{core}(G))$ into* $\text{core}(G)$;
(ii) $N(\text{core}(G)) = \bigcap \{V(G) - S : S \in \Omega(G)\} = V(G) - \text{corona}(G)$.

The *deficiency* $def(G)$ is the number of non-saturated vertices relative to a maximum matching, i.e., $def(G) = n(G) - 2\mu(G)$ [28]. A proof of a conjecture of Graffiti.pc [7] yields a new characterization of König-Egerváry graphs: these are exactly the graphs, where there exists a critical maximum independent set [15].

Theorem 3. *[19] For a König-Egerváry graph G the following equalities hold*

$$d(G) = |\text{core}(G)| - |N(\text{core}(G))| = \alpha(G) - \mu(G) = def(G).$$

Using this finding, we have strengthened a characterization of König-Egerváry graphs given in [15].

Theorem 4. *[19] G is a König-Egerváry graph if and only if each of its maximum independent sets is critical.*

For a graph G, let us denote

- $\ker(G) = \bigcap \{A : A \textit{ is a critical independent set}\}$ [18],
- $\text{MaxCritIndep}(G) = \{S : S \textit{ is a maximum critical independent set}\}$,
- $\text{nucleus}(G) = \bigcap \text{MaxCritIndep}(G)$ [12],
- $\text{diadem}(G) = \bigcup \text{MaxCritIndep}(G)$ [24].

Clearly, $\text{isol}(G) \subseteq \ker(G) \subseteq \text{nucleus}(G)$ and, according to Theorem 1, the inclusion $\text{diadem}(G) \subseteq \text{corona}(G)$ is true for every graph G.

In this paper we present several properties of $\ker(G)$, in relation with $\text{core}(G)$, $\text{corona}(G)$, $\text{diadem}(G)$, and $\text{nucleus}(G)$.

2 Preliminaries

Theorem 5. *[18] For a graph G, the following assertions are true:*

(i) *the function d is supermodular, i.e., $d(A \cup B) + d(A \cap B) \geq d(A) + d(B)$ for every $A, B \subseteq V(G)$;*
(ii) *if A and B are critical in G, then $A \cup B$ and $A \cap B$ are critical as well;*
(iii) *G has a unique minimal independent critical set, namely, $\ker(G)$.*

As a consequence, we have the following.

Corollary 1. *For every graph G, $\mathrm{diadem}(G)$ is a critical set.*

For instance, $\mathrm{diadem}(G) = \{v_1, v_2, v_3, v_4, v_6, v_7, v_8, v_{10}\}$ is critical, but not independent, where the graph G is from Fig. 1.

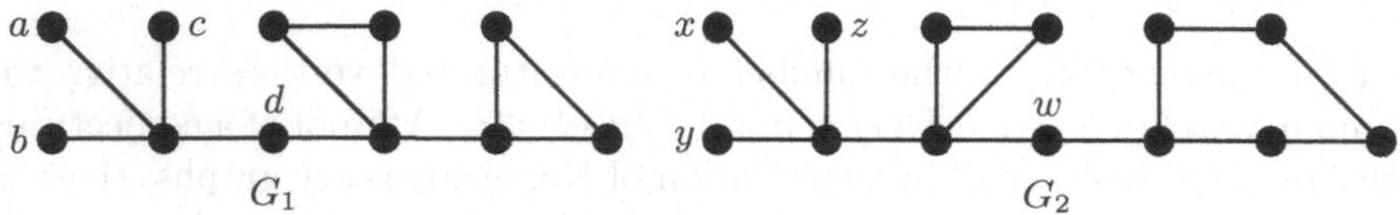

Fig. 2. Both G_1 and G_2 are not König-Egerváry graphs.

Combining Theorems 4 and 5*(ii)*, we deduce the following.

Corollary 2. *If G is a König-Egerváry graph, then both $\mathrm{core}(G)$ and $\mathrm{corona}(G)$ are critical sets.*

Let us consider the graphs G_1 and G_2 from Fig. 2: $\mathrm{core}(G_1) = \{a, b, c, d\}$ and it is a critical set, while $\mathrm{core}(G_2) = \{x, y, z, w\}$ and it is not critical.

Theorem 6. *[14] All inclusion maximal critical independent sets are of the same size. In other words,*

$$\{A : A \text{ is an inclusion maximal critical independent set}\} = \mathrm{MaxCritIndep}(G).$$

By Theorems 1, 6, $\mathrm{corona}(G) \supseteq \mathrm{diadem}(G)$ for every graph.

Theorem 7. *[12] If $\mathrm{core}(G)$ is a critical set, then $\mathrm{core}(G) \subseteq \mathrm{nucleus}(G)$. If, in addition, $\mathrm{diadem}(G) = \mathrm{corona}(G)$, then $\mathrm{core}(G) \subseteq \mathrm{nucleus}(G)$.*

3 Structural Properties of ker (G)

Deleting a vertex from a graph may change its critical difference. For instance, $d(G - v_1) = d(G) - 1$, $d(G - v_{13}) = d(G)$, while $d(G - v_3) = d(G) + 1$, where G is the graph of Fig. 1.

Proposition 1. *[21] For a vertex v in a graph G, the following assertions hold:*

(i) $d(G-v) = d(G) - 1$ *if and only if* $v \in \ker(G)$;
(ii) *if* $v \in \ker(G)$, *then* $\ker(G-v) \subseteq \ker(G) - \{v\}$.

Note that $\ker(G-v)$ may differ from $\ker(G) - \{v\}$. For example, $\ker(K_{3,2})$ is equal to the partite set of size 3, but $\ker(K_{3,2} - v) = \emptyset$ whenever v is in that set. Also, if $G = C_4$, then $\ker(G) - \{v\} = \emptyset - \{v\} = \emptyset$, while $\ker(G-v) = N_G(v)$ for every $v \in V(G)$.

Since $d(G)$ is polynomially computable [34], Proposition 1 implies the following.

Corollary 3. *[18] The set* $\ker(G)$ *can be computed by an algorithm of polynomial complexity.*

It seems interesting to find even better polynomial approximations of $\mathrm{core}(G)$.

Theorem 8. *[14] There is a matching from $N(S)$ into S for every critical independent set S.*

In the graph G of Fig. 1, let $S = \{v_1, v_2, v_3\}$. By Theorem 8, there is a matching from $N(S)$ into $S = \{v_1, v_2, v_3\}$, for instance, $M = \{v_2v_5, v_3v_4\}$, since S is critical independent. On the other hand, there is no matching from $N(S)$ into $S - v_3$.

Theorem 9. *[21] For a critical independent set A in a graph G, the following statements are equivalent:*

(i) $A = \ker(G)$;
(ii) *there is no set* $B \subseteq N(A), B \neq \emptyset$ *such that* $|N(B) \cap A| = |B|$;
(iii) *for each* $v \in A$ *there exists a matching from* $N(A)$ *into* $A - v$.

The graphs G_1 and G_2 in Fig. 3 satisfy $\ker(G_1) = \mathrm{core}(G_1)$, $\ker(G_2) = \{x, y, z\} \subset \mathrm{core}(G_2)$, and both $\mathrm{core}(G_1)$ and $\mathrm{core}(G_2)$ are critical sets of maximum size. The graph G_3 in Fig. 3 has $\ker(G_3) = \{u, v\}$, the set $\{t, u, v\}$ as a critical independent set of maximum size, while $\mathrm{core}(G_3) = \{t, u, v, w\}$ is not a critical set.

An independent set S is *inclusion minimal with* $d(S) > 0$ if no proper subset of S has positive difference. For example, in Fig. 3 one can see that $\ker(G_1)$ is an inclusion minimal independent set with positive difference, while for the graph

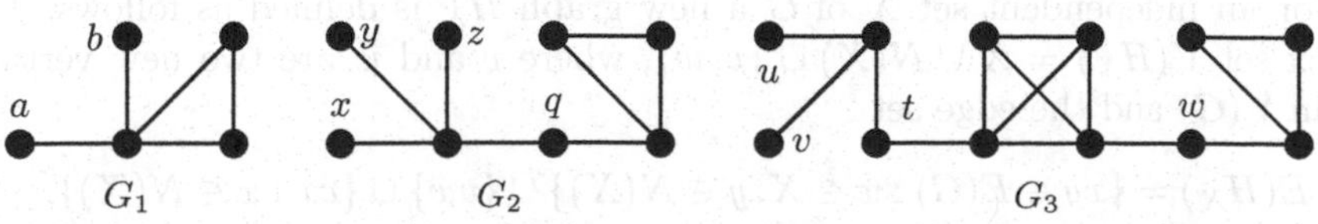

Fig. 3. $\mathrm{core}(G_1) = \{a, b\}$, $\mathrm{core}(G_2) = \{q, x, y, z\}$, $\mathrm{core}(G_3) = \{t, u, v, w\}$.

G_2 the sets $\{x, y\}, \{x, z\}, \{y, z\}$ are inclusion minimal independent with positive difference, and $\ker(G_2) = \{x, y\} \cup \{x, z\} \cup \{y, z\}$.

Actually, all inclusion minimal independent sets S with $d(S) > 0$ are of the same difference.

Proposition 2. *[21] If S_0 is an inclusion minimal independent set such that $d(S_0) > 0$, then $d(S_0) = 1$. In other words,*

$$\{S_0 : S_0 \textit{ is an inclusion minimal independent set with } d(S_0) > 0\} =$$
$$\{S_0 : S_0 \textit{ is an inclusion minimal independent set with } d(S_0) = 1\}.$$

The converse of Proposition 2 is not true. For instance, $S = \{x, y, u\}$ is independent in the graph G of Fig. 4 and $d(S) = 1$, but S is not minimal with this property.

Theorem 10. *[21] If $\ker(G) \neq \emptyset$, then*

$$\ker(G) = \bigcup \{S_0 : S_0 \textit{ is inclusion minimal independent with } d(S_0) = 1\}$$
$$= \bigcup \{S_0 : S_0 \textit{ is inclusion minimal independent with } d(S_0) > 0\}.$$

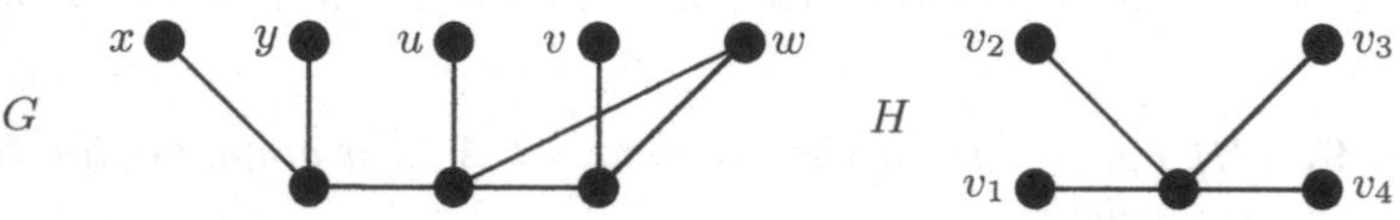

Fig. 4. Both $S_1 = \{x, y\}$ and $S_2 = \{u, v, w\}$ are inclusion minimal independent sets satisfying $d(S) > 0$. The same is true for each pair of leaves in H.

In a graph G, the union of all minimum cardinality independent sets S with $d(S) > 0$ may be a proper subset of $\ker(G)$. For example, consider the graph G in Fig. 4, where $\{x, y\} \subset \ker(G) = \{x, y, u, v, w\}$.

Proposition 3. *[21]* $\min\{|S_0| : d(S_0) > 0, S_0 \in \mathrm{Ind}(G)\} \leq |\ker(G)| - d(G) + 1$ *is true for every graph G.*

Conjecture 1. [21] The number of inclusion minimal independent set S such that $d(S) > 0$ is greater than or equal to $d(G)$.

For an independent set X of G a new graph H_X is defined as follows. The vertex set $V(H_X) = X \cup N(X) \cup \{v, w\}$, where v and w are two new vertices not in $V(G)$ and the edge set

$$E(H_X) = \{xy \in E(G) : x \in X, y \in N(X)\} \cup \{vw\} \cup \{vx : x \in N(X)\}.$$

Note that if G is a connected graph with $|V(G)| > 1$, then H_X is a connected bipartite graph. Also observe that for all $Y \subset X$, $d_{H_X}(Y) = d_G(Y)$.

Theorem 11. *[3] If X is an independent set of G with $d(X) > 0$ such that $d(Y) < d(X)$ for all proper subsets Y of X, then* $\ker(H_X) = X \subset \ker(G)$.

Proposition 4. *[3] A set S with $d(S) > 0$ such that no proper subset of S has positive difference must be an independent set.*

Theorem 12. *[3] Let $X \in \text{Ind}(G)$ with $d(X) = k > 0$. If $d(Y) < k$ for all proper subsets Y of X, then X can be expressed as a union of k distinct inclusion minimal sets with positive difference.*

Putting $X = \ker(G)$ in Theorem 12, one may conclude that there exist $d(\ker(G)) = d(G)$ inclusion minimal independent sets, which validates Conjecture 1.

4 Relationships Between ker (G) and Core(G)

Let us consider again the graph G_2 from Fig. 2: $\text{core}(G_2) = \{x, y, z, w\}$ and it is not critical, but $\ker(G_2) = \{x, y, z\} \subsetneq \text{core}(G_2)$. Clearly, the same inclusion holds for G_1, whose $\text{core}(G_1)$ is a critical set.

Theorem 13. *[18] For every graph G,* $\ker(G) \subseteq \text{core}(G)$.

Let I_c be a maximum critical independent set of G, and $X = I_c \cup N(I_c)$. In [30] it is proved that $\text{core}(G[X]) \subseteq \text{core}(G)$. Moreover, in [18], we showed that the chain of relationships $\ker(G) = \ker(G[X]) \subseteq \text{core}(G[X]) \subseteq \text{core}(G)$ holds for every graph G. Theorem 13 allows an alternative proof of the following finding due to Lorentzen.

Corollary 4. *[18, 27, 29] The inequality $d(G) \geq \alpha(G) - \mu(G)$ holds for every graph.*

The following lemma will be used further to give an alternative proof for the assertion that $\ker(G) = \text{core}(G)$ holds for every bipartite graph G.

Lemma 1. *If $G = (A, B, E)$ is a bipartite graph with a perfect matching, say M, $S \in \Omega(G)$, $X \in \text{Ind}(G)$, $X \subseteq V(G) - S$, and $G[X \cup M(X)]$ is connected, then*

$$X^1 = X \cup M((N(X) \cap S) - M(X))$$

is an independent set, and $G[X^1 \cup M(X^1)]$ is connected.

Proof. Let us show that the set $M((N(X) \cap S) - M(X))$ is independent. Suppose, to the contrary, that there exist $v_1, v_2 \in M((N(X) \cap S) - M(X))$ such that $v_1 v_2 \in E(G)$. Hence $M(v_1), M(v_2) \in (N(X) \cap S) - M(X)$.

If $M(v_1)$ and $M(v_2)$ have a common neighbor $w \in X$, then the set of vertices $\{v_1, v_2, M(v_2), w, M(v_1)\}$ spans C_5, which is forbidden for bipartite graphs.

Otherwise, let $w_1, w_2 \in X$ be neighbors of $M(v_1)$ and $M(v_2)$, respectively. Since $G[X \cup M(X)]$ is connected, there is a path with an even number of edges connecting w_1 and w_2. Together with $\{w_1, M(v_1), v_1, v_2, M(v_2), w_2\}$ this path produces a cycle of odd length in contradiction with the hypothesis on G being a bipartite graph.

To complete the proof of independence of the set

$$X^1 = X \cup M\left(\left(N\left(X\right) \cap S\right) - M\left(X\right)\right)$$

it is enough to demonstrate that there are no edges connecting vertices of X and $M\left(\left(N\left(X\right) \cap S\right) - M\left(X\right)\right)$.

Assume, to the contrary, that there is an edge $vw \in E$, such that $v \in M\left(\left(N\left(X\right) \cap S\right) - M\left(X\right)\right)$ and $w \in X$. Since $M\left(v\right) \in \left(N\left(X\right) \cap S\right) - M\left(X\right)$ and $G\left[X \cup M\left(X\right)\right]$ is connected, it follows that there exists a path with an odd number of edges connecting $M\left(v\right)$ to w. This path together with the edges vw and $vM\left(v\right)$ produces cycle of odd length, in contradiction with the bipartiteness of G.

Finally, since $G\left[X \cup M\left(X\right)\right]$ is connected, $G\left[X^1 \cup M\left(X^1\right)\right]$ is connected as well, by definitions of set functions N and M (Fig. 5).

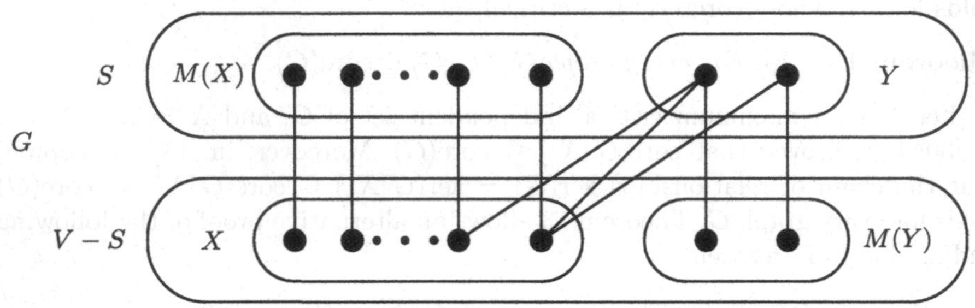

Fig. 5. $S \in \Omega(G)$, $Y = \left(N\left(X\right) \cap S\right) - M\left(X\right)$ and $X^1 = X \cup M\left(Y\right)$.

Theorem 13 claims that $\ker(G) \subseteq \mathrm{core}(G)$ for every graph.

Theorem 14. *[20] If G is a bipartite graph, then* $\ker(G) = \mathrm{core}(G)$.

Proof (Alternative Proof). The assertions are clearly true, whenever $\mathrm{core}(G) = \emptyset$, i.e., for G having a perfect matching. Assume that $\mathrm{core}(G) \neq \emptyset$.

Let $S \in \Omega\left(G\right)$ and M be a maximum matching. By Theorem 2*(i)*, M matches $V\left(G\right) - S$ into S, and $N(\mathrm{core}(G))$ into $\mathrm{core}(G)$.

According to Theorem 9*(ii)*, it is sufficient to show that there is no set $Z \subseteq N\left(\mathrm{core}(G)\right)$, $Z \neq \emptyset$, such that $\left|N\left(Z\right) \cap \mathrm{core}(G)\right| = \left|Z\right|$.

Suppose, to the contrary, that there exists a non-empty set $Z \subseteq N\left(\mathrm{core}(G)\right)$ such that $\left|N\left(Z\right) \cap \mathrm{core}(G)\right| = \left|Z\right|$. Let Z_0 be a minimal non-empty subset of $N\left(\mathrm{core}(G)\right)$ enjoying this equality.

Clearly, $H = G\left[Z_0 \cup M\left(Z_0\right)\right]$ is bipartite, because it is a subgraph of a bipartite graph. Moreover, the restriction of M on H is a perfect matching.

Claim 1. Z_0 is independent.

Since H is a bipartite graph with a perfect matching, it has two maximum independent sets at least. Hence there exists $W \in \Omega(H)$ different from $M(Z_0)$. Thus $W \cap Z_0 \neq \emptyset$. Therefore, $N(W \cap Z_0) \cap \mathrm{core}(G) = M(W \cap Z_0)$. Consequently,

$$|N(W \cap Z_0) \cap \mathrm{core}(G)| = |M(W \cap Z_0)| = |W \cap Z_0|.$$

Finally, $W \cap Z_0 = Z_0$, because Z_0 has been chosen as a minimal subset of $N(\mathrm{core}(G))$ such that $|N(Z_0) \cap \mathrm{core}(G)| = |Z_0|$. Since $|Z_0| = \alpha(H) = |W|$ we conclude with $W = Z_0$, which means, in particular, that Z_0 is independent.

Claim 2. H is a connected graph.

Otherwise, for any connected component of H, say $\tilde{H}$, the set $V\left(\tilde{H}\right) \cap Z_0$ contradicts the minimality property of Z_0.

Claim 3. $Z_0 \cup (\mathrm{core}(G) - M(Z_0))$ is independent.

By Claim 1, Z_0 is independent. The equality $|N(Z_0) \cap \mathrm{core}(G)| = |Z_0|$ implies $N(Z_0) \cap \mathrm{core}(G) = M(Z_0)$, which means that there are no edges connecting Z_0 and $\mathrm{core}(G) - M(Z_0)$. Consequently, $Z_0 \cup (\mathrm{core}(G) - M(Z_0))$ is independent.

Claim 4. $Z_0 \cup (\mathrm{core}(G) - M(Z_0))$ is included in a maximum independent set.

Let $Z_i = M((N(Z_{i-1}) \cap S) - M(Z_{i-1})), 1 \leq i < \infty$. By Lemma 1 all the sets $Z^i = \bigcup\limits_{0 \leq j \leq i} Z_j, 1 \leq i < \infty$ are independent. Define

$$Z^{\infty} = \bigcup_{0 \leq i \leq \infty} Z_i,$$

which is, actually, the largest set in the sequence $\{Z^i, 1 \leq i < \infty\}$ (Fig. 6).

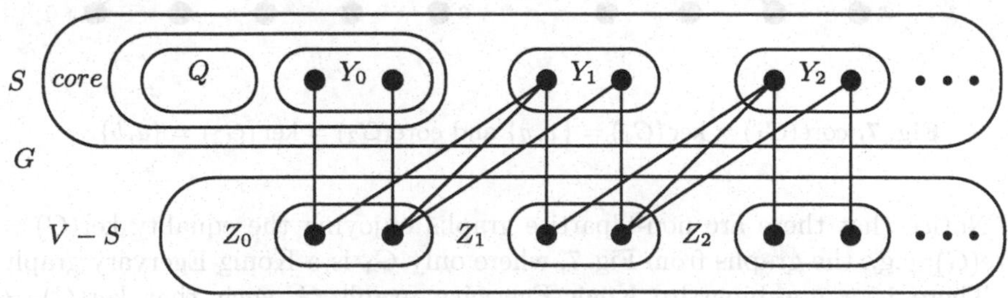

Fig. 6. $S \in \Omega(G)$, $Q = \mathrm{core}(G) - M(Z_0)$, $Y_0 = M(Z_0)$, $Y_1 = (N(Z_0) - M(Z_0)) \cap S$, $Y_2 = \ldots$, and $Z_i = M(Y_i), i = 1, 2, \ldots$.

The inclusion

$$Z_0 \cup (\text{core}(G) - M(Z_0)) \subseteq (S - M(Z^\infty)) \cup Z^\infty$$

is justified by the definition of Z^∞.

Since $|M(Z^\infty)| = |Z^\infty|$ we obtain $|(S - M(Z^\infty)) \cup Z^\infty| = |S|$. According to the definition of Z^∞ the set

$$(N(Z^\infty) \cap S) - M(Z^\infty)$$

is empty. In other words, the set $(S - M(Z^\infty)) \cup Z^\infty$ is independent. Therefore, we arrive at

$$(S - M(Z^\infty)) \cup Z^\infty \in \Omega(G).$$

Thus $(S - M(Z^\infty)) \cup Z^\infty$ is a desired enlargement of the set $Z_0 \cup (\text{core}(G) - M(Z_0))$.

Claim 5. $\text{core}(G) \cap ((S - M(Z^\infty)) \cup Z^\infty) = \text{core}(G) - M(Z_0)$.

The only part of $(S - M(Z^\infty)) \cup Z^\infty$ that interacts with core(G) is the subset

$$Z_0 \cup (\text{core}(G) - M(Z_0)).$$

Hence we obtain

$$\begin{aligned} &\text{core}(G) \cap ((S - M(Z^\infty)) \cup Z^\infty) = \\ &\text{core}(G) \cap (Z_0 \cup (\text{core}(G) - M(Z_0))) = \text{core}(G) - M(Z_0). \end{aligned}$$

Since Z_0 is non-empty, by Claim 5 we arrive at the following contradiction

$$\text{core}(G) \textit{ is not a subset of } (S - M(Z^\infty)) \cup Z^\infty \in \Omega(G).$$

Finally, we conclude with the fact there is no set $Z \subseteq N(\text{core}(G)), Z \neq \emptyset$ such that $|N(Z) \cap \text{core}(G)| = |Z|$, which, by Theorem 9, means that core(G) and ker(G) coincide.

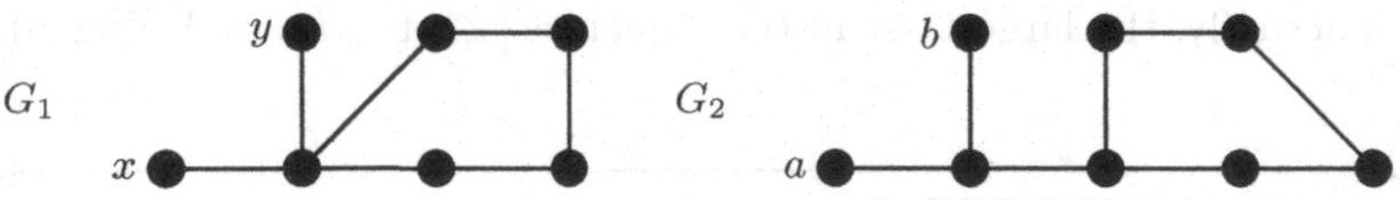

Fig. 7. $\text{core}(G_1) = \ker(G_1) = \{x, y\}$ and $\text{core}(G_2) = \ker(G_2) = \{a, b\}$.

Notice that there are non-bipartite graphs enjoying the equality $\ker(G) = \text{core}(G)$; e.g., the graphs from Fig. 7, where only G_1 is a König-Egerváry graph.

There is a non-bipartite König-Egerváry graph G, such that $\ker(G) \neq \text{core}(G)$. For instance, the graph G_1 from Fig. 8 has $\ker(G_1) = \{x, y\}$, while $\text{core}(G_1) = \{x, y, u, v\}$. The graph G_2 from Fig. 8 has $\ker(G_2) = \emptyset$, while $\text{core}(G_2) = \{w\}$.

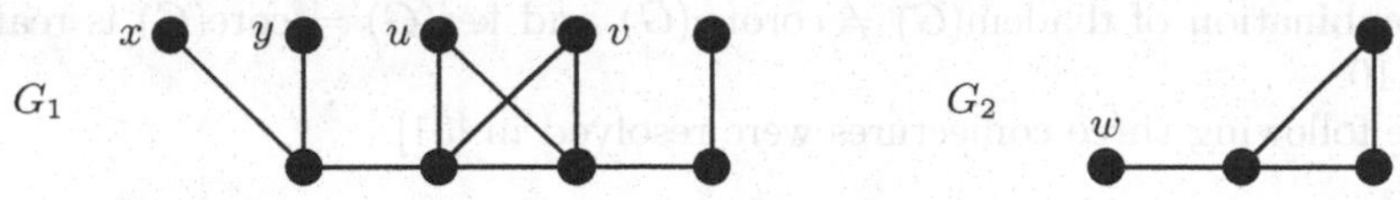

Fig. 8. Both G_1 and G_2 are König-Egerváry graphs. Only G_2 has a perfect matching.

5 Interrelationships Between ker (G), Nucleus(G), Diadem(G) and Corona(G)

There is a non-König-Egerváry graph G with $V(G) = N(\text{core}(G)) \cup \text{corona}(G)$; e.g., the graph G from Fig. 9.

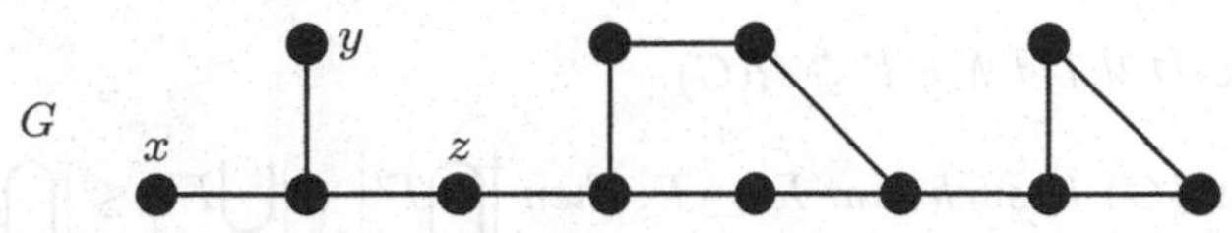

Fig. 9. G is not a König-Egerváry graph, and $\text{core}(G) = \{x, y, z\}$.

Theorem 15. *If G is a König-Egerváry graph, then*

(i) *[23]* $|\text{corona}(G)| + |\text{core}(G)| = 2\alpha(G)$;
(ii) *[25]* $\text{diadem}(G) = \text{corona}(G)$, *while* $\text{diadem}(G) \subseteq \text{corona}(G)$ *is true for every graph;*
(iii) *[25]* $|\ker(G)| + |\text{diadem}(G)| \leq 2\alpha(G)$.

Notice that the graph from Fig. 9 has $|\text{corona}(G)| + |\text{core}(G)| > 2\alpha(G)$. For a König-Egerváry graph with $|\ker(G)| + |\text{diadem}(G)| < 2\alpha(G)$, see Fig. 8. Figure 9 shows that a graph may have $\text{diadem}(G) \neq \text{corona}(G)$ and $\ker(G) \neq \text{core}(G)$.

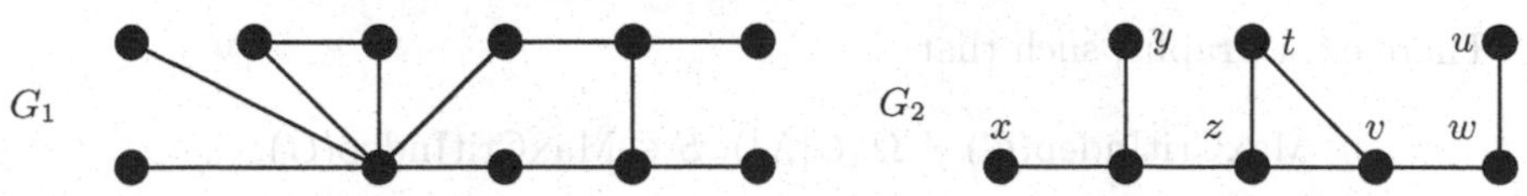

Fig. 10. G_1 is a non-bipartite König-Egerváry graph, such that $\ker(G_1) = \text{core}(G_1)$ and $\text{diadem}(G_1) = \text{corona}(G_1)$; G_2 is a non-König-Egerváry graph, such that $\ker(G_2) = \text{core}(G_2) = \{x, y\}$; $\text{diadem}(G_2) \cup \{z, t, v, w\} = \text{corona}(G_2)$.

The combination of $\text{diadem}(G) \neq \text{corona}(G)$ and $\ker(G) = \text{core}(G)$ is realized in Fig. 10.

The following three conjectures were resolved in [31].

Conjecture 2. [11,22] $|\ker(G)| + |\text{diadem}(G)| \leq 2\alpha(G)$ for every graph G.

Conjecture 3. [12] If $|\text{nucleus}(G)| + |\text{diadem}(G)| = 2\alpha(G)$, then G is a König-Egerváry graph.

Conjecture 4. [11] If $|\text{diadem}(G)| = |\text{corona}(G)|$, then G is a König-Egerváry graph.

Actually, all these conjectures are involved in a more general framework, where they appear as corollaries.

If Γ, Γ' are two set collections, we write $\Gamma' \preceq \Gamma$ if $\bigcup \Gamma' \subseteq \bigcup \Gamma$ and $\bigcap \Gamma \subseteq \bigcap \Gamma'$ [12].

Theorem 16. *[12] Let $\emptyset \neq \Gamma \subseteq \Omega(G)$.*

(i) *If $\Gamma' \subseteq \text{Ind}(G)$ is such that $\Gamma' \preceq \Gamma$, then* $\left|\bigcap \Gamma'\right| + \left|\bigcup \Gamma'\right| \leq \left|\bigcap \Gamma\right| + \left|\bigcup \Gamma\right|$.

(ii) $2\alpha(G) \leq \left|\bigcap \Gamma\right| + \left|\bigcup \Gamma\right|$.

(iii) *If, in addition, G is a König-Egerváry graph, then* $\left|\bigcap \Gamma\right| + \left|\bigcup \Gamma\right| = 2\alpha(G)$, *and, in particular,* $|\text{corona}(G)| + |\text{core}(G)| = 2\alpha(G)$.

Notice that if $S \in \text{Ind}(G)$, then $G[N[S]]$ is not necessarily a König-Egerváry graph.

Theorem 17. *[15] For every graph G, there is some $X \subseteq V(G)$, such that:*

(i) $X = N[S]$ *for every* $S \in \text{MaxCritIndep}(G)$*;*
(ii) $G[X]$ *is a König-Egerváry graph.*

In other words, Theorem 17*(i)* claims that $X = N[S]$ does not depend on the choice of $S \in \text{MaxCritIndep}(G)$.

Lemma 2. *If* $S \in \text{MaxCritIndep}(G)$ *and* $X = N[S]$, *then* $\text{MaxCritIndep}(G) \preceq \Omega(G[X])$.

There exist graphs, such that

$$\text{MaxCritIndep}(G) \neq \Omega(G[X]), S \in \text{MaxCritIndep}(G),$$

and $X = N[S]$.

Corollary 5. *[31] If* $S \in \text{MaxCritIndep}(G)$ *and* $X = N[S]$, *then*

$$\text{diadem}(G) \subseteq \text{diadem}(G[X]) \text{ and } \text{nucleus}(G[X]) \subseteq \text{nucleus}(G).$$

The *critical independence number* is

$$\alpha'(G) = \max\{|S| : S \in \text{MaxCritIndep}(G)\}$$

[15].

Lemma 3. *[26] If* $\emptyset \neq \Gamma' \subseteq \text{MaxCritIndep}(G)$ *and* $\emptyset \neq \Gamma \subseteq \Omega(G)$, *then*

$$\left|\bigcap \Gamma'\right| + \left|\bigcup \Gamma'\right| \leq 2\alpha'(G) \leq 2\alpha(G) \leq \left|\bigcap \Gamma\right| + \left|\bigcup \Gamma\right|.$$

If $\Gamma' = \text{MaxCritIndep}(G)$ and $\Gamma = \Omega(G)$, Lemma 3 implies the following.

Corollary 6. *[31]* $|\text{nucleus}(G)| + |\text{diadem}(G)| \leq 2\alpha(G)$ *for every graph* G.

Since $\ker(G) \subseteq \text{nucleus}(G)$, Corollary 6 validates Conjecture 2. An alternative proof of Conjecture 2 may be found in [3].

A family $\Gamma \subseteq \text{Ind}(G)$ is a *König-Egerváry collection* if $\left|\bigcap \Gamma\right| + \left|\bigcup \Gamma\right| = 2\alpha(G)$ [12]. It is worth mentioning that $\Omega(G)$ may be a König-Egerváry collection, while G is not a König-Egerváry graph.

Theorem 18. *[26] For a graph* G, *the following assertions are equivalent:*

(i) *G is a König-Egerváry graph;*
(ii) *every non-empty family of maximum critical independent sets of G is a König-Egerváry collection;*
(iii) *there is a König-Egerváry collection of maximum critical independent sets of G.*

Since $|\text{nucleus}(G)| + |\text{diadem}(G)| = 2\alpha(G)$ means that $\text{MaxCritIndep}(G)$ is a König-Egerváry collection, Theorem 18 implies the validity of Conjecture 3.

Corollary 7. *[31] If* $|\text{nucleus}(G)| + |\text{diadem}(G)| = 2\alpha(G)$, *then G is a König-Egerváry graph.*

If $\emptyset \neq \Gamma \subseteq \Omega(G)$, then none of $\bigcap \Gamma$ and $\bigcup \Gamma$ is necessarily critical.

Proposition 5. *[26] Let* $\Gamma \subseteq \Omega(G)$ *and* $\emptyset \neq \Gamma' \subseteq \text{MaxCritIndep}(G)$ *be such that for every* $A \in \Gamma'$ *there exists* $S \in \Gamma$ *such that* $A \subseteq S$. *If* $\bigcup \Gamma' = \bigcup \Gamma$, *then G is a König-Egerváry graph.*

If $\Gamma' = \text{MaxCritIndep}(G)$ and $\Gamma = \Omega(G)$, Proposition 5 immediately implies the validity of Conjecture 4.

Corollary 8. *[31] If* $\text{diadem}(G) = \text{corona}(G)$, *then G is a König-Egerváry graph.*

6 Conclusions

Theorem 14 claims that the equality $\ker(G) = \text{core}(G)$ is true for bipartite graphs.

Problem 1. [20] Characterize graphs with $\ker(G) = \text{core}(G)$.

By Corollary 2, $\text{core}(G)$ is critical for every König-Egerváry graph.

Problem 2. [11] Characterize graphs, where $\text{core}(G)$ is a critical set.

Conjecture 5. If $\text{core}(G)$ is a critical set, then $\text{core}(G) = \text{nucleus}(G)$.

By Theorem 4, for König-Egerváry graphs $\text{core}(G) = \text{nucleus}(G)$.

Problem 3. [12] Characterize graphs with $\text{core}(G) = \text{nucleus}(G)$.

For König-Egerváry graphs $\text{corona}(G)$ is critical in accordance with Theorem 4 and Theorem 5*(ii)*.

Problem 4. [25] Characterize graphs such that $\text{corona}(G)$ is a critical set.

By Theorem 16, every subcollection of a König-Egerváry collection of maximum independent sets is König-Egerváry as well.

Problem 5. [12] Characterize the graphs such that every collection of maximum independent sets is König-Egerváry. In other words, characterize the graphs such that $|\text{corona}(G)| + |\text{core}(G)| = 2\alpha(G)$.

Theorem 3 says that $d(G) = \alpha(G) - \mu(G)$ for König-Egerváry graphs.

Problem 6. Characterize graphs satisfying $d(G) = \alpha(G) - \mu(G)$.

Acknowledgments. The first author would like to thank the organizers of the Mathematical Optimization Theory and Operations Research Conference - MOTOR2019 for an opportunity to deliver an invited lecture on critical independent sets.

References

1. Ageev, A.A.: On finding critical independent and vertex sets. SIAM J. Discret. Math. **7**, 293–295 (1994)
2. Beckenbach, I., Borndörfer, R.: Hall's and König's theorem in graphs and hypergraphs. Discret. Math. **341**, 2753–2761 (2018)
3. Bhattacharya, A., Mondal, A., Murthy, T.S.: Problems on matchings and independent sets of a graph. Discret. Math. **341**, 1561–1572 (2018)
4. Bonomo, F., Dourado, M.C., Durán, G., Faria, L., Grippo, L.N., Safe, M.D.: Forbidden subgraphs and the König-Egerváry property. Discret. Appl. Math. **161**, 2380–2388 (2013)
5. Boros, E., Golumbic, M.C., Levit, V.E.: On the number of vertices belonging to all maximum stable sets of a graph. Discret. Appl. Math. **124**, 17–25 (2002)

6. Butenko, S., Trukhanov, S.: Using critical sets to solve the maximum independent set problem. Oper. Res. Lett. **35**, 519–524 (2007)
7. DeLaVina, E.: Written on the Wall II, Conjectures of Graffiti.pc. http://cms.dt.uh.edu/faculty/delavinae/research/wowII/
8. DeLaVina, E., Larson, C.E.: A parallel algorithm for computing the critical independence number and related sets. Ars Math. Contemp. **6**, 237–245 (2013)
9. Deming, R.W.: Independence numbers of graphs - an extension of the König-Egerváry theorem. Discret. Math. **27**, 23–33 (1979)
10. Garey, M., Johnson, D.: Computers and Intractability, 1st edn. W. H. Freeman and Company, New York (1979)
11. Jarden, A., Levit, V.E., Mandrescu, E.: Critical and maximum independent sets of a graph. Discret. Appl. Math. **247**, 127–134 (2018)
12. Jarden, A., Levit, V.E., Mandrescu, E.: Monotonic properties of collections of maximum independent sets of a graph, Order (2018). https://link.springer.com/article/10.1007/s11083-018-9461-8
13. Korach, E., Nguyen, T., Peis B.: Subgraph characterization of red/blue-split graphs and König-Egerváry graphs. In: Proceedings of the Seventeenth Annual ACM-SIAM Symposium on Discrete Algorithms, pp. 842–850. ACM Press (2006)
14. Larson, C.E.: A note on critical independence reductions. Bull. Inst. Comb. Appl. **5**, 34–46 (2007)
15. Larson, C.E.: The critical independence number and an independence decomposition. Eur. J. Comb. **32**, 294–300 (2011)
16. Levit, V.E., Mandrescu, E.: Combinatorial properties of the family of maximum stable sets of a graph. Discret. Appl. Math. **117**, 149–161 (2002)
17. Levit, V.E., Mandrescu, E.: On α^{+}-stable König-Egerváry graphs. Discret. Math. **263**, 179–190 (2003)
18. Levit, V.E., Mandrescu, E.: Vertices belonging to all critical independent sets of a graph. SIAM J. Discret. Math. **26**, 399–403 (2012)
19. Levit, V.E., Mandrescu, E.: Critical independent sets and König-Egerváry graphs. Graphs Comb. **28**, 243–250 (2012)
20. Levit, V.E., Mandrescu, E.: Critical sets in bipartite graphs. Ann. Comb. **17**, 543–548 (2013)
21. Levit, V.E., Mandrescu, E.: On the structure of the minimum critical independent set of a graph. Discret. Math. **313**, 605–610 (2013)
22. Levit, V.E., Mandrescu, E.: Critical independent sets in a graph. In: 3rd International Conference on Discrete Mathematics, 10–14 June 2013, Karnatak University, Dharwad, India (2013)
23. Levit, V.E., Mandrescu, E.: A set and collection lemma. Electron. J. Comb. **21**, #P1.40 (2014)
24. Levit, V.E., Mandrescu, E.: Critical independent sets of a graph. arXiv:1407.7368 [cs.DM], 15 p. (2014)
25. Levit, V.E., Mandrescu, E.: Intersections and unions of critical independent sets in bipartite graphs. Bulletin mathématique de la Société des Sciences Mathématiques de Roumanie **57**, 257–260 (2016)
26. Levit, V.E., Mandrescu, E.: On König-Egerváry collections of maximum critical independent sets. Art Discret. Appl. Math. **2**, #P1.02 (2019)
27. Lorentzen, L.C.: Notes on covering of arcs by nodes in an undirected graph, Technical report ORC 66-16, Operations Research Center, University of California, Berkeley, California (1966)
28. Lovász, L., Plummer, M.D.: Matching Theory, Annals of Discrete Mathematics, vol. 29, North-Holland, Amsterdam (1986)

29. Schrijver, A.: Combinatorial Optimization. Springer, Berlin (2003)
30. Short, T.M.: KE Theory & the number of vertices belonging to all maximum independent sets in a graph, M.Sc. thesis, Virginia Commonwealth University (2011)
31. Short, T.M.: On some conjectures concerning critical independent sets of a graph. Electron. J. Comb. **23**, #P2.43 (2016)
32. Sterboul, F.: A characterization of the graphs in which the transversal number equals the matching number. J. Comb. Theory B **27**, 228–229 (1979)
33. Trukhanov, S.: Novel approaches for solving large-scale optimization problems on graphs, Ph.D. thesis, University of Texas (2008)
34. Zhang, C.Q.: Finding critical independent sets and critical vertex subsets are polynomial problems. SIAM J. Discret. Math. **3**, 431–438 (1990)

Mathematical Programming

On Generating Nonconvex Optimization Test Problems

Maria V. Barkova(✉)

Matrosov Institute for System Dynamics and Control Theory of SB of RAS, Irkutsk 664033, Russia

mbarkova@icc.ru

Abstract. This paper addresses a technique for generating two types of nonconvex test problems. We study quadratic problems with d.c. inequality constraints and sum-of-ratios programs where both numerators and denominators are quadratic functions. Based on the idea of P. Calamai and L. Vicente, we propose the procedures for constructing nonconvex test problems with quadratic functions of any dimension, where global and local solutions are known. The implementation of the procedures does not require any complicated operations and solving auxiliary problems, except for elementary operations with matrices and vectors.

Keywords: Nonconvex test problems · Quadratic function · Fractional program · d.c. functions

1 Introduction

Test problems play an important role in computational testing of numerical methods. They help verify the efficiency of algorithm and allow us to compare it with other methods. Test problems often come from two sources: (pseudo)random generators and existing test collections (libraries). As for randomly generated problems, we usually do not know the properties of problems, such as the number of stationary points, the number of local and global solutions. Moreover, we often do not even know if these solutions actually exist. The libraries of test problems offer us specific classes of problems of given dimension. For instance, there are Floudas and Pardalos' collection [3], COCONUT Benchmark [14], DEGEN collection, etc. But it is rather difficult to find test instances of required dimension for minimizing a quadratic function with quadratic inequality constraints in these collections. Let alone instances with nonconvex functions both in the objective function and in the constraints. The situation remains the same with test problems for fractional optimization. There are some examples with affine functions [8] or with quadratic functions, but with the small number of fractions [6].

Thus, this paper was motivated by the necessity to have a test problem collection for nonconvex problems with quadratic functions. We consider a method of generating nonconvex test problems based on the technique proposed by Calamai

M. Khachay et al. (Eds.): MOTOR 2019, LNCS 11548, pp. 21–33, 2019.
https://doi.org/10.1007/978-3-030-22629-9_2

and Vicente [1,2,13]. The idea of their method is to construct a "big" problem of the desired dimension by combining a finite number of low-dimensional problems. These so-called kernel problems are rather simple, and we can find all their local and global solutions. In addition, this technique does not require solving auxiliary problems or systems of equations. Thus, we can construct test instances of any dimension with needed properties and known local and global solutions.

In the paper we propose a method for generating nonconvex quadratic problems in the following form:

$$(\mathcal{P}): \qquad \begin{cases} f_0(x) := \langle x, Q^0 x\rangle + \langle b^0, x\rangle + d_0 \downarrow \min\limits_x, \ x \in S, \\ f_i(x) := \langle x, Q^i x\rangle + \langle b^i, x\rangle + d_i \leq 0, \ \ i \in \mathcal{I} = \{1, ..., N\}, \end{cases}$$

where $S \subset I\!R^n$ is a closed convex set, $Q^i \in I\!R^{n\times n}$ are indefinite, symmetric matrices, and $x, b^i \in I\!R^n$, $d_i \in I\!R$, $i = 0, ..., N$.

Furthermore, we propose a method for constructing fractional programs in the following form:

$$(\mathcal{FP}): \qquad \sum_{i=1}^{m} \frac{\psi_i(x)}{\varphi_i(x)} = \sum_{i=1}^{m} \frac{\langle x, A_i x\rangle + \langle p^i, x\rangle + q_i}{\langle x, B_i x\rangle + \langle c^i, x\rangle + t_i} \downarrow \min_x, \ \ x \in S,$$

where $A_i, B_i \in I\!R^{n\times n}$, $x, p^i, c^i \in I\!R^n$, $q_i, t_i \in I\!R$, $i = 1, \ldots, m$, and ψ_i, φ_i such that

$$(\mathcal{H}_0): \qquad \psi_i(x) > 0, \ \varphi_i(x) > 0 \ \ \forall x \in S, \ \ i = 1, \ldots, m.$$

In contrast to the techniques for generating test quadratic programs proposed in [1] or in [10,11], we construct problems with nonconvex quadratic inequality constraints $(\mathcal{P})$. Moreover, due to some easy additional operations, we can generate fractional test problems $(\mathcal{FP})$.

2 Quadratic Program Generating Scheme

The proposed method of the generation of the nonconvex quadratic test problem $(\mathcal{P})$ consists of three stages [1]. The first one includes the construction of low-dimensional kernel problems and the analytical search for all local and global solutions of these problems. At the second stage, a separable problem of the required dimension is constructed by merging a finite number of kernel problems having different properties. Finally, the separable problem is transformed in order to eliminate the separability of constructed problem. Each of these stages is described below.

2.1 Kernel Problems

In this section, we describe the classes of nonconvex quadratic kernel problems. Each class possesses its own properties. These classes will be combined to generate a "big" separable problem.

Define kernel problems in the following way:

$$(\mathcal{P}_k): \quad \begin{cases} f_0(x) = -p_k x_1(x_1 - 3) - 0.5 + x_2 \downarrow \min\limits_x, \\ f_1(x) = -x_1^2 - x_2 \le 0, \\ f_2(x) = (x_1 - 1)^2 - x_2 - 2.5 \le 0, \end{cases}$$

where $x \in I\!R^2, p_k \in \{p_1, p_2, p_3\}$.

The problems $(\mathcal{P}_k)$ are nonconvex quadratic problems in the space $I\!R^2$ with parameters p_k, $k = 1, 2, 3$, in the objective function. All stationary points of problems $(\mathcal{P}_k)$ were obtained analytically using KKT-theorem. They are illustrated in Fig. 1.

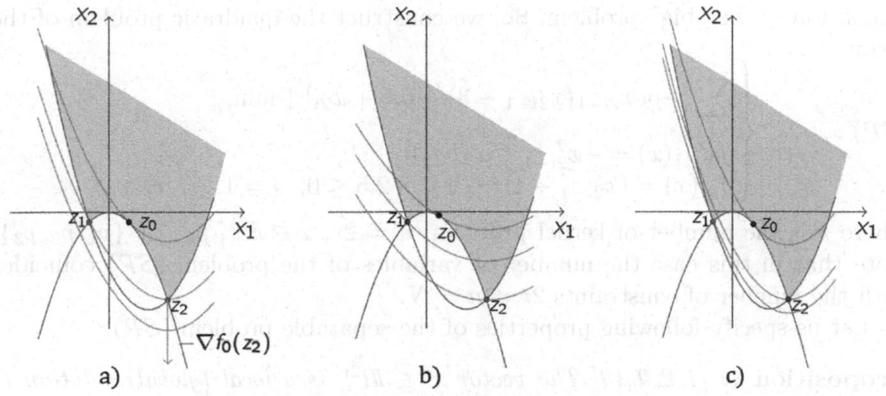

Fig. 1. Three classes of kernel problems

Analytical solutions for all types of kernel problem are provided below. There are three following classes to consider.

Class 1 ($p_1 = 0.5$):

$$z_0 = (1/2, -1/4), \quad f_0(z_0) = -0.125;$$

$$z_1 = (-1/2, -1/4), \quad f_0(z_1) = -1.625;$$

$$z_2 = (3/2, -9/4), \quad f_0(z_2) = -1.625.$$

Points z_1, z_2 are the global solutions, z_0 is the stationary point (see Fig. 1(a)).

Class 2 ($p_2 = 0.25$):

$$z_0 = (3/10, -9/100), \quad f_0(z_0) = -1.7375;$$

$$z_1 = (-1/2, -1/4), \quad f_0(z_1) = -1.1875;$$

$$z_2 = (3/2, -9/4), \quad f_0(z_2) = -2.1875.$$

Point z_2 is the global solution, z_1, z_0 are the stationary ones (see Fig. 1(b)).

$$
\begin{aligned}
&\text{Class 3 } (p_3 = 0.6):\\
z_0 &= (9/16, -81/256), \quad f_0(z_0) = -0.3164;\\
z_1 &= (-1/2, -1/4), \quad f_0(z_1) = -1.8;\\
z_2 &= (3/2, -9/4), \quad f_0(z_2) = -1.4.
\end{aligned}
$$

Point z_1 is the global solution, z_2, z_0 are the stationary points (see Fig. 1(c)).

Further, these kernel problems will be used to construct a separable problem.

2.2 Separable Quadratic Problem

At the second stage of the method, a separable problem is generated by combining a finite number of different classes of kernel problems. Let n is desired dimension of the "big" problem. So, we construct the quadratic problem of the form:

$$
(\mathcal{SP}): \quad \begin{cases} \sum\limits_{i=1}^{r} [-p_k x_{2i-1}(x_{2i-1} - 3) - 0.5 + x_{2i}] \downarrow \min\limits_{x}, \\ f_{2i-1}(x) = -x_{2i-1}^2 - x_{2i} \leq 0, \\ f_{2i}(x) = (x_{2i-1} - 1)^2 - x_{2i} - 2.5 \leq 0, \quad i = 1, \ldots, r, \end{cases}
$$

where r is the number of kernel problem, $n = 2r$, $x \in I\!R^n$, $p_k \in \{p_1, p_2, p_3\}$. Note that in this case the number of variables of the problem $(\mathcal{SP})$ coincides with the number of constraints $2r = n = N$.

Let us specify following properties of the separable problem $(\mathcal{SP})$.

Proposition 1. *[1,2,9,13] The vector $x^* \in I\!R^{2r}$ is a local (global) solution to the problem $(\mathcal{SP})$ if and only if its components are local (global) solutions to the problems $(\mathcal{P}_k)$, $k = 1, 2, 3$.*

Proposition 2. *[1,2,9,13] The problem $(\mathcal{SP})$ that includes r_1 kernel problems of the class 1, r_2 kernel problems of the class 2, and r_3 kernel problems of the class 3, so that $r_1 + r_2 + r_3 = r$, has $3^{r_1+r_2+r_3}$ stationary points among which 2^{r_1} solutions are global solutions to this problem.*

The problem $(\mathcal{SP})$ can be reformulated as follows:

$$
(\mathcal{QP}): \quad \begin{cases} f_0(x) = \langle x, Q_0 x \rangle + \langle b_0, x \rangle + d_0 \downarrow \min\limits_{x}, \\ f_i(x) = \langle x, Q_i x \rangle + \langle b_i, x \rangle + d_i \leq 0, \quad i \in \mathcal{I} = \{1, ..., 2r\}, \end{cases}
$$

where

$$
Q_0 = \begin{pmatrix} -p_k & & & & & \\ & 0 & & & (0) & \\ & & -p_k & & & \\ & & & 0 & & \\ & & & & \ddots & \\ & (0) & & & -p_k & \\ & & & & & 0 \end{pmatrix}, \quad b_0 = \begin{pmatrix} 3p_k \\ 1 \\ 3p_k \\ 1 \\ \vdots \\ 3p_k \\ 1 \end{pmatrix}, \quad d_0 = -0.5r;
$$

and for odd indexes i, $i \in \mathcal{I}$:

$$Q_i = \begin{pmatrix} 0 & & & & \\ & \ddots & & (0) & \\ & & -1 & & \\ & (0) & & \ddots & \\ & & & & 0 \end{pmatrix}_i, \quad b_i = \begin{pmatrix} 0 \\ \vdots \\ -1 \\ \vdots \\ 0 \end{pmatrix}_i, \quad d_i = 0;$$

for even indexes i, $i \in \mathcal{I}$:

$$Q_i = \begin{pmatrix} 0 & & & & \\ & \ddots & & (0) & \\ & & 1 & & \\ & (0) & & \ddots & \\ & & & & 0 \end{pmatrix}_i, \quad b_i = \begin{pmatrix} 0 \\ \vdots \\ -2 \\ -1 \\ \vdots \\ 0 \end{pmatrix}\begin{matrix} \\ \\ i-1 \\ i \\ \\ \\ \end{matrix}, \quad d_i = -1.5;$$

$Q_i \in I\!R^{2r \times 2r}, \;\; b_i \in I\!R^{2r}, \; d_i \in I\!R, \; i = 0, \ldots, 2r.$

2.3 Transformation of the Separable Problem

At the final stage of the method, it is necessary to get rid of the separability to expand the variety of the constructed problem. For this purpose, we use the substitution

$$x = M \cdot M^{-1} x,$$

where $M \in I\!R^{2r \times 2r} : detM \neq 0$, and replacement of a variable:

$$z = M^{-1} x.$$

Applying the above-described transformation to an arbitrary problem of the class $(\mathcal{QP})$, we obtain the following problem:

$$(\mathcal{QP}'): \quad \begin{cases} f_0(z) = \langle z, M^T Q_0 M z \rangle + \langle M^T b_0, z \rangle + d_0 \downarrow \min\limits_{z}, \\ f_i(z) = \langle z, M^T Q_i M z \rangle + \langle M^T b_i, z \rangle + d_i \leq 0, \;\; i \in \mathcal{I}. \end{cases}$$

Using the standard definitions of the linear algebra [4], one can easily show that the matrices Q_i remain indefinite after the transformation $M^T Q_i M$, $i = 0, ..., 2r$. Therefore, the problem $(\mathcal{QP}')$ remains in the class of nonconvex quadratic problems.

Let us describe one of the methods of constructing the matrix M. First, calculate a random Householder matrix [7] satisfying

$$H = E - \frac{2}{\langle h, h \rangle} hh^T, \tag{1}$$

where E is the identity matrix, $h \in I\!R^{2r}$ is an arbitrary nonzero vector.

And then the matrix M is in the following form:

$$M = \Lambda \cdot H, \tag{2}$$

where Λ is random positive defined diagonal matrix $(2r \times 2r)$.

Therefore, inverse matrices with respect to M can be obtained by

$$M^{-1} = H \cdot \Lambda^{-1} = W.$$

Proposition 3. *[1] The problem $(\mathcal{P})$ in the variables $z \in I\!R^{2r}$ is equivalent to the problem $(\mathcal{Q}P')$ in the variables $\bar{z} \in I\!R^{2r}$ under the nonsingular transformation $\bar{z} = Wz$.*

2.4 A Simple Example

Let us construct a low-dimension example to demonstrate how the method can be used to generate nonconvex quadratic problems.

Suppose that $n = 4$, $r = 2$, $r_1 = 1$, $r_2 = 1$ and $r_3 = 0$, i.e. we choose one problem from the first class of kernel problem and another one from the second class. This corresponds to the following separable quadratic problem:

$$\left.\begin{aligned} f_0(x) = -0.5x_1^2 + 1.5x_1 + x_2 - 0.25x_3^2 + 0.75x_3 + x_4 - 1 \downarrow \min_x,\\ f_1(x) = -x_1^2 - x_2 \le 0,\\ f_2(x) = (x_1 - 1)^2 - x_2 - 2.5 \le 0,\\ f_3(x) = -x_3^2 - x_4 \le 0,\\ f_4(x) = (x_3 - 1)^2 - x_4 - 2.5 \le 0. \end{aligned}\right\} \tag{3}$$

The problem (3) is a nonconvex quadratic 4-dimension problem with 2 nonconvex and 2 convex quadratic constrains. Since $r = 2$, $r_1 = 1$, $r_2 = 1$, $r_3 = 0$, the problem (3) has $3^{r_1+r_2+r_3} = 3^2 = 9$ stationary points including $2^{r_1} = 2^1 = 2$ global solutions, namely $z_1^* = (-1/2, -1/4, 3/2, -9/4)$ and $z_2^* = (3/2, -9/4, 3/2, -9/4)$.

Then after using transformations (1)–(2) with the following parameters

$$\Lambda = \begin{pmatrix} 9 & 0 & 0 & 0 \\ 0 & 3 & 0 & 0 \\ 0 & 0 & 10 & 0 \\ 0 & 0 & 0 & 4 \end{pmatrix}, \quad h = \begin{pmatrix} -6 \\ -5 \\ 2 \\ -1 \end{pmatrix}, \quad M = \begin{pmatrix} -0.82 & -8.18 & 3.27 & -1.64 \\ -2.73 & 0.73 & 0.91 & -0.45 \\ 3.64 & 3.03 & 8.79 & 0.61 \\ -0.73 & -0.61 & 0.24 & 3.88 \end{pmatrix},$$

the transformed nonconvex quadratic problem ($\mathcal{QP}'$) would be:

$$f_0(\bar{x}) = \begin{pmatrix}\bar{x}_1\\ \bar{x}_2\\ \bar{x}_3\\ \bar{x}_4\end{pmatrix}^T \begin{pmatrix} -3.64 & -6.10 & -6.65 & -1.22\\ -6.10 & -35.76 & 6.73 & -7.15\\ -6.65 & 6.73 & -24.66 & 1.34\\ -1.22 & -7.15 & 1.34 & -1.43\end{pmatrix} \begin{pmatrix}\bar{x}_1\\ \bar{x}_2\\ \bar{x}_3\\ \bar{x}_4\end{pmatrix} - \begin{pmatrix}-1.95\\ -9.87\\ 12.65\\ 1.42\end{pmatrix}^T \begin{pmatrix}\bar{x}_1\\ \bar{x}_2\\ \bar{x}_3\\ \bar{x}_4\end{pmatrix} - 1 \downarrow \min_{\bar{x}},$$

$$f_1(\bar{x}) = \begin{pmatrix}\bar{x}_1\\ \bar{x}_2\\ \bar{x}_3\\ \bar{x}_4\end{pmatrix}^T \begin{pmatrix} -0.66 & -6.69 & 2.67 & -1.33\\ -6.69 & -66.94 & 26.77 & -13.38\\ 2.67 & 26.77 & -10.71 & 5.35\\ -1.33 & -13.38 & 5.35 & -2.67\end{pmatrix} \begin{pmatrix}\bar{x}_1\\ \bar{x}_2\\ \bar{x}_3\\ \bar{x}_4\end{pmatrix} - \begin{pmatrix}2.72\\ -0.72\\ -0.90\\ 0.45\end{pmatrix}^T \begin{pmatrix}\bar{x}_1\\ \bar{x}_2\\ \bar{x}_3\\ \bar{x}_4\end{pmatrix} \leq 0,$$

$$f_2(\bar{x}) = \begin{pmatrix}\bar{x}_1\\ \bar{x}_2\\ \bar{x}_3\\ \bar{x}_4\end{pmatrix}^T \begin{pmatrix} 0.66 & 6.69 & -2.67 & 1.33\\ 6.69 & 66.94 & -26.77 & 13.38\\ -2.67 & -26.77 & 10.71 & -5.35\\ 1.33 & 13.38 & -5.35 & 2.67\end{pmatrix} \begin{pmatrix}\bar{x}_1\\ \bar{x}_2\\ \bar{x}_3\\ \bar{x}_4\end{pmatrix} - \begin{pmatrix}4.36\\ 15.63\\ -7.45\\ 3.72\end{pmatrix}^T \begin{pmatrix}\bar{x}_1\\ \bar{x}_2\\ \bar{x}_3\\ \bar{x}_4\end{pmatrix} \leq 1.5,$$

$$f_3(\bar{x}) = \begin{pmatrix}\bar{x}_1\\ \bar{x}_2\\ \bar{x}_3\\ \bar{x}_4\end{pmatrix}^T \begin{pmatrix} -13.22 & -11.01 & -31.95 & -2.20\\ -11.01 & -9.18 & -26.62 & -1.83\\ -31.95 & -26.62 & -77.22 & -5.32\\ -2.20 & -1.83 & -5.32 & -0.36\end{pmatrix} \begin{pmatrix}\bar{x}_1\\ \bar{x}_2\\ \bar{x}_3\\ \bar{x}_4\end{pmatrix} - \begin{pmatrix}0.72\\ 0.60\\ -0.24\\ -3.87\end{pmatrix}^T \begin{pmatrix}\bar{x}_1\\ \bar{x}_2\\ \bar{x}_3\\ \bar{x}_4\end{pmatrix} \leq 0,$$

$$f_4(\bar{x}) = \begin{pmatrix}\bar{x}_1\\ \bar{x}_2\\ \bar{x}_3\\ \bar{x}_4\end{pmatrix}^T \begin{pmatrix} 13.22 & 11.01 & 31.95 & 2.20\\ 11.01 & 9.18 & 26.62 & 1.83\\ 31.95 & 26.62 & 77.22 & 5.32\\ 2.20 & 1.83 & 5.32 & 0.36\end{pmatrix} \begin{pmatrix}\bar{x}_1\\ \bar{x}_2\\ \bar{x}_3\\ \bar{x}_4\end{pmatrix} - \begin{pmatrix}-6.54\\ -5.45\\ -17.81\\ -5.09\end{pmatrix}^T \begin{pmatrix}\bar{x}_1\\ \bar{x}_2\\ \bar{x}_3\\ \bar{x}_4\end{pmatrix} \leq 1.5.$$

As we can see, we have obtained a nonconvex quadratic problem with nonconvex inequality constraints. All transformed matrices are indefinite and dense.

The two global minima for this problem are:

$$\bar{z}_1 = (0.238, 0.161, 0.052, -0.514),$$
$$\bar{z}_2 = (0.823, -0.203, -0.069, -0.453),$$

with value $f_0(\bar{z}_1) = f_0(\bar{z}_2) = -3.8125$.

3 Fractional Programming Test Problem

For describing the technique of construction fractional programming problem it is necessary to recall reduction theorem which shows the relations between fractional and d.c. minimization problem. This reduction will be need to produce the fractional problem ($\mathcal{FP}$) from the quadratic in the test problem generation scheme.

In that purpose we consider the following d.c. minimization problem with vector parameter α

$$(\mathcal{P}_\alpha): \qquad \Phi(x,\alpha) := \sum_{i=1}^{m}[\psi_i(x) - \alpha_i\varphi_i(x)] \downarrow \min_x, \quad x \in S,$$

where $\alpha = (\alpha_1, \ldots, \alpha_m)^\top \in I\!R^m_+$.

Suppose, that the data of the problem $(\mathcal{FP})$ satisfies "the nonnegativity condition", so that the following inequalities hold

$$(\mathcal{H}_\alpha): \qquad \psi_i(x) - \alpha_i\varphi_i(x) \geq 0 \;\; \forall x \in S, \;\; i = 1, \ldots, m.$$

In addition, suppose that the following assumptions are fulfilled:

$$(\mathcal{H}_1): \quad \begin{cases} (a)\ \mathcal{V}(\alpha) > -\infty \;\; \forall \alpha \in \mathcal{K}, \text{where } \mathcal{K} \text{ is a convex set from } I\!R^m; \\ (b)\ \forall \alpha \in \mathcal{K} \subset I\!R^m \text{ there exists a solution } z = z(\alpha) \text{ to Problem } (\mathcal{P}_\alpha). \end{cases}$$

Theorem 1. *[5] Suppose that the assumptions $(\mathcal{H}_0)$, $(\mathcal{H}_1)$ are satisfied in the problem $(\mathcal{FP})$.*

In addition, let there exist a vector $\alpha_0 = (\alpha_{01}, \ldots, \alpha_{0m})^\top \in \mathcal{K} \subset I\!R^m$ at which "the nonnegativity condition" $(\mathcal{H}_{\alpha_0})$ holds.

Besides, suppose that in Problem $(\mathcal{P}_{\alpha_0})$ the following equality takes place

$$\mathcal{V}(\alpha_0) \stackrel{\triangle}{=} \min_x \left\{ \sum_{i=1}^{m}[\psi_i(x) - \alpha_{0i}\varphi_i(x)] \;:\; x \in S \right\} = 0. \tag{4}$$

Then, any solution $z = z(\alpha_0)$ to the problem $(\mathcal{P}_{\alpha_0})$ is a solution to Problem $(\mathcal{FP})$, so that $z \in Sol(\mathcal{P}_{\alpha_0}) \subset Sol(\mathcal{FP})$.

Let us turn to the generation scheme (Fig. 2). In order to generate one ratio of the problem $(\mathcal{FP})$, first of all, we should construct the quadratic problem of the required dimension with linear or box constraints by the scheme described in Sect. 2. It is well known that all quadratic functions can be represented as a

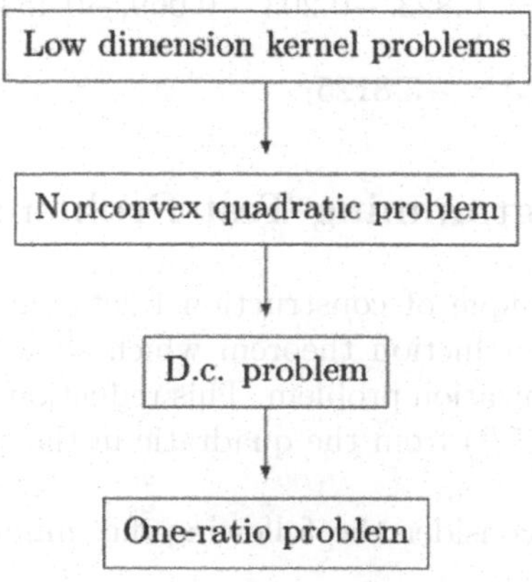

Fig. 2. Generation scheme

difference of two convex functions, i.e. d.c. functions [12]. Therefore, after that, we rewrite the constructed quadratic problem in a d.c. form. Then, in view of the reduction Theorem 1, we obtain one ratio with quadratic functions. By repeating this procedure m times we generate the sum-of-ratios problem $(\mathcal{FP})$ with m fractions.

3.1 Kernel Problems

Let us construct the following kernel problems:

$$(\mathcal{KP}_k) \qquad \begin{cases} f_k(x) \downarrow \min\limits_x, \\ x \in S_k, \quad k = 1, 2, 3, \end{cases}$$

$$\text{where } f_1(x) = -\frac{2}{3}x^2 + \frac{14}{3}x - 4, \quad S_1 = \{x \in I\!R \mid -2x + 2 \leq 0, x - 5.5 \leq 0\},$$

$$f_2(x) = -\frac{1}{2}x^2 + 3x - \frac{5}{2}, \quad S_2 = \{x \in I\!R \mid -2x + 2 \leq 0, x - 4.5 \leq 0\},$$

$$f_3(x) = -\frac{1}{4}x^2 + \frac{9}{4}x - 2, \quad S_3 = \{x \in I\!R \mid -2x + 2 \leq 0, x - 7.5 \leq 0\}.$$

These problems are nonconvex quadratic minimization problems with linear constraints, and their solutions are illustrated by Fig. 3.

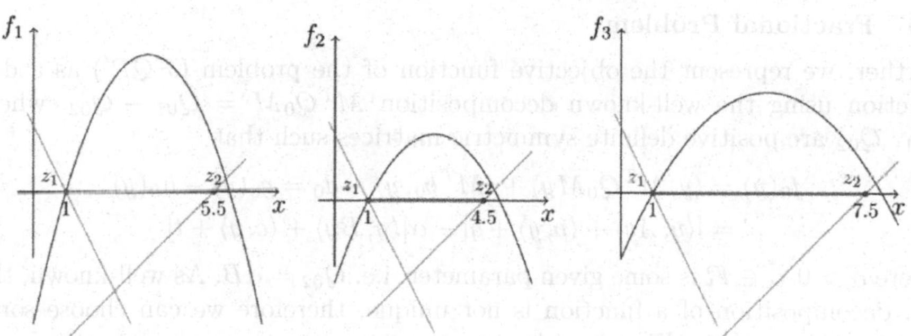

Fig. 3. Kernel problems

The bold line in Fig. 3 shows the feasible sets of the problems $(\mathcal{KP}_k)$, $k = 1, 2, 3$. The point z_1 is the global solution for each kernel problem and the point z_2 is the local one. It should be noted that, in order for the "nonnegative condition" (H_α) to be fulfilled, the optimal values of all problems are equal to zero and their global solutions coincide.

3.2 Separable Problem and Its Transformation

At the second stage of the method we combine a finite number of kernel problems to construct the following separable problem

$$(\mathcal{FSP}) \qquad \begin{cases} \sum_{i=1}^{l} f_k(x_i) \downarrow \min_x, \\ -2x_i + 2 \le 0, \\ x_i - p_k \le 0, \quad i = 1, \dots, l, \end{cases}$$

where l is the number of classes of kernel problems, $x \in I\!R^l$, $f_k \in \{f_1, f_2, f_3\}$, $p_k \in \{5.5, 4.5, 7.5\}$.

Then we reformulate problem $(\mathcal{FSP})$ as follows:

$$(\mathcal{FQP}) \qquad \begin{cases} f_0(x) = \langle x, Q_0 x\rangle + \langle b_0, x\rangle + d_0 \downarrow \min_x, \\ \mathcal{A}x \le \mathcal{B}, \end{cases}$$

where $Q_0 \in I\!R^{l\times l}$, $b_0 \in I\!R^l$, $d_0 \in I\!R$, $\mathcal{A} \in I\!R^{2l\times l}$, $\mathcal{B} \in I\!R^{2l}$.

In order to increase the variety of constructed problems, it is necessary to eliminate separateness. For this purpose, we use the substitution $x = M \cdot M^{-1}x$, where $M \in I\!R^{l\times l} : detM \neq 0$, and a change of the variable: $y = M^{-1}x$.

We obtain the following problem:

$$(\mathcal{FQP}') \qquad \begin{cases} f_0(y) = \langle yM, Q_0 My\rangle + \langle M^T b_0, y\rangle + d_0 \downarrow \min_y, \\ \mathcal{A}My \le \mathcal{B}. \end{cases}$$

Thus, we have generated nonconvex quadratic problems of the given dimension.

3.3 Fractional Problem

Further, we represent the objective function of the problem $(\mathcal{FQP}')$ as a d.c. function using the well-known decomposition $M^T Q_0 M = Q_{01} - Q_{02}$, where Q_{01}, Q_{02} are positive definite symmetric matrices such that

$$\begin{aligned} f_0(y) &= \langle y, M^T Q_0 My\rangle + \langle M^T b_0, y\rangle + d_0 = g_0(y) - h_0(y) \\ &= [\langle y, Ay\rangle + \langle p, y\rangle + q] - \alpha[\langle y, By\rangle + \langle c, y\rangle + t] \end{aligned}$$

where $\alpha > 0, \alpha \in I\!R$ is some given parameter, i.e. $Q_{02} = \alpha B$. As well-known, the d.c. decomposition of a function is not unique, therefore we can choose some positive parameter α. For example, we can set $\alpha = 1$.

Further, due to the reduction Theorem 1 $f_0(\cdot) = 0$ (see equality (4) for $m = 1$), we get one ratio $\alpha = \dfrac{\langle y, Ay\rangle + \langle p, y\rangle + q}{\langle y, By\rangle + \langle c, y\rangle + t}$, and the corresponding fractional program

$$\frac{\langle y, Ay\rangle + \langle p, y\rangle + q}{\langle y, By\rangle + \langle c, y\rangle + t} \downarrow \min_y, \quad \mathcal{A}My \le \mathcal{B}. \tag{5}$$

Repeating this procedure m times and summing ratios, we generate an objective function of the test Problem $(\mathcal{FP})$. The feasible set of $(\mathcal{FP})$ is obviously constructed by adding all constraints of all problems (5).

3.4 A Simple Example

The following example demonstrates how the method can be used to generate a fractional problem with quadratic functions in 2 ratios.

First of all, let us generate a nonconvex quadratic problem for the first fraction. Suppose that number of kernel problems of first class is $l_1 = 1$, the second class is $l_2 = 1$, and the third one is $l_3 = 0$. So, the total number of kernel problem is $l = l_1 + l_2 + l_3 = 2$. This corresponds to the following separable quadratic problem:

$$\left.\begin{array}{l} f_0(x) = -\frac{2}{3}x_1^2 + \frac{14}{3}x_1 - \frac{1}{2}x_2^2 + 3x_2 - \frac{13}{2} \downarrow \min_x, \\ \qquad -2x_1^2 + 2 \leq 0, \\ \qquad x_1 - 5.5 \leq 0, \\ \qquad -2x_2^2 + 2 \leq 0, \\ \qquad x_2 - 4.5 \leq 0. \end{array}\right\} \tag{6}$$

Then, let us generate a nonconvex quadratic problem for the second fraction with parameters: $l_1 = 0$, $l_2 = 1$, $l_3 = 1$, $l = 2$.

$$\left.\begin{array}{l} f_0(x) = -\frac{1}{2}x_2^2 + 3x_2 - \frac{1}{4}x_3^2 + \frac{9}{4}x_3 - \frac{9}{2} \downarrow \min_x, \\ \qquad -2x_2^2 + 2 \leq 0, \\ \qquad x_2 - 4.5 \leq 0, \\ \qquad -2x_3^2 + 2 \leq 0, \\ \qquad x_3 - 7.5 \leq 0. \end{array}\right\} \tag{7}$$

Problems (6)–(7) are nonconvex quadratic 3-dimension problems each of which has $2^{l_1+l_2+l_3} = 2^2 = 4$ local solutions and one global solution. These global solutions coincide $y_1^* = y_2^* = (1, 1, 1)$ and the optimal values of the problems are equal to zero.

Now suppose the following data were used in the transformation:

$$\Lambda = \begin{pmatrix} 8\,0\,0 \\ 0\,3\,0 \\ 0\,0\,6 \end{pmatrix}, \quad h = \begin{pmatrix} 4 \\ 8 \\ 10 \end{pmatrix}, \quad M = \begin{pmatrix} 6.58 & -2.84 & -3.56 \\ -1.07 & 0.87 & -2.67 \\ -2.67 & -5.33 & -0.67 \end{pmatrix}.$$

This yield the following quadratic programming problems ($\mathcal{FQP}'$): for the first ratio

$$f_0^1(y) = \begin{pmatrix} y_1 \\ y_2 \\ y_3 \end{pmatrix}^T \begin{pmatrix} -29.41 & 12.94 & 14.17 \\ 12.94 & -5.77 & -5.59 \\ 14.17 & -5.59 & -11.98 \end{pmatrix} \begin{pmatrix} y_1 \\ y_2 \\ y_3 \end{pmatrix} - \begin{pmatrix} 27.49 \\ -10.67 \\ -24.59 \end{pmatrix}^T \begin{pmatrix} y_1 \\ y_2 \\ y_3 \end{pmatrix} - 6.5 \downarrow \min_y,$$

subject to

$$\begin{array}{r} -13.15y_1 + 5.69y_2 + 7.11y_3 \leq -2, \\ 6.57y_1 - 2.84y_2 - 3.56y_3 \leq 5.5, \\ 2.13y_1 - 1.73y_2 + 5.33y_3 \leq -2, \\ -1.07y_1 + 0.87y_2 - 2.67y_3 \leq 4.5, \\ 5.33y_1 + 10.67y_2 + 1.33y_3 \leq -2, \\ -2.67y_1 - 5.33y_2 - 0.67y_3 \leq 7.5, \end{array} \tag{8}$$

and for the second one:

$$f_0^2(y) = \begin{pmatrix} y_1 \\ y_2 \\ y_3 \end{pmatrix}^T \begin{pmatrix} -2.35 & -3.09 & -1.87 \\ -3.09 & -7.49 & 0.27 \\ -1.87 & 0.27 & -3.67 \end{pmatrix} \begin{pmatrix} y_1 \\ y_2 \\ y_3 \end{pmatrix} - \begin{pmatrix} -9.2 \\ -9.4 \\ -9.5 \end{pmatrix}^T \begin{pmatrix} y_1 \\ y_2 \\ y_3 \end{pmatrix} - 4.5 \downarrow \min_y,$$

subject to (8).

Further, we represent objective functions f_0^1, f_0^2 of transformed problems as d.c. functions using the well-known decomposition [12]:

$$f_0^1(y) = [\langle y, A^1 y\rangle + \langle p^1, y\rangle + q_1] - \alpha[\langle y, B^1 y\rangle + \langle c^1, y\rangle + t_1],$$

and

$$f_0^2(y) = [\langle y, A^2 y\rangle + \langle p^2, y\rangle + q_2] - \alpha[\langle y, B^2 y\rangle + \langle c^2, y\rangle + t_2],$$

where

$$A^1 = \begin{pmatrix} 27.11 & 12.94 & 14.17 \\ 12.94 & 12.94 & 0 \\ 14.17 & 0 & 14.17 \end{pmatrix}, \quad p^1 = \begin{pmatrix} 27.49 \\ -10.67 \\ -24.59 \end{pmatrix}, \quad q_1 = -6.5,$$

$$B^1 = \begin{pmatrix} 56.52 & 0 & 0 \\ 0 & 18.71 & 5.59 \\ 0 & 5.59 & 26.15 \end{pmatrix}, \quad c^1 = \begin{pmatrix} 0 \\ 0 \\ 0 \end{pmatrix}, \quad t_1 = 0,$$

$$A^2 = \begin{pmatrix} 2.61 & 0 & 0 \\ 0 & 0.27 & 0.27 \\ 0 & 0.27 & 0.27 \end{pmatrix}, \quad p^2 = \begin{pmatrix} -9.2 \\ -9.4 \\ -9.5 \end{pmatrix}, \quad q_2 = -6.5,$$

$$B^2 = \begin{pmatrix} 4.96 & 3.09 & 1.87 \\ 3.09 & 7.75 & 0 \\ 1.87 & 0 & 3.93 \end{pmatrix}, \quad c^2 = \begin{pmatrix} 0 \\ 0 \\ 0 \end{pmatrix}, \quad t_2 = 0, \quad \alpha = 1.$$

Finally, due to the reduction Theorem 1, we obtain 2 ratios. Summing them, we generate an objective function of the test problem. Its feasible set is obviously constructed by adding all constraints.

Thus, we have the following sum-of-ratios test problem:

$$f(y) = \frac{\langle y, A^1 y\rangle + \langle p^1, y\rangle + q_1}{\langle y, B^1 y\rangle + \langle c^1, y\rangle + t_1} + \frac{\langle y, A^2 y\rangle + \langle p^2, y\rangle + q_2}{\langle y, B^2 y\rangle + \langle c^2, y\rangle + t_2} \downarrow \min_y, \qquad (9)$$

with linear inequality constraints (8).

The global minimum for this problem is:

$$\bar{y} = (-0.089, -0.096, -0.370),$$

with an optimal value $f_0(\bar{y}) = 2$. It should be noted that, since we have chosen the parameter $\alpha = 1$, the global value of the problem (9) coincides with the number of ratios.

4 Conclusion

To conclude, it can be said that the described technique for generating nonconvex quadratic problems and fractional problems allows us to construct various test problems of the desired dimension with known local and global solutions. Moreover, this technique does not require any complicated operations and solving auxiliary problems. Therefore, our approach looks quite promising and beneficial.

References

1. Calamai, P.H., Vicente, L.N., Judice, J.J.: A new technique for generating quadratic programming test problems. Math. Program. **61**, 215–231 (1993)
2. Calamai, P., Vicente, L.: Generating quadratic bilevel programming test problems. ACM Trans. Math. Softw. **20**, 103–119 (1994)
3. Floudas, C.A., Pardalos, P.M.: Handbook of Test Problems in Local and Global Optimization. Kluwer Academic Publishers, Boston (1999)
4. Gantmacher, F.R.: The Theory of Matrices. Fizmatlit, Moscow (2010). (in Russian)
5. Gruzdeva, T.V., Strekalovskiy, A.S.: On solving the sum-of-ratios problem. Appl. Math. Comput. **318**, 260–269 (2018)
6. Jong, Y.-C.: An efficient global optimization algorithm for nonlinear sum-of-ratios problem (2012). http://www.optimization-online.org/DB_FILE/2012/08/3586.pdf
7. Kalitkin, N.N.: Chislennye metody (Numerical Methods). Nauka, Moscow (1978). (in Russian)
8. Ma, B., Geng, L., Yin, J., Fan, L.: An effective algorithm for globally solving a class of linear fractional programming problem. J. Softw. **8**(1), 118–125 (2013)
9. Orlov, A.V., Malyshev, A.V.: Test problem generation for quadratic-linear pessimistic bilevel optimization. Numer. Anal. Appl. **7**(3), 204–214 (2014)
10. Pardalos, P.M.: Construction of test problems in quadratic bivalent programming. ACM Trans. Math. Softw. **17**(1), 74–87 (1991)
11. Pardalos, P.M.: Generation of large-scale quadratic programs for use as global optimization test problems. ACM Trans. Math. Softw. **13**(2), 133–137 (1987)
12. Strekalovsky, A.S.: Elements of Nonconvex Optimization. Nauka, Novosibirsk (2003). (in Russian)
13. Vicente, L.N., Calamai, P.H., Judice, J.J.: Generation of disjointly constrained bilinear programming test problems. Comput. Optim. Appl. **1**(3), 299–306 (1992)
14. The COCONUT Benchmark. https://www.mat.univie.ac.at/ neum/glopt/coconut/Benchmark/Benchmark.html. Accessed 21 Feb 2019

Non-Convex Quadratic Programming Problems in Short Wave Antenna Array Optimization

Anton V. Eremeev[1,2(✉)], Nikolay N. Tyunin[1], and Alexander S. Yurkov[3]

[1] Sobolev Institute of Mathematics, Omsk, Russia
nik.tyunin.92@mail.ru
[2] Dostoevsky Omsk State University, Omsk, Russia
eremeev@ofim.oscsbras.ru
[3] Institute of Radiophysics and Physical Electronics Omsk Scientific Center SB RAS, Omsk, Russia
fitec@mail.ru

Abstract. In this paper, we describe a non-convex constrained quadratic programming problem arising in short wave transmitting antenna array synthesis and provide preliminary computational results. We consider problem instances for three different antenna designs including up to 25 radiators. In the computational experiments, BARON package is compared to the gradient optimization method, applied to the unconstrained problem formulation using the penalty function method. Global optimality of the obtained solutions is established using BARON package the smallest instances of 4 radiators. On small instances, both methods have demonstrated similar results, while on larger instances significant difference has been observed. The set of local optima is studied experimentally. It is established that even though the problem instances have numerous local optima, the objective function in many local optima has the same value.

Keywords: Quadratic programming · Local optima · Antenna array · Gradient optimization · Computational experiment

1 Introduction

Phased antenna arrays (PAA) are regular arrays of radiators, connected to devices that provide the required distribution of phases and amplitudes. PAA are widely used in super-high frequency (SHF) band to obtain directional radiation (see, for example, [8]). On high frequency band (HF), which corresponds to the short waves (SW), such systems are not widely used. However, the possibilities to increase the energy of the communication channel of the HF-band, and to reduce the occupied space through the use of phased antenna arrays attract the attention to such systems [14,20,21].

M. Khachay et al. (Eds.): MOTOR 2019, LNCS 11548, pp. 34–45, 2019.
https://doi.org/10.1007/978-3-030-22629-9_3

On the SHF band, mutual influence of radiators is usually weak and this allows to assume their independence. However, on HF, where other designs of radiators are commonly used, such an assumption is inappropriate. This leads to more complex optimization problems for such antenna arrays.

Various optimization methods have been used to solve the PAA optimization problems. In particular, it was reported in the literature about methods based on semi-definite relaxation [7], gradient optimization methods [13], metaheuristics [1,3,13,19], methods of linear algebra [23,24], approximation theory [15] etc. In some cases, it is possible to solve the original problem approximately, using convex programming methods [6].

Nevertheless, the optimization problems for PAA in HF-band with strong mutual influence of radiators are lacking consideration. In this paper, we study the applicability of a gradient-based algorithm to such a problem (penalty function method is used to take into account the constrains) and compare its results with the results of the BARON solver built into GAMS package. Besides that, we study the properties of the local optima found in multiple restarts of the gradient-based algorithm and discuss the directions for further research.

2 Basic Notations and Problem Formulation

In this paper, similarly to the works [22,24], we consider HF phased antenna arrays consisting of broadband vertical monopoles (BVM), see Fig. 1a, broadband vertical dipoles (BVD), see Fig. 1b, and whip antennae. Each BVM consists of 8 wires that make up a "thick" vertical radiator, fed against the counterpoise system. The counterpoise system of each radiator consists of 6 wires, located parallel to the ground. BVD is designed similarly to BVM with the only difference being that instead of the counterpoise system, another "thick" radiator is attached, pointing in the opposite direction. The whip antenna has a standard dipole configuration. In principle, any other designes of the radiators may be considered, provided that relevant mathematical models may be constructed for them.

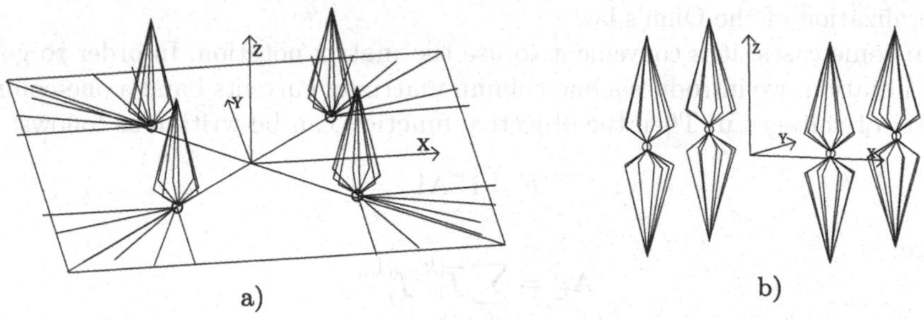

Fig. 1. Broadband vertical monopoles (a) and broadband vertical dipoles (b)

Our goal is to maximize radiation of the antenna array in a given direction (i.e. to maximize the antenna system gain) under constraints on the power input to the antenna system. This problem can be formulated as follows (see [23,24] for details). Let k be the index of a directional vector component: $k = 1$ for the angle of the horizontal direction and $k = 2$ for the vertical angle (the distance can be considered a sufficiently large constant, which can be omitted in what follows). The total electromagnetic field $f_\Sigma^{(k)}$ is given by

$$f_\Sigma^{(k)} = \sum_{i=1}^{N} I_i f_i^{(k)}, \tag{1}$$

where I_i is the complex current in i-th feeding point; $f_i^{(k)}$ is the partial field that is radiated if a unit current flows at the i-th feeding point of the radiating system while the current at all other feeding points of the radiating system is zero. This formula follows by the linearity, so that the total field $f_\Sigma^{(k)}$ is a superposition of the partial fields from the currents at each power point of the radiating system. The values $f_i^{(k)}$ and $f_\Sigma^{(k)}$ are functions of direction and frequency. These values may be computed using some antenna modelling system (in this study NEC-2 system [4] will be used).

Let $\overline{f}$ denote a complex conjugate number to f. As mentioned above, the goal is to maximize the nonnormalized energy flux in the given direction, therefore

$$F = \sum_{k} \overline{f}_\Sigma^{(k)} f_\Sigma^{(k)} \tag{2}$$

is the objective function. It is quite obvious that there are restrictions imposed on the currents I_i due to the fact that the power of sources of these currents is limited. To find the power of source i, we express the corresponding complex voltage U_i as follows:

$$U_i = \sum_{i=1}^{N} \mathbf{Z}_{ij} I_j, \tag{3}$$

where $\mathbf{Z}_{ij}$ are matrix elements of the impedance matrix $\mathbf{Z}$. Equation (3) is a generalization of the Ohm's law.

In some cases, it is convenient to use the matrix notation. In order to go to such notation, we introduce a one-column matrix of currents $\mathbf{i}$ and a one-column matrix of voltages $\mathbf{u}$. Then the objective function can be written as follows:

$$F = \mathbf{i}^{+} \mathbf{A} \mathbf{i}, \tag{4}$$

where

$$\mathbf{A}_{ij} = \sum_{k} \overline{f}_i^{(k)} f_j^{(k)}. \tag{5}$$

Similarly, the relationship between currents and voltages can be written as follows:

$$\mathbf{u} = \mathbf{Z} \mathbf{i}. \tag{6}$$

There are various forms of restrictions that correspond to a different antenna systems. For example, the total power at all feeding points P can be restricted. In this case, the optimization problem can be written as follows:

$$\begin{cases} \mathbf{i}^{+}\mathbf{A}\mathbf{i} \rightarrow \max, \\ \mathbf{i}^{+}\mathbf{B}\mathbf{i} = 1, \end{cases} \tag{7}$$

where

$$\mathbf{B} = \frac{1}{4P}(\mathbf{Z} + \mathbf{Z}^{+}), \tag{8}$$

the superscript + means the Hermitian conjugation. Such a problem can be solved analytically [23].

The optimization problem is more difficult in the case when the power is restricted at each feeding point. This problem can be written as follows:

$$\begin{cases} \mathbf{i}^{+}\mathbf{A}\mathbf{i} \rightarrow \max, \\ 0 \leq \mathbf{i}^{+}\mathbf{B}^{(1)}\mathbf{i} \leq 1, \\ \ldots \\ 0 \leq \mathbf{i}^{+}\mathbf{B}^{(n)}\mathbf{i} \leq 1, \\ \mathbf{i} \in \mathbb{C}^{n}, \end{cases} \tag{9}$$

where

$$\mathbf{B}^{(k)} = \frac{1}{4P_m^{(k)}}(\mathbf{Z}^{+}\mathcal{P}^{(k)} + \mathcal{P}^{(k)}\mathbf{Z}), \tag{10}$$

$P_m^{(k)}$ is the maximum admissible power at k-th feeding point, $\mathcal{P}^{(k)}$ is the matrix projector having a single nonzero matrix element $\mathcal{P}_{kk}^{(k)} = 1$. It can be proved [23] that:

1. All matrices $\mathbf{B}^{(k)}$ have no more than two nonzero eigenvalues. One of them is positive, the other one is negative or zero.
2. Matrices $\mathbf{A}$ and $\mathbf{B}^{(k)}$ are Hermitian, i.e. $\mathfrak{a}_{ij} = \bar{\mathfrak{a}}_{ji}\ \forall ij$.
3. Matrix $\mathbf{A}$ is positive semi-definite.

Apparently, the problem (9) can be solved only by numerical methods. For the development of solution algorithms, it is convenient to reformulate it in terms of real numbers. Let us denote the corresponding real matrices: $\mathbf{G}$ for the objective function and $\mathbf{H}^{(k)} \in \mathbb{R}^{(2n)^2}; 0 \leq k \leq n$ for each of the constraints. Let $\mathbf{y} \in \mathbb{C}^n$, $\mathbf{A} \in \mathbb{C}^{n^2}$, and let $\mathbf{x} \in \mathbb{R}^{2n}$ be a vector, where the first n components are the real parts of the corresponding components of the vector $\mathbf{y}$ while the rest of the components are imaginary, i. e.

$$\mathbf{y}_i \in \mathbb{C} \longleftrightarrow (\mathbf{x}_i, \mathbf{x}_{n+i}),\ \mathbf{x}_i = Re(\mathbf{y}_i),\ \mathbf{x}_{n+i} = Im(\mathbf{y}_i).$$

Let $\mathbf{G} \in \mathbb{R}^{(2n)^2}$ denote a matrix of the following form:

$$\left(\begin{array}{c|c} Re(\mathbf{A}) & -Im(\mathbf{A}) \\ \hline Im(\mathbf{A}) & Re(\mathbf{A}) \end{array}\right) \tag{11}$$

Then

$$\mathbf{Ay} = \mathbf{Gx}\,. \tag{12}$$

Indeed

$$\mathbf{Gx} = \begin{pmatrix} \dfrac{Re(\mathbf{A})Re(\mathbf{y}) - Im(\mathbf{A})Im(\mathbf{y})}{Im(\mathbf{A})Im(\mathbf{y}) + Re(\mathbf{A})Re(\mathbf{y})} \end{pmatrix} = \begin{pmatrix} b \\ c \end{pmatrix}$$

$$\mathbf{Gx} = \mathbf{b} + i\mathbf{c} = Re(\mathbf{A})Re(\mathbf{y}) - Im(\mathbf{A})Im(\mathbf{y}) + Im(\mathbf{A})Im(\mathbf{y}) + Re(\mathbf{A})Re(\mathbf{y})$$

$$\mathbf{Ay} = (Re(\mathbf{A}) + iIm(\mathbf{A})(Re(\mathbf{y}) + iIm(\mathbf{y})) =$$

$$Re(\mathbf{A})Re(\mathbf{y}) - Im(\mathbf{A})Im(\mathbf{y}) + Im(\mathbf{A})Im(\mathbf{y}) + Re(\mathbf{A})Re(\mathbf{y})$$

The fact that the matrix $\mathbf{A}$ is Hermitian leads to the symmetry of matrix $\mathbf{G}$. Indeed, since the matrix $\mathbf{A}$ is Hermitian, this implies symmetry of $Re(\mathbf{A})$ and a skew-symmetry of $Im(\mathbf{G})$. This means that

$$\mathbf{G}^T = \left(\begin{array}{c|c} Re(\mathbf{A}) & (Im(\mathbf{A}))^T \\ \hline (-Im\mathbf{A})^T & Re(\mathbf{A}) \end{array}\right) = \left(\begin{array}{c|c} Re(\mathbf{A}) & -Im(\mathbf{A}) \\ \hline Im(\mathbf{A}) & Re(\mathbf{A}) \end{array}\right) = \mathbf{G}\,.$$

Thus, $\mathbf{G}$ is a symmetric matrix. The same applies to all matrices of the constraints $\mathbf{H}^{(k)} \in \mathbb{R}^{(2n)^2}; 0 \le k \le n$. In real numbers, the optimization problem (9) has the following formulation:

$$\begin{cases} \mathbf{x}^T\mathbf{Gx} \to \max, \\ 0 \le \mathbf{x}^T\mathbf{H}^{(1)}\mathbf{x} \le 1, \\ \ldots \\ 0 \le \mathbf{x}^T\mathbf{H}^{(n)}\mathbf{x} \le 1, \\ \mathbf{x} \in \mathbb{R}^{2n}. \end{cases} \tag{13}$$

To summarize, the proposed mathematical programming problem (13) for short wave antenna array optimization has the objective function defined by a quadratic form with symmetric positive semi-definite matrix $\mathbf{G}$. Every constraint is given by a quadratic form, defined by a symmetric matrix $\mathbf{H}^{(k)}$ with two identical positive eigenvalues and two identical non-negative eigenvalues, the rest of the eigenvalues are equal to zero. Globally optimal solutions to the non-convex mathematical programming problems of this type may be found using branch and bound methods [12,18] or DC programming approaches [11,17]. If global optima are hard to find, then at least the locally optimal solutions may be found by means of the gradient-based optimization or Newton's algorithm [9]. In case of numerous local optima, different metaheuristic algorithms may be helpful as well (see e.g. [5,16]).

2.1 Bounding the Feasible Area

In computational experiments, it will be helpful to reduce the size of the set of feasible solutions to problem (13), still keeping at least one globally optimal solution in this area.

First of all, note that problem (13) in complex numbers has an obvious symmetry with respect to the transformation $\mathbf{i} \to e^{i\phi}\mathbf{i}$ of all complex coordinates (up to an arbitrary angle ϕ). In physical terms, this symmetry corresponds to the phase-independence of the radiation power flux (symmetry of the objective function) and phase-independence of real power flow in each feeder of the antenna system (symmetry of the set of feasible solutions).

The symmetry w.r.t. the transformation $\mathbf{i} \to e^{i\phi}\mathbf{i}$ can be utilized to reduce the search space dimensionality by one, e.g. by fixing $Im(y_n) = 0$, which is equivalent to adding a constraint $x_{2n} = 0$ to problem (13).

Now we can also impose a bound on the set of feasible solutions in terms of the Euclidean distance to the origin. Note that if $\mathbf{x}$ satisfies all constraints of problem (13), this implies

$$\sum_{k=1}^{n} \mathbf{x}^T \mathbf{H}^{(k)} \mathbf{x} \leq n.$$

Denote $\mathbf{H}_{\text{sum}} := \sum_{k=1}^{n} \mathbf{H}^{(k)}$. By physical properties of the problem, we can assume that the minimum eigenvalue of $\mathbf{H}_{\text{sum}}$ (denoted $\lambda_{\min}$) is positive. Then, in view of the fact that

$$\min\{\mathbf{z}^T \mathbf{H}_{\text{sum}} \mathbf{z} \ : \ \mathbf{z} \in \mathbb{R}^{2n}, \ ||\mathbf{z}|| = 1\} = \lambda_{\min},$$

(see e.g. [10], Chap. 1, § 1.0.2), it holds that

$$\mathbf{x}^T \mathbf{H}_{\text{sum}} \mathbf{x} \geq ||\mathbf{x}||^2 \lambda_{\min}$$

and

$$||\mathbf{x}|| \leq \sqrt{\frac{n}{\lambda_{\min}}}. \tag{14}$$

This upper bound will be helpful to establish global optimality to some problem instances by means of BARON solver in the next section.

3 Computational Experiment

The general procedure for solving the antenna array optimization problem, when power is restricted at each feeding point, is as follows:

1. For each radiator in the array, calculate the partial field components.
2. Compute matrices $\mathbf{G}$ and $\mathbf{H}^{(k)}$, $k = 1, \ldots, n$, by the above formulas.
3. Solve problem (13).

As the most basic optimization method, we consider a gradient-based maximization method (a maximization version of the steepest descent) with quadratic approximation algorithm as a line search procedure [9]. In this case the constraint optimization problem was reduced to an unconstraint optimization problem by the external point method [2] as follows:

$$\mathbf{x}^T \mathbf{G} \mathbf{x} - r \cdot \sum_{k=1}^{n} \left(\min\left(0, \mathbf{x}^T \mathbf{H}^{(k)} \mathbf{x}\right) + \min\left(0, 1 - \mathbf{x}^T \mathbf{H}^{(k)} \mathbf{x}\right) \right)^4 \to \max, \tag{15}$$

where r is the penalty parameter. An optimal solution obtained for this unconstraint optimization problem may be infeasible w.r.t. problem (13), but the greater penalty parameter r is chosen, the smaller violations of constraints will be. Besides that, given an infeasible solution $\mathbf{x}$, such that only the inequalities on the right-hand side of the constraints are violated in problem (13), one can easily convert it into a feasible solution $\mathbf{x}' := \alpha(\mathbf{x})^{-1/2}\mathbf{x}$, where $\alpha(\mathbf{x}) := \max_{k=1,\ldots,n} \mathbf{x}^T\mathbf{H}^{(k)}\mathbf{x}$. The objective function value $\mathbf{x}^T\mathbf{G}\mathbf{x}$ will be reduced only by a factor of $\alpha(\mathbf{x})$. In what follows, we will refer to the results of gradient optimization method, including this feasibility restoring post-processing procedure. The gradient optimization algorithm was performed repeatedly, using a randomly generated vector $X \in \mathbb{R}^{2n}$ as a starting point. The distribution of X is described in Subsect. 3.2.

In order to establish the global optimality by means of the BARON solver, based on branch and bound approach [18], it is necessary to provide a bounding box or an upper bound on the norm of feasible solutions. To this end, we use the inequality (14). Unfortunately, the values of $\lambda_{\min}$ turned out to be practically applicable only for three arrays. The BARON solver was tested in terms of global optimality proof only for these instances (denoted by sign "*" in Tables 1, 2 and 4 below).

3.1 Problem Instances

The computational experiments were carried out on three types of regular phased antenna arrays: grids of broadband vertical monopoles (BVM) (Fig. 1a), arrays of broadband vertical dipoles (BVD) (Fig. 1b), and arrays of vertical whip dipoles (VWD). The array sizes are 2×2, 3×3 and 5×5, but only the whip antennae array of size 5×5 are considered because NEC system could not handle 5×5 BVM and 5×5 BVD due to large complexity of these models.

BVM arrays are modelled as placed above the real ground at a height of 0.2 m (the conductivity of the ground equals 0.01 S/m and its dielectric constant equals 10), while BVD and VWD are modelled in free space. In the case of BVM and BVD, the distance between the neighboring radiators is 20 m. The height of each BVM radiator is 15 m. The distance between the endpoints of dipoles in each BVD radiator is 30 m. In the case of whip antennae, the length of the whole radiator is 10 m, the distance between the neighboring radiators is 5 m. The frequency of the signal is 5 MHz in all instances. The target direction for maximization of radiation is given as 45 degrees in the horizontal plane and 70° vertical.

3.2 Results of Computational Experiments

In this subsection, we compare the results of the simple gradient method and the BARON solver in its default mode. In all experiments described below, the overall time limit was set to 1000 sec. of Intel i7 processor. If a gradient optimization converged (termination by minimal admissible step size 10^{-4}), then it was restarted again, until the overall time budget was used. At each initialization of the gradient method, the random starting point X was chosen independently with the uniform distribution in the cube $[0, 2000]^{2n}$. The best found solution is considered as a final result. The penalty weight parameter r in the gradient optimization method is set to 10^6 in all runs. The BARON solver of version 18.5.8 was given the same amount of CPU time, when it was compared to the gradient method (row "BARON") and practically unlimited time when the global optimality was tested (row "BARON*"). In all tables, the column "Time" provides the time till the reported solution was found the first time or the global optimality was established (denoted by "*").

Tables 1, 2 and 3 compare the simple gradient method and the BARON solver, applied to the 2×2 arrays.

Table 1. CPU times and objective function values for BVM 2×2

Solver	Time (sec)	Objective function
Gradient	0.058	138.2
BARON	0.12	139.2
BARON*	0.69*	139.2*

Table 2. CPU times and objective function values for BVD 2×2

Solver	Time (sec)	Objective function
Gradient	0.14	459.7
BARON	0.27	463.6
BARON*	1.14*	463.6*

Table 3. CPU times and objective function values for VWD 2×2

Solver	Time (sec)	Goal function
Gradient	3.3	303.0
BARON	26.62	306.2

As an illustration of physical properties of the obtained solutions, in Figs. 2 and 3 we show a horizontal plan of the beam pattern for the PAA using the solutions obtained by the gradient method (the solutions found by BARON solver look very similar). Figure 2 contains the beam pattern for the 2×2 array of BVD elements. As it can be seen from this figure, the maximal radiation is attained in the given direction (45°), but 2×2 array does not form a sharp main lobe. Figure 3 contains the beam pattern for the 3×3 array of BVD elements, given the same target direction. Comparison of this figure to Fig. 2 shows that the 3×3 array allows to form a sharper main lobe (Tables 5 and 6).

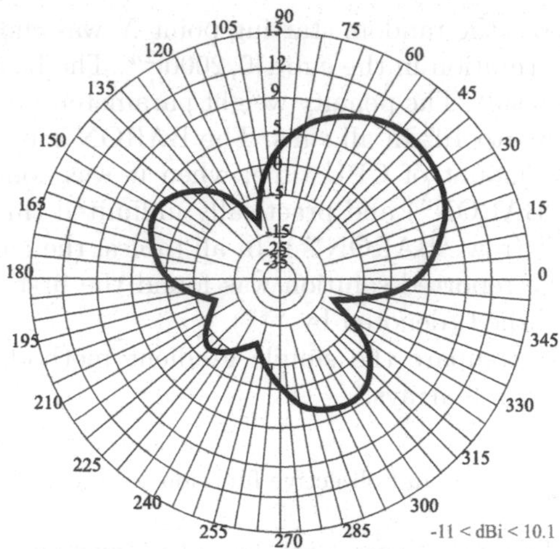

Fig. 2. Horizontal plan of beam pattern for 2×2 array of BVD elements

Table 4. CPU times and objective function values for BVM 3×3

Solver	Time (sec)	Goal function
Gradient	0.68	575.7
BARON	0.34	580.6
BARON*	1873.83*	580.6*

Table 5. CPU times and objective function values for BVD 3×3

Solver	Time (sec)	Goal function
Gradient	40.0	1954.8
BARON	0.56	1980.7

3.3 Statistics on the Number of Different Local Optima Found

Table 7 shows the statistics on the number of different local optima found by the multistart procedure during 1000 s CPU time. In Table 7, M is a number of restarts made, M_{ne} is the number of non-equivalent local optima, and M_f is the number of local optima with identical objective function value (up to a error tolerance below the last digit reported in tables above). In the case of 5×5 VWD the gradient method found just one solution in 1000 s but this time was not enough to find a local optimum, so this instance is not included into Table 7.

It can be observed from Table 7 that the local optima corresponding to the problem instances under consideration are not unique. The column M_f suggests that value of the objective function of all local optima computed for and BVM, BVD was identical for each of the problems, except for BVD 3×3 where local

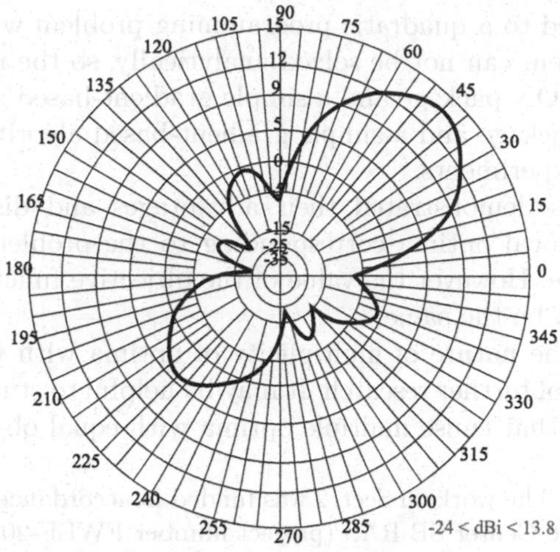

Fig. 3. Horizontal plan of beam pattern for 3×3 array of BVD elements

Table 6. CPU times and objective function values for VWD 5×5

Solver	Time (sec)	Goal function
Gradient	1000	1382.7
BARON	217.94	33.5

Table 7. The number of local optima found

PAA	M	M_{ne}	M_f
BVM 2×2	31691	340	1
BVD 2×2	9531	194	1
VWD 2×2	305	302	24
BVM 3×3	1551	94	1
BVD 3×3	94	52	52

optima with 52 different objective function values were found. In VWD 2×2, a total of 24 local optima with different objective function values were found. The results for BVM 2×2, BVD 2×2 and BVM 3×3 suggest that it is plausible that for many instances of problem (13) for BVM and BVD arrays, all local optima are in fact global solutions.

4 Conclusion

In this paper, we have investigated the problem of optimizing an antenna array in conditions where power restrictions are imposed at each feeding point. Such a problem is reduced to a quadratic programming problem with quadratic constraints. This problem can not be solved analytically, so the numerical solution methods from BARON package and a simple gradient-based algorithm are considered. BARON package and a simple gradient-based algorithm are compared in computational experiments.

Both algorithms demonstrated their advantages and disadvantages. It is observed that the local optima corresponding to the problem under consideration are not unique. However the value of the objective function in many local optima turns out to be the same.

Elucidation of the nature of multiplicity of optima with identical objective value is the subject of further research. It may be helpful to study the symmetries of the feasible area that cause multiple optima with equal objective.

Acknowledgment. The work on Sect. 2 was funded in accordance with the state task of the Omsk Scientific Center SB RAS (project number FWEF-2019-0006).

References

1. Akdagli, A., Guney, K.: Shaped-beam pattern synthesis of equally and unequally spaced linear antenna arrays using a modified tabu search algorithm. Microwave Opt. Technol. Lett. **36**(1), 16–20 (2003)
2. Aoki, M.: Introduction to optimization techniques. fundamentals and applications of nonlinear programming. Technical report, California Univ Los Angeles Dept of System Science (1971)
3. Boriskin, A.V., Balaban, M.V., Galan, O.Y., Sauleau, R.: Efficient approach for fast synthesis of phased arrays with the aid of a hybrid genetic algorithm and a smart feed representation. In: 2010 IEEE International Symposium on Phased Array Systems and Technology, pp. 827–832. IEEE (2010)
4. Burke, G.J., Poggio, A.J., Logan, J.C., Rockway, J.W.: Numerical electromagnetic code (NEC). In: 1979 IEEE International Symposium on Electromagnetic Compatibility, pp. 1–3. IEEE (1979)
5. Eberhart, R., Kennedy, J.: Particle swarm optimization. In: Proceedings of the IEEE international Conference on Neural Networks, vol. 4, pp. 1942–1948. IEEE (1995)
6. Echeveste, J.I., de Aza, M.A.G., Zapata, J.: Shaped beam synthesis of real antenna arrays via finite-element method, floquet modal analysis, and convex programming. IEEE Trans. Antennas Propag. **64**(4), 1279–1286 (2016)
7. Fuchs, B.: Application of convex relaxation to array synthesis problems. IEEE Trans. Antennas Propag. **62**(2), 634–640 (2014)
8. Hansen, R.C.: Phased Array Antennas, vol. 213. Wiley, Hoboken (2009)
9. Himmelblau, D.M.: Applied Nonlinear Programming. McGraw-Hill Companies, New York (1972)
10. Horn, R.A., Horn, R.A., Johnson, C.R.: Matrix Analysis. Cambridge University Press, Cambridge (1990)
11. Horst, R., Pardalos, P.M.: Handbook of Global Optimization, vol. 2. Springer, Dordrecht (2013)
12. Horst, R., Tuy, H.: Global Optimization: Deterministic Approaches. Springer, New York (2013)

13. Indenbom, M., Izhutkin, V., Sharapov, A., Zonov, A.: Synthesis of conical phased antenna arrays optimization of amplitude distribution parameters. DEStech Transactions on Computer Science and Engineering (optim) (2018)
14. Kudzin, V.P., Lozovsky, V.N., Shlyk, N.I.: The compact linear antenna array system of the short-wave band consisting of "butterfly" radiators. In: 2013 IX Internatioal Conference on Antenna Theory and Techniques, pp. 252–253. IEEE (2013)
15. Obukhovets, V.A.: Antenna array iterative synthesis algorithm. In: 2017 Radiation and Scattering of Electromagnetic Waves (RSEMW), pp. 58–60. IEEE (2017)
16. Storn, R., Price, K.: Differential evolution-a simple and efficient heuristic for global optimization over continuous spaces. J. Global Optim. **11**(4), 341–359 (1997)
17. Strekalovsky, A.S.: Global optimality conditions in nonconvex optimization. J. Optim. Theory Appl. **173**(3), 770–792 (2017)
18. Tawarmalani, M., Sahinidis, N.V.: Global optimization of mixed-integer nonlinear programs: a theoretical and computational study. Math. Program. **99**(3), 563–591 (2004)
19. Villegas, F.J.: Parallel genetic-algorithm optimization of shaped beam coverage areas using planar 2-D phased arrays. IEEE Trans. Antennas Propag. **55**(6), 1745–1753 (2007)
20. Wilensky, R.: High-power, broad-bandwidth HF dipole curtain array with extensive vertical and azimuthal beam control. IEEE Trans. Broadcast. **34**(2), 201–209 (1988)
21. Yin, Y., Deng, J.: Design of short wave communication system with phased array antenna. Electronic Eng. **33**(9), 31–33 (2007). in Chinese
22. Yurkov, A.S.: O vliyanii poter v zemle na rabotu chetyrehelementnoi FAR KV diapazona. Tehnika radiosvyazi **1**, 78–81 (2014). in Russian
23. Yurkov, A.S.: Optimizatsiya vozbuzhdeniya peredayushih fazirovannyh antennyh reshotok dekametrovogo diapazona dlin voln. ONIIP, Omsk (2014). in Russian
24. Yurkov, A.S.: Directivity maximization of the short wave band phased antenna array. Tehnika radiosvyazi **2**, 46–53 (2016). in Russian

Splitting Method with Adaptive Step-Size

Igor Konnov[1] and Olga Pinyagina[2(✉)]

[1] Institute of Computational Mathematics and Information Technologies, Department of System Analysis and Information Technologies, Kazan Federal University, Kazan, Russia
konn-igor@ya.ru

[2] Institute of Computational Mathematics and Information Technologies, Department of Data Mining and Operations Research, Kazan Federal University, Kazan, Russia
Olga.Piniaguina@kpfu.ru

Abstract. We suggest the modified splitting method for mixed variational inequalities and prove its convergence under rather mild assumptions. This method maintains the basic convergence properties but does not require any iterative step-size search procedure. It involves a simple adaptive step-size choice, which takes into account the problem behavior along the iterative sequence. The key element of this approach is a given majorant step-size sequence converging to zero. The next decreased value of step-size is taken only when the current iterate does not give a sufficient descent of the objective function. This descent value is estimated with the help of an Armijo-type condition, similar to the rule used in the inexact step-size linesearch. If the current iterate gives a sufficient descent, we can even take an increasing step-size value at the next iterate. Preliminary results of computational experiments confirm the efficiency of the proposed modification in comparison with the ordinary splitting method using the inexact step-size linesearch procedure.

Keywords: Splitting method · Forward-backward method · Adaptive step-size choice · Mixed variational inequality · Nonsmooth optimization problem

1 Introduction

The so-called forward-backward splitting method was proposed first in [1] for the sum of two monotone mappings. It allows one to take into account the peculiarities of each problem under solution for more efficient implementation. Its convergence was established in [2], but required rather restrictive additional assumptions. These assumptions can be removed in the integrable case where a

The results of the first author in this work were obtained within the state assignment of the Ministry of Science and Education of Russia, project No. 1.460.2016/1.4. In this work, the first author was also supported by the RFBR grant, project No. 19-01-00431.

M. Khachay et al. (Eds.): MOTOR 2019, LNCS 11548, pp. 46–58, 2019.
https://doi.org/10.1007/978-3-030-22629-9_4

linesearch procedure with respect to a cost function is inserted. The first splitting method with linesearch was proposed in [3] and further developed in [4–10]. This method is suitable for decomposable problems such as mixed variational inequalities or optimization problems of a special structure. The objective function of such optimization problems can be split into two parts, where the first part is differentiable but can be non-convex, and the second one is convex but non-differentiable in general.

On the other hand, a general scheme of simple adaptive step-size choice was recently proposed for iterative optimization methods in [11,12]. In the present paper, we apply this scheme to the splitting method for solving mixed variational inequalities with potential cost mappings and prove its convergence under rather mild assumptions. This scheme takes into account the behavior of the problem along the iterative sequence. The key element of this approach is a given majorant step-size sequence converging to zero. In accordance with this majorant, the next decreased value of the step-size is taken only when the current iterate does not give a sufficient descent, which is estimated with the help of an Armijo-type condition. If the current iterate gives a sufficient descent, we can even take an increasing step-size value at the next iterate.

The rest of the paper is organized as follows. In Sect. 2 we recall a general scheme of the splitting method for solving mixed variational inequalities. Section 3 contains the main result of the paper, it describes the splitting method with the adaptive step-size choice and the proof of its convergence. Preliminary numerical tests are presented in Sect. 4.

2 The General Scheme of the Splitting Method

Let $F : R^n \to R^n$ be a continuous mapping, $h : R^n \to R$ be a convex but not necessarily differentiable function, $D \in R^n$ be a nonempty convex closed feasible set. A mixed variational inequality is the problem of finding a point $x^* \in D$ such that

$$\langle F(x^*), x - x^* \rangle + h(x) - h(x^*) \geq 0 \quad \forall x \in D. \tag{1}$$

We denote by D^0 the solution set of this problem and assume that it is nonempty. This is the case if, for example, the set D is bounded. If the mapping F is strongly monotone or the function h is strongly convex, then D^0 is a singleton.

Now we recall the general scheme of the splitting method for problem (1) (see, for example, [5]). Let us be given a current iterative point $x^k \in D$. Then the next iterative point $x^{k+1} \in D$ can be defined as a solution of the following mixed variational inequality:

$$\langle F(x^k) + \theta^{-1}(x^{k+1} - x^k), y - x^{k+1} \rangle + h(y) - h(x^{k+1}) \geq 0 \quad \forall y \in D. \tag{2}$$

On the one hand, if the function h is constant, then algorithm (2) becomes the well known projection method

$$x^{k+1} = \pi_D \left[x^k - \theta F(x^k) \right],$$

where $\pi_D : R^n \to D$ is the projection operator onto the set D. On the other hand, if $F \equiv 0$, we obtain a pure implicit (proximal) process.

Under the given assumptions, there exists a unique solution to the optimization problem

$$\min_{y \in D} \to \{h(y) + (2\theta)^{-1}\|y - x\|^2\} \tag{3}$$

for any $\theta > 0$ and any point $x \in R^n$.

We denote this solution by $P_\theta(x)$, therefore we define the proximal continuous mapping $x \to P_\theta(x)$. It follows from the optimality condition of problem (3) that the point $P_\theta(x) \in D$ is a solution to the variational inequality

$$\exists d \in \partial h(P_\theta(x)), \quad \langle d + \theta^{-1}(P_\theta(x) - x), y - P_\theta(x)\rangle \geq 0 \quad \forall y \in D.$$

Then splitting method (2) can equivalently be defined by the following formula

$$x^{k+1} = P_\theta[x^k - \theta F(x^k)], \quad \theta > 0. \tag{4}$$

If we define the mapping $x \to \bar{P}_\theta(x)$ as follows

$$\bar{P}_\theta(x) = P_\theta(x - \theta F(x)), \tag{5}$$

then $\bar{P}_\theta(x)$ is a solution to the problem

$$\min_{y \in D} \longrightarrow \left\{\langle F(x), y\rangle + h(y) + (2\theta)^{-1}\|y - x\|^2\right\}.$$

or the following variational inequality

$$\exists d \in \partial h(\bar{P}_\theta(x)), \quad \langle F(x) + d + \theta^{-1}(\bar{P}_\theta(x) - x), y - \bar{P}_\theta(x)\rangle \geq 0 \quad \forall y \in D. \tag{6}$$

Evidently, the mappings $x \to P_\theta(x)$ and $x \to \bar{P}_\theta(x)$ can equivalently be used.

Let us remind some important properties of the mapping $x \to \bar{P}_\theta(x)$ under the given assumptions (see also [5]).

Proposition 1. *The mapping $x \to \bar{P}_\theta(x)$ has the following properties:*
(a) it is continuous;
(b) if $\bar{x} = \bar{P}_\theta(\bar{x})$ for some $\bar{x} \in D$, then $\bar{x} \in D^0$;
(c) $\exists d \in \partial h(x)$, $\langle F(x) + d, \bar{P}_\theta(x) - x\rangle \leq -\theta^{-1}\|\bar{P}_\theta(x) - x\|^2$ $\forall x \in D$.

Proof. To prove property (a), we arbitrarily fix two points x', $x'' \in D$ and sum inequality (6) with $x = x'$, $y = \bar{P}_\theta(x'')$ and the same inequality with $x = x''$ and $y = \bar{P}_\theta(x')$. Then we obtain

$$\langle F(x') - F(x'') + \theta^{-1}(x'' - x'), \bar{P}_\theta(x'') - \bar{P}_\theta(x')\rangle \geq \theta^{-1}\|\bar{P}_\theta(x'') - \bar{P}_\theta(x')\|^2.$$

It follows that

$$\|F(x') - F(x'')\| + \theta^{-1}\|x'' - x'\| \geq \theta^{-1}\|\bar{P}_\theta(x'') - P_\theta(x')\|.$$

We conclude that the mapping $x \to \bar{P}_\theta(x)$ is continuous, assertion (a) holds true.

Now let $\bar{x} = \bar{P}_\theta(\bar{x})$. Then by setting $x^k = \bar{x}$ in (2) and taking into account (4) and (5) we obtain that $\bar{x}$ solves problem (1). Conversely, let $\bar{x}$ solves problem (1). Then combining inequalities (1) with $y = \bar{P}_\theta(\bar{x})$ and (2) with $y = \bar{x}$, $x^k = \bar{x}$ we have that $-\theta^{-1}\|\bar{P}_\theta(\bar{x}) - \bar{x}\| \geq 0$, therefore $\bar{x} = \bar{P}_\theta(\bar{x})$. Assertion (b) was proven.

Further, using assertion (6) with $y = x$, we have

$$\begin{aligned}-\theta^{-1}\|\bar{P}_\theta(x) - x\|^2 &\geq \langle F(x), \bar{P}_\theta(x) - x\rangle + h(\bar{P}_\theta(x)) - h(x)\\ &\geq \langle F(x) + d, \bar{P}_\theta(x) - x\rangle\end{aligned}$$

for all $x \in D$ and certain $d \in \partial h(x)$. Hence, assertion (c) is also true.

We note that the general scheme (2) requires choosing some procedure for constructing the step size θ. On the one hand, one can use constant step sizes, but they are usually tightly connected with the initial problem properties such as Lipschitz constants or strong monotonicity constants. On the other hand, procedure (2) yields a descent direction $\bar{P}_\theta(x) - x$ and then one can use this direction in certain iterative procedures of exact or inexact step-size search.

In the next section, we describe a variant of the splitting method with adaptive step size choice, which is independent of such problem constants and does not require iterative line-search procedures.

3 The Adaptive Step-Size Choice in the Splitting Method

In what follows, we consider mixed variational inequalities with potential mappings, i.e., we assume that there exists a function $f : R^n \to R$ such that $f'(x) = F(x)\ \forall x \in R^n$. We denote $\varphi(x) = f(x) + h(x)$. Then variational inequality (1) is equivalent to the following optimization problem

$$\min_{x \in D} \longrightarrow \varphi(x). \tag{7}$$

We denote by D^* the solution set of problem (7) and by φ^* the optimal value of its objective function.

We also use a general coercivity condition, which is necessary for the convergence of the method, if the feasible set of the initial problem is unbounded.

(A1) There exists a number $\gamma > \varphi^*$ such that the set

$$D_\gamma = \{x \in D : \varphi(x) \leq \gamma\}$$

is bounded.

Let us apply the general scheme of adaptive step size choice from [12] and describe the modified splitting method, which solves mixed variational inequality (1) and can also solve optimization problem (7) if the function f is convex.

The Splitting Method with Adaptive Step-Size (SMA)

Step 0. Choose an initial point $x^0 \in D_\gamma$, a coefficient $\beta \in (0, 1)$, and a majorant sequence $\{\tau_l\} \to 0$, $\tau_l \in (0, 1)$. Set $k = 0$, $l = 0$, $u^0 = x^0$, choose an initial step size $\lambda_0 \in (0, \tau_0]$.

Step 1. Take a point $y^k = \bar{P}_\theta(x^k)$. If $y^k = x^k$, then stop. Otherwise set $d^k = y^k - x^k$ and $z^{k+1} = x^k + \lambda_k d^k$.

Step 2. If

$$\varphi(z^{k+1}) - \varphi(x^k) \leq -\beta\lambda_k \|d^k\|^2, \tag{8}$$

then take $\lambda_{k+1} \in [\lambda_k, \tau_l]$, set $x^{k+1} = z^{k+1}$, and go to Step 4.

Step 3. Set $\lambda'_{k+1} = \min\{\lambda_k, \tau_{l+1}\}$, $l = l + 1$, and take $\lambda_{k+1} \in (0, \lambda'_{k+1}]$. If $\varphi(z^{k+1}) \leq \gamma$, set $x^{k+1} = z^{k+1}$ and go to Step 4. Otherwise set $x^{k+1} = u^k$, $u^{k+1} = u^k$, $k = k + 1$, and go to Step 1.

Step 4. If $\varphi(x^{k+1}) < \varphi(u^k)$, set $u^{k+1} = x^{k+1}$. Set $k = k+1$ and go to Step 1.

Here we apply a very simple rule of step choice, which follows the approach from [12]. We use condition (8) similar to conditions used in the Armijo-type line-search. Even if this condition is violated at the current iteration and the objective function does not sufficiently decrease (but it does not exceed the threshold γ), we *do* this step, but we take the next value of step-size for using at the *next* iteration.

We note that the auxiliary sequence $\{u^k\}$ contains the best current points of the iterative sequence $\{x^k\}$, in other words,

$$\varphi(u^k) = \min_{0 \leq i \leq k} \varphi(x^i).$$

If the method stops at Step 1 at some point $\bar{x}^k$, then $\bar{x}^k = \bar{P}_\theta(\bar{x}^k)$ and $\bar{x}^k$ is the exact solution to the initial problem due to property (b) of Proposition 1. Hence, in what follows we consider the case when the iterative sequence $\{x^k\}$ is infinite. The proof of the next theorem is similar to that of Theorem 1 from [12], but involves certain differences.

Theorem 1. *Let the assumption* **(A1)** *be fulfilled and* $\beta < \theta^{-1}$. *Then the following assertions hold true.*

(i) The iterative sequence $\{x^k\}$ *generated by SMA has a limit point, which belongs to* D^0.

(ii) If $D^* = D^0$, *then all the limit points of the iterative sequence* $\{x^k\}$ *belongs to the set* D^* *and*

$$\lim_{k\to\infty} \varphi(x^k) = \varphi^*. \tag{9}$$

Proof. We note that the iterative sequence $\{x^k\}$ is contained in a bounded set D_γ, therefore it has limit points. Due to property (a) of Proposition 1, so are sequences $\{y^k\}$ and $\{d^k\}$. We take a subsequence of indices $\{i_s\}$ such that

$$\varphi(z^{i_s+1}) > \varphi(x^{i_s}) - \beta\lambda_{i_s} \|d^{i_s}\|^2, \tag{10}$$

$$\varphi(z^{i_s+1}) > \gamma, \ \varphi(x^{i_s}) \leq \gamma. \tag{11}$$

In other words, i_s are indices of such iterations, which do not give a sufficient descent and the step value will decrease at the next iterate. In addition, we do not take z^{k+1} as the next iterative point, because the objective function value is too large at z^{k+1}.

Then several cases are possible.

Case 1. The subsequence $\{i_s\}$ is infinite.
Take an arbitrary limit point x' of subsequence $\{x^{i_s}\}$. Without loss of generality we can assume that

$$x' = \lim_{s\to\infty} x^{i_s}, \quad y' = \lim_{s\to\infty} y^{i_s},$$

where $y' = \bar{P}_\theta(x')$, because the mapping $x \to \bar{P}_\theta(x)$ is continuous due to assertion (a) of Proposition 1. We also note that we take the next value of step majorant for the next iterate at each i_s, i.e.,

$$\lambda_{i_s} \in (0, \tau_{l_s}], \quad \lambda_{i_s+1} \in (0, \tau_{l_s+1}],$$

for some infinite subsequence of indices l_s, where $\lim_{s\to\infty} \tau_{l_s} = 0$. Therefore $\lim_{s\to\infty} \lambda_{i_s} = 0$. Earlier we noted that the sequence $\{d^{i_s}\}$ is bounded, then by construction of $\{z^{i_s+1}\}$ the limit points of subsequences $\{z^{i_s+1}\}$ and $\{x^{i_s}\}$ coincide. Hence from (11) we obtain

$$\varphi(x') = \gamma > \varphi^*. \tag{12}$$

Since the function h is convex by definition, from assumption (10) we obtain

$$f(x^{i_s} + \lambda_{i_s} d^{i_s}) - f(x^{i_s}) + \lambda_{i_s} \langle g(z^{i_s+1}), d^{i_s} \rangle > -\beta\lambda_{i_s} \|d^{i_s}\|^2$$

for some subgradient $g(z^{i_s+1}) \in \partial h(z^{i_s+1})$. Taking the limit in the previous inequality as $s \to \infty$ yields

$$\langle F(x') + g(x'), y' - x' \rangle \geq -\beta \|y' - x'\|^2$$

for some subgradient $g(x') \in \partial h(x')$. Using assertion (c) from Proposition 1, we obtain

$$\beta \|y' - x'\|^2 \geq \theta^{-1} \|y' - x'\|^2.$$

Therefore, $x' = \bar{P}_\theta(x')$ and due to assertion (b) of Proposition 1,

$$x' \in D^0. \tag{13}$$

Assertion (i) is proven for this case.

Case 2: The subsequence $\{i_s\}$ is finite.
We assumed that the sequence $\{x^k\}$ is infinite. Then $z^k = x^k$ for sufficiently large k. The further proof depends on the properties of the sequence λ_k.

Case 2a: The number of changes of the index l is finite.
Then we have $\lambda_k \geq \bar{\lambda} > 0$ for numbers k large enough, therefore we obtain from condition (8)

$$\varphi(x^{k+1}) \leq \varphi(x^k) - \beta\lambda_k \|d^k\|^2 \leq \varphi(x^k) - \beta\bar{\lambda}\|d^k\|^2$$

for k large enough. Since $\varphi(x^k) \geq \varphi^* > -\infty$, we obtain

$$\lim_{k\to\infty} \varphi(x^k) = \mu \tag{14}$$

and

$$\lim_{k\to\infty} \|y^k - x^k\| = 0. \tag{15}$$

Let x' be an arbitrary limit point of the sequence $\{x^k\}$. From (15) we have

$$\bar{P}_\theta(x') = x',$$

which gives $x' \in D^0$ due to assertion (b) of Proposition 1. Hence all the limit points of the iterative sequence $\{x^k\}$ belong to the set D^0. Therefore, assertion (i) is also true in this case.

Case 2b: The number of changes of the index l is infinite.
In this case, there exists an infinite subsequence of indices $\{k_l\}$ such that $x^{k_l+1} = x^{k_l} + \lambda_{k_l} d^{k_l}$ and condition (8) is violated:

$$\varphi(x^{k_l} + \lambda_{k_l} d^{k_l}) - \varphi(x^{k_l}) = \varphi(x^{k_l+1}) - \varphi(x^{k_l}) > -\beta\lambda_{k_l}\|d^{k_l}\|^2, \tag{16}$$

in addition,

$$\lambda_{k_l} \in (0, \tau_l], \quad \lambda_{k_l+1} \in (0, \tau_{l+1}],$$

and $\lim\limits_{l\to\infty} \tau_l = 0$. Therefore, $\lim\limits_{l\to\infty} \lambda_{k_l} = 0$. Note that since the subsequence $\{d^{k_l}\}$ is bounded, the limit points of the subsequences $\{x^{k_l+1}\}$ and $\{x^{k_l}\}$ coincide. Let us take an arbitrary limit point x' of this subsequence $\{x^{k_l}\}$. Without loss of generality we can assume that

$$x' = \lim_{s\to\infty} x^{k_l}, \quad y' = \lim_{s\to\infty} y^{k_l},$$

where $y' = \bar{P}_\theta(x')$. Since the function h is convex by definition, we obtain from assumption (16)

$$f(x^{k_l} + \lambda_{k_l} d^{k_l}) - f(x^{k_l}) + \lambda_{k_l}\langle g(x^{k_l+1}), d^{k_l}\rangle > -\beta\lambda_{k_l}\|d^{k_k}|^2.$$

for some subgradient $g(x^{k_l+1}) \in \partial h(x^{k_l+1})$. Taking the limit in the previous inequality as $l \to \infty$ yields

$$\langle F(x') + g(x'), y' - x'\rangle \geq -\beta\|y' - x'\|^2$$

for some subgradient $g(x') \in \partial h(x')$. Using assertion (c) from Proposition 1, we obtain

$$\beta\|y' - x'\|^2 \geq \theta^{-1}\|y' - x'\|^2.$$

i.e., $x' = \bar{P}_\theta(x')$ and $x' \in D^0$. Therefore, all the limit points of the subsequence $\{x^{k_l}\}$ belong to the set D^0. Assertion (i) is also proven for this case.

Now we assume that the solution sets of problems (1) and (7) coincide, i.e., $D^* = D^0$. Note that relations (12) and (13) are inconsistent and Case 1 is impossible. This means that the subsequence $\{x^{i_s}\}$ cannot be infinite. In Case

2a we have $\mu = \varphi^*$ in (14) which gives (9). We conclude that assertion (i) holds true in this case.

In Case 2b, as we showed above, the limit points of the subsequences $\{x^{k_l}\}$ and $\{x^{k_l+1}\}$ coincide and all they now belong to the set D^*. For any index k we define the index $m(k)$ as follows

$$m(k) = \max\{j : j \leq k,\ \varphi(x^j) - \varphi(x^{j-1}) > -\beta\lambda_{j-1}\|d^{j-1}\|^2\},$$

i.e., $m(k)$ is the closest to k but not greater index from the subsequence $\{x^{k_l+1}\}$. This means that $m(k) = k$ if $\varphi(x^k) - \varphi(x^{k-1}) > -\beta\lambda_{k-1}\|d^{k-1}\|^2$. By definition, we have

$$\varphi(x^k) \leq \varphi(x^{m(k)}). \tag{17}$$

Now let us take an arbitrary limit point x^* of the sequence $\{x^k\}$, i.e., $\lim_{s\to\infty} x^{t_s} = x^*$. Construct the corresponding infinite subsequence $\{x^{m(t_s)}\}$. From condition (17) we have $\varphi^* \leq \varphi(x^{t_s}) \leq \varphi(x^{m(t_s)})$, but all the limit points of the sequence $\{x^{m(t_s)}\}$ belong to the set D^* because it is contained in the sequence $\{x^{k_s+1}\}$. Choose any limit point $\bar{x}$ of $\{x^{m(t_s)}\}$. Then, taking a subsequence if necessary, we obtain

$$\varphi^* \leq \varphi(x^*) \leq \varphi(\bar{x}) = \varphi^*,$$

therefore $x^* \in D$. This means that all the limit points of the iterative sequence $\{x^k\}$ belong to the set D^* and condition (9) is fulfilled. We conclude that assertion (ii) also holds true. The proof is complete.

The proposed method can be simplified in the case when the feasible set D is bounded. Then we can set $\gamma = +\infty$ and remove all the calculations of the sequence $\{u^k\}$ of best current points. It is easy to verify that all the assertions of Theorem 1 remain true.

4 Preliminary Computational Results

We compared the proposed version of the splitting method with adaptive step-size (SMA) with the ordinary version of this method (SMI), using the inexact line-search procedure.

The Splitting Method with Inexact Step-Size Line-Search (SMI)

Step 0. Choose an initial point $x^0 \in D$, coefficients $\beta \in (0,1)$, $\alpha \in (0,1)$. Set $k = 0$.

Step 1. Take a point $y^k = \bar{P}_\theta(x^k)$. If $y^k = x^k$, then stop. Otherwise set $d^k = y^k - x^k$.

Step 2. Find the smallest nonnegative number m such that

$$\varphi(x^k + \alpha^m d^k) - \varphi(x^k) \leq -\beta\alpha^m\|d^k\|^2,$$

set $\lambda_k = \alpha^m$, $x^{k+1} = x^k + \lambda_k d^k$, $k = k+1$, and go to Step 1.

The computational results are presented in tables, which have the following structure. The first column contains the dimensions of problems. Each row presents the aggregate results of 100 problem instances: mean values (mean val.) and standard deviations (st.dev.) of calculation time and iterations numbers. All randomly generated data are uniformly distributed.

The smooth part of the objective function has the form

$$f(x) = 1/2\langle Ax, x\rangle - \langle b, x\rangle. \tag{18}$$

Here $A = B^T B$. Coefficients b_{ij} and b_i are randomly generated numbers from the segment $[-1, 1]$, $i, j = 1, \ldots, n$. The stopping criterion is $\|d^k\| < 0.001$. The coefficients of methods are $\alpha = 0.5$, $\beta = 0.5$, $\theta = 1$, $\tau_0 = 1$, $\tau_{k+1} = 0.5\tau_k$.

We remind that if the current iterate gives a sufficient descent, we can even take an increasing step-size value at the next iterate. At each 20-th iterate, we increase the step-size value $\lambda_{k+1} = \lambda_k/0.5$.

Example 1. The first series of experiments contains simple nonsmooth functions, which are defined as follows

$$h(x) = \sum_{i=1,\ldots,n} |x_i|. \tag{19}$$

For the sake of simplicity we consider the unconditional optimization problem, i.e., $D = R^n$. The computational results for Example 1 are presented in Table 1.

Table 1. Results for Example 1.

n	*SMI*				*SMA*			
	Time (s)		Iterations		Time (s)		Iterations	
	mean val.	st.dev.	mean val.	st.dev.	mean val.	st.dev.	mean val.	st.dev.
50	0.005	0.007	35	8	0.002	0.005	99	9
100	0.025	0.008	54	8	0.018	0.005	124	17
150	0.076	0.016	67	13	0.052	0.009	163	20
200	0.183	0.031	83	13	0.107	0.016	185	22
250	0.294	0.048	84	14	0.186	0.014	210	12
300	0.534	0.065	102	13	0.307	0.020	244	13
350	0.868	0.099	118	14	0.451	0.044	264	26
400	1.273	0.136	129	14	0.644	0.051	287	20
450	1.508	0.214	121	18	0.810	0.084	286	30

Example 2. The functions f and h are defined in (18) and (19), and the feasible set is a parallelepiped

$$D = \{x \in R^n : d_i \leq x_i \leq e_i, \quad i = 1, \ldots, n\}, \tag{20}$$

where the coefficients d_i, e_i are randomly generated numbers from the segment $[-10, 10]$, $i = 1, \ldots, n$, taking into account that $d_i \leq e_i$ for all $i = 1, \ldots, n$. The computational results for Example 2 are presented in Table 2.

Table 2. Results for Example 2.

n	*SMI*				*SMA*			
	Time (s)		Iterations		Time (s)		Iterations	
	mean val.	st.dev.	mean val.	st.dev.	mean val.	st.dev.	mean val.	st.dev.
50	0.016	0.012	142	81	0.008	0.008	163	30
100	0.241	0.075	466	130	0.053	0.010	341	49
150	0.481	0.133	406	112	0.168	0.024	502	67
200	2.523	0.430	1 096	173	0.413	0.045	685	76
250	4.854	0.551	1 305	144	0.771	0.088	826	95
300	4.412	1.245	822	233	1.447	0.181	1 065	135
350	10.068	1.706	1 340	213	2.315	0.267	1 261	146
400	25.893	3.091	2 441	278	3.421	0.481	1 406	187
450	44.228	3.772	3 123	137	4.851	0.684	1 539	201

Table 3. Results for Example 3.

n	*SMI*				*SMA*			
	Time (s)		Iterations		Time (s)		Iterations	
	mean val.	st.dev.	mean val.	st.dev.	mean val.	st.dev.	mean val.	st.dev.
50	0.016	0.012	132	82	0.006	0.008	155	28
100	0.205	0.076	408	141	0.049	0.010	322	47
150	0.438	0.132	368	112	0.167	0.024	487	64
200	2.372	0.438	1 034	174	0.395	0.048	655	78
250	4.813	0.539	1295	140	0.765	0.104	811	102
300	4.523	1.269	848	236	1.321	0.180	981	132
350	9.852	2.012	1 310	251	2.222	0.297	1 207	160
400	24.450	3.399	2317	271	3.233	0.510	1 312	170
450	42.609	2.519	3 104	144	4.522	0.622	1 459	192

Example 3. Now we consider conditional optimization problems with nonsmooth functions

$$h(x) = \sum_{i=1,\ldots,n} \alpha_i |x_i|,$$

where α_i are randomly generated numbers from the segment $[0, 10]$, $i = 1, \ldots, n$. The functions f are defined in (18) and D is given in (20). The computational results for Example 3 are presented in Table 3.

Example 4. Now we consider problems with nonsmooth functions

$$h(x) = \sum_{i=1,\ldots,n} |x_i - c_i|,$$

Table 4. Results for Example 4.

n	SMI				SMA			
	Time (s)		Iterations		Time (s)		Iterations	
	mean val.	st.dev.	mean val.	st.dev.	mean val.	st.dev.	mean val.	st.dev.
50	0.019	0.013	156	97	0.007	0.008	165	24
100	0.229	0.080	451	143	0.051	0.009	339	45
150	0.424	0.131	361	110	0.176	0.025	517	68
200	2.507	0.418	1 089	165	0.406	0.040	668	64
250	5.002	0.544	1 295	134	0.820	0.116	843	99
300	4.528	1.432	813	246	1.443	0.218	1 035	128
350	10.907	2.675	1 391	291	2.377	0.295	1 248	151
400	24.822	2.992	2 420	262	3.362	0.394	1 400	165
450	43.486	2.644	3 212	157	4.757	0.618	1 545	174

where c_i are randomly generated numbers from the segment $[-10, 10]$, $i = 1, \ldots, n$. The functions f are defined in (18) and D is given in (20). The computational results for Example 4 are presented in Table 4.

Example 5. In conclusion, we consider problems with nonsmooth functions

$$h(x) = \sum_{i=1,\ldots,n} \alpha_i |x_i - c_i|,$$

where α_i are randomly generated numbers from the segment $[0, 10]$, $i = 1, \ldots, n$, c_i are randomly generated numbers from the segment $[-10, 10]$, $i = 1, \ldots, n$. The

Table 5. Results for Example 5.

n	SMI				SMA			
	Time (s)		Iterations		Time (s)		Iterations	
	mean val.	st.dev.	mean val.	st.dev.	mean val.	st.dev.	mean val.	st.dev.
50	0.015	0.012	133	85	0.007	0.005	160	26
100	0.226	0.076	449	137	0.047	0.007	326	34
150	0.457	0.143	387	119	0.162	0.022	503	67
200	2.481	0.473	1072	177	0.385	0.050	674	85
250	4.695	0.432	1276	115	0.732	0.073	839	82
300	4.108	1.270	772	237	1.316	0.172	1038	136
350	10.161	1.996	1338	246	2.150	0.230	1221	127
400	24.757	2.788	2410	247	3.050	0.350	1348	151
450	43.283	2.629	3185	148	4.501	0.687	1544	211

functions f are defined in (18) and D is given in (20). The computational results for Example 5 are presented in Table 5.

These preliminary results of computational tests show the efficiency of the proposed method. It is more flexible in the choice of parameters in comparison with the original version using inexact line-search. In our opinion, this approach is promising for further investigations.

The program was written in Visual C# with double precision, tested on an Intel i3-4170 CPU at 3.7 GHz, 4 Gb, running under Windows 7.

5 Conclusion

In the present work, we propose the modified splitting method for mixed variational inequalities and prove its convergence under rather mild assumptions. This method maintains the basic convergence properties but does not require any iterative step-size search procedure. It involves a simple adaptive step-size choice, which takes into account the behavior of the problem along the iterative sequence. The key element of this approach is a given majorant step-size sequence converging to zero. The next decreased value of step-size is taken only when the current iterate does not give a sufficient descent of the objective function. This descent value is estimated with the help of an Armijo-type condition, similar to the rule used in the inexact step-size linesearch. If the current iterate gives a sufficient descent, we can even take an increasing step-size value at the next iterate.

Preliminary results of computational experiments confirm the efficiency of the proposed modification in comparison with the ordinary splitting method using the inexact step-size line-search procedure.

References

1. Lions, P.L., Mercier, B.: Splitting algorithms for the sum of two monotone operators. SIAM. J. Num. Anal. **16**(6), 964–979 (1979)
2. Gabay, D.: Application of the method of multipliers to variational inequalities. In: Fortin, M., Glowinski, R. (eds.) Augmented Lagrangian Methods: Applications to the Numerical Solution of Boundary-Value Problems, pp. 299–331. North-Holland, Amsterdam (1983)
3. Fukushima, M., Mine, H.: A generalized proximal point algorithm for certain nonconvex minimization problems. Int. J. Syst. Sci. **12**, 989–1000 (1981)
4. Patriksson, M.: Cost approximations: a unified framework of descent algorithms for nonlinear programs. SIAM J. Optim. **8**(2), 561–582 (1998)
5. Patriksson, M.: Nonlinear Programming and Variational Inequality Problems: A Unified Approach. Kluwer, Dordrecht (1999)
6. Konnov, I.V., Kum, S.: Descent methods for mixed variational inequalities in a Hilbert space. Nonlinear Anal. Theory Methods Appl. **47**(1), 561–572 (2001)
7. Konnov, I.V.: Iterative solution methods for mixed equilibrium problems and variational inequalities with non-smooth functions. In: Haugen, I.N., Nilsen, A.S. (eds.) Game Theory: Strategies, Equilibria, and Theorems, pp. 117–160. NOVA, Hauppauge (2008)

8. Konnov, I.V.: Descent methods for mixed variational inequalities with non-smooth mappings. In: Reich, S., Zaslavski, A.J. (eds.) Optimization Theory and Related Topics. Contemporary Mathematics, vol. 568, pp. 121–138. American Mathematical Society, Providence (2012)
9. Konnov, I.V.: Sequential threshold control in descent splitting methods for decomposal optimization problems. Optim. Methods Softw. **30**(6), 1238–1254 (2015)
10. Konnov, I.V., Salahuddin: Two-level iterative method for non-stationary mixed variational inequalities. Russ. Mathem. (Iz. VUZ) **61**(10), 44–53 (2017)
11. Konnov, I.: Conditional gradient method without line-search. Russ. Math. **62**(1), 82–85 (2018)
12. Konnov, I.: A simple adaptive step-size choice for iterative optimization methods. Adv. Model. Optim. **20**(2), 353–369 (2018)

A Dynamic Algorithm for Constructing the Dual Representation of a Polyhedral Cone

Sergey O. Semenov and Nikolai Yu. Zolotykh(✉)

Lobachevsky State University of Nizhny Novgorod,
Gagarin Avenue 23, Nizhny Novgorod 603600, Russia
{sergey.semenov,nikolai.zolotykh}@itmm.unn.ru

Abstract. We propose a dynamic version of the double description method for generating the extreme rays of a polyhedral cone. The dynamic version of the algorithm supports online input of inequalities. Some modifications of the method were implemented and the results of computational experiments are presented. On a series of problems, our implementation of the algorithm showed higher performance results in comparison with the known analogues.

Keywords: System of linear inequalities · Convex hull · Cone · Polyhedron · Double description method

1 Introduction

It is well-known that any convex polyhedron $P \subseteq F^d$, where F is an ordered field, can be represented in any of the following two ways:

(1) as the set $P = \{x \in F^d : \; Ax \leq b\}$ of solutions to a system of linear inequalities, where $A \in F^{m \times d}$, $b \in F^m$ (*facet description*);
(2) as the sum of the conical hull of a set of vectors $v_1, \ldots, v_s$ in F^d and the convex hull of a set of points $w_1, \ldots, w_n$ in F^d (*vertex description*).

The problem of finding the representation (1) given the representation (2) is called *the convex hull problem*. According to the classical theorem of Weyl this problem is equivalent (dual) to the problem of constructing the representation (2) given the representation (1). These two problems are referred to as *finding the dual representation of a polyhedron*.

The problem of constructing the dual representation of a convex polyhedron plays a central role in the theory of systems of linear inequalities and computational geometry [11,21]. For some applications, the representation (1) is convenient, while in other cases the representation (2) is more usable, therefore, it is important to quickly move from one description to another. The importance

This work was supported by the Russian Science Foundation Grant No. 17-11-01336.

M. Khachay et al. (Eds.): MOTOR 2019, LNCS 11548, pp. 59–69, 2019.
https://doi.org/10.1007/978-3-030-22629-9_5

of studying this problem is also emphasized by the fact that it has a variety of applications, the most known of which are linear and integer programming [18], combinatorial optimization [19,21] and global optimization [15]. Some of the newer applications are biological kinetics [20], analysis and verification of software and hardware [5], identification of dynamic systems [13] and computer algebra [17,22].

From an algorithmic or theoretical points of view it is convenient to consider these problems only for polyhedral cones. Any polyhedral cone $C \subseteq F^d$ can be represented in two equivalent ways:

(1) as the set $C = \{x \in F^d : Ax \geq 0\}$ of solutions to a homogeneous system of linear inequalities, where $A \in F^{m \times d}$, or
(2) as the conical hull of a set of vectors $v_1, \ldots, v_s$.

There is a standard method for reducing the problem of finding the dual representation for convex polyhedra to the corresponding one for polyhedral cones. For example, in order to find representation (2) for a polyhedron $P = \{x \in F^d : Ax \leq b\}$ it is sufficient to solve the corresponding problem for the polyhedral cone $\{x = (x_0, x_1, \ldots, x_d) \in F^{d+1} : bx_0 - Ax \geq 0,\ x_0 \geq 0\}$, and then set $x_0 = 1$ (see Section 1.5 in [21]).

There are several known algorithms for solving problems of finding the dual representation. One of the most popular ones is *the double description method* (DDM) [16], also known as the Motzkin–Burger algorithm [11] or Chernikova's algorithm [12]. The double description method generally outperforms the other algorithms when applied to degenerate inputs and/or outputs [3].

There are multiple known programs implementing various modifications of the double description method. Among the most well-known are:

- cdd [14] (www.inf.ethz.ch/personal/fukudak/cdd_home);
- SKELETON [23] (www.uic.unn.ru/~zny/skeleton);
- QSkeleton [9] (github.com/sbastrakov/qskeleton);
- Parma Polyhedra Library [5] (bugseng.com/products/ppl).

Implementations of other algorithms solving the given problem should also be noted:

- QHull [6] (www.qhull.org);
- lrs [2,4] (cgm.cs.mcgill.ca/ avis/C/lrs.html);
- pd [10] (www.cs.unb.ca/~bremner/software/pd).

In this paper we consider the *dynamic* problem of finding the dual representation. This problem appears in many of the applications listed above. For definiteness, we will deal with the problem of constructing a description (2) if the description (1) is given. In dynamic problem the full list of constraints is not known in advance and the constraints come to the input of the algorithm online as the computation proceeds, and the dual description must be computed at each iteration for the current system of linear inequalities. Such framework does not allow the use of many of the heuristics proposed by various authors (see the references above) to accelerate the algorithm. Nevertheless, we propose such an algorithm (as a version of the double description method) for the dynamic problem, and our algorithm is usually not inferior in performance to other offline algorithms.

The problem of efficient removal of constraints from the double description is considered in [1,7]. Other issues concerning the dynamic problem are studied in [8].

2 Preliminaries

The material in this section is based on [11,18,21]. A *polyhedral cone*, or simply a *cone*, in F^d is defined as a set

$$C = \{x \in F^d : Ax \geq 0\},$$

where F is an ordered field, $A \in F^{m \times d}$. The system of linear inequalities $Ax \geq 0$ is said to *define* the cone C. A cone is called *pointed*, if it contains no zero subspaces. It is well-known that for a cone to be pointed it is necessary and sufficient that rank A $= d$, where rank A denotes the rank of matrix A. Any polyhedral cone C can be defined as the conical hull of a finite set of vectors $v_1, v_2, \ldots, v_s$ in F^d, i.e.

$$C = \{x = \alpha_1 v_1 + \alpha_2 v_2 + \cdots + \alpha_s v_s : \alpha_i \geq 0 \ (i = 1, \ldots, s)\}.$$

By writing vectors $v_1, v_2, \ldots, v_s$ as rows of matrix $V \in F^{s \times d}$ the conical hull can be defined as

$$C = \{x = \alpha V, \ \alpha \in F^s, \ \alpha \geq 0\},$$

where α is a row vector. The set of vectors $v_1, \ldots, v_s$ are said to *generate* the cone C.

A non-zero vector $u \in C$ is referred to as a *ray* of the cone C. Two rays u and v are *equal* (written as $u \simeq v$) if for some $\alpha > 0$ it is true that $u = \alpha v$. A ray $u \in C$ is said to be *extreme* if the condition $u = \alpha v + \beta w$, where $\alpha \geq 0$, $\beta \geq 0$ and $v, w \in C$ implies $u \simeq v \simeq w$. Suppose that P is a convex subset of F^d, and for some $a \in F^d$, $\alpha \in F$, it holds that $P \subseteq \{x : ax \leq \alpha\}$. Then $P \cap \{x : ax = \alpha\}$ is called a *face* of the set P. Two different extreme rays u and v of a pointed cone C are said to be *adjacent*, if no minimal face containing both rays contains any other extreme rays of the cone C.

The problem of constructing the set of vectors generating polyhedral cone $C = \{x \in F^d : Ax \geq 0\}$ is reduced to finding the extreme generators of a pointed cone by transition to the orthogonal complement $L^{\perp}$ of the maximal subspace $L = \{x \in F^d : Ax = 0\}$ contained inside C. Unfortunately, in our study we cannot take advantage of this fact (as is often the case), since in the process of adding new inequalities, the space L may change.

Notations: $I_{d \times d}$ is the identity matrix of size $d \times d$, $O_{s \times m}$ is the zero matrix of size $s \times m$, A_i is the i-th row of matrix A, A_{ij} is the element from the i-th row and j-th column of matrix A.

3 Dynamic Double Description Method

The dynamic variant of the double description method is based on the regular double description method [11]. The procedure takes a matrix $A \in F^{m\times d}$ as its input. The output is matrices $U \in F^{t\times d}$ and $V \in F^{s\times d}$, the rows of which compose a basis of maximal subspace $L = \{x \in F^d : Ax = 0\}$ and an irreducible set of vectors generating the cone $C = \{x \in F^d : Ax \geq 0\}$ respectively.

```
procedure DDM-DYN(A)
    Input: A ∈ F^{m×d}
    Output: the basis U of maximal subspace L = {x ∈ F^d : Ax = 0}
        and the set of vectors V generating the cone C = {x ∈ F^d : Ax ≥ 0}
    U ← I_{d×d}
    V ← O_{0×d}
    Q ← O_{0×d}
    for i = 1, 2, ..., m
        p ← U · A_i^T
        q ← V · A_i^T
        if p = 0
            insert the column BOOL(q) into Q
            J_+ ← {j : q_j > 0}
            J_- ← {j : q_j < 0}
            V_new ← O_{0×d}
            Q_new ← O_{0×m}
            E ← ADJACENT(Q, J_+, J_-)
            for {j_1, j_2} ∈ E
                append the row NORMALIZE (q_{j_1} V_{j_2} − q_{j_2} V_{j_1}) to V_new
                append the row NORMALIZE (Q_{j_1} ∨ Q_{j_2}) to Q_new
            append the zero column to Q_new
            remove the rows J_- from V and from Q
            append the rows of V_new to V and the rows of Q_new to Q
        else
            find j_0 with p_{j_0} ≠ 0
            if p_{j_0} < 0
                U_{j_0} ← −U_{j_0}
                p_{j_0} ← −p_{j_0}
            for j = 1, 2, ..., d
                if j ≠ j_0 and p_j ≠ 0
                    U_j ← NORMALIZE (p_{j_0} U_j − p_j U_{j_0})
            for j = 1, 2, ..., s
                if q_j ≠ 0
                    V_j ← NORMALIZE (p_{j_0} V_j − q_j U_{j_0})
            insert the row U_{j_0} into V
            remove the j_0-th row from U
            insert the zero column into Q
            insert the (0, 0, ..., 0, 1) row into Q
```

The normalization function NORMALIZE used in the algorithm is necessary to prevent unlimited growth of elements. Its implementation may vary, the one used in this study for integer calculations divides each element of the input vector by their greatest common divisor. The ADJACENT function checks a pair of extreme rays for adjacency.

Typically modifications of the double description method alter some of the following three parameters:

(1) the order in which the rows of the primal representation are considered
(2) the test for adjacency of two extreme rays
(3) the moment when the extreme rays are tested for adjacency.

Since the dynamic version of the double description method has to support online input of inequalities, only modifications of the latter two aspects are applicable to this algorithm. Two versions of adjacency test were implemented: the combinatorial test [16], and the graph test [23].

function ADJACENT.COMBINATORIAL(Q, J_1, J_2)
Input: $Q \in F^{s \times m}$
$J_1, J_2 \subseteq \{1, 2, \ldots, s\}$
Output: $E = \{\{j_1, j_2\} : j_1 \in J_1, j_2 \in J_2, j_1$ and j_2 - adjacent rays$\}$
$E = \emptyset$
for $j_1 \in J_1$
  for $j_2 \in J_2$
    $Z \leftarrow \{\ell : Q_{j_1\ell} = 0 \wedge Q_{j_2\ell} = 0\}$
    if $|Z| \geq r - 2$
      if $\forall\, k = 1, 2, \ldots, s,\ k \neq j_1 \wedge k \neq j_2\ \exists\, \ell \in Z : Q_{k\ell} = 1$
        $E \leftarrow E \cup \{\{j_1, j_2\}\}$
return E

function ADJACENT.GRAPH(Q, J_1, J_2)
Input: $Q \in F^{s \times m}$
$J_1, J_2 \subseteq \{1, 2, \ldots, s\}$
Output: $E = \{\{j_1, j_2\} : j_1 \in J_1, j_2 \in J_2,\ j_1$ and j_2 are adjacent rays$\}$
$E \leftarrow \emptyset$
for $j_1 \in J_1$
  $D \leftarrow \emptyset$
  for $j \in \{1, 2 \ldots, m\}$
    $Z \leftarrow \{\ell : Q_{j_1\ell} = 0 \wedge Q_{j\ell} = 0\}$
    if $|Z| \geq r - 2$
      $D \leftarrow D \cup \{j\}$
  for $j_2 \in D \cap J_2$
    if $\forall\, k = 1, 2, \ldots, s,\ k \neq j_1 \wedge k \neq j_2\ \exists\, \ell \in Z : Q_{k\ell} = 1$
      $E \leftarrow E \cup \{\{j_1, j_2\}\}$
return E

Some modifications of the double description method were proposed (e.g. [14]) where the set of adjacent extreme rays is maintained and rebuilt immediately after the list of extreme rays is updated. Instead of iterating over all pairs of rays (u, v) for which $uA_i^\top \cdot vA_i^\top < 0$ to generate new extreme rays the algorithm iterates over all pairs of adjacent extreme rays. Below is an adaptation of such a modification for the dynamic version of the algorithm which will be referred to as M1.

```
procedure DDM-DYN.M1(A)
    Input: A ∈ F^{m×d}
    Output: the basis U of maximal subspace L = {x ∈ F^d : Ax = 0}
        and the set of vectors V generating the cone C = {x ∈ F^d : Ax ≥ 0}
    U ← I_{d×d}
    V ← O_{0×d}
    Q ← O_{0×d}
    E ← ∅
    for i = 1, 2, ..., m
        p ← U · A_i^T
        q ← V · A_i^T
        if p = 0
            insert the column BOOL(q) into Q
            E_new ← ∅
            E_old ← ∅
            for {j_1, j_2} ∈ E
                if q_{j_1} > 0 ∧ q_{j_2} < 0
                    append row NORMALIZE(q_{j_1} V_{j_2} − q_{j_2} V_{j_1}) to V_new
                    append row NORMALIZE(Q_{j_1} ∨ Q_{j_2}) to Q_new
                    j ← index of the new rows after their insertion
                    E_new ← E_new ∪ {{j_1, j}}
                    E_old ← E_old ∪ {{j_1, j_2}}
                else if q_{j_1} < 0 ∧ q_{j_2} > 0
                    append row NORMALIZE(q_{j_2} V_{j_1} − q_{j_1} V_{j_2}) to V_new
                    append row NORMALIZE(Q_{j_1} ∨ Q_{j_2}) to Q_new
                    j ←index of the new rows after their insertion
                    E_new ← E_new ∪ {{j, j_2}}
                    E_old ← E_old ∪ {{j_1, j_2}}
                else if q_{j_1} ≤ 0 ∧ q_{j_2} ≤ 0
                    E_old ← E_old ∪ {{j_1, j_2}}
            E ← E ∪ E_new \ E_old
            J_− ← {j : q_j < 0}
            J_± ← {j : q_j = 0}
            remove rows J_− from V and from Q
            update row indices in E and J_±
            append rows of V_new to V and insert their new indices into J_±
            append rows of Q_new to Q
            E ← E ∪ ADJACENT.M1(Q, J_±)
```

```
else
    find $j_0$ with $p_{j_0} \neq 0$
    if $p_{j_0} < 0$
        $U_{j_0} \leftarrow -U_{j_0}$
        $p_{j_0} \leftarrow -p_{j_0}$
    for $j = 1, 2, \ldots, d$
        if $j \neq j_0$ and $p_j \neq 0$
            $U_j \leftarrow$ NORMALIZE $(p_{j_0} U_j - p_j U_{j_0})$
    for $j = 1, 2, \ldots, s$
        if $q_j \neq 0$
            $V_j \leftarrow$ NORMALIZE $(p_{j_0} V_j - q_j U_{j_0})$
    append row $U_{j_0}$ to $V$, $j \leftarrow$ its new index
    $E \leftarrow E \cup \{\{j', j\} : j' \in \{1, 2 \ldots, j-1\}\}$
    remove row $j_0$ from $U$
    append the zero column to $Q$
    append the $(0, 0, \ldots, 0, 1)$ row into $Q$
```

ADJACENT.M1$(Q, J_\pm)$ is a simple modification of ADJACENT(Q, J_1, J_2) that iterates over $\{\{j_1, j_2\} : j_1 \in J_\pm, j_2 \in J_\pm, j_1 \neq j_2\}$ rather than $\{(j_1, j_2) : j_1 \in J_1, j_2 \in J_2\}$.

4 Computational Results

A C++ implementation of the dynamic double description method and its modifications presented above has been developed. The computational experiments were performed on a computer with Intel(R) Core(TM) i7-8700K CPU at 3.70 GHz, Microsoft Windows 10 operating system, using the Microsoft Visual Studio 2017 compiler. The experiments were run using the problem instances described in [14].

Tables 1 and 2 present the performance comparison of DDM-DYN, its modification DDM-DYN.M1 and SKELETON [23], with/without its PlusPlus modification. Since computation time depends significantly on the order in which the rows of the primal representation are considered and which is fixed in the case of the dynamic algorithm, SKELETON was used with the minindex order of consideration. Note that the PlusPlus modification of SKELETON reduces the number of adjacency checks by relying on the fact that the entire primal representation is known ahead of time and, therefore, it cannot be adopted for use with online input of inequalities.

Figures 1, 2 and 3 demonstrate the dependence of the number of adjacency checks on the number of iteration made by DDM-DYN and DDM-DYN.M1 on cube18, mit729-9 and ccc7 problems. The number of the checks varies from one problem instance to another and heavily impacts total computation time.

Table 1. Performance comparison of DDM-DYN and SKELETON, with combinatorial adjacency test (s)

Problem	Input	Output	DDM-DYN	DDM -DYN.M1	SKELETON	SKELETON, PlusPlus
cube16	32×17	65536×17	3.512	7.210	12.199	**3.354**
cube18	36×19	262144×19	45.184	103.514	207.037	**27.069**
mit729-9	729×9	4862×9	164.727	**120.745**	274.795	255.393
ccc7	63×22	38780×22	14793.694	15413.77	16016.7	**3437.64**

Table 2. Performance comparison of DDM-DYN and SKELETON, with graph adjacency test (s)

Problem	Input	Output	DDM-DYN	DDM -DYN.M1	SKELETON	SKELETON, PlusPlus
cube16	32×17	65536×17	**4.316**	4.961	4.777	4.489
cube18	36×19	262144×19	**48.215**	54.552	50.698	48.498
mit729-9	729×9	4862×9	874.575	**106.253**	289.037	258.991
ccc7	63×22	38780×22	>5 h	2988.16	3041.86	**1854.63**

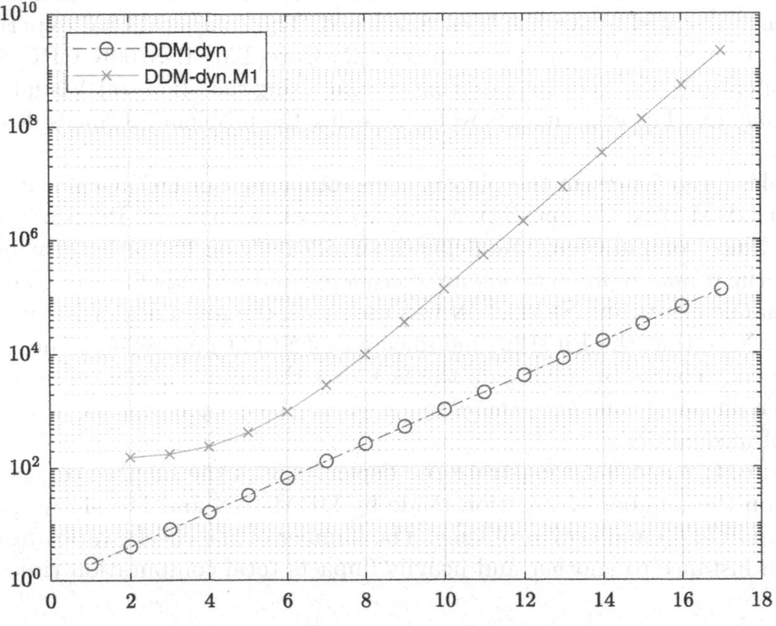

Fig. 1. Number of adjacency checks on cube18

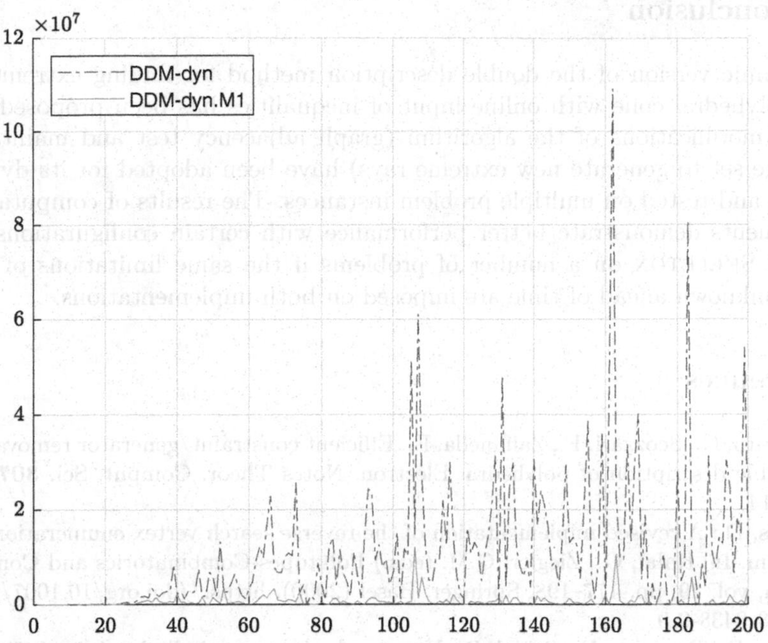

Fig. 2. Number of adjacency checks on mit729-9 (first 200 iterations)

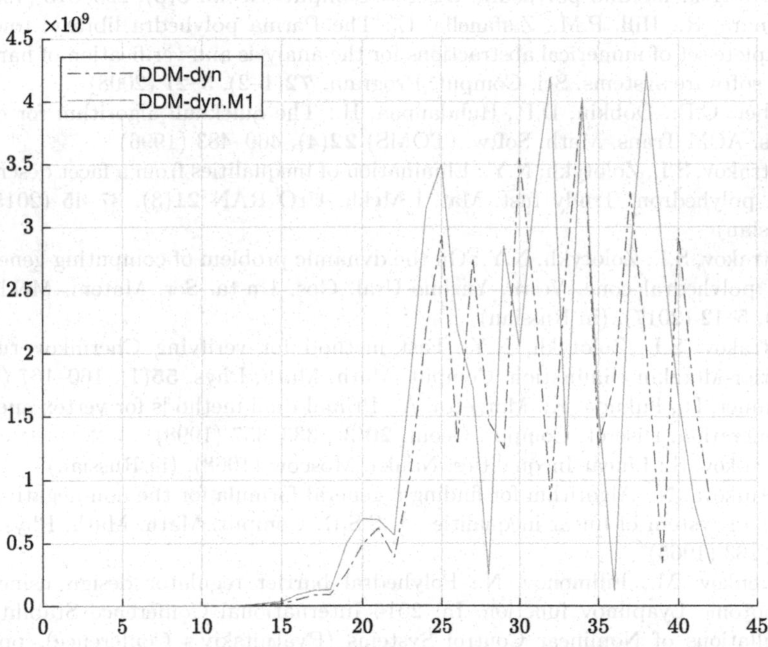

Fig. 3. Number of adjacency checks on ccc7

5 Conclusion

A dynamic version of the double description method for finding extreme rays of a polyhedral cone with online input of inequalities has been proposed. Two known modifications of the algorithm (graph adjacency test and maintaining the edge set to generate new extreme rays) have been adopted for its dynamic variant and tested on multiple problem instances. The results of computational experiments demonstrate better performance with certain configurations than that of SKELETON on a number of problems if the same limitations of input being unknown ahead of time are imposed on both implementations.

References

1. Amato, G., Scozzari, F., Zaffanella, E.: Efficient constraint/generator removal from double description of polyhedra. Electron. Notes Theor. Comput. Sci. **307**, 3–15 (2014)
2. Avis, D.: A revised implementation of the reverse search vertex enumeration algorithm. In: Kalai, G., Ziegler, G.M. (eds.) Polytopes-Combinatorics and Computation, vol. 29, pp. 177–198. Springer, Basel (2000). https://doi.org/10.1007/978-3-0348-8438-9_9
3. Avis, D., Bremner, D., Seidel, R.: How good are convex hull algorithms? Comput. Geom. **7**(5–6), 265–301 (1997)
4. Avis, D., Fukuda, K.: A pivoting algorithm for convex hulls and vertex enumeration of arrangements and polyhedra. Discret. Comput. Geom. **8**(3), 295–313 (1992)
5. Bagnara, R., Hill, P.M., Zaffanella, E.: The Parma polyhedra library: toward a complete set of numerical abstractions for the analysis and verification of hardware and software systems. Sci. Comput. Program. **72**(1–2), 3–21 (2008)
6. Barber, C.B., Dobkin, D.P., Huhdanpaa, H.: The quickhull algorithm for convex hulls. ACM Trans. Math. Softw. (TOMS) **22**(4), 469–483 (1996)
7. Bastrakov, S.I., Zolotykh, N.Y.: Elimination of inequalities from a facet description of a polyhedron. Trudy Inst. Mat. i Mekh. UrO RAN **21**(3), 37–45 (2015). (in Russian)
8. Bastrakov, S.I., Zolotykh, N.Y.: On the dynamic problem of computing generators of a polyhedral cone. Vestn. Yuzhno-Ural. Gos. Un-ta. Ser. Matem. Mekh. Fiz. **9**(1), 5–12 (2017). (in Russian)
9. Bastrakov, S.I., Zolotykh, N.Y.: Fast method for verifying Chernikov rules in Fourier-Motzkin elimination. Comput. Math. Math. Phys. **55**(1), 160–167 (2015)
10. Bremner, D., Fukuda, K., Marzetta, A.: Primal-dual methods for vertex and facet enumeration. Discret. Comput. Geom. **20**(3), 333–357 (1998)
11. Chernikov, S.: Linear Inequalities. Nauka, Moscow (1968). (in Russian)
12. Chernikova, N.: Algorithm for finding a general formula for the non-negative solutions of system of linear inequalities. U.S.S.R. Comput. Math. Math. Phys. **5**(2), 228–233 (1965)
13. Demenkov, M., Filimonov, N.: Polyhedral barrier regulator design using non-monotonic Lyapunov function. In: 2016 International Conference Stability and Oscillations of Nonlinear Control Systems (Pyatnitskiy's Conference), pp. 1–3. IEEE (2016)

14. Fukuda, K., Prodon, A.: Double description method revisited. In: Deza, M., Euler, R., Manoussakis, I. (eds.) CCS 1995. LNCS, vol. 1120, pp. 91–111. Springer, Heidelberg (1996). https://doi.org/10.1007/3-540-61576-8_77
15. Horst, R., Pardalos, P.M., Van Thoai, N.: Introduction to Global Optimization. Springer, Dordrecht (2000)
16. Motzkin, T., Raiffa, H., Thompson, G., Thrall, R.: The double description method. In: Kuhn, H., Tucker, A.W. (eds.) Contributions to Theory of Games, vol. 2. Princeton University Press, Princeton (1953)
17. Perry, J.: Exploring the dynamic Buchberger algorithm. In: Proceedings of the 2017 ACM on International Symposium on Symbolic and Algebraic Computation, pp. 365–372. ACM (2017)
18. Schrijver, A.: Theory of Linear and Integer Programming. Wiley, New York (1998)
19. Schrijver, A.: Combinatorial Optimization: Polyhedra and Efficiency, vol. 24. Springer, Heidelberg (2003)
20. Terzer, M., Stelling, J.: Large-scale computation of elementary flux modes with bit pattern trees. Bioinformatics **24**(19), 2229–2235 (2008)
21. Ziegler, G.M.: Lectures on Polytopes, vol. 152. Springer, Heidelberg (2012)
22. Zolotykh, N.Y., Kubarev, V.K., Lyalin, S.S.: Double description method over the field of algebraic numbers. Vestn. Udmurtsk. Univ. Mat. Mekh. Komp. Nauki **28**(2), 161–175 (2018). (in Russian)
23. Zolotykh, N.: New modification of the double description method for constructing the skeleton of a polyhedral cone. Comput. Math. Math. Phys. **52**(1), 146–156 (2012)

Comparison of Several Stochastic and Deterministic Derivative-Free Global Optimization Algorithms

Vladislav Sovrasov(✉)

Lobachevsky State University of Nizhni Novgorod, Nizhni Novgorod, Russia
sovrasov.vlad@gmail.com

Abstract. In this paper popular open-source solvers are compared against Globalizer solver, which is developed at the Lobachevsky State University. The Globalizer is designed to solve problems with black-box objective function satisfying the Lipschitz condition and shows competitive performance with other similar solvers. The comparison is done on several sets of challenging multi-extremal benchmark functions. Also this work considers a method of heuristic hyperparameters control for the Globalizer allowing to reduce amount of initial tuning before optimization. The proposed scheme allows substantially increase convergence speed of the Globalizer by switching between "local" and "global" search phases in runtime.

Keywords: Deterministic global optimization · Stochastic global optimization · Algorithms comparison · Derivative-free algorithms · Black-box optimization · Multi-extremal problems

1 Introduction

The problem of finding the global minima of the nonlinear nonconvex functions is considered to be one of the most difficult mathematical programming problems traditionally. Often, it appears to be more complex than the local optimization in an essentially multidimensional space. For the latter, the application of the simplest gradient descent method or of the pattern search algorithms may appear to be sufficient [26] whereas in order to *guarantee* the finding of the global optimum, the optimization methods have to accumulate the information on the behavior of the objective function in the whole search domain [3,10,17,25]. Recently, various stochastic global optimization algorithms, first of all, the evolution ones [12,20,23] became popular. These ones have rather simple structure and allow solving the problems of large dimensionality. However, these methods provide the global convergence in the probabilistic meaning only.

The study was supported by the Russian Science Foundation, project No. 16-11-10150.

M. Khachay et al. (Eds.): MOTOR 2019, LNCS 11548, pp. 70–81, 2019.
https://doi.org/10.1007/978-3-030-22629-9_6

In the present work, the open-source implementations of the eight different global optimization methods included into the NLOpt library [8] and the SciPY package [9] are considered. All algorithms were tested on a set of 900 essentially multiextremal functions, which has been generated with the use of special problem generators [5,7].

2 Related Work

Earlier, the comparison of the stochastic global optimization algorithms [1,16] as well as of the deterministic ones [13,14,18] between each other has been considered in the literature. In these works, most of modern methods have been studied in details. In the majority of works, the sets of well-known test problems (for example, the Rastrigin function, Ackley function, etc.) were taken as the sets of test functions. The sizes of such sets don't exceed 100 different functions usually, some of which can be the single-extremal ones (such as the Rosenbrock function).

In [2], some general principles were formulated, which, in the author's opinion, should be obeyed when comparing the optimization methods. In particular, the authors say about the advantages of the problem generators allowing generating the large sets of problems thus minimizing the random effects when comparing the methods. At the same time, the use of a single generator can appear to be not enough for a comprehensive comparison of the methods. In order to overcome this problem in part, the authors of the paper [2] advise to use several generators of various nature and to create the sets of problems of various complexity.

Taking into account the experience of the preceding works in the field of comparison of the optimization methods, two generators of the test problems of different nature will be used in the present work. Using these ones, 9 sets of 100 problems of various complexity with the dimensionality varying from 2 to 5 were generated.

3 Statement of Multidimensional Global Optimization Problem

In this paper, the core class of optimization problems, which can be solved using global optimization methods, is formulated. This class involves the multidimensional global optimization problems without constraints, which can be defined in the following way:

$$\begin{gathered}\varphi(y^*) = \min\{\varphi(y) : y \in D\}, \\ D = \{y \in \mathbb{R}^N : a_i \leq y_i \leq b_i, 1 \leq i \leq N\}\end{gathered} \tag{1}$$

with the given boundary vectors a and b. It is supposed, that the objective function $\varphi(y)$ satisfies the Lipschitz condition

$$|\varphi(y_1) - \varphi(y_2)| \leq L\|y_1 - y_2\|, y_1, y_2 \in D, \tag{2}$$

where $L > 0$ is the Lipschitz constant, and $||\cdot||$ denotes the norm in $\mathbb{R}^N$ space.

Usually, the objective function $\varphi(y)$ is defined as a computational procedure, according to which the value $\varphi(y)$ can be calculated for any vector $y \in D$ (let us further call such a calculation *a trial*). It is supposed that this procedure is time-consuming.

4 Review of Considered Optimization Methods

4.1 Algorithm of Global Search

Dimension Reduction with Space-Filling Curves. Within the framework of the information-statistical global optimization theory, the Peano space-filling curves (or *evolvents*) $y(x)$ mapping the interval $[0, 1]$ onto an N-dimensional hypercube D unambiguously are used for the dimensionality reduction [22,24,25].

As a result of the reduction, the initial multidimensional global optimization problem (1) is reduced to the following one-dimensional problem:

$$\varphi(y(x^*)) = \min\{\varphi(y(x)) : x \in [0, 1]\}. \tag{3}$$

It is important to note that this dimensionality reduction scheme transforms the Lipschitzian function from (1) to the corresponding one-dimensional function $\varphi(y(x))$, which satisfies the uniform Hölder condition, i. e.

$$|\varphi(y(x_1)) - \varphi(y(x_2))| \leq H|x_1 - x_2|^{\frac{1}{N}}, x_1, x_2 \in [0, 1], \tag{4}$$

where the constant H is defined by the relation $H = 2L\sqrt{N+3}$, L is the Lipschitz constant from (2), and N is the dimensionality of the optimization problem (1).

The algorithms for the numerical construction of the Peano curve approximations are given in [25].

The computational scheme obtained as a result of the dimensionality reduction consists of the following:

- The optimization algorithm performs the minimization of the reduced one-dimensional function $\varphi(y(x))$ from (3),
- After determining the next trial point x, a multidimensional image y is calculated by using the mapping $y(x)$,
- The value of the initial multidimensional function $\varphi(y)$ is calculated at the point $y \in D$,
- The calculated value $z = \varphi(y)$ is used further as the value of the reduced one-dimensional function $\varphi(y(x))$ at the point x.

Optimization method applied in Globalizer [6] to solve the reduced problem (3) is based on the AGS method, which can be presented as follows—see [24], [25].

The algorithm considered for solving the stated problem implies generating a sequence of points x_k, in which the values of the minimized function $z_k = f(x_k)$ are computed. Let us call the process of computating the function value (including calculating an image $y^k = y(x^k)$) a trial, and the pair (x^k, z^k)—the result of the trial. A set of the pairs $\{(x^k, z^k)\}, 1 \leqslant k \leqslant n$ makes up the search information accumulated by the method after executing n steps.

The initial iteration of the algorithm is performed at an arbitrary point $x^1 \in (0, 1)$. Then, let us suppose that k, $k \geq 1$, optimization iterations have been completed already. The selection of the trial point x^{k+1} for the next iteration is performed according to the following rules.

Step 1. Renumber the points in the set $X_k = \{x^1, \ldots, x^k\} \cup \{0\} \cup \{1\}$, which includes the boundary points of the interval $[0, 1]$ as well as the points of preceding trials, by the lower indices in order of increasing coordinate values i.e.

$$0 = x_0 < x_1 < \ldots < x_{k+1} = 1$$

Step 2. Assuming $z_i = f(x_i), 1 \leqslant i \leqslant k$, compute the values

$$\mu = \max_{1 \leqslant i \leqslant k} \frac{|z_i - z_{i-1}|}{\Delta_i}, M = \begin{cases} r\mu, \mu > 0 \\ 1, \mu = 0 \end{cases} \tag{5}$$

where $r > 1$ is a predefined parameter for the method, and $\Delta_i = (x_i - x_{i-1})^{\frac{1}{N}}$.

Step 3. For each interval $(x_{i-1}, x_i), 1 \leqslant i \leqslant k+1$, compute the characteristics according to the formulae

$$R(1) = 2\Delta_1 - 4\frac{z_1}{M}, R(k+1) = 2\Delta_{k+1} - 4\frac{z_k}{M}, \tag{6}$$

$$R(i) = \Delta_i + \frac{(z_i - z_{i-1})^2}{M^2 \Delta_i} - 2\frac{z_i + z_{i-1}}{M}, 1 < i < k+1. \tag{7}$$

Step 4. Determine the interval with the maximum characteristic (x_{t-1}, x_t), $t = \operatorname{argmax}_{1 \leqslant i \leqslant k+1} R(i)$

Step 5. Execute a new trial at point x_{k+1} computed according to the formula

$$x_{k+1} = \frac{x_t + x_{t-1}}{2}, t = 1, t = k+1,$$

$$x_{k+1} = \frac{x_t + x_{t-1}}{2} - \operatorname{sign}(z_t - z_{t-1}) \frac{1}{2r} \left[\frac{|z_t - z_{t-1}|}{\mu} \right]^N, 1 < t < k+1. \tag{8}$$

The stopping condition, which terminated the trials, is defined by the inequality $\Delta_t \leqslant \varepsilon$ for the interval with the maximum characteristics from Step 4 and $\varepsilon > 0$ is the predefined accuracy of the optimization problem solution. If the stopping condition is not satisfied, the index k is incremented by 1, and the new global optimization iteration is executed.

The convergence conditions of the described algorithm are given, for example, in [25].

Hyperparameters Control in AGS. The parameter r from (5) affects the global convergence of AGS directly (see [25], Chap. 8): at high enough value of r, the method converges to all global minima of the objective function with guarantee. At the same time, according to (7) and (8), at the infinitely high value of r, AGS turns into the brute force search method on a uniform grid.

In the ideal case, in order to provide the highest convergence speed, the estimate of the Lipschitz constant from (5) should not be too overestimated, but in practice the actual value of L from (2) in unknown, and one has either to take an obviously overestimated value of r or to execute several runs of AGS with different parameters. In order to resolve the problem of choosing r to some extent, let us use the following scheme:

- execute q iterations of AGS with $r = r_{max}$;
- execute q iterations of AGS with $r = r_{min}$;
- repeat the above steps either until convergence or until the allowed number of iterations are exhausted.

In the above algorithm, $r_{min} < r_{max}$, $q > 1$. Instead one parameter r, now 3 ones should be selected. However, according to the results of the numerical experiments, it is easier than to find the optimal value of r. Intuitively, the practical efficiency of the proposed scheme can be explained by the fact that now the operation of the method takes place in two modes: the global search with $r = r_{max}$ and the local one with $r = r_{min}$. If during the global search phase, the method approached the global minimum whereas during the next phase, the estimate of the global minimum would be refined rapidly. If two phases are not enough, the process is continued. This way, a better trade-off between the exploration and the exploitation is achieved. Further, we will denote the method utilizing the scheme described above as AGS-AR.

4.2 Other Optimization Methods

- **Multi Level Single Linkage** [11]. MLSL is an improved multistart algorithm. It samples low-discrepancy starting points and does local optimizations from them. In contrast to the dummy multistart schemes MLSL uses some clustering heuristics to avoid multiple local descents to already explored local minima.
- **DIRECT** [10]. The algorithm is deterministic and recursively divides the search space and forms a tree of hyper-rectangles (boxes). DIRECT uses the objective function values and the Lipschitz condition (2) to estimate promising boxes.
- **Locally-biased DIRECT (DIRECT*l*)** [4]. It's a variation of DIRECT which pays less attention to non-promising boxes and therefore has less exploration power: it can converge faster on problems with few local minima, but lost the global one in complicated cases.

- **Dual Simulated Annealing** [27]. This stochastic method is a combination of the Classical Simulated Annealing and the Fast Simulated Annealing coupled to a strategy for applying a local search on accepted locations. It converges much faster than both parent algorithms, CSA and FSA.
- **Differential Evolution** [23]. DE is an adaptation of the original genetic algorithm to the continuous search domain.
- **Controlled Random Search** [19]. The CRS starts with a set of random points and then defines the next trial point in relation to a simplex chosen randomly from a stored configuration of points. CRS in not an evolutional algorithm, although stores something like population and performs transformation resembling a mutation.
- **StoGO** [15]. StoGO is dividing the search space into smaller hyper- rectangles via a branch-and-bound approach, and searching them by a local-search algorithm, optionally including some randomness.

All the mentioned algorithms are available in source codes as parts of widespread optimization packages. DIRECT, DIRECT*l*, CRS, MLSL and StoGO are part of the NLOpt library [8]. Differential Evolution and DSA can be found in the latest version of the SciPy [9] package for Python.

5 Tools for Comparison of Global Optimization Algorithms

The use of the sets of test problems with known solutions generated by some random mechanisms is one of commonly accepted approaches to comparing the optimization algorithms [2]. In the present work, we will use two generators of test problems generating the problems of different nature [5,7][1].

Let us denote the problem set obtained with the use of the first generator from [7] as F_{GR}. The mechanism of generation of the problems F_{GR} doesn't provide the control of the problem complexity and of the number of local optima. However, the generated functions are known to be the multiextremal ones essentially. Besides, the problems generated by F_{GR} are the two-dimensional ones. In the present work, we will use 100 functions from the class F_{GR} generated randomly.

The GKLS generator [5] allows obtaining the problems of given dimensionality with given number of extrema. Moreover, GKLS allows adjusting the complexity of the problems by decreasing or increasing the size of the global minimum attractor. In [21] the parameters of the generator allowing generating the sets of 100 problems each of two levels of complexity (Simple and Hard) of the dimensionality equal to 2, 3, 4, and 5 are given. Following the authors of the GKLS generator, we will use the parameters proposed by them and, this way, add 800 more problems of various dimensionalities and complexity into the test problem set.

[1] Software implementations of these generators are available in source codes at the page https://github.com/sovrasov/global-optimization-test-problems.

Let us suppose a test problem to be solved if the optimization method executes the scheduled trial y^k in a δ-vicinity of the global minimum y^*, i.e. $\|y^k - y^*\| \leq \delta = \alpha \|b - a\|$, where a and b are the left and the right boundaries of the hypercube from (1), α is relative precision. If this relation is not fulfilled before the expiration of the limit of the number of trials, the problem was considered to be unsolved. The limit of the number of trials and α were set for each problem class according to the dimensionality and complexity (see Table 1).

Table 1. Trials limits and relative precision for the test problem classes

Problems class	Trials limit	α
F_{GR}	5000	0.01
GKLS 2d Simple	8000	0.01
GKLS 2d Hard	9000	0.01
GKLS 3d Simple	15000	0.01
GKLS 3d Hard	25000	0.01
GKLS 4d Simple	150000	$\sqrt[4]{10^{-6}}$
GKLS 4d Hard	250000	$\sqrt[4]{10^{-6}}$
GKLS 5d Simple	350000	$\sqrt[5]{10^{-7}}$
GKLS 5d Hard	600000	$\sqrt[5]{10^{-7}}$

Let us consider the averaged number of trials executed to solve a single problem and the number of solved problems as the characteristics of the optimization method on each class. The less the number of trials, the faster the method converges to a solution, hence the less times it turns to a potentially computation-costly procedure of computing the objective function. The number of solved problems evidences the reliability of the method at given parameters on the class of test problems being solved. In order to make independent the quantities featuring the reliability and the speed of convergence, averaged number of trials always was calculated taking into account solved problems only.

The average number of trials doesn't represent the real behavior of an optimization method on a problems set in some cases. For an instance, if a method performs well on the most problems and spends too much trials to solve the least several problems, we wouldn't catch such case looking at the average number of trials only. As an advanced measure of performance we will use the operating characteristic [7]. It's defined by a set of points on the (K, P) plane where K is the average number of search trials conducted before satisfying the termination condition when minimizing a function from a given class, and P is the proportion of problems solved successfully. If at a given K, the operating characteristic of a method goes higher than one from another method, it means that at fixed search costs, the former method has a greater probability of finding the solution. If some value of P is fixed, and the characteristic of a method goes to the left from that of another method, the former method requires fewer resources to achieve the same reliability.

6 Results of Numerical Experiments

The results of various algorithms on different problem classes depend on the adjustments of algorithms directly. In most cases, the authors of software implementations are oriented onto the problems of medium difficulty. In order to obtain a satisfactory result when solving the essentially multiextremal problems, a correction of some parameters is required. When conducting the comparison, the following parameters for the methods were employed:

- in the AGS-AR method, the parameter of alternation the global and local stages q was set to be equal to $50 \cdot \log_2(N-1) \cdot N^2$, also $r_{min} = 3$, $r_{max} = 2 \cdot r_{min}$;
- in the DIRECT and DIRECTl methods, the parameter $\epsilon = 10^{-4}$;
- in the SDA method, the parameter $visit = 2.72$.

The rest parameters were varied subject to the problem class (see Table 2). For the AGS the value of the r parameter, such that the method solves all problems and performs the minimum amount of trials, was estimated by brute force on the uniform grid with step 0.1.

Table 2. Class-specific parameters of the optimization algorithms

	AGS	CRS	DE
F_{GR}	$r = 3$	popsize = 150	mutation = (1.1, 1.9), popsize = 60
GKLS 2d Simple	$r = 4.6$	popsize = 200	mutation = (1.1, 1.9), popsize = 60
GKLS 2d Hard	$r = 6.5$	popsize = 400	mutation = (1.1, 1.9), popsize = 60
GKLS 3d Simple	$r = 3.7$	popsize = 1000	mutation = (1.1, 1.9), popsize = 70
GKLS 3d Hard	$r = 4.4$	popsize = 2000	mutation = (1.1, 1.9), popsize = 80
GKLS 4d Simple	$r = 4.7$	popsize = 8000	mutation = (1.1, 1.9), popsize = 90
GKLS 4d Hard	$r = 4.9$	popsize = 16000	mutation = (1.1, 1.9), popsize = 100
GKLS 5d Simple	$r = 4$	popsize = 25000	mutation = (1.1, 1.9), popsize = 120
GKLS 5d Hard	$r = 4$	popsize = 30000	mutation = (1.1, 1.9), popsize = 140

The results of running the optimization methods on the considered problem classes are presented in Tables 3 and 4. The DIRECT, AGS and AGS-AR methods have demonstrated the best convergence speed on all classes, at that AGS-AR inferior to DIRECT on the 2d problems from the Simple classes and has an advantage on the problems of the Hard classes. As one can see from Table 4, the deterministic methods (AGS, AGS-AR, DIRECT, and DIRECTl)

Table 3. Averaged number of trials executed by optimization methods for solving the test optimization problems

	AGS	AGS-AR	CRS	DIRECT	DIRECT*l*	MLSL	SDA	DE	StoGO
F_{GR}	193.1	248.3	400.3	**182.2**	214.9	947.2	691.2	1257.3	1336.8
GKLS 2d Simple	254.9	221.6	510.6	**189.0**	255.2	556.8	356.3	952.2	1251.5
GKLS 2d Hard	**728.7**	785.0	844.7	985.4	1126.7	1042.5	1637.9	1041.1	2532.2
GKLS 3d Simple	1372.1	1169.5	4145.8	**973.6**	1477.8	4609.2	2706.5	5956.9	3856.1
GKLS 3d Hard	3636.1	**1952.1**	6787.0	2298.7	3553.3	5640.1	4708.4	6914.3	7843.2
GKLS 4d Simple	5729.8	**4919.1**	19883.6	7328.8	15010.0	41484.8	22066.0	6271.2	29359.2
GKLS 4d Hard	13113.4	**12860.1**	27137.4	22884.4	55596.1	80220.1	68048.0	12487.6	58925.5
GKLS 5d Simple	**5821.5**	6241.3	62921.7	5966.1	10795.5	52609.2	34208.8	20859.4	69206.8
GKLS 5d Hard	**17008.6**	21555.1	87563.9	61657.3	148637.8	138011.8	115634.6	26850.0	141886.5

Table 4. Number of test optimization problems solved by the methods

	AGS	AGS-AR	CRS	DIRECT	DIRECT*l*	MLSL	SDA	DE	StoGO
F_{GR}	100	100	76	100	100	97	96	96	67
GKLS 2d Simple	100	100	85	100	100	100	100	98	90
GKLS 2d Hard	100	97	74	100	100	100	93	85	77
GKLS 3d Simple	100	100	75	100	100	100	89	86	44
GKLS 3d Hard	100	100	72	100	99	100	88	77	43
GKLS 4d Simple	100	100	74	100	100	94	82	68	72
GKLS 4d Hard	100	100	60	99	99	94	73	55	69
GKLS 5d Simple	100	100	86	100	100	98	100	88	82
GKLS 5d Hard	100	100	77	100	93	79	86	77	78

were the most reliable. Among the stochastic methods, MLSL and SDA have demonstrated the highest reliability.

Operating characteristic of the methods (Figs. 1a–d) demonstrates that AGS and AGS-AR faster than the other methods achieve 100% success rate. Also on GKLS 5d Simple the DIRECT generally has the best performance, but there are several hard problems that affect it's average number of trials metric.

Robustness of AGS and AGS-AR to the Hyperparameters Choice. In order to investigate the influence of hyperparameters to the convergence speed of the AGS and AGS-AR, experiments with the following settings were conducted on the problems from GKLS 5d Simple class:

- AGS with $r = 4$ (like in the Table 2);
- AGS with $r = 6$;
- AGS-AR with parameters from the beginning of the Sect. 6 ($q = 50 \cdot \log_2(4) \cdot 25 = 2500$, $r_{min} = 3$, $r_{max} = 2 \cdot r_{min}$);
- AGS-AR with $r_{max} = 8$ and other parameters from the previous experiment;
- AGS-AR with $q = 1000$ and other parameters from the beginning of the Sect. 6;

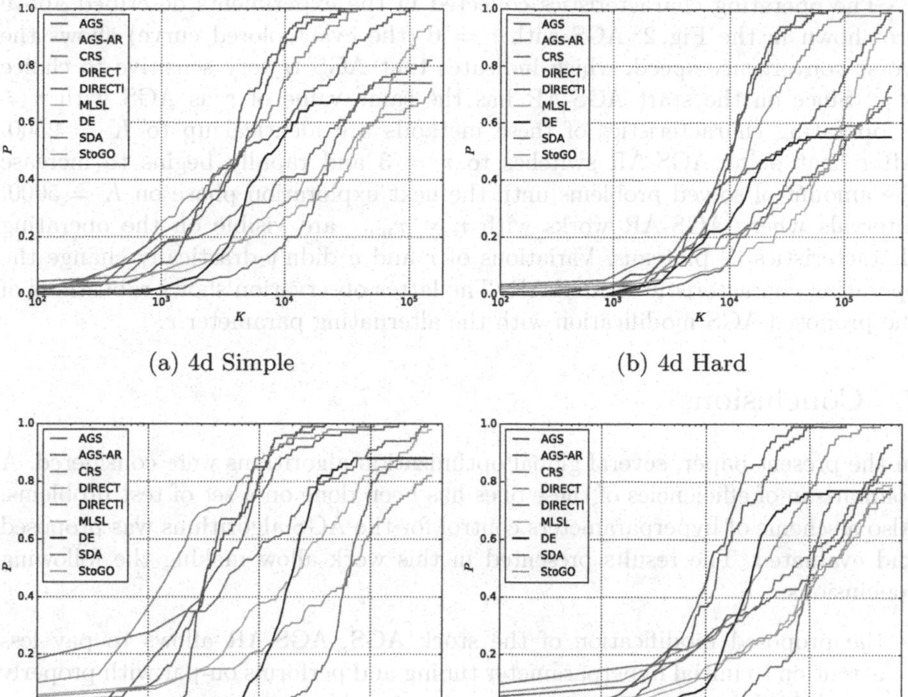

(a) 4d Simple

(b) 4d Hard

(c) 5d Simple

(d) 5d Hard

Fig. 1. Operating characteristics of the algorithms when solving problems from the GKLS 4d and 5d classes. Best viewed in color.

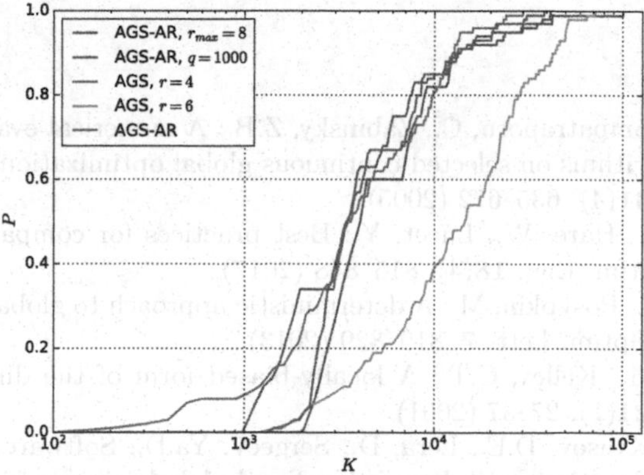

Fig. 2. Operating characteristics of AGS and AGS-AR with different hyperparameters when solving problems from the GKLS 5d Simple classes. Best viewed in color. (Color figure online)

The operating characteristics collected in the experiments described above are shown in the Fig. 2. AGS with $r = 6$ (the cyan-colored curve) shows the worst convergence speed, which indicates that AGS is very sensitive to choice of r. Since on the start AGS-AR has the same value of r as AGS with $r = 6$, operating characteristics of these methods are identical up to $K = 2500$. After that point AGS-AR switches to $r = 3$ and rapidly begins to increase the amount of solved problems until the next exploration phase on $K = 5000$. Intervals where AGS-AR works with $r = r_{max}$ are visible on the operating characteristics as plateaus. Variations of r and q didn't drastically change the operating characteristic of AGS-AR. The latter observation shows robustness of the proposed AGS modification with the alternating parameter r.

7 Conclusions

In the present paper, several global optimization algorithms were considered. A comparison of efficiencies of these ones has been done on a set of test problems. Also a scheme of hyperparameters control for the AGS algorithms was proposed and evaluated. The results presented in this work allow making the following conclusions:

- the proposed modification of the stock AGS, AGS-AR allows to pay less attention to initial hyperparameter tuning and performs on-par with properly tuned AGS;
- AGS-AR method has demonstrated the convergence speed and reliability at the level of DIRECT and exceeds many other algorithms, the open-source implementations of which are available;
- the stochastic optimization methods inferior to the deterministic ones in the convergence speed and in reliability. It is manifested especially strongly on more complex multiextremal problems.

References

1. Ali, M.M., Khompatraporn, C., Zabinsky, Z.B.: A numerical evaluation of several stochastic algorithms on selected continuous global optimization test problems. J. Glob. Optim. **31**(4), 635–672 (2005)
2. Beiranvand, V., Hare, W., Lucet, Y.: Best practices for comparing optimization algorithms. Optim. Eng. **18**(4), 815–848 (2017)
3. Evtushenko, Y., Posypkin, M.: A deterministic approach to global box-constrained optimization. Optim. Lett. **7**, 819–829 (2013)
4. Gablonsky, J.M., Kelley, C.T.: A locally-biased form of the direct algorithm. J. Glob. Optim. **21**(1), 27–37 (2001)
5. Gaviano, M., Kvasov, D.E., Lera, D., Sergeev, Ya.D.: Software for generation of classes of test functions with known local and global minima for global optimization. ACM Trans. Math. Softw. **29**(4), 469–480 (2003)
6. Gergel, V.P., Barkalov, K.A., Sysoyev, A.V.: A novel supercomputer software system for solving time-consuming global optimization problems. Numer. Algebr. Control. Optim. **8**(1), 47–62 (2018)

7. Grishagin, V.: Operating characteristics of some global search algorithms. Probl. Stoch. Search **7**, 198–206 (1978). (In Russian)
8. Johnson, S.G.: The NLOpt nonlinear-optimization package. http://ab-initio.mit.edu/nlopt. Accessed 24 Dec 2018
9. Jones, E., Oliphant, T., Peterson, P., et al.: SciPy: open source scientific tools for Python (2001–). http://www.scipy.org/. Accessed 24 Dec 2018
10. Jones, D.R.: The direct global optimization algorithm. In: Floudas, C., Pardalos, P. (eds.) The Encyclopedia of Optimization, pp. 725–735. Springer, Boston (2009). https://doi.org/10.1007/978-0-387-74759-0_128
11. Kan, A.H.G.R., Timmer, G.T.: Stochastic global optimization methods Part II: multi level methods. Math. Program. **39**, 57–78 (1987)
12. Kennedy, J., Eberhart, R.: Particle swarm optimization. In: Proceedings of ICNN 1995 - International Conference on Neural Networks, vol. 4, pp. 1942–1948 (1995)
13. Kvasov, D.E., Mukhametzhanov, M.S.: Metaheuristic vs. deterministic global optimization algorithms: the univariate case. Appl. Math. Comput. **318**, 245–259 (2018)
14. Liberti, L., Kucherenko, S.: Comparison of deterministic and stochastic approaches to global optimization. Int. Trans. Oper. Res. **12**, 263–285 (2005)
15. Madsen, K., Zertchaninov, S.: A new branch-and-bound method for global optimization (1998)
16. Mullen, K.: Continuous global optimization in R. J. Stat. Softw. **60**(6), 1–45 (2014)
17. Paulavivcius, R., Zilinskas, J., Grothey, A.: Parallel branch and bound for global optimization with combination of Lipschitz bounds. Optim. Methods Softw. **26**(3), 487–498 (1997)
18. Pošík, P., Huyer, W., Pál, L.: A comparison of global search algorithms for continuous black box optimization. Evol. Comput. **20**(4), 509–541 (2012)
19. Price, W.L.: Global optimization by controlled random search. J. Optim. Theory Appl. **40**(3), 333–348 (1983)
20. Schluter, M., Egea, J.A., Banga, J.R.: Extended ant colony optimization for non-convex mixed integer nonlinear programming. Comput. Oper. Res. **36**(7), 2217–2229 (2009)
21. Sergeyev, Y., Kvasov, D.: Global search based on efficient diagonal partitions and a set of Lipschitz constants. SIAM J. Optim. **16**(3), 910–937 (2006)
22. Sergeyev, Y.D., Strongin, R.G., Lera, D.: Introduction to Global Optimization Exploiting Space-Filling Curves. Springer Briefs in Optimization. Springer, New York (2013). https://doi.org/10.1007/978-1-4614-8042-6
23. Storn, R., Price, K.: Differential evolution - a simple and efficient heuristic for global optimization over continuous spaces. J. Glob. Optim. **11**(4), 341–359 (1997)
24. Strongin, R.: Numerical Methods in Multiextremal Problems (Information-Statistical Algorithms). Nauka, Moscow (1978). (In Russian)
25. Strongin R.G., Sergeyev Ya.D.: Global Optimization with Non-convex Constraints: Sequential and Parallel Algorithms. Kluwer Academic Publishers, Dordrecht (2000)
26. Torczon, V.: On the convergence of pattern search algorithms. SIAM J. Optim. **9**(1), 1–25 (1997)
27. Xiang, Y., Sun, D., Fan, W., Gong, X.: Generalized simulated annealing algorithm and its application to the thomson model. Phys. Lett. A **233**(3), 216–220 (1997)

On Some Methods for Strongly Convex Optimization Problems with One Functional Constraint

Fedor S. Stonyakin[1(✉)], Mohammad S. Alkousa[2], Alexander A. Titov[2], and Victoria V. Piskunova[1]

[1] V.I. Vernadsky Crimean Federal University, Simferopol, Russia
fedyor@mail.ru, viktoryapiskunova@yandex.ru

[2] Moscow Institute of Physics and Technology, Moscow, Russia
{mohammad.alkousa,a.a.titov}@phystech.edu

Abstract. We consider the classical optimization problem of minimizing a strongly convex, non-smooth, Lipschitz-continuous function with one Lipschitz-continuous constraint. We develop the approach in [10] and propose two methods for the considered problem with adaptive stopping rules. The main idea of the methods is using the dichotomy method and solving an auxiliary one-dimensional problem at each iteration. Theoretical estimates for the proposed methods are obtained. Partially, for smooth functions, we prove the linear rate of convergence of the methods. We also consider theoretical estimates in the case of non-smooth functions. The results for some examples of numerical experiments illustrating the advantages of the proposed methods and the comparison with some adaptive optimal method for non-smooth strongly convex functions are also given.

Keywords: Optimization with functional constraint · Adaptive method · Lipschitz-continuous function · Lipschitz-continuous gradient · Strongly convex objective function · Dichotomy method

1 Introduction

The optimization of non-smooth functions with constraints attracts wide interest in large-scale optimization and its applications [4,14]. There are a lot of methods of solving such kind of optimization problems. Some examples of these methods, to name but a few, are: bundle-level method [13], penalty method

The authors are very grateful to Alexander V. Gasnikov and Anastasiya S. Ivanova for fruitful discussions. The research of F. Stonyakin in Algorithm 1, Theorem 2 and Lemma 2 was supported by Russian Foundation for Basic Research according to the project 18-29-03071 mk. The research of F. Stonyakin in Subsects. 3.4 and 3.5 was supported by Russian Science Foundation grant 18-71-10044.

M. Khachay et al. (Eds.): MOTOR 2019, LNCS 11548, pp. 82–96, 2019.
https://doi.org/10.1007/978-3-030-22629-9_7

[15] and Lagrange multipliers method [5]. Recently in [2], some adaptive Mirror Descent methods were proposed for optimization problems of convex and strongly convex functions with non-smooth constraints

$$\min\{f(x): \quad x \in Q \subset E, \quad g(x) \leq 0\}, \tag{1}$$

where Q is a convex and compact subset of a finite-dimensional real vector space E, $f: Q \to \mathbb{R}$ and $g: E \to \mathbb{R}$ are convex Lipschitz-continuous functions. In the case of several strongly convex non-smooth constraints, we consider one max-type constraint which is also strongly convex.

Methods in [2] are optimal from the point of view of lower oracle bounds and guarantee achieving acceptable precision ε with complexity $O\left(\varepsilon^{-1}\right)$ for strongly convex, Lipschitz-continuous objective f and convex Lipschitz-continuous constraint g.

In this paper, we develop the approach in [10] and propose an alternative approach for the problem (1) with a strongly convex Lipschitz-continuous objective f and a convex Lipschitz-continuous constraint g. Our approach is based on the transition to a strongly convex dual problem. In this case, the dual function depends on one dual variable $\lambda \geq 0$. When the Slater conditions for the problem (1) hold, all possible values of the dual variable are limited to a certain segment. This allows us to apply the dichotomy method similarly to [10] to search for the value of the dual variable λ, which is close to the appropriate λ_*, for which

$$\lambda_* \cdot g(x(\lambda_*)) = 0. \tag{2}$$

We propose two algorithms with adaptive stopping criterion that meet the necessary condition (2) in the general situation $\lambda_* \geq 0$ (Algorithm 1), as well as under the stronger assumption of the existence of $\lambda_* > 0$ (Algorithm 2). Partially, the last condition holds for the economic problem considered in [10].

It turns out that, with the possibility of a relatively quick solution of auxiliary problems, due to the proposed adaptive stopping criterion, Algorithms 1 and 2 may work faster than the optimal schemes in [2]. In proposed Algorithms 1 and 2 strong convexity of g is not required, and there is also no need to know the value of the strong convexity parameter of f.

The paper consists of an Introduction and four main sections. In Sect. 2 we consider the problem statement and some basic information concerning the necessary conditions of the extremum. In Sect. 3 we describe two main algorithms and give some estimates of the rate of convergence for them. Section 4 is devoted to basic information for optimal Mirror Descent Algorithms in the class of non-smooth strongly convex functions [2]. In Sect. 5 we make a comparison between the proposed algorithms and Mirror Descent Algorithm [2].

Thus, in the paper, we propose two methods for solving the problem (1) with the following types of assumptions:

$$|f(x) - f(y)| \leqslant M_f||x-y||_2, \quad |g(x) - g(y)| \leqslant M_g||x-y||_2 \tag{3}$$

or

$$||\nabla f(x) - \nabla f(y)||_2 \leqslant L_f||x-y||_2, \quad ||\nabla g(x) - \nabla g(y)||_2 \leqslant L_g||x-y||_2 \tag{4}$$

for all $x, y \in Q$, and for some real positive numbers M_f, M_g, L_f, L_g.

The contributions of this paper can be summarized as follows.

- With assumptions (4), the proposed methods have complexity

$$O\left(\log_2^2 \frac{1}{\varepsilon}\right), \tag{5}$$

i.e. the linear rate of convergence. Note that we assume the strong convexity for the objective f only. The functional constraint g may not be strongly convex.

- With assumptions (3) we obtain complexity $O\left(\frac{1}{\varepsilon^2}\log_2 \frac{1}{\varepsilon}\right)$, which is generally not optimal. However, due to the adaptivity of Algorithms 1 and 2, these methods can work faster than the optimal ones in [2] (see Sect. 5 below). Note that, unlike ([2], Subection 3.2), we require the strong convexity only of the objective functional f. In this case, the functional g, in general, may not be strongly convex.
- Also, a class of non-smooth functionals is considered, for which Algorithms 1 and 2 have complexity (5) (see Subsect. 3.4 below).

2 Problem Statement

Let $(E, ||\cdot||_2)$ be a normed finite-dimensional vector space with inner product $\langle\cdot,\cdot\rangle$ and norm $||x||_2 = \sqrt{\langle x, x\rangle}$. In this paper we consider the following optimization problem

$$f(x) \to \min_{\substack{g(x)\leqslant 0\\ x\in Q}}, \tag{6}$$

where f is a μ_f-strongly convex function with respect to the 2-norm, i.e.

$$f(\alpha x + (1-\alpha)y)) \leq \alpha f(x) + (1-\alpha)f(y) - \alpha(1-\alpha)\frac{\mu_f}{2}\|x-y\|_2^2$$

for $\alpha \in [0,1]$ and for all $x, y \in Q$. Assume that f and g are Lipschitz-continuous:

$$|f(y) - f(x)| \leqslant M_f\|y-x\|_2,\ \ \forall x, y \in Q,$$

$$|g(y) - g(x)| \leqslant M_g\|y-x\|_2,\ \ \forall x, y \in Q.$$

Let us introduce a dual factor $\lambda \geqslant 0$ and consider the dual problem to (6).

$$\min_{\substack{g(x)\leqslant 0\\ x\in Q}} f(x) = \min_{x\in Q}\left\{f(x) + \max_{\lambda\geqslant 0}(\lambda g(x))\right\} = \max_{\lambda\geqslant 0}\left\{\underbrace{\min_{x\in Q}\left(f(x) + \lambda g(x)\right)}_{=\varphi(\lambda)}\right\}.$$

Then the dual problem to the problem (6) is:

$$\varphi(\lambda) = f(x(\lambda)) + \lambda g(x(\lambda)) \to \max_{\lambda\geqslant 0}, \tag{7}$$

where

$$x(\lambda) = \arg\min_{x \in Q} \{f(x) + \lambda g(x)\}. \quad (8)$$

Let us mention the following important well-known Demyanov-Danskin-Rubinov Theorem, see [7,8].

Theorem 1. *Let* $\varphi(\lambda) = \min_{x \in X} F(x, \lambda)$ *for all* $\lambda \geqslant 0$, *where* $F(x, \lambda)$ *is a smooth convex function with respect to* λ *and* $x(\lambda)$ *is the only maximum point. Then*

$$\varphi'(\lambda) = F'_{\lambda}(x(\lambda), \lambda).$$

For the problem (7) Theorem 1 means that:

$$\varphi'(\lambda) = g(x(\lambda)). \quad (9)$$

Let λ^* be a solution of the dual problem (7). Then, according to the necessary condition of the extremum, the following equality must be satisfied for λ^*:

$$\lambda^* g(x(\lambda^*)) = 0, \ \lambda^* \geqslant 0,$$

which, by using (9), can be modified as follows:

$$\lambda^* \varphi'(\lambda^*) = 0, \ \lambda^* \geqslant 0. \quad (10)$$

3 Algorithms and Estimates of the Accuracy of Solutions and the Rate of Convergence

To solve the above-mentioned optimization problem (6), we proposed two algorithms. The main idea of the proposed algorithms is using the dichotomy method to solve the dual problem and solving an auxiliary one-dimensional problem at each iteration of the algorithms. Note that stopping criteria are the only difference between these algorithms.

Algorithm 1

Require: convex function f; initial localization interval $[\lambda^0_{min}, \lambda^0_{max}]$ of the dual variable; accuracy δ for auxiliary problems; accuracy ε.

$N := 0$
repeat
  $\lambda^N := \frac{\lambda^N_{min} + \lambda^N_{max}}{2}$;
  $x_\delta(\lambda^N) = \arg\min_{x \in Q}\{f(x) + \lambda^N g(x)\}$;
  $\varphi'(\lambda^N) = g(x_\delta(\lambda^N))$;
  if $\varphi'(\lambda^N) < 0$ **then** $\lambda^{N+1}_{max} := \frac{\lambda^N_{min} + \lambda^N_{max}}{2}$;
  if $\varphi'(\lambda^N) > 0$ **then** $\lambda^{N+1}_{min} := \frac{\lambda^N_{min} + \lambda^N_{max}}{2}$;
  $N := N + 1$;
until $\lambda^N |g(x_\delta(\lambda^N))| \leq \varepsilon$.

Ensure: λ^N, with $\lambda^N |g(x_\delta(\lambda^N))| \leq \varepsilon$; $x_\delta(\lambda^N)$.

Algorithm 2

Require: convex function f; initial localization interval $[\lambda^0_{min}, \lambda^0_{max}]$ of the dual variable; accuracy δ for auxiliary problems; accuracy ε.

$N := 0$
repeat
  $\lambda^N := \frac{\lambda^N_{min}+\lambda^N_{max}}{2}$;
  $x_\delta(\lambda^N) = \arg\min_{x\in Q}\{f(x) + \lambda^N g(x)\}$;
  $\varphi'(\lambda^N) = g(x_\delta(\lambda^N))$;
  if $\varphi'(\lambda^N) < 0$ **then** $\lambda^{N+1}_{max} := \frac{\lambda^N_{min}+\lambda^N_{max}}{2}$;
  if $\varphi'(\lambda^N) > 0$ **then** $\lambda^{N+1}_{min} := \frac{\lambda^N_{min}+\lambda^N_{max}}{2}$;
  $N := N + 1$;
until $|g(x_\delta(\lambda^N))| \leq \varepsilon$.

Ensure: λ^N, with $|g(x_\delta(\lambda^N))| \leq \varepsilon$; $x_\delta(\lambda^N)$.

Remark 1. Note that the stopping criterion of Algorithm 1 is necessarily reached due to the assumption that there exists such $k \in \mathbb{N}$, $\lambda^k = 0$. However, we need an additional assumption to guarantee that the Algorithm 2 stops. Suppose there exists a point $\overline{x} \in Q$, such that $g'(\overline{x}) = 0$.

3.1 Slater Condition

In order to use the dichotomy method and solve the dual problem, it is necessary to compactify the dual variable. So, the initial interval of the localization of the dual variable must be determined. As the dual variable reflects namely the inequality constraint, we can take zero as the lower bound, that means

$$\lambda_{min} = 0.$$

To determine the upper bound, we need to use the Slater condition.

Lemma 1. *Consider the problem of convex optimization*

$$f(x) \to \min_{\substack{g(x)\leqslant 0 \\ x\in Q}}.$$

Suppose the Slater condition is satisfied, so there is such a point $\overline{x} \in Q$ that $g(\overline{x}) < 0$, i.e. there exists $\gamma > 0$ such that $g(\overline{x}) = -\gamma < 0$. Then the following estimate holds

$$\lambda^* \leqslant \frac{1}{\gamma}(f(\overline{x}) - \min_{x\in Q} f(x)), \tag{11}$$

where λ^* is a solution of the dual problem $\varphi(\lambda) \to \max_{\lambda\geqslant 0}$.

Proof. Note the following inequality

$$\min_{x\in Q} f(x) = \min_{x\in Q}\left\{ f(x) + \underbrace{\lambda}_{=0} g(x)\right\} \leqslant \max_{\lambda\geqslant 0}\min_{x\in Q}\{f(x)+\lambda g(x)\}$$
$$= \min_{x\in Q}\{f(x)+\lambda^* g(x)\} \leqslant f(\bar{x}) + \lambda^* g(\bar{x}) = f(\bar{x}) + \lambda^*\gamma.$$

Using this inequality one can get

$$\lambda^*\gamma \leqslant f(\overline{x}) - \min_{x\in Q} f(x).$$

□

Thus, by using lemma (1), we can take the upper bound for the dual variable λ as follows:

$$\lambda_{max} = \frac{1}{\gamma}\Big(f(\overline{x}) - \min_{x\in Q} f(x)\Big).$$

3.2 An Estimate of the Accuracy of Solutions for the Proposed Algorithms

To estimate the rate of convergence of the previous Algorithms 1 and 2, we need the following analogue of Theorem 1 from [11].

Theorem 2. *Let $f(x)$ be a μ_f-strongly convex function, the function $g(x)$ satisfies the Lipschitz condition with a constant M_g. Then the function $\varphi(\lambda)$, defined in (7), where $x(\lambda)$ is determined by the condition (8), is an M_g^2/μ_f-smooth function, i.e. the derivative of the function $\varphi(\lambda)$ satisfies the following Lipschitz condition*

$$|\varphi'(\lambda_2) - \varphi'(\lambda_1)| \leqslant L_\varphi\,|\lambda_2 - \lambda_1|, \tag{12}$$

with a constant $L_\varphi = M_g^2/\mu_f$.

Proof. Let $\lambda_1,\ \lambda_2 \in [\lambda_{min}, \lambda_{max}]$. Define

$$x_1 = \arg\min_{x\in Q}\{f(x)+\lambda_1 g(x)\},\quad x_2 = \arg\min_{x\in Q}\{f(x)+\lambda_2 g(x)\}.$$

Since x_1 and x_2 are unique due to the strong convexity of the function f and by using (9), one can get

$$\varphi'(\lambda_1) = g(x_1),\ \varphi'(\lambda_2) = g(x_2).$$

Recall the necessary optimality conditions are

$$\langle\nabla f(x_1) + \lambda_1\nabla g(x_1),\, x_1 - x_2\rangle \leqslant 0,\quad \langle\nabla f(x_2) + \lambda_2\nabla g(x_2),\, x_2 - x_1\rangle \leqslant 0.$$

Summing these inequalities, we get

$$\langle\nabla f(x_1) - \nabla f(x_2),\, x_2 - x_2\rangle \leqslant \langle\lambda_1\nabla g(x_1) - \lambda_2\nabla g(x_2),\, x_2 - x_1\rangle.$$

Due to the strong convexity of $f(x)$, we obtain the following inequality

$$\langle \nabla f(x_2) - \nabla f(x_1), x_2 - x_1 \rangle \geqslant \mu_f ||x_2 - x_1||_2^2.$$

Then

$$\begin{aligned}
\mu_f ||x_2 - x_1||_2^2 &\leqslant \langle \lambda_1 \nabla g(x_1) - \lambda_2 \nabla g(x_2), x_2 - x_1 \rangle \\
&= \underbrace{\lambda_1}_{\geqslant 0} \underbrace{\langle \nabla g(x_1) - \nabla g(x_2), x_2 - x_1 \rangle}_{\leqslant 0} + (\lambda_1 - \lambda_2) \langle \nabla g(x_2), x_2 - x_1 \rangle \\
&\leqslant |\lambda_1 - \lambda_2| \langle \nabla g(x_2), x_2 - x_1 \rangle \leqslant |\lambda_1 - \lambda_2| \, ||\nabla g(x_2)||_2 \, ||x_2 - x_1||_2 \\
&\leqslant M_g |\lambda_1 - \lambda_2| \, ||x_2 - x_1||_2,
\end{aligned}$$

where $||\nabla g(x_2)||_2 \leqslant M_g$ since g satisfies Lipschitz condition (3).
Thus, for $x_1 \neq x_2$ we get

$$\mu_f ||x_2 - x_1||_2 \leqslant M_g |\lambda_2 - \lambda_1|.$$

As a result, the following estimate holds

$$|\varphi'(\lambda_2) - \varphi'(\lambda_1)| = |g(x_2) - g(x_1)| \leqslant M_g ||x_2 - x_1||_2 \leqslant \frac{M_g^2}{\mu_f} |\lambda_2 - \lambda_1|.$$

□

In order to estimate the accuracy of solutions of the proposed Algorithms 1 and 2, we set the following two lemmas.

Lemma 2. *Suppose the stopping criterion of Algorithm 1 holds for $\lambda = \lambda^N$, then the following inequalities hold*

$$f(x_\delta(\lambda)) - f(x^*) \leqslant \varepsilon + \delta, \quad g(x_\delta(\lambda)) \leqslant \frac{\varepsilon}{\lambda}.$$

For the case $\delta = \varepsilon$ we get

$$f(x_\delta(\lambda)) - f(x^*) \leqslant 2\varepsilon, \quad g(x_\delta(\lambda)) \leqslant \frac{\varepsilon}{\lambda}.$$

Proof. Let λ^* be a solution of the dual problem (7). Denote $x^* = x(\lambda^*)$. Then we get the following relation

$$\begin{aligned}
f(x_\delta(\lambda)) + \lambda g(x_\delta(\lambda)) &\leqslant f(x(\lambda)) + \lambda g(x(\lambda)) + \delta = \varphi(\lambda) + \delta \\
&\leqslant \varphi(\lambda^*) + \delta = f(x^*) + \lambda^* \underbrace{g(x^*)}_{\leqslant 0} + \delta \leqslant f(x^*) + \delta.
\end{aligned}$$

Consequently,

$$f(x_\delta(\lambda)) - f(x^*) \leqslant -\lambda g(x_\delta(\lambda)) + \delta \leqslant \varepsilon + \delta$$

due to the stopping criterion of Algorithm 1, as required. The inequality $g(x_\delta(\lambda)) \leqslant \frac{\varepsilon}{\lambda}$ follows from the stopping criterion of Algorithm 1 (see item 9).

□

Also an analogue of Lemma 2 takes place.

Lemma 3. *Suppose the stopping criterion of Algorithm 2 holds for $\lambda = \lambda^N$, then the following inequalities hold*

$$f(x_\delta(\lambda)) - f(x^*) \leqslant \lambda\varepsilon + \delta, \quad g(x_\delta(\lambda)) \leqslant \varepsilon.$$

For the case $\delta = \varepsilon$ we get

$$f(x_\delta(\lambda)) - f(x^*) \leqslant (\lambda + 1)\varepsilon, \quad g(x_\delta(\lambda)) \leqslant \varepsilon.$$

Remark 2. Let us analyze Lemmas 2 and 3. Algorithm 1 (Lemma 2) guarantees the desirable accuracy of the solution with respect to the objective function, but, possibly, unsatisfactory accuracy of the solution with respect to the constraint, as the estimate is huge in case λ is small. Algorithm 2 (Lemma 3) provides the desirable accuracy of the solution with respect to the constraint and, possibly, unsatisfactory accuracy of the solution with respect to the objective function in case λ is huge. So one of the Algorithms 1, 2 surely guarantees the desirable accuracy with respect to both the objective function and the constraint.

3.3 Estimates of the Rate of Convergence for Lipschitz-Continuous Functionals

The idea of the proposed methods is the consistent decrease of the localization interval of the values of the dual variable λ. At each iteration of Algorithms 1 and 2, this interval decreases by 2 times and every time contains λ_*, for which $\lambda_* g(x(\lambda_*)) = 0$ (for Algorithm 1)

$$\lambda_* g(x(\lambda_*)) = \lambda_* \varphi'(\lambda_*) = 0$$

or $g(x(\lambda_*)) = 0$ (for Algorithm 2)

$$g(x(\lambda_*)) = \varphi'(\lambda_*) = 0.$$

By Theorem 2 for all $\lambda_1, \lambda_2 \in [0; \lambda_{\max}]$

$$|\varphi'(\lambda_2) - \varphi'(\lambda_1)| \leqslant \frac{M_g^2}{\mu_f}|\lambda_2 - \lambda_1|, \tag{13}$$

whence

$$|\lambda_2\varphi'(\lambda_2) - \lambda_1\varphi'(\lambda_1)| \leqslant \left(|\varphi'(0)| + \frac{M_g^2\lambda_{\max}}{\mu_f}\right)|\lambda_2 - \lambda_1| = C|\lambda_2 - \lambda_1|, \tag{14}$$

where $C = |\varphi'(0)| + \frac{M_g^2\lambda_{\max}}{\mu_f}$. Therefore, the achievement of the stopping criterion for Algorithm 2 (item 9) is possible with

$$\lambda_{\max}^N - \lambda_{\min}^N = \frac{\lambda_{\max}}{2^N} \leqslant \frac{\varepsilon}{2C},$$

i.e.

$$N \geqslant \log_2 \frac{2C\lambda_{\max}}{\varepsilon}.$$

So, Algorithm 1 stops after no more than

$$O\left(\log_2 \frac{M_g^2 \lambda_{\max}^2}{\varepsilon \mu_f}\right)$$

iterations. Similarly, if there is $\lambda_* : \varphi'(\lambda_*) = 0$, then (14) means that Algorithm 2 stops after no more than

$$O\left(\log_2 \frac{M_g^2 \lambda_{\max}}{\varepsilon \mu_f}\right)$$

iterations.

Let us analyze the rate of convergence of proposed Algorithms 1 and 2. We need some results from [2] concerning a strongly convex objective function.

Method which guarantees an optimal rate of convergence for the problem (6) is an algorithm based on the restarting of another Adaptive Mirror Descent Algorithm. Information concerning the ordinary Adaptive Mirror Descent Algorithm and the algorithm with its restart can be found in Sect. 4 (Algorithms 3 and 4 respectively). In each iteration of Algorithms 1 and 2 the auxiliary problem

$$x_\delta(\lambda) = \arg\min_{x \in Q} \{f(x) + \lambda g(x)\}$$

is being solved inexactly with the accuracy δ, which means

$$f(x_\delta(\lambda)) + \lambda g(x_\delta(\lambda)) - f(x^*(\lambda)) + \lambda g(x^*(\lambda)) \leqslant \delta,$$

where the function $f(x) + \lambda g(x)$ is strongly convex and satisfies the Lipschitz condition for any fixed λ due to the properties of the functions $f(x)$ and $g(x)$.

To solve the auxiliary problem of minimization of the functional $F_\lambda(x) = f(x) + \lambda g(x)$, we use the standard gradient method. Let us note an important statement [1]. After k iterations of the standard projected subgradient method the following inequality holds

$$F_\lambda(x^k) - F_\lambda(x^*) \leqslant \frac{2M_{F_\lambda}^2}{k \cdot \mu_f},$$

where $M_{F_\lambda} = \max\{M_f, \lambda \cdot M_g\}$. Due to the strong convexity of f we have

$$F_\lambda(x) \geqslant F_\lambda(x^*) + \langle \nabla F_\lambda(x^*), x - x^* \rangle + \frac{\mu_f}{2}\|x - x^*\|_2^2 \geqslant F_\lambda(x^*) + \frac{\mu_f}{2}\|x - x^*\|_2^2.$$

So,

$$\|x - x^*\|_2^2 \leqslant \frac{2}{\mu_f}\left(F_\lambda(x) - F_\lambda(x^*)\right).$$

Taking $x = x^k$ the following estimate holds

$$\|x - x^*\|_2^2 \leqslant \frac{4M_F^2}{k \cdot \mu_f^2} \leqslant \delta^2.$$

Thus, the required number of iterations does not exceed

$$k = \frac{4M_F^2}{\mu_f^2 \delta^2}.$$

Now by using Theorem 2 and taking into account the complexity $O\left(\log_2(\frac{1}{\varepsilon})\right)$ of the dichotomy in Algorithms 1 and 2, the general complexity is

$$O\left(\frac{1}{\delta^2} log_2 \frac{1}{\varepsilon}\right).$$

Remark 3. If $\delta = \varepsilon$ then the general complexity of Algorithms 1 and 2:

$$O\left(\frac{1}{\varepsilon^2} log_2 \frac{1}{\varepsilon}\right).$$

3.4 Estimate for Composite Formulation

Let us emphasize an important remark. Let f have a Lipschitz-continuous gradient, with a constant L_f

$$||\nabla f(x) - \nabla f(y)||_2 \leqslant L_f ||x - y||_2 \forall x, y \in Q,$$

and g be a so-called simple function, i.e. g is a non-smooth convex function of a simple structure. The latter means that Lebesgue sets

$$\Lambda_y = \{x \in Q : g(x) < y\} \tag{15}$$

have a simple structure. For example, to such problems can be attributed the LASSO problem [3,9,12]:

$$\frac{1}{2}||Ax - b||_2^2 + \lambda ||x||_1 \to \min_{x \in \mathbb{R}^n}, \tag{16}$$

where A is a matrix of $(m \times n)$ dimension, $b \in \mathbb{R}^m$, λ is a regularization parameter and $||\cdot||_1$ denotes the standard l_1-norm.

Then we can use the following gradient-type procedure

$$x^{k+1} = \arg\min_{x \in Q} \left\{ \langle \nabla f(x^k), x - x^k \rangle + \lambda g(x) + \frac{L_f}{2} ||x - x^k||_2^2 \right\}. \tag{17}$$

For the method (17) we can achieve $||x - x(\delta)||_2 \leqslant \varepsilon$ after

$$\sqrt{\frac{L_f}{\mu}} \log_2 \frac{1}{\delta}$$

iterations of the method (17) [9]. In such a case, the general complexity of Algorithms 1 and 2:

$$O\left(log_2\frac{1}{\delta}log_2\frac{1}{\varepsilon}\right). \quad (18)$$

The convergence rate is similar in the case when g is a smooth convex function of a simple structure (see (15)). Let g have a Lipschitz-continuous gradient, with a constant L_g

$$||\nabla g(x) - \nabla g(y)||_2 \leqslant L_g||x-y||_2 \forall x, y \in Q$$

and f be a non-smooth convex function. Then we can use the following gradient-type procedure

$$x^{k+1} = \arg\min_{x \in Q}\left\{\langle \lambda \nabla g(x^k), x - x^k\rangle + f(x) + \frac{\lambda L_g}{2}||x-x^k||_2^2\right\}. \quad (19)$$

For the method (19) we can achieve $||x - x(\delta)||_2 \leqslant \varepsilon$ after

$$\sqrt{\frac{\lambda L_g}{\mu_f}}\log_2\frac{1}{\delta}$$

iterations of the method (19) and the general complexity (18) for Algorithms 1 and 2.

3.5 The Case of Smooth Functionals

Suppose functions f and g are smooth, i.e. there exist some L_f, L_g such that

$$||\nabla f(x) - \nabla f(y)||_2 \leqslant L_f||x-y||_2 \ \forall x, y \in Q,$$
$$||\nabla g(x) - \nabla g(y)||_2 \leqslant L_g||x-y||_2 \ \forall x, y \in Q.$$

Then the auxiliary problem

$$\arg\min_{x \in Q} F_\lambda(x),$$

where $F_\lambda(x) = f(x) + \lambda g(x)$, is also smooth and it can be solved, for example, with Gradient Descent [9]

$$x^{k+1} = x^k - \alpha \nabla F_\lambda(x^k).$$

Note that F_λ is a μ_f-strongly convex function.

In such a case, the following estimate for the rate of convergence holds ([6], [9])

$$||x^k - x(\delta)||_2^2 \leqslant ||x^0 - x(\delta)||_2^2\left(1 - \frac{\mu_f}{\max\{L_f, \lambda L_g\}}\right)^k.$$

It means that the complexity of Algorithms 1 and 2 is (18). For $\delta = \varepsilon$ the estimate (18) is

$$O\left(log_2^2\frac{1}{\varepsilon}\right).$$

4 Comparison with Mirror Descent Algorithms

In this section, we compare the proposed methods with two variants of the Mirror Descent Algorithm. These are the classical variant and the one based on the restart method. Let us, according to [2], present basic information concerning Mirror Descent Algorithms. Assume that there exists a constant $\varTheta_0 > 0$, that $\frac{1}{2}\|x - x^*\|_2^2 \leq \varTheta_0^2$. If there is a set of solutions of the problem $\{x_i^*\}$, assume that

$$\min_{x^* \in \{x_i^*\}} \frac{1}{2}\|x - x^*\|_2^2 \leq \varTheta_0^2.$$

The standard definition of the mirror descent operator with Euclidean proximal setup is defined as

$$Mirr_x(p) = \arg\min_{v \in Q}\left\{\langle p, v\rangle + \frac{1}{2}\|x - v\|_2^2\right\} \text{ for each } x \in Q \text{ and } p \in E^*,$$

and assume that it is easily computable.

Algorithm 3. Adaptive Mirror Descent Algorithm.

Require: $\varepsilon > 0, \varTheta_0$ s.t. $\frac{1}{2}\|x - x^*\|_2^2 \leqslant \varTheta_0^2$.
$x^0 = argmin_{x \in Q} \frac{1}{2}\|x - x^*\|_2^2$
$I =: \emptyset$
$N \leftarrow 0$
repeat
  if $g(x^N) \leqslant \varepsilon$ **then**
    $M_N = \|\nabla f(x^N)\|_2$, $h_N = \frac{\varepsilon}{M_N^2}$
    $x^{N+1} = Mirr_{x^N}(h_N \nabla f(x^N))$ *"productive step"*
    $N \to I$
  else
    $M_N = \|\nabla g(x^N)\|_2$, $h_N = \frac{\varepsilon}{M_N^2}$
    $x^{N+1} = Mirr_{x^N}(h_N \nabla g(x^N))$ *"non-productive step"*
  end if
  $N \leftarrow N + 1$
until $\sum_{j=0}^{N-1} \frac{1}{M_j^2} \geqslant 2\frac{\varTheta_0^2}{\varepsilon^2}$
Ensure: $\bar{x}^N := \frac{\sum_{k \in I} x^k h_k}{\sum_{k \in I} h_k}$

Theorem 3. *Let the functionals f and g satisfy the Lipschitz condition with constants M_f and M_g respectively. Then Algorithm 3 works no more than*

$$N = \left\lceil \frac{2\max\{M_f^2, M_g^2\}\varTheta_0^2}{\varepsilon^2} \right\rceil$$

iterations, and the point $\overline{x}^N$ *is a* ε*-solution of* (6). *It means that*

$$f(\overline{x}^k) - f(x^*) \leq \varepsilon, \quad g(\overline{x}^k) \leq \varepsilon. \tag{20}$$

Consider the case of μ-strong convex f and g. We need to modify some proposed assumptions. Assume that

$$x_0 = \arg\min_{x \in Q} \frac{1}{2}\|x - x^*\|_2^2, \quad \frac{1}{2}\|x - x^*\|_2^2 \leq \frac{\Omega}{2} \quad \forall x \in Q : \|x\|_2 \leq 1,$$

where Ω is some known constant. Suppose that there exists some initial starting point $x_0 \in Q$ and a number $R_0 > 0$ such that $\|x_0 - x^*\|_2^2 \leq R_0^2$.

Algorithm 4. Adaptive Mirror Descent Algorithm for Strongly Convex Functions (with restart of Algorithm 3).

Require: accuracy $\varepsilon > 0$; starting point x_0; Ω s.t. $\frac{1}{2}\|x - x^*\|_2^2 \leq \frac{\Omega}{2}$ $\forall x \in Q : \|x\|_2 \leq 1$; strong convexity parameter μ; R_0 s.t. $\|x_0 - x^*\|_2^2 \leq R_0^2$.

Set $d_0(x) = \frac{1}{2}\|\left(\frac{x - x_0}{R_0}\right) - x^*\|_2^2$.
Set $p = 1$.
repeat
  Set $R_p^2 = R_0^2 \cdot 2^{-p}$.
  Set $\varepsilon_p = \frac{\mu R_p^2}{2}$.
  Set x_p as the output of Algorithm 3 with accuracy ε_p, prox-function $d_{p-1}(\cdot)$ and $\frac{\Omega}{2}$ as Θ_0^2.
  $d_p(x) \leftarrow \frac{1}{2}\|\left(\frac{x - x_p}{R_p}\right) - x^*\|_2^2$.
  Set $p = p + 1$.
until $p > \log_2 \frac{\mu R_0^2}{2\varepsilon}$.

Ensure: x^p.

Theorem 4. *Assume that* f *and* g *satisfy the Lipschitz condition with constants* M_f *and* M_g *respectively. Then solving the* μ*-strongly convex problem* (6), *Algorithm 4 works no more than*

$$k = \left\lceil log_2 \frac{\mu R_0^2}{2\varepsilon} \right\rceil + \frac{32\Omega \max\{M_f^2, M_g^2\}}{\mu\varepsilon}$$

iterations. The output point x_p *of Algorithm 4 is satisfied to* (20) *and the following inequality holds*

$$\|x_p - x^*\|_2^2 \leq \frac{2\varepsilon}{\mu}.$$

5 Numerical Experiments

To compare Algorithms 1, 2 and 4, a series of numerical experiments were carried out. Consider three different examples of strongly convex, Lipschitz-continuous objective functions, as follows

Example 1.

$$f(x) = x_1^2 + \sum_{i=1}^{n} i x_i^2 + \frac{1}{100} \sum_{i=1}^{n} \left(\sum_{j=1}^{i} x_j \right)^2 .$$

Example 2.

$$f(x) = \sum_{i=1}^{n-1} i x_i^2 + \sum_{i=1}^{n-2} (x_i + x_{i+1} + x_{i+2})^2 .$$

Example 3.

$$f(x) = \sum_{i=1}^{n} i x_i^4 + \frac{1}{2} \|x\|_2^2.$$

The functional constraint has the next form: $g(x) = \max_{1 \leq i \leq m} \{g_i(x)\}$, where

$$g_i((x_1, \ldots, x_n)) = \langle a_i x, x \rangle - 5,$$

a_i^T $(i = 1, \ldots, m)$ are the rows in the matrix $A \in \mathbb{R}^{m \times n}$ with entries drawn from the discrete uniform distribution in the half open interval $[1, 6)$.

Let us choose the set $Q = \{x = (x_1, x_2, ..., x_n) \in \mathbb{R}^n \,;\, x_1^2 + x_2^2 + ... + x_n^2 \leq 1\}$. For Algorithms 1 and 2, we choose $\lambda_{min} = 0, \lambda_{max} = \frac{f(\bar{x})}{-g(\bar{x})}$, where $\bar{x}$ is an arbitrary point such that $g(\bar{x}) < 0$. For Algorithm 4 we choose standard Euclidean proximal setup as prox-function, starting point $x_0 = \frac{(1,\ldots,1)}{\sqrt{n}}$, $\Theta_0 = \sqrt{2}$ (i.e. $\Omega = 4$) and $R_0 = 1$.

For $\varepsilon = \frac{1}{2}, \frac{1}{4}, \frac{1}{8}, \frac{1}{16}, \frac{1}{32}$ the results of the work of Algorithms 1, 2 and 4, for Examples 1 and 2, when $n = 200, m = 100$, are represented in Figs. 1 and 2 below. For Example 3, when $n = 1000$ and $m = 100$, they are represented in Fig. 3. These results demonstrate the comparison of the running time (in seconds) for each algorithm, with different accuracy ε.

All experiments were implemented in Python 3.4, on a computer fitted with Intel(R) Core(TM) i7-8550U CPU @ 1.80GHz, 1992 Mhz, 4 Core(s), 8 Logical Processor(s). The RAM of the computer is 8 GB.

In general, from all experiments conducted, we can see that Algorithm 1 is the best algorithm, the efficiency of this algorithm is represented by its very high execution speed, where by this algorithm one needs a few seconds to achieve the solution and to reach its stopping criterion. In some details, from Fig. 1 and Fig. 2, for Examples 1 and 2 when $n = 200, m = 100$, one can see that, according to the running time of each algorithm, Algorithm 1 works better than Algorithm 2, which works better than Algorithm 4. We note that the running time of Algorithm 4 is very long compared with the running time of Algorithms 1 and 2. Therefore, for the objective functions in Examples 1 and 2 (quadratic functions), we can see that Algorithm 4 works badly, unlike Algorithm 1. For Example 3 when $n = 1000, m = 100$, from Fig. 3, we can see that Algorithm 1 is still the best, but now Algorithm 4 works better than Algorithm 2. We note that the difference between the running time of Algorithms 1 and 4 is very small, but it is very long compared with the running time of Algorithm 2.

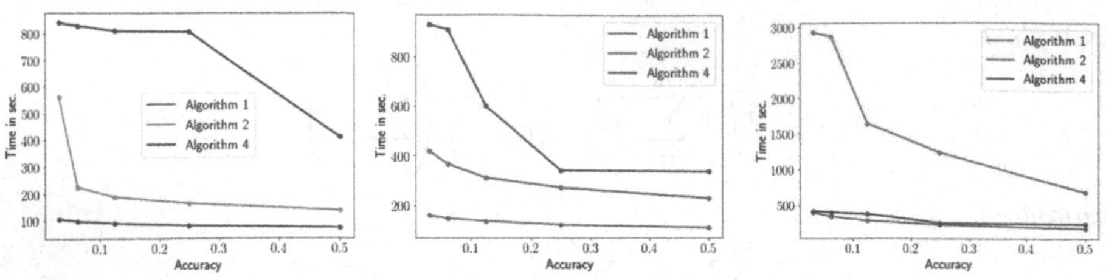

Fig. 1. Example 1, $n = 200$. **Fig. 2.** Example 2, $n = 200$. **Fig. 3.** Example 3, $n = 1000$.

References

1. Aravkin A.Y., Burke J.V., Drusvyatskiy D.: Convex Analysis and Nonsmooth Optimization (2017). https://sites.math.washington.edu/~burke/crs/516/notes/graduate-nco.pdf
2. Bayandina, A., Dvurechensky, P., Gasnikov, A., Stonyakin, F., Titov, A.: Mirror descent and convex optimization problems with non-smooth inequality constraints. In: Giselsson, P., Rantzer, A. (eds.) Large-Scale and Distributed Optimization. LNM, vol. 2227, pp. 181–213. Springer, Cham (2018). https://doi.org/10.1007/978-3-319-97478-1_8
3. Beck, A., Teboulle, M.: A fast iterative shrinkage-thresholding algorithm for linear inverse problems. SIAM J. Imaging Sci. **2**(1), 183–202 (2009)
4. Ben-Tal, A., Nemirovski, A.: Robust truss topology design via semidefinite programming. SIAM J. Optim. **7**(4), 991–1016 (1997)
5. Boyd, S., Vandenberghe, L.: Convex Optimization. Cambridge University Press, New York (2004)
6. Bubeck, S.: Convex optimization: algorithms and complexity. Found. Trends Mach. Learn. **8**(3–4), 231–357 (2015). https://arxiv.org/pdf/1405.4980.pd
7. Danskin, J.M.: The theory of Max-Min, with applications. J. SIAM Appl. Math. **14**(4) (1966)
8. Demyanov, V.F., Malozemov, V.N.: Introduction to Minimax. Nauka, Moscow (1972). (in Russian)
9. Gasnikov, A.V.: Modern numerical optimization methods. The method of universal gradient descent (2018). (in Russian). https://arxiv.org/ftp/arxiv/papers/1711/1711.00394.pdf
10. Ivanova, A., Gasnikov, A., Nurminski, E., Vorontsova, E.: Walrasian equilibrium and centralized distributed optimization from the point of view of modern convex optimization methods on the example of resource allocation problem (2019). (in Russian). https://arxiv.org/pdf/1806.09071.pdf
11. Nesterov, Y.: Smooth minimization of non-smooth functions. Math. Program. **103**(1), 127–152 (2005)
12. Nesterov, Y.: Gradient methods for minimizing composite functions. Math. Program. **140**(1), 125–161 (2013)
13. Nesterov, Y.: Introductory Lectures on Convex Optimization: A Basic Course. Kluwer Academic Publishers, Massachusetts (2004)
14. Shpirko, S., Nesterov, Y.: Primal-dual subgradient methods for huge-scale linear conic problem. SIAM J. Optim. **24**(3), 1444–1457 (2014)
15. Vasilyev, F.: Optimization Methods. Fizmatlit, Moscow (2002). (in Russian)

Gradient Methods for Problems with Inexact Model of the Objective

Fedor S. Stonyakin[1,4(✉)], Darina Dvinskikh[2,3], Pavel Dvurechensky[2,3], Alexey Kroshnin[3,4], Olesya Kuznetsova[4], Artem Agafonov[4], Alexander Gasnikov[3,4,5], Alexander Tyurin[5], César A. Uribe[6], Dmitry Pasechnyuk[7], and Sergei Artamonov[5]

[1] V.I. Vernadsky Crimean Federal University, Simferopol, Russia
fedyor@mail.ru

[2] Weierstrass Institute for Applied Analysis and Stochastics, Berlin, Germany
{darina.dvinskikh,pavel.dvurechensky}@wias-berlin.de

[3] Institute for Information Transmission Problems RAS, Moscow, Russia
gasnikov@yandex.ru

[4] Moscow Institute of Physics and Technologies, Moscow, Russia
{kroshnin,kuznetsova.oa,agafonov.ad}@phystech.edu

[5] National Research University Higher School of Economics, Moscow, Russia
alexandertiurin@gmail.com, sartamonov@hse.ru

[6] Massachusetts Institute of Technology, Cambridge, USA
cauribe@mit.edu

[7] 239-th School of St. Petersburg, Saint Petersburg, Russia
pasechnyuk2004@gmail.com

Abstract. We consider optimization methods for convex minimization problems under inexact information on the objective function. We introduce inexact model of the objective, which as a particular cases includes inexact oracle [16] and relative smoothness condition [36]. We analyze gradient method which uses this inexact model and obtain convergence rates for convex and strongly convex problems. To show potential applications of our general framework we consider three particular problems. The first one is clustering by electorial model introduced in [41]. The second one is approximating optimal transport distance, for which we propose a Proximal Sinkhorn algorithm. The third one is devoted to approximating optimal transport barycenter and we propose a Proximal Iterative Bregman Projections algorithm. We also illustrate the practical performance of our algorithms by numerical experiments.

Keywords: Gradient method · Inexact oracle · Strong convexity · Relative smoothness · Bregman divergence

1 Introduction

In this paper we consider optimization methods for convex problems under inexact information on the objective function. This information is given by an object,

M. Khachay et al. (Eds.): MOTOR 2019, LNCS 11548, pp. 97–114, 2019.
https://doi.org/10.1007/978-3-030-22629-9_8

which we call *inexact model.* Inexact model generalizes the inexact oracle introduced in [16], where inexactness is assumed to be present in the objective value and its gradient. The authors show that, based on these two objects, it is possible to construct a linear function, which is a lower approximation and, up to a quadratic term, an upper approximation of the objective, and these two approximations are enough to obtain convergence rates for gradient method and accelerated gradient method. We go beyond and assume that the approximations of the objective are given through some function, which is not necessarily linear.

This allows us to construct general gradient-type method which is applicable in for different problem classes and allows to obtain convergence rates in these situations as a corollary of our general theorem. Besides convex problems we focus also on strongly convex objectives and illustrate the application of our general theory by two examples. The first example is data clustering by electoral model [41]. The second example relates to Wasserstein distance and barycenter, which are widely used in data analysis [12,13].

Many optimization methods use some model of the objective function to define a step by minimization of this model. Usually the model is constructed using exact first-order [18,39,43], second-order [42], or higher-order information [9,40] information on the objective. The influence of inexactness on the convergence of gradient-type methods have being studied at least since [46]. Accelerated first-order methods with inexact oracle are studied in [11,14,16,21,37]. Some recent works study also non-convex problems in this context [8,19]. Randomized methods with inexact oracle are also studied in the literature, e.g. coordinate descent in [27,53], random gradient-free methods and random directional derivative methods in [22,23]. A method with inexact oracle for variational inequalities can be found in [26].

The contributions of this paper can be summarized as follows.

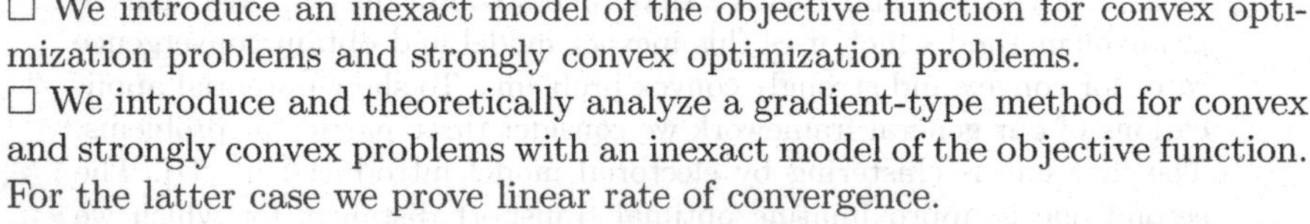

- ☐ We introduce an inexact model of the objective function for convex optimization problems and strongly convex optimization problems.
- ☐ We introduce and theoretically analyze a gradient-type method for convex and strongly convex problems with an inexact model of the objective function. For the latter case we prove linear rate of convergence.
- ☐ We apply our method to, generally speaking, non-convex optimization problem which arises in clustering model introduced in [41]. To do this we construct an inexact model and apply our general algorithms and convergence theorems.
- ☐ We apply our general framework for Wasserstein distance and barycenter problems and show that it allows to construct a proximal á la [10] version of the Sinkhorn's algorithm [49] and Iterative Bregman Projection algorithm [5].

Notation. We define $\mathbf{1} = (1, ..., 1)^T \in \mathbb{R}^n$, $KL(z|t)$ to be the Kullback-Leibler divergence: $KL(z|t) = \sum\limits_{k=1}^{n} z_k \ln(z_k/t_k)$, $\forall z, t \in S_n(1)$, where $S_n(1)$ is the standard simplex in $\mathbb{R}^n$. We also denote by $\odot$ the entrywise product of two matrices.

2 Gradient Methods with Inexact Model of the Objective

Consider the convex optimization problem

$$f(x) \to \min_{x \in Q}, \tag{1}$$

where function f is convex and $Q \subseteq \mathbb{R}^n$ is a simple convex compact set. Moreover, assume that $\min_{x \in Q} f(x) = f(x_*)$ for some $x_* \in Q$.

To solve this problem, we introduce a norm $\|\cdot\|$ on $\mathbb{R}^n$ and a prox-function $d(x)$ which is continuous and convex. We underline that, unlike most of the literature, we do not require d to be strongly convex.

Without loss of generality, we assume that $\min\limits_{x \in \mathbb{R}^n} d(x) = 0$. Further, we define *Bregman divergence* $V[y](x) := d(x) - d(y) - \langle \nabla d(y), x - y \rangle$. Next we define the inexact model of the objective function, which generalizes the inexact oracle of [16] (see also [8,21,24,29,52,54]).

Definition 1. *Let function $\psi_\delta(x, y)$ be convex in $x \in Q$ and satisfy $\psi_\delta(x, x) = 0$ for all $x \in Q$.*

(i) We say that $\psi_\delta(x, y)$ is a (δ, L)-model of the function f at a given point y with respect to $V[y](x)$ iff, for all $x \in Q$, the inequality

$$0 \le f(x) - (f(y) + \psi_\delta(x, y)) \le LV[y](x) + \delta$$

holds for some $L, \delta > 0$.

(ii) We say that $\psi_\delta(x, y)$ is a (δ, L, μ)-model of the function f at a given point y with respect to $V[y](x)$ iff, for all $x \in Q$, the inequality

$$\mu V[y](x) \le f(x) - (f(y) + \psi_\delta(x, y)) \le LV[y](x) + \delta \tag{2}$$

Note that we allow L to depend on δ. We refer to the case (i) as convex case and to the case (ii) as strongly convex case.

Remark 1. In the particular case of function f possessing (δ, L)-oracle [16] at a given point y, one has

$$0 \le f(x) - f(y) - \langle g_\delta(y), x - y \rangle \le \frac{L}{2} \|x - y\|^2 + \delta$$

and $\psi_\delta(x, y) = \langle g_\delta(y), x - y \rangle$. In the same way, if function f is equipped with (δ, L, μ)-oracle [17], i.e.,

$$\frac{\mu}{2} \|x - y\|^2 \le f(x) - f(y) - \langle g_{\delta,L,\mu}(y), x - y \rangle \le \frac{L}{2} \|x - y\|^2 + \delta \quad \forall x \in Q,$$

we have $\psi_\delta(x, y) = \langle g_{\delta,L,\mu}(y), x - y \rangle$.

The algorithms we develop are based on solving auxiliary simple problems on each iteration. We assume that these problems can be solved inexactly and, following [4] introduce a definition of inexact solution of a problem.

Definition 2. *Consider a convex minimization problem*

$$\phi(x) \to \min_{x \in Q \subseteq \mathbb{R}^n}. \tag{3}$$

If ϕ is smooth, we say that we solve it with $\widetilde{\delta}$-'precision' ($\widetilde{\delta} \geq 0$) if we find $\tilde{x}$ s.t. $\max_{x \in Q} \langle \nabla \phi(\tilde{x}), \tilde{x} - x \rangle = \widetilde{\delta}$. If ϕ is general convex, we say that we solve this problem with $\widetilde{\delta}$-'precision' if we find $\tilde{x}$ s.t. $\exists h \in \partial \phi(\widetilde{x}),\ \langle h, x_* - \widetilde{x} \rangle \geq -\widetilde{\delta}$. In both cases we denote this $\tilde{x}$ as $\operatorname{argmin}^{\widetilde{\delta}}_{x \in Q} \phi(x)$.

We notice that the case $\widetilde{\delta} = 0$ corresponds to the case when $\tilde{x}$ is an exact solution of convex optimization problem (3) [4,39].

The connection of Definition 2 with standard definitions of inexact solution, e.g. in terms of the objective residual, can be found in Appendix G of the full version of the paper [51].

2.1 Convex Case

In this subsection we describe a gradient-type method for problems with (δ, L)-model of the objective. This algorithm is a natural extension of gradient method, see [29,52,54].

Algorithm 1. Gradient method with (δ, L)-model of the objective.

Input: x_0 is the starting point, $L > 0$ and $\delta, \widetilde{\delta} > 0$.
for $k \geq 0$ **do**

$$\phi_{k+1}(x) := \psi_\delta(x, x_k) + LV[x_k](x), \quad x_{k+1} := \arg\min_{x \in Q}{}^{\widetilde{\delta}} \phi_{k+1}(x).$$

end for
Output: $\bar{x}_N = \frac{1}{N} \sum_{k=0}^{N-1} x_{k+1}$

Theorem 1. *Let $V[x_0](x_*) \leq R^2$, where x_0 is the starting point, and x_* is the nearest minimum point to the point x_0 in the sense of Bregman divergence $V[y](x)$. Then, for the sequence, generated by Algorithm 1 the following inequality holds:*

$$f(\bar{x}_N) - f(x_*) \leq \frac{LR^2}{N} + \widetilde{\delta} + \delta,$$

In Appendix A of the full version of the paper [51] we prove this theorem and provide an adaptive version of Algorithm 1, which does not require knowledge of the constant L.

2.2 Strongly Convex Case

In this subsection we consider problem (1) with (δ, L, μ)-model of the objective function satisfying (2). This more strong assumption allows us to obtain linear rate of convergence of the proposed algorithm. Our algorithm is listed as Algorithm 2 and it is a version of Algorithm 1, which is adaptive to possibly unknown constant L.

Algorithm 2. Adaptive gradient method with an oracle using the (δ, L, μ)-model

---

Input: x_0 is the starting point, $\mu > 0$ $L_0 \geq 2\mu$ and δ.
Set $S_0 := 0$
for $k \geq 0$ **do**
  Find the smallest $i_k \geq 0$ such that

$$f(x_{k+1}) \leq f(x_k) + \psi_\delta(x_{k+1}, x_k) + L_{k+1} V[x_k](x_{k+1}) + \delta,$$

  where $L_{k+1} = 2^{i_k - 1} L_k$ for $L_k \geq 2\mu$ and $L_{k+1} = 2^{i_k} L_k$ for $L_k < 2\mu$,
  $\alpha_{k+1} := \frac{1}{L_{k+1}}$, $S_{k+1} := S_k + \alpha_{k+1}$.

$$\phi_{k+1}(x) := \psi_\delta(x, x_k) + L_{k+1} V[x_k](x), \quad x_{k+1} := \arg\min_{x \in Q}{}^{\widetilde{\delta}} \phi_{k+1}(x).$$

end for
Output: $\bar{x}_N = \frac{1}{S_N} \sum_{k=0}^{N-1} \frac{x_{k+1}}{L_{k+1}}$

Let's introduce average parameter $\hat{L}$:

$$1 - \frac{\mu}{\hat{L}} = \sqrt[k+1]{\left(1 - \frac{\mu}{L_{k+1}}\right)\left(1 - \frac{\mu}{L_k}\right) \cdots \left(1 - \frac{\mu}{L_1}\right)}.$$

Note that by $L_i \geq \mu$ $(i = 1, 2, \ldots)$

$$\min_{1 \leq i \leq k+1} L_i \leq \hat{L} \leq \max_{1 \leq i \leq k+1} L_i \leq 2L.$$

The following result holds.

Theorem 2. *Let $\psi_\delta(x, y)$ is a (δ, L, μ)-model for f w.r.t. $V[y](x)$. Then, after k iterations of Algorithm 2, we have*

$$V[x_{k+1}](x_*) \leq \frac{2L(\delta + \widetilde{\delta})}{\mu^2} \left(1 - \left(1 - \frac{\mu}{2L}\right)^{k+1}\right) + \left(1 - \frac{\mu}{\hat{L}}\right)^{k+1} V[x_0](x_*),$$

$$f(x_{k+1}) - f(x_*) \leq \frac{4L^2(\delta + \widetilde{\delta})}{\mu^2} \left(1 - \left(1 - \frac{\mu}{2L}\right)^{k+1}\right) + 2L \left(1 - \frac{\mu}{\hat{L}}\right)^{k+1} V[x_0](x_*).$$

The details of proof can be found in Appendix B of the full version of the paper [51]. Note that Algorithm 1 also has linear convergence rate for the strongly convex case. The details can be found in Appendix C of the full version of the paper [51]. The benefit of Algorithm 1 is that there is no need to know the strong convexity parameter μ for the algorithm to work. On the other hand, this parameter is needed for assessing the quality of the solution returned by the algorithm. The benefit of the adaptive version is that it does not require to know the value of the parameter L and adapts to it. Moreover, the parameter L can be different for the model at different points and the algorithm adapts also for the local value of this parameter.

3 Clustering by Electorial Model

In this section we consider clustering model introduced in [41]. In this model voters (data points) choose a party (cluster) in an iterative manner by alternative minimization of the following function.

$$f_{\mu_1,\mu_2}(x=(z,p)) = g(x) + \mu_1 \sum_{k=1}^{n} z_k \ln z_k + \frac{\mu_2}{2}\|p\|_2^2 \to \min_{z \in S_n(1), p \in \mathbb{R}_+^m}, \quad (4)$$

where $\mathbb{R}_+^m$ is a non-negative orthant and $S_n(1)$ is the standard n-dimensional simplex in $\mathbb{R}^n$.

The vector z contains probabilities with which voters choose the considered party, and vector p describes the position of the party in the space of voter opinions. The minimized potential is the result of combining two optimization problems into one: voters choose the party whose position is closest to their personal opinion and the party adjusts its position minimizing dispersion and trying not to go too far from its initial position. Yu. Nesterov in [41] used sequential elections process to show that under some natural assumptions the process convergence and gives the clustering of the data-points. This was done for a particular choice of the function g which has limited interpretability. We show, how our framework of inexact model of the objective allows to construct a gradient-type method for the case of general function g, which is not necessarily convex.

Assume that $g(x)$ (generally, non-convex) is a function with L_g-Lipschitz continuous gradient:

$$\|\nabla g(x) - \nabla g(y)\|_* \leq L_g\|x-y\| \quad \forall x, y \in S_n(1) \times \mathbb{R}_+^m,$$

and, following [41], the numbers μ_1, μ_2 are chosen such that $L_g \leq \mu_1$ and $L_g \leq \mu_2$.

The norm $\|\cdot\|$ in $S_n(1) \times \mathbb{R}_+^m$ is defined as $\|(z,p)\|^2 = \|z\|_1^2 + \|p\|_2^2$, where $\|z\|_1 = \sum_{k=1}^{n} z_k$ and $\|p\|_2 = \sqrt{\sum_{k=1}^{m} p_k^2}$. The correctness of this definition is proven in Appendix I of the full version of the paper [51].

It can be shown that

$$\psi_\delta(x,y) = \langle \nabla g(y), x-y\rangle - L_g \cdot KL(z_x|z_y) - \frac{L_g}{2}\|p_x - p_y\|_2^2 + \mu_1(KL(z_x|1) - KL(z_y|1)) + \frac{\mu_2}{2}(\|p_x\|_2^2 - \|p_y\|_2^2)$$

is a $(0, 2L_g)$-model of $f_{\mu_1,\mu_2}(x)$ in x with respect to the following Bregman divergence

$$V[y](x) = KL(z_x|z_y) + \frac{1}{2}\|p_x - p_y\|_2^2.$$

The proof is detailed in Appendix I of the full version of the paper [51].

Further, for the case $\min\{\mu_1,\mu_2\} > L_g$ $\psi_\delta(x,y)$ is a strongly convex w.r.t. $V[y](x)$:

$$\psi_\delta(x,y) = \psi_\delta^{lin}(x,y) + (\mu_1 - L_g)\cdot KL(z_x|z_y) + \frac{\mu_2 - L_g}{2}\|p_x - p_y\|_2^2 \geq (\min\{\mu_1,\mu_2\} - L_g)\cdot V[y](x), \tag{5}$$

where

$$\psi_\delta^{lin}(x,y) = \langle \nabla g(y), x-y\rangle + \mu_1\langle \nabla KL(z_y|1), z_x - z_y\rangle + \mu_2\langle p_y, p_x - p_y\rangle$$

is linear in y. The proof of (5) is given in Appendix I of the full version of the paper [51].

Thus, $\psi_\delta^{lin}(x,y)$ is a $(0, \max\{\mu_1,\mu_2\} + L_g, \min\{\mu_1,\mu_2\} - L_g)$-model of the function f_{μ_1,μ_2}:

$$f_{\mu_1,\mu_2}(y) + \psi_\delta^{lin}(x,y) + (\min\{\mu_1,\mu_2\} - L_g)V[y](x) \leq f_{\mu_1,\mu_2}(x)$$

and

$$f_{\mu_1,\mu_2}(x) \leq f_{\mu_1,\mu_2}(y) + \psi_\delta^{lin}(x,y) + (\max\{\mu_1,\mu_2\} + L_g)V[y](x).$$

So, we can apply our Algorithms 1 and 2 to the problem (4).

4 Proximal Sinkhorn Algorithm for Optimal Transport

In this section we consider the problem of approximating an optimal transport (OT) distance. Recently optimal transport distances has gained a lot of interest in machine learning and statistical applications [3,6,15,28,34,45,50]. To state the OT problem, assume that we are given two discrete probability measures $p, q \in S_n(1)$ and ground cost matrix $C \in \mathbb{R}_+^{n\times n}$, then the optimal transport problem is

$$\langle C, \pi\rangle \to \min_{\pi\in\mathcal{U}(p,q)}, \quad \mathcal{U}(p,q) = \{\pi \in \mathbb{R}_+^{n\times n} : \pi\mathbf{1} = p, \pi^T\mathbf{1} = q\} \tag{6}$$

where $\langle \cdot, \cdot \rangle$ denotes Frobenius dot product of matrices, π is a transportation plan.

The above optimal transport problem is the Kantorovich [31] linear program (LP) formulation of the problem, which goes back to the Monge's problem [38].

The best known theoretical complexity for this linear program is[1] $\widetilde{O}(n^{2.5})$, see [35]. However, there is no known practical implementation of this algorithm. In practice, the simplex method gives complexity $O(n^3 \ln n)$ [44]. We follow the alternative approach based on entropic regularization of the OT problem [12]. We show how our general framework of inexact model of the objective allows to construct Proximal Sinkhorn algorithm with better computational stability in comparison with the standard Sinkhorn algorithm.

For any optimization problem (1), $\psi_\delta(x, y) = f(x) - f(y)$ satisfies Definition 1 with any $L \geq 0$. In this case, our Algorithm 1 becomes inexact *Bregman proximal gradient method*

$$x_{k+1} = \arg\min_{x \in Q}{}^{\tilde{\delta}} \left\{ f(x) + LV[x_k](x) \right\}.$$

Our idea is to apply this proximal method for the OT problem and approximately find the next iterate x_{k+1} by Sinkhorn's algorithm [2,12,25,49]. The latter is made possible by the choice of V as KL divergence, which makes the problem of finding the point x_{k+1} to be an entropy-regularized OT problem, which, in turn, is efficiently solvable by the Sinkhorn algorithm.

Consider the iterates

$$\begin{aligned} \pi^0 = pq^T \in \mathcal{U}(p, q), \quad \pi^{k+1} &= \arg \min_{\pi \in \mathcal{U}(p,q)} {}^{\varepsilon/2} \left\{ \langle C, \pi \rangle + L \cdot KL(\pi | \pi^k) \right\} \\ &= \arg \min_{\pi \in \mathcal{U}(p,q)} {}^{\varepsilon/2} KL\left(\pi \,\middle|\, \pi^k \odot \exp\left(-\frac{C}{L} \right) \right), \end{aligned} \tag{7}$$

which we call outer iterations. On each outer iteration we use Sinkhorn's algorithm 3, which solves the minimization problem in (7) with accuracy $\tilde{\varepsilon}$ in terms of its objective residual. Notice that here ε' differs from the one from [2,25] as we need approximated solution to the regularized problem. Moreover, unlike [25] we use a slightly refined theoretical bounds for the Sinkhorn's algorithm not depending on vectors p, q[2].

Theorem 3. *Let* $\bar{\pi}^N = \frac{1}{N} \sum_{k=1}^N \pi^k$, *where* π^k *are the iterates of* (7). *Then, after* $N = \frac{4L \ln n}{\varepsilon}$ *iterations, it holds that* $\langle C, \bar{\pi}^N \rangle \leq \min_{\pi \in \mathcal{U}(p,q)} \langle C, \pi \rangle + \varepsilon$. *Moreover, the accuracy* $\tilde{\varepsilon}$ *for the solution of* (7) *is sufficient to be set as* $\widetilde{O}(\varepsilon^4/(Ln^4))$ *and the complexity of Sinkhorn's Algorithm on* k*-th iteration is bounded as*

$$n^2 \widetilde{O}\left(\min\left\{ \exp\left(\frac{\bar{c}_k}{L} \right) \left(\frac{\bar{c}_k}{L} + \ln \frac{\bar{c}_k}{\tilde{\varepsilon}} \right), \frac{\bar{c}_k^2}{L\tilde{\varepsilon}} \right\} \right), \tag{8}$$

[1] Here and below for all (large) n: $\widetilde{O}(g(n)) \leq \tilde{C} \cdot (\ln n)^r g(n)$ with some constants $\tilde{C} > 0$ and $r \geq 0$. Typically, $r = 1$, but not in this particular case. If $r = 0$, then $\widetilde{O}(\cdot) = O(\cdot)$.

[2] One can find the proof in Appendix E of the full version of the paper [51].

Algorithm 3. Sinkhorn's Algorithm

Input: Accuracy $\tilde{\varepsilon}$, matrix $K = e^{-C/\gamma}$, marginals $p, q \in S_n(1)$.

Set $t = 0$, $u^0 = \ln p$, $v^0 = \ln q$, $\varepsilon' = \frac{\tilde{\varepsilon}}{4}\left(\max_{i,j} C_{ij} - \min_{i,j} C_{ij} + 2\gamma \ln\left(\frac{4\gamma n^2}{\tilde{\varepsilon}}\right)\right)^{-1}$.
repeat
  if $t \bmod 2 = 0$ **then**
    $u^{t+1} = u^t + \ln p - \ln(B(u^t, v^t)\mathbb{1})$, where $B(u, v) := \mathrm{diag}(e^u) K \,\mathrm{diag}(e^v)$
    $v^{t+1} = v^t$
  else
    $v^{t+1} = v^t + \ln q - \ln(B(u^t, v^t)^T \mathbb{1})$
    $u^{t+1} = u^t$
  end if
  $t = t + 1$
until $\|B(u^t, v^t)\mathbb{1} - p\|_1 + \|B(u^t, v^t)^T\mathbb{1} - q\|_1 \le \varepsilon'$
Find $\hat{\pi}$ as the projection of $B(u^t, v^t)$ on $\mathcal{U}(p, q)$ by Algorithm 2 in [2].

Output: $\hat{\pi}$.

where[3]

$$\bar{c}_k = \|C\|_\infty + L \ln\left(\frac{\max_{i,j} \pi^k_{ij}}{\min_{i,j} \pi^k_{ij}}\right).$$

Remark 2. The standard Sinkhorn's method can be seen as a particular case of our algorithm (7) with only one step. To obtain an ε-approximate solution of (6), the regularization parameter L needs to be chosen $O(\varepsilon/\ln n)$ [2,25,30]. This can lead to instability of the Sinkhorn's algorithm [48]. On the opposite, our Proximal Sinkhorn algorithm allows to run Sinkhorn's algorithm with larger regularization parameter. This parameter can be chosen by minimization of the theoretical bound (8), which gives $L = \widetilde{O}(\|C\|_\infty)$. In practice one can choose this constant adaptively since we have a (δ, L)-model for any L and can vary L from iteration to iteration. First, the inner problem (7) is solved with overestimated L. Then, we set $L := L/2$ and the problem is solved with the updated value of the parameter and so on until a significant increase (e.g. 10 times) in the complexity of the auxiliary entropy-linear programming problem in comparison with the initial complexity

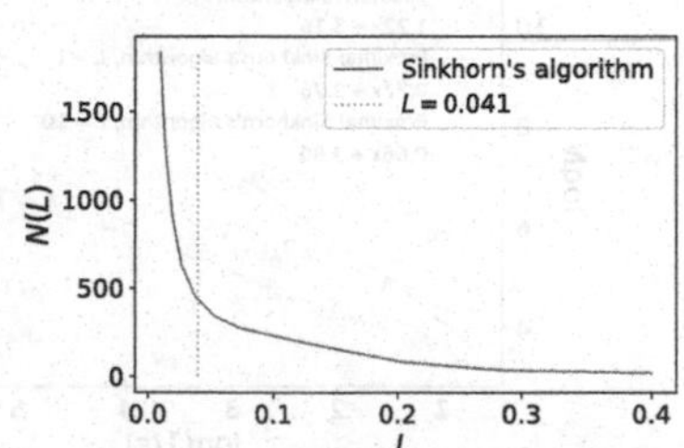

Fig. 1. Adaptive choice of L

[3] This bound is rough and typically $\bar{c}_k$ is smaller in practice. By proper rounding of π^k one can guarantee (without loss of generality) that $\pi^k_{ij} \ge \varepsilon/(2n^2\|C\|_\infty)$, which gives

$$\frac{\bar{c}_k}{L} = \frac{\|C\|_\infty}{L} + \ln\left(\frac{2n^2\|C\|_\infty}{\varepsilon}\right).$$

But, in practice there often is no need to make 'rounding' after each outer iteration.

is detected, see Fig. 1, where $N(L)$ is a number of required iterations of Sinkhorn algorithm to solve the inner problem with accuracy ε.

From the Theorem 3 and Remark 2 one can roughly estimate the total complexity of Proximal Sinkhorn algorithm as[4] $\widetilde{O}(n^4/\varepsilon^2)$.

We also mention several recent complexity bounds[5] for the OT problem $\tilde{O}(n^2/\varepsilon^3)$ [2], $\tilde{O}(n^2/\varepsilon^2)$ and $\tilde{O}(n^{2.5}/\varepsilon)$ [25], $\tilde{O}(n^2/\varepsilon)$ [7,47], $\tilde{O}(n/\varepsilon^{3+d})$, $d \geq 1$ [1].

4.1 Numerical Illustration

In this subsection we provide numerical illustration of the Proximal Sinkhorn algorithm.[6] In the experiments we use a standard MNIST dataset with images scaled to a size 10×10. The vectors p and q contain the pixel intensities of the first and second images respectively. The value of c_{ij} is equal to the Euclidean distance between the i-th pixel from the vector p and the j-th pixel from the vector q on the image pixel grid. For experiments with varying number of pixels n the images are resized to be images of $10 \cdot m \times 10 \cdot m$ pixels, where $m \in \mathbb{N}$. We replace all the zero elements in p and q with 10^{-3} and, then, normalize these vectors.

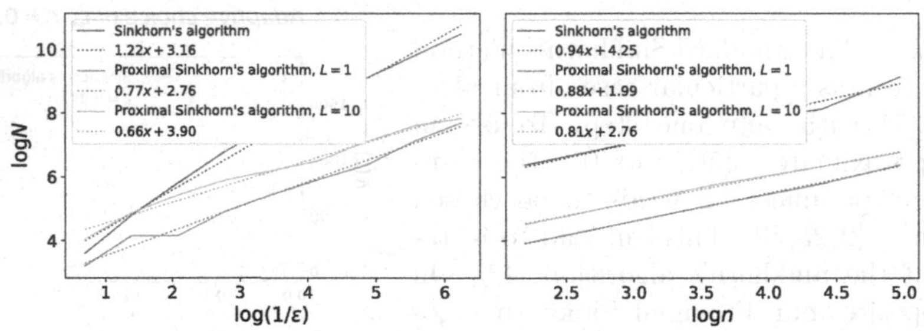

Fig. 2. Comparison of iteration number of Sinkhorn's algorithm and total number of Sinkhorn steps in Proximal Sinkhorn's algorithm for different L.

Figure 2 shows that the growth rate of the iteration number with increasing accuracy or size of the problem for the Sinkhorn's algorithm is greater than for

[4] Our experiments on MNIST data set show (see Figs. 2, 3) that in practice the bound is better.

[5] Strictly speaking for the moment we can not verify all the details of the proof of estimate $\tilde{O}(n^2/\varepsilon)$. Also the proposed in [7,47] methods are mainly theoretical, like Lee–Sidford's method for OT problem with the complexity $\tilde{O}(n^{2.5})$ [35]. For the moment it is hardly possible to implement these methods such that theirs practical efficiencies correspond to the theoretical ones.

[6] The code is available at https://github.com/dmivilensky/Proximal-Sinkhorn-algorithm.

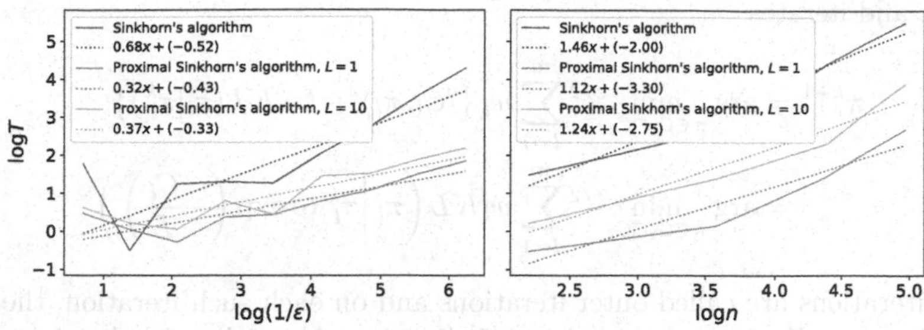

Fig. 3. Comparison of working time of Sinkhorn's algorithm and Proximal Sinkhorn's algorithm with different L.

the Proximal Sinkhorn's method. At the same time, with a higher value of L in proximal method, the iteration number is greater, and the growth rates with some precision are equal. The same type of dependence on the accuracy and the size of the problem can be seen for the working time (Fig. 3).

More experiments can be found in the full version of this paper [51], in particular, on the mean number of inner iterations[7].

5 Proximal IBP Algorithm for Wasserstein Barycenter

In this section we consider a more complicated problem of approximating an OT barycenter. OT barycenter is a natural definition of a mean in a space endowed with an OT distance. Such barycenters are used in the analysis of data with geometric structure, e.g. images, and other machine learning applications [5,13,32,33,45].

For a set of probability measures $\{p_1, \dots, p_m\}$, cost matrices $C_1, \dots, C_m \in \mathbb{R}_+^{n\times n}$, and $w \in S_n(1)$, the weighted barycenter of these measures is defined as a solution of the following convex optimization problem

$$\sum_{l=1}^{m} w_l \min_{\pi_l \in \mathcal{U}(p_l,q)} \langle C_l, \pi_l\rangle \to \min_{q\in S_n(1)} \Longleftrightarrow \sum_{l=1}^{m} w_l \langle C_l, \pi_l\rangle \to \min_{\pi\in\mathcal{C}_1\cap\mathcal{C}_2},$$

$$\mathcal{C}_1 = \{\pi = [\pi_1, \dots, \pi_m] : \forall l\ \pi_l \mathbb{1} = p_l\}, \quad \mathcal{C}_2 = \left\{\pi = [\pi_1, \dots, \pi_m] : \pi_1^T \mathbb{1} = \dots = \pi_m^T \mathbb{1}\right\}.$$

The idea is similar to the one in Sect. 4, namely, we use our framework to define a Proximal Iterative Bregman Projections algorithm.

[7] Figures 5–8 are given in the more complete version of the text by link https://arxiv.org/abs/1902.09001

The algorithm starts from the point $\boldsymbol{\pi}$ s.t. $\pi_l^0 = \frac{1}{n} p_l \mathbb{1}^T \in \mathcal{U}(p_l, \mathbb{1}/n)$, $l = 1, ..., m$ and iterates

$$\begin{aligned} \boldsymbol{\pi}^{k+1} &= \arg \min_{\boldsymbol{\pi} \in \mathcal{C}_1 \cap \mathcal{C}_2}{}^{\varepsilon/2} \sum_{l=1}^{m} w_l \left\{ \langle C_l, \pi_l \rangle + L \cdot KL(\pi_l | \pi_l^k) \right\} \\ &= \arg \min_{\boldsymbol{\pi} \in \mathcal{C}_1 \cap \mathcal{C}_2}{}^{\varepsilon/2} \sum_{l=1}^{m} w_l KL \left(\pi_l \middle| \pi_l^k \odot \exp\left(-\frac{C_l}{L}\right) \right). \end{aligned} \tag{9}$$

These iterations are called outer iterations and on each such iteration, the Iterative Bregman Projections algorithm [5] listed as Algorithm 4 below is used to solve the auxiliary minimization problem.

Algorithm 4. Iterative Bregman Projection

Input: $C_1, \ldots, C_m$, $p_1, \ldots, p_m$, $L > 0$, $\tilde{\varepsilon} > 0$

1: $u_l^0 := 0$, $v_l^0 := 0$, $K_l := \exp\left(-\frac{C_l}{L}\right)$, $l = 1, \ldots, m$
2: **repeat**
3: $\quad v_l^{t+1} := \sum_{k=1}^m w_k \ln K_k^T e^{u_k^t} - \ln K_l^T e^{u_l^t}$, $\quad \mathbf{u}^{t+1} := \mathbf{u}^t$
4: $\quad t := t + 1$
5: $\quad u_l^{t+1} := \ln p_l - \ln K_l e^{v_l^t}$, $\quad \mathbf{v}^{t+1} := \mathbf{v}^t$
6: $\quad t := t + 1$
7: **until** $\sum_{l=1}^m w_l \left\| B_l^T(u_l^t, v_l^t)\mathbb{1} - \bar{q}^t \right\|_1 \leq \frac{\tilde{\varepsilon}}{4 \max_l \|C_l\|_\infty}$, where $B_l(u_l, v_l) = \operatorname{diag}(e^{u_l}) K_l \operatorname{diag}(e^{v_l})$, $\bar{q}^t := \sum_{l=1}^m w_l B_l^T(u_l^t, v_l^t)\mathbb{1}$
8: $q := \frac{1}{\sum_{l=1}^m w_l \langle \mathbb{1}, B_l \mathbb{1} \rangle} \sum_{l=1}^m w_l B_l^T \mathbb{1}$
9: Calculate $\hat{\pi}_1, \ldots, \hat{\pi}_m$ by Algorithm 2 from [2] s.t. $\hat{\pi}_l \in \mathcal{U}(p_l, q)$, $\|\hat{\pi}_l - B_l\|_1 \leq \|B_l \mathbb{1} - p_l\|_1 + \|B_l^T \mathbb{1} - q\|_1$.

Output: q, $\hat{\boldsymbol{\pi}} = [\hat{\pi}_1, \ldots, \hat{\pi}_m]$.

Theorem 4. *Let* $\bar{\boldsymbol{\pi}}^N = \frac{1}{N} \sum_{k=1}^N \boldsymbol{\pi}^k$, *where* $\boldsymbol{\pi}^k$ *are the iterates of* (9). *Then, after* $N = \frac{4Lm \ln n}{\varepsilon}$ *iterations, it holds that*

$$\sum_{l=1}^{m} w_l \langle C_l, \bar{\pi}_l^N \rangle \leq \min_{\boldsymbol{\pi} \in \mathcal{C}_1 \cap \mathcal{C}_2} \sum_{l=1}^{m} w_l \langle C_l, \pi_l \rangle + \varepsilon.$$

Moreover, the accuracy $\tilde{\varepsilon}$ *for the solution of* (9) *is sufficient to be set as* $\tilde{\varepsilon} = \widetilde{O}(\varepsilon^2/(mn^3))$ *and the complexity of IBP on k-th iteration is bounded as*

$$mn^2 \widetilde{O} \left(\min \left\{ \exp\left(\frac{\bar{c}_k}{L}\right) \ln \frac{\bar{c}_k}{\tilde{\varepsilon}}, \frac{\bar{c}_k^2}{L\tilde{\varepsilon}} \right\} \right),$$

$$\bar{c}_k = O\left(\max_{l=1,\ldots,m} \left[\|C_l\|_\infty + L \ln \left(\frac{\max_{i,j} [\pi_l^k]_{ij}}{\min_{i,j} [\pi_l^k]_{ij}} \right) \right] \right).$$

The proof of Theorem 4 is based on Theorem 1 and [32]. All the remarks from Sect. 4 for Proximal Sinkhorn algorithm also hold for Proximal IBP.

In [32] it was shown that complexity of IBP is $\widetilde{O}\left(n^2/\varepsilon^2\right)$. Despite the theoretical complexity of Proximal IBP is worse than this bound, we show in the next section that in practice Proximal IBP beats the standard IBP algorithm. As an alternative to the IBP algorithm we mention primal-dual accelerated gradient descent [20,55].

5.1 Numerical Illustration

In this section, we present preliminary computational results for the numerical performance analysis of the Proximal Iterative Bregman Projection (ProxIBP) method discussed above asthe iterates (9).

Initially, we show the results for the computation of a non-regularized Wasserstein barycenter of a set of 10 truncated Gaussian distributions with finite support. For the finite support $x = [-5, -4.9, -4.8, \ldots, -0.1, 0, 0.1, \ldots, 4.8, 4.9, 5]$, we set the finite distribution p_l such that $p_l(i) = \mathcal{N}(x_i; \mu_i, \sigma_i)$, that is, the value at coordinate i of the distribution p_l, for $1 \leq l \leq m$, is the value of the Normal distribution with mean μ_i and standard deviation σ_i. The values $\{\mu_i\} \sim \text{Uniform}[-5, 5]$, are uniformly chosen in the line segment $[-5, 5]$, and the values are selected as $\{\mu_i\} \sim \text{Uniform}[0.25, 1.25]$. For simplicity of exposition, we select uniform weighting for all distributions, i.e., $w_l = 1/m$.

Figure 4 shows the numerical results for a number of comparative scenarios between the Iterative Bregman Projection (IBP) algorithm proposed in [5] and its Proximal variant in (9). For both algorithms, we show the function values achieved by the generated iterates, and the final approximated barycenter. The results for the IBP algorithm are shown in Fig. 4(a) and (b). Figure 4(a) shows the weighted distance between the generated barycenter and the original distributions for three different desired accuracy values.

It is clear that a bigger ε generates a faster convergence, but the final cost is slightly higher than in other cases. Figure 4(b) shows the resulting barycenter for the three values of the accuracy parameter. For higher accuracy, the effects of the regularization constant are smaller and thus we obtain a "spikier" barycenter. Figure 4(c) and (d) shows a similar analysis for the proposed Proximal IBP in (9), in Fig. 4(c) we observe the function value of the generated barycenter, for a fixed number of inner loop iterations, and changing values of L, note that here L is not a regularization parameter but the weight on the Bregman function. For larger values of L, the inner loop problem is easier to solve, requires less iterations to achieve certain accuracy, with the price in a larger number of iterations in the outer loop. For the particular problem studied, 200 iterations in the outer loop are sufficient to achieve good performance even with relatively smaller values of L. Figure 4(c) shows the generated barycenters for the Proximal IBP algorithm. Finally, Fig. 4(e) and (f) show the results, for the analogous adaptive stopping condition described in Line 11 of Algorithm 3 with $\varepsilon = 1 \cdot 10^{-10}$. We test two different values of the parameter L, namely 1 and 0.1. Additionally, we explore the suggested adaptive search procedure, where one decreases the value of the parameter L at each iteration, until the inner problem has become particularly hard to solve. This last approach is shown a fast convergence as

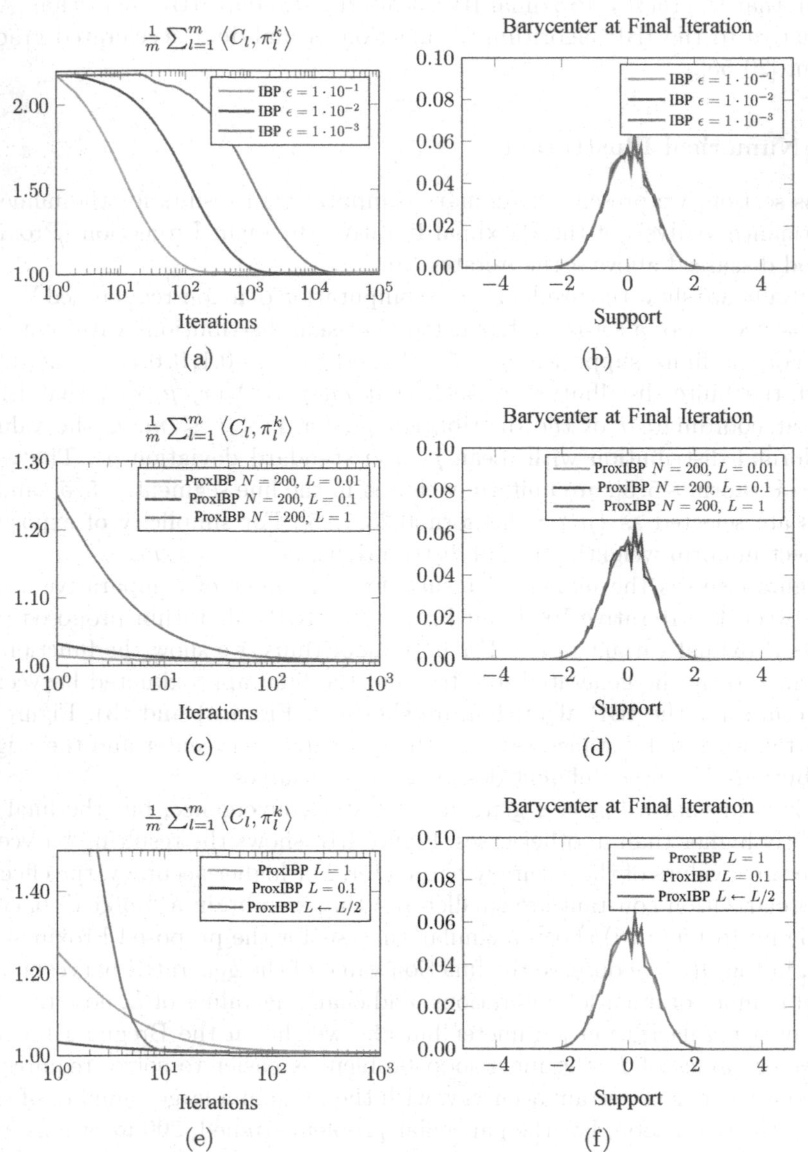

Fig. 4. Numerical results for the computation of the barycenter of 10 truncated Gaussian random variables with finite support for the IBP Algorithm and the Proximal IBP algorithm. Both function value and final resulting barycenter are shown for an number of simulation scenarios.

it reaches a comparable value in around 10 iterations. Figure 4(f) shows the resulting barycenters.

Again, we refer to the full version [51] for additional experiments e.g. on computing Wasserstein barycenters of images from MNIST dataset[8].

6 Conclusions

In this paper we consider gradient methods with inexact information of the objective given by inexact model of this objective. We analyze a gradient-type method for this type of problems and provide its convergence rate. To illustrate the applications, we consider optimization problems in optimal transport and a clustering model. Notably, our framework allows to solve non-convex problems which have a convex inexact model, which is illustrated in the section devoted to clustering model.

Acknowledgments. The work in Sects. 4 and 5 was funded by Russian Science Foundation (project 18-71-10108). The work in Subsect. 2.1 and Sect. 3 was supported by Russian Foundation for Basic Research 18-31-20005 mol_a_ved. The work of F. Stonyakin on Algorithm 2 and Theorem 2 was supported by Russian Science Foundation (project 18-71-00048). The work of A. Gasnikov in Sect. 2 was supported within the framework of the HSE University Basic Research Program and funded by the Russian Academic Excellence Project "5-100". The work of A. Kroshnin in Sect. 3 was supported within the framework of the HSE University Basic Research Program and funded by the Russian Academic Excellence Project "5-100". The work of S. Artamonov in Sect. 3 was supported by Academic Fund Program at the National Research University Higher School of Economics (HSE) in 2019–2020 (grant No 19-01-024) and by the Russian Academic Excellence Project "5-100".

References

1. Altschuler, J., Bach, F., Rudi, A., Weed, J.: Approximating the quadratic transportation metric in near-linear time. arXiv:1810.10046 (2018)
2. Altschuler, J., Weed, J., Rigollet, P.: Near-linear time approxfimation algorithms for optimal transport via sinkhorn iteration. In: Guyon, I., et al. (eds.) Advances in Neural Information Processing Systems 30, pp. 1961–1971. Curran Associates, Inc. (2017)
3. Arjovsky, M., Chintala, S., Bottou, L.: Wasserstein GAN. arXiv:1701.07875 (2017)
4. Ben-Tal, A., Nemirovski, A.: Lectures on modern convex optimization (lecture notes). Personal web-page of A. Nemirovski (2015). http://www2.isye.gatech.edu/~nemirovs/Lect_ModConvOpt.pdf
5. Benamou, J.D., Carlier, G., Cuturi, M., Nenna, L., Peyré, G.: Iterative bregman projections for regularized transportation problems. SIAM J. Sci. Comput. **37**(2), A1111–A1138 (2015)
6. Bigot, J., Klein, T., et al.: Consistent estimation of a population barycenter in the Wasserstein space. arXiv:1212.2562 (2012)

[8] Figures 5–8 are given in the more complete version of the text by link https://arxiv.org/abs/1902.09001

7. Blanchet, J., Jambulapati, A., Kent, C., Sidford, A.: Towards optimal running times for optimal transport. arXiv:1810.07717 (2018)
8. Bogolubsky, L., et al.: Learning supervised PageRank with gradient-based and gradient-free optimization methods. In: NIPS 2016 (2016). http://papers.nips.cc/paper/6565-learning-supervised-pagerank-with-gradient-based-and-gradient-free-optimization-methods.pdf
9. Cartis, C., Gould, N.I.M., Toint, P.L.: Improved second-order evaluation complexity for unconstrained nonlinear optimization using high-order regularized models. arXiv:1708.04044 (2018)
10. Chen, G., Teboulle, M.: Convergence analysis of a proximal-like minimization algorithm using bregman functions. SIAM J. Optim. **3**(3), 538–543 (1993)
11. Cohen, M.B., Diakonikolas, J., Orecchia, L.: On acceleration with noise-corrupted gradients. arXiv:1805.12591 (2018)
12. Cuturi, M.: Sinkhorn distances: lightspeed computation of optimal transport. In: Burges, C.J.C., Bottou, L., Welling, M., Ghahramani, Z., Weinberger, K.Q. (eds.) Advances in Neural Information Processing Systems 26, pp. 2292–2300. Curran Associates, Inc. (2013)
13. Cuturi, M., Doucet, A.: Fast computation of Wasserstein barycenters. In: Xing, E.P., Jebara, T. (eds.) Proceedings of the 31st International Conference on Machine Learning. Proceedings of Machine Learning Research, vol. 32, Bejing, China, 22–24 June 2014, pp. 685–693. PMLR (2014). http://proceedings.mlr.press/v32/cuturi14.html
14. d'Aspremont, A.: Smooth optimization with approximate gradient. SIAM J. Optim. **19**(3), 1171–1183 (2008). https://doi.org/10.1137/060676386
15. Del Barrio, E., Lescornel, H., Loubes, J.M.: A statistical analysis of a deformation model with Wasserstein barycenters: estimation procedure and goodness of fit test. arXiv:1508.06465 (2015)
16. Devolder, O., Glineur, F., Nesterov, Y.: First-order methods of smooth convex optimization with inexact oracle. Math. Program. **146**(1), 37–75 (2014). https://doi.org/10.1007/s10107-013-0677-5
17. Devolder, O., Glineur, F., Nesterov, Y., et al.: First-order methods with inexact oracle: the strongly convex case. CORE Discussion Papers **2013016** (2013)
18. Drusvyatskiy, D., Ioffe, A.D., Lewis, A.S.: Nonsmooth optimization using taylor-like models: error bounds, convergence, and termination criteria. arXiv:1610.03446 (2016)
19. Dvurechensky, P.: Gradient method with inexact oracle for composite non-convex optimization. arXiv:1703.09180 (2017)
20. Dvurechensky, P., Dvinskikh, D., Gasnikov, A., Uribe, C.A., Nedić, A.: Decentralize and randomize: faster algorithm for Wasserstein barycenters. In: Bengio, S., Wallach, H., Larochelle, H., Grauman, K., Cesa-Bianchi, N., Garnett, R. (eds.) Advances in Neural Information Processing Systems 31, pp. 10783–10793. NeurIPS 2018, Curran Associates, Inc. (2018). arXiv:1802.04367
21. Dvurechensky, P., Gasnikov, A.: Stochastic intermediate gradient method for convex problems with stochastic inexact oracle. J. Optim. Theory Appl. **171**(1), 121–145 (2016). https://doi.org/10.1007/s10957-016-0999-6
22. Dvurechensky, P., Gasnikov, A., Gorbunov, E.: An accelerated directional derivative method for smooth stochastic convex optimization. arXiv:1804.02394 (2018)
23. Dvurechensky, P., Gasnikov, A., Gorbunov, E.: An accelerated method for derivative-free smooth stochastic convex optimization. arXiv:1802.09022 (2018)
24. Dvurechensky, P., Gasnikov, A., Kamzolov, D.: Universal intermediate gradient method for convex problems with inexact oracle. arXiv:1712.06036 (2017)

25. Dvurechensky, P., Gasnikov, A., Kroshnin, A.: Computational optimal transport: complexity by accelerated gradient descent is better than by Sinkhorn's algorithm. In: Dy, J., Krause, A. (eds.) Proceedings of the 35th International Conference on Machine Learning. Proceedings of Machine Learning Research, vol. 80, pp. 1367–1376 (2018). arXiv:1802.04367
26. Dvurechensky, P., Gasnikov, A., Stonyakin, F., Titov, A.: Generalized Mirror Prox: Solving variational inequalities with monotone operator, inexact oracle, and unknown Hölder parameters (2018). https://arxiv.org/abs/1806.05140
27. Dvurechensky, P., Gasnikov, A., Tiurin, A.: Randomized similar triangles method: a unifying framework for accelerated randomized optimization methods (coordinate descent, directional search, derivative-free method) (2017). https://arxiv.org/abs/1707.08486
28. Ebert, J., Spokoiny, V., Suvorikova, A.: Construction of non-asymptotic confidence sets in 2-Wasserstein space (2017). https://arxiv.org/abs/1703.03658
29. Gasnikov, A.: Universal gradient descent (2017). https://arxiv.org/abs/1711.00394
30. Gasnikov, A., et al.: Universal method with inexact oracle and its applications for searching equillibriums in multistage transport problems (2015). https://arxiv.org/abs/1506.00292
31. Kantorovich, L.: On the translocation of masses. Doklady Acad. Sci. USSR (N.S.) **37**(7–8), 227–229 (1942)
32. Kroshnin, A., Dvinskikh, D., Dvurechensky, P., Gasnikov, A., Tupitsa, N., Uribe, C.: On the complexity of approximating Wasserstein barycenter (2019). https://arxiv.org/abs/1901.08686
33. Kroshnin, A., Spokoiny, V., Suvorikova, A.: Statistical inference for bures-Wasserstein barycenters (2019). https://arxiv.org/abs/1901.00226
34. Le Gouic, T., Loubes, J.M.: Existence and consistency of Wasserstein barycenters. Probab. Theory Relat. Fields **168**(3–4), 901–917 (2017)
35. Lee, Y.T., Sidford, A.: Path finding methods for linear programming: solving linear programs in o (vrank) iterations and faster algorithms for maximum flow. In: 2014 IEEE 55th Annual Symposium on Foundations of Computer Science Foundations of Computer Science (FOCS), pp. 424–433 (2014)
36. Lu, H., Freund, R.M., Nesterov, Y.: Relatively smooth convex optimization by first-order methods, and applications. SIAM J. Optim. **28**(1), 333–354 (2018)
37. Mairal, J.: Optimization with first-order surrogate functions. In: International Conference on Machine Learning, pp. 783–791 (2013)
38. Monge, G.: Mémoire sur la théorie des déblais et des remblais. Histoire de l'Académie Royale des Sciences de Paris (1781)
39. Nesterov, Y.: Introductory Lectures on Convex Optimization: A Basic Course. Kluwer Academic Publishers, Massachusetts (2004)
40. Nesterov, Y.: Implementable tensor methods in unconstrained convex optimization. CORE Discussion Papers 2018005, Université catholique de Louvain, Center for Operations Research and Econometrics (CORE), March 2018. https://ideas.repec.org/p/cor/louvco/2018005.html
41. Nesterov, Y.: Soft clustering by convex electoral model. CORE Discussion Papers 2018001, Université catholique de Louvain, Center for Operations Research and Econometrics (CORE), January 2018. https://ideas.repec.org/p/cor/louvco/2018001.html
42. Nesterov, Y., Polyak, B.: Cubic regularization of Newton method and its global performance. Math. Program. **108**(1), 177–205 (2006)
43. Ochs, P., Fadili, J., Brox, T.: Non-smooth non-convex bregman minimization: unification and new algorithms. J. Optim. Theory Appl. **181**(1), 244–278 (2019)

44. Pele, O., Werman, M.: Fast and robust earth mover's distances. In: 2009 IEEE 12th International Conference on Computer Vision, pp. 460–467 (2009)
45. Peyré, G., Cuturi, M.: Computational optimal transport. Found. Trends Mach. Learn. **11**(5–6), 355–607 (2019)
46. Polyak, B.: Introduction to Optimization. Optimization Software, New York (1987)
47. Quanrud, K.: Approximating optimal transport with linear programs. In: 2nd Symposium on Simplicity in Algorithms (SOSA 2019), vol. 69, pp. 6:1–6:9. Schloss Dagstuhl-Leibniz-Zentrum fuer Informatik, Dagstuhl, Germany (2018)
48. Schmitzer, B.: Stabilized Sparse Scaling Algorithms for Entropy Regularized Transport Problems (2016). https://arxiv.org/abs/1610.06519
49. Sinkhorn, R.: Diagonal equivalence to matrices with prescribed row and column sums. II. Proc. Amer. Math. Soc. **45**(2), 195–198 (1974)
50. Solomon, J., Rustamov, R.M., Guibas, L., Butscher, A.: wasserstein propagation for semi-supervised learning. In: Proceedings of the 31st International Conference on International Conference on Machine Learning, vol. 32, pp. 306–314. PMLR (2014)
51. Stonyakin, F., et al.: Gradient methods for problems with inexact model of the objective. arXiv:1902.09001 (2019)
52. Stonyakin, F., et al.: Inexact Model: A Framework for Optimization and Variational Inequalities (2019). https://arxiv.org/abs/1902.00990
53. Tappenden, R., Richtárik, P., Gondzio, J.: Inexact coordinate descent: complexity and preconditioning. J. Optim. Theory Appl. **170**(1), 144–176 (2016)
54. Tyurin, A., Gasnikov, A.: Fast gradient descent method for convex optimization problems with an oracle that generates a (δ, L)-model of a function in a requested point. Comput. Math. Math. Phys. (2019, accepted). https://arxiv.org/abs/1711.02747
55. Uribe, C.A., Dvinskikh, D., Dvurechensky, P., Gasnikov, A., Nedić, A.: Distributed computation of Wasserstein barycenters over networks. In: 2018 IEEE Conference on Decision and Control (CDC), pp. 6544–6549 (2018)

A Variant of the Simplex Method for Second-Order Cone Programming

Vitaly Zhadan[1,2(✉)]

[1] Dorodnicyn Computing Centre, FRC "Computer Science and Control" of RAS, 40, Vavilova Street, Moscow 119333, Russia
zhadan@ccas.ru
[2] Moscow Institute of Physics and Technology (State Research University), 9 Institutskiy per., Dolgoprudny, Moscow Region 141701, Russia

Abstract. The linear second-order cone programming problem is considered. For its solution a variant of the primal simplex-type method is proposed. This variant is a generalization on the cone programming of the standard simplex method for linear programming. At each iteration the dual variable and dual slack are defined, and the move from the given extreme point to another one is realized. Finite and infinite convergence of the method to the solution of the problem having a special form is discussed.

Keywords: Second-order cone programming · Simplex-type method · Finite and infinite convergence

1 Introduction

Cone programming is more general setting with respect to linear programming (LP). In cone programming, the requirement that variables must be non-negative is replaced by belonging them to convex cones. The second-order cone programming (SOCP) is the very important special case of cone programming, in which the linear goal function is minimized over the intersection of a linear manifold with direct product of second-order cones [1,2]. Many optimization problems, including, in particular, quadratically constrained convex quadratic problems, robust optimization and combinatorial optimization problems, may be formulated as SOCP [2,3].

The most popular methods of solving SOCP are primal-dual interior point techniques, which were developed for LP and were extended for cone programming [4,5]. The simplex-type algorithms for SOCP are developed essentially less. There are only a few simplex-type methods for SOCP. This situation with simplex-type methods is explained by the presence of infinitely many extreme

This work was supported partially by the Russian Foundation for Basic Research (project no. 17-07-00510).

M. Khachay et al. (Eds.): MOTOR 2019, LNCS 11548, pp. 115–129, 2019.
https://doi.org/10.1007/978-3-030-22629-9_9

points in feasible sets. The general approach for constructing simplex-type methods for cone programming was proposed in [6]. In [7] the SOCP problem of special structure with the single second-order cone and other nonnegative variables was considered and the simplex-type algorithm for its solution was developed. The other variant of a simplex-type algorithm for the general linear SOCP problem had been worked out in [8]. This algorithm is based on the reformulation of the SOCP problem as a linear semi-infinite programme and on the consequent application of the dual-simplex primal-exchange method from [9] for solving the reformulated problem. As it is mentioned by authors, their algorithm is more advantage, when it solves the SOCP problems with similar structure.

In the present paper, the general SOCP problem is considered. For its solution, a variant of the primal simplex-type method is proposed. This variant can be treated as the simple extension of the well-known simplex method for LP. In pivoting procedures all variables, belonging to the second-order cone, are taken in the form of a single variable. The method can be interpreted as a special way of solving the system of optimality conditions. The primal feasibility and complementarity between primal variables and dual slack variables are preserved in the course of iterations. The dual slack variable (dual slack) is estimated at each iteration in order to define the primal variable, which must enter the list of basic variables. The similar way of constructing the primal simplex-type method for solving linear semidefinite programming problems had been used in [10].

The paper is organized as follows. In Sect. 2, the statement of SOCP is given. Here we also introduce some notions and notations. Among them definitions of regular and irregular extreme points of the feasible set are rather important. In Sect. 3, the approach to updating regular extreme points is described. Finally, in Sect. 4, the partial case of the SOCP problem is considered. It is shown that the sequence, generated by the proposed algorithm, converges to the solution of the problem.

2 The Problem Statement and Basic Definitions

Let K^n denote the second-order (Lorentz) cone in $\mathbb{R}^n$. By its definition

$$K^n = \left\{[x^0; \bar{x}] \in \mathbb{R} \times \mathbb{R}^{n-1}: \; x^0 \geq \|\bar{x}\|\right\},$$

where $\|\cdot\|$ refers to the standard Euclidean norm and n is the dimension of K^n. Here and in what follows we use ";" for adjoining vectors or components of a vector in a column. The cone K^n is self-dual, and it induces a partial order in $\mathbb{R}^n$, namely: $x_1 \succeq_{K^n} x_2$, if $x_1 - x_2 \in K^n$.

Consider the cone programming problem

$$\begin{array}{c} \min \; \sum_{i=1}^r \langle c_i, x_i \rangle, \\ \sum_{i=1}^r A_i x_i = b, \quad x_1 \succeq_{K^{n_1}} 0_{n_1}, \; \ldots, \; x_r \succeq_{K^{n_r}} 0_{n_r}. \end{array} \tag{1}$$

Here, $c_i \in \mathbb{R}^n_i$, $1 \leq i \leq r$, and $b \in \mathbb{R}^m$. Matrices A_i are of dimensions $m \times n_i$, $1 \leq i \leq r$, and 0_{n_i} is a zero vector of dimension n_i. The angle brackets denote the usual Euclidean scalar product in $\mathbb{R}^{n_i}$.

The dual problem to (1) has the following form

$$\max \langle b, u \rangle, \quad A_i^T u + y_i = c_i, \ 1 \le i \le r; \quad y_1 \succeq_{K^{n_1}} 0_{n_1}, \ \ldots, \ y_r \succeq_{K^{n_r}} 0_{n_r}, \tag{2}$$

in which $u \in \mathbb{R}^m$.

Let $n = n_1 + \cdots + n_r$. Denoting

$$c = [c_1; \ldots; c_r] \in \mathbb{R}^n, \quad x = [x_1; \ldots; x_r] \in \mathbb{R}^n; \quad y = [y_1; \ldots; y_r] \in \mathbb{R}^n,$$

and $\mathcal{A} = [A_1, \ldots A_r]$, $\mathcal{K} = K^{n_1} \times \cdots \times K^{n_r}$, it is possible to rewrite the pair of problems (1) and (2) as

$$\min \langle c, x \rangle, \quad \mathcal{A}x = b, \quad x \succeq_{\mathcal{K}} 0_n, \tag{3}$$

$$\max \langle b, u \rangle, \quad \mathcal{A}^T u + y = c, \quad y \succeq_{\mathcal{K}} 0_n. \tag{4}$$

We assume that solutions of both problems (3) and (4) exist. Moreover, we assume that $m < n$ and rows of the matrix $\mathcal{A}$ are linear independent. The feasible set in problem (3) is denoted by $\mathcal{F}_P$. Observe, that LP is a special case of (3), (4) with the nonnegative orthant $\mathbb{R}^n_+$ as $\mathcal{K}$.

In order that both problems (3) and (4) have solutions it is necessary that the following system of equalities and inclusions

$$\langle x, y \rangle = 0, \quad \mathcal{A}x = b, \quad \mathcal{A}^T u + y = c, \quad x \in \mathcal{K}, \quad y \in \mathcal{K} \tag{5}$$

be solvable. The simplex-method under consideration is one of the possible ways for solving this system.

Let $x \in \mathcal{K}$. We split all components x_i, composed the vector x, onto zero and nonzero components. In addition, we split nonzero components onto internal components x_i, belonging to interior of the cone K^{n_i}, and onto boundary components, belonging to the boundary of the cone K^{n_i} (more exactly, to a nonzero face of K^{n_i}). From boundary, internal and zero components of the vector x it is possible to compose three blocks of components: x_F, x_I, x_N. Without loss of generality, we assume that these blocks are located in the mentioned order, i.e.

$$x = [x_F; x_I; x_N]. \tag{6}$$

We suppose also that

$$x_F = [x_1; \ldots; x_{r_F}], \quad x_I = [x_{r_F+1}; \ldots; x_{r_F+r_I}], \quad x_N = [x_{r_F+r_I+1}; \ldots; x_r]. \tag{7}$$

Thus, the first block of components x_F consists of $r_F = r_F(x)$ components x_i. Respectively, the second and the third blocks consist of $r_I = r_I(x)$ and $r_N = r_N(x)$ components x_i, respectively. Some blocks may be empty, then the corresponding numbers r_F, r_I or r_N are equal to zero. We have $r_F + r_I + r_N = r$.

Let $J^r = [1 : r]$. The following partition of the index set J^r onto three subsets

$$J^r_F(x) = [1, \ldots, r_F], \quad J^r_I(x) = [r_F+1, \ldots, r_F+r_I], \quad J^r_N(x) = [r_F+r_I+1, \ldots, r]$$

corresponds to the introduced splitting of x onto blocks. Below we use also notations:

$$r_B = r_B(x) = r_F(x) + r_I(x), \qquad J^r_B(x) = J^r_F(x) \cup J^r_I(x).$$

For any nonzero component x_i, $i \in J^r_B(x)$, the following spectral decomposition

$$x_i = \eta_{i,1}\, \mathbf{d}_{i,1} + \eta_{i,n_i}\, \mathbf{d}_{i,n_i} \tag{8}$$

takes place (see [2]). In (8) the pair $\{\mathbf{d}_{i,1}, \mathbf{d}_{i,n_i}\}$ is the "so-called" Jordan frame. The frame vectors $\mathbf{d}_{i,1}$ and $\mathbf{d}_{i,n_i}$ have the forms

$$\mathbf{d}_{i,1} = \frac{1}{\sqrt{2}}\left[1; \frac{\bar{x}_i}{\|\bar{x}_i\|}\right], \qquad \mathbf{d}_{i,n_i} = \frac{1}{\sqrt{2}}\left[1; -\frac{\bar{x}_i}{\|\bar{x}_i\|}\right],$$

and

$$\eta_{i,1} = \frac{1}{\sqrt{2}}\left(x_i^0 + \|\bar{x}_i\|\right), \qquad \eta_{i,n_i} = \frac{1}{\sqrt{2}}\left(x_i^0 - \|\bar{x}_i\|\right).$$

Both vectors $\mathbf{d}_{i,1}$ and $\mathbf{d}_{i,n_i}$ are unit vectors and are orthogonal with each other. If $x_i \in K^{n_i}$, then $\eta_{i,1} \geq 0$, $\eta_{i,n_i} \geq 0$.

Introduce in $\mathbb{R}^{n_i}$ the system of coordinates associated with the current point x_i. For this purpose we set

$$\mathbf{g}_{i,0} = [1; 0; \ldots; 0], \quad \mathbf{g}_{i,1} = \left[0; \frac{\bar{x}_i}{\|\bar{x}_i\|}\right].$$

Both vectors $\mathbf{g}_{i,0}$, $\mathbf{g}_{i,1}$ are unite vectors and

$$\mathbf{d}_{i,1} = \frac{1}{\sqrt{2}}\left(\mathbf{g}_{i,0} + \mathbf{g}_{i,1}\right), \qquad \mathbf{d}_{i,n_i} = \frac{1}{\sqrt{2}}\left(\mathbf{g}_{i,0} - \mathbf{g}_{i,1}\right).$$

The vector $\mathbf{g}_{i,0}$ coincides with the basis vector in $\mathbb{R}^{n_i}$ corresponding to the component with zero index. Furthermore, in the subspace

$$\mathbb{R}^{n_i}_0 = \{x_i = [x_i^0; \bar{x}_i] \in \mathbb{R}^{n_i} :\ x_i^0 = 0\}$$

we take arbitrary unit vectors $\mathbf{g}_{i,2}, \ldots, \mathbf{g}_{i,n_i-1}$, which are orthogonal to each others and orthogonal to the vector $\mathbf{g}_{i,1}$ too. Then the vectors $\mathbf{g}_{i,j}$, $1 \leq j \leq n_i - 1$, form the orthonormal basis in $\mathbb{R}^{n_i}_0$, and jointly with $\mathbf{g}_{i,0}$—the orthonormal basis in $\mathbb{R}^{n_i}$.

Let $\mathbf{G}_i, i \in J^r_B(x)$, denote the orthogonal matrix $\mathbf{G}_i = \left[\mathbf{g}_{i,0},\ \mathbf{g}_{i,1},\ \ldots,\ \mathbf{g}_{i,n_i-1}\right]$ of order n_i. For $i \in J^r_N(x)$ the intrinsic basis $\mathbb{R}^{n_i}$ is taken as $\mathbf{G}_i$. Then for any point $x_i \in \mathbb{R}^{n_i}$, $i \in J^r$, the representation $x_i = \mathbf{G}_i \nu_i$ is valid, where $\nu_i = [\nu_{i,0}; \nu_{i,1}; \ldots; \nu_{i,n_i-1}] \in \mathbb{R}^{n_i}$ and $\nu_i = \mathbf{G}_i^T x_i$.

Introduce additionally $n_i \times (n_i - 1)$ matrix

$$\mathbf{\Lambda}_i = \begin{bmatrix} \nu^0_{i,1}\ \nu^0_{i,2} \cdots \nu^0_{i,n_i-1} \\ \nu_{i,1}\ \ 0\ \ \ldots\ \ 0 \\ 0\ \ \nu_{i,2}\ \ 0 \ldots\ \ 0 \\ \ldots \\ 0\ \ \ldots\ \ 0\ \ \nu_{i,n_i-1} \end{bmatrix}, \qquad i \in J^r,$$

and denote $\lambda_{i,j} = \mathbf{\Lambda}_i e_{i,j} \in \mathbb{R}^{n_i}$, where $e_{i,j}$ is the jth unit orth in $\mathbb{R}^{n_i}$. Then for components x_i, $i \in J_B^r(x)$, with representations (8), the following equality

$$x_i = \mathbf{G}_i \lambda_{i,1} = \nu_{i,1}^0 \mathbf{g}_{i,0} + \nu_{i,1} \mathbf{g}_{i,1}, \tag{9}$$

takes place. Moreover, if $i \in J_F^r(x)$, the equality $\nu_{i,0} = \nu_{i,1} = \nu_{i,1}^0 = x_i^0$ holds. All other components $\lambda_{i,j}$, $2 \leq j \leq n_i - 1$, are zero vectors. For $i \in J_I^r(x)$ we have: $\nu_{i,1}^0 = x_i^0, \nu_{i,1} = \|\bar{x}_i\|$ and $\nu_{i,1}^0 > \nu_{i,1}$. If we set $[\nu_{i,1}^0; \nu_{i,1}] = [0; 0]$, then, formally, the representation (9) is valid for x_i, when $i \in J_N^r(x)$.

Denote by $\mathbf{G}$ and $\mathbf{\Lambda}$ block diagonal matrices

$$\mathbf{G} = \text{Diag}\,(\mathbf{G}_1, \ldots, \mathbf{G}_r)\,, \qquad \mathbf{\Lambda} = \text{Diag}\,(\mathbf{\Lambda}_1, \ldots, \mathbf{\Lambda}_r)\,.$$

Moreover, denote by $\mathbf{e}_1$ the n-dimensional vector $\mathbf{e}_1 = [e_{1,1}; e_{2,1}; \ldots; e_{r,1}]$. Then, for the vector $x = [x_F; x_I; x_N] \in \mathcal{F}_P$ with components x_i, $i \in J^r$ we obtain $\mathcal{A}x = \mathcal{A}^{\mathbf{G}} \mathbf{\Lambda} \mathbf{e}_1 = b$, where $\mathcal{A}^{\mathbf{G}} = \mathcal{A}\mathbf{G}$. The matrix $\mathcal{A}^{\mathbf{G}}$ together with the matrix $\mathcal{A}$ has full rank equal to m.

Consider the sets

$$\mathbf{S}_{i,j} = \{x_i \in \mathbb{R}^{n_i} : \; x_i = \mathbf{G}_i \mathbf{\Lambda}_i e_{i,j}\}\,, \quad 1 \leq j \leq n_i - 1.$$

By $\mathbf{S}_{i,j}^+$ we denote the following subset of the set $\mathbf{S}_{i,j}$:

$$\mathbf{S}_{i,j}^+ = \{x_i \in \mathbb{R}^{n_i} : \; x_i = \mathbf{G}_i \mathbf{\Lambda}_i e_{i,j}, \, \nu_{i,j}^0 \geq |\nu_{i,j}|\}\,, \quad 1 \leq j \leq n_i - 1.$$

The set $\mathbf{S}_{i,j}^+$, being a two-dimensional second-order cone, is the section of the cone K^{n_i}.

Let $x_{i,j} = \mathbf{G}_i \lambda_{i,j} \in \mathbf{S}_{i,j}$, $1 \leq j \leq n_i - 1$. The cone K^{n_i} is convex, therefore, $x_i = \sum_{j=1}^{n_i-1} x_{i,j} \in K^{n_i}$, if $x_{i,j} \in \mathbf{S}_{i,j}^+$, $1 \leq j \leq n_i - 1$. From the other hand, if $x \in K_2^{n_i}$, then x can be represented as the sum of the vectors $x_{i,j} \in \mathbf{S}_{i,j}$, $1 \leq j \leq n_i - 1$, but at nonunique way.

In what follows, we will need in extreme rays of the cone $\mathbf{S}_{i,j}^+$, which are the sets

$$\mathbf{l}_{i,j}^+ = \{x_i = \mathbf{G}_i \Lambda_i e_{i,j} \in \mathbf{S}_{i,j}^+ : \; \nu_{i,j}^0 = \nu_{i,j}\}\,,$$
$$\mathbf{l}_{i,j}^- = \{x_i = \mathbf{G}_i \Lambda_i e_{i,j} \in \mathbf{S}_{i,j}^+ : \; \nu_{i,j}^0 = -\nu_{i,j}\}\,.$$

Both rays $\mathbf{l}_{i,j}^+$ and $\mathbf{l}_{i,j}^-$ belong to the boundary of the cone $\mathbf{S}_{i,j}^+$, and, consequently, belong to the boundary of the cone K^{n_i}.

Definition 1. *A point $x_{i,j} = \mathbf{G}_i \lambda_{i,j} \in \mathbf{S}_{i,j}^+$ is called interior point of the cone $\mathbf{S}_{i,j}^+$, if the pair $[\nu_{i,j}^0; \nu_{i,j}]$ is such, that $\nu_{i,j}^0 > |\nu_{i,j}|$.*

Definition 2. *A point $x_{i,j} = \mathbf{G}_i \lambda_{i,j} \in \mathbf{S}_{i,j}^+$ is called nonzero boundary point of the cone $S_{i,j}^+$, if the pair $[\nu_{i,j}^0; \nu_{i,j}]$ is such, that $\nu_{i,j}^0 = |\nu_{i,j}| > 0$.*

Proposition 1. *Let $x_i = \sum_{j=1}^{n_i-1} x_{i,j}$, where $x_{i,j} \in \mathbf{S}_{i,j}^+$, $1 \leq j \leq n_i - 1$. Let also at least one point $x_{i,j}$ be an interior point of the cone $\mathbf{S}_{i,j}^+$. Then x_i is the interior point of the cone K^{n_i}.*

Proposition 2. *Let $x_i = x_{i,1}$, where $x_{i,1}$ is an interior point of the cone $\mathbf{S}^+_{i,1}$. Let, in addition, $\Delta x_i = \sum_{j=1}^{n_i-1} \Delta x_{i,j}$, where $\Delta x_{i,j} \in \mathbf{S}_{i,j}$, $1 \leq j \leq n_i - 1$. Then there is $\alpha_* > 0$, such that $x_i + \alpha \Delta x_i \in K^{n_i}$ for any $0 \leq \alpha \leq \alpha_*$.*

Denote by $F_{\min}(x|\mathcal{K})$ the minimal face of the cone $\mathcal{K}$, containing the point $x \in \mathcal{K}$. Denote also by $\mathcal{N}(\mathcal{A})$ the null space of the matrix $\mathcal{A}$. According to [6] the vector $x \in \mathcal{F}_P$ is *an extreme* point of the set $\mathcal{F}_P$, if

$$\mathrm{lin}\,(F_{\min}(x \,|\, \mathcal{K})) \cap \mathcal{N}(\mathcal{A}) = \{0_n\},$$

where $\mathrm{lin}\,(F_{\min}(x|\mathcal{K}))$ is a linear hull of the face $F_{\min}(x|\mathcal{K})$. Moreover, the following inequality $\dim F_{\min}(x \,|\, \mathcal{K}) \leq m$ must hold.

We have

$$\dim F_{\min}(x_i \,|\, K^{n_i}) = \begin{cases} 1, & i \in J^r_F(x), \\ n_i, & i \in J^r_I(x). \end{cases}$$

Hence, for the dimension of a minimal face, containing the extreme point x, the inequality $\dim F_{\min}(x \,|\, \mathcal{F}_P) \leq m$ is fulfilled, where

$$\dim F_{\min}(x \,|\, \mathcal{F}_P) = r_F + n_I, \qquad n_I = n_I(x) = \sum_{i \in J^r_I(x)} n_i.$$

We call an extreme point $x \in \mathcal{F}_P$ *regular*, if $\dim F_{\min}(x \,|\, \mathcal{F}_P) = m$. In the case, where $\dim F_{\min}(x \,|\, \mathcal{F}_P) < m$, we call an extreme point $x \in \mathcal{F}_P$ *irregular*.

3 Updating of Regular Extreme Point

Let x be a regular extreme point of the feasible set $\mathcal{F}_P$. We want to move from x to another extreme point $\hat{x} \in \mathcal{F}_P$ next to it. Moreover, we want to make this move in such a manner, that the value of the objective function at the updated point $\hat{x}$ is less than at x. To this end, we will determine at first the slack dual variable $y \in \mathbb{R}^n$, satisfying to all equalities from (5). In addition, we require that these equalities be reserved during the move from x to $\hat{x}$.

As a preliminary, we determine the dual vector $u \in \mathbb{R}^m$ in order to determine y. Assume that the extreme point $x \in \mathcal{F}_P$ is regular.

Partition (6) of the vector x onto components x_F, x_I and x_N generates the partition of the vector y onto corresponding components y_F, y_I and y_N, where

$$y_F = [y_1; \ldots; y_{r_F}], \quad y_I = [y_{r_F+1}; \ldots; y_{r_B}], \quad y_N = [y_{r_B+1}; \ldots; y_{r_B+r_N}].$$

By analogy with x each component y_i, $i \in J^r$, may be represented as $y_i = \mathbf{G}_i \sigma_i$ with $\sigma_i = [\sigma_{i,0}; \sigma_{i,1}; \ldots; \sigma_{i,n_i-1}] \in \mathbb{R}^{n_i}$. Moreover, for any vector $y = [y_1; \ldots; y_r] \in \mathbb{R}^n$ the equality $y = \mathbf{G}\sigma$ is valid, where $\sigma = [\sigma_1; \ldots; \sigma_r] \in \mathbb{R}^n$. Thus, $\sigma = \mathbf{G}^T y$.

The complementary condition $\langle x, y \rangle = 0$ from (5) may be rewritten as

$$\langle x, y \rangle = \sum_{i \in J^r_B(x)} \langle x_i, y_i \rangle = 0.$$

It follows from here that this condition is fulfilled, if the following equalities

$$\langle x_i, y_i\rangle = 0, \qquad i \in J^r_B(x), \tag{10}$$

hold. In the case, where $i \in J^r_I(x)$, equality (10) is fulfilled, if, for example, $y_i = 0_{n_i}$. For $i \in J^r_F(x)$, i.e. when $x_i = \eta_{i,1}\mathbf{d}_{i,1}$ is a boundary point of K^{n_i}, we may take $y_i = \mathbf{d}_{i,n_i}$. Since $\mathbf{d}_{i,1} \perp \mathbf{d}_{i,n_i}$, the equality (10) is also fulfilled in this case.

Take into account that $y_i = c_i - A_i^T u$. Then conditions $y_i = 0_{n_i}$, $i \in J^r_I(x)$, may be written as the system of linear equations with respect to the dual variable u, that is:

$$\left(A_i^{\mathbf{G}_i}\right)^T u = c_i^{\mathbf{G}_i}, \qquad i \in J^r_I(x). \tag{11}$$

Here and in what follows, $A_i^{\mathbf{G}_i} = A_i\mathbf{G}_i$, $c_i^{\mathbf{G}_i} = \mathbf{G}_i^T c_i$.

Consider now the case, where $i \in J^r_F(x)$. In this case $x_i = \eta_{i,1}\mathbf{d}_{i,1}$ and $\eta_{i,1} = \sqrt{2}x_i^0$. From here we have

$$\nu_i = \mathbf{G}_i^T x_i = \eta_{i,1}\mathbf{G}_i^T\mathbf{d}_{i,1} = x_i^0\mathbf{G}_i^T\left(\mathbf{g}_{i,0} + \mathbf{g}_{i,1}\right) = x_i^0\left[1;1;0;\ldots;0\right].$$

Hence, for $y_i = \mathbf{d}_{i,n_i}$ we obtain

$$\sigma_i = \mathbf{G}_i^T y_i = \frac{1}{\sqrt{2}}\mathbf{G}_i^T\left(\mathbf{g}_{i,0} - \mathbf{g}_{i,1}\right) = \frac{1}{\sqrt{2}}\left[1;-1;0;\ldots;0\right].$$

Thus, it is sufficient to require $\sigma_{i,0} = -\sigma_{i,1}$ in order to satisfy the equality $\langle x_i, y_i\rangle = 0$.

Let $i \in J^r_F(x)$. Denote by $c_{i,0}^{\mathbf{G}_i}$ and $c_{i,1}^{\mathbf{G}_i}$ the zero and the first components of the vector $c_i^{\mathbf{G}_i}$, respectively. Denote also by $A_{i,0}^{\mathbf{G}_i}$ and $A_{i,1}^{\mathbf{G}_i}$ the zero and the first columns of the matrix $A_i^{\mathbf{G}_i}$. We set $\widetilde{c}_{i,1}^{\mathbf{G}_i} = c_{i,0}^{\mathbf{G}_i} + c_{i,1}^{\mathbf{G}_i}$, $\widetilde{A}_{i,1}^{\mathbf{G}_i} = A_{i,0}^{\mathbf{G}_i} + A_{i,1}^{\mathbf{G}_i}$. Since $\sigma_i = \mathbf{G}_i^T y_i = c_i^{\mathbf{G}_i} - \left(A_i^{\mathbf{G}_i}\right)^T u$, we derive from here and (11) the following system of linear equations

$$\left(\widetilde{A}_{i,1}^{\mathbf{G}_i}\right)^T u = \widetilde{c}_{i,1}^{\mathbf{G}_i}, \quad i \in J^r_F(x); \qquad \left(A_i^{\mathbf{G}_i}\right)^T u = c_i^{\mathbf{G}_i}, \quad i \in J^r_I(x). \tag{12}$$

At the regular point $x \in \mathcal{F}_P$ the system (12) consists of m equations, the number of variables (components of the vector u) is also equal m.

Denoting by $\mathcal{A}_B^{\mathbf{G}}$ the square matrix of the order m

$$\mathcal{A}_B^{\mathbf{G}} = \left[\widetilde{A}_{1,1}^{\mathbf{G}_i},\ \ldots,\ \widetilde{A}_{r_F,1}^{\mathbf{G}_i},\ A_{r_F+1}^{\mathbf{G}_{r_F+1}},\ \ldots,\ A_{r_F+r_I}^{\mathbf{G}_{r_F+r_I}}\right],$$

and denoting by $c_B^{\mathbf{G}}$ the m-dimensional vector

$$c_B^{\mathbf{G}} = \left[\widetilde{c}_{1,1}^{\mathbf{G}_i};\ \ldots;\ \widetilde{c}_{r_F,1}^{\mathbf{G}_i};\ c_{r_F+1}^{\mathbf{G}_{r_F+1}};\ \ldots;\ c_{r_F+r_I}^{\mathbf{G}_{r_F+r_I}}\right],$$

rewrite the system of equations (12) as

$$\left(\mathcal{A}_B^{\mathbf{G}}\right)^T u = c_B^{\mathbf{G}}. \tag{13}$$

If the matrix of this system is nonsingular, then, solving (13), we obtain

$$u = \left(\mathcal{A}_B^{\mathbf{G}}\right)^{-T} c_B^{\mathbf{G}}.$$

Here and in what follows the notation M^{-T} is used instead of $(M^T)^{-1}$.

Let us give the definition of the non-degenerate point $x \in \mathcal{F}_P$ from [2].

Definition 3. *The point $x \in \mathcal{F}_P$ is called non-degenerate, if $\mathcal{T}_\mathcal{K}(x) + \mathcal{N}_\mathcal{A} = \mathbb{R}^n$, where $\mathcal{T}_\mathcal{K}(x)$ is the tangent space to the cone $\mathcal{K}$ at x, and $\mathcal{N}_\mathcal{A}$ is the null-space of the matrix $\mathcal{A}$.*

Denote by $\bar{A}_i^{\mathbf{G}_i}$, $i \in J_F^r(x)$, the matrix

$$\bar{A}_i^{\mathbf{G}_i} = \left[A_{i,0}^{\mathbf{G}_i} + A_{i,1}^{\mathbf{G}_i}, A_{i,2}^{\mathbf{G}_i}, \ \ldots, \ A_{i,n_i-1}^{\mathbf{G}_i}\right],$$

and by $\bar{\mathcal{A}}_B^{\mathbf{G}}$—the matrix

$$\bar{\mathcal{A}}_B^{\mathbf{G}} = \left[\bar{A}_1^{\mathbf{G}_1}, \ \ldots, \ \bar{A}_{r_F}^{\mathbf{G}_{r_F}}, \ A_{r_F+1}^{\mathbf{G}_{r_F+1}}, \ \ldots, \ A_{r_F+r_I}^{\mathbf{G}_{r_F+r_I}}\right].$$

The matrix $\bar{\mathcal{A}}_B^{\mathbf{G}}$ has the dimension $m \times (n_B - r_F)$, where $n_B = n_B(x) = \sum_{i \in J_B^r(x)} n_i$.

Proposition 3 (Non-degeneracy criterion). *The point $x = [x_F; x_I; x_N] \in \mathcal{F}_P$ is non-degenerate if and only if the rows of the matrix $\bar{\mathcal{A}}_B^{\mathbf{G}}$ are linear independent.*

According to Proposition 3, the inequality $m + r_F \leq n_B$ must hold at the non-degenerate point $x \in \mathcal{F}_P$.

Now, let us give the criterion of extreme point $x \in \mathcal{F}_P$ (see [6]).

Proposition 4 (Criterion of an extreme point). *The point $x \in \mathcal{F}_P$ is an extreme point of the set $\mathcal{F}_P$, if and only if columns of the matrix $\mathcal{A}_B^{\mathbf{G}}$ are linear independent.*

By Proposition 4 the inequality $r_F + n_I \leq m$ must hold at any extreme point of $\mathcal{F}_P$.

Proposition 5. *Let $x \in \mathcal{F}_P$ be a regular extreme point. Then x is a non-degenerate point.*

Proof. Since all columns of the matrix $\mathcal{A}_B^{\mathbf{G}}$ are contained in the matrix $\bar{\mathcal{A}}_B^{\mathbf{G}}$, and since the row rank of the matrix $\bar{\mathcal{A}}_B^{\mathbf{G}}$ coincides with its column rank, all m rows of $\bar{\mathcal{A}}_B^{\mathbf{G}}$ are linear independent. Taking into account Proposition 3, we come to conclusion, that any regular extreme point $x \in \mathcal{F}_P$ is a non-degenerate point. □

Theorem 1. *Let x be a regular extreme point of $\mathcal{F}_P$. Then the matrix $\mathcal{A}_B^{\mathbf{G}}$ is non-singular.*

Proof. The matrix $\mathcal{A}_B^{\mathbf{G}}$ consists of m columns. But the extreme point x is regular. Hence, according to the assertion of Proposition 4 these columns are linear independent. We obtain that the square matrix $\mathcal{A}_B^{\mathbf{G}}$ is nonsingular. □

By Theorem 1, columns of the matrix $\mathcal{A}_B^{\mathbf{G}}$ are linear independent, and we may regard $\mathcal{A}_B^{\mathbf{G}}$ as the *matrix of basis* at the regular extreme point $x \in \mathcal{F}_P$. In addition, we may regard x_i, $i \in J_B^r(x)$, as *basic variables*, and we may regard x_i, $i \in J_N^r(x)$, as *non-basic* variables. What is more, we call the basic variables x_i, $i \in J_F^r(x)$, by *facet basic variables*, and we call the basic variables x_i, $i \in J_I^r(x)$, by *interior basic variables*.

Further, we take the obtained dual variable u and define the dual slack

$$y = c - \mathcal{A}^T u = c - \mathcal{A}^T \left(\mathcal{A}_B^{\mathbf{G}}\right)^{-T} c_B^{\mathbf{G}}.$$

For the vector of coefficients $\sigma = [\sigma_1; \ldots; \sigma_r] \in \mathbb{R}^n$, we obtain respectively

$$\sigma = [\sigma_1; \ldots; \sigma_r] = c^{\mathbf{G}} - (\mathcal{A}^{\mathbf{G}})^T \left(\mathcal{A}_B^{\mathbf{G}}\right)^{-T} c_B^{\mathbf{G}}.$$

In the case, where $y \in \mathcal{K}$, the point x is a solution of problem (3), and $[u, y]$ is a solution of problem (4).

In what follows, we assume that the inclusion $y \in \mathcal{K}$ is violated, that is $y_i \notin K^{n_i}$ for at least one index $1 \le i \le r$. Since u satisfies equations (13), we have $\sigma_i = 0_{n_i}$, $i \in J_I^r(x)$. Therefore, $y_i = 0_{n_i}$, when $i \in J_I^r(x)$. Hence, the inclusion $y_i \in K^{n_i}$ may be broken only in cases, where $i \in J_N^r(x)$ or $i \in J_F^r(x)$.

Consider firstly the case, where there exists the index $k \in J_N^r(x)$ such that $y_k \notin K^{n_i}$. We take this y_k and make the spectral decomposition

$$y_k = \theta_{k,1}\, \mathbf{f}_{k,1} + \theta_{k,n_k}\, \mathbf{f}_{k,n_k}, \tag{14}$$

where

$$\theta_{k,1} = \frac{1}{\sqrt{2}}\left(y_k^0 + \|\bar{y}_k\|\right), \qquad \theta_{k,n_k} = \frac{1}{\sqrt{2}}\left(y_k^0 - \|\bar{y}_k\|\right),$$

and $\mathbf{f}_{k,1}$, $\mathbf{f}_{k,n_k}$ are frame vectors:

$$\mathbf{f}_{k,1} = \frac{1}{\sqrt{2}}\left[1; \frac{\bar{y}_k}{\|\bar{y}_k\|}\right], \quad \mathbf{f}_{k,n_k} = \frac{1}{\sqrt{2}}\left[1; -\frac{\bar{y}_k}{\|\bar{y}_k\|}\right]. \tag{15}$$

Both vectors (3) are unit vectors. Since $y_k \notin K^{n_k}$, at least one of two coefficients $\theta_{k,1}$ or θ_{k,n_k} is negative. We suppose for definiteness, that $\theta_{k,1} < 0$.

Change components of the vector x, setting

$$\hat{x}_i = \hat{x}_i(\alpha) = x_i + \alpha \Delta x_i, \quad 1 \le i \le r, \tag{16}$$

where $\alpha > 0$ is a step length. The vectors Δx_i, $i \in J^r$, are defined by different ways, depending on the case, to which set $i \in J_N^r(x)$, $i \in J_I^r(x)$ or $i \in J_F^r(x)$ the index i belongs. First of all, we set

$$\Delta x_i = \begin{cases} \mathbf{f}_{k,1}, & i = k, \\ 0_{n_i}, & i \ne k. \end{cases}, \qquad i \in J_N^r(x). \tag{17}$$

Thus, $\hat{x}_i = x_i = 0_{n_i}$, when $i \in J_N^r(x)$ and $i \neq k$. For index k the updated point $\hat{x}_k$ is defined as $\hat{x}_k = \alpha \mathbf{f}_{k,1}$. The vectors Δx_i, $i \in J_I^r(x)$, are arbitrary from the space $\mathbb{R}^{n_i}$.

For $i \in J_F^r(x)$ we take Δx_i in the form

$$\Delta x_i = [\Delta x_{i,0}; \Delta x_{i,1}; 0; \ldots ; 0] \tag{18}$$

under the additional condition, that $\Delta x_{i,0} = \Delta x_{i,1}$. The vector Δx_i in this case belongs to the linear hull of the ray $\mathbf{l}_{i,1}^+$.

In order to satisfy the equality $\mathcal{A}\hat{x} = b$, we require that

$$\sum_{i \in J_F^r(x)} \widetilde{A}_{i,1} \Delta x_{i,1} + \sum_{i \in J_I^r(x)} A_i \Delta x_i + A_k \mathbf{f}_{k,1} = 0_m. \tag{19}$$

Replacing Δx_i, $i \in J_B^r(x)$, by its coefficients $\Delta \nu_i = \mathbf{G}_i^T \Delta x_i$, we obtain

$$\sum_{i \in J_F^r(x)} \left(\widetilde{A}_{i,1}^{\mathbf{G}_i} \right) \Delta \nu_{i,1} + \sum_{i \in J_I^r(x)} A_i^{\mathbf{G}_i} \Delta \nu_i + A_k^{\mathbf{G}_k} \Delta \sigma_{k,1} = 0_m, \tag{20}$$

where $\Delta \sigma_{k,1} = \mathbf{G}_k^T \mathbf{f}_{k,1}$.

Let

$$\Delta \nu_B = [\Delta \nu_{1,1}; \ldots ; \Delta \nu_{r_F,1};\ \Delta \nu_{r_F+1}; \ldots ; \Delta \nu_{r_F+r_I}] \in \mathbb{R}^m.$$

Then the system (20) may be rewritten as

$$\mathcal{A}_B^{\mathbf{G}} \Delta \nu_B + A_k^{\mathbf{G}_k} \Delta \sigma_{k,1} = 0_m. \tag{21}$$

According to Theorem 1, the matrix $\mathcal{A}_B^{\mathbf{G}}$ of this system is nonsingular. Solving the system (21), we get

$$\Delta \nu_B = -\left(\mathcal{A}_B^{\mathbf{G}} \right)^{-1} A_k^{\mathbf{G}_k} \Delta \sigma_{k,1}. \tag{22}$$

Analyze now, is it possible the case, when $k \in J_F^r(x)$.

Proposition 6. *The index k can not belong to the set $J_F^r(x)$.*

Proof. Under the assumption that $k \in J_F^r(x)$, we must set $\Delta x_i = 0_{n_i}$, $i \in J_N^r(x)$. Moreover, as in the case $k \in J_N^r(x)$, we must take Δx_i, $i \in J_I^r(x)$, arbitrary from the space $\mathbb{R}^{n_i}$.

Consider now situations with $i \in J_F^r(x)$. If $i \neq k$, then we take Δx_i in previous form (18). For $i = k$ we must set

$$\Delta x_k = [\Delta x_{k,0}; \Delta x_{k,1}; 0; \ldots ; 0] + \mathbf{f}_{k,1},$$

where $\Delta x_{k,0} = \Delta x_{k,1}$. Respectively, in the space of coefficients ν we have

$$\Delta \nu_k = [\Delta \nu_{k,0}; \Delta \nu_{k,1}; 0; \ldots , 0] + \Delta \sigma_{k,1}, \qquad \Delta \nu_{k,0} = \Delta \nu_{k,1}. \tag{23}$$

Let the following representation $\Delta \sigma_{k,1} = [\rho_{k,0}; \rho_{k,1}; \ldots ; \rho_{k,n_k-1}]$ hold. Note, that $y_k = \mathbf{d}_{k,n_k}$, because of from just this condition and from similar conditions for other dual slacks y_i, $i \in J^r_F(x)$, all these dual slacks, including y_k, are chosen. Hence, $\rho_{k,0} = 1$, $\rho_{k,1} = -1$. All other coefficient $\rho_{k,j}$, $2 \le j \le n_k - 1$, are zeros. Thus, (23) can be rewritten as

$$\Delta\nu_k = [\Delta\nu_{k,0} + 1; \Delta\nu_{k,1} - 1; 0; \ldots ; 0]. \tag{24}$$

The matrix $A_k^{\mathbf{G}_k}$ is contained as $\widetilde{A}_{k,1}^{\mathbf{G}_k}$ in more general matrix $\mathcal{A}_B^{\mathbf{G}}$. Therefore, dropping zero components in (24) and substituting the reduced vector

$$\Delta\nu_k = [\Delta\nu_{k,0} + 1; \Delta\nu_{k,1} - 1]$$

in general vector $\Delta\nu_B$, we obtain the homogeneous system of linear equation

$$\mathcal{A}_B^{\mathbf{G}} \Delta\nu_B = 0_m \tag{25}$$

with the nonsingular matrix. The solution of the system (25) is $\Delta\nu_B = 0_m$. Therefore, all components $\Delta\nu_i$, with the exception of $\Delta\nu_k$, are zeros. For $\Delta\nu_k$ we have $\Delta\nu_{k,0} = -1$ and $\Delta\nu_{k,1} = 1$. This contradicts to equality: $\Delta\nu_{k,0} = \Delta\nu_{k,1}$. □

From Proposition 6 we come to conclusion, that index k must belong only to the set $J^r_N(x)$.

Denote by $\mathcal{C}_{\mathcal{K}}(x)$ a cone of feasible directions with respect of $\mathcal{K}$ at the point $x \in \mathcal{K}$. The cone $\mathcal{C}_{\mathcal{K}}(x)$ is the direct product of cones of feasible directions $\mathcal{C}_{K^{n_i}}(x_i)$ at points $x_i \in K^{n_i}$, $i \in J^r$, that is

$$\mathcal{C}_{\mathcal{K}}(x) = \mathcal{C}_{K^{n_1}}(x_1) \times \cdots \times \mathcal{C}_{K^{n_r}}(x_r).$$

According to Lemma 3.2.1 from [11], the direction $h \in \mathbb{R}^{n_i}$ belongs to $\mathcal{C}_{K^{n_i}}(x_i)$ if and only if $h = h_1 + h_2$, where $h_1 \in lin\,(F_{\min}(x_i|K^{n_i}))$ and $h_2 \in K^{n_i}$. The vector h belongs to cone of feasible directions with respect to the set $\mathcal{F}_P$ at point $x \in \mathcal{F}_P$, if h is a feasible direction with respect to the cone $\mathcal{K}$ and $\mathcal{A}h = 0_m$.

The following result is valid.

Proposition 7. *The direction* Δx, *defined by* (17), (18) *and* (19), *is a feasible direction with respect to the set* $\mathcal{F}_P$.

Proof. Observe that according to (19) the equality $\mathcal{A}\Delta x = 0_m$ holds. Observe also that the vector $\mathbf{f}_{k,1}$ belongs to the cone K^{n_k} (more exactly, to the boundary of K^{n_k}). Hence, $\Delta x_i \in K^{n_k}$, if $i \in J^r_N(x)$ and $i = k$.

For $i \in J^r_I(x)$ the point x_i is an interior point of the cone K^{n_i}. Therefore, the cone of feasible directions at this point with respect to K^{n_i} coincides with the space $\mathbb{R}^{n_i}$. At last, the vector Δx_i, when $i \in J^r_F(x)$, belongs to the linear hull of the minimal face $F_{\min}(x_i|K^{n_i})$, which is defined by the frame vector $\mathbf{d}_{i,1}$.

Thus, the assertion, that Δx belongs to the cone of feasible directions with respect to the set $\mathcal{F}_P$, follows from the representation of this cone. □

Proposition 8. *Let $x \in \mathcal{F}_P$ be a regular extreme non-optimal point. Let also $k \in J_N^r(x)$ be such that $y_k \notin K^{n_k}$. Then,*

$$\langle c, \Delta x\rangle = \theta_{k,1} < 0, \tag{26}$$

where $\theta_{k,1} < 0$ is taken from the decomposition (14).

Proof. We have due to (22)

$$\begin{aligned}\langle c, \Delta x\rangle &= \textstyle\sum_{i\in J_B^r(x)}\langle c_i, \Delta x_i\rangle + \langle c_k, \Delta x_k\rangle \\ &= \textstyle\sum_{i\in J_B^r(x)}\langle c_i^{\mathbf{G}_i}, \Delta \nu_i\rangle + \langle c_k^{\mathbf{G}_k}, \Delta\sigma_{k,1}\rangle = \langle c_B^{\mathbf{G}}, \Delta\nu_B\rangle + \langle c_k^{\mathbf{G}_k}, \Delta\sigma_{k,1}\rangle \\ &= \langle c_k^{\mathbf{G}_k}, \Delta\sigma_{k,1}\rangle - \langle (\mathcal{A}_B^{\mathbf{G}})^{-T} c_B^{\mathbf{G}}, A_k^{\mathbf{G}_k}\Delta\sigma_{k,1}\rangle \\ &= \langle c_k^{\mathbf{G}_k} - \left(A_k^{\mathbf{G}_k}\right)^T u, \Delta\sigma_{k,1}\rangle = \langle \sigma_k, \Delta\sigma_{k,1}\rangle \\ &= \langle \theta_{k,1}\mathbf{f}_{k,1} + \theta_{k,n_k}\mathbf{f}_{k,n_k}, \mathbf{f}_{k,1}\rangle = \theta_{k,1}\|\mathbf{f}_{k,1}\|^2 = \theta_{k,1}.\end{aligned}$$

Hence, the inequality (26) is correct. □

The step length α is chosen as large as possible under the condition that the updated point $\hat{x}$ belongs to the feasible set $\mathcal{F}_P$. Since $\mathcal{A}\Delta x = 0_m$, the step length α is defined as minimal among maximal step lengths, satisfying to conditions: $\hat{x}_i(\alpha) \in K^{n_i}$ for all cones K^{n_i}, $i \in J_B^r(x)$.

Proposition 9. *Let index $k \in J_N^r(x)$ be such that $\Delta x_i \in K^{n_i}$ for all $i \in J_B^r(x)$. Then the set $\mathcal{F}_P$ is unbounded and $\langle c, \hat{x}(\alpha)\rangle \to -\infty$, when $\alpha \to +\infty$.*

If the assertion of Proposition 9 is not realized, the step length α is finite and it is possible to make the move from the extreme point x to another feasible point $\hat{x}(\alpha) \in \mathcal{F}_P$ with decreasing the value of goal function.

Proposition 10. *Let x be a regular extreme point of $\mathcal{F}_P$, and let the step length α be finite. Then the updated point $\hat{x}(\alpha)$ is an extreme point of $\mathcal{F}_P$ too.*

Proof. Since the step length α is finite, there are only two situations, when it is possible:

(1) There is the index $s \in J_F^r(x)$ such that at updated point $\hat{x}$ this index s belongs to the set $J_N^r(\hat{x})$. In other words, the facet basic variable becomes a non-basic variable.
(2) There is the index $s \in J_I^r(x)$ such that at the updated point $\hat{x}$ this index s belongs either to the set $J_F^r(\hat{x})$ or to $J_N^r(\hat{x})$. In other words, the interior basic variable becomes either a non-basic variable or a facet basic variable.

In principle, the cases are possible, when each of these situations or both situations happen simultaneously.

Denote $n_B(x) = r_F(x) + n_I(x)$. If x is an extreme point of $\mathcal{F}_P$, then $n_B(x) \leq m$. Regardless of the way, how α is defined, we obtain that at the updated point $\hat{x}$ the following inequality $n_B(\hat{x}) \leq n_B(x)$ holds. Here we take into account that the non-basis variable x_k becomes the facet basic variable at the updated point. The inequality $n_B(\hat{x}) \leq n_B(x)$ is necessary for $\hat{x}$ be an extreme point.

Let us show that Proposition 4 is fulfilled at the point $\hat{x}$. Since x is an extreme point, columns of the matrix $\mathcal{A}_B^{\mathbf{G}}$ are linear independent. Moreover, at regular extreme point x the number of these linear independent columns exactly equals to m. The following representation is valid for the right hand side vector

$$b = \sum_{i \in J_F^r(x)} \widetilde{A}_{i,1}^{\mathbf{G}_i} \nu_{i,1} + \sum_{i \in J_I^r(x)} A_i^{\mathbf{G}_i} \nu_i, \tag{27}$$

and more, in (27) all $\nu_{i,1} > 0$, $i \in J_F^r(x)$, and all $\nu_i \in K_i^n$, $i \in J_I^r(x)$.

Denote

$$\text{pos}_{K^{n_i}} A_i^{\mathbf{G}_i} = \left\{ z_i \in \mathbb{R}^m : \ z_i = A_i^{\mathbf{G}_i} \nu_i, \ \nu_i \in K^{n_i} \right\}.$$

The set $\text{pos}_{K^{n_i}} A_i^{\mathbf{G}_i}$ is an image of the convex cone K^{n_i} under the linear mapping. Thus, $\text{pos}_{K^{n_i}} A_i^{\mathbf{G}_i}$ is a convex cone in $\mathbb{R}^m$. Observe, that in (27) $\nu_i \in \text{int} K^{n_i}$, where $\text{int} K^{n_i}$ is an interior of the cone K^{n_i}. Hence, $z_i = A_i^{\mathbf{G}_i} \nu_i$ is an interior point of the convex cone $\text{pos}_{K^{n_i}} A_i^{\mathbf{G}_i}$.

Let $\mathcal{W}_i = \text{cone} \widetilde{A}_{i,1}^{\mathbf{G}_i}$, $i \in J_F^r(x)$, be a cone hull of the vector $\widetilde{A}_{i,1}^{\mathbf{G}_i}$. In other words, it is the ray generated by $\widetilde{A}_{i,1}^{\mathbf{G}_i}$. Let also $\mathcal{W}_i = \text{pos}_{K^{n_i}} A_i^{\mathbf{G}_i}$, $i \in J_I^r(x)$. All these sets are convex cones in $\mathbb{R}^m$, which don't intersect between themselves. We take the sum of these cones

$$\mathcal{W} = \mathcal{W}_1 + \ldots + \mathcal{W}_{r_F} + \mathcal{W}_{r_F+1} + \ldots + \mathcal{W}_{r_F+r_I}.$$

Since columns of the matrix $\mathcal{A}_B^{\mathbf{G}}$ are linear independent, the cone $\mathcal{W}$ has non-empty interior. The point $\nu_B = [\nu_{1,1}; \cdots; \nu_{r_F,1}; \ \nu_{r_F+1}; \cdots; \nu_{r_F+r_I}] \in \mathbb{R}^m$ belongs to the interior of $\mathcal{W}$.

In the case, where facet basic variable x_s, $s \in J_F^r(x)$ becomes nonbasic, the column $\widetilde{A}_{s,1}^{\mathbf{G}_i}$ is taken out from the decomposition (27), and the column $\widetilde{A}_{k,1}^{\mathbf{G}_k}$ is introduced. If this column $\widetilde{A}_{k,1}^{\mathbf{G}_k}$ together with the rest columns are linear dependent, it means that the vector b belongs to the boundary of the cone $\mathcal{W}$, which is impossible. The same conclusion is valid, when an interior basic variable becomes the facet basic variable.

Thus, the assertion of Proposition 4 is fulfilled at the updated point $\hat{x}$. Therefore, $\hat{x}$ is an extreme point of $\mathcal{F}_P$. □

By Propositions 8 and 10, we may construct *the simplex-type iterative algorithm*, in which all points are extreme points of the feasible set, and values of the goal function monotonically decreases from iteration to iteration.

4 Partial Case of SOCP Problem

Consider the partial case of problem (3), when it is known in advance, that the solution of (3) is an extreme point of $\mathcal{F}_P$ with all basic variables being facet basic variables. Then it is possible to take an extreme point $x^0 \in \mathcal{F}_P$ with only facet basic variables as a starting point. The move from any extreme point to another one turns out to be such that only facet basic variables are at all these extreme points.

In what follows, we call the sequence of points $\{x^l\}$, generated by the algorithm, *regular*, if all these points x^l are regular extreme points of $\mathcal{F}_P$. Moreover, denote by $\text{ext}_F(\mathcal{F}_P)$ the subset of all extreme points from $\mathcal{F}_P$ with all basic variables being of facet type. We say that the problem (3) is non-degenerate with respect to $\text{ext}_F(\mathcal{F}_P)$, if all points from this set are non-degenerate.

Proposition 11. *Let $x^* \in \text{ext}_F(\mathcal{F}_P)$ be regular unique solution of the problem* (3). *Let also the starting extreme point $x^0 \in \text{ext}_F(\mathcal{F}_P)$ be such that the sequence $\{x^l\}$ is regular. If $\{x^l\}$ is finite, then the last point of this sequence coincides with x^*.*

Theorem 2. *Let Problem* (3) *be non-degenerate with respect to the set $\text{ext}_F(\mathcal{F}_P)$. Let also all assumptions of Proposition 11 be fulfilled, except the assumption that $\{x^l\}$ is a finite sequence. Suppose additionally that the starting point x^0 is such that the set*

$$\mathcal{F}_P(x^0) = \{x \in \mathcal{F}_P : \langle c, x\rangle \leq \langle c, x^0\rangle\}$$

is bounded. Then the sequence $\{x^l\}$ converges to x^*.

Proof. Since the sequence $\{x^l\}$ is bounded, there exist limit points of $\{x^l\}$. Let $\{x^{l_s}\}$ be a convergent subsequence of $\{x^l\}$, and let $x^{l_s} \to \bar{x}$. All points of $\{x^{l_s}\}$ are regular extreme points of $\mathcal{F}_P$. Moreover, $x^{l_s} \in \text{ext}_F(\mathcal{F}_P)$ for $s \geq 1$. The point $\bar{x}$ is also an extreme point of $\mathcal{F}_P$, and more: $\bar{x} \in \text{ext}_F(\mathcal{F}_P)$.

The number of all possible sets $J^r_F(x)$, consisting of m indices, is finite. Therefore, we may assume without loss of generality, that sets $J^r_F(x^{l_s})$ are the same for all s. Denote this set by $\bar{J}^r_F$. Because of continuity we have $J^r_F(\bar{x}) \subseteq \bar{J}^r_F$. Moreover, the matrices $\mathcal{A}^{\mathbf{G}}_B$ converge to a certain matrix $\bar{\mathcal{A}}^{\mathbf{G}}_B$, and vectors $c^{\mathbf{G}}_B$ converge to a certain vector $\bar{c}^{\mathbf{G}}_B$. As a matter of fact, $\bar{\mathcal{A}}^{\mathbf{G}}_B$ and $\bar{c}^{\mathbf{G}}_B$ are the matrix and the vector, defined at the extreme point $\bar{x}$. Since $\bar{x} \in \text{ext}_F(\mathcal{F}_P)$, we have that $\bar{x}$ is a non-degenerate point. From here we derive that $\bar{x}$ is a regular extreme point. Hence, the matrix $\bar{\mathcal{A}}^{\mathbf{G}}_B$ is nonsingular.

Let $\bar{u}$ be a dual variable, satisfying the system of linear Eq. (13) with $\bar{\mathcal{A}}^{\mathbf{G}}_B$ and $\bar{c}^{\mathbf{G}}_B$. Let also $\bar{y}$ be the corresponding dual slack. Since $\bar{x}$ is not an optimal point, the coefficient $\bar{\theta}$ in the decomposition of $\bar{y}$ is such that $\bar{\theta}_{1,k} < 0$. Dual variables u^{l_s}, being solutions of system (13) at points x^{l_s}, converge to $\bar{u}$. Dual slacks y^{l_s} converge to $\bar{y}$ too. We obtain by Proposition 8 that $\langle c, x^{l_s+1}\rangle \leq \langle c, x^{l_s}\rangle + \alpha^{l_s}\theta^{l_s}_{1,k} < \langle c, x^{l_s}\rangle$. As Δx^{l_s} are bounded for s sufficiently large, the step lengths α^{l_s} don't tend to zero. Thus, we obtain at some iteration that $\langle c, x^{l_s+1}\rangle < \langle c, \bar{x}\rangle$. This contradicts to monotone decreasing of values of the objective function at each iteration and convergence of the sequence $\{x^{l_s}\}$ to $\bar{x}$. Hence, $\bar{x}$ may be only the optimal point. □

5 Conclusion

We presented a variant of the simplex method for SOCP problems. The main attention has been given to updating of regular extreme points. In principle, it is possible to develop an approach for updating irregular extreme points. However, this approach is more complicated compared with the regular case.

References

1. Anjos, M.F., Lasserre, J.B. (eds.): Handbook of Semidefinite, Cone and Polynomial Optimizatin: Theory, Algorithms, Software and Applications. Springer, New York (2011). https://doi.org/10.1007/978-1-4614-0769-0
2. Alizadeh, F., Goldfarb, D.: Second-order cone programming. Math. Program. Ser. B. **95**, 3–51 (2003)
3. Lobo, M.S., Vandenberghe, L., Boyd, S., Lebret, H.: Applications of second order cone programming. Linear Algebra Appl. **284**, 193–228 (1998)
4. Nesterov, YuE, Todd, M.J.: Primal-dual interior-point methods for self-scaled cones. SIAM J. Optim. **8**, 324–364 (1998)
5. Monteiro, R.D.C., Tsuchiya, T.: Polynomial convergence of primal-dual algorithms for second-order cone program based on the MZ-family of directions. Math. Program. **88**(1), 61–83 (2000)
6. Pataki, G.: Cone-LP's and semidefinite programs: geometry and a simplex-type method. In: Cunningham, W.H., McCormick, S.T., Queyranne, M. (eds.) IPCO 1996. LNCS, vol. 1084, pp. 162–174. Springer, Heidelberg (1996). https://doi.org/10.1007/3-540-61310-2_13
7. Muramatsu, M.: A pivoting procedure for a class of second-order cone programming. Optim. Methods Softw. **21**(2), 295–314 (2006)
8. Hayashi, Sh., Okuno, T., Ito, Y.: Simplex-type algorithm for second-order cone programming via semi-infinite programming reformulation. Optim. Methods Softw. **31**(4–6), 1272–1297 (2016)
9. Goberna, M.A., Lopez, M.A.: Linear Semi-Infinite Optimization. Wiley, New York (1998)
10. Zhadan V.G.: Two-phase simplex method for linear semidefinite optimization. Optim. Lett. 1–16 (2018). https://doi.org/10:1007/s11590-018-133-z
11. Wolkowicz, H., Saigal, R., Vandenberghe, L. (eds.): Handbook of Semidefinite Programming. Kluwer Academic Publishers, Dordrecht (2000)

Bilevel Optimization

The Competitive Hub Location Under the Price War

Dimitrije D. Čvokić[1,4(✉)], Yury A. Kochetov[2,3], Aleksandr V. Plyasunov[2,3], and Aleksandar Savić[1,4]

[1] Faculty of Natural Sciences and Mathematics, University of Banja Luka, Mladena Stojanovića 2, 78000 Banja Luka, Srpska Republic, Bosnia and Herzegovina
dimitrije.cvokic@pmf.unibl.org

[2] Sobolev Institute of Mathematics, pr. Akademika Koptyuga 4, Novosibirsk 630090, Russia
{jkochet,apljas}@math.nsc.ru

[3] Novosibirsk State University, 1, Pirogova str., Novosibirsk 630090, Russia

[4] Faculty of Mathematics, University of Belgrade, Studentski trg 16, 11000 Belgrade, Serbia
aleks3rd@gmail.com
https://matinf.pmf.unibl.org
http://www.math.nsc.ru/LBRT/k5/lab.html
http://tc.nsu.ru
http://matf.bg.ac.rs

Abstract. Two transportation companies want to enter the market and they are aware of each other. The objective for the both of competitors is to maximize their respective profits by finding the best hub and spoke networks and price structures. One company wants to establish r hubs and the other wants to establish p hubs. It is assumed that the customers choose the route by price and the logistic regression based model is used to estimate how the demand is shared. After setting their networks, the competing companies engage in the price war. We propose a new model for finding a Stackelberg strategy that includes a price game, as bi-level nonlinear mixed-integer program, called the $(r|p)$ hub-centroid problem under the price war. It is shown that there is a unique finite Bertrand-Nash price equilibrium. On the basis of this result, we show the solution existence, propose a new equations for the best response pricing, and address the computational complexity of the problem. Finally, we discuss some possible future research directions that concern the solution approach and some other competitive scenarios that involve pricing.

Keywords: Hub location · Logit model · Stackelberg strategy · Bertrand price equilibrium · Complexity

The research of the second author was supported by the Russian Foundation for Basic Research, grant 18-07-00599. The research of the third author was supported by the program of fundamental scientific researches of the SB RAS, project 0314-2019-0014.

M. Khachay et al. (Eds.): MOTOR 2019, LNCS 11548, pp. 133–146, 2019.
https://doi.org/10.1007/978-3-030-22629-9_10

1 Introduction

In the location theory, the important part of literature is devoted to the hub location problems (HLP). Roughly speaking, the goal is to find the optimal locations of hubs (nodes in a graph) and allocations of non-hub nodes to hubs (called spokes), regarding a given objective. Hubs serve as concentration points through which the flows are routed between the origin and destination pairs (O-D). This concept should provide several benefits like reducing the number of spokes and empowering the economies of scale. The HLPs can be classified according to: the source that determines the number of hubs, the set of possible hub locations is discrete or not, the number of allocations of a non-hub node to hubs, whether the hubs are capacitated or uncapacitated, the hub location and allocation cost, etc. An interested reader is referred to the papers [1,2] for a better and deeper introduction into the topic.

In the current HLP research, the work is focused on the solution approaches that are able to handle real-life instances and on the extension of classical HLP models in order to grasp more reality. Several extensions of the classic hub location problems have been used quite successfully. Lüer-Villagra and Marianov [3] have argued that the location or route opening decisions can be very dependent on the revenues that a company can obtain. In turn, revenues depend on the price structure. The paper of Sasaki and Fukushima [4] addresses a *continuous* Stackelberg competition in which the incumbent competes with several entrants for profit maximization. For every route, only one hub was allowed. Mahmutogullari and Kara [5] addressed the competitive bi-level HLP in which the goal is *the market share* maximization. In their paper, the demand is divided among the competitors by the "winner-takes-it-all" rule: a competitor with lower route cost gets the whole demand for a given O-D pair. Čvokić et al. [6] have formulated, in a sketchy way, the model of Stackelberg competition that includes a price game. The results were primarily announced, i.e., only the sketches of proofs are given. More general review of facility location and pricing problems can be found in [7]. A good review concerning the equilibrium solutions for the location games is the paper of Karakitsiou and Migdalas [8]. In this study we focus on the Stackelberg strategy for a scenario in which two competing transportation companies intend to enter the market and both of them aspire to maximize their respective profits by finding the best hub and spoke networks and price structures, assuming the price war scenario.

The organization of this paper is given as follows. In Sect. 2, we describe a new bi-level problem for finding the Stackelberg strategy as an $(r|p)$ hub-centroid problem under the price war ($(r|p)$HCPuPW) and formulate the corresponding bi-level mathematical model. The existence of Bertrand-Nash price equilibrium for this model is shown in Sect. 3, alongside with the derivation of transcendent price equations. On the basis of this result, the solution existence is proved. The complexity of this problem is separately investigated in Sect. 4. Finally, in Sect. 5, the concluding remarks on this research are given, alongside with some proposals for the future work.

2 The $(r|p)$ Hub-Centroid Problem Under the Price War

The two competing transportation companies intend to enter the market. They are aware of each other. The both of them aspire to maximize their respective profits by finding the best hub and spoke networks and price structures. One company wants to locate p hubs and the other wants to locate r hubs. After setting their networks, it is expected that the competing companies will engage in the price war, which assumes responding to the current opponent's pricing with the more competitive one. The solution of the price war, if it exists, is a Bertrand-Nash equilibrium, in which none of the competitors have incentive to change their price decisions, unilaterally. This setting implies that the leader does not have the ability to impose prices to the follower. The pricing is a result of the follower's entrance to the market.

The basic setting for the problem is a complete digraph $G = G(N, A)$, where N is the non-empty node set and A is the set of arcs. A hub can only be established at the node $k \in N$, it can be shared, and there are no capacity constraints. All hubs should be mutually interconnected. Establishing a hub is considered as a strategic decision. For every arc $(i, j) \in A$ there is a transportation cost per unit of flow $c_{ij} \geq 0$. For each O-D pair $(i, j) \in N^2$, only one route can be established. Multiple allocations of non-hub nodes to hubs are allowed. The transportation factors $\aleph$, α and δ are already known for the market and they correspond to flow consolidation in collection (origin to hub), transfer between hubs, and distribution (hub to destination), respectively. Concatenation of arcs composes a route, where hubs are located at the joints. At most two hubs are allowed to be on a single route, i.e., at most two stops are permitted. Transportation cost on a route $i \rightarrow k \rightarrow l \rightarrow j$ is given as $c_{ij,kl} = \aleph c_{ij} + \alpha c_{kl} + \delta c_{lj}$, for all $i, j, k, l \in N$. It is assumed that the customers choose routes according to the prices. Both competitors are using the mill pricing, i.e., the customers are paying their respective expenses. The multinomial logit model (MNL) is used to resolve the issue with the discrete choice, as in [3,6,9]. The MNL is essentially a rule that determines what fraction of the flow is going to be captured. It has a sensitivity parameter $\Theta \geq 0$ with an already known non-negative value assigned. A higher Θ means that customers are very sensitive to price differences, so they will mostly choose less expensive routes. On the other hand, a smaller Θ means that the clients are less sensitive to price differences. The demand $w_{ij} \geq 0$ for every O-D pair is taken to be perfectly inelastic. Every customer must be served by one of the competitors. Following the specific reasoning presented in [6,10] we require that both companies have to cover all nodes. Sometimes, we will use different pronouns for the competitors: "she" for the leader and "he" for the follower.

The following variables are used to describe the choices made by the leader and follower:

- $x_k = 1$ if the leader has established a hub at node $k \in N$, and 0 otherwise
- $\rho_{ij,kl} = 1$ if the leader has established a transportation route $i \rightarrow k \rightarrow l \rightarrow j$ from i to j, and 0 otherwise
- $t_{ij,kl}$ is the price charged by the leader on a route $i \rightarrow k \rightarrow l \rightarrow j$

- $y_k = 1$ if the follower has established a hub at node $k \in N$, and 0 otherwise
- $\varsigma_{ij} = 1$ if the follower has established a transportation route $i \to k \to l \to j$ from i to j, and 0 otherwise
- $q_{ij,kl}$ is the price charged by the follower on a route $i \to k \to l \to j$

In order to represent the sequence of variables, we will use a more compact notation: $c = (c_{ij,kl})_{i,j,k,l \in N}$, $x = (x_k)_{k \in N}$, $\rho = (\rho_{ij,kl})_{i,j,k,l \in N}$, $t = (t_{ij,kl})_{i,j,k,l \in N}$, $y = (y_k)_{k \in N}$, $\varsigma = (\varsigma_{ij,kl})_{i,j,kl \in N}$, and $q = (q_{ij,kl})_{i,j,k,l \in N}$. The set of follower's solutions for a given leader's solution is shortly denoted as $\mathcal{F}(x, \rho)$. The optimal solutions are denoted with the asterisk, as usual.

Bitran and Ferrer in [11] provided the closed form expression for the optimal response price q over a cost c, when the opponent's price p is known:

$$q^* = c + \frac{1}{\Theta}\left(1 + W_0\left(e^{-\Theta(c-t)-1}\right)\right) \tag{1}$$

Here, W_0 is the principal branch of the Lambert W function. Lüer-Villagra and Marianov [3] have generalized this expression considering HLP with multiple routes of the same O-D pair. This motivates us to introduce the function to represent the optimal price response. Following the result in [3], this new function $\lambda_{ij,kl} : \mathbb{N} \times \mathbb{R}_+^{4|N|^4} \longrightarrow \mathbb{R}$ is defined as

$$\lambda_{ij,kl}(N, c, \rho, p, \varsigma) = c_{ij,kl} + \frac{1}{\Theta}\left(1 + W_0\left(\frac{\sum_{u,v \in N} e^{-\Theta c_{ij,uv} - 1}\varsigma_{ij,uv}}{\sum_{u,v \in N} e^{-\Theta t_{ij,uv}}\rho_{ij,uv}}\right)\right) \tag{2}$$

Usually two scenarios are considered in the literature: simultaneous and sequential entrance to the market. In the first scenario, the price war is a natural assumption. The issue is that we could expect multiple Nash equilibria to exists, when it comes to the competitors' hub and spoke topologies. Finding payoff-dominant equilibrium could be a daunting task. Moreover, the payoff-dominant equilibrium does not need to be in pure strategies, i.e., it can be characterized by a cycle of the best responses. The standard interpretation of Nash equilibrium in mixed-strategies is not acceptable. The company will not flip a coin to choose the network. In the second scenario, the price war is not assumed, i.e., the first competitor that enters the market is committed to its location and price decisions. But the existence of finite Stackelberg price solution implies the existence of feasible Bertrand-Nash price equilibrium, which opens the door to a cooperative price game with transferable utilities.

Taking all this into the account, we find it interesting in this study to consider an intermediate variant, i.e., a Stackelberg competition under the price war, where one company, usually called the leader, enters the market as the first competitor, anticipating the entrance of the other company, incidentally called the follower. The prices are set according to the solution of the price war. In other words, the leader is setting the prices, so that she does not have the incentive to deviate after the followers move. This setting is equivalent to the search for the Stackelberg strategy if the game is simultaneous, and it can be related to the search for the price *status quo* point when a cooperative pricing is considered.

The $(r|p)$HCPuPW can be represented as a bi-level mix-integer non-linear mathematical program. For the leader, we propose the following model.

$$\max \sum_{i,j,k,l \in N} w_{ij}(t_{ij,kl} - c_{ij,kl}) \frac{\rho_{ij,kl} e^{-\Theta t_{ij,kl}}}{\sum_{u,v \in N} \rho_{ij,uv} e^{-\Theta t_{ij,uv}} + \sum_{u,v \in N} \varsigma^*_{ij,uv} e^{-\Theta q^*_{ij,uv}}} \quad (3)$$

$$t_{ij,kl} = \lambda_{ij,kl}(N, c, q^*, \varsigma^*, \rho), \qquad \forall i,j,k,l \in N \quad (4)$$

$$\sum_{k \in N} \rho_{ij,kl} \le x_l, \qquad \forall i,j,l \in N \quad (5)$$

$$\sum_{l \in N} \rho_{ij,kl} \le x_k, \qquad \forall i,j,k \in N \quad (6)$$

$$\sum_{k,l \in N} \rho_{ij,kl} = 1, \qquad \forall i,j \in N \quad (7)$$

$$\sum_{k \in N} x_k = p \quad (8)$$

$$(q^*, y^*, \varsigma^*) \in \mathcal{F}^*(x, \rho) \quad (9)$$

$$x_k \in \{0,1\}, \qquad \forall k \in N \quad (10)$$

$$\rho_{ij,kl} \in \{0,1\}, \quad \forall i,j,k,l \in N \quad (11)$$

The leader's profit (3) is calculated as a sum of all net incomes. The leader's price is the best response to the anticipated follower's optimal price, according to the Eq. (4). We note that (4) is not a constraint set, but a parameter definition. The constraints (5)–(6) require that the nodes can be allocated solely to hubs. Only one route can be established per O-D pair and that is imposed with the constraint set (7). The number of hubs to locate is exogenous and specified with Eq. (8). Constraint (9) denotes that for a given leader's solution only optimal follower's solutions are considered. The domain of decision variables is stated in (10)–(11).

Recalling the terminology for the bi-level problems, a solution $((x, \rho), (q, y, \varsigma))$ is called *semi-feasible* if (x, ρ) satisfies (5)–(8), (10)–(11) (i.e., without (9)) and $(q, y, \varsigma) \in F(x, \rho)$. In other words, the optimality for the follower's solution is not required. For a solution to be *feasible*, it is required that (q, y, ς) is optimal [12,13].

Now, for the follower's problem, we propose the following multi-objective mixed-integer non-linear program, for which the preferred solutions are obtained by an *a priori* lexicographic method. It is assumed that the follower's behavior is altruistic, i.e., the leader has optimistic expectations concerning the follower's attitude.

$$\max \sum_{i,j,k,l \in N} w_{ij}(q_{ij,kl} - c_{ij,kl}) \frac{\varsigma_{ij,kl} e^{-\Theta q_{ij,kl}}}{\sum_{u,v \in N} \rho_{ij,uv} e^{-\Theta t_{ij,uv}} + \sum_{u,v \in N} \varsigma_{ij,uv} e^{-\Theta q_{ij,uv}}} \quad (12)$$

$$\max \sum_{i,j,k,l \in N} w_{ij}(t_{ij,kl} - c_{ij,kl}) \frac{\rho_{ij,kl} e^{-\Theta t_{ij,kl}}}{\sum_{s,t \in N} \rho_{ij,uv} e^{-\Theta t_{ij,uv}} + \sum_{s,t \in N} \varsigma_{ij,uv} e^{-\Theta q_{ij,uv}}} \quad (13)$$

$$t_{ij,kl} = \lambda_{ij,kl}(N, c, q, \varsigma, \rho), \qquad \forall i,j,k,l \in N \tag{14}$$

$$\sum_{k \in N} \varsigma_{ij,kl} \le y_l, \qquad \forall i,j,l \in N \tag{15}$$

$$\sum_{l \in N} \varsigma_{ij,kl} \le y_k, \qquad \forall i,j,k \in N \tag{16}$$

$$\sum_{k,l \in N} \varsigma_{ij,kl} = 1, \qquad \forall i,j \in N \tag{17}$$

$$\sum_{k \in N} y_k = r \tag{18}$$

$$q_{ij,kl} = \lambda_{ij,kl}(N, c, \rho, (\lambda_{ij,uv}(N, c, \varsigma, q, \rho))_{i,j,u,v \in N}, \varsigma), \qquad \forall i,j,k,l \in N \tag{19}$$

$$q_{ij,kl} \ge 0, \qquad \forall i,j,k,l \in N \tag{20}$$

$$y_k \in \{0,1\}, \qquad \forall k \in N \tag{21}$$

$$\varsigma_{ij,kl} \in \{0,1\}, \qquad \forall i,j,k,l \in N \tag{22}$$

The follower's profit (12) is calculated as a sum of all net incomes. The behavior of follower as the altruistic competitor is defined by (13). As in the previous model, we note that (14) is not a constraint set, but a parameter definition. The difference is that here the value of $t_{ij,kl}$ is not based on the optimal follower's solution (the star notation is omitted). The constraints (15)–(16) require that the nodes can be allocated solely to hubs. Only one route can be established per O-D pair and that is imposed with the constraint set (17). The number of hubs to locate is exogenous and specified with Eq. (18). The follower is setting the equilibrium prices (19). Basically, the follower sets his prices so that he does not have an incentive to change his own price decisions, after the leader's best price response. The domains of price and network variables are stated in (20)–(22).

Remark 1. Changing the second objective from maximization to minimization we obtain the model for the selfish follower's behavior. In that case, the leader would have pessimistic expectations concerning the follower's attitude.

Summing the leader's objective function with the follower's first objective one will not result in a constant. Therefore, we can not claim that this is a zero-sum game. This is the main reason behind the bi-objective formulation. Otherwise, the problem could be ill-posed.

The lower level model, regarding solely the first objective, is concerned with finding a medianoid affected by the price war, for which the leader's set of hubs H_L is fixed. Therefore, we will call it the $(r|H_p)$ *hub-medianoid problem under the price war.* The lower level model, regarding the second objective, is usually in the literature denoted as *the auxiliary model* [12,13].

Remark 2. If the Bertrand-Nash price equilibrium exists, then the equilibrium equation holds for the leader's prices, too. In other words,

$$t_{ij,kl} = \lambda_{ij,kl}(N, c, \varsigma^*, (\lambda_{ij,uv}(N, c, \rho, t, \varsigma^*))_{i,j,u,v \in N}, \rho). \tag{23}$$

3 The Solution Existence

If a finite Bertrand-Nash price equilibrium does not exists, then obviously the set of feasible solutions is empty for both competitors. On the other hand, the existence of multiple price equilibria could make problem ill-posed. In Čvokić et al. [6], the result concerning the Bertrand-Nash price equilibrium is primarily announced, i.e., only the sketch of proof is given. Here, we provide the complete rigorous proof with the appropriate discussion. We find that not only the statement itself is important, but also the way how it is proved and interpreted.

Theorem 1. *For given hub and spoke networks in the (r|p) hub-centroid problem under the price war, there is a unique finite Bertrand-Nash price equilibrium.*

Proof. The objective function for both competitors are separable by O-D pairs. Taking into account that the networks are already given, we know which routes are established. Thus, we can focus on a particular O-D pair in our analysis and neglect the indexes entirely. Because each competitor can establish only one route per O-D pair, the best response price constraints are reduced to (1).

The derived closed form expression for the best response in terms of margins for the competitors are given as follows:

- the leader's best response margin $r_L(r_F) = \frac{1}{\Theta}\left(1 + W_0\left(Qe^{\Theta r_F - 1}\right)\right)$
- the follower's best response margin $r_F(r_L) = \frac{1}{\Theta}\left(1 + W_0\left(\frac{e^{\Theta r_L - 1}}{Q}\right)\right)$

where $Q = \frac{e^{-\Theta c_{h_L}}}{e^{-\Theta c_{h_F}}}$. The margins of best responses are bijective functions (continuous, monotone increasing) from a domain of non-negative real numbers, to corresponding co-domains, and vice versa for the inverses. We need to prove that the finite stable point, i.e., a Bertrand-Nash price equilibrium, always exists. In other words, when it comes to the margins, we need to solve the following equation $r_L^* = r_L(r_F^*) = r_L(r_F(r_L^*))$, which is reduced to the system

$$\tau = W_0\left(Qe^{W_0\left(\frac{e^{\tau}}{Q}\right)}\right) \tag{24}$$

$$r_L^* = \frac{\tau + 1}{\Theta} \tag{25}$$

Algebra can also be done for the other player, in the same fashion. The principal branch of Lambert W function can be represented by an infinitely nested logarithm as $W_0(x) = \ln\left(\frac{x}{W_0(x)}\right)$. Using this, we can transform Eq. (24) into $W_0\left(Qe^{W_0\left(\frac{e^{\tau}}{Q}\right)}\right)e^{\tau} = Qe^{W_0\left(\frac{e^{\tau}}{Q}\right)}$. After multiplication of both sides by $W_0\left(\frac{e^{\tau}}{Q}\right)$ and simplifying the equation, we obtain the next system of equations with their corresponding constraints

$$W_0\left(Qe^{\xi}\right) = \frac{1}{\xi} \wedge \xi > 0 \tag{26}$$

$$\xi = W_0\left(\frac{e^{\tau}}{Q}\right) \tag{27}$$

$$r_L^* = \frac{\tau + 1}{\Theta} \wedge r_L^* \geq 0 \tag{28}$$

The first equation always has a solution on $(0, \infty)$. What remains is to check if the solution is feasible, i.e., if $r_L^* \geq 0$. The last two equations result in $e^{\tau} = Q\xi e^{\xi} \wedge \xi > 0 \wedge \tau \geq -1 \iff \xi \geq W_0\left(\frac{1}{Qe}\right)$. So, we need to prove that $W_0\left(Qe^{W_0((Qe)^{-1})}\right) \leq \frac{1}{W_0((Qe)^{-1})}$ for all $Q > 0$. To do that, we will analyze the function $f(Q) = W_0\left((Qe)^{-1}\right) W_0\left(Qe^{W_0((Qe)^{-1})}\right)$.

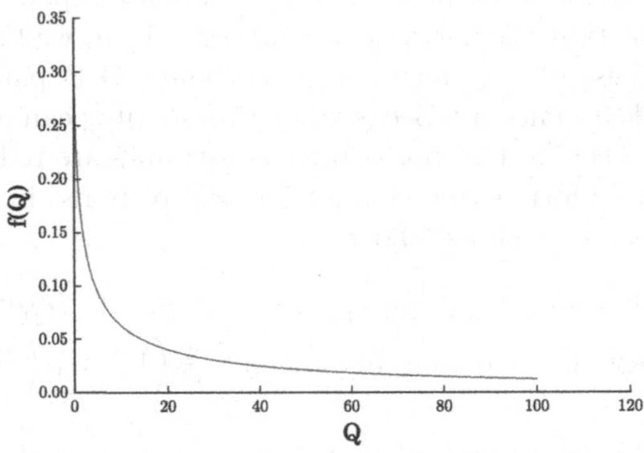

Fig. 1. The graph of the function $f(Q)$ when $Q \in (0, \infty)$. The limit of $f(Q)$ when $Q \to 0+$ is $\frac{1}{e}$, and the limit when $Q \to \infty$ is 0.

We observe that $\lim_{Q\to\infty} f(Q) = 0$, which can be seen through the series expansion at $x = \infty$. Next, $W_0\left(\frac{1}{Qe}\right) W_0\left(Qe^{W_0\left(\frac{1}{Qe}\right)}\right)^2 = 0$ is representing the first order condition for $f(Q)$, which doesn't have a solution on $(0, \infty)$. At the end, $\lim_{Q\to 0+} f(Q) = \frac{1}{e}$, because $W_0((Qe)^{-1}) \to \infty$ when $Q \to 0+$, and $\lim_{x\to\infty} xW_0\left(\frac{a}{x}\right) = a$ for some real a, which can again be seen from the series expansion at $x = \infty$. In our case $a = \frac{1}{e}$. The graph of the function $f(Q)$ is presented in Fig. 1. □

On a plot, the Bertrand-Nash price equilibrium can be represented as an intersection of the best response curves, as it is done in Fig. 2.

Remark 3. The pair (∞, ∞) is also the equilibrium, but it is not feasible.

Remark 4. The logit model and possibly different route costs yield a Bertrand-Nash price equilibrium that is not *a perfect competition.*

Proposition 1 (The follower's Bertrand-Nash equilibrium pricing). *For a given leader's network, the follower's Bertrand-Nash equilibrium price $q^*_{ij,kl}$ on a route $i \to k \to l \to j$ is given by the following equations*

$$\tau_{ij,kl} = W_0\left(\frac{e^{\tau_{ij,kl}}}{W_0\left(e^{\tau_{ij,kl}+\Theta(c_{ij,kl}-c_{ij,uv})}\right)}\right) \tag{29}$$

$$q^*_{ij,kl} = c_{ij,kl} + \frac{\tau_{ij,kl}+1}{\Theta} \tag{30}$$

where (u,v) is the pair of hubs connecting the route established by the leader for the O-D pair (i,j).

Proof. The statements follows from the proof of the Theorem 1, when the networks are fixed and equations are derived from the follower's point of view. To obtain Eq. (29) from Eq. (24) we exploit the identity $e^{W_0(x)} = \frac{x}{W_0(x)}$. □

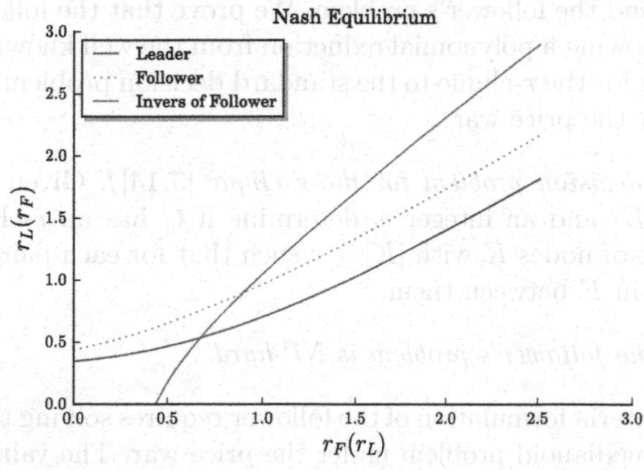

Fig. 2. The Nash pricing equilibrium for $\Theta = 3.35$ and $Q = 0.5$, presented as the intersection of the best response curves: the leader's (blue) and the follower's (red). Doted line represents the follower's best response in the same coordinate system as the leader's one (just for comparison). (Color figure online)

Remark 5. The new Bertrand-Nash equilibrium follower's pricing does not take into account the leader's price — only the route costs.

Remark 6. The proof of Theorem 1 gives the starting point for computing the follower's price as $\max\left\{W_0\left(e^{-\Theta(c_{ij,uv}-c_{ij,kl})-1}\right), \frac{1}{\Theta}\left(1+W_0\left(e^{-\Theta(c_{ij,uv}-c_{ij,kl})-1}\right)\right)\right\}$.

Remark 7. The existence of Bertrand-Nash price equilibrium does not depend on the behavior of the follower.

Thus, the search for the optimal leader's solution can be based on finding the best feasible follower's hub and spoke topology for which the prices are computed by (29)–(30).

From the Theorem 1 we obtain the following result about the solution existence.

Theorem 2. *The optimal solution exists for the* $(r|p)$ *hub-centroid problem under the price war.*

Proof. The number of possible hub and spoke networks for both players in the market is finite. For each pair of networks, there is a unique finite Bertrand-Nash price equilibrium. Therefore, there exists the optimal solution for the problem (3)–(11). The same reasoning can be applied in the case of pessimistic leader's expectations about the follower's behavior. □

4 The Computational Complexity

In this section we address some questions about the computational complexity for the leader's and the follower's problem. We prove that the follower's problem is NP-hard by showing a polynomial reduction from the well known NP-complete decision problem for the r-clique to the standard decision problem of $(r|H_p)$ hub-medianoid under the price war.

Problem 1 (the decision problem for the r*-clique* [5,14]*).* Given an undirected graph $G = (N, E)$ and an integer r, determine if G has an r-clique, i.e., that there exists a set of nodes K with $|K| \geq r$ such that for each pair of nodes in K there is an edge in E between them.

Theorem 3. *The follower's problem is NP-hard.*

Proof. The bi-criteria formulation of the follower requires solving the corresponding $(r|H_p)$ hub-medianoid problem under the price war. The values of variables $q_{ij,kl}$ $(\forall i, j, k, l \in N)$ can be precomputed, i.e., we can consider them as constants. From the constraint sets (15)–(17) we know that only one route can be established per O-D pair. If $\varsigma_{ij,kl} = 1$, for some $k, l \in N$, then for all other $k', l' \in N \wedge (k', l') \neq (k, l)$ we have that $\varsigma_{ij,k'l'} = 0$, and vice versa. This means that instead of $\sum_{u,v \in N} \varsigma_{ij,uv} e^{-\Theta q_{ij,uv}}$, we can write just $e^{-\Theta q_{ij,uv}}$ in the denominator, which leads to the following reformulation of the first objective:

$$\max \sum_{i,j,k,l \in N} w_{ij} Q_{ij,kl} \varsigma_{ij,kl}. \tag{31}$$

where $Q_{ij,kl}$ is computed as $(q_{ij,kl} - c_{ij,kl}) \frac{e^{-\Theta q_{ij,kl}}}{\sum_{s,t \in N} \rho_{ij,st} e^{-\Theta t_{ij,uv}} + e^{-\Theta q_{ij,kl}}}$. As we have already said, if we show a polynomial reduction from the r-clique decision problem to the standard decision problem of $(r|H_p)$ hub-medianoid and pricing, then the problem of follower is NP-hard.

Consider an r-clique instance $G(N, E)$, where N is the set of nodes and E is the set of edges. We can construct a (digraph) network $G'(N', A')$ where $N' = N$ and $(i, j) \in A' \subseteq N^2$ if $(i, j) \in E$. Note that for the $(r|H_L)$ hub-medianoid problem under the price war, the opponent's network does not need to satisfy the constraints (5)–(8), nor the route costs have to be computed in the same way. It just needs to be a valid hub and spoke topology with non-negative finite values for the route costs.

Now, assume that in the opponent's network all established routes are spokes $i \to i \to j \to j$, for all available O-D pairs $(i, j) \in A'$. Furthermore, take that $\Theta \geq 0$, $\alpha = 1$, and $w_{ij} = \frac{2\Theta}{1+\tau}$ $(\forall i, j \in N)$, where τ is the solution of equation $\tau = W_0(e^{W_0(e^{\tau})})$.

If the r-clique exists, then there is a network (y, ς) in which the hub backbone corresponds to this r-clique. For each inter-hub spoke, both competitors have the same route costs, which implies the same equilibrium prices with margin $\frac{1+\tau}{\Theta}$ and equal market share. Therefore, the profit obtained just on the hub backbone is greater than $\frac{r(r-1)}{2}$.

On the other hand, the solution existence itself implies the existence of r-clique, because of constraint (18). If the solution with objective value greater then or equal to $\frac{r(r-1)}{2}$ does not exists, then we have two cases:

(1) the objective value of every feasible solution is strictly less than $\frac{r(r-1)}{2}$
(2) the set of feasible solutions is empty.

In the first case, we easily reach the contradiction. In the second case, we infer that there is no r-clique.

The situation when $\alpha \in [0, 1)$ is analyzed in the similar fashion. We just need to assume that the opponent's route costs are all discounted. □

Although, the follower's problem is NP-hard, the corresponding allocation problem is easier to solve.

Theorem 4. *The linear relaxation of the corresponding allocation problem for the follower has an integral solution.*

Proof. Regarding the follower's first objective, for an O-D pair $(i, j) \in N^2$, the highest coefficient of variable $\varsigma_{ij,kl}$ (calculated form (31)) determines the optimal route. Moreover, a linear relaxation where $\varsigma_{ij,kl} \in [0, 1]$ must have an integral optimal solution. If we assume that the fractional optimal solution exists for the linear relaxation, we can easily see that the corresponding first objective function (31) will have a linear deviation.

Regarding the second follower's objective, we know that solution should always be from the set of optimal solutions concerning the first objective. Observe that the higher values of $e^{-\Theta q_{ij,uv}}$ $(u, v \in N)$ (i.e. the lower values of $q_{ij,uv}$) are more preferable, for a given O-D pair $(i, j) \in N^2$. If we assume that the fractional optimal solution exists for this linear relaxation, we can easily see that the corresponding second objective function will have a linear deviation, too. □

From the proof of this statement, the following two corollaries follow.

Corollary 1. *The allocation problem of the $(r|H_p)$ hub-medianoid under the price war is polynomially solvable.*

Corollary 2. *The allocation problem of the auxiliary problem is polynomially solvable.*

Because finding the optimal solution for the leader requires solving the NP-hard problem of the follower, we could assess that the leader's problem is NP-hard. The proof of statement about the leader's computational complexity is based on the vertex cover decision problem.

Problem 2 (the decision problem for the vertex cover [5,14]*).* Given an undirected graph $G = (N, E)$ and an integer p, determine if G has a vertex cover, i.e., if there is a set of vertices C with $|C| \leq p$ such that for each edge $(i, j) \in E$, either i or j is in C.

Theorem 5. *The leader's problem is NP-hard.*

Proof. Given an instance of vertex cover problem on an undirected graph $G(N, E)$, we can construct a digraph $G'(N', A')$ where $N' = N$ and $A' = N' \times N'$. Let $r > p$, and

$$w_{ij} = \begin{cases} 1, & \text{if } (i,j) \in E \\ 0, & \text{otherwise.} \end{cases} \tag{32}$$

We need to show that there exists a vertex cover C with $|C| \leq p$ if and only if there exists a set of p nodes H_p on G', such that the follower's network will coincide with the leader's one on edges (i, j), for which $w_{ij} = 1$. We know from the expression (1) what are the margins for both competitors, if their profits are equal. Therefore, we are able to know exactly the leader's profit, from which we can derive the corresponding standard decision problem.

($\rightarrow$) Assume that the vertex cover problem has a solution $C \subsetneq N$ and $|C| \leq p$. We can let $H_p \supseteq C$ and observe that if $i \in H_L$ or $j \in H_p$ then the unit flow w_{ij} could get value 1, depending on the membership to E. In all other cases w_{ij} is always equal to 0. In other words, the pricing can be important only for those O-D pairs (i, j) that have at least i or j in H_p. Therefore, the leader and the follower could compete for the profit only on those routes in which each flow is routed via one (single) spoke that has at least one end in C (subset of H_p). In this situation, the follower can not choose strictly better routes for O-D pairs than those the leader is already using.

($\leftarrow$) Suppose that H_p on G' is such that the best follower's response is to copy of the leader's solution on all O-D pairs for which $w_{ij} = 1$. If H_p does not contain as a subset the vertex cover of G, then there exists an edge $(i, j) \in E$ and $i \notin H_p$ or $j \notin H_p$ (otherwise H_p would be a vertex cover). Then, on that particular O-D pair (i, j) the follower can profit more than the leader, if his hub backbone includes i or j. In this situation, the follower is offering a non-stop (direct) route to the customers, while the leader has to route the flow through the intermediate hubs. This contradicts to the assumption of follower using the same solution as the leader, as his best one.

Hence, we can conclude that the standard decision problem for the leader is polynomially equivalent to the decision problem for the vertex cover. □

5 Concluding Remarks

This study introduces an intermediate variant of hub location and pricing problem in which competitors are sequentially entering into the market (a leader-follower scenario), but the pricing is resolved as in the Bertrand price game. Involving pricing is a more realistic scenario than relaying solely on the costs and demands. The leader and the follower are intending to establish p and r hubs, respectively. The setting for hub location and establishing routes is derived from the classic uncapacitated multiple allocation hub location problem. Multiple allocations are allowed and there are no limits on hub capacities. Only one route can be established per O-D pair. The demand is perfectly inelastic and split between the competitors according to the logit model, which is also a more realistic assumption. The objective for both companies is the profit maximization, contrary to the usual viewpoint in which the company is interested in the minimization of its costs. The problem is denoted as the $(r|p)$ hub-centroid problem under the price war. Compared to some other bi-level problems in the literature, here we need to properly define the behavior of follower.

The existence of finite Bertrand-Nash price equilibrium for perfectly inelastic demand is shown, which further implied the existence of optimal solution to the problem itself. The new price equations are proposed for the follower, and they could be seen as a game theoretic generalization of the expression given by Bitran and Ferrer [11]. The logit model and possibly different route costs yield a Bertrand-Nash price equilibrium that is not a perfect competition. Besides the pricing related statements, we have addressed the computational complexity for the leader's and follower's problems. It is shown that the follower's problem is NP-hard, but on the other hand the derived allocation problem is in fact polynomially solvable. Also, it is shown, as one could asses, that the leader's problem is NP-hard, too.

The finite Bertrand-Nash price equilibrium, indicates that it would be reasonable to consider the cooperative price game with transferable utilities. In this setting, the leader announces the strategy profile in terms of prices and has an incentive to find the position on the market that would ensure the best credible threat strategy, or the *status quo* point, depending whether the side payment is allowed or not. Also, one could address some robust variants of this problem (or a similar one), i.e., to explore situations in which the demand or cost can be affected by some uncertain factors.

The solution approach is another line of research for the future work. It is a tough proposition to design the exact solution method for the bi-level optimization problems with non-linear objectives and where the follower's behavior must be properly defined. As a matter of fact, finding a good heuristic approaches for these kind of problems is not an easy piece of work, too. But on the other hand, the location patterns and managerial insights provide the stepping stone for the further development and refinement of models.

An interesting research direction is the investigation of relationships concerning the polynomial and approximation hierarchies, similarly as it was done in [15].

References

1. Campbell, J., O'Kelly, M.: Twenty-five years of hub location research. Transp. Sci. **46**(2), 153–169 (2012)
2. Farahani, R., Hekmatfar, M., Arabani, A., Nikbakhsh, E.: Hub location problems: a review of models, classification, solution techniques, and applications. Comput. Ind. Eng. **64**(4), 1096–1109 (2013)
3. Lüer-Villagra, A., Marianov, V.: A competitive hub location and pricing problem. Eur. J. Oper. Res. **231**(3), 734–744 (2013)
4. Sasaki, M., Fukushima, M.: Stackelberg hub location problem. J. Oper. Res. Soc. Japan **44**(4), 390–405 (2001)
5. Mahmutogullari, I., Kara, B.: Hub location under competition. Eur. J. Oper. Res. **250**(1), 214–225 (2016)
6. Čvokić, D., Kochetov, Y., Plyasunov, A.: The existance of equilibria in the leader-follower hub location and pricing problem. In: Dörner, K., Ljubic, I., Pflug, G., Tragler, G. (eds.) Operations Research Proceedings 2015. Springer, Cham (2017). https://doi.org/10.1007/978-3-319-42902-1_73
7. Karakitsiou, A.: Modeling Discrete Competitive Facility Location. Springer, Cham (2015). https://doi.org/10.1007/978-3-319-21341-5
8. Karakitsiou, A., Migdalas, A.: Nash type games in competitive facilities location. Int. J. Decis. Support. Syst. **2**(1/2/3), 4–12 (2016)
9. Čvokić, D.D., Kochetov, Y.A., Plyasunov, A.V.: A leader-follower hub location problem under fixed markups. In: Kochetov, Y., Khachay, M., Beresnev, V., Nurminski, E., Pardalos, P. (eds.) DOOR 2016. LNCS, vol. 9869, pp. 350–363. Springer, Cham (2016). https://doi.org/10.1007/978-3-319-44914-2_28
10. Sasaki, M.: Hub network design model in a competitive environment with flow threshold. J. Oper. Res. Soc. Jpn. **48**, 158–71 (2005)
11. Bitran, G., Ferrer, J.C.: On pricing and composition of bundles. Prod. Oper. Manag. **16**, 93–108 (2007)
12. Dempe, S.: Foundations of Bilevel Programming, pp. 1–6. Kluwer Academic, Dordrecht (2002)
13. Alekseeva, E., Kochetov, Y.: Matheuristics and exact methods for the discrete $(r|p)$-centroid problem. In: El-Ghazali, T. (ed.) Metaheuristics for Bi-level Optimization. SCI, vol. 482, pp. 189–219. Springer, Heidelberg (2013). https://doi.org/10.1007/978-3-642-37838-6_7
14. Garey, M., Johnson, D.: Computers and Intractability: A Guide to the Theory of NP-Completeness. Series of Books in the Mathematical Sciences. W. H. Freeman and Company, New York (1979)
15. Kononov, A.A., Panin, A.A., Plyasunov, V.A.: A bilevel competitive location and pricing model with non-uniform split of the demand. J. Appl. Ind. Math. **13**, 519 (2019)

Computing Locally Optimal Solutions of the Bilevel Optimization Problem Using the KKT Approach

Stephan Dempe[(✉)]

Institute of Numerical Mathematics and Optimization,
TU Bergakademie Freiberg, Freiberg, Germany
dempe@tu-freiberg.de
http://www.mathe.tu-freiberg.de/ dempe/

Abstract. If the lower-level problem in a bilevel optimization problem is replaced by its Karush-Kuhn-Tucker conditions, a mathematical program with complementarity constraints is obtained. Solving this nonconvex optimization problem, locally optimal solutions are computed which do in general not correspond to locally optimal solutions of the bilevel problem. Using a relaxation of this problem in two constraints it can be shown that a sequence of locally optimal solutions of the relaxed problems converges to a point which is related to a locally optimal solution of the bilevel optimization problem. If the lower-level problem is a linear one, relaxation of only the complementarity constraint is sufficient.

Keywords: Optimistic bilevel optimization · KKT transformation · Locally optimal solutions

1 Introduction

Bilevel optimization problems are hierarchical optimization problems of two levels. The lower-level problem is

$$\min_y \{f(x,y) : g(x,y) \leq 0\} \tag{1}$$

depending on the upper-level variable x, where $f : \mathbb{R}^n \times \mathbb{R}^m \to \mathbb{R}$ as well as $g : \mathbb{R}^n \times \mathbb{R}^m \to \mathbb{R}^p$. We can add equality constraints by adjusting the constraint qualification. Let

$$\varphi(x) := \min_y \{f(x,y) : g(x,y) \leq 0\} : \mathbb{R}^n \to \mathbb{R}$$

denote the optimal value function of problem (1) and

$$\Psi(x) := \operatorname{Argmin}_y \{f(x,y) : g(x,y) \leq 0\} : \mathbb{R}^n \to 2^{\mathbb{R}^m}$$

Supported by Deutsche Forschungsgemeinschaft.

M. Khachay et al. (Eds.): MOTOR 2019, LNCS 11548, pp. 147–157, 2019.
https://doi.org/10.1007/978-3-030-22629-9_11

its solution set mapping. If

$$\mathbf{gph}\,\Psi := \{(x,y) \in \mathbb{R}^n \times \mathbb{R}^m : y \in \Psi(x)\}$$

denotes the graph of the solution set mapping, the upper-level optimization problem can be formulated as

$$\min_{x,y}\{F(x,y) : G(x) \leq 0, (x,y) \in \mathbf{gph}\,\Psi\}, \tag{2}$$

where $F : \mathbb{R}^n \times \mathbb{R}^m \to \mathbb{R}$ and $G : \mathbb{R}^n \to \mathbb{R}^q$. We do not want to consider upper-level constraints depending on both x and y since this implies some ambiguity if the solution set $\Psi(x)$ does not reduce to a singleton for some x. Problem (2) is the so-called optimistic version of the bilevel optimization problem. For the pessimistic formulation where the upper-level decision maker has to bound the damage resulting from an unwelcome decision of the lower-level decision maker, see e.g. [12]. All functions F, f, g_i, G_i are assumed to be sufficiently smooth.

A point $(\overline{x}, \overline{y}) \in \mathbb{R}^n \times \mathbb{R}^m$ is called *feasible* for (1), (2) if $G(\overline{x}) \leq 0$, $\overline{y} \in \Psi(\overline{x})$. It is a *locally optimal solution* provided there exists an open neighborhood U of $(\overline{x}, \overline{y})$ such that

$$F(x,y) \geq F(\overline{x}, \overline{y}) \ \forall \text{ feasible points } (x,y) \in U.$$

Finally, it is a *globally optimal solution* if $U = \mathbb{R}^n \times \mathbb{R}^m$ can be used.

Problem (1), (2) has been investigated at least in three monographs [2,5,11], it has many applications, see [6]. It is a nonconvex optimization problem with, moreover, a feasible set which is not given in an explicit form using equality or inequality constraints.

To investigate it and to find possible solution algorithms, it needs to be replaced by a single-level optimization problem. For that we have different possibilities:

1. We can use the optimal value function of the lower-level problem and bound its objective function from above. This idea goes back to [20] and results in the fully equivalent, nonconvex optimization problem

$$\min_{x,y}\{F(x,y) : G(x) \leq 0,\ f(x,y) \leq \varphi(x),\ g(x,y) \leq 0\}. \tag{3}$$

   Problem (3) is irregular in the sense that the nonsmooth Mangasarian-Fromovitz constraint qualification is violated at every feasible point and the function $\varphi(\cdot)$ is not differentiable even if the lower-level problem (1) is a parametric linear optimization problem. For more information about (3) see e.g. [8,9].
2. If the functions $y \mapsto f(x,y)$, $y \mapsto g_i(x,y)$, $i = 1, \ldots, p$, are differentiable and a regularity condition is satisfied at all points (x,y), problem (1) can be replaced by its Karush-Kuhn-Tucker conditions resulting in

$$\begin{aligned}\min_{x,y,u}\{F(x,y) : G(x) \leq 0, \nabla_y L(x,y,u) = 0,\\ g(x,y) \leq 0,\ u \geq 0,\ u^\top g(x,y) = 0\},\end{aligned} \tag{4}$$

where $L(x, y, u) = f(x, y) + u^\top g(x, y)$ is the Lagrange function of (1). This is the *KKT transformation* which can only be used if the lower-level problem (1) is a convex optimization problem, see [18]. Problem (4) is a co-called mathematical program with complementarity constraints (MPCC), it is a nonconvex optimization problem for which the Mangasarian-Fromovitz constraint qualification is violated at every feasible point [23]. This problem has been the topic of a large number of monographs and articles, see e.g. [19]. If the regularity condition (Slater's condition) is violated for the lower-level problem at some points (x, y) the F.-John conditions can be used to replace (1), see [1], or the feasible set of problem (4) is perhaps no longer closed.

3. Under some restrictive assumptions, the optimal solution $y(x)$ of the lower-level problem is strongly stable [14] and can be inserted directly into (2) resulting is a nondifferentiable single level optimization problem

$$\min_x \{f(x, y(x)) : G(x) \leq 0\}, \tag{5}$$

which, under certain assumptions, is a Lipschitz continuous optimization problem, see [4].

Topic of the article is the solution of problem (4). This problem has often be used to solve the bilevel optimization problem (1), (2). From now on assume that $y \mapsto f(x, y)$ and $y \mapsto g_i(x, y)$, $i = 1, \ldots, p$, are convex functions. Problem (4) is a nonconvex optimization problem. Globally optimal solutions of this problem can easily be shown to be related to globally optimal solutions of the bilevel optimization problem [7]. This is in general not the case for locally optimal solutions, see Sect. 2. Section 3 is devoted to an algorithm for computing local solutions of the bilevel optimization problem. After that, the special case of a linear bilevel optimization problem is investigated in Sect. 4.

2 Relations Between the Bilevel Optimization Problem and Its KKT-transformation

Problem (4) is often used for solving the bilevel optimization problem. The following example from [7] shows that this is not so easy.

Example 1 (Dempe and Dutta [7]*).* Consider the linear lower-level problem

$$\min_y \{-y : x + y \leq 1,\ -x + y \leq 1\} \tag{6}$$

and the upper-level problem

$$\min\{(x-1)^2 + (y-1)^2 : (x, y) \in \mathbf{gph}\,\Psi\}. \tag{7}$$

This problem has the unique optimal solution $(\overline{x}, \overline{y}) = (0.5, 0.5)$ and no other locally optimal solutions.

Consider the point $(x^0, y^0) = (0, 1)$. Here, $\Lambda(x^0, y^0) = \text{conv}\{(1, 0), (0, 1)\}$ is the set of regular Lagrange multipliers, where conv$A$ denotes the convex hull of the set A.

Then, the point $(x^0, y^0, u^0) = (0, 1, (0, 1))$ can be shown to be a locally optimal solution of the KKT-reformulation (4) of (6), (7). $\diamond$

Since solution algorithms solving nonconvex, irregular MPCCs can often only be shown to compute stationary solutions and these do not need to be related to stationary solutions of the bilevel optimization problem, this approach is not without difficulties.

Remark 1. If the bilevel optimization problem is a linear one, i.e. if the functions f, F, g_i, G_j are all affine linear, the feasible set of problem (4) is polyhedral: it is the union of a finite number of convex polyhedral sets. Hence, stationary points of (4) are at least local minima.

Theorem 1 ([7]). *Let Slater's condition be satisfied for (1) for all x with $G(x) \leq 0$, i.e., $\exists\ w(x)$ satisfying $g_i(x, w(x)) < 0\ \forall\ i = 1, \ldots, p$. Then, a feasible point $(\overline{x}, \overline{y})$ of (1), (2) is a locally optimal solution if and only if the point $(\overline{x}, \overline{y}, u)$ is a locally optimal solution for (4) for all*

$$u \in \Lambda(\overline{x}, \overline{y}) := \{w \in \mathbb{R}^p : w \geq 0,\ w^\top g(\overline{x}, \overline{y}) = 0,\ \nabla_y L(\overline{x}, \overline{y}, w) = 0\}.$$

The feasible point $(\overline{x}, \overline{y})$ is a globally optimal solution of (1), (2) if and only if $(\overline{x}, \overline{y}, u)$ is a globally optimal solution of (4) for some $u \in \Lambda(\overline{x}, \overline{y})$.

The point $(x^0, y^0, u) = (0, 1, (1, 0))$ is not a locally optimal solution of (4) in Example 1.

3 Relaxation of the KKT Transformation

Since the Mangasarian-Fromovitz constraint qualification is violated at every feasible point of the problem (4), see [23], and standard solution algorithms for optimization problems suppose a constraint qualification for sure convergence to a solution, special approaches need to be used. One often used approach is a relaxation approach [13]. Within the relaxation approaches, on the basis of a numerical comparison with other relaxation methods, the one by Scholtes [24] is favored in [13]. The idea here is to replace (4) by

$$\min_{x,y,u} F(x, y) \tag{8}$$

$$s.t.\ G(x) \leq 0, \tag{9}$$

$$\nabla_y L(x, y, u) = 0, \tag{10}$$

$$g(x, y) \leq 0,\ u \geq 0,\ -u^\top g(x, y) \leq \varepsilon, \tag{11}$$

where the last equation in (4) is relaxed, see (11). This problem is solved for different relaxation parameters $\varepsilon > 0$ tending to zero.

Theorem 2 ([13]). *Let $\{\varepsilon_k\} \downarrow 0$ and let (x^k, y^k, u^k) be a stationary point of (8)–(11) for $\varepsilon = \varepsilon_k$ with $\{(x^k, y^k, u^k)\}$ converging to $(\overline{x}, \overline{y}, \overline{u})$ such that MPEC-MFCQ holds at $(\overline{x}, \overline{y}, \overline{u})$. Then $(\overline{x}, \overline{y}, \overline{u})$ is a C-stationary point of (4).*

Here, MPEC-MFCQ is the Mangasarian-Fromovitz constraint qualification adapted to (4): The complementarity constraint is not considered and some of the inequality constraints are replaced by equations. C-stationarity means that conditions similar to the Karush-Kuhn-Tucker conditions are satisfied where the multipliers to the biactive constraints ($\overline{u}_i = 0$ and $g_i(\overline{x}, \overline{y}) = 0$) do not need to be nonnegative (as in the Karush-Kuhn-Tucker conditions) but satisfy a weaker condition: they have the same sign, their product is not negative.

Theorem 2 shows that, if locally optimal solutions of problem (8)–(11) are computed, the algorithm will in general not converge to a locally optimal solution which corresponds to a locally optimal solution of the bilevel optimization problem (cf. Theorem 1). Consider the following relaxation of (4) where additionally to the last equation in (11) also the Eq. (10) is relaxed:

$$\min_{x,y,u} F(x, y) \tag{12}$$

$$\text{s.t. } G(x) \leq 0, \tag{13}$$

$$\|\nabla_y L(x, y, u)\| \leq \varepsilon_1, \tag{14}$$

$$g(x, y) \leq 0,\ u \geq 0,\ -u^\top g(x, y) \leq \varepsilon_2. \tag{15}$$

Here, $\|\cdot\|$ is an arbitrary norm in $\mathbb{R}^m$. We prefer to use the Chebyshev norm $\|a\| = \max_i |a_i|$ meaning that (14) reduces to a set of $2m$ inequalities.

The use of problem (12)–(15) for decreasing $\varepsilon_j, j = 1, 2$, for solving the bilevel optimization problem has originally been suggested in [16], see also [17]. There, the following convergence result has been shown:

Theorem 3 ([17]). *Let $\{(x^k, y^k, u^k)\}$ be a sequence of locally optimal solutions of problem (12)–(15) for $\{\varepsilon^k\}$ tending to zero converging to $(\overline{x}, \overline{y}, \overline{u})$. If Slater's condition, the constant rank constraint qualification (CRCQ) and the strong sufficient optimality condition of second order (SSOSC) are satisfied for the lower-level problem (1) and the Mangasarian-Fromovitz constraint qualification is satisfied for the upper-level constraints then, $(\overline{x}, \overline{y})$ is a Bouligand stationary point for the bilevel optimization problem.*

The assumptions (MFCQ) (or Slater's condition), (CRCQ) and (SSOSC) for the lower-level problem imply that the (globally by convexity) optimal solution $y(x)$ of (1) is a $\mathcal{PC}^1$ function (i.e. a continuous and piecewise continuously differentiable function) which is locally Lipschitz continuous and directionally differentiable [21]:

$$y'(\overline{x} : d) := \lim_{t \downarrow 0} \frac{y(\overline{x} + td) - y(\overline{x})}{t}$$

exists and is finite for all directions d. It is a strongly stable function $y(x)$ by [14]. Hence, problem (1)–(2) can locally be replaced by

$$\min_x \{F(x, y(x)) : G(x) \leq 0\}. \tag{16}$$

The point $(\overline{x}, \overline{y})$ with $\overline{y} = y(\overline{x})$ is Bouligand stationary if the necessary optimality condition

$$\frac{\partial}{\partial x} F(\overline{x}, \overline{y})d + \frac{\partial}{\partial y} F(\overline{x}, \overline{y})y'(\overline{x}; d) \geq 0 \ \forall \ d \in T(\overline{x}),$$

where $T(\overline{x}) := \{d : \nabla G_i(\overline{x})d \leq 0$ for all $i \in I(\overline{x}) := \{j : G_j(\overline{x}) = 0\}\}$ is satisfied at $\overline{x}$, see [4]. The proof of Theorem 3 uses an algorithm of feasible directions where, in each iteration, a descent direction d is computed. For that, a bilevel optimization problem needs to be solved (the directional derivative $y'(\overline{x} : d)$ is the optimal solution of some quadratic optimization problem, see [21]). Hence, Theorem 3 is theoretically interesting but very difficult to be applied in solving the bilevel optimization problem.

The topic of this article is to show that the sequence of problems (12)–(15) can be used even if the assumptions of Theorem 3 are violated and that this problem can be solved using other methods than an algorithm of feasible directions.

Since the feasible set of problem (12)–(15) contains the feasible set of (4) the following theorem is clear due to Theorem 1.

Theorem 4. *Let Slater's condition be satisfied for the lower-level problem at all x with $G(x) \leq 0$ and let $\{(x^k, y^k, u^k)\}$ be a sequence of globally optimal solutions of problem (12)–(15) for $\{\varepsilon^k\} \subset \mathbb{R}^2_+$ converging to zero. Then, any accumulation point $(\overline{x}, \overline{y}, \overline{u})$ of $\{(x^k, y^k, u^k)\}$ corresponds to a globally optimal solution $(\overline{x}, \overline{y})$ of (1), (2).*

To abbreviate notation, let $M(\varepsilon)$ denote the feasible set of problem (12)–(15) for fixed $\varepsilon_i > 0, \ i = 1, 2$.

Theorem 5. *Assume that Slater's condition is satisfied for (1) for all x with $G(x) \leq 0$ and let $\{(x^k, y^k, u^k)\}$ be a sequence of locally optimal solutions of problem (12)–(15) for $\varepsilon_i^k > 0, \ i = 1, 2$ tending to zero. Assume that there exists $\delta > 0$ such that $F(x, y) \geq F(x^k, y^k)$ for all $(x, y, u) \in M(\varepsilon_1^k, \varepsilon_2^k)$ with $\|(x, y) - (x^k, y^k)\|_2 \leq \delta$ and all k. Then, each accumulation point $(\overline{x}, \overline{y}, \overline{u})$ of $\{(x^k, y^k, u^k)\}$ corresponds to a locally optimal solution $(\overline{x}, \overline{y})$ of (1), (2).*

Proof. Assume without loss of generality that $\{(x^k, y^k, u^k)\}$ converges to $(\overline{x}, \overline{y}, \overline{u})$ and that $(\overline{x}, \overline{y})$ is not a locally optimal solution of (1), (2). Then, by Theorem 1 there exists $\widetilde{u} \in \Lambda(\overline{x}, \overline{y})$ such that $(\overline{x}, \overline{y}, \widetilde{u})$ is not a local minimum of (4). Hence, there exists a sequence $\{(\overline{x}^t, \overline{y}^t, \overline{u}^t)\}$ of feasible points to (4) converging to $(\overline{x}, \overline{y}, \widetilde{u})$ with

$$F(\overline{x}^t, \overline{y}^t) < F(\overline{x}, \overline{y}) \ \forall t.$$

Consider the sequence $\{(\overline{x}^t, \overline{y}^t, \widetilde{u})\}$. For sufficiently large t and $\varepsilon_i^k > 0, \ i = 1, 2$, $(\overline{x}^t, \overline{y}^t, \widetilde{u}) \in M(\varepsilon^k)$ and $\|(\overline{x}^t, \overline{y}^t) - (x^k, y^k)\|_2 \leq \delta$. Hence,

$$F(x^k, y^k) \leq F(\overline{x}^t, \overline{y}^t) < F(\overline{x}, \overline{y}) \ \forall k \text{ and sufficiently large } t.$$

Let $(x^0, y^0) \in \mathbf{gph}\,\Psi$ with $G(x^0) \leq 0$ be arbitrarily chosen such that

$$\|(x^0, y^0) - (\overline{x}, \overline{y})\|_2 \leq \delta/2.$$

Take $u^0 \in \Lambda(x^0, y^0)$. Then, (x^0, y^0, u^0) is feasible for (4) and also for (12)–(15) for all $\varepsilon^k > 0$, $\forall\, k$. For sufficiently large k we have $\|(x^k, y^k) - (\overline{x}, \overline{y})\|_2 \leq \delta/2$ and hence $\|(x^k, y^k) - (x^0, y^0)\|_2 \leq \delta$ implying $F(x^k, y^k) \leq F(x^0, y^0)$ by our assumption. Passing to the limit we derive

$$F(\overline{x}, \overline{y}) \leq F(x^0, y^0).$$

Note, that we can also take $(x^0, y^0) = (\overline{x}^t, \overline{y}^t)$ which contradicts now our assumption that $(\overline{x}, \overline{y}, \widetilde{u})$ is not locally optimal for (4). □

Note that we used a neighborhood of feasible points of (4) not depending on the Lagrange multiplier in the lower-level problem. This has also been done in [15,25]. This is justified here since the objective function of the upper-level problem does not depend on u.

Example 2. Consider Example 1 again at the point $x^0 = 0$, $y^0 = 1$, $u^0 = (0, 1)$. Problem (12)–(15) reads as

$$\min_{x,yx,y,u} \; (x-1)^2 + (y-1)^2 \tag{17}$$

$$s.t.\, x + y \leq 1, \; -x + y \leq 1 \tag{18}$$

$$|u_1 + u_2 - 1| \leq \varepsilon_1 \tag{19}$$

$$u_1 \geq 0, \; x + y - 1 \leq 0, \; u_1(1 - x - y) \leq \varepsilon_2 \tag{20}$$

$$u_2 \geq 0, \; -x + y - 1 \leq 0, \; u_2(1 + x - y) \leq \varepsilon_2 \tag{21}$$

It is easy to see that, for $\varepsilon > 0$, the point $x = t$, $y = 1 - t$, $u_1 = t$, $u_2 = 1 - t$ has a smaller function value than (x^0, y^0, u^0) for sufficient small $t > 0$ and is feasible: Objective function value is $F(t, 1-t) = 2t^2 - 2t + 1 < 1$ for $0 < t < 1$, $u_1 + u_2 = 1$, $x + y = 1$, $-x + y = 1 - 2t < 1$, $u_2(1 + x - y) = (1-t)(2t) = 2t - 2t^2 \leq \varepsilon_2$ for $0 < t \leq 0.5 - \sqrt{0.25 - \varepsilon_2}$. Hence, the point (x^0, y^0, u^0) is no longer locally optimal.

Note that this is correct also for $\varepsilon_1 = 0$ since the lower-level problem is a linear optimization problem with right-hand-side perturbations.

4 The Linear Bilevel Optimization Problem

Now, let the lower-level problem be a right-hand-side perturbed linear optimization problem:

$$\Psi_L(x) := \operatorname*{Argmin}_y \{c^\top y : Ay \leq x\} \tag{22}$$

and consider the upper-level problem (2) with $\Psi(x) = \Psi_L(x)$ and $G(x) = Gx - b$ for some matrix G and vector b of appropriate dimension. The KKT transformation of this problem is

$$\min_{x,y,u} F(x, y), \tag{23}$$

$$s.t.\, Gx \leq b, \tag{24}$$

$$A^\top u = c, \tag{25}$$

$$u \geq 0, \; Ay - x \leq 0, \; u^\top(Ay - x) = 0. \tag{26}$$

Consider now an algorithm which solves the relaxation in the sense of Scholtes [24] of problem (23)–(26):

$$\min_{x,y,u} F(x,y), \tag{27}$$

$$s.t.\, Gx \le b, \tag{28}$$

$$A^\top u = c, \tag{29}$$

$$u \ge 0,\ Ay - x \le 0,\ u^\top(x - Ay) \le \varepsilon \tag{30}$$

for $\varepsilon \downarrow 0$. Again we see easily that accumulation points of a sequence of globally optimal solutions of these problems for $\varepsilon \downarrow 0$ correspond to globally optimal solutions of the bilevel optimization problem (2), (22).

Theorem 6. *Let $\{(x^k, y^k, u^k)\}$ be a sequence of locally optimal solutions of problem (27)–(30) for $\{\varepsilon^k\}$ tending to zero and let $(\overline{x}, \overline{y}, \overline{u})$ be an accumulation point of this sequence. Then, $(\overline{x}, \overline{y})$ is a locally optimal solution of (2), (22).*

Proof. Without loss of generality assume that $\lim_{k\to\infty} (x^k, y^k, u^k) = (\overline{x}, \overline{y}, \overline{u})$ and assume that this point does not correspond to a local minimum of (2), (22). Since the set

$$K := \{(x,y,u) : A^\top u = c,\ u \ge 0,\ Ay - x \le 0,\ u^\top(x - Ay) = 0\}$$

is polyhedral there exist convex polyhedra K_j and index sets I_j, J_j such that

$$\begin{aligned} K_j := \{(x,y,u) : A^\top u = c,\ u \ge 0,\ u_i = 0,\ i \in I_j, \\ Ay - x \le 0,\ (Ay - x)_i = 0,\ i \in J_j\}, \end{aligned}$$

$I_j \cup J_j = \{1, \ldots, p\}$, where p is the number of inequalities in the lower-level problem (22), and $K = \cup_j K_j$. Since $(\overline{x}, \overline{y})$ is assumed to be not a local minimum of (2), (22), by Theorem 1, there exists

$$\widetilde{u} \in \Lambda(\overline{x}, \overline{y}) := \{u : u \ge 0,\ A^\top u = c,\ u^\top(Ay - x) = 0\}$$

such that $(\overline{x}, \overline{y}, \widetilde{u})$ is not a local minimum of (23)–(26). Hence, since the feasible set $\{x \times \{0\} \times \{0\} : Gx \le b\} \cap K$ is polyhedral, there exists a direction (d_x, d_y, d_u) of descent:

$$\nabla_x F(\overline{x}, \overline{y}) d_x + \nabla_y F(\overline{x}, \overline{y}) d_y < 0$$

and

$$(\overline{x}, \overline{y}, \widetilde{u}) + t(d_x, d_y, d_u) \in K_j,\ G(\overline{x} + t d_x) \le b \text{ for sufficiently small } t > 0,$$

and some index sets $I_j \subseteq \{i : \widetilde{u}_i = 0\}$, $J_j \subseteq \{i : (A\overline{y} - \overline{x})_i = 0\}$ and $I_j \cup J_j = \{1, \ldots, p\}$. Hence,

$$A^\top d_u = 0,\ (Ad_y - d_x)_i = 0,\ \forall\, i \in J_j,\ (d_u)_i = 0,\ \forall\, i \in I_j,$$

$$(Gd_x)_i \le 0,\ \forall\, i : (Gx - b)_i = 0.$$

For $i \notin I_j$ we have $i \in J_j$ and hence, by feasibility of (x^k, y^k),

$$(A(y^k + td_y) - x^k - td_x)_i \leq 0 \text{ for small } t > 0.$$

For $i \notin J_j$ but $(A\overline{y} - \overline{x})_i = 0$ we have

$$(Ad_y - d_x)_i \leq 0 \text{ and thus } (A(y^k + td_y) - x^k - td_x)_i \leq 0 \text{ for small } t > 0.$$

In the last case, if $(A\overline{y} - \overline{x})_i < 0$, $(A(y^k + td_y) - x^k - td_x)_i \leq 0$ follows from continuity. Hence, $A(y^k + td_y) - x^k - td_x \leq 0$ for sufficiently small $t > 0$. We similarly have $G(x^k + td_x) \leq b$.

Consequently, $(x^k + td_x, y^k + ty_d, \widetilde{u}) \in K_j$ is feasible for (27)–(30) and has a smaller objective function value than (x^k, y^k, u^k).

Since $\widetilde{u} \in \Lambda(\overline{x}, \overline{y})$ and $u^k \geq 0$ we have $u^k + \alpha(\widetilde{u} - u^k) \geq 0$ for $0 \leq \alpha \leq 1$ and $A^\top(u^k + \alpha(\widetilde{u} - u^k)) = c$. Moreover, by $\{u^{k\top}(x^k - Ay^k)\}$ converging to zero there exists $k' \geq k$ such that

$$u^{k'\top}(x^{k'} - Ay^{k'}) < \varepsilon^k$$

and thus

$$(u^{k'} + \alpha(\widetilde{u} - u^{k'}))^\top(x^{k'} - Ay^{k'}) \leq \varepsilon^k \text{ for } \alpha > 0 \text{ sufficiently small.}$$

Setting without loss of generality $k = k'$ we see that $(d_x, d_y, \widetilde{u} - u^k)$ is a direction of descent in (x^k, y^k, u^k) which is, thus, not a local minimum. Since this contradicts our assumption, the proof is completed.

□

Example 2 illustrates this result.

5 Conclusion

The bilevel optimization problem is often transformed into a single level optimization problem using the Karush-Kuhn-Tucker conditions of the lower-level problem resulting in an MPCC. This is a nonconvex optimization problem and, solving it, locally optimal solutions or even stationary points are obtained. Locally optimal solutions of the MPCC do in general not correspond to locally optimal solutions of the bilevel optimization problem. To overcome this unpleasant situation, a relaxation approach is suggested in which two constraints of the MPCC, namely the complementarity constraint and the condition that the gradient of the Lagrange function of the lower-level problem with respect to the lower-level variable vanishes, are relaxed. Then, it can be shown that a sequence of locally optimal solutions of the relaxed problems converges to a solution which corresponds to a locally optimal solution of the bilevel optimization problem. A similar result has been obtained in the article [10]. If the lower-level problem is a linear optimization problem parameterized in the right-hand side, the relaxation of the complementarity constraint as suggested in [24] is sufficient for deriving the same result, see [3]. Numerical test runs in [22] have shown yet that the solution of linear bilevel optimization problems using the first, more general, approach leads often to better locally optimal solutions than applying the second one.

References

1. Allende, G.B., Still, G.: Solving bilevel programs with the KKT-approach. Math. Program. **138**, 309–332 (2013)
2. Bard, J.: Practical Bilevel Optimization: Algorithms and Applications. Kluwer Academic Publishers, Dordrecht (1998)
3. Burtscheidt, J., Claus, M., Dempe, S.: Risk-averse models in bilevel stochastic linear programming. arXiv preprint arXiv:1901.11349 (2019)
4. Dempe, S.: A necessary and a sufficient optimality condition for bilevel programming problems. Optimization **25**, 341–354 (1992)
5. Dempe, S.: Foundations of Bilevel Programming. Kluwer Academic Publishers, Dordrecht (2002)
6. Dempe, S.: Bilevel optimization: theory, algorithms and applications (2018). Optimization http://www.optimization-online.org/DB_HTML/2018/08/6773.html
7. Dempe, S., Dutta, J.: Is bilevel programming a special case of a mathematical program with complementarity constraints? Math. Program. **131**, 37–48 (2012)
8. Dempe, S., Franke, S.: Solution algorithm for an optimistic linear Stackelberg problem. Comput. Oper. Res. **41**, 277–281 (2014)
9. Dempe, S., Franke, S.: On the solution of convex bilevel optimization problems. Comput. Optim. Appl. **63**, 685–703 (2016)
10. Dempe, S., Franke, S.: Solution of bilevel optimization problems using the KKT approach. Optimization 1–19 (2019). online first publication
11. Dempe, S., Kalashnikov, V., Pérez-Valdés, G., Kalashnykova, N.: Bilevel Programming Problems: Theory, Algorithms and Application to Energy Networks. Springer, Heidelberg (2015). https://doi.org/10.1007/978-3-662-45827-3
12. Dempe, S., Luo, G., Franke, S.: Pessimistic bilevel linear optimization. J. Nepal Math. Soc. **1**, 1–10 (2018)
13. Hoheisel, T., Kanzow, C., Schwartz, A.: Theoretical and numerical comparison of relaxation methods for mathematical programs with complementarity constraints. Math. Program. **137**(1–2), 257–288 (2013)
14. Kojima, M.: Strongly stable stationary solutions in nonlinear programs. In: Robinson, S. (ed.) Analysis and Computation of Fixed Points, pp. 93–138. Academic Press, New York (1980)
15. Lampariello, L., Sagratella, S.: A bridge between bilevel programs and Nash games. J. Optim. Theory Appl. **174**(2), 613–635 (2017). https://doi.org/10.1007/s10957-017-1109-0
16. Mersha, A.: Solution Methods for Bilevel Programming Problems. Ph.D. thesis, TU Bergakademie Freiberg (2008)
17. Mersha, A., Dempe, S.: Feasible direction method for bilevel programming problem. Optimization **61**(4–6), 597–616 (2012)
18. Mirrlees, J.: The theory of moral hazard and unobservable bevaviour: part I. Rev. Econ. Stud. **66**, 3–21 (1999)
19. Outrata, J., Kočvara, M., Zowe, J.: Nonsmooth Approach to Optimization Problems with Equilibrium Constraints. Kluwer Academic Publishers, Dordrecht (1998)
20. Outrata, J.: On the numerical solution of a class of Stackelberg problems. ZOR - Math. Meth. Oper. Res. **34**, 255–277 (1990)
21. Ralph, D., Dempe, S.: Directional derivatives of the solution of a parametric nonlinear program. Math. Program. **70**, 159–172 (1995)
22. Rog, R.: Lösungsalgorithmen für die KKT-Transformation von Zwei-Ebenen-Optimierungsaufgaben. Master's thesis, TU Bergakademie Freiberg, Fakultät für Mathematik und Informatik (2017)

23. Scheel, H., Scholtes, S.: Mathematical programs with equilibrium constraints: stationarity, optimality, and sensitivity. Math. Oper. Res. **25**, 1–22 (2000)
24. Scholtes, S.: Convergence properties of a regularization scheme for mathematical programs with complementarity constraints. SIAM J. Optim. **11**, 918–936 (2001)
25. Ye, J., Zhu, D.: New necessary optimality conditions for bilevel programs by combining the MPEC and value function approaches. SIAM J. Optim. **20**(4), 1885–1905 (2010)

Stackelberg Model and Public-Private Partnerships in the Natural Resources Sector of Russia

Sergey Lavlinskii[1,2,3(✉)], Artem Panin[1,2], and Aleksandr V. Plyasunov[1,2]

[1] Sobolev Institute of Mathematics, Novosibirsk, Russia
{lavlin,apljas}@math.nsc.ru, aapanin1988@gmail.com
[2] Novosibirsk State University, Novosibirsk, Russia
[3] Zabaikalsky State University, Chita, Russia

Abstract. A comparative analysis is conducted of the efficiency of different partnership models in the natural resources sector of Russia. The first one is a classic public-private partnership (PPP) model used in developed countries, whereby a private company builds an object of public property and transfers it to the government either immediately after the construction or after a certain period of operation of the object. The second model represents for the government a costly alternative of the former and is used in Russia in underdeveloped regions. This model assumes that the government supports the investor in infrastructure development and, in part, in the implementation of mandatory environmental protection measures and can also provide tax incentives. In practical terms, this work aims to look into possible ways of transforming the current Russian PPP model towards the classic forms of partnership. To conduct the comparative analysis of the PPP models, Stackelberg models are formulated and original iterative algorithms are developed for solving the corresponding bilevel Boolean programming problems based on probabilistic local search. The properties of the equilibrium solutions are studied using real data for the Transbaikal krai. Based on the modeling results, the different partnership models are compared to find out the conditions under which the private investor would choose to invest in publicly owned industrial infrastructure facilities in Russia.

Keywords: Stackelberg game ·
Bilevel mathematical programming problems ·
Mineral resources development program ·
Probabilistic local search algorithm

1 Introduction

The public-private partnership (PPP) scheme has gained popularity worldwide, offering an effective tool to reach a compromise of interests in various sectors of economy. Here, the classical partnership model has gained foothold,

M. Khachay et al. (Eds.): MOTOR 2019, LNCS 11548, pp. 158–171, 2019.
https://doi.org/10.1007/978-3-030-22629-9_12

whereby a private company builds an object of public property and transfers it to the government either immediately after the construction or after a certain period of operation of the object [1–4]. Developed nations apply this model in the natural resource sector, where it substantially expands the sources of project financing and encourages subsoil users to develop new deposits in hard-to-reach areas.

Russia's natural resource sector is only beginning to institutionalize the PPP scheme. To date, a few PPP-based projects have been implemented in underdeveloped regions, which apply a model, in a sense, alternative to the classical one. In this model, the government supports the investor in infrastructure development and, in part, in the implementation of mandatory environmental protection measures and can also provide tax incentives.

Practical experience in implementing Russian PPP projects shows that this partnership model, firstly, puts a financial burden on the government. Secondly, it requires a well-calibrated decision-making methodology. That is why all the attempts undertaken by the government of Russia to encourage various partnership schemes with private businesses were not accompanied by economically sound managerial decisions [1,5–7].

This article continues our research into cooperation between public and private investors in the natural resource sector [8–10]. This work aims to analyze and compare the classical and Russian partnership models in terms of efficiency, using the game-theoretical Stackelberg model. This way we can explore possible ways to transform the Russian PPP model, using the resources of the Investment Fund of Russia, towards the classical forms of partnership. This is important for addressing a whole range of issues related to the strategic management of the natural resource sector in Russia.

2 Mathematical Models

The choice of the Stackelberg model is dictated by the features of the hierarchy of interactions between the government and the private investor in the natural resource sector. Although the private investor usually initiates the development of a new field, the model assigns the leadership role to the government. Until the government makes critical decisions such as selling a license, choosing infrastructure projects, approving environmental measures, etc., the investor cannot decide to implement the project. Thus, we assume that the government makes the first move.

Classical PPP Model. A private company builds an object of public property and transfers it to the government either immediately after the construction or after a certain period of operation of the object. In underdeveloped regions rich in natural resources, this model in the natural resource sector develops according to the following scenario. The investor cannot launch its field development projects due to the lack of the necessary infrastructure. Therefore, the investor negotiates with the government a list of infrastructure projects that "open up" the target development projects and implements these infrastructure projects at its own expense. The government compensates the investor for the costs as soon as the budget receives the taxes from the extraction of natural resources by the private investor.

The government can compensate the investor for the infrastructure costs in two ways. In the first case, the investor does not trust the government and demands the reimbursement regardless of the overall results of the development program. For the government, this situation necessitates such a schedule of payments within the budget constraints, starting from the time when the taxes from mining operations come into the budget, that will compensate the investor for the infrastructure s costs with the discount. In the second scheme of mutual settlements, the investor shows substantially greater trust in the government. This scheme implies a coordinated assessment of the integral effect for the investor, taking into account the infrastructure costs and the compensation payments from the government that guarantee a positive final net present value for the investor.

Thus, the following information serves as input data in the PPP model:

- A schedule of compensation payments to the investor for infrastructure development.
- A set of industrial projects to develop natural resource sites, implemented by the private investor.
- A set of infrastructure development projects.
- A list of environmental projects necessary to compensate for the environmental damage caused by the implementation of the investment projects.

The output of the model is a natural resource development program, i.e., a set of infrastructure, environmental, and industrial projects implemented by the private investor.

A formal description of the classical PPP model can be presented as follows. We use the following notation:

T is a planning horizon; I is a set of investment projects; J is a set of infrastructure development projects; K is a set of environmental projects;

Investment project i in year t:

CFP_i^t is the cashflow (the difference between the incomes and expenses of all kinds, taking into account a transaction costs, constructive borrowed from [11]);

EPP_i^t is the environmental damage from the implementation of project;

DBP_i^t is the government revenue from the implementation of project;

Infrastructure development project j in year t:

ZI_j^t is the costs of implementation of project;

EPI_j^t is the environmental damage from the implementation of project;

VDI_j^t is the government revenue from local economic development as a result of the implementation of project;

Environmental project k in year t: ZE_k^t is the costs of implementation of project;

The matrices μ and ν define the relationship between the projects, where μ_{ij} is a coherence indicator for the infrastructure and investment projects, $i \in I$,

$j \in J$, and ν_{ij} is a coherence indicator for the environmental and investment projects, $i \in I$, $k \in K$:

$$\mu_{ij} = \begin{cases} 1, & \text{if the implementation of investment project } i \\ & \text{requires the implementation of infrastructure development project } j, \\ 0 & \text{otherwise;} \end{cases}$$

$$\nu_{ik} = \begin{cases} 1, & \text{if the implementation of investment project } i \\ & \text{requires the implementation of environmental project } k, \\ 0 & \text{otherwise.} \end{cases}$$

The discounts of the government and the investor:

DG is the discount of the government; DI is the discount of the investor;

The budget constraints:

b_t^G is the government budget in year t; b_t^O is the investor budget in year t.

We use the following variables:

$$v_j = \begin{cases} 1, & \text{if the investor launches infrastructure development project } j, \\ 0 & \text{otherwise;} \end{cases}$$

$$z_i = \begin{cases} 1, & \text{if the investor launches investment project } i, \\ 0 & \text{otherwise;} \end{cases}$$

$$u_k = \begin{cases} 1, & \text{if the investor launches environmental project } k, \\ 0 & \text{otherwise.} \end{cases}$$

W_t, $\bar{W}_t$ is the schedule of compensation payments for infrastructure development in year t, which was proposed by the government and used by the investor.

The government problem $\mathcal{PS}$:

$$\sum_{t\in T}\Big(\sum_{i\in I}(DBP_i^t - EPP_i^t)z_i + \sum_{j\in J}(VDI_j^t - EPI_j^t)v_j - W_t\Big)/(1+DG)^t \to \max_{W,v,z} \quad (1)$$

subject to:

$$\sum_{1\leq t\leq\omega}\bar{W}_t \leq \sum_{1\leq t\leq\omega} b_t^G; \omega \in T; \quad (2)$$

$$\bar{W}_t \geq 0; t \in T; \quad (3)$$

$$(z,v) \in \mathcal{F}^*(\bar{W}); \quad (4)$$

The set $\mathcal{F}^*$ is a set of optimal solutions of the low-level parametric investor problem.

The investor problem $\widetilde{\mathcal{PI}}(\bar{W})$:

$$\sum_{t\in T}\Big(\sum_{i\in I} CFP_i^t z_i - \sum_{k\in K} ZE_k^t u_k - \sum_{j\in J} ZI_j^t v_j - W_t\Big)/(1+DI)^t \to \max_{z,u,v} \quad (5)$$

subject to:

$$\sum_{t\in T}\Big(W_t - \sum_{j\in J} ZI_j^t v_j\Big)/(1+DI)^t \geq 0; \quad (6)$$

$$\sum_{k\in K} ZE_k^t u_k + \sum_{j\in J} ZI_j^t v_j - \sum_{i\in I} CFP_i^t z_i - W_t \le b_t^O; t \in T; \tag{7}$$

$$v_j \ge \mu_{ij} z_i; i \in I, j \in J; \tag{8}$$

$$u_k \ge \nu_{ik} z_i; i \in I, k \in K; \tag{9}$$

$$\sum_{i\in I} \nu_{ik} z_i \ge u_k; k \in K; \tag{10}$$

$$\sum_{t\in T}\Big(\sum_{i\in I}(DBP_i^t - EPP_i^t)z_i - W_t\Big)/(1+DG)^t \ge 0; \tag{11}$$

$$W_t \le \bar{W}_t; t \in T; \tag{12}$$

$$v_j, z_i, u_k \in \{0,1\}; i \in I, k \in K, j \in J. \tag{13}$$

The objective function of the government is the part of the natural resource rent received by the government in the form of taxes, taking into account the compensation payments to the investor for infrastructure development. Budget constraint (2) is soft, i.e., financial resources that are not spent in the current year go to the next year. Constraints (8)–(9) capture the interrelations between the industrial, infrastructure, and environmental projects. Each environmental project must be necessary for the implementation of some industrial project (10). Constraint (11) blocks industrial programs that do not provide a positive balance between budget revenues and compensation payments, taking into account the discounts and environmental losses.

Problem (1)–(13) describes the behavior of an investor who does not trust the government and demands reimbursement regardless of the overall results of the development program. The model of coordinated partnership with a substantially higher level of trust in the government on the part of the investor lacks constraint (6): this constraint formalizes the requirement of unconditional reimbursement of the investor's costs.

Russian PPP Model. While the classical PPP model implies that the investor helps the government, the Russian version does things the other way round: the government supports the investor on underdeveloped territories in building the infrastructure and implementing the necessary environmental measures and, in some cases, provides tax incentives. The full version of this type of model has been described by the authors in [10]. Here, we use a simplified version of the model, which, however, can be used to compare the properties of the two PPP models based on the analysis of the properties of equilibrium solutions.

The Russian PPP version can be formalized as the following Stackelberg model. We use the following variables additionally:

TP^t_{im} is a tax incentive of level for project;

M is a set of tax incentive levels;

$$x_j = \begin{cases} 1, \text{ if the government launches infrastructure development project } j, \\ 0 \text{ otherwise;} \end{cases}$$

$$\bar{y}_k = \begin{cases} 1, \text{ if the government is prepared to launch environmental project } k \\ \quad \text{(the government has included it into the budget expenses),} \\ 0 \text{ otherwise;} \end{cases}$$

$$y_k = \begin{cases} 1, \text{ if the government launches environmental project } k \\ \quad \text{as agreed with the investor,} \\ 0 \text{ otherwise;} \end{cases}$$

$$\bar{\varphi}_{im} = \begin{cases} 1, \text{ if the government is prepared to provide the investor with a tax} \\ \quad \text{incentive of level } m \text{ for investment project } i, \\ 0 \text{ otherwise;} \end{cases}$$

$$\varphi_{im} = \begin{cases} 1, \text{ if the government provides the investor with a tax incentive} \\ \quad \text{of level } m \text{ for investment project } i, \\ 0 \text{ otherwise;} \end{cases}$$

The government problem $\mathcal{PS}$:

$$\sum_{t \in T} \Big(\sum_{i \in I} (DBP^t_i - EPP^t_i) z_i - \sum_{i \in I} \sum_{m \in M} TP^t_{im} \varphi_{im} + \sum_{j \in J} (VDI^t_j - EPI^t_j - ZI^t_j) x_j - \sum_{k \in K} ZE^t_k \, y_k \Big) / (1 + DG)^t \to \max_{x, \bar{y}, \bar{\varphi}, \varphi, y, z, u} \tag{14}$$

subject to:

$$\sum_{j \in J} ZI^t_j \, x_j + \sum_{k \in K} ZE^t_k \, \bar{y}_k \le b^G_t; t \in T; \tag{15}$$

$$\sum_{m \in M} \bar{\varphi}_{im} \le 1; i \in I; \tag{16}$$

$$(y, z, u, \varphi) \in \mathcal{F}^*(x, \bar{y}, \bar{\varphi}); \tag{17}$$

$$\bar{\varphi}_{im}, x_j, \bar{y}_k, \in \{0, 1\}; j \in J, k \in K, m \in M. \tag{18}$$

The objective function of the government represents the net present value received by the government (14). Constraints (15) guarantee that the government expenses on infrastructure and environmental protection stay within the budget. Constraint (16) forbids the government to provide several tax incentives within one project. Constraint (17) means that the investor acts in an optimal way, which implies solving the low-level problem (the investor problem). The set $\mathcal{F}^*(x, \bar{y}, \bar{\varphi})$ is a set of optimal solutions of the low-level parametric problem.

The investor problem $\mathcal{PI}(x, \bar{y}, \bar{\varphi})$:

$$\sum_{t \in T}\Big(\sum_{i \in I}(CFP_i^t z_i + \sum_{m \in M} TP_{im}^t \varphi_{im}) - \sum_{k \in K} ZE_k^t u_k\Big)/(1+DI)^t \to \max_{z,u,y,\varphi} \quad (19)$$

subject to:

$$\sum_{k \in K} ZE_k^t u_k - \sum_{i \in I}(CFP_i^t z_i - \sum_{m \in M} TP_{im}^t \varphi_{im}) \le b_t^O; t \in T; \quad (20)$$

$$x_j \ge \mu_{ij} z_i; i \in I, j \in J; \quad (21)$$

$$y_k + u_k \le 1; k \in K; \quad (22)$$

$$y_k + u_k \ge \nu_{ik} z_i; i \in I, k \in K; \quad (23)$$

$$y_k \le \bar{y}_k; k \in K; \quad (24)$$

$$\varphi_{im} \le \bar{\varphi}_{im}; i \in I; m \in M; \quad (25)$$

$$\sum_{m \in M} \varphi_{im} \le z_i; i \in I; \quad (26)$$

$$y_k, z_i, u_k, \varphi_{im} \in \{0,1\}; i \in I, k \in K, m \in M. \quad (27)$$

The income of the private investor is determined by objective function (19). From (20) it follows that the investor's costs in each year do not exceed the budget, considering the income received from the investment projects and the tax incentives. Constraints (21)–(23) ensure technological coherence of the projects and prevent the situation when the investor and the government implement the same environmental projects simultaneously. Constraint (24) guarantees that the government implements only included environmental projects.

In solving the PPP planning problems, we applied an approximate hybrid algorithm based on the ideas of local descent and the CPLEX package [12–16]. The latter is applied for solving both the one-level problem, where the government decides for the investor, and the investor problem. The local descent is used to search for a good approximate solution for the government.

First, we describe the schema of the algorithm for the Russian PPP model. Since the problem being studied has two levels and an arbitrary feasible solution $(x, \bar{y}, \varphi, \bar{\varphi}, y, z, u)$ contains the optimal solution (y, z, u, φ) of the parametric investor problem with the parameters x, $\bar{y}$ and $\bar{\varphi}$, and, we call the solution $(x, \bar{y}, \bar{\varphi})$ an almost feasible solution if it satisfies constraints (15), (16), and (18) and the investor problem with the parameters $(x, \bar{y}, \bar{\varphi})$ is solvable.

Algorithm parameters:

mIter is the maximum number of iterations in the algorithm for finding the initial solution (Step 2);

cfBound is the coefficient of constraint relaxation by the value of objective function (14) in solving the auxiliary problem in Step 2.3.

Hybrid algorithm:

Step 1. Calculate the upper bound *Bound* by solving the government problem with the low-level constraints (i.e., the problem with objective function (14) and constraints (15), (16), (18) and (20)–(27)), using CPLEX.

Step 2. Find a feasible solution $(x^0, \bar{y}^0, \bar{\varphi}^0)$ (which will later be used as an initial solution in the local search algorithm):

Step 2.1. $iter := 1$.

Step 2.2: If $iter \leq mIter$, then solve the investor problem with the government's variables and constraints (i.e., the problem with objective function (19) and constraints (15), (16), (18) and (20)–(27)) and the additional constraint that the government's objective function (14) is no less than $(Bound - 1)/iter$ by CPLEX. Otherwise, proceed to Step 3.

Step 2.3: If the problem in the previous step is solvable and $(x, \bar{y}, \varphi, \bar{\varphi}, y, z, u)$ is an optimal solution, then calculate the value f of objective function (14) by solving the problem $\mathcal{PI}(x, \bar{y}, \bar{\varphi})$ using CPLEX. If $f < (Bound - 1)/(iter * cfBound)$ or the problem in the previous step has no solution, then assume that $iter := iter + 1$ and proceed to Step 2.2; otherwise, assume $x^0 := x$, $\bar{y}^0 := \bar{y}$, $\bar{\varphi}^0 := \bar{\varphi}$, and $f^0 := f$ and proceed to Step 3.

Step 3: If we could not find a feasible solution in Step 2, then we use a zero solution as a feasible one; i.e., we assume that $x^0 := 0$, $\bar{y}^0 := 0$ and $\bar{\varphi}^0 := 0$, and calculate the value f^0 of objective function (14) by solving the problem $\mathcal{PI}(x^0, \bar{y}^0, \bar{\varphi}^0)$ using CPLEX. Then we apply the local search algorithm:

Step 3.1: Take $(x, \bar{y}, \bar{\varphi}) := (x^0, \bar{y}^0, \bar{\varphi}^0)$ as a staring solution and $f := f^0$ as a record value.

Step 3.2: Find the best neighbor $(x^*, \bar{y}^*, \bar{\varphi}^*)$ in the neighborhood of the solution $(x, \bar{y}, \bar{\varphi})$.

Step 3.3: If the value of the objective function $f(x^*, \bar{y}^*, \bar{\varphi}^*) > f$, then assume that $x := x^*$, $\bar{y} := \bar{y}^*$, $\bar{\varphi} := \bar{\varphi}^*$ and $f := f(x^*, \bar{y}^*, \bar{\varphi}^*)$ and proceed to Step 3.2; otherwise, stop the algorithm.

We used as a neighborhood in the local search algorithm the following randomized neighborhood with precisely one neighbor. The randomized neighborhood of the solution $(x, \bar{y}, \bar{\varphi})$ has precisely one solution $(x', \bar{y}', \bar{\varphi}')$, which was obtained as follows. Each component of the vector x' is a random value which is equal with a probability of $1 - 1/|J|$ to the corresponding component of the vector x and with a probability of $1/|J|$ to the corresponding component of the vector $1 - x$. The situation with the vectors $\bar{y}'$ and $\bar{y}$ is the same, except that the probabilities are $1 - 1/|K|$ and $1/|K|$, respectively. For the tax incentives, the probabilities are $1 - 1/(|I||M|)$ and $1/(|I||M|)$, respectively. The result of the algorithm is the best found solution.

For the classical PPP model, the hybrid algorithm changes as follows. In the Step 3 we find the best $\bar{W}$ using the local search algorithm with the following randomized neighborhood. The neighborhood consists of preciously one neighbor. It is obtained when we change any of the vector $\bar{W}$ with a probability of $1/T$. The value $\bar{W}_t$ is selected randomly from the segment $[0, b_t^O]$.

The stopping criterion for the local search algorithm with the randomized neighborhood was a limit on the number of iterations. The time of the local descent was limited to 5000 iterations, and the parameters $mIter$ and $cfBound$ were 30 and 3, respectively.

3 Numerical Experiment

To demonstrate the methodology of application of the described tools, we designed a special model test site, whose prototype was a set of 50 polymetallic ore deposits in the Transbaikal krai. For this model test site, we composed a set of 10 infrastructure projects, some of which are being implemented (railroads and powerlines) while others make up for the infrastructure that is currently missing but is necessary for the deposit development projects (powerlines and highways). For each of the deposits, there are 5 levels of tax incentives and a set of compensating environmental activities integrated into the relevant environmental project.

Thus, the model test site captures the specificity of the object being modeled: long timeframe of investment processes, nonstationary market conditions, and well-established technology of natural resource management. The methodology for studying the properties of the Stackelberg equilibrium draws upon an analysis of how sensitive the solutions of the corresponding bilevel Boolean programming problem are to changes in the key parameters of the model. This is a critical issue, primarily because for many of the model parameters, we know only the operational ranges of their values. Likewise, when designing a subsoil development program, an expert has access only to the project-related data and can only use approximate estimates for many parameters such as the discounts of the partnership participants, environmental costs and losses, etc.

The figures below show the results of the calculations that analyzed the sensitivity of the solution to changes in the key model parameters, i.e., the discounts of the investor and the government. In the calculations, we used a single information base to compare four models: the classical model of the "distrustful" investor (Classical1), classical model of coordinated partnership (Classical2), and Russian PPP models without (Russian1) and with tax incentives (Russian2).

Figure 1 shows the dependence of the government's objective function on the discounts of the PPP participants. The upper panel corresponds to the case where the investor builds the infrastructure. Here, we see that among the classical models, the coordinated partnership model yields at small values of the investor's discount almost twice-as-good results. However, with the increase in the investor's discount, the value of the government's objective function falls at a significantly higher rate than in the distrustful-investor model.

The alternative partnership scheme, whereby the government builds the infrastructure, provides it with higher values of the functional, compared with the classical version. Here, we see a greater resistance to the growth of the investor's discount in Russian1, and the introduction of tax incentives removes the hollow on the surface of Russian1, ensuring even greater stability of the partnership results in Russian2.

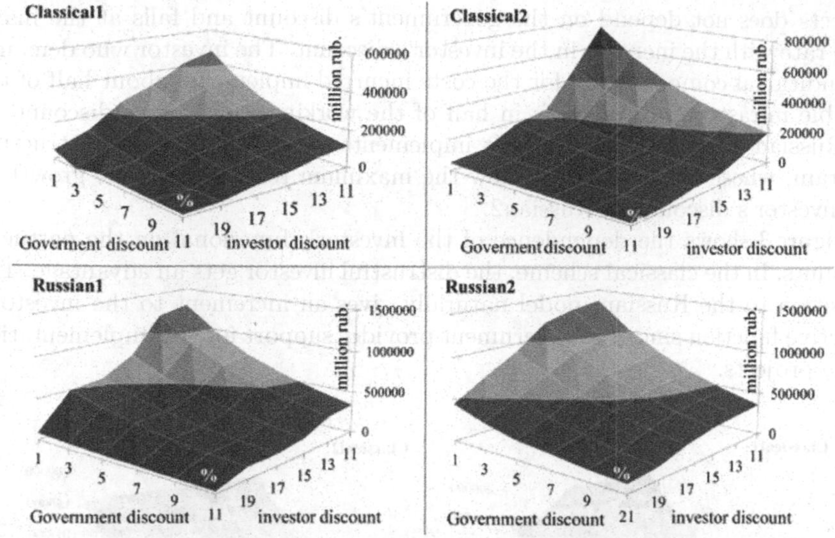

Fig. 1. The government objective function and the partner discounts

Figure 2 explains why the objective function of the government behaves in such a way. This figure shows the dependence of the intensity of infrastructure development on the ratio between the partner's discounts. In the classical coordinated partnership model Classic2, the number of implemented infrastructure

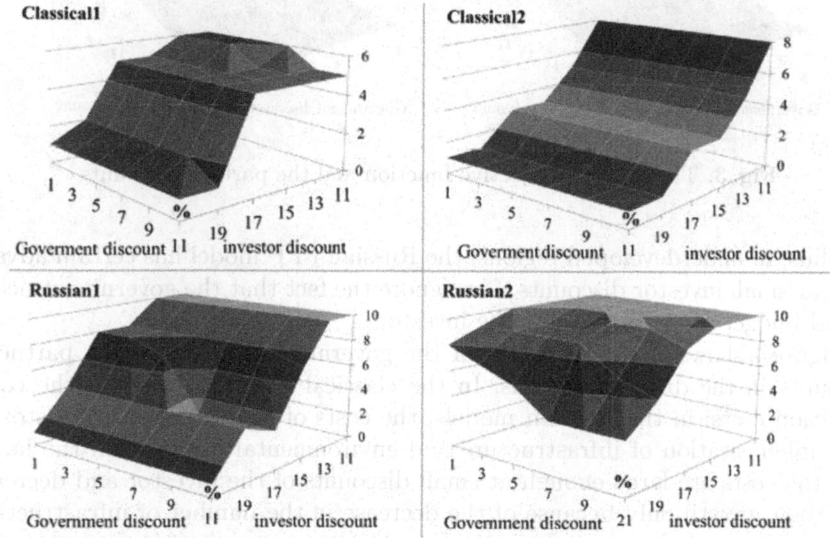

Fig. 2. Number of implemented infrastructure projects

projects does not depend on the government's discount and falls at the maximum rate with the increase in the investor's discount. The investor who demands unconditional compensation for the costs incurred implements about half of the possible infrastructure projects in half of the working range of its discount. In the Russian model, the government implements a more intensive infrastructure program, whose characteristics show the maximum resistance to the growth of the investor's discount in Russian2.

Figure 3 shows the dependence of the investor's functional on the partner's discounts. In the classical scheme, the distrustful investor gets an advantage. The transition to the Russian model naturally gives an increment to the investor's objective function since the government provides support for the implementation of the projects.

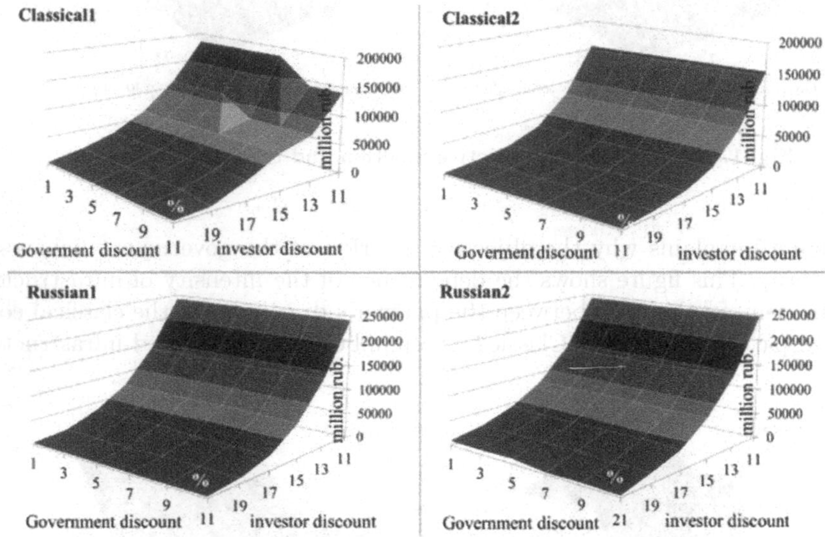

Fig. 3. The investor objective function and the partner discounts

Thus, in underdeveloped regions, the Russian PPP model has certain advantages at small investor discounts if we ignore the fact that the government needs to find budget funds to support the investor.

Figure 4 shows the dependence of the government's costs on the partner's discounts in the different models. In the classical scheme, these are the compensation costs; in the Russian models, the costs of supporting the investor in the implementation of infrastructure and environmental projects. In the latter case, the costs are large enough at small discounts of the investor and decrease with their growth only because of the decrease in the number of infrastructure development projects.

The costs in the classical models differ by an order of magnitude and increase with increasing discount of the investor although the infrastructure program

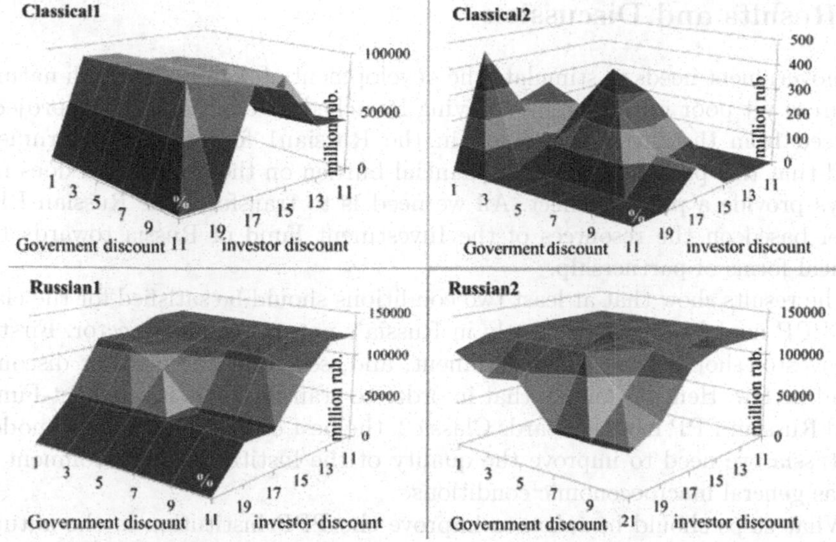

Fig. 4. Government expenses on the compensation payments

shrinks to a minimum. What is the relationship between the total compensation costs of the government and the costs of implemented infrastructure projects in the different models?

Figure 5 shows the surfaces reflecting the balance between the costs and the acquired production assets for the government (costs of new infrastructure minus the total compensation costs) in the different classical models. We see that in the distrustful- investor model, the amount of compensation payments from the government significantly exceeds the actual cost of the implemented infrastructure program: it would be more cost-efficient for the government to finance the infrastructure development itself. The coordinated partnership model is a different story. Here, we see a positive balance at all the discounts of the partners.

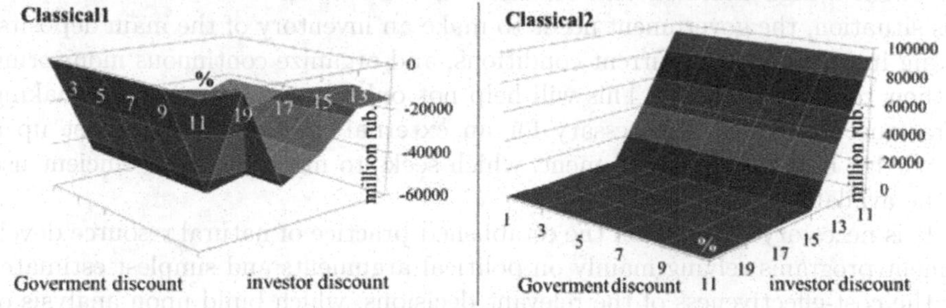

Fig. 5. The balance between the costs and the acquired production assets for the government

4 Results and Discussion

The government needs to stimulate the development of regions with rich natural resources yet poor infrastructure. Having launched several partnership projects financed from the Investment Fund in the Russian1 format, the government found that this path puts a heavy financial burden on the budget and does not always provide a positive effect. All we need is to transform the Russian PPP model based on the resources of the Investment Fund of Russia towards the classical forms of partnership.

The results show that at least two conditions should be satisfied for the classical PPP model to work efficiently in Russia's natural resource sector. Firstly, the investor should trust the government, and, secondly, the investor discount should be low. Hence it follows that in order to transform the Investment-Fund-based Russian PPP model towards Classic2, the best of the classical PPP models for Russia, we need to improve the quality of the institutional environment as well as general macroeconomic conditions.

What steps should be taken to improve the PPP institution in the natural resource sector of Russia?

The government decision-makers need to gain a sufficiently detailed understanding of the natural resource sites and the respective development projects. This is possible if the decision-making on subsoil use is supported by the relevant government institutions that evaluate subsoil development projects from the perspective of the government and society as a whole. Institutions that examined long-term natural resource development projects were lost during the Russian economic reforms. It is necessary to restore the institutional infrastructure of natural resource development planning, which is capable of working out professional solutions for long-term goals of sustainable development in Russia.

The potential investor needs, apart from macroeconomic stability, a comprehensive cost assessment of subsoil deposits, i.e., detailed, up-to-date information on economically viable development projects as potential investment targets. Such an assessment cannot rely on the previous expert assessments of deposit development projects because the situation in raw materials markets undergoes substantial changes over time, and so do the price ratios in a national economy with a high inflation and an unsteady rate of the national currency. In this situation, the government needs to make an inventory of the main deposits, taking into account the current conditions, and organize continuous monitoring of their rental assessment. This will help not only develop the decision-making infrastructure, which is necessary for an external investor, but also set up a knowledge base for the government, which seeks to make the most efficient use of the available natural resources.

It is necessary to abandon the established practice of natural resource development programs relying mainly on political arguments and simplest estimates for the cost-effectiveness of the relevant decisions, which build upon analysis of technological projects and current prices of raw materials. A priori confidence that PPP always brings a positive result is groundless if we set the task of safeguarding the interests not only of private business but society as a whole.

The search for options for reconciling these interests represents a separate problem, and a very complicated one, solving which might benefit from the use of the Stackelberg model.

Acknowledgements. This work was financially supported by the Russian Science Foundation (project No. 16-18-00073).

References

1. Reznichenko, N.V.: Public-private partnership models. Bull. St. Petersburg Univ. Ser. 8 Manag. **4**, 58–83 (2010). (in Russian)
2. Quiggin, J.: Risk, PPPs and the public sector comparator. Aust. Account. Rev. **14**(33), 51–61 (2004)
3. Grimsey, D., Levis, M.K.: Public private partnerships: the worldwide revolution in infrastructure provision and project finance. Edward Elgar, Cheltenham (2004)
4. Bennett, J., Iossa, E.: Delegation of contracting in the private provision of public services. Rev. Ind. Organ. **29**(1), 75–92 (2006)
5. Lavlinskii, S.M.: Public-private partnership in a natural resource region: ecological problems, models, and prospects. Stud. Russ. Econ. Dev. **21**(1), 71–79 (2010). https://doi.org/10.1134/S1075700710010089
6. Glazyrina, I.P., Kalgina, I.S., Lavlinskii, S.M.: Problems in the development of the mineral and raw material base of Russia's far east and prospects for the modernization of the region's economy in the framework of Russian-Chinese cooperation. Reg. Res. Russ. **3**(4), 21–29 (2013). https://doi.org/10.1134/S2079970514010055
7. Glazyrina, I.P., Lavlinskii, S.M., Kalgina, I.S.: Public-private partnership in the mineral resources complex of Zabaikalskii krai: problems and prospects. Geogr. Nat. Resour. **35**(4), 359–364 (2014). https://doi.org/10.1134/S1875372814040088
8. Lavlinskii, S., Panin, A., Pliasunov, A.: A two-level planning model for public-private partnership. Autom. Remote Control. **11**, 89–103 (2015). https://doi.org/10.1134/S0005117915110077
9. Lavlinskii, S., Panin, A., Pliasunov, A.: Comparison of models of planning the public-private partnership. J. Appl. Ind. Math. **10**(3), 1–17 (2016). https://doi.org/10.1134/S1990478916030017
10. Lavlinskii, S., Panin, A.A., Plyasunov, A.V.: Public-private partnership models with tax incentives: numerical analysis of solutions. In: Eremeev, A., Khachay, M., Kochetov, Y., Pardalos, P. (eds.) OPTA 2018. CCIS, vol. 871, pp. 220–234. Springer, Cham (2018). https://doi.org/10.1007/978-3-319-93800-4_18
11. Glazyrina, I., Lavlinskii, S.: Transaction costs and problems in the development of the mineral and raw-material base of the resource region. J. New Econ. Assoc. New Econ. Assoc. **38**(2), 121–143 (2018)
12. Dempe, S.J.: Foundations of Bilevel Programming. Kluwer Academic Publishers, Dordrecht (2002)
13. Davydov, I., Kochetov, Yu., Plyasunov, A.: On the complexity of the (r|p)-centroid problem in the plane. TOP **22**(2), 614–623 (2014)
14. Plyasunov, A.V., Panin, A.A.: The pricing problem, Part I: exact and approximate algorithms. J. Appl. Ind. Math. **7**(2), 1–14 (2013)
15. Plyasunov, A.V., Panin, A.A.: The pricing problem, Part II: computational complexity. J. Appl. Ind. Math. **7**(3), 1–13 (2013)
16. Kononov, A.V., Kochetov, Yu. A., Plyasunov, A.V.: Competitive facility location models. Comput. Math. Math. Phys. **49**(6), 994–1009 (2009)

The Local and Global Searches in Bilevel Problems with a Matrix Game at the Lower Level

Andrei V. Orlov(✉) and Tatiana V. Gruzdeva

Matrosov Institute for System Dynamics and Control Theory of SB of RAS, Irkutsk, Russia
{anor,gruzdeva}@icc.ru

Abstract. This work addresses the simplest class of the bilevel optimization problems (BOPs) with equilibrium at the lower level. We study linear BOPs with a matrix game at the lower level in their optimistic statement. First, we transform this problem to a single-level nonconvex optimization problem with the help of the optimality conditions for the lower level problem. Then we apply the special Global Search Theory (GST) for general d.c. optimization problems to the reduced problem. Following this theory, the methods of local and global searches in this problem are constructed. These methods take into account the structure of the problem in question.

Keywords: Bilevel optimization · Bilevel problems with equilibrium at the lower level · Matrix game · Optimistic solution · Reduction theorem · D.C. constraint problem · Global Search Theory · Local search · Global search

1 Introduction

The study of optimization problems with hierarchical structure is now at the front edge of the recent advances in Operations Research [1,2]. Such problems arise in modeling of complex systems, which are characterized by unequal status of the participants. Hierarchical problems have many applications in control, economy, transport, networks, energy, etc. [3–5]. Over the last 50 years, a great number of scientists have been focusing their research on hierarchical problems [3,4], which are much more difficult than the classical one-level formulations because they imply taking into account the interests of different levels of the hierarchy. Therefore, their study is often bounded at a bilevel structure [1].

Usually in a hierarchical bilevel problem, the upper level depends on the lower level through the objective function and/or the feasible set, and the lower level depends on the upper one in the same way. It is assumed that the upper level makes the first move [1]. Additionally, note that bilevel optimization problems (BOPs) are closely connected with the so-called Mathematical Programs with Equilibrium Constraints (MPECs) [6,7].

M. Khachay et al. (Eds.): MOTOR 2019, LNCS 11548, pp. 172–183, 2019.
https://doi.org/10.1007/978-3-030-22629-9_13

Using the optimization problem at the lower level, either one or several players depending on the upper level can be modeled. In the latter case, it is necessarily assumed that these players are independent of each other (and make decisions simultaneously and independently). Then we can assume that actually one aggregated player acts on the lower level. On the one hand, such a model makes it possible to investigate cases when the upper level player controls several players at the lower level. Such situations prevail in practice (for example, a corporation usually has several branches). On the other hand, the assumption about the independence of players may reduce the adequacy of the model. Therefore, the study of bilevel problems with several players at the lower (Single-Leader-Multi-Follower-Problem (SLMFP)) or at the upper level (Multi-Leader-Single-Follower-Problem (MLSFP)), which help model more complex systems, is gaining popularity [8–12]. Some researches even attempted to study Multi-Leader-Follower-Problems (MLFPs) [8–12]. The research in this area is mostly motivated by practical applications, and at the moment there is no general approach to developing numerical methods for such problems.

At the same time, according to Pang [2] for example, hierarchy and equilibrium are the promising paradigms in mathematical optimization in the 21st century. Therefore, development of the efficient numerical methods even for the simplest classes of BOPs with an equilibrium is a challenge of modern Operations Research. In this connection, the present paper addresses a new approach to a special type of BOPs with the simplest equilibrium at the lower level, where one leader is connected with two followers. We consider a bilevel problem with a parametric matrix game [13–15] at the lower level and with a linear objective function subject to linear constraints at the upper level.

In order to elaborate numerical methods for solving BOPs with a matrix game at the lower level, we reformulate it as a single-level optimization problem using a reduction theorem. This auxiliary problem turns out to be a global optimization problem with a nonconvex feasible set (see, e.g., [16–18]). It is well known that classical convex optimization methods do not provide global solutions for nonconvex optimization problems [18–20]. Therefore, to solve the obtained single-level problem, we apply a special Global Search Theory (GST) developed by A.S. Strekalovsky for optimization problems with d.c. functions [18,21–24]. In the recent years, the GST has proved to be an efficient tool for numerical solution of different nonconvex problems of Operations Research (including problems with hierarchical and equilibrium structures) [18,21,25–34].

2 Problem Statement and Reduction

Let us formulate the BOPs with a matrix game at the lower level in the following way:

$$\langle c, x\rangle + \langle d_1, y\rangle + \langle d_2, z\rangle \uparrow \max_{x,y,z}, \quad x \in X, \ (y, z) \in C(\Gamma M(x)), \qquad (\mathcal{BP}_{\Gamma M})$$

where $X = \{x \in I\!R^m \mid Ax \leq a,\ x \geq 0,\ \langle b_1, x\rangle + \langle b_2, x\rangle = 1\}$, $C(\Gamma M(x))$ is a set of saddle points [13–15] of the game

$$\left.\begin{array}{ll} \langle y, Bz\rangle \uparrow \max\limits_{y}, & y \in Y(x) = \{y \mid y \geq 0,\ \ \langle e_{n_1}, y\rangle = \langle b_1, x\rangle\}, \\ \langle y, Bz\rangle \downarrow \min\limits_{z}, & z \in Z(x) = \{z \mid z \geq 0,\ \ \langle e_{n_2}, z\rangle = \langle b_2, x\rangle\}; \end{array}\right\} \qquad (\Gamma M(x))$$

$c, b_1, b_2 \in I\!R^m$; $y, d_1 \in I\!R^{n_1}$; $z, d_2 \in I\!R^{n_2}$; $a \in I\!R^{m_1}$; $b_1 \geq 0,\ b_1 \neq 0,\ \ b_2 \geq 0,\ b_2 \neq 0$; A, B are matrices and $e_{n_1} = (1, ..., 1)$, $e_{n_2} = (1, ..., 1)$ are vectors of appropriate dimension.

It can be readily seen that at the lower level we formulate the special matrix game with mixed strategies, which is considered on simplexes depending on the upper level variable x, instead of canonical simplexes. The expression $\langle b_1, x\rangle + \langle b_2, x\rangle = 1$ can be interpreted as some resource, which should be distributed by the leader among the followers.

Additionally, the problem $(\mathcal{BP}_{\Gamma M})$ is formulated in the so-called optimistic statement, when interests of the upper level can be coordinated with the actions of the lower level [1].

In order to elaborate numerical methods for solving the bilevel problem $(\mathcal{BP}_{\Gamma M})$, we need to transform it to a single-level problem.

Let us set $\xi_1 := \langle b_1, x\rangle$, $\xi_2 := \langle b_2, x\rangle$ (x is fixed) and formulate the so-called non-normalized matrix game with parameters ξ_1, ξ_2:

$$\left.\begin{array}{ll} \langle y, Bz\rangle \uparrow \max\limits_{y}, & y \in Y = \{y \mid y \geq 0, \langle e_{n_1}, y\rangle = \xi_1 > 0\}, \\ \langle y, Bz\rangle \downarrow \min\limits_{z}, & z \in Z = \{z \mid z \geq 0, \langle e_{n_2}, z\rangle = \xi_2 > 0\}. \end{array}\right\} \qquad (\Gamma M)$$

Further we present the optimality conditions for the non-normalized matrix game (ΓM). These conditions are a generalization of classical optimality conditions for a classical matrix game [13–15]. Note that if we use an expression like yB, we mean that y is a row vector here, but if we write Bz, we mean that z is a column vector.

Definition 1. The situation $(y^*, z^*) \in Y \times Z$ satisfying the inequalities

$$\forall y \in Y \quad \langle y, Bz^*\rangle \leq v_* \leq \langle y^*, Bz\rangle \quad \forall z \in Z, \tag{1}$$

where $v_* \triangleq \langle y^*, Bz^*\rangle$ is an optimal value of the game (ΓM), is said to be the *saddle point* of the game (ΓM) $((y^*, z^*) \in C(\Gamma M))$.

Theorem 1. *The tuple $(y^*, z^*) \in C(\Gamma M)$, if and only if there exists a number v_*, such that the following system is fulfilled:*

$$\left.\begin{array}{lll} \xi_1(Bz^*) \leq v_* e_{n_1}, & z^* \geq 0, & \langle e_{n_2}, z^*\rangle = \xi_2; \\ \xi_2(y^* B) \geq v_* e_{n_2}, & y^* \geq 0, & \langle e_{n_1}, y^*\rangle = \xi_1. \end{array}\right\} \tag{2}$$

Proof. *Necessity.* Set $y_i = (0, ..., \overset{i}{\xi_1}, ..., 0) \in Y\ \ \forall i = 1, ..., n_1$ and $z_j = (0, ..., \overset{j}{\xi_2}, ..., 0) \in Z\ \ \forall j = 1, ..., n_2$ in (1). Then we obtain (2).

Sufficiency. Scalarly multiplying the first inequality in (2) by y^*, we obtain:

$$\xi_1\langle y^*, Bz^*\rangle \le v_* \sum_{i=1}^{n_1} y_i^*.$$

Similarly, multiplying the first inequality in the second line of the (2) by z^*, we get:

$$v_* \sum_{j=1}^{n_2} z_j^* \le \xi_2\langle y^*, Bz^*\rangle.$$

Hence, $v_* = \langle y^*, Bz^*\rangle$.

Now, multiplying the first inequality in (2) by an arbitrary $y \in Y$ and the first inequality in the second line of (2) by an arbitrary $z \in Z$, we obtain:

$$\xi_1\langle y, Bz^*\rangle \le v_* \sum_{i=1}^{n_1} y_i^* \quad \forall y \in Y; \quad v_* \sum_{j=1}^{n_2} z_j^* \le \xi_2\langle y^*, Bz\rangle \quad \forall z \in Z.$$

So, we arrive at (1). □

Note that conditions (2) represent finite numbers of equalities and inequalities. Now we can replace a game at the lower level by its optimality conditions. Hence, for the bilevel problem $(\mathcal{BP}_{\Gamma M})$, it is possible to formulate the following equivalent single-level problem:

$$\left.\begin{array}{c} -f_0(x,y,z) \stackrel{\triangle}{=} \langle c, x\rangle + \langle d_1, y\rangle + \langle d_2, z\rangle \uparrow \max\limits_{x,y,z,v}, \\ (x,y,z) \in S \stackrel{\triangle}{=} \{x,y,z \mid Ax \le a,\ x \ge 0,\ \langle b_1, x\rangle + \langle b_2, x\rangle = 1, \\ y \ge 0,\ \langle e_{n_1}, y\rangle = \langle b_1, x\rangle, \quad z \ge 0,\ \langle e_{n_2}, z\rangle = \langle b_2, x\rangle\}, \\ \langle b_1, x\rangle(Bz) \le v e_{n_1}, \\ -\langle b_2, x\rangle(yB) \le -v e_{n_2}. \end{array}\right\} \quad (\mathcal{PM})$$

More precisely, the following theorem takes place.

Theorem 2. *The triplet (x^*, y^*, z^*) is a global optimistic solution of the bilevel problem $(\mathcal{BP}_{\Gamma M})$ $((x^*, y^*, z^*) \in \mathrm{Sol}(\mathcal{BP}_{\Gamma M}))$, if and only if there exist a number v_* such that the 4-tuple (x^*, y^*, z^*, v_*) is a global solution of the problem $(\mathcal{PM})$.*

Proof. *Necessity.* Let the triplet $(x^*, y^*, z^*) \in \mathrm{Sol}(\mathcal{BP}_{\Gamma M})$. Then $(y^*, z^*) \in C(\Gamma M(x^*))$ and Theorem 1 is fulfilled. Therefore, there exists v_*: the conditions (2) hold under $\xi_1 = \langle b_1, x^*\rangle$ and $\xi_2 = \langle b_2, x^*\rangle$. Since, in addition, $x_* \in X$, then the 4-tuple (x^*, y^*, z^*, v_*) is feasible in the problem $(\mathcal{PM})$.

Let on the contrary $(x^*, y^*, z^*, v_*) \notin \mathrm{Sol}(\mathcal{PM})$. Then there exists a feasible in the problem $(\mathcal{PM})$ 4-tuple $(\bar{x}, \bar{y}, \bar{z}, \bar{v})$, such that

$$\langle c, \bar{x}\rangle + \langle d_1, \bar{y}\rangle + \langle d_2, \bar{z}\rangle > \langle c, x^*\rangle + \langle d_1, y^*\rangle + \langle d_2, z^*\rangle. \quad (3)$$

At the same time, the conditions (2) are fulfilled for the 4-tuple $(\bar{x}, \bar{y}, \bar{z}, \bar{v})$ (under $\xi_1 = \langle b_1, \bar{x}\rangle$, $\xi_2 = \langle b_2, \bar{x}\rangle$, and $(y^*, z^*, v_*) = (\bar{y}, \bar{z}, \bar{v})$) because of

$(\bar{x}, \bar{y}, \bar{z}, \bar{v})$ is feasible in the problem $(\mathcal{PM})$. Therefore, according to Theorem 1 $(\bar{y}, \bar{z}) \in C(\Gamma M(\bar{x}))$, and the triplet $(\bar{x}, \bar{y}, \bar{z})$ is feasible in the problem $(\mathcal{BP}_{\Gamma M})$, because $\bar{x} \in X$. Since the objective functions of the problems $(\mathcal{BP}_{\Gamma M})$ and $(\mathcal{PM})$ coincide, the inequality (3) contradicts $(x^*, y^*, z^*) \in \text{Sol}(\mathcal{BP}_{\Gamma M})$.

Sufficiency. Now let the 4-tuple $(x^*, y^*, z^*, v_*) \in \text{Sol}(\mathcal{PM})$. Then $x_* \in X$ and the conditions (2) hold for the 4-tuple (x^*, y^*, z^*, v_*) (under $\xi_1 = \langle b_1, x^*\rangle$ and $\xi_2 = \langle b_2, x^*\rangle$). Therefore, according to Theorem 1, $(y^*, z^*) \in C(\Gamma M(x^*))$ and the triplet (x^*, y^*, z^*) is feasible in the problem $(\mathcal{BP}_{\Gamma M})$.

Further let there exist a feasible in the problem $(\mathcal{BP}_{\Gamma M})$ 3-tuple $(\tilde{x}, \tilde{y}, \tilde{z})$, such that

$$\langle c, \tilde{x}\rangle + \langle d_1, \tilde{y}\rangle + \langle d_2, \tilde{z}\rangle > \langle c, x^*\rangle + \langle d_1, y^*\rangle + \langle d_2, z^*\rangle. \tag{4}$$

According to Theorem 1, again there exists a number $\tilde{v}$: the conditions (2) are fulfilled for the 4-tuple $(\tilde{x}, \tilde{y}, \tilde{z}, \tilde{v})$ (under $\xi_1 = \langle b_1, \tilde{x}\rangle$, $\xi_2 = \langle b_2, \tilde{x}\rangle$, and $(y^*, z^*, v_*) = (\tilde{y}, \tilde{z}, \tilde{v})$). Then this 4-tuple is feasible in the problem $(\mathcal{PM})$ and the inequality (4) holds. As above, this contradicts $(x^*, y^*, z^*, v_*) \in \text{Sol}(\mathcal{PM})$. □

It can be readily seen that the problem $(\mathcal{PM})$ is a global optimization problem with a nonconvex feasible set (see, e.g., [16–18]). A nonconvexity in the problem $(\mathcal{PM})$ is generated by the two last vector constraints (two groups of $(n_1 + n_2)$ bilinear constraints in total). These constraints have arisen from optimality conditions for a non-normalized matrix game at the lower level of the bilevel problem $(\mathcal{BP}_{\Gamma M})$. It is known that a bilinear function can be represented as a difference of two convex functions (i.e. a bilinear function is a d.c. function) [14,26]. Therefore, the problem $(\mathcal{PM})$ belongs to the class of nonconvex optimization problems with d.c. constraints [18,22–24] and we can apply the GST to solving this class of nonconvex problems.

3 D.C. Decomposition

The first stage of the application of the Global Search Theory to the problem under scrutiny is an explicit decomposition of nonconvex functions from the problem statement as a difference of two convex functions. As noted above, $(n_1 + n_2)$ bilinear constraints generate the basic nonconvexity in the problem $(\mathcal{PM})$.

Therefore, we should find a decomposition of the following functions by the difference of two convex functions. First, let us obtain an explicit d.c. representation of the i-th scalar constraint in the first group:

$$f_i(x, z, v) = \langle b_1, x\rangle\langle (B)_i, z\rangle - v \le 0, \quad i = 1, \ldots, n_1, \tag{5}$$

where $(B)_i$ is an i-th row of the matrix B.

Introduce the denotation $Q_i^T = (b_1^{(1)}(B)_i;\ b_1^{(2)}(B)_i;\ \ldots;\ b_1^{(m)}(B)_i)$, where $b_1^{(1)}, b_1^{(2)} \ldots, b_1^{(m)}$ are components of the vector b_1. Hence, we can reduce (5) to a standard bilinear form $f_i(x, z, v) = \langle xQ_i, z\rangle - v \le 0$, $i = 1, \ldots, n_1$. And we can use here the known d.c. representation based on the property of a scalar product [14,26]:

$$f_i(x,z,v) = g_i(x,z,v) - h_i(x,z), \tag{6}$$

where $g_i(x,z,v) = \frac{1}{4}\|xQ_i + z\|^2 - v$, $h_i(x,z) = \frac{1}{4}\|xQ_i - z\|^2$.

Similarly, if we introduce the matrix $R_j^T = (b_2^{(1)}(B)_j;\ b_2^{(2)}(B)_j;\ \ldots;\ b_2^{(m)}(B)_j)$ ($(B)_j$ is a j-th column of the matrix B), we obtain a d.c. representation of the constraints in the second group:

$$f_j(x,y,v) = -\langle b_2, x\rangle\langle y, (B)_j\rangle + v = g_j(x,y,v) - h_j(x,y),\ j = 1,...,n_2, \tag{7}$$

where $g_j(x,y,v) = \frac{1}{4}\|xR_j - y\|^2 + v$, $h_i(x,z) = \frac{1}{4}\|xR_j + y\|^2$.

In total, we obtain the following problem with $(n_1 + n_2)$ d.c. constraints:

$$\left.\begin{array}{c} f_0(x,y,z) \downarrow \min\limits_{x,y,z,v}, \quad (x,y,z) \in S, \\ f_i(x,z,v) := g_i(x,z,v) - h_i(x,z) \le 0, \quad i = 1,\ldots,n_1, \\ f_j(x,y,v) := g_j(x,y,v) - h_j(x,y) \le 0, \quad j = 1,\ldots,n_2, \end{array}\right\} \tag{$\mathcal{P}$}$$

where the functions f_0, g_i, h_i $\forall i \in I = \{1,...,n_1\}$, and g_j, h_j $\forall j \in J = \{1,...,n_2\}$ as well as the set

$$S = \{x,y,z \ge 0 \mid Ax \le a,\ \langle b_1,x\rangle + \langle b_2,x\rangle = 1,\ \langle e_{n_1}, y\rangle = \langle b_1,x\rangle,\ \langle e_{n_2}, z\rangle = \langle b_2,x\rangle\},$$

are convex.

And now we are ready to apply the GST to the problem $(\mathcal{P})$.

4 Local Search

As has been mentioned, for the purpose of solving the d.c. constraint problem $(\mathcal{P})$, we develop the Global Search Algorithm based on the Global Search Theory (GST) [18,23,24] using the d.c. decomposition constructed above. According to the GST, the algorithm for solving the problem $(\mathcal{P})$ should consist of two principal stages:

(1) a special Local Search Method (LSM), which takes into account the structure of the problem in question [18,22];

(2) the procedure based on the Global Optimality Conditions (GOCs) and the Global Search Algorithm based on the Global Search Theory (GST) [18,23,24], which allows us to improve the point provided by the LSM [18,22].

In order to develop the LSM for the problem $(\mathcal{P})$, we suppose that the feasible set $D := \{(x,y,z,v) \mid (x,y,z) \in S;\ f_i(x,z,v) \le 0,\ i \in I;\ f_j(x,y,v) \le 0,\ j \in J\}$ of the problem $(\mathcal{P})$ is not empty and the optimal value $\mathcal{V}(\mathcal{P}) := \inf\{f_0(x,y,z) \mid (x,y,z,v) \in D\}$ of the problem $(\mathcal{P})$ is finite: $\mathcal{V}(\mathcal{P}) > -\infty$.

Furthermore, let us denote $w := (x,y,z,v) \in I\!R^{m+n_1+n_2+1}$ and assume that a feasible starting point $w^0 \in D$ is given and, in addition, after several iterations it has derived the current iterate $w^s \in D$, $s \in Z_+ = \{0,1,2,\ldots\}$.

In order to propose an LSM for the problem $(\mathcal{P})$, let us apply a classical idea of linearization with respect to the basic nonconvexity of the problem (i.e. with respect to $h_i(\cdot)$, $i \in I$ and $h_j(\cdot)$, $j \in J$) at the point w^s [18–20,22]. Thus, we obtain the following linearized problem:

$$\left.\begin{array}{c} f_0(x,y,z) \downarrow \min\limits_{x,y,z,v}, \quad (x,y,z) \in S, \\ \varphi_{is}(x,z,v) := g_i(x,z,v) - \langle \nabla h_i(x^s,z^s),(x,z)-(x^s,z^s)\rangle \\ -h_i(x^s,z^s) \le 0,\ i=1,\ldots,n_1, \\ \varphi_{js}(x,y,v) := g_j(x,y,v) - \langle \nabla h_j(x^s,y^s),(x,y)-(x^s,y^s)\rangle \\ -h_j(x^s,y^s) \le 0,\ j=1,\ldots,n_2, \end{array}\right\} \qquad (\mathcal{PL}_s)$$

where

$$\begin{array}{c} \nabla_x h_i(x^s,z^s) = \frac{1}{2}\Big(Q_i(x^sQ_i - z^s)\Big),\ \ \nabla_z h_i(x^s,z^s) = -\frac{1}{2}\Big(x^sQ_i - z^s\Big), \\ i \in I; \\ \nabla_x h_j(x^s,y^s) = \frac{1}{2}\Big(R_j(x^sR_j + y^s)\Big),\ \ \nabla_y h_j(x^s,y^s) = \frac{1}{2}\Big(x^sR_j + y^s\Big), \\ j \in J. \end{array} \qquad (8)$$

Suppose that the point w^{s+1} is provided by the approximate solution to the problem $(\mathcal{PL}_s)$, so that

$$\begin{array}{c} w^{s+1} \in D_s = \{(x,y,z,v) \mid (x,y,z) \in S; \\ \varphi_{is}(x,z,v) \le 0,\ i \in I;\ \ \varphi_{js}(x,y,v) \le 0,\ j \in J\} \end{array}$$

and the inequality

$$f_0(x^{s+1},y^{s+1},z^{s+1}) \le \mathcal{V}(\mathcal{PL}_s) + \delta_s$$

holds. Here $\mathcal{V}(\mathcal{PL}_s)$ is the optimal value to the problem $(\mathcal{PL}_s)$:

$$\begin{array}{c} \mathcal{V}(\mathcal{PL}_s) \triangleq \inf\limits_{w}\{f_0(x,y,z) \mid (x,y,z) \in S,\ \varphi_{is}(x,z,v) \le 0,\ i \in I, \\ \varphi_{js}(x,y,v) \le 0,\ j \in J\}, \end{array}$$

and the sequence $\{\delta_s\}$ satisfies the following condition: $\sum\limits_{s=0}^{\infty} \delta_s < +\infty$.

Therefore, the LSM generates the sequence $\{w^s\}$, $w^s \in D_s$, $s \in Z_+$ of solutions to the problems $(\mathcal{PL}_s)$. As it was proven in [22], the cluster point w_* of the sequence $\{w^s\}$ is a solution to the linearized problem $(\mathcal{PL}_*)$ (which is the problem $(\mathcal{PL}_s)$ with w_* instead of w^s), and w_* can be called the critical point with respect to the LSM. Thus, the algorithm constructed in this way provides critical points by employing suitable convex optimization methods with any given accuracy τ.

The inequalities

$$f_0(x^s,y^s,z^s) - f_0(x^{s+1},y^{s+1},z^{s+1}) \le \frac{\tau}{2}, \quad \delta_s \le \frac{\tau}{2},$$

can be chosen as a stopping criterion for the LSM [22].

Computational simulations (see, e.g., [30]) confirm the efficiency of the LSM developed, the performance of which naturally depends on the choice of the method or the software employed to solve auxiliary problems (we can use, for example, IBM ILOG CPLEX or other efficient software for solving convex optimization problems). Thus, the LSM can be applied in future implementations of the global search algorithm for solving the bilevel problem $(\mathcal{BP}_{\Gamma M})$ which is formulated as an equivalent single-level problem with d.c. constraints $(\mathcal{PM})$.

5 Global Search

The second part of the global search procedure can be viewed as the most important and even the crucial one, because we have to address the issue of escaping a critical point (the non-global solution provided by a local search). Such a procedure is substantiated by the theoretical basis produced with the help of the Global Search Theory (GST) developed by Strekalovsky [18,23,24] for a problem with d.c. constraints. Furthermore, we intend to solve the problem $(\mathcal{P})$ using the exact penalization approach to the d.c. optimization developed in [23,24]. Therefore, we introduce the penalized problem in the following way

$$\begin{array}{c}\theta_\sigma(x,y,z,v) = f_0(x,y,z) + \sigma \max\{0; f_i(x,z,v), i=1,\dots n_1; \\ f_j(x,y,v), j=1,\dots n_2\} \downarrow \min, \quad (x,y,z) \in S.\end{array} \qquad (\mathcal{P}_\sigma)$$

It can be readily seen that the penalized function $\theta_\sigma(\cdot)$ is a d.c. function, because the functions $f_i(\cdot) = g_i(\cdot) - h_i(\cdot)$, $i \in I$, $f_j(\cdot) = g_j(\cdot) - h_j(\cdot)$, $j \in J$ are d.c. functions and the objective function $f_0(\cdot)$ of the problem $(\mathcal{P})$ (see also single-level problem $(\mathcal{PM})$) is an affine function.

Actually, since $\sigma > 0$, $\theta_\sigma(x,y,z,v) = G_\sigma(x,y,z,v) - H_\sigma(x,y,z)$,

$$\begin{array}{c}H_\sigma(x,y,z) := \sigma\left[\sum\limits_{i=1}^{n_1} h_i(x,z) + \sum\limits_{j=1}^{n_2} h_j(x,y)\right], \\ G_\sigma(x,y,z,v) := \theta_\sigma(x,y,z,v) + H_\sigma(x,y,z) \\ = f_0(x,y,z) + \sigma \max\left\{\sum\limits_{i=1}^{n_1} h_i(x,z) + \sum\limits_{j=1}^{n_2} h_j(x,y);\right. \\ \left[g_i(x,z,v) + \sum\limits_{\substack{l=1\\ l\neq i}}^{n_1} h_l(x,z) + \sum\limits_{j=1}^{n_2} h_j(x,y)\right], i \in I; \\ \left.\left[g_j(x,y,v) + \sum\limits_{\substack{l=1\\ l\neq j}}^{n_2} h_l(x,y) + \sum\limits_{i=1}^{n_1} h_i(x,z)\right], j \in J\right\}.\end{array} \qquad (9)$$

It is clear that $G_\sigma(\cdot)$ and $H_\sigma(\cdot)$ are convex functions.

It is well-known that if for some σ the 4-tuple $(x(\sigma), y(\sigma), z(\sigma), v(\sigma)) \in \text{Sol}(\mathcal{P}_\sigma)$, and $(x(\sigma), y(\sigma), z(\sigma), v(\sigma))$ is feasible in the problem $(\mathcal{P})$, then $(x(\sigma), y(\sigma), z(\sigma), v(\sigma))$ is a global solution to the problem $(\mathcal{P})$ [19,20,24]. Moreover, this situation remains the same when the value of σ grows [19,20,24].

Hence, the key point for using exact penalization theory here is the existence of a threshold value $\hat{\sigma} > 0$ of the penalty parameter σ such that $\text{Sol}(\mathcal{P}_\sigma) \subset \text{Sol}(\mathcal{P})$ $\forall \sigma \geq \hat{\sigma}$. In other words, for $\sigma \geq \hat{\sigma}$ the problems $(\mathcal{P})$ and $(\mathcal{P}_\sigma)$ turn out to be equivalent in the sense that $\text{Sol}(\mathcal{P}) = \text{Sol}(\mathcal{P}_\sigma)$ [24]. Due to the fact that the objective function f_0 of the problem $(\mathcal{P})$ satisfies the Lipschitz property [19,20] with respect to all variables, such threshold value $\hat{\sigma}$ exists [24], and we can use the following Global Optimality Conditions in order to characterize a global solution to this problem [24].

Theorem 3. *[24] Let a feasible point* $(x, y, z, v) \in D$, $\zeta := f_0(x, y, z)$, *be a solution to the problem* $(\mathcal{P})$ *and* $\sigma \geq \sigma_* > 0$, *where* $\sigma_* \geq 0$ *is a threshold value of the penalty parameter such that* $\text{Sol}(\mathcal{P}) = \text{Sol}(\mathcal{P}_\sigma)$ $\forall \sigma \geq \sigma_*$.

Then, for every pair $(q, \beta) \in I\!R^{m+n_1+n_2} \times I\!R$ *such that*

$$H_\sigma(q) = \beta - \zeta, \tag{10}$$

the following inequality holds

$$G_\sigma(x, y, z, v) - \beta \geq \langle \nabla H_\sigma(q), (x, y, z) - q \rangle \quad \forall (x, y, z) \in S. \tag{11}$$

It is not difficult to notice that Theorem 3 reduces the solution of the nonconvex problem $(\mathcal{P}_\sigma)$ to an investigation of the family of the convex (linearized) problems

$$G_\sigma(x, y, z, v) - \langle \nabla H_\sigma(q), (x, y, z) \rangle \downarrow \min_{(x,y,z,v)}, \quad (x, y, z) \in S, \qquad (\mathcal{P}_\sigma L(q))$$

depending on the pairs $(q, \beta) \in I\!R^{m+n_1+n_2} \times I\!R$, which fulfill the Eq. (10).

It is worth noting that the linearization is carried out here with respect to the "unified" nonconvexity of the problem $(\mathcal{P})$ accumulated by the function $H_\sigma(\cdot)$ from the d.c. representation (9) and $\nabla H_\sigma(\cdot)$ can be constructed in the following way

$$\nabla_x H_\sigma(x, y, z) := \sigma \left[\sum_{i=1}^{n_1} \nabla_x h_i(x, z) + \sum_{j=1}^{n_2} \nabla_x h_j(x, y) \right],$$
$$\nabla_y H_\sigma(x, y) := \sigma \sum_{j=1}^{n_2} \nabla_y h_j(x, y),$$
$$\nabla_z H_\sigma(x, z) := \sigma \sum_{i=1}^{n_1} \nabla_z h_i(x, z),$$

where $\nabla_x h_i(x, z)$, $\nabla_z h_i(x, z)$, $i \in I$, $\nabla_x h_j(x, y)$, $\nabla_y h_j(x, y)$, $j \in J$, given by formulas (8).

Hence, the verification of the principal inequality (11) can be performed by solving the linearized problems $(\mathcal{P}_\sigma L(q))$ and varying the parameters (q, β) satisfying (10).

Let the Lagrange multipliers, associated with the constraints and corresponding to the point $w^k \in I\!R^{m+n_1+n_2+1}$, $k \in \{0, 1, 2, ...\}$, be denoted by $\lambda^k := (\lambda_1, \ldots, \lambda_{n_1+n_2}) \in I\!R^{n_1+n_2}$.

Basing on the relations and connections (see [24]) between the conditions (10), (11) of Theorem 3 and the Classical Optimality Conditions [19,20], we can present the following Global Search Scheme in the problem $(\mathcal{P}_\sigma)$.

Let there be given a starting point $(x_0, y_0, z_0) \in S$, numerical sequences $\{\tau_k\}$, $\{\delta_k\}$ $(\tau_k,\ \delta_k > 0,\ k = 0, 1, 2, ...;\ \tau_k \downarrow 0,\ \delta_k \downarrow 0 (k \to \infty))$.

Global Search Scheme

Step 0. Set $k := 0$.

Step 1. Using the LSM from Sect. 4, find a τ_k-critical point $w^k = (x^k, y^k, z^k, v_k)$ in the problem $(\mathcal{P})$.

Step 2. Set $\sigma_k := \sum\limits_{l=1}^{n_1+n_2} \lambda_l^k$.

Choose a number $\beta :\ \inf(G_\sigma, S) \leq \beta \leq \sup(G_\sigma, S)$.
Choose an initial $\beta_0 = G_\sigma(w^k)$, $\zeta_k = \theta_\sigma(w^k)$.

Step 3. Construct a finite approximation

$$\mathcal{A}_k(\beta) = \{q^1, \ldots, q^{N_k} \mid H_\sigma(q^p) = \beta - \zeta_k,\ p = 1, \ldots, N_k,\ N_k = N_k(\beta)\}$$

of the level surface $\mathcal{U}(\zeta_k) = \{(x, y, z) \mid H_\sigma(x, y, z) = \beta - \zeta_k\}$ of the function $H_\sigma(\cdot)$.

Step 4. Find a δ_k-solution $\bar{u}^p \in I\!R^{m+n_1+n_2+1}$ to the following linearized problem:

$$G_\sigma(x, y, z, v) - \langle \nabla H_\sigma(q^p), (x, y, z) \rangle \downarrow \min_{(x,y,z,v)},\ (x, y, z) \in S. \qquad (P_\sigma L(q^p))$$

Step 5. Starting from the point $\bar{u}^p$, find a τ_k-critical point $\bar{w}^p = (\bar{x}^p, \bar{y}^p, \bar{z}^p, \bar{v}^p)$ by the LSM from Sect. 4.

Step 6. Choose the point $(\hat{x}, \hat{y}, \hat{z})$:

$$f_0(\hat{x}, \hat{y}, \hat{z}) \leq \min\{f_0(\bar{x}^p, \bar{y}^p, \bar{z}^p),\ p = 1, ..., N_k\}.$$

Step 7. If $f_0(\hat{x}, \hat{y}, \hat{z}) < f_0(x^k, y^k, z^k)$, then set $w_{k+1} = (\hat{x}, \hat{y}, \hat{z}, \hat{v})$, $k = k + 1$ and go to Step 2.

Step 8. Otherwise, choose a new value of β and go to Step 3. □

It can be readily seen that the Global Search Scheme is not an algorithm in the conventional sense, because some of its steps are not specified. For example, we do not know precisely how to construct a starting point and the approximation of the level surface of the function, how to implement a local search and solve the problem $(P_\sigma L(q^p))$ etc. The answer to these questions depends on the real data of numerical instances and uses the previous computational experience concerning the nonconvex problems solution [18,21,25–34].

6 Concluding Remarks

In the present paper, we developed a new approach to finding optimistic solutions to a special bilevel problem with an equilibrium (with a parametric matrix game) at the lower level. These methods are based on the original Global Search Theory for nonconvex (d.c.) optimization by A.S. Strekalovsky.

We described in detail the reduction of the original bilevel problem to a nonconvex single-level problem and showed how to develop special local and global search methods that take into account properties of the problem in question.

In our future research we will increase the complexity of the bilevel model (we are going to study problems with bimatrix game at the lower level) and carry out numerical testing of the developed methods on specially generated examples. Based on our previous computational experience (see, for example, results on solution of other bilevel problems [27,28]), we hope that the algorithms developed can also be used for efficient numerical solution of bilevel problems with an equilibrium at the lower level.

References

1. Dempe, S.: Foundations of Bilevel Programming. Kluwer Academic Publishers, Dordrecht (2002)
2. Pang, J.-S.: Three modeling paradigms in mathematical programming. Math. Program. Ser. B **125**, 297–323 (2010)
3. Colson, B., Marcotte, P., Savard, G.: An overview of bilevel optimization. Ann. Oper. Res. **153**, 235–256 (2007)
4. Dempe, S.: Bilevel programming. In: Audet, C., Hansen, P., Savard, G. (eds.) Essays and Surveys in Global Optimization, pp. 165–193. Springer, Boston (2005). https://doi.org/10.1007/b135610
5. Dempe, S., Kalashnikov, V.V., Perez-Valdes, G.A., Kalashnykova, N.: Bilevel Programming Problems: Theory, Algorithms and Applications to Energy Networks. Springer, Heidelberg (2015). https://doi.org/10.1007/978-3-662-45827-3
6. Dempe, S., Dutta, J.: Is bilevel programming a special case of a mathematical program with complementarity constraints? Math. Program. Ser. A **131**, 37–48 (2012)
7. Luo, Z.-Q., Pang, J.-S., Ralph, D.: Mathematical Programs with Equilibrium Constraints. Cambridge University Press, Cambridge (1996)
8. Hu, M., Fukushima, M.: Existence, uniqueness, and computation of robust Nash equilibria in a class of multi-leader-follower games. SIAM J. Optim. **23**(2), 894–916 (2013)
9. Ang, J., Fukushima, M., Meng, F., Noda, T., Sun, J.: Establishing Nash equilibrium of the manufacturer-supplier game in supply chain management. J. Glob. Optim. **56**, 1297–1312 (2013)
10. Zhang, H., Bennis, M., Da Silva, L.A., Han, Z.: Multi-leader multi-follower Stackelberg game among Wi-Fi, small cell and macrocell networks. In: 2014 IEEE Global Communications Conference, pp. 4520–4524 (2014)
11. Ramos, M., Boix, M., Aussel, D., Montastruc, L., Domenech, S.: Water integration in eco-industrial parks using a multi-leader-follower approach. Comput. Chem. Eng. **87**, 190–207 (2016)
12. Yang, Zh, Ju, Y.: Existence and generic stability of cooperative equilibria for multi-leader-multi-follower games. J. Glob. Optim. **65**, 563–573 (2016)
13. Owen, G.: Game Theory. Academic Press, New York (1995)

14. Strekalovsky, A.S., Orlov, A.V.: Bimatrix Games and Bilinear Programming. FizMatLit, Moscow (2007). (in Russian)
15. Mazalov, V.: Mathematical Game Theory and Applications. Wiley, New York (2014)
16. Horst, R., Tuy, H.: Global Optimization: Deterministic Approaches. Springer, Berlin (1993). https://doi.org/10.1007/978-3-662-03199-5
17. Strongin, R.G., Sergeyev, Ya.D: Global Optimization with Non-convex Constraints: Sequential and Parallel Algorithms. Springer, New York (2000). https://doi.org/10.1007/978-1-4615-4677-1
18. Strekalovsky, A.S.: Elements of Nonconvex Optimization. Nauka, Novosibirsk (2003). (in Russian)
19. Nocedal, J., Wright, S.J.: Numerical Optimization. Springer, New York, Berlin, Heidelberg (2000). https://doi.org/10.1007/978-0-387-40065-5
20. Bonnans, J.-F., Gilbert, J.C., Lemarechal, C., Sagastizabal, C.A.: Numerical Optimization: Theoretical and Practical Aspects. Springer, Heidelberg (2006). https://doi.org/10.1007/978-3-540-35447-5
21. Strekalovsky, A.S.: On solving optimization problems with hidden nonconvex structures. In: Rassias, T.M., Floudas, C.A., Butenko, S. (eds.) Optimization in Science and Engineering, pp. 465–502. Springer, New York (2014). https://doi.org/10.1007/978-1-4939-0808-0_23
22. Strekalovsky, A.S.: On local search in d.c. optimization problems. Appl. Math. Comput. **255**, 73–83 (2015)
23. Strekalovsky, A.S.: Global optimality conditions in nonconvex optimization. J. Optim. Theory Appl. **173**(3), 770–792 (2017)
24. Strekalovsky, A.S.: Global optimality conditions and exact penalization. Optim. Lett. https://doi.org/10.1007/s11590-017-1214-x (published online)
25. Orlov, A.V., Strekalovsky, A.S.: Numerical search for equilibria in bimatrix games. Comput. Math. Math. Phys. **45**, 947–960 (2005)
26. Orlov, A.V.: Numerical solution of bilinear programming problems. Comput. Math. Math. Phys. **48**, 225–241 (2008)
27. Gruzdeva, T.V., Petrova, E.G.: Numerical solution of a linear bilevel problem. Comput. Math. Math. Phys. **50**, 1631–1641 (2010)
28. Strekalovsky, A.S., Orlov, A.V., Malyshev, A.V.: On computational search for optimistic solutions in bilevel problems. J. Glob. Optim. **48**(1), 159–172 (2010)
29. Orlov, A.V., Strekalovsky, A.S., Batbileg, S.: On computational search for Nash equilibrium in hexamatrix games. Optim. Lett. **10**(2), 369–381 (2016)
30. Gruzdeva, T., Strekalovsky, A.: An approach to fractional programming via D.C. constraints problem: local search. In: Kochetov, Yu., Khachay, M., Beresnev, V., Nurminski, E., Pardalos, P. (eds.) DOOR 2016. LNCS, vol. 9869, pp. 404–417. Springer, Cham (2016). https://doi.org/10.1007/978-3-319-44914-2_32
31. Gruzdeva, T.V.: On a continuous approach for the maximum weighted clique problem. J. Glob. Optim. **56**, 971–981 (2013)
32. Gaudioso, M., Gruzdeva, T.V., Strekalovsky, A.S.: On numerical solving the spherical separability problem. J. Glob. Optim. **66**(1), 21–34 (2016)
33. Enkhbat, R., Gruzdeva, T.V., Barkova, M.V.: DC programming approach for solving an applied ore-processing problem. J. Ind. Manag. Optim. **14**(2), 613–623 (2018)
34. Gruzdeva, T.V., Strekalovskiy, A.S.: On solving the sum-of-ratios problem. Appl. Math. Comput. **318**, 260–269 (2018)

Integer Programming

How the Difference in Travel Times Affects the Optima Localization for the Routing Open Shop

Ilya Chernykh[1,2,3](✉) and Ekaterina Lgotina[2]

[1] Sobolev Institute of Mathematics, Novosibirsk, Russia
idchern@math.nsc.ru
[2] Novosibirsk State University, Novosibirsk, Russia
kate.lgotina@gmail.com
[3] Novosibirsk State Technical University, Novosibirsk, Russia

Abstract. The routing open shop problem, being a generalization of the metric TSP and the open shop scheduling problem, is known to be NP-hard even in case of two machines with a transportation network consisting of two nodes only. We consider a generalization of this problem with unrelated travel times of each machine. We determine a tight optima localization interval for the two-machine problem in the case when the transportation network consists of at most three nodes. As a byproduct of our research, we present a linear time $\frac{5}{4}$-approximation algorithm for the same problem. We prove that the algorithm has the best theoretically possible approximation ratio with respect to the standard lower bound.

Keywords: Scheduling · Routing open shop · Unrelated travel times · Optima localization · Approximation algorithm

1 Introduction

The idea of determining the *tight optima localization interval* for scheduling problems was introduced more than 20 years ago in [14]. It can be described as follows. Consider some minimization problem $f(x) \to \min$ with a lower bound LB on the optimum. Then the tight optima localization interval (subject to LB) is an interval of type $[LB, \rho \cdot LB]$ with the smallest possible value of ρ guaranteed to contain an optimum value for any problem instance. The systematic search for such tight intervals $[LB, \rho \cdot LB]$ is useful for the following reasons. First, it proves that the approximation factor ρ of an algorithm guaranteed to find a solution with $f(x) \leqslant \rho LB$ is as good as possible with respect to LB. Second, it also allows the quality of the lower bound LB to be estimated: any approximation algorithm with worst-case ratio performance guarantee less than ρ has to be based either on another lower bound (more precise than LB) or on the optimum

This research was supported by the Russian Foundation for Basic Research, projects 17-01-00170, 17-07-00513 and 18-01-00747.

M. Khachay et al. (Eds.): MOTOR 2019, LNCS 11548, pp. 187–201, 2019.
https://doi.org/10.1007/978-3-030-22629-9_14

itself. Third, as it typically turns out for scheduling problems, one can compute the upper and lower bounds in linear time. Thus, essentially, optima localization can yield linear-time approximation algorithms that are provably as good as possible with respect to a certain lower bound.

An example of such a research for the classic open shop scheduling problem [9] with three machines with respect to the standard lower bound was implemented in [14]. The approach used includes an *instance reduction* preserving the lower bound, and intellectual *enumeration of subsets of instances* in the branch-and-bound manner. Altogether that approach can be referred to as an *algorithm of proving* and can be used to obtain similar results for a huge variety of scheduling problems. The proof described in [14] involved massive enumeration of subsets and therefore was a computer-aided one. In this paper we illustrate the application of the same approach to a special case of some relatively new scheduling model being a generalization of the open shop and the metric traveling salesman problem. Note that the proof in our case is compact and not computer-aided due to the following reasons. First, we consider a two-machine case (note that our case is still a generalization of a known NP-hard problem). Second, but not least, we fine-tuned the approach allowing a more efficient instance reduction. The main goal of the paper is to describe the approach in details and to obtain optima localization results for a new scheduling problem.

The problem under consideration is a certain generalization of the routing open shop model. Routing open shop was first introduced in [1,2]. In this model, the sets of jobs $\mathcal{J}$ and machines $\mathcal{M}$ are given, and machines have to perform operations of each job (with given processing times) in an arbitrary order similar to the classic open shop scheduling problem: different operations of the same job cannot be processed simultaneously. Jobs are distributed among the nodes of some transportation network represented by an edge-weighted graph G. The weight $\mathbf{dist}(v, u)$ of an edge $e = [v, u]$ represents the travel distance between the nodes v and u. All the machines are initially located at a predefined node referred to as *the depot*. Machines have to travel with unit speed between the nodes of the transportation network to process their operations and to return to the depot. Machines are allowed to use shortest paths between the nodes, therefore we may assume that travel times satisfy the triangle inequality.

Such a problem can appear in a production, where mobile machines have to perform technological operations on some large unmovable parts, located in different workshops.

For any schedule S, the value $R_{\max}(S)$ denotes the *makespan* of S which is the moment when the last machine returns to the depot after processing all the operations (see Sect. 2 for details). The goal is to minimize the makespan. Following the traditional three-field notation (see [12] for example) the m-machine routing open shop problem is denoted as $ROm||R_{\max}$ or $ROm|G = X|R_{\max}$ in case we want to specify the structure X of a transportation network. In the latter case we use standard graph theory notation, like $G = K_p$ for a complete graph with p nodes, or $G = tree$.

Note that in the open shop environment, due to the duality of sets $\mathcal{J}$ and $\mathcal{M}$, the routing open shop problem is very similar to an open shop with sequence-dependent transportation delays (see, e.g., [3]). However, traditionally researchers consider problems with a constant number of machines, hence the focus of the research differs depending on which model we use.

We assume that each node (with the possible exception of the depot) contains at least one job. This makes it necessary for each machine to visit each node at least once. Therefore, the routing open shop with a single machine is equivalent to the classic metric TSP which is well-known to be NP-hard in the strong sense. On the other hand, a single-node routing open shop is just a plain open shop problem and is NP-hard for three and more machines while being polynomially solvable in the two-machine case [9]. Surprisingly, the combination of those classic problems remains NP-hard even in the case with two machines on a link ($RO2|G = K_2|R_{\max}$), as proved in [2]. A fully polynomial time approximation scheme for $RO2|G = K_2|R_{\max}$ is described in [11].

Let us give a brief review of the routing open shop problem focusing on a case of two machines $RO2||R_{\max}$. A first $\frac{7}{4}$-approximation algorithm was proposed in [2]. It was further improved in [5] where a $\frac{13}{8}$-approximation algorithm is described. This improvement is relatively significant due to the following remark. Note that the $RO2||R_{\max}$ problem includes the metric TSP as a special case. Since the best known up to date approximation algorithm for the metric TSP is the $\frac{3}{2}$-approximation algorithm due to Christofides [8] and Serdyukov [13] we cannot hope to achieve better than $\frac{3}{2}$-approximation for $RO2||R_{\max}$ until a better approximation for the metric TSP is found. On the other hand, the *easy-TSP* version of the $RO2||R_{\max}$ problem (the case when an optimal solution for the underlying TSP is known, or the time complexity of its solving is not taken into account) admits a $\frac{4}{3}$-approximation algorithm described in [5].

The standard lower bound $\bar{R}$ for $RO||R_{\max}$ was introduced in [1] (see Sect. 2). All the approximation algorithms mentioned in the previous paragraph use $\bar{R}$ to justify their performance guarantees: ρ-approximation algorithms actually obtain a schedule with makespan belonging to an interval $[\bar{R}, \rho\bar{R}]$. Hence the search for the tight optima localization interval for $RO2||R_{\max}$ with respect to $\bar{R}$ is an important step in the design of such approximation algorithms.

Such an interval is known only for a few special cases of the $RO||R_{\max}$ problem. The problem with 2 nodes $RO2|G = K_2|R_{\max}$ was thoroughly investigated in [1]. It was shown that the optimal makespan for any instance does not exceed $\frac{6}{5}\bar{R}$ and this upper bound is tight. A few years ago that result was generalized to a problem with 3 nodes $RO2|G = K_3|R_{\max}$ [6]: it was shown that the same optima localization interval holds for the triangular transportation network. The tight upper bound of the optima localization interval for $RO2||R_{\max}$ is still an open question. The properties of the $\frac{4}{3}$-approximation algorithm from [5] imply that the upper bound of the optima localization interval for $RO2||R_{\max}$ does not exceed $\frac{4}{3}\bar{R}$. Therefore, the exact value of the upper bound of the optima localization interval for the two-machine problem lies between $\frac{6}{5}\bar{R}$ and $\frac{4}{3}\bar{R}$.

In this paper we consider the following generalization of the routing open shop problem introduced in [4]. In this model, travel times are specific for each machine. Let $\mathbf{dist}_i(v,u)$ stand for the travel times of machine $M_i \in \mathcal{M}$ between nodes v and u. We assume distances to be symmetrical: $\mathbf{dist}_i(v,u) = \mathbf{dist}_i(u,v)$ for any nodes u, v. Consider the following hierarchy of the travel time models, introduced in [4]:

- $RO|Ptt|R_{\max}$: $\mathbf{dist}_i(v,u) = \mathbf{dist}(v,u)$ (*identical* travel times, equivalent to $RO||R_{\max}$);
- $RO|Qtt|R_{\max}$: $\mathbf{dist}_i(v,u) = \dfrac{\mathbf{dist}(v,u)}{\sigma_i}$ (*uniform* travel times, σ_i represents the travel speed of machine M_i);
- $RO|Rtt|R_{\max}$: $\mathbf{dist}_i(v,u)$ are individual for each machine (*unrelated* travel times).

The usage of P, Q and R for notation is inspired by a well-known notation for scheduling problems with parallel machines (see [12]). Note that in Rtt environment it is possible to model a situation when each machine has to visit only some of the nodes of the transportation network, i.e., in a case when each job has to be processed only by a subset of $\mathcal{M}$.

In this paper we consider the following special cases with uniform and unrelated travel times: $RO2|Qtt, G = K_2|R_{\max}$, $RO2|Qtt, G = K_3|R_{\max}$, $RO2|Rtt, G = K_2|R_{\max}$ and $RO2|Rtt, G = K_3|R_{\max}$. Obviously, $RO2|Qtt, G = K_2|R_{\max}$ can be considered as a subcase of both $RO2|Qtt, G = K_3|R_{\max}$ and $RO2|Rtt, G = K_2|R_{\max}$ problems while the latter problems are special cases of $RO2|Rtt, G = K_3|R_{\max}$ (see Fig. 1).

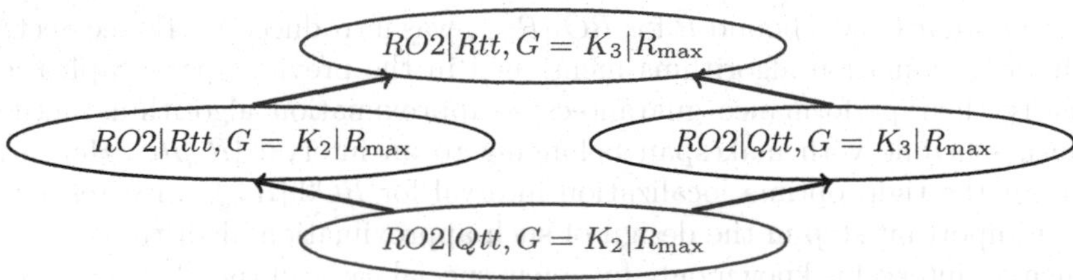

Fig. 1. Reduction graph for routing open shop problems with individual travel times.

First, we show that there exists a series of instances I_ε with $\varepsilon \in (0,1)$ for $RO2|Qtt, G = K_2|R_{\max}$ such that $\lim\limits_{\varepsilon\to 0} \frac{R^*_{\max}(I_\varepsilon)}{\bar{R}(I_\varepsilon)} = \frac{5}{4}$. Second, we constructively prove that for any instance I of $RO2|Rtt, G = K_3|R_{\max}$ there exists a schedule S such that $R_{\max}(S) \in [\bar{R}(I), \frac{5}{4}\bar{R}(I)]$ with the standard lower bound $\bar{R}(I)$ being adapted for individual travel times. Based on the reducibility discussed in the previous paragraph we derive that the interval $[\bar{R}(I), \frac{5}{4}\bar{R}(I)]$ is the tight optima localization interval for all four problem cases considered.

The structure of the paper is as follows. Section 2 contains a formal description of the problem under consideration, the necessary notation, some preliminary results and the description of the series I_ε of instances. In Sect. 3, we provide the proof of the main result for three important special cases. The final proof and the description of the $\frac{5}{4}$-approximation algorithm for the $RO2|Rtt, G = K_3|R_{\max}$ problem, as well as some concluding remarks, are given in Sect. 4.

2 Preliminary Notes

Let us give a formal description of $RO2|Qtt|R_{\max}$ and $RO2|Rtt|R_{\max}$ problems.

Sets $\mathcal{J} = \{J_1, \dots, J_n\}$ of jobs and $\mathcal{M} = \{M_1, M_2\}$ of machines are given. Each job J_j consists of two operations a_j and b_j to be processed by machines M_1 and M_2, respectively, in an arbitrary order. We use the same notation (a_j, b_j) for operations' processing times. An undirected transportation network is described by a graph $G = \langle V, E\rangle$, a node $v_0 \in V$ is referred to as the *depot*. Jobs are distributed among the nodes of G, $\mathcal{J}(v)$ denotes the set of jobs located in the node $v \in V$. (We also use notation $\mathcal{J}(I; v)$ in case we want to specify a problem instance I.) We assume that each non-depot node contains at least one job. Two weights $\mathbf{dist}_1(v, u)$ and $\mathbf{dist}_2(v, u)$ are associated with each edge $e = [v, u] \in E$ and represent travel times over e for M_1 and M_2, respectively. Machines can visit nodes without job processing and can visit each node multiple times. Any number of machines can travel over the same edge simultaneously in any direction. For each machine M_i, travel times $\mathbf{dist}_i(v, u)$ satisfy the triangle inequality. In the case of Qtt, for each $e \in E$, travel times are related by $\mathbf{dist}_2(v, u) = \sigma\mathbf{dist}_1(v, u)$ with σ being the travel speed of machine M_1 (in this case, without loss of generality, we assume that M_2 travels with unit speed). Note that for $RO2||R_{\max}$, travel times of both machines are equal: $\mathbf{dist}_1(v, u) = \mathbf{dist}_2(v, u) = \mathbf{dist}(v, u)$.

Machines are initially located at the depot and have to travel over G to process jobs in arbitrary order. Jobs cannot be processed by two machines simultaneously (however two jobs from the same location can be processed at the same time). The goal is to perform all the operations and to return to the depot as soon as possible. As preemption is not allowed, any schedule S can be described by specifying the starting times $s_{j1}(S)$ and $s_{j2}(S)$ for operations a_j and b_j of each job J_j. The completion times of the operations of job J_j and machine M_i can be defined as $c_{j1}(S) = s_{j1}(S) + a_j$, $c_{j2}(S) = s_{j2}(S) + b_j$. We also use notation $s(a_j), s(b_j), c(a_j), c(b_j)$ if schedule S is specified.

Any feasible schedule S has to agree with the following conditions. If machine M_i processes job $J_j \in \mathcal{J}(v)$ before job $J_{j'} \in \mathcal{J}(u)$, then $s_{j'i}(S) \geq c_{ji}(S) + \mathbf{dist}_i(u, v)$. If job $J_j \in \mathcal{J}(v)$ is the first job processed by machine M_i in schedule S, then $s_{ji}(S) \geq \mathbf{dist}_i(v_0, v)$.

Let job $J_l \in \mathcal{J}(v)$ be the last job processed by machine M_i in schedule S. Then the *return time* of M_i in S is

$$R_i(S) = \max_j c_{ji}(S) + \mathbf{dist}_i(v_0, v) = c_{li}(S) + \mathbf{dist}_i(v_0, v).$$

The goal is to minimize the *makespan* $R_{\max}(S) = \max_i R_i(S)$.

For any problem instance I we use the following notation.

- $\ell_1 = \sum_{j=1}^{n} a_j$ and $\ell_2 = \sum_{j=1}^{n} b_j$ are *machine loads* of M_1, M_2, $\ell_{\max} = \max_i \ell_i$,
- $d_j = a_j + b_j$ is the *job length* of J_j, $d_{\max}(v) = \max_{J_j \in \mathcal{J}(v)} d_j$,
- $\Delta(v) = \sum_{J_j \in \mathcal{J}(v)} d_j$ is the *load* of the node v, $\Delta = \sum_{v \in V} \Delta(v)$,
- T_i^* is the length of an optimal tour on G for machine M_i (TSP optimum for distances $\mathbf{dist}_i$ in G),
- $R^*_{\max}(I)$ is the optimal makespan.

The standard lower bound on the makespan for $RO||R_{\max}$ is described in [1]:

$$R^*_{\max} \geqslant \max\left\{\ell_{\max} + T^*, \max_{v \in V}\big(d_{\max}(v) + 2\mathbf{dist}(v_0, v)\big)\right\}.$$

(In this case $T^* = T_1^* = T_2^*$.)

A similar lower bound for problems with individual travel times was introduced in [4] and has the following form

$$\bar{R} \doteq \max\left\{\max_i(\ell_i + T_i^*), \max_{v \in V}\big(d_{\max}(v) + \mathbf{dist}_1(v_0, v) + \mathbf{dist}_2(v_0, v)\big)\right\}. \quad (1)$$

Note that (1) implies

$$\Delta = \ell_1 + \ell_2 \leqslant 2\bar{R} - T_1^* - T_2^*. \quad (2)$$

Note that $\bar{R}$ can be easily computed in linear time if we do know the optimal solution of the underlying TSPs (otherwise it is not polynomially computable).

We denote the sets of all instances with nonzero standard lower bound for $RO2|G = X|R_{\max}$, $RO2|Qtt, G = X|R_{\max}$ and $RO2|Rtt, G = X|R_{\max}$ by $\mathcal{I}_2^X$, $\mathcal{I}_{Q2}^X$ and $\mathcal{I}_{R2}^X$, respectively. If the graph structure is not restricted, we omit the notation X.

We use the following terminology (consistent with one introduced in [10]).

Definition 1. *A feasible schedule S for an instance $I \in \mathcal{I}_2 \cup \mathcal{I}_{Q2} \cup \mathcal{I}_{R2}$ is* normal *if $R_{\max}(S) = \bar{R}(I)$. An instance I is referred to as* normal *if it admits constructing a normal schedule.*

Definition 2. *The* abnormality *of instance I is $\alpha(I) = \dfrac{R^*_{\max}(I)}{\bar{R}(I)}$.*

The abnormality for some class of instances $\mathcal{K}$ is defined as $\alpha(\mathcal{K}) = \sup_{I \in \mathcal{K}} \alpha(I)$.

Obviously, the tight optima localization interval for a class $\mathcal{K}$ coincides with $[\bar{R}, \alpha(\mathcal{K})\bar{R}]$, therefore the search for such an interval is equivalent to the search for an instance $I \in \mathcal{K}$ with maximal abnormality (if any).

We know that $\alpha\left(\mathcal{I}_2^{K_2}\right) = \alpha\left(\mathcal{I}_2^{K_3}\right) = \frac{6}{5}$ [1,6]. In this paper we prove that

$$\alpha\left(\mathcal{I}_{Q2}^{K_2}\right) = \alpha\left(\mathcal{I}_{Q2}^{K_3}\right) = \alpha\left(\mathcal{I}_{R2}^{K_2}\right) = \alpha\left(\mathcal{I}_{R2}^{K_3}\right) = \frac{5}{4}.$$

Lemma 1. $\alpha\left(\mathcal{I}_{Q2}^{K_2}\right) \geqslant \frac{5}{4}$.

Proof. Consider the following instance $I_\varepsilon \in \mathcal{I}_{Q2}^{K_2}$ for any $0 < \varepsilon \leqslant 1$. The depot v_0 contains a single job J_0, with $a_0 = 0$ and $b_0 = 4$. There are two identical jobs J_1, J_2 in v_1, with $a_1 = a_2 = 4-\varepsilon$ and $b_1 = b_2 = 1$. Machine M_2 travels with unit speed, $\tau_2 = \tau_{01}^2 = 1$. Travel speed of M_1 is $\sigma = \varepsilon^{-1}$, hence $\tau_1 = \tau_{01}^1 = \varepsilon$. Note that $\bar{R}(I_\varepsilon) = 8$. Let us prove that the schedule for I_ε given in Fig. 2 is optimal, and $R^*_{\max}(I_\varepsilon) = 10 - \varepsilon$.

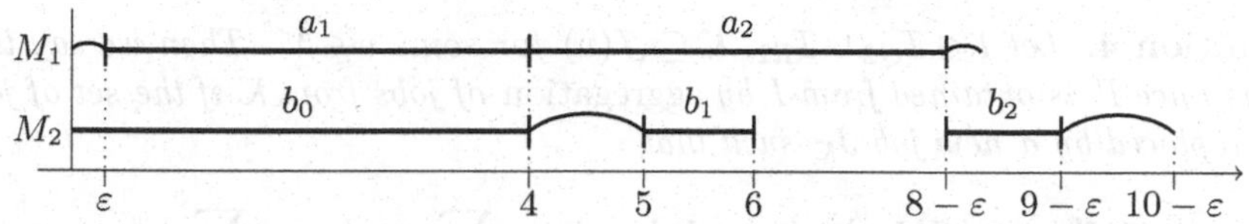

Fig. 2. Optimal schedule for I_ε.

In any optimal schedule, as soon as $R^*_{\max}(I_\varepsilon) \leqslant 10 - \varepsilon$, the machine M_2 travels only once (otherwise $R_2 \geqslant \ell_2 + 4\tau_2 = 10$). Without loss of generality, M_2 performs operations in the order b_0, b_1, b_2: jobs J_1 and J_2 are identical, J_0 cannot be processed between them because the machine makes a single trip, and the order b_2, b_1, b_0 can be transformed into b_0, b_1, b_2 by reversing the direction of time. Therefore, $c(b_1) \geqslant b_0 + \tau_2 + b_1 = 6$. Hence operation a_1 precedes b_1 (otherwise $R_1 \geqslant c(a_1) + \tau_1 \geqslant 10$), a_2 precedes b_2 by similar reasoning and a_1 precedes a_2 (otherwise $c(b_1) \geqslant 9-\varepsilon$ and $R_2 \geqslant 11-\varepsilon$). Therefore, $s(b_2) \geqslant \tau_1 + a_1 + a_2 = 8-\varepsilon$ and $R_2 \geqslant 10 - \varepsilon$, hence $R^*_{\max}(I_\varepsilon) = 10 - \varepsilon$.

Since $\lim_{\varepsilon\to 0} \alpha(I_\varepsilon) = \lim_{\varepsilon\to 0} \dfrac{10-\varepsilon}{8} = \dfrac{5}{4}$, this proves the Lemma. □

Note that for the problem $RO2|G = K_2, Rtt|R_{\max}$ it is sufficient to consider a single instance I_0 (with $\varepsilon = 0$) to confirm that $\alpha\left(\mathcal{I}_{R2}^{K_2}\right) \geqslant \frac{5}{4}$.

The proof of the main result is based on the following job aggregation procedure similar to one described in detail in [6].

2.1 Job Aggregation

Definition 3. *Let $I \in \mathcal{I}_2 \cup \mathcal{I}_{Q2} \cup \mathcal{I}_{R2}$. A node $v \in V$ is called* overloaded *if*

$$\Delta(v) + \mathbf{dist}_1(v_0, v) + \mathbf{dist}_2(v_0, v) > \bar{R},$$

otherwise the node v is underloaded.

Proposition 1. *Any $I \in \mathcal{I}_2 \cup \mathcal{I}_{Q2} \cup \mathcal{I}_{R2}$ contains at most one overloaded node.*

Proof. Suppose we have two overloaded nodes v and u:

$$\Delta(v) + \mathbf{dist}_1(v_0, v) + \mathbf{dist}_2(v_0, v) > \bar{R},\ \Delta(u) + \mathbf{dist}_1(v_0, u) + \mathbf{dist}_2(v_0, u) > \bar{R}.$$

Then we have

$$\begin{aligned}\Delta \geqslant \Delta(v)+\Delta(u) > 2\bar{R} - (\mathbf{dist}_1(v_0, v)+\mathbf{dist}_1(v_0, u)) - (\mathbf{dist}_2(v_0, v)+\mathbf{dist}_2(v_0, u))\\ \geqslant 2\bar{R} - T_1^* - T_2^*,\end{aligned}$$

which contradicts (2). □

Definition 4. *Let $I \in \mathcal{I}_{Q2} \cup \mathcal{I}_{R2}$, $\mathcal{K} \subseteq \mathcal{J}(v)$ for some $v \in V$. Then we say that an instance I' is obtained from I by* aggregation *of jobs from $\mathcal{K}$ if the set of jobs $\mathcal{K}$ is replaced by a new job $J_{\mathcal{K}}$ such that*

$$\mathcal{J}(I'; v) = \mathcal{J}(I; v) \setminus \mathcal{K} \cup \{J_{\mathcal{K}}\},\ a_{\mathcal{K}} = \sum_{J_j \in \mathcal{K}} a_j,\ b_{\mathcal{K}} = \sum_{J_j \in \mathcal{K}} b_j,$$

$$\forall u \neq v\ \mathcal{J}(I'; u) = \mathcal{J}(I; u).$$

The instance $\tilde{I}$ obtained from I by a series of job aggregations will be referred to as a modification *of I.*

Any feasible schedule for some modification of I can be treated as a feasible schedule for I with the same makespan. Therefore, the optimum of any modification of I is greater or equal to $R^*_{\max}(I)$.

Note that machine loads and node loads are preserved by any job aggregation operation, but the standard lower bound $\bar{R}$ might grow if $d_{\mathcal{K}}$ is large enough. In order to preserve $\bar{R}$ we may only aggregate such sets $\mathcal{K}$ that

$$\sum_{J_j \in \mathcal{K}} d_j \leqslant \bar{R} - \mathbf{dist}_1(v_0, v) - \mathbf{dist}_2(v_0, v). \tag{3}$$

Proposition 2. *For every instance $I \in \mathcal{I}_2 \cup \mathcal{I}_{Q2} \cup \mathcal{I}_{R2}$ there exists its modification $\tilde{I}$ such that*

1. *$\bar{R}(\tilde{I}) = \bar{R}(I)$,*
2. *each underloaded node in $\tilde{I}$ contains exactly one job, the only overloaded node (if any) contains at most three jobs.*

Proof. For any underloaded node v, set $\mathcal{K} = \mathcal{J}(v)$ satisfy (3) by Definition 3, and therefore we can aggregate such sets preserving $\bar{R}$. Now we need to prove that all jobs from an overloaded node (if any) can be aggregated into at most three jobs without alteration of $\bar{R}$.

Let v be overloaded and $\mathcal{J}(v) = \{J_1, \ldots, J_p\}$. Let j be the maximal number such that $\sum_{t=1}^{j} d_t \leqslant \bar{R} - \mathbf{dist}_1(v_0, v) - \mathbf{dist}_2(v_0, v)$. Note that $j < p$, as v is overloaded. Lets aggregate all jobs from the set $\mathcal{K} = \{J_1, \ldots, J_j\}$. Due to the choice of j, we have $d_{\mathcal{K}} + d_{j+1} > \bar{R} - \mathbf{dist}_1(v_0, v) - \mathbf{dist}_2(v_0, v)$. Suppose that $j+1 < p$ (otherwise we have two jobs at after the aggregation v and the claim holds). Let $\mathcal{K}' = \{J_{j+2}, \dots, J_p\}$. From (2) we have

$$\sum_{J_t \in \mathcal{K}'} d_t \leqslant \Delta - d_{\mathcal{K}} - d_{j+1} < 2\bar{R} - T_1^* - T_2^* - (\bar{R} - \mathbf{dist}_1(v_0, v) - \mathbf{dist}_2(v_0, v))$$
$$\leqslant \bar{R} - \mathbf{dist}_1(v_0, v) - \mathbf{dist}_2(v_0, v),$$

therefore aggregating the set $\mathcal{K}'$ does not increase $\bar{R}$. Thus, the modification claimed to exist is achieved by aggregating all jobs at each underloaded node, then of jobs in $\mathcal{K}$, and finally in $\mathcal{K}'$. □

Note that, for any instance I, such a modification $\tilde{I}$ can be found in time $O(n)$.

Let $\tilde{I}$ be a modification of I, with $\bar{R}(\tilde{I}) = \bar{R}(I) = \bar{R}$. If there exists a schedule S for $\tilde{I}$, such that $R_{\max}(S) \leqslant \rho\bar{R}$, then $R^*_{\max}(I) \leqslant \rho\bar{R}$. Hence, there exists an instance with maximal abnormality, which has at most three jobs at an overloaded node (if any) due to Proposition 2. It is sufficient to consider only *irreducible* modifications, for which no further reduction (preserving $\bar{R}$) is possible. In the next section we consider all the three possibilities for an irreducible instance:

1. Instance has an overloaded node with exactly three jobs. (Following [7] we refer to such a node as *superoverloaded*.)
2. Instance has an overloaded node with exactly two jobs.
3. Instance has only underloaded nodes.

3 Optima Localization for Irreducible Instances

According to Proposition 2 any irreducible instance contains at most five jobs. We describe schedules for such an instance in the following manner. We specify the order of the operations for each job and each machine. Such an ordering is represented by a weighted digraph referred to as a *scheme* of a schedule. Each node's weight is the corresponding operation's processing time, and arc weights are travel times. For clarity, we add the source S and the sink F (both with zero weight) to each scheme. Denote the set of paths connecting nodes x and y by $\mathcal{P}_{x,y}$. For some path $P \in \mathcal{P}_{x,y}$ its *length* $|P|$ is the total weight of all nodes and arcs belonging to P. The *early completion time* of operation x in the scheme is the maximal length over all the paths from $\mathcal{P}_{S,x}$: $\hat{c}(x) = \max\limits_{P \in \mathcal{P}_{S,x}} |P|$. Any operation's processing in any feasible schedule cannot be completed earlier than its early completion time. For any scheme $\mathcal{H}$ we consider an *early schedule* $S_{\mathcal{H}}$, in which each operation x completes at $\hat{c}(x)$. The makespan of early schedule $R_{\max}(S_{\mathcal{H}}) = \max\limits_{P \in \mathcal{P}_{S,F}} |P| = |\hat{P}|$, there $\hat{P}$ is referred to as a *critical path* in schedule $S_{\mathcal{H}}$.

The weights in $\mathcal{H}$ depend on an instance, so we cannot tell in advance which path is critical. However, one can easily describe the set $\mathcal{P}_{S,F}$ of *complete* paths and consider the length of any $P \in \mathcal{P}_{S,F}$ as a total sum of corresponding variable processing times and travel times. In our analysis we have to consider each *non-trivial* complete path as a candidate for a critical one. We will not consider trivial paths with the total length not exceeding $\bar{R}$ for any problem instance.

For any schedules $S_1, \ldots, S_k$, the best among them is denoted by $S_1 \vee \cdots \vee S_k$.

3.1 One of the Nodes is Superoverloaded

It was proved in [7] that, unless $P = NP$, one cannot test in polynomial time whether a node is superoverloaded. However, if we have an irreducible modification with three jobs in some node, then the node is definitely superoverloaded. The next Theorem concerns exactly that special case.

Theorem 1. *Let $I \in \mathcal{I}_{R2}^{K_3}$ and one of the nodes in I is superoverloaded. Then*

$$R^*_{\max}(I) \leqslant \frac{7}{6}\bar{R}(I).$$

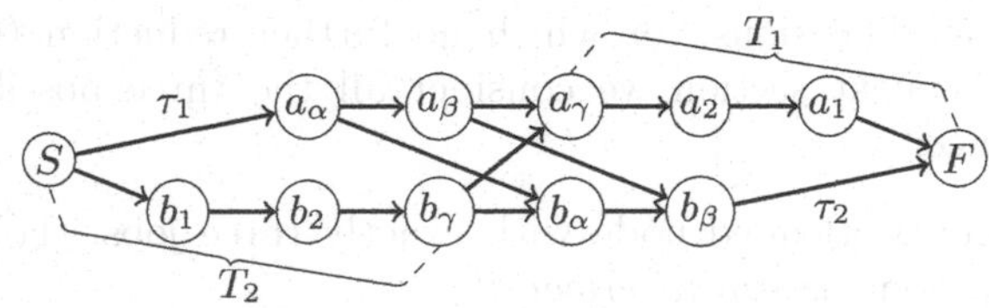

Fig. 3. Scheme $\mathcal{H}'$.

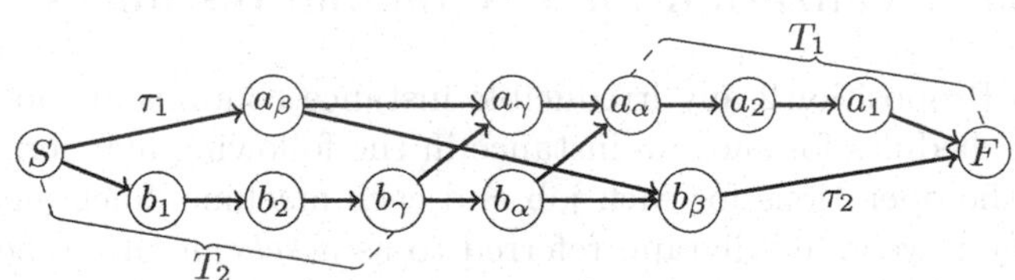

Fig. 4. Scheme $\mathcal{H}''$.

Proof. Apply the job aggregation procedure preserving $\bar{R}$, obtaining a single job in each underloaded node and exactly three jobs $J_\alpha, J_\beta, J_\gamma$ in the superoverloaded node v. Let τ_1 and τ_2 be travel times from the depot to v for M_1 and M_2, respectively. Then for any $p, q \in \{\alpha, \beta, \gamma\}$ with $p \neq q$ we have

$$d_p + d_q > \bar{R} - \tau_1 - \tau_2. \tag{4}$$

Without loss of generality, we assume

$$a_\alpha + \tau_1 = \min\bigl\{\min\{a_\alpha, a_\beta, a_\gamma\} + \tau_1, \min\{b_\alpha, b_\beta, b_\gamma\} + \tau_2\bigr\}. \tag{5}$$

Under this assumption

$$a_\alpha \leqslant \frac{1}{3}\ell_1 \leqslant \frac{1}{3}\bar{R}. \tag{6}$$

Denote single jobs in underloaded nodes by J_1 and J_2, where J_1 belongs to the depot in the case $v \neq v_0$.

Consider an early schedule $S_1 = S_{\mathcal{H}'}$ (see Fig. 3).

Note that the travel time of M_2 along the path from S to b_γ is exactly $T_2 = T_2^* - \tau_2 \geqslant \tau_2$, and the travel time of M_1 from a_γ to F is $T_1 = T_1^* - \tau_1$. Hence, by (5), $c(b_\gamma) \geqslant c(a_\alpha)$. Let us prove that $R_1 \leqslant \bar{R}$. Indeed,

$$R_1 = \max\{c(a_\beta), c(b_\gamma)\} + a_\gamma + a_2 + a_1 + T_1 = \max\{\ell_1 + T_1^*, d_1 + d_2 + d_\gamma + T_2 + T_1\}.$$

Using (1), (2) and (4) we have $R_1 \leqslant \max\{\bar{R}, \Delta + T_1^* + T_2^* - (d_\alpha + d_\beta + \tau_1 + \tau_2)\} \leqslant \bar{R}$. If S_1 is not normal, then

$$R_{\max}(S_1) = R_2 = \tau_1 + a_\alpha + a_\beta + b_\beta + \tau_2. \tag{7}$$

In this case consider an early schedule $S_2 = S_{\mathcal{H}''}$ (Fig. 4).

If S_2 is not normal, then

$$R_{\max}(S_2) = T_2 + b_1 + b_2 + b_\gamma + \max\{b_\alpha, a_\gamma\} + a_\alpha + a_2 + a_1 + T_1. \tag{8}$$

Let $S = S_1 \vee S_2$. Then, by (7), (8) and (6),

$$2R_{\max}(S) \leqslant R_{\max}(S_1) + R_{\max}(S_2) = \ell_1 + T_1^* + \ell_2 + T_2^* - \min\{b_\alpha, a_\gamma\} + a_\alpha \leqslant \frac{7}{3}\bar{R},$$

therefore $R_{\max}(S) \leqslant \frac{7}{6}\bar{R}$.

Note that, according to Proposition 2, such a schedule can be constructed in time $O(n)$.

According to Lemma 1 and Theorem 1, an instance $I \in \mathcal{I}_{R2}^{K_3}$ with maximal abnormality does not have a superoverloaded node (although one of the nodes can be overloaded). Those cases are considered in the next two subsections.

3.2 One of the Nodes is Overloaded

The next lemma is concerned with the case when the irreducible instance contains exactly two jobs in an overloaded node. Although, in this case, that node might still be superoverloaded in the initial instance (and this can be shown by some other reduction), this need not hold in general. The following lemma completes the consideration of the general case with an overloaded node.

Lemma 2. *Let an instance $I \in \mathcal{I}_{R2}^{K_3}$ have a single job at each node except v, which is overloaded and contains two jobs. Then there exists a feasible schedule S for I such that $R_{\max}(S) \leqslant \frac{5}{4}\bar{R}$.*

Proof. We use notation similar to that in the proof of Theorem 1. Let $\mathcal{J}(v) = \{J_\alpha, J_\beta\}$, and let τ_1 and τ_2 be travel times from the depot to v for M_1 and M_2, respectively. The remaining travel time for each machine is denoted by $T_i = T_i^* - \tau_i$, for $i = 1, 2$. Denote single jobs from the underloaded nodes by J_1, J_2, where J_1 belongs to the depot in the case $v \neq v_0$.

Since v is overloaded, we have $d_\alpha + d_\beta > \bar{R} - \tau_1 - \tau_2$. Together with (2) that implies

$$d_1 + d_2 + T_1 + T_2 < \bar{R}. \tag{9}$$

Consider schedules $S_1 = S_{\mathcal{H}_1}$, $S_2 = S_{\mathcal{H}_2}$, $S_3 = S_{\mathcal{H}_3}$, and $S_4 = S_{\mathcal{H}_4}$ (Fig. 5).

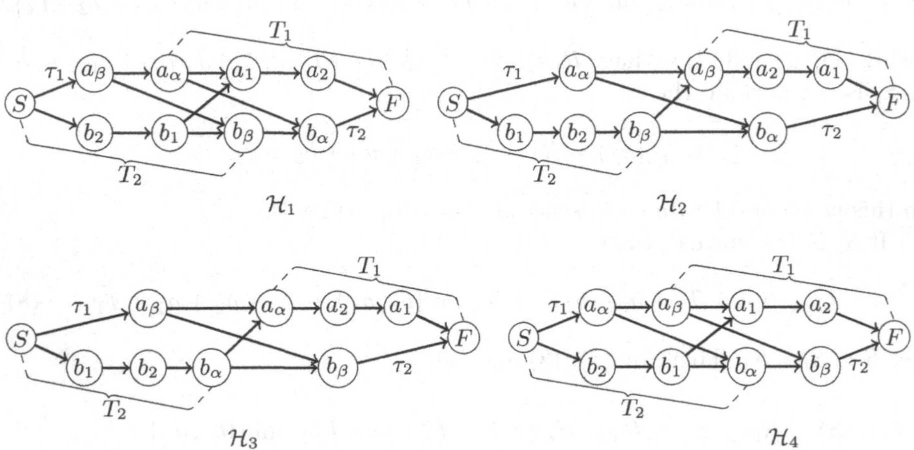

Fig. 5. Schemes $\mathcal{H}_1$, $\mathcal{H}_2$, $\mathcal{H}_3$ and $\mathcal{H}_4$ for the case with overloaded node.

Note that both $\mathcal{H}_1$ and $\mathcal{H}_4$ contain a complete path $S \to b_2 \to b_1 \to a_1 \to a_2 \to F$ of length not greater than $d_1 + d_2 + T_1 + T_2$; by (9) its length is less than $\bar{R}$ and it is therefore not critical. Schemes $\mathcal{H}_2$ and $\mathcal{H}_3$ have a single non-trivial complete path each. Assuming that each of these schedules is not normal, we obtain

$$\begin{aligned} R_{\max}(S_1) &= \tau_1 + \tau_2 + a_\beta + b_\alpha + \max\{a_\alpha, b_\beta\}, \\ R_{\max}(S_2) &= b_1 + b_2 + b_\beta + a_\beta + a_2 + a_1 + T_1 + T_2, \\ R_{\max}(S_3) &= b_1 + b_2 + b_\alpha + a_\alpha + a_2 + a_1 + T_1 + T_2, \\ R_{\max}(S_4) &= \tau_1 + \tau_2 + a_\alpha + b_\beta + \max\{a_\beta, b_\alpha\}. \end{aligned}$$

For the schedule $S = S_1 \vee S_2 \vee S_3 \vee S_4$ we have

$$4R_{\max}(S) \leqslant R_{\max}(S_1) + R_{\max}(S_2) + R_{\max}(S_3) + R_{\max}(S_4) =$$
$$2(\ell_1 + T_1^* + \ell_2 + T_2^*) + (\max\{a_\alpha, b_\beta\} + \max\{a_\beta, b_\alpha\}) \leqslant 4\bar{R} + \max\{\ell_1, \ell_2, d_\alpha, d_\beta\} \leqslant 5\bar{R},$$

therefore $R_{\max}(S) \leqslant \frac{5}{4}\bar{R}$.

3.3 Each Node is Underloaded

Lemma 3. *Let an instance $I \in \mathcal{I}_{R2}^{K_3}$ contain a single job at each node. Then there exists a feasible schedule S for I such that $R_{\max}(S) \leqslant \frac{5}{4}\bar{R}(I)$.*

Proof. Let $\mathcal{J}(v_k) = \{J_k\}$, $k = 0, 1, 2$. We also use the following notation for travel times:

$$\tau_i = \mathbf{dist}_i(v_0, v_1), \nu_i = \mathbf{dist}_i(v_0, v_2), \mu_i = \mathbf{dist}_i(v_1, v_2), T_i^* = \tau_i + \mu_i + \nu_i, \ i = 1, 2.$$

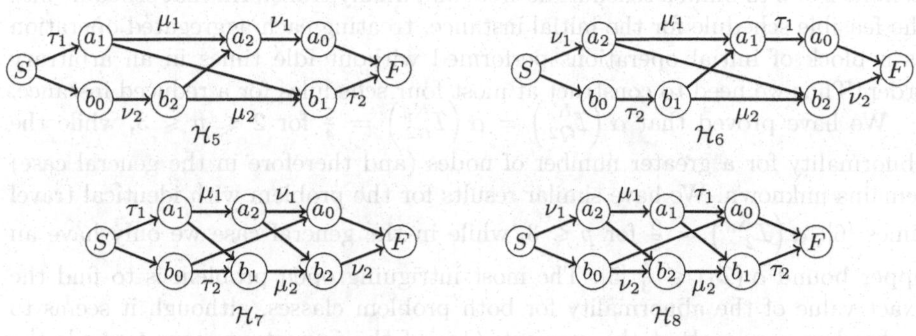

Fig. 6. Schemes $\mathcal{H}_5$, $\mathcal{H}_6$, $\mathcal{H}_7$ and $\mathcal{H}_8$ for the case with underloaded nodes.

Consider schedules $S_1 = S_{\mathcal{H}_5}$, $S_2 = S_{\mathcal{H}_6}$, $S_3 = S_{\mathcal{H}_7}$, and $S_4 = S_{\mathcal{H}_8}$ (Fig. 6). Assuming that none of these schedules is normal, we have

$$R_{\max}(S_1) = b_0 + b_2 + a_2 + a_0 + \nu_1 + \nu_2,$$

$$R_{\max}(S_2) = b_0 + b_1 + a_1 + a_0 + \tau_1 + \tau_2,$$

$$R_{\max}(S_3) = \tau_1 + \nu_2 + a_1 + b_2 + \max\{\mu_1 + a_2, \mu_2 + b_1\},$$

$$R_{\max}(S_4) = \nu_1 + \tau_2 + a_2 + b_1 + \max\{\mu_1 + a_1, \mu_2 + b_2\}.$$

For $S = S_1 \vee S_2 \vee S_3 \vee S_4$ we have

$$4R_{\max}(S) \leqslant R_{\max}(S_1) + R_{\max}(S_2) + R_{\max}(S_3) + R_{\max}(S_4)$$
$$\leqslant 2\ell_1 + 2\ell_2 + 2(\tau_1 + \tau_2 + \nu_1 + \nu_2 + \max\{\mu_1, \mu_2\}) + (\max\{a_2, b_1\} + \max\{a_1, b_2\})$$
$$\leqslant 2(\ell_1 + T_1^*) + 2(\ell_2 + T_2^*) + \max\{\ell_1, \ell_2, d_1, d_2\} \leqslant 5\bar{R},$$

therefore $R_{\max}(S) \leqslant \frac{5}{4}\bar{R}$.

4 Conclusion

Proposition 2, Theorem 1, Lemmas 2 and 3 imply the following

Theorem 2. *For any instance* $I \in \mathcal{I}_{R2}^{K_3}$ *there exists a feasible schedule* S *with makespan in the interval* $[\bar{R}, \frac{5}{4}\bar{R}]$. *Such a schedule can be built in linear time.*

Indeed, we just need to perform the job aggregation procedure to obtain an irreducible modification $\tilde{I}$ discussed in Sect. 2. If $\tilde{I}$ has a node with exactly three jobs, then we build a schedule according to the proof of Theorem 1, with makespan not exceeding $\frac{7}{6}\bar{R}$. In other cases, we use the proof of the corresponding Lemma 2 or 3 to build a schedule desired, and finally transform that schedule into the feasible schedule for the initial instance, treating each aggregated operation as a block of initial operations performed without idle times in an arbitrary order. Thus we need to construct at most four schedules for a reduced instance.

We have proved that $\alpha\left(\mathcal{I}_{Q2}^{K_p}\right) = \alpha\left(\mathcal{I}_{R2}^{K_p}\right) = \frac{5}{4}$ for $2 \leqslant p \leqslant 3$, while the abnormality for a greater number of nodes (and therefore in the general case) remains unknown. We have similar results for the problem with identical travel times [6]: $\alpha\left(\mathcal{I}_{2}^{K_p}\right) = \frac{6}{5}$ for $p \leqslant 3$, while in the general case we only have an upper bound $\alpha(\mathcal{I}_2) \leqslant \frac{4}{3}$ [5]. The most intriguing open problem is to find the exact value of the abnormality for both problem classes, although it seems to be hard to approach at the moment. One of the important steps towards the solution of that problem is to find an instance with a greater abnormality (if any). We have no knowledge of an existence of such an instance and would like to propose the following

Conjecture 1. $\alpha(\mathcal{I}_2) = \frac{6}{5}$, $\alpha(\mathcal{I}_{Q2}) = \alpha(\mathcal{I}_{R2}) = \frac{5}{4}$.

We suggest the following open questions for future research.

Question 1. If Conjecture 1 is not true, what is the smallest number of nodes p such that $\alpha\left(\mathcal{I}_{2}^{K_p}\right) > \frac{6}{5}$? $\alpha\left(\mathcal{I}_{Q2}^{K_p}\right) > \frac{5}{4}$? $\alpha\left(\mathcal{I}_{R2}^{K_p}\right) > \frac{5}{4}$? $\alpha\left(\mathcal{I}_{Q2}^{K_p}\right) < \alpha\left(\mathcal{I}_{R2}^{K_p}\right)$?

This question is a significant motivation for studying special cases with a small transportation network ([2,6] and this paper).

Question 2. It was proved in [7] that any instance of $RO2|G = K_2|R_{\max}$ with a superoverloaded node is normal. This result can be easily extended for $G = K_3$. The claim of Theorem 1 is significantly weaker (although sufficient for our purposes). The question is, does a similar result on the normality of an instance with a superoverloaded node hold for $RO2|Qtt, G = K_3|R_{\max}$ and $RO2|Rtt, G = K_3|R_{\max}$?

Question 3. It is known [1] that $\alpha\left(\mathcal{I}_{2}^{K_2}\right) = \frac{6}{5}$, although different travel times can increase that value up to $\frac{5}{4}$ (Lemma 1). We probably need an unbounded machine's travel speed σ_1 to achieve that abnormality. Suppose we have an instance $I \in \mathcal{I}_{Q2}^{K_2}$ with travel speeds $\sigma_2 = 1$ and $\sigma_1 \in [1, x]$. What is the maximal abnormality of such an instance as a function of x?

References

1. Averbakh, I., Berman, O., Chernykh, I.: A 6/5-approximation algorithm for the two-machine routing open shop problem on a 2-node network. Eur. J. Oper. Res. **166**(1), 3–24 (2005). https://doi.org/10.1016/j.ejor.2003.06.050
2. Averbakh, I., Berman, O., Chernykh, I.: The routing open-shop problem on a network: complexity and approximation. Eur. J. Oper. Res. **173**(2), 521–539 (2006). https://doi.org/10.1016/j.ejor.2005.01.034
3. Brucker, P., Knust, S., Edwin Cheng, T.C., Shakhlevich, N.: Complexity results for flow-shop and open-shop scheduling problems with transportation delays. Ann. Oper. Res. **129**, 81–106 (2004). https://doi.org/10.1023/b:anor.0000030683.64615.c8
4. Chernykh, I.: Routing open shop with unrelated travel times. In: Kochetov, Y., Khachay, M., Beresnev, V., Nurminski, E., Pardalos, P. (eds.) DOOR 2016. LNCS, vol. 9869, pp. 272–283. Springer, Cham (2016). https://doi.org/10.1007/978-3-319-44914-2_22
5. Chernykh, I., Kononov, A., Sevastyanov, S.: Efficient approximation algorithms for the routing open shop problem. Comput. Oper. Res. **40**(3), 841–847 (2013). https://doi.org/10.1016/j.cor.2012.01.006
6. Chernykh, I., Lgotina, E.: The 2-machine routing open shop on a triangular transportation network. In: Kochetov, Y., Khachay, M., Beresnev, V., Nurminski, E., Pardalos, P. (eds.) DOOR 2016. LNCS, vol. 9869, pp. 284–297. Springer, Cham (2016). https://doi.org/10.1007/978-3-319-44914-2_23
7. Chernykh, I., Pyatkin, A.: Refinement of the optima localization for the two-machine routing open shop. In: Proceedings of the 8th International Conference on Optimization and Applications (OPTIMA 2017), vol. 1987, pp. 131–138. CEUR Workshop Proceedings (2017)
8. Christofides, N.: Worst-case analysis of a new heuristic for the travelling salesman problem. Report 388, Graduate School of Industrial Administration, Carnegie-Mellon University, Pittsburg, PA (1976)
9. Gonzalez, T., Sahni, S.: Open shop scheduling to minimize finish time. J. Assoc. Comput. Mach. **23**, 665–679 (1976). https://doi.org/10.1145/321978.321985
10. Kononov, A., Sevastianov, S., Tchernykh, I.: When difference in machine loads leads to efficient scheduling in open shops. Ann. Oper. Res. **92**, 211–239 (1999). https://doi.org/10.1023/a:1018986731638
11. Kononov, A.: On the routing open shop problem with two machines on a two-vertex network. J. Appl. Ind. Math. **6**(3), 318–331 (2012). https://doi.org/10.1134/s1990478912030064
12. Lawler, E.L., Lenstra, J.K., Kan, A.H.G.R., Shmoys, G.B.: Sequencing and scheduling: algorithms and complexity. In: Graves, S.S., Rinnooy-Kan, A.H.G., Zipkin, P. (eds.) Logistics of Production and Inventory. Elsevier, Amsterdam (1993)
13. Serdyukov, A.: On some extremal routes in graphs. Upravlyaemye Sistemy **17**, 76–79 (1978). (in Russian)
14. Sevastianov, S.V., Tchernykh, I.D.: Computer-aided way to prove theorems in scheduling. In: Bilardi, G., Italiano, G.F., Pietracaprina, A., Pucci, G. (eds.) ESA 1998. LNCS, vol. 1461, pp. 502–513. Springer, Heidelberg (1998). https://doi.org/10.1007/3-540-68530-8_42

Inland Waterway Efficiency Through Skipper Collaboration and Joint Speed Optimization

Christof Defryn[1], Julian Golak[1], Alexander Grigoriev[1](✉), and Veerle Timmermans[2]

[1] Maastricht University School of Business and Economics, P.O.Box 616, 6200 MD Maastricht, The Netherlands
{c.defryn,j.golak,a.grigoriev}@maastrichtuniversity.nl

[2] RWTH Aachen, Department of Management Science, Kackertstrasse 7, 52072 Aachen, Germany
veerle.tantimmermans@oms.rwth-aachen.de

Abstract. We address the problem of minimizing the aggregated fuel consumption by the vessels in an inland waterway (a river) with a single lock. The fuel consumption of a vessel depends on its velocity and the slower it moves, the less fuel it consumes. Given entry times of the vessels into the waterway and the deadlines before which they need to leave the waterway, we decide on optimal velocities of the vessels that minimize their private fuel consumption. Presence of the lock and possible congestions on the waterway make the problem computationally challenging. First, we prove that in general Nash equilibria might not exist, i.e., if there is no supervision on the vessels velocities, there might not exist a strategy profile from which no vessel can unilaterally deviate to decrease its private fuel consumption. Next, we introduce simple supervision methods to guarantee existence of Nash equilibria. Unfortunately, though a Nash equilibrium can be computed, the aggregated fuel consumption of such a stable solution is high compared to the consumption in a social optimum, where the total fuel consumption is minimized. Therefore, we propose a mechanism involving payments between vessels, guaranteeing Nash equilibria while minimizing the fuel consumption. This mechanism is studied for both the offline setting, where all information is known beforehand, and online setting, where we only know the entry time and deadline of a vessel when it enters the waterway.

Keywords: Lock scheduling · Congestions · Social welfare · Mechanism design · Online scheduling

1 Introduction

The high fuel prices, a congested road network and the increasing demand for transport due to globalization put a high pressure on the existing transportation

M. Khachay et al. (Eds.): MOTOR 2019, LNCS 11548, pp. 202–217, 2019.
https://doi.org/10.1007/978-3-030-22629-9_15

network, especially road transport. The growing sense of resource scarcity and climate change motivates companies to rethink their logistical operations and, if possible shift towards a more sustainable transport mode. In comparison to other transportation modes, the use of barges is more sustainable (less greenhouse gas emission) and relatively cheap (due to economies of scale). Moreover, as a single barge can replace over 100 trucks, increased use of the water network is likely to reduce congestion and the number of accidents on the road network. The Netherlands, located around the mouth of multiple important European rivers, has a dense network of over 4600 km of navigable inland waterways [2], on which 36% of all freight transport (in tonne-kilometre) takes place [3].

Besides longer travel times, mainly due to the relatively low density of the network, the high uncertainty in arrival time is one the major drawbacks of freight transport over inland waterways. This uncertainty is caused by the presence of many river obstacles, such as low bridges, narrow river segments, harbors and locks, which gives rise to unexpected congestion and waiting time. This requires the skipper, the person in charge of the boat, to increase the speed afterwards to guarantee an on-time arrival at the destination. However, the operational cost for the skipper is largely determined by the fuel consumption, which is related directly to the required power and, therefore, the speed of the vessel. The required speeding up results therefore in a direct increase of operational costs for the skipper.

In this paper, we investigate how coordination and scheduling of all movement around these river obstacles can help to reduce congestion and waiting times, and therefore increase the efficiency of inland waterway transport. Moreover, by optimizing a recommended speed for each barge between two consecutive obstacles, one can control the arrival times of the vessels at each obstacle, guaranteeing the minimal throughput time and at the same time the minimal total fuel consumption. For a single lock, the strategy of reducing the speed of the vessel to avoid waiting time has resulted in significant economic benefits [14].

2 Literature Review

Lock Scheduling. Existing research on the optimization of river obstacles is mainly focused on lock scheduling. In a single lock scheduling problem, the operating times of a single lock are optimized for a set of vessels with given arriving time at the lock. By batching the vessels together and determining the optimal service time for each batch, the goal is to reduce overall waiting time at the lock.

Passchyn et al. [8] provide a polynomial time algorithm to optimally solve the single lock scheduling problem, given the arrival times of the boats and the capacity of the lock. Passchyn, Briskorn and Spieksma present in [6] a complexity analysis of this problem and provide a polynomial time algorithm that applies to special cases for the single lock scheduling problem with multiple parallel chambers. The problem of physically placing vessels inside the chamber of the lock has been addressed by Verstichel et al. [16,17]. The joint optimization of multiple sequential locks on the river is considered by Passchyn et al. [7] and Prandtstetter et al. [10]. Here, Prandtstetter et al. propose a Variable Neighborhood Search for solving the problem. Passchyn et al. propose an MILP to find an exact solution, which is also used in the current work. In all the contributions above the vessel speeds and arrival times in the river segment are deterministic and given. In an optimal lock schedule the aim is always to minimize the aggregated fuel cost or emissions, and selfish behavior of skippers is not addressed.

There are also multiple case studies conducted for the lock scheduling problem, focused on specific lock sequences on important waterways in the world. Petersen et al. [9] consider the *Welland Canal* in North America for which they provide a heuristic that employs optimal dynamic programming submodels for scheduling individual locks in order to determine operating schedules for the lock sequence. Smith et al. [13] present a simulation model to evaluate the quality of different heuristics on lock operations on the *Upper Mississippi River* in the US. This research has been extended by Smit et al. [12]. Here, the authors propose a MIP model to solve the lock scheduling problem with sequence-dependent setup- and processing times. On the same river segment, Nauss [5] incorporated the malfunctioning of locks in order to efficiently resolve a queue of vessels that might arise due to the malfunctioning. Also, a model for the lock scheduling problem with multiple parallel chambers for this river layout has been investigated by Ting and Schonfeld [15]. Finally, the *Kiel Canal* is considered by Günter, Lübbecke and Möring [4]. They incorporate collision of ships in their model and provide a heuristic to determine a routing and scheduling to fleet of ships in a collision-free manner.

In contrast to the previous literature, only Passchyn et al. [7] take into account that skippers can choose the speed of their boat, and hence influence the time in which they arrive at the lock. They minimize overall CO_2 emissions by optimizing the speed at which vessels have to approach the locks using a MILP formulation. Although this approach is closely related to the problem addressed in this paper, the authors of [7] focus on minimizing the aggregated emissions without considering the fact that each skipper is mainly interested in minimizing his personal fuel cost and emissions. As a consequence, skippers might deviate from the proposed solution and increase their individual utility. In this paper, we view this problem from a game-theoretic point of view, and propose a schedule in which no skipper can profitably deviate from the proposed solution.

Fuel Reduction. Academic literature on fuel savings has been extensive in the context of ocean vessels. We refer to [11] for a more detailed survey. Though, inland waterways are significantly different compared to the ocean, as there are no 'river' obstacles in the ocean. Research on fuel consumption in inland waterways is sparse. Ting et al. [14] found that the strategy of reducing vessels speed to avoid idle time has resulted in significant economic benefits for a single lock. This may been seen as a key observation for the motivation of the current work. The fact that fuel consumption grows non-linearly in the vehicle's speed is corroborated by Bialystockia and Konovessis [1].

Our Contributions. Previous research on the lock scheduling is based on the assumption that lock operators have the full power to determine the operating schedule for the lock. In practice, this schedule is typically determined using the first come first serve (FIFO) principle based on the order at which vessels arrive at the lock. Skippers that know this have the incentive to speed up when approaching a lock in order to pass their predecessors and get served first. This action leads to longer waiting times before the locks, and increases the operational cost for these skippers due to the higher fuel consumption that is caused by maintaining a higher speed.

In this paper, we aim to minimize the aggregated fuel consumption by the vessels in the river, while keeping in mind that each skipper is a rational individual with the sole goal of minimizing his personal fuel cost or emissions. In the solutions we present, we determine an optimal speed for each individual boat and for each river segment. The positive relation between vessel speed and fuel consumption leads to the observation that maintaining the slowest speed—yet meeting the arrival deadline at the destination harbour—minimizes the total fuel consumption of a single vessel. Unfortunately, even a single lock on the river becomes a source of congestion and the speeds of the vessels have to be adjusted accordingly.

The paper is structured as follows. In Sect. 3, we model the problem as a non-cooperative game and discuss a variety of priority rules that can be used by the lock operators in case multiple vessels approach the lock (possibly in the opposite directions). Moreover, we discuss the existence of *Nash equilibria*—situations in which no skipper can unilaterally deviate from the proposed solution and decrease its individual cost. In Sect. 4, we introduce a *cooperative game* perspective on the traffic optimization problem at hand. We assume that binding contracts between different skippers are possible and propose a mechanism based on monetary payments. This situation will give rise to new Nash equilibria. We design an algorithm that computes these Nash equilibria while minimizing total fuel consumption on the river. Finally, in Sect. 5, we extend this algorithm to comply with an online setting.

3 Non-cooperative Game for Traffic Optimization at River Obstacles

3.1 Mathematical Notation of the System

Without loss of generality, we assume a waterway with a single lock L. Let this lock be defined by its capacity C, i.e., the number of boats that can be leveled up or down simultaneously, and its current state P, indicating whether the level of the water is high (equal to the upstream level) or low (equal to the downstream level). Let T be the time to change the lock state from high to low or vice versa. If a batch of vessels is processed, an additional T_i times units are required for each vessel i in the batch. That time represents the loading and unloading of vessels and varies across different types and sizes of vessels [13].

The total processing time of a batch of vessels is the sum of lockage time T and the individual processing times T_i for every vessel i in the batch. Moreover, let L_u and L_d be the distances between the upstream and downstream end points of the waterway respectively and the lock. From the moment that a vessel is within that distance from the lock, we consider it to be in the system. The complete system is, therefore, determined by the tuple $L = \{C, P, T, L_u, L_d\}$.

Now, let U and D be sets of vessels, that sail upstream or downstream respectively and let $S = U \cup D$ be the set of all vessels. Let $n = |S|$ be the size of the entire fleet. For each vessel $i \in U$, we are given an arrival time at the upstream end point of the river, denoted by a_i, and a deadline d_i, the latest time when the vessel has to reach the downstream end point of the waterway. Similarly, a_j and d_j are defined for each vessel $j \in D$, sailing in the opposite direction. Furthermore, we assume that vessels in set S are ordered according to their arrival times and that between any two sequential vessel arrivals at least ε time elapses. Finally, let $v_{i,p}$ denote the speed of vessel i along river segment $p \in \{u, d\}$, where u and d represent the upstream and downstream segments respectively. We assume the minimum and the maximum speed for any vessel is bounded by $v_{\min}$ and $v_{\max}$.

3.2 Model Definition

In the game, each vessel $i \in U \cup D$ decides on $v_{i,d}$ and $v_{i,u} \in [v_{\min}, v_{\max}]$, such that $v_i = (v_{i,d}, v_{i,u})$. Furthermore, let v_{-i} denote the strategy profile of every player in the game except for i and let $\mathbf{v} = (v_i, v_{-i})$. Note that only constant speeds have been specified for both, upstream and downstream, waterway segments. Due to the convexity of the cost function, defined below, skippers will have no incentive to alter their speed midway of the segments. The assumption of constant speeds is relaxed, when an online setting of the game is considered, in Sect. 5. To illustrate the game, consider the following example.

Example 1. Assume three vessels (see also Fig. 1): 1 and 2 sailing upstream and 3 sailing downstream. The waterway is 20 km long, and the lock is placed in the middle of the waterway. As a result, $L_u = L_d = 10$. The lock has an infinite capacity and $T = T_1 = T_2 = T_3 = 0.5$. The entry/arrival times of the vessels are as follows: $a_1 = 0$, $a_2 = \varepsilon$ and $a_3 = 2\varepsilon$. Moreover, we know that $(v_{1,u}, v_{1,d}) = (5, 5)$, $(v_{2,u}, v_{2,d}) = (10, 5)$ and $(v_{3,u}, v_{3,d}) = (5, 10)$. Given the current speeds, vessel 1 arrives at the lock at time 2, vessel 2 at time $1+\varepsilon$ and vessel 3 is expected to arrive at the lock at time $1 + 2\varepsilon$.

The total fuel consumption is given by the function $E(v)$, where v represents the speed of the vessel. The function is measured in tons per kilometer. We assume that fuel consumption is equal to zero if the vessel is not moving, i.e., its speed is equal to zero, and vessels are only standing still inside the lock. Following the conventions from the related literature, we assume convexity of $E(v)$, $v > 0$ (see [7]).

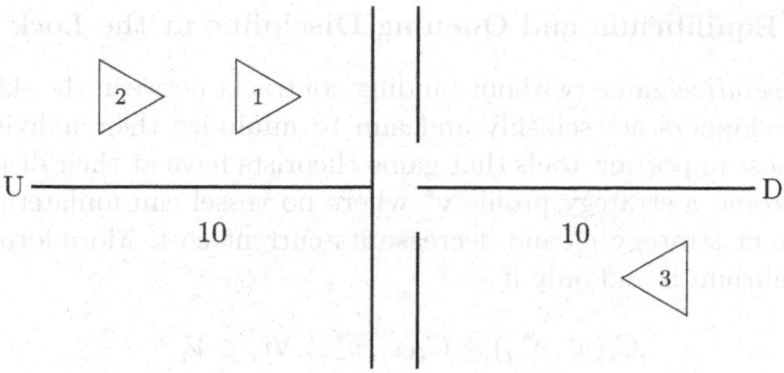

Fig. 1. The setup of locks and vessel for Example 1

To further simplify notations, and without loss of generalization, we consider the fuel consumption function to be the same for every vessel and equal to

$$E_i(v_i) = L_u E(v_{i,u}) + L_d E(v_{i,d}). \tag{1}$$

The fuel consumption of the entire fleet can therefore be written as

$$E_{tot}(\mathbf{v}) = \sum_{i \in S} E_i(v_i). \tag{2}$$

Each skipper i aims to minimize its total fuel consumption $E_i(v_i)$, given its deadline (denoted as d_i) on the arrival time at the destination. This is considered a hard constraint. Arriving at the destination after the predefined deadline is considered infeasible, represented by an infinite penalty cost. In case the deadline is unrestrictive for the vessel, it will sail at the minimum speed $v_{\min}$. Therefore, we define the cost function for skipper $i \in S$ by

$$C_i(\mathbf{v}) = \begin{cases} E_i(v_i) & \text{if } a_i + L_u/v_{i,u} + L_d/v_{i,d} + q_i(\mathbf{v}) \leq d_i; \\ \infty & \text{otherwise ,} \end{cases} \tag{3}$$

where $q_i(\mathbf{v})$ is the total processing time of vessel i at the lock, i.e., waiting time before entering the lock plus the lock re-level time T and the individual loading times. This waiting time depends on the congestion induced by the strategy profile, i.e., individual speeds of all vessels in the system.

We now define the *social cost* $C(\mathbf{v})$ of a strategy profile $\mathbf{v}$ as the aggregated cost of all players, defined as

$$C(\mathbf{v}) = \sum_{i \in S} C_i(\mathbf{v}). \tag{4}$$

The strategy profile $\mathbf{v}$ that minimizes the social cost is called the *social optimum*, and has a social cost of

$$C_{opt} = \min_{\mathbf{v}} C(\mathbf{v}). \tag{5}$$

3.3 Nash Equilibrium and Queuing Discipline at the Lock

In a *non-cooperative* game (without binding contracts between the skippers), we assume that skippers act selfishly and aim to minimize their individual costs. One of the most important tools that game theorists have at their disposal is the *Nash equilibrium*: a strategy profile $\mathbf{v}^*$ where no vessel can unilaterally deviate from its current strategy v_i^* and decrease its current cost. More formally, $\mathbf{v}^*$ is a Nash equilibrium if and only if

$$C_i(v_i^*, v_{-i}^*) \leq C_i(v_i, v_{-i}^*), \forall v_i \in \mathcal{V}_i. \tag{6}$$

The importance of the Nash equilibrium comes from the natural observation that agents/players/skippers are rather interested in selfishly minimizing their individual costs than reducing the social cost, i.e., the total cost of the entire fleet. The Nash equilibrium is calculated by minimizing the regret of the individual players, where regret is defined as the cost they could have saved by altering the strategy.

The existence of the Nash Equilibrium is dependent on the waiting time of vessels in front of the locks. In turn, this waiting time is subject to the *queuing discipline* of the lock. This queuing discipline dictates the order in which vessels are served by the lock operator. As the waiting time impacts the optimal (required) speed after the lock, the queuing discipline directly affects the cost of each skipper. Therefore, different lock mechanisms yield different characteristics of the game. We consider the following three simple lock mechanisms:

Mechanism 1: Lock FIFO. For any $i, j \in U \cup D$, vessel i is served by the lock before vessel j if i arrives at the lock before j. If vessels i and j arrive at the lock at the same time, i will be served first if $a_i < a_j$.

Mechanism 2: System FIFO. For any $i, j \in U \cup D$, vessel i is served by the lock before vessel j if $a_i < a_j$.

Mechanism 3: System FIFO with filling idle time. Consider vessel $i \in U \cup D$. Assume that skippers choose strategies sequentially and all $(v_j)_{j=1,\ldots,i-1}$ are given. For any $i, j \in U \cup D$ such that $j < i$, vessel i is served before j if it does not affect the time of departure of vessel j determined by the strategy profile $(v_j)_{j=1,\ldots,i-1}$.

The following example illustrates how these three mechanisms work and how they affect the payoff of a strategy profile.

Example 2. Consider again the setup of Example 1. Let us remind that the entry/arrival times of the vessels were $a_1 = 0$, $a_2 = \varepsilon$ and $a_3 = 2\varepsilon$. Furthermore, given the current speeds of the vessels, the arrival times at the locks are 2, $1 + \varepsilon$ and $1 + 2\varepsilon$, for vessels 1, 2 and 3, respectively.

First, if the lock operates under Mechanism 1, only the arrival times at the lock are relevant. Note that vessel 2 arrives at the lock first, vessel 3 second and vessel 1 is the last one. As vessels are processed in order of arrival time, the waiting times under the strategy profile are $2+\varepsilon, 1, 2-\varepsilon$ for vessel 1, 2 and 3 respectively.

Second, under mechanism 2, only the arrival times into the system are relevant. Note that vessel 1 arrives first in the system, vessel 2 second and vessel 3 last. The waiting times are $1, 2-\varepsilon, 3-2\varepsilon$ for vessel 1, 2 and 3 respectively.

Lastly, when Mechanism 3 is applied, the arrival times into the system and at the locks are relevant. Note that if vessel 2 or 3 is served before vessel 1, the exit from the lock of vessel 1 would be delayed. Since vessel 1 arrives first into the system, it has priority and therefore it is processed first. Once vessel 1 is processed, the lock is open to the downstream side and vessels 2 and 3 are waiting on the upstream and downstream segments, respectively. Vessel 2 arrives first into the system, therefore it has priority. However, when vessel 1 has been processed, the lock is on the side of vessel 3. Thus, serving vessel 3 does not affect the waiting time of vessel 2. Therefore, under this mechanism, vessel 3 is processed second and vessel 2 is processed last. The waiting times are now equal to $1, 4-\varepsilon, 3-2\varepsilon$ for vessel 1, 2 and 3 respectively.

Since the choice of a lock mechanism influences the behavior of vessels, it also influences the existence of equilibria. Under the assumption of Mechanism 1, where the priority of vessels is determined by the arrival of vessels at the lock, equilibria might not exist, which is shown in the following example.

Example 3 (Mechanism 1). Assume there are two vessels: vessel 1 sailing upstream and vessel 2 sailing downstream. The complete river segment is again 20 km long, and the lock is placed in the middle of the waterway, hence, $L_u = L_d = 10$. The lock has capacity of 1 (though, any positive capacity will do) and its duration T and loading times T_1 and T_2 are set to 0.5. We assume that the fuel consumption function $E(v)$ is convex, non-negative and increasing in speeds $v_{i,p} \in [5, 10]$, $p \in \{u, d\}$. We assume that the lock starts on the upstream side, but can switch to the downstream side in time whenever vessel 2 is the first one to arrive at the lock. We assume the arrival times in the system are given by $a_1 = 0$ and $a_2 = \epsilon$ and the deadlines are $d_1 = 4$ and $d_2 = 4 + \epsilon$. Note that whenever a vessel has decided on its speed up to the lock, there is a unique speed after the lock that minimizes the fuel consumption such that the deadline, if possible, will not be exceeded. Therefore, the strategy of the vessels can be expressed in their speed before the lock (denoted by v_1 for vessel 1, and v_2 for vessel 2). We divide all possible speed scenarios into six cases, presented in Table 1. We see that in every strategy profile, there is a skipper that can decrease its fuel consumption by changing its speed. Hence, there does not exist a Nash equilibrium. Note that for this example, $v_{opt} = 6.\bar{6}$, i.e., the optimal speed for each vessel if it would be the only vessel on this waterway segment.

Under lock operating mechanisms 2 and 3, however, the Nash equilibrium does exist as the order in which the vessels enter the lock is determined solely by the order in which they arrive into the system. Hence, it cannot occur that vessels race each other to the lock, which is the main idea behind our previous example. Under these two mechanisms, vessels cannot affect the costs of vessels

Table 1. Speed scenarios for example 3.

Scenario	v_1	v_2	Improving move
1	10	$[5, v_{opt}]$	Player 1 should decrease v_1 to v_{opt}
2	10	$(v_{opt}, 10]$	Player 2 should decrease v_2 to 5
3	$(5, 10)$	$v_2 \leq v_1$	Player 2 should increase v_2 to 10
4	$(5, 10)$	$v_2 \geq v_1$	Player 1 should increase v_1 to 10
5	5	$(v_{opt}, 10]$	Player 2 should decrease v_2 to v_{opt}
6	5	$[5, v_{opt}]$	Player 1 should increase v_1 to v_{opt}

that entered the river section earlier. This implies that vessels can sequentially choose a best response, taking into account the arrival times of the previous vessels. We prove this statement more formally in the next theorem.

Theorem 1. *Consider the single lock scheduling problem, where the lock operates under Mechanism 2 or 3. Then, each game possesses at least one Nash equilibrium.*

Proof. We provide a generic construction of a strategy profile and show that this strategy profile constitutes a Nash equilibrium. Observe that under both Mechanism 2 and 3, for any speed v_i, the waiting time of vessel i, $q_i(\mathbf{v})$, only depends on the vessels arriving earlier in the system than vessel i. Consequently, knowing the strategies $v_1, \ldots, v_{i-1}$ is sufficient to determine optimal strategy v_i.

By construction of the strategy profile, it is apparent that each vessel i chooses its best possible strategy with respect to the early arriving vessels. Also, strategies of vessels that arrive later cannot influence the costs experienced by vessel i. Hence, vessel i can not decrease its private cost and therefore the resulting strategy profile is a Nash equilibrium.

Note, that the difference between the two mechanisms occurs in the individual optimization of strategies: under Mechanism 3 the waiting times caused by profile $\mathbf{v}$ might be different from the waiting times under Mechanism 2 using the same vector $\mathbf{v}$. However, the implications and the arguments stay the same: the cost for vessel i is only affected by the strategies of the first $i - 1$ vessels. □

A central authority could guarantee the existence of a Nash equilibrium by forcing the lock operators to use Mechanism 2 or Mechanism 3. However, the fact that a Nash equilibrium exists does not tell us anything about its cost efficiency. Selfish decision making may lead to a Nash equilibrium with a high social cost, which then leads to a waste of resources and high pollution on rivers. In Mechanism 2 and 3, individual costs highly depend on the strategies taken by the previous vessels. Therefore, selfish decision making may lead to the scenario in which later vessels are unable to cross the river segment before their deadline, resulting in a Nash equilibrium with an infinitely high social cost. Such scenario indicates that the *price of anarchy* of this game (the ratio between the highest social cost of any Nash equilibrium and the minimal social cost) is unbounded. This becomes apparent in the following example.

Example 4. We consider the same instance as in Example 3. However, this time we assume that the lock operates under Mechanism 2. We construct a Nash equilibrium with the procedure described in the proof of Theorem 1. This implies that $v_1^* = (20/3, 20/3)$. Note that there is no strategy in the strategy space of vessel 2, such that it passes the river segment before its deadline. Thus the social cost of this instance is infinitely high.

There is a strategy profile such that both vessels cross the river before their deadlines. More precisely, $\mathbf{v} = (5, 10)$ leads to a finite costs for both vessels. Because of this, the price of anarchy of the game at hand is unbounded. Note that the same results hold, when the lock is assumed to operate under Mechanism 3.

The goal of this section was to show that, though the concept of a Nash equilibrium seems appealing, in the non-cooperative setting it might not exist or it might be extremely inefficient compared to a socially optimal strategy profile. In the next section, we review the problem from a cooperative game point of view as we introduce the possibility to make binding contracts between the vessels.

4 Cooperative Game for Traffic Optimization at River Obstacles

We now assume that the vessels can make binding contracts and allow *payments* between skippers. As a result, the agents/skippers can incentivize their counter-agents to adapt their speeds by reimbursing their extra costs. We aim to find a solution concept that is cost optimal while making sure that no player can profit from a unilateral deviation from the social optimum. More precisely, we introduce a payment system that fulfills two criteria:

1. By participating in the payment system, the cost of a player can never be higher than when he/she did not participate.
2. The payment system should give a vessel an incentive to behave as in the social optimum.

In this section, we consider full information about the lock, river segments and vessels that will enter the system to be known in advance. An online variant of this problem is presented in Sect. 5, in which only the information about the river segment and the lock are publicly known while information about the vessels becomes only available when a vessel physically enters the waterway. First, we propose an algorithm that returns for each vessel a speed v_i, and the payment scheme $P_{i,j}$ indicating payment of skipper i to skipper j for the requested velocity adjustment. Second, we prove that the solution proposed by the algorithm satisfies the two criteria mentioned above.

4.1 Iterative Payment Scheme Algorithm

The algorithm sequentially determines optimal speeds and payments in the order of vessels arrival by considering all vessels 1 through i, denoted by the set $\bar{S}_i$. In the first iteration, only vessel 1 is considered and its optimal speed is determined. Let ζ_1 be the operating cost associated with this strategy such that $\zeta_1 = C_1(v_1)$. During future iterations, it will be ensured that the cost for this skipper will not go above the cost of this benchmark situation. To do this, other skippers should fully reimburse any cost increase that results from changing the strategy for the skipper.

Now, let $P^*_{j,j'}$ be the payment scheme for all $j' < j < i$ at iteration i. Moreover, all guaranteed costs ζ_j are considered to be known for all $j < i$. To determine the speeds v_j for all $j \in \bar{S}_i$ and payments $P_{i,j}$ for all $j < i$, we solve the following optimization problem: determine new velocities of the vessels from $\bar{S}_i$ such that the sum of the costs and payments for vessel i is minimized, while the total cost of each vessel $j < i$ is at most ζ_j. Then, we compute the value of the guaranteed cost ζ_i of player i. More formally, we define following relations.

$$C_{opt,k}(\bar{S}_i) := C_k\big((v^*_j)_{j\in\bar{S}_i}\big) \quad k \in \bar{S}_i, \tag{7}$$

$$P^*_{i,k} := C_{opt,k}(\bar{S}_i) - \zeta_k - \sum_{j\in\bar{S}_{i-1}:j>k} P^*_{jk} \quad k \in \bar{S}_{i-1}, \tag{8}$$

$$\zeta_i := C_{opt,i}(\bar{S}_i), \tag{9}$$

where v^* and P^* are the solutions to the following optimization problem. For a given vessel $i > 1$, having computed all optimal values P^* for all $i' < i$, the mathematical program reads

$$\min_{(v_j)_{j\in\bar{S}_i};P_{i,j}} \left(C_i\big((v_j)_{j\in\bar{S}_i}\big) + \sum_{k\in\bar{S}_{i-1}} P_{i,k} \right) \tag{10}$$

$$C_k\big((v_j)_{j\in\bar{S}_i}\big) - P_{i,k} - \sum_{\substack{j\in\bar{S}_{i-1}\\ j>k}} P^*_{j,k} \leq \zeta_k, \qquad k \in \bar{S}_{i-1}. \tag{11}$$

Algorithm 1 represents the payment system which outputs both optimal speeds and payments for all skippers. Note that the optimization problem has been replaced by a computation of the social optimal speeds. This is a valid substitution due to Theorem 2 below.

Input: $(L := (C, T, P, L_u, L_d), U, D, (a_i, d_i, v_{\min}, v_{\max})_{i \in U \cup D})$
Output: Optimal set of speeds and payments.
$\bar{S}_1 = \{1\}$;
$\zeta_1 = C_{opt}(S_i)$;
for *i from 2 to n* **do**
  $\bar{S}_i = \bar{S}_{i-1} \cup \{i\}$;
  Compute $C_{opt}(\bar{S}_i)$ and let $(v_j^*)_{j \in \bar{S}_i}$ be the optimal parameters;
  $C_{opt,k}(\bar{S}_i) := C_k\big((v_j^*)_{j \in \bar{S}_i}\big) \ \forall k \in \bar{S}_i$;
  $P_{i,k}^* := C_{opt,k}(\bar{S}_i) - \zeta_k - \sum_{j \in \bar{S}_{i-1} : j > k} P_{jk}^* \ \forall\, k \in \bar{S}_{i-1}$;
  $\zeta_i := C_{opt,k}(\bar{S}_i)$;
end
return $((v_j^*)_{j \in S}, (P_{ij}^*)_{i,j \in S})$

Algorithm 1. Payment mechanism

The subroutine computing of $C_{opt}(\bar{S}_i)$ can be implemented in various ways. In the Appendix, we provide a MIP-formulation to solve the lock scheduling problem to optimality. This formulation is based on the model in [7] and has been adjusted to comply with our problem statement. Moreover, we show that the problem is NP-complete in the strong sense, this way motivating design of MIP-formulations and approximation algorithms for the problem. Regarding existence of good approximation algorithms, we leave this question open but stress that any α-approximation algorithm directly leads to an α-approximate Nash equilibrium.

Given a solution to the optimization problem above, in Theorem 2, we show that the optimal speeds in that problem are equivalent to the speeds in the social optimum computed on vessels in the set $\bar{S}_i$.

Theorem 2. $(v_j^*)_{j \in \bar{S}_i} = \operatorname{argmin}_{(v_j)_{j \in \bar{S}_i}} \sum_{k \in \bar{S}_i} C_k(v_j)_{j \in \bar{S}_i}$.

Proof. For each $k \in \bar{S}_{i-1}$,

$$P_{i,k} = C_k\big((v_j)_{j \in \bar{S}_i}\big) - \zeta_k - \sum_{\substack{j \in \bar{S}_{i-1} \\ j > k}} P_{j,k}^* \tag{12}$$

and therefore the optimization problem can be written as

$$(v_j^*)_{j \in \bar{S}_i} = \operatorname*{argmin}_{(v_j)_{j \in \bar{S}_i}} C_i\big((v_j)_{j \in \bar{S}_i}\big) + \sum_{k \in \bar{S}_{i-1}} P_{i,k} \tag{13}$$

$$= \operatorname*{argmin}_{(v_j)_{j \in \bar{S}_i}} C_i\big((v_j)_{j \in \bar{S}_i}\big) + \sum_{k \in \bar{S}_{i-1}} \Big(C_k\big((v_j)_{j \in \bar{S}_i}\big) - \zeta_k - \sum_{\substack{j \in \bar{S}_{i-1} \\ j > k}} P_{j,k}^*\Big) \tag{14}$$

$$= \operatorname*{argmin}_{(v_j)_{j \in \bar{S}_i}} \sum_{k \in \bar{S}_i} C_k\big((v_j)_{j \in \bar{S}_i}\big) - \sum_{k \in \bar{S}_{i-1}} \Big(\zeta_k + \sum_{\substack{j \in \bar{S}_{i-1} \\ j>k}} P^*_{j,k}\Big) \tag{15}$$

$$= \operatorname*{argmin}_{(v_j)_{j \in \bar{S}_i}} \sum_{k \in \bar{S}_i} C_k\big((v_j)_{j \in \bar{S}_i}\big), \tag{16}$$

□

Lastly, in Theorem 3, we show that in the i-th iteration of the algorithm the best response for skipper i is to obey the payment mechanism. This means that the guaranteed cost of vessel i plus the payments this skipper has to pay to all other skipper is lower than the cost of any strategy not involving the payments.

Theorem 3. *In Algorithm 1 for each* $\bar{S}_i$*, it holds that*

$$\zeta_i + \sum_{k \in \bar{S}_{i-1}} P_{i,k} \leq C_i(v_i, (v^*_j)_{j \in \bar{S}_{i-1}}) \text{ for all } v_i \in V_i. \tag{17}$$

Proof. Note that after every iteration i, it holds for every $k \in \bar{S}_{i-1}$

$$P^*_{i,k} = C_{opt,k}(\bar{S}_i) - \zeta_k - \sum_{\substack{j \in \bar{S}_{i-1} \\ j>k}} P^*_{jk} \tag{18}$$

$$C_{opt,k}(\bar{S}_i) = \zeta_k + \sum_{\substack{j \in \bar{S}_i \\ j>k}} P^*_{jk} \tag{19}$$

This leads to the following equality.

$$\zeta_i + \sum_{k \in \bar{S}_{i-1}} P^*_{ik} = C_{opt,i}(\bar{S}_i) + \sum_{k \in \bar{S}_{i-1}} \Big(C_{opt,k}(\bar{S}_i) - \zeta_k - \sum_{\substack{j \in \bar{S}_{i-1} \\ j>k}} P^*_{jk}\Big) \tag{20}$$

$$= \sum_{k \in \bar{S}_i} C_{opt,k}(\bar{S}_i) - \sum_{k \in \bar{S}_{i-1}} (\zeta_k + \sum_{\substack{j \in \bar{S}_{i-1} \\ j>k}} P^*_{jk}) \tag{21}$$

$$= \sum_{k \in \bar{S}_i} C_{opt,k}(\bar{S}_i) - \sum_{k \in \bar{S}_{i-1}} C_{opt,k}(\bar{S}_{i-1}) \tag{22}$$

$$= C_{opt}(\bar{S}_i) - C_{opt}(\bar{S}_{i-1}) \tag{23}$$

Furthermore, we know that

$$C_{opt}(\bar{S}_i) \leq C_i(v_i, (v^*_j)_{j \in \bar{S}_{i-1}}) + C_{opt}(\bar{S}_{i-1}) \text{ for all } v_i \in \mathcal{V}_i \tag{24}$$

$$C_{opt}(\bar{S}_i) - C_{opt}(\bar{S}_{i-1}) \leq C_i(v_i, (v^*_j)_{j \in \bar{S}_{i-1}}) \text{ for all } v_i \in \mathcal{V}_i \tag{25}$$

$$\zeta_i + \sum_{k \in \bar{S}_{i-1}} P^*_{ik} \leq C_i(v_i, (v^*_j)_{j \in \bar{S}_{i-1}}) \text{ for all } v_i \in \mathcal{V}_i \tag{26}$$

□

From Theorems 2 and 3, it follows that the stated criteria for an efficient payment mechanism are fulfilled by Algorithm 1.

5 Online Setting

The assumption of perfect information on arrival times is likely to be violated in real-life. That is, there is no information prior to the arrival of the vessels at the boundaries of the system. Each time a vessel enters, the optimal speed and payments are recomputed taking into account the location of the vessels already present on the waterway. Note that the definition of a social optimum and a best response of a player are dependent on the information setting of the game. Therefore, we have to dynamically redefine/adjust these quantities in an online setting.

Let the distance between vessel i and the exit of the waterway at time t be denoted by h_i^t. The best response of vessel i, given the strategies of the other vessels, is defined as the strategy that minimizes the cost of vessel i conditional on the position of the other vessels at time t. The cost of vessel i under strategy profile $\mathbf{v}$ conditional on the position of all vessels in set $\bar{S}$ at time t is denoted as $C_i\big((v_j)_{j\in\bar{S}}|(h_j^t)_{j\in\bar{S}}\big)$. The social optimum is defined as a strategy profile, which provides the lowest possible cost given the positions of vessels in $\bar{S}$ at time t.

Similar to the offline setting, the algorithm sequentially determines optimal speeds and payments at the arrival of each vessel. In each iteration, a set $\bar{S}_i$ is constructed containing all vessels currently in the system. Assume that vessel i arrives and vessels in $\bar{S} = \{k, k+1, \ldots, i\}$ have not left the waterway yet. Assume that the payments $P^*_{j,j'}$ for all $k \leq j' < j < i$ and the guaranteed costs ζ_j for all $k \leq j < i$ are given. Since each vessel is at a different position, payments and costs are normalized to units per kilometers. Therefore, we solve the following optimization problem: determine new velocities of the vessels from $\bar{S}_i$ such that the sum of the costs and payments for vessel i is minimized, while the normalized total cost of each vessel $k \leq j < i$ is at most the normalized guaranteed cost. Given the following relations

$$C^{a_i}_{opt,k}(\bar{S}_i) = C_k\big((v^*_j)_{j\in\bar{S}_i}|(h^{a_i}_j)_{j\in\bar{S}_i}\big) \ \ \forall\, k \in \bar{S}_i, \tag{27}$$

$$P^*_{i,k} := C^{a_i}_{opt,k}(\bar{S}_i) - h^{a_i}_k\Bigg(\sum_{\substack{j\in\bar{S}_i\setminus\{i\}\\ j>k}}\left(\frac{P^*_{j,k}}{h^{a_j}_k}\right) + \frac{\zeta_k}{l_d+l_u}\Bigg) \ \ \forall\, k \in \bar{S}_i\setminus\{i\}, \tag{28}$$

$$\zeta_i := C^{a_i}_{opt,i}(\bar{S}_i) \tag{29}$$

we define the online optimization problem as

$$(v^*_j)_{j\in\bar{S}_i} \in \underset{(v_j)_{j\in\bar{S}_i}:P_{i,j}}{\operatorname{argmin}}\left(C_i\big((v_j)_{j\in\bar{S}_i}|(h^{a_i}_j)_{j\in\bar{S}_i}\big) + \sum_{k\in\bar{S}_i\setminus\{i\}} P_{i,k}\right) \tag{30}$$

s.t

$$\frac{C_k\big((v_j)_{j\in\bar{S}_i}|(h^{a_i}_j)_{j\in\bar{S}_i}\big)}{h^{a_i}_k} \frac{P_{i,k}}{h^{a_i}_k} - \sum_{\substack{j\in\bar{S}_i\setminus\{i\}\\ j>k}} \frac{P^*_{j,k}}{h^{a_j}_k} \leq \frac{\zeta_k}{l_d+l_u} \quad k \in \bar{S}_i\setminus\{i\}. \tag{31}$$

Again, it can be shown that the two conditions for an efficient payment mechanism are fulfilled in the online setting. The proof is similar to the one discussed in the offline case. The resulting algorithm is given in Algorithm 2.

Input: $(L := (C, T, P, L_u, L_d), U, D, (a_i, d_i, v_{\min}, v_{\max})_{i \in U \cup D})$
Output: Optimal set of speeds and payments.
vessel i arrives in the system at time a_i:
For each vessel currently present in the waterway, update the distance to the destination;
Let $\bar{S}_i$ be the set of vessels in the waterway at time a_i;
if $\bar{S}_i \neq \emptyset$ **then**
Compute $C^{a_i}_{opt}(\bar{S}_i)$ and let $(v^*_j)_{j \in \bar{S}_i}$ be the optimal parameters;

$$C^{a_i}_{opt,k}(\bar{S}_i) = C_k\big((v^*_j)_{j \in \bar{S}_i} | (h^{a_i}_j)_{j \in \bar{S}_i}\big) \ \ \forall\, k \in \bar{S}_i;$$

$$P^*_{i,k} := C^{a_i}_{opt,k}(\bar{S}_i) - h^{a_i}_k \Big(\sum_{\substack{j \in \bar{S}_i \setminus \{i\} \\ j > k}} \left(\frac{P^*_{j,k}}{h^{a_j}_k} \right) + \frac{\zeta_k}{l_d + l_u} \Big) \ \ \forall\, k \in \bar{S}_i \setminus \{i\};$$

$$\zeta_i := C^{a_i}_{opt,i}(\bar{S}_i);$$

else

$$\zeta_i = \min_{v_i} C_i(v_i);$$

end

Algorithm 2. Payment mechanism Online Setting

Note that whenever a vessel enters the lock, its total fuel cost and payments to the other vessels are known, and will not change anymore. Hence, the lock operator can also operate as a bank: whenever a vessel crosses the lock, it pays (or receives) the payments. This implies that the lock operator needs a cash reserve, as it is likely that the first vessels entering the lock receive money from the vessels that did not arrive at the lock yet. Clearly, this cash reserve needs to be at most the cost of an optimal profile minus the minimum fuel cost of all earlier vessels. In the journal version of the paper we give a simple and insightful example where the cash reserve is actually completely needed.

References

1. Bialystockia, N., Konovessis, K.: On the estimation of vessel's fuel consumption and speed curve: a statistical approach. J. Ocean Eng. Sci. **1**(2), 157–166 (2016)
2. Eurostat: Navigable inland waterways, by horizontal dimensions of vessels and pushed convoys (2016). http://appsso.eurostat.ec.europa.eu/nui/show.do?dataset=iww_if_hordim&lang=en. Accessed 1 Apr 2019

3. Inland Navigation in Europe, Market Observation. Central commission for the navigation of the Rhine, annual report (2017). https://www.inland-navigation-market.org/wp-content/uploads/2017/09/CCNR_annual_report_EN_Q2_2017_BD_-1.pdf. Accessed 1 Apr 2019
4. Günther, E., Lübbecke, M.E., Möhring, R.H.: Vessel traffic optimization for the Kiel canal. TRISTAN VII Book of Extended Abstracts 104 (2010)
5. Nauss, R.M.: Optimal sequencing in the presence of setup times for tow/barge traffic through a river lock. Eur. J. Oper. Res. **187**(3), 1268–1281 (2008)
6. Passchyn, W., Briskorn, D., and Spieksma, F.C.R.: No-wait scheduling for locks. Technical Report KBI_1605, KU Leuven, Research group Operations Research and Business Statistics, Leuven, Belgium (2016)
7. Passchyn, W., Briskorn, D., Spieksma, F.C.R.: Mathematical programming models for lock scheduling with an emission objective. Eur. J. Oper. Res. **248**(3), 802–814 (2016)
8. Passchyn, W., Coene, S., Briskorn, D., Hurink, J.L., Spieksma, F.C.R., Vanden Berghe, G.: The lockmaster's problem. Eur. J. Oper. Res. **251**(2), 432–441 (2016)
9. Petersen, E.R., Taylor, A.J.: An optimal scheduling system for the Welland Canal. Transp. Sci. **22**(3), 173–185 (1988)
10. Prandtstetter, M., Ritzinger, U., Schmidt, P., Ruthmair, M.: A variable neighborhood search approach for the interdependent lock scheduling problem. In: Ochoa, G., Chicano, F. (eds.) EvoCOP 2015. LNCS, vol. 9026, pp. 36–47. Springer, Cham (2015). https://doi.org/10.1007/978-3-319-16468-7_4
11. Psaraftis, H.N., Kontovas, C.A.: Speed models for energy-efficient maritime transportation: a taxonomy and survey. Transp. Res. Part C: Emerg. Technol. **26**, 331–351 (2013)
12. Smith, L.D., Nauss, R.M., Mattfeld, D.C., Li, J., Ehmke, J.F., Reindl, M.: Scheduling operations at system choke points with sequence-dependent delays and processing times. Transp. Res. Part E: Logistics Transp. Rev. **47**(5), 669–680 (2011)
13. Smith, L.D., Sweeney, D.C., Campbell, J.F.: Simulation of alternative approaches to relieving congestion at locks in a river transportion system. J. Oper. Res. Soc. **60**(4), 519–533 (2009)
14. Ching-Jung, T., Schonfeld, P.: Effects of speed control on tow travel costs. J. Waterw. Port Coastal Ocean Eng. **125**(4), 203–206 (1999)
15. Ching-Jung, T., Schonfeld, P.: Control alternatives at a waterway lock. J. Waterw. Port Coastal Ocean Eng. **127**(2), 89–96 (2001)
16. Verstichel, J., De Causmaecker, P., Spieksma, F.C.R., Vanden Berghe, G.: Exact and heuristic methods for placing vessels in locks. Eur. J. Oper. Res. **235**(2), 387–398 (2014)
17. Verstichel, J., De Causmaecker, P., Spieksma, F.C.R., Vanden Berghe, G.: The generalized lock scheduling problem: an exact approach. Transp. Res. Part E: Logistics Transp. Rev. **65**, 16–34 (2014)

Integer Conic Function Minimization Based on the Comparison Oracle

Dmitriy V. Gribanov[1,2(✉)] and Dmitriy S. Malyshev[1,2]

[1] Lobachevsky State University of Nizhny Novgorod,
23 Gagarina Avenue, Nizhny Novgorod 603950, Russian Federation
dimitry.gribanov@gmail.com

[2] National Research University Higher School of Economics, 25/12 Bolshaja Pecherskaja Ulitsa, Nizhny Novgorod 603155, Russian Federation
dsmalyshev@rambler.ru

Abstract. Let $f : \mathbb{R}^n \to \mathbb{R}$ be a conic function and $x_0 \in \mathbb{R}^n$. In this note, we show that the shallow separation oracle for the set $K = \{x \in \mathbb{R}^n : f(x) \le f(x_0)\}$ can be polynomially reduced to the comparison oracle of the function f. Combining these results with known results of D. Dadush et al., we give an algorithm with $(O(n))^n \log R$ calls to the comparison oracle for checking the non-emptiness of the set $K \cap \mathbb{Z}^n$, where K is included to the Euclidean ball of a radius R. Additionally, we give a randomized algorithm with the expected oracle complexity $(O(n))^n \log R$ for the problem to find an integral vector that minimizes values of f on an Euclidean ball of a radius R. It is known that the classes of convex, strictly quasiconvex functions, and quasiconvex polynomials are included into the class of conic functions. Since any system of conic functions can be represented by a single conic function, the last facts give us an opportunity to check the feasibility of any system of convex, strictly quasiconvex functions, and quasiconvex polynomials by an algorithm with $(O(n))^n \log R$ calls to the comparison oracle of the functions. It is also possible to solve a constraint minimization problem with the considered classes of functions by a randomized algorithm with $(O(n))^n \log R$ expected oracle calls.

Keywords: Nonlinear integer programming · Conic function · Convex function · Quasiconvex function · Comparison oracle · Separation oracle · Membership oracle · Convex set · Integral lattice

1 Introduction

Let K be a convex set included into the Euclidean ball $a + R \cdot B_2^n$ with a center $a \in \mathbb{Q}^n$ and a radius $R \in \mathbb{Q}_+$. Let, additionally, $\Lambda = \Lambda(B)$ be the lattice induced

This research is supported by Russian Foundation for Basic Research (Project 18-31-20001-mol-a-ved).

M. Khachay et al. (Eds.): MOTOR 2019, LNCS 11548, pp. 218–231, 2019.
https://doi.org/10.1007/978-3-030-22629-9_16

by the columns of a matrix $B \in \mathbb{Q}^{n \times n}$. In this note, we consider the following two problems. The first problem is to check the non-emptiness of a set

$$K \cap \Lambda \tag{1}$$

and return $x \in K \cap \Lambda$, if it is not empty. The second one is to find an exact minimizer of the following minimization problem

$$\begin{aligned} f(x) &\to \min \\ x &\in K \cap \Lambda, \end{aligned} \tag{2}$$

where $f : K \to \mathbb{R}$ is a quasiconvex function, or assert that $K \cap \Lambda$ is empty.

The integer minimization problem of (quasi)convex functions under (quasi) convex constraints is a well-known and intensively studied generalization of the integer linear programming problem [1,3,4,6–8,10–15,18–20].

The goal function and constraints can be defined explicitly or by an oracle. In the papers [7,20], for the non-emptiness (feasibility) problem (1), it was proposed a polynomial-time on $\log R$ algorithm, for a fixed dimension n, where K is given by the separation hyperplane oracle. A modification of these results and a good survey can be found in [6], where a $(O(n))^n$ poly$(\log R)$-oracle time algorithm has been presented. Additionally, in [6] a randomized $(O(n))^n$ poly$(\log R)$-expected oracle time algorithm for the minimization problem (2) has been presented, where K is given by the separation hyperplane oracle and f is convex and given by the subgradient oracle. A novel approach in integer convex optimization, based on the centerpoint concept, is proposed in [2,18].

The main disadvantage of the oracles mentioned above is that they are hard to implement. A more convenient way is to use the comparison oracle and the 0-th order oracle that computes function values. For any two points $x, y \in \operatorname{dom}(f)$, the comparison oracle allows to decide which of the two possibilities $f(x) \le f(y)$ or $f(x) > f(y)$ holds. It was shown in [5] that the minimization problem (2), for an arbitrary quasiconvex function f and for $K = R \cdot B_2^n$, can not be solved by an algorithm with a polynomial on $\log R$ number of calls to the comparison oracle. Additionally, it means that the problem (2), for the separation hyperplane oracle, can not be polynomially reduced to the same problem for the comparison oracle.

The integer minimization problem of convex (and closed to them) functions, given by the comparison oracle or by the 0-th order oracle, was considered in [4,5,21].

In the paper [4], an algorithm was developed for minimization of integer strictly quasiconvex function for $n = 2$ with the number of comparison oracle calls at most $2\log_2^2 R + O(\log R)$. In the paper [21], it was considered the symmetric version of the problem for $n = 2$ with the 0-th order oracle and the number of oracle calls at most $4 \log_2 R + O(1)$.

In the paper [5], it was considered the question about a possibility of narrowing the class of quasiconvex functions, given by the comparison oracle, for which the integer optimization problem, for a fixed dimension, can be solved in polynomial on $\log R$ time. In particular, some classes of functions were introduced there, called conic and discretely conic. The class of conic functions contains the classes of convex, strictly quasiconvex functions and the class quasiconvex polynomials. For the minimization problem (2), where $K = R \cdot B_2^n$ and f is a conic function, a deterministic algorithm with the oracle complexity $(O(n))^{2n} \log R$ was presented in [5]. Additionally in [5], the lower oracle complexity bound $3^{n-1} \log(2R - 1)$ was obtained.

1.1 Results of This Article

Let $g : \mathbb{R}^n \to \mathbb{R}$ be a conic function and $x_0 \in \mathbb{R}^n$. We show that the shallow separation oracle for the set $K = \{x \in \mathbb{R}^n : g(x) \leq g(x_0)\}$ can be polynomially reduced to the comparison oracle of the function g. Using the shallow cut ellipsoid method [9,17], introduced by A. Nemirovsky and D. Yudin, we then can find an ellipsoid E with a center $a \in \mathbb{Q}^n$, such that K is well-sandwiched by E:

$$\frac{1}{2(n+1)\sqrt{n}} E \subseteq K - a \subseteq E.$$

The presented result gives us an opportunity to use results of Dadush et al. [6]. The work [6] contains an algorithm with the oracle complexity $(O(n))^n \log R$ for the non-emptiness problem (1), where K is defined by the strict separation hyperplane oracle. A detailed consideration of the algorithm from [6] shows that it also works for the set K, equipped by the membership oracle, for which it is possible to find an ellipsoid E in $2^{O(n)}$-time, such that K is well-sandwiched by E. All together, it gives an algorithm for the non-emptiness (feasibility) problem (1), where $K = \{x \in \mathbb{R}^n : g(x) \leq g(x_0)\}$ with $(O(n))^n \log R$ calls to the comparison oracle for g. In a similar way, a randomized algorithm with the expected number $(O(n))^n \log R$ of calls to the oracle is presented for the minimization problem (2), where $K = \{x \in \mathbb{R}^n : g(x) \leq g(x_0)\}$ and f is conic.

2 Definitions, Notation and Some Preliminary Results

Let B_p^n be the n-dimensional unit ball related to the norm l_p. In other words,

$$B_p^n = \{x \in \mathbb{R}^n : ||x||_p \leq 1\}.$$

For a matrix $B \in \mathbb{R}^{m \times n}$, $\operatorname{cone}(B) = \{Bt : t \in \mathbb{R}_+^n\}$ is the *cone spanned by columns of* B, $\operatorname{conv.hull}(B) = \{Bt : t \in \mathbb{R}_+^n, \sum_{i=1}^n t_i = 1\}$ is the *convex hull spanned by columns of* B, $\operatorname{affine}(B) = \{Bt : t \in \mathbb{R}^n, \sum_{i=1}^n t_i = 1\}$ is the *affine hull spanned by columns of* B, and $\Lambda(B) = \{Bt : t \in \mathbb{Z}^n\}$ is the *lattice spanned by columns of* B.

For points $x^{(1)}, x^{(2)}, \ldots, x^{(k)} \in \mathbb{R}^n$, the set

$$x^{(k)} + \operatorname{cone}(x^{(k)} - x^{(1)}, \ldots, x^{(k)} - x^{(k-1)}) \tag{3}$$

is denoted as $\operatorname{cone}(x^{(1)}, x^{(2)}, \ldots, x^{(k-1)} | x^{(k)})$.

For a set $D \subseteq \mathbb{R}^n$, $\text{int}(D)$ and $\text{br}(D)$ are the sets of *interior* and *boundary points* of D, respectively. The sets of *interior* and *boundary points related to* $\text{affine}(D)$ are denoted by $\text{rel.int}(D)$ and $\text{rel.br}(D)$, respectively. The closure of D is denoted by $cl(D)$.

For a vector $x \in \mathbb{R}^n$, x_i is the i-th component of x. The set of integer values, started from i and ended in j, is denoted by $i : j = \{i, i+1, \dots, j\}$. The interval between points $y, z \in \mathbb{R}^n$ is denoted by

$$[y, z] = \{x = ty + (1-t)z : 0 \leq t \leq 1\}.$$

We will use the symbol (y, z) to define an open interval. The set D is said to be *convex* if $\forall x, y \in D$ we have $[x, y] \subseteq D$. For a function f, $\text{dom}(f)$ is the domain of f. For any $y \in \text{dom}(f)$, $H_f^{\leq}(y)$ is the set of contour lines for f. In other words,

$$H_f^{\leq}(y) = \{x \in \text{dom}(f) : \; f(x) \leq f(y)\}.$$

The set $H_f^{=}(y)$ is defined in a similar way.

For any symmetric positive definite matrix $A \in \mathbb{R}^n$ and vector $a \in \mathbb{R}^n$, $\text{E}(A, a)$ is the ellipsoid $\text{E}(A, a) = \{x \in \mathbb{R}^n : (x-a)^\top A^{-1}(x-a) \leq 1\}$.

2.1 Classes of Functions

Let us consider the set of functions $f : \text{dom}(f) \to \mathbb{R}$, such that $\text{dom}(f) \subseteq \mathbb{R}^n$ is convex. A function f is said to be *quasiconvex* if

$$\forall x, y \in \text{dom}(f), \forall z \in (x, \, y) \quad f(z) \leq \max\{f(x), \, f(y)\}.$$

A function f is said to be *strictly quasiconvex* if

$$\forall x, y \in \text{dom}(f), \forall z \in (x, \, y) \quad f(z) < \max\{f(x), \, f(y)\}.$$

A function f is said to be *convex* if

$$\forall x, y \in \text{dom}(f), \forall t \in (0, 1) \quad f(tx + (1-t)y) \leq tf(x) + (1-t)f(y).$$

We will denote these classes by the symbols $QConv_n$, $SQConv_n$, and $Conv_n$, respectively. Additionally, we denote by $QCPoly_n$ the class of quasiconvex polynomials with real coefficients.

Definition 1. *Let D be a set, equipped by a linear (total) order $\preceq$. Let f : $\text{dom}(f) \to D$, where $\text{dom}(f)$ is convex.*

The function f is conic *if $\forall y, z \in \text{dom}(f)$ and $\forall t \geq 0$, such that $f(y) \preceq f(z)$ and $z + t(z-y) \in \text{dom}(f)$, we have*

$$f(z + t(z-y)) \succeq f(z).$$

In our work, we mainly use $D = \mathbb{R}$ with the standard ordering, but the results of the work are valid for the general case too. Additionally, In Section "Algorithms for the optimization problem" we need to use $D = \mathbb{R}^2$ with the lexicographical ordering to reduce a constraint minimization problem to an unconstrained variant.

Clearly, the class $Conic_n$ of conic functions is a subclass of the quasiconvex functions class, that is $Conic_n \subset QConv_n$. The inclusion is strict, a counterexample is the quasiconvex function $\text{sgn}(x_1)$.

The next theorem from [5] gives two additional ways to define the class of conic functions.

Theorem 1. *Let $f : \text{dom}(f) \to D$, where $\text{dom}(f) \subseteq \mathbb{R}^n$ is convex, and D be a set, equipped by a linear (total) order $\preceq$. The following definitions are equivalent:*

1. *For any pair of points $y, z \in \text{dom}(f)$ and $\forall t \geq 0$, such that $f(y) \preceq f(z)$ and $z + t(z - y) \in \text{dom}(f)$, we have*
$$f(z + t(z - y)) \succeq f(z).$$
2. *For any set of points $x^{(1)}, x^{(2)}, \dots, x^{(k)}$, $y \in \text{dom}(f)$, such that*
$$f(x^{(1)}) \preceq f(x^{(2)}) \preceq \dots \preceq f(x^{(k)}) \text{ and}$$
$$y \in \text{cone}(x^{(1)}, x^{(2)}, \dots, x^{(k-1)} | x^{(k)}),$$
the inequality $f(y) \succeq f(x^{(k)})$ holds. Furthermore, we can assume that the points $x^{(1)}, x^{(2)}, \dots, x^{(k)}$ are in general position, i.e. no hyperplane contains more than n of them.
3. *For any $x \in \text{dom}(f)$, the set $H_f^{\leq}(x)$ is convex (which is equivalent to the quasiconvexity of the function f) and*
$$\forall x \in \text{dom}(f) \setminus M \quad H_f^{=}(x) \subseteq \text{rel.\,br}(H_f^{\leq}(x)),$$
where $M = \arg\min_{x \in \text{dom}(f)} f(x)$. If the set M is not defined, we will put it to be empty.

The next theorems from [5] show that the class $Conic_n$ contains some important subclasses.

Theorem 2. *The following strict inclusions hold:*

1. *$SQConv_n \subset Conic_n \subset QConv_n$,*
2. *$QCPoly_n \subset Conic_n$,*
3. *$Conv_n \subset Conic_n$.*

Theorem 3. *The class $Conic_n$ is closed with respect to the following operations.*

1. *Let $f_i \in Conic_n$ be real-valued functions and $w_i \in \mathbb{R}_+$, for any $i \in 1 : k$. Then the function $g(x) = \max_{i \in 1:k}\{w_i f_i(x)\}$ belongs to the class $Conic_n$, where $\text{dom}(g) = \bigcap_{i \in 1:k} \text{dom}(f_i)$.*

2. *Let $f \in Conic_n$ be real-valued function and $h : \mathbb{R} \to \mathbb{R}$ be a non-decreasing function. Then the function $g = h \cdot f$ belongs to the class $Conic_n$.*
3. *Let $f \in Conic_m$, $A \in \mathbb{R}^{m \times n}$, and $b \in \mathbb{R}^m$. Then the affine image $g(x) = f(Ax + b)$ belongs to the class $Conic_n$.*

2.2 Computational Model

Let us present types of oracles that we will need in this paper. With some slight modifications, we adopt the terminology from [6,9].

Let $K \subseteq \mathbb{R}^n$ be a convex set. We say that K is *(a, R)-circumscribed* if $K \subseteq a + R \cdot B_2^n$ for some $a \in \mathbb{Q}^n$ and $R \in \mathbb{Q}_+$. If the set K is $(0, R)$-circumscribed, then we simply call it *R-circumscribed.* For $\epsilon \in \mathbb{Q}_+$, we define

$$K^{\epsilon} = K + \epsilon B_2^n \qquad \text{and} \qquad K^{-\epsilon} = \{x \in K : x + \epsilon B_2^n \subseteq K\}.$$

For $A \in \mathbb{Q}^{m \times n}$ we define $\text{size}(A)$ as the length of the binary encoding of A.

Definition 2. *The* membership oracle O_K *for K is a function, which takes as an input a point $x \in \mathbb{Q}^n$ and returns $O_K(x) = [x \in K]$.*

Definition 3. *The* strong separation oracle $SSEP_K$ *on an input $y \in \mathbb{Q}^n$ either returns YES if $y \in K$ or some $c \in \mathbb{Q}^n$, such that $c^\top x < c^\top y$ for any $x \in K$.*

Definition 4. *The* weak separation oracle $WSEP_K$ *on an input $y \in \mathbb{Q}^n$ and a rational $\epsilon > 0$, either*

1. *asserts that $y \in K^{\epsilon}$ or*
2. *finds a vector $c \in \mathbb{Q}^n$ with $||c||_\infty = 1$, such that $c^\top x \leq c^\top y + \epsilon$ for every $x \in K^{-\epsilon}$*

When working with separation oracles, we assume that there is a polynomial ϕ, such that on an input y and ϵ as above, the outputs of $SSEP_K$ and $WSEP_K$ have a size bounded by $\phi(\text{size}(y) + \text{size}(\epsilon))$. The running times of algorithms using $SSEP_K$ and $WSEP_K$ will therefore depend on ϕ. Clearly, the membership oracle can be easily derived from the strong separation oracle.

Definition 5. *Let $f : K \to \mathbb{R}$ be a convex function with $\text{dom}(f) = K$. We refer that the function f is equipped by the* subgradient oracle, *if we have a query access to a subgradient $v \in \delta f(x)$ and to the value $f(x)$ for any $x \in K$.*

Definition 6. *The* shallow separation oracle, *for a convex set $K \subseteq \mathbb{R}^n$, is an oracle, whose input is an ellipsoid $\mathrm{E}(A, a)$, described by a positive definite matrix $A \in \mathbb{Q}^{n \times n}$ and a vector $a \in \mathbb{Q}^n$. The shallow separation oracle, denoted by $SHALL_K(A, a)$, can output one of the following possible answers:*

1. *a vector $c \in \mathbb{Q}^n \setminus \{0\}$, so that the halfspace $H = \{x \in \mathbb{R}^n : c^\top x \leq c^\top a + (n + 1)^{-1}\sqrt{c^\top A c}\}$ contains $K \cap \mathrm{E}(A, a)$ (a vector c with this property is called a* shallow cut *for K and $\mathrm{E}(A, a)$),*
2. *the assertion that $\mathrm{E}(A, a)$ is* tough.

As above, we assume the existence of a polynomial ϕ, such that the size of the output of $SHALL_K(A, a)$ is bounded by $\phi(\text{size}(A) + \text{size}(a))$. "Toughness" is a parameter left open, and in every instance of the shallow separation oracle the particular meaning of "tough" has to be specified. In our work, a tough ellipsoid is an ellipsoid $\mathrm{E}(A, a)$, such that $\mathrm{E}(\frac{1}{4n(n+1)^2} A, a) \subseteq K$.

Theorem 4. (Shallow cut ellipsoid method [17], [9, **pp. 94–102**]**).** *There exists an oracle-polynomial time algorithm, called* the shallow cut ellipsoid method, *so that for any rational number $\epsilon > 0$ and for any R-circumscribed closed convex set, given by the shallow separation oracle $SHALL_K$, finds a positive definite matrix $A \in \mathbb{Q}^{n \times n}$ and a point $a \in \mathbb{Q}^n$, such that one of the following holds:*

1. $\mathrm{E}(A, a)$ *has been declared tough by the oracle,*
2. $K \subseteq \mathrm{E}(A, a)$ *and* $\mathrm{vol}(\mathrm{E}(A, a)) \leq \epsilon$.

Here we give the definitions of strong and weak versions of comparison oracles. Our definitions are slightly stronger then the standard ones, because they allow to compare the function values with zero. We need this possibility, because we want to work with sets of the type $\{x \in \mathbb{R}^n : f(x) \leq 0\}$ instead of $\{x \in \mathbb{R}^n : f(x) \leq f(x_0)\}$. It makes sense clearer, and the constraint is not crucial from our point of view.

Definition 7. *The* comparison oracle $COMP_f$ *for $f : \mathbb{R}^n \to \mathbb{R}$ is a function, which takes as an input a pair of points $x, y \in \mathbb{Q}^n$, and returns $COMP_f(x, y) = [f(x) \leq f(y)]$.*

It also possible to compare the value of the function f in the point $x \in \mathbb{Q}^n$ with zero. In this case, we assume that $COMP_f$ takes as an input x, and returns $COMP_f(x) = [f(x) \leq 0]$.

Definition 8. *The* weak comparison oracle $WCOMP_f$ *for $f : \mathbb{R}^n \to \mathbb{R}$ is a function, which takes as an input a pair of points $x, y \in \mathbb{Q}^n$ and a rational ϵ, and returns $WCOMP_f(x, y, \epsilon) = [f(x) \leq f(y)]$ if $|f(x) - f(y)| > \epsilon$, where any answer is possible if $|f(x) - f(y)| \leq \epsilon$.*

It also possible to compare the value of the function f in the point $x \in \mathbb{Q}^n$ with zero. In this case, we assume that $WCOMP_f$ takes as an input x and a rational $\epsilon > 0$, and returns $WCOMP_f(x, \epsilon) = [f(x) \leq 0]$ if $|f(x)| > \epsilon$, where any answer is possible if $|f(x)| \leq \epsilon$.

The next theorems give some knowledge about the power of comparison oracles for convex functions.

Theorem 5. *Let an R-circumscribed convex set K be given by the strong separation hyperplane oracle $SSEP_K$. Then, there exists a conic function $f : \mathbb{R}^n \to \mathbb{R}$, equipped by the weak comparison oracle $WCOMP_f$, such that $K = \{x \in \mathbb{R}^n : f(x) \leq 0\}$. The computation of $WCOMP_f$ needs a polynomial number of calls to $SSEP_K$ on any input.*

Proof. Let $f : \mathbb{R}^n \to \mathbb{R}$ be the function given by the formula

$$f(x) = \begin{cases} 0, & \text{for } x \in K \\ 1 + d_K(x), & \text{for } x \notin K, \end{cases}$$

where $d_K(x) = \inf_{y \in K} ||x - y||_2$ is the distance from the point x to K. Clearly, $K = \{x \in \mathbb{R}^n : f(x) \leq 0\}$ and the function f is conic by the third item of Theorem 1. Let us show how to implement $WCOMP_f(y, z, \epsilon)$ for $y, z \in \mathbb{Q}^n$ and a rational ϵ, assuming that $|d_K(y) - d_K(z)| > \epsilon$. The separation hyperplane oracle gives an opportunity to check the statements $y \in K$ and $z \in K$. In case $y \in K$ we put $WCOMP_f(y, z, \epsilon) = 1$. If $y \notin K$ and $z \in K$ we put $WCOMP_f(y, z, \epsilon) = 0$. In the case, when $y \notin K$ and $z \notin K$ we need to compute the distances $d_K(y)$ and $d_K(z)$ and compare them. The problem to compute the value of $d_K(y)$ can be formulated as the following convex optimization problem

$$\min_{x \in K} ||y - x||_2.$$

The results, described in [9, pp. 56, 105–107], state that this problem for an R-circumscribed convex set K, equipped by the weak separation hyperplane oracle $WSEP_K$, can be solved with any accuracy $\delta > 0$ in time polynomial on $\text{size}(y) + \text{size}(R) + \text{size}(\delta)$. Finally, we can compute $d_1 \approx d_K(y)$ and $d_2 \approx d_K(z)$ with the accuracy $\delta = \epsilon/2$ and return $WCOMP_f(y, z, \epsilon) = [d_1 \leq d_2]$.

The next theorem states that a polynomial-time reduction from the weak separation oracle to the weak comparison oracle is also possible in the case, when K is additionally closed.

Theorem 6. *Let a closed R-circumscribed convex set K be given by the weak separation hyperplane oracle $WSEP_K$. Then, there exists a conic function $f : \mathbb{R}^n \to \mathbb{R}$, equipped by the weak comparison oracle $WCOMP_f$, such that $K = \{x \in \mathbb{R}^n : f(x) \leq 0\}$. The computation of $WCOMP_f$ needs a polynomial number of calls to $WSEP_K$ on any input.*

Proof. Let $f = d_K(x)$, where $d_K(x) = \inf_{y \in K} ||x - y||_2$ is the distance from the point x to K. Clearly, $cl(K) = \{x \in \mathbb{R}^n : f(x) \leq 0\}$ and the function f is conic by the third item of Theorem 1. Since K is closed, we have $K = \{x \in \mathbb{R}^n : f(x) \leq 0\}$. Assuming that $|d_K(y) - d_K(z)| > \epsilon$, the oracle $WCOMP_f(y, z, \epsilon)$, for $y, z \in \mathbb{Q}^n$ and a rational ϵ, can be implemented by the following way. In the previous proof, it has already been mentioned that the value of $d_K(y)$ can be computed with any accuracy δ in time polynomial on $\text{size}(y) + \text{size}(R) + \text{size}(\delta)$. Hence, we can compute $d_1 \approx d_K(y)$ and $d_2 \approx d_K(z)$ with the accuracy $\delta = \epsilon/2$ and return $WCOMP_f(y, z, \epsilon) = [d_1 \leq d_2]$.

3 Algorithms for the Feasibility Problem

The goal of this section is to design an algorithm for the non-emptiness problem (1), when $K = \{x \in \mathbb{R}^n : g_i(x) \leq 0, \text{ for } i \in 1 : m\}$, where $g_i : \mathbb{R}^n \to \mathbb{R}$

are conic functions equipped by the comparison oracle. By the first property of conic functions from Theorem 3, the function $f(x) = \max_i g_i(x)$ is also conic and the comparison oracle for f can be easily derived by comparison oracles of the functions g_i. Moreover, $K = \{x \in \mathbb{R}^n : f(x) \leq 0\}$. Hence, the considered problem is equivalent to the non-emptiness problem with only one function f. Finally, by the third property from Theorem 3, the superposition of a conic function with an affine map is a conic function, and we can assume that $\Lambda = \mathbb{Z}^n$.

Theorem 7. *Let $f : \mathbb{R}^n \to \mathbb{R}$ be a conic function, equipped by the comparison oracle $COMP_f$, and $K = \{x \in \mathbb{R}^n : f(x) \leq 0\}$. Then the shallow separation oracle $SHALL_K$ can be polynomially reduced to $COMP_f$.*

Proof. We shall only describe the underlying simple geometric idea of the algorithm, supposing that all calculations with real numbers can be carried out exactly. The necessary rounding can be done by standard methods, see for example [9, pp. 86–102]. We will give the detailed analysis in the extended version of the work.

Let $A \in \mathbb{Q}^{n \times n}$, $a \in \mathbb{Q}^n$ and an ellipsoid $\mathrm{E}(A, a)$ be an input of $SHALL_D(A, a)$. Since A is a symmetric positive definite matrix, then there exists the unique symmetric positive definite matrix $A^{1/2}$, such that $A = A^{1/2} A^{1/2}$. The affine map $x \to A^{1/2} x + a$ transforms the ellipsoid $\mathrm{E}(A, a)$ to the unit Euclidean ball. By the third property of Theorem 3, the function $f(Ax + a)$ is conic and the comparison oracle for $f(Ax + a)$ can be easily derived from $COMP_f$. Hence, we can assume that the input of $SHALL_D$ is the unit Euclidean ball $\mathrm{E}(I, 0) = B_2^n$.

Let $\gamma = \frac{1}{n+1}$, $B = B_2^n$ and $B(t) = tB$, for $t \in (0, \gamma)$. Let $r \in \mathbb{R}^n$ and $||r||_2 = 1$, then the rotation cone around a ray r with an angle ϕ is denoted by the symbol

$$C(r, \phi) = \{x \in \mathbb{R}^n : (x, r) \geq ||x||_2 \cos \phi\}, \text{ for } 0 \leq \phi \leq \frac{\pi}{2}.$$

The hyperplane $H(r) = \{x \in \mathbb{R}^n : r^\top x = \gamma\}$ supports $B(\gamma)$ in the point γr. Clearly, there exists an angle $\phi(t)$, such that the cone $tr + C(r, \phi(t))$ intersects the bound of the ball B along the hyperplane $H(r)$. In other words

$$H(r) \cap \operatorname{br} B = (tr + C(r, \phi)) \cap \operatorname{br} B, \text{ for } \phi = \phi(t).$$

Let us show, that

$$\cos \phi(t) = \frac{\gamma - t}{\sqrt{t^2 - 2\gamma t + 1}}. \tag{4}$$

Without loss of generality, we can assume that $r = e_1$. The intersection of the hyperplane $H(r)$ with $\operatorname{br} B$ has the equation $x_2^2 + x_3^2 + \cdots + x_n^2 = 1 - \gamma^2$ and contains the point $y = \gamma e_1 + \sqrt{1 - \gamma^2} e_2$. Clearly, the $\phi(t)$ is an angle between the vectors r and $y - tr$. We have that $r^\top (y - tr) = \gamma - t$ and $||y - tr||_2 = \sqrt{(\gamma - t)^2 + (1 - \gamma^2)} = \sqrt{t^2 - 2\gamma t + 1}$. So the formula (4) for $\cos \phi(t)$ is correct.

The reduction algorithm is following. Firstly, using the comparison oracle $COMP_f$, we compute the point $p = \arg\max_{i \in 1:n} f(\pm\frac{1}{2}\gamma e_i)$. If $f(p) \le 0$, then we can declare that B is tough, because

$$\frac{1}{2\sqrt{n}}\gamma B \subseteq \text{conv. hull}_{i\in 1:n}(\pm\frac{1}{2}\gamma e_i) \subseteq K.$$

Suppose, that $f(p) > 0$, so $p \notin K$.

The paper [5, p. 26] gives an algorithm that, for a given angle $0 < \alpha < \pi/2$ can build a sequence of points

$$x^{(1)}, x^{(2)}, \ldots, x^{(n)}, x^{(n+1)} \in \frac{1}{2}\gamma B,$$

such that the cone $D = \text{cone}(x^{(1)}, x^{(2)}, \ldots, x^{(n)} | x^{(n+1)})$ has the following properties:

1. $x^{(n+1)} + C(\frac{x^{(n+1)}}{||x^{(n+1)}||_2}, \alpha) \subseteq D$,
2. $f(x) \ge f(p)$ for $x \in D$.

The algorithm complexity is polynomial by size$(\cos\phi)$. Additionally, we need to note that the considered algorithm was firstly developed by Nemirovsky and Yudin in [16] (see also [17, p. 345]) for convex optimization with the 0-th order oracle. In [5, p. 26], the reformulation of this algorithm for convex functions has been given.

Next, we run this algorithm for the function f in the ball $\frac{1}{2}\gamma B$ with the parameter $\alpha = \phi(1/2\gamma)$. Let $w = x^{(n+1)}$. By the definition of cone D, we have

$$K \subseteq \{x \in \mathbb{R}^n : w^\top x \le \gamma||w||_2\} \cap B,$$

and the corresponding hyperplane supports γB, so the vector w is a shallow cut.

Theorem 7, together with the shallow cut ellipsoid method from Theorem 4, gives us an algorithm that outputs an ellipsoid $E(A, a)$, such that either

$$E(\frac{1}{4(n+1)^2 n}A, a) \subseteq K \subseteq E(A, a), \text{ for } K = \{x \in \mathbb{R}^n : f(x) \le 0\}, \quad (5)$$

or $K \subseteq \mathrm{E}(A, a)$ and $\mathrm{vol}(\mathrm{E}(A, a)) \le \epsilon$.

The following theorem was proved in [6, p. 220].

Theorem 8. *Let an R-circumscribed convex set K be given by the separation hyperplane oracle $SSEP_K$ and $\Lambda = \Lambda(B)$, for $B \in \mathbb{Q}^{n\times n}$. Then there is an algorithm with the oracle complexity $(O(n))^n$ poly$(\log R)$ that finds a point $x \in K \cap \Lambda$ or asserts that $K \cap \Lambda = \emptyset$.*

The algorithm presented in [6, p. 240] that solves the problem from the previous theorem is called "An Improved Kannan Type Algorithm". A detailed consideration of this algorithm shows that the separation oracle $SSEP_K$ is only needed for deriving an ellipsoid with the property (5). Everywhere in other parts of the algorithm the strong membership oracle O_K may only be used. Combining all facts together and using the Dadush's algorithm, we obtain a solution for the problem defined in beginning of this section.

Theorem 9. *Let a R-circumscribed convex set K be given by the formula $K = \{x \in \mathbb{R}^n : f(x) \leq 0\}$, where $f : \mathbb{R}^n \to \mathbb{R}$ is a conic function, equipped by the comparison oracle $COMP_f$. Then there is an algorithm with the oracle complexity $(O(n))^n \operatorname{poly}(\log R)$ that finds a point $x \in K \cap \mathbb{Z}^n$ or asserts that $K \cap \mathbb{Z}^n = \emptyset$.*

4 Algorithms for the Optimization Problem

The goal of this section is to design an algorithm for the minimization problem (2), when $K = \{x \in \mathbb{R}^n : g_i(x) \leq 0, \text{ for } i \in 1 : m\}$, where $g_i : \mathbb{R}^n \to \mathbb{R}$ and f are conic functions, equipped by the comparison oracle. By the first and second properties of conic functions from Theorem 3, the function $g(x) = \max_i\{(g_i(x))_+\}$ is also conic, where $(x)_+ = [x \geq 0]\, x$ is a positive part of x, and the comparison oracle for g can be easily derived by comparison oracles of the functions g_i. Moreover, $\arg \min_{x \in \mathbb{R}^n} g(x) = K$. Now, consider the function $h : \mathbb{R}^n \to \mathbb{R}^2$, given by the formula $h(x) = \begin{pmatrix} g(x) \\ f(x) \end{pmatrix}$. Clearly, the function h is conic and the lexicographical comparison oracle for h can be easily derived from comparison oracles of g and f. Clearly,

$$\arg \min_{x \in K \cap \Lambda} f(x) = \arg \min_{x \in \Lambda} h(x). \tag{6}$$

Finally, by Theorem 3, the superposition of a conic function with an affine map is a conic function. Hence, the constrained minimization problem 2 on the lattice $\Lambda = \Lambda(B)$ is equivalent to the problem $\min_{x \in \mathbb{Z}^n} h(Bx)$.

The following theorem was proved in [6, p. 247].

Theorem 10. *Let an R-circumscribed convex set K be given by the separation hyperplane oracle $SSEP_K$ and $\Lambda = \Lambda(B)$, for $B \in \mathbb{Q}^{n \times n}$. Let, additionally, $f : K \to \mathbb{R}$ be a convex function, equipped by the subgradient oracle. Then there is a randomized algorithm with the expected number of oracle calls $(O(n))^n \operatorname{poly}(\log R)$ that finds an exact minimizer of the problem $\min_{x \in K \cap \Lambda} f(x)$ or asserts that $K \cap \Lambda = \emptyset$.*

The algorithm presented in [6, p. 247] that solves the problem from the previous theorem is called "Convex Integer Minimization". A detailed consideration of this algorithm shows that the separation oracle $SSEP_K$ is only needed for deriving an ellipsoid with the property (5). Everywhere in other parts of the algorithm the strong membership oracle O_K may only be used. Now we will show how to avoid usage of the subgradient oracle for f by changing it to the comparison oracle. The main idea of Dadush's algorithm is to show that if the convex set K is sufficiently wide, then K contains a deep lattice point, and in the opposite case, a dimension can be reduced.

Denote by $y \in K \cap \Lambda$ this lattice point. Clearly, an optimal point is contained in the set $K' = K \cap \{x \in \mathbb{R}^n : f(x) \leq f(y)\}$. Dadush showed that if there exists a hyperplane, supporting K' in the point y, then $\mathrm{vol}_k(K') \leq (1 - \frac{1}{2}10^{-k}) \mathrm{vol}_k(K)$, where $k = \dim(K)$. He used a subgradient of f in y as the normal vector of the supporting hyperplane. The next step is defining the separation oracle for K', and the process is repeated. This approach leads to exponential decreasing of the volume of K' that gives its flatness sooner or later.

Thus, if we want to apply the Dadush's result, we need to show the existence of an hyperplane that supports $K' = K \cap \{x \in \mathbb{R}^n : f(x) \leq f(y)\}$ in the point $y \in K \cap \Lambda$, and to show how to implement the comparison oracle $COMP_{f'}$ for a conic function f', such that $K' = \{x \in \mathbb{R}^n : f'(x) \leq 0\}$. Let us answer the first question. By the third item of Theorem 1, the point y is a boundary point of the set $\{x \in \mathbb{R}^n : f(x) \leq f(y)\}$. Hence, y is also a boundary point of the set K'. By the definition of a convex set, there exists a hyperplane supporting K' in the point y. Now, the conic function f' and its comparison oracle can be easily derived using the properties of conic functions from Theorem 3. See the example marked by (6). Combining all facts together and using the Dadush's algorithm, we obtain a solution for the problem, defined in beginning of this section.

Theorem 11. *Let $f : \mathbb{R}^n \to \mathbb{R}$ be a conic function, equipped by the comparison oracle $COMP_f$, with the property that*

$$\arg\min_{x \in \mathbb{Z}^n} f(x) \subseteq a + R \cdot B_2^n.$$

Then there is a randomized algorithm with the expected number of oracle calls $(O(n))^n \operatorname{poly}(\log R)$ that finds an exact minimizer of the problem $\min_{x \in \mathbb{Z}^n} f(x)$.

5 Conclusion: Future Work and Remarks

It has already been mentioned that the class of conic functions is sufficiently wide [5]. It was shown in [5] that any constraint minimization problem for conic functions can be reduced to an unconstrained minimization problem with only one conic function. Moreover, it has been shown in Subsection "Computational model" (see Theorems 5 and 6) that any convex set K, equipped by the strong separation hyperplane oracle, can by represented as $K = \{x \in \mathbb{R}^n : f(x) \leq 0\}$, where $f : \mathbb{R}^n \to \mathbb{R}$ is a conic function, equipped by the weak comparison oracle. If, additionally, the set K is closed, then the reduction from the weak separation hyperplane oracle to the weak comparison oracle is also possible. This reductions, from separation to weak comparison, can be done by a polynomial-time algorithm. But, the existence of a polynomial-time reduction between the strong separation hyperplane oracle and the strong comparison oracle is an open question. Another open question is the possibility of an inverse polynomial-time reduction. Probably, the answer is "No", since the known algorithms for continuous conic function minimization only guaranty decreasing of the distance to an optimum point and not guaranty decreasing of the function value. But, the question needs additional consideration. We are planning to answer these questions in the extended version of this article.

References

1. Ahmadi, A., Olshevsky, A., Parrilo, P., Tsitsiklis, J.: NP-hardness of deciding convexity of quadratic polynomials and related problems. Math. Program. **137**(1–2), 453–476 (2013). https://doi.org/10.1007/s10107-011-0499-2
2. Basu, A., Oertel, T.: Centerpoints: a link between optimization and convex geometry. SIAM J. Optim. **27**(2), 866–889 (2017). https://doi.org/10.1007/978-3-319-33461-5_2
3. Bredereck, R., Faliszewski, P., Niedermeier, R., Skowron, P., Talmon, N.: Mixed integer programming with convex/concave constraints: fixed-parameter tractability and applications to multicovering and voting. CoRR, https://arxiv.org/abs/1709.02850 (2017)
4. Chirkov, A.: Minimization of quasi-convex function on two-dimensional integer lattice. Vestn. Nizhegorod. Univ. N. I. Lobachevskogo, Mat. Model. Optim. Upr. **1**, 227–238 (2003). (in Russian)
5. Chirkov, A., Gribanov, D., Malyshev, D., Pardalos, P., Veselov, S., Zolotykh, A.: On the complexity of quasiconvex integer minimization problem. J. Glob. Optim. **73**(4), 761–788 (2019). https://doi.org/10.1007/s10898-018-0729-8
6. Dadush, D.: Integer programming, lattice algorithms, and deterministic volume estimation. ProQuest LLC, Ann Arbor, MI. thesis (Ph.D.), Georgia Institute of Technology (2012)
7. Dadush, D., Peikert, C., Vempala, S.: Enumerative lattice algorithms in any norm via M-ellipsoid coverings. In: Proceedings of the 52nd Annual IEEE Symposium on Foundations of Computer Science (FOCS 11), pp. 580–589 (2011). https://doi.org/10.1109/FOCS.2011.31
8. De Loera, J., Hemmecke, R., Koppe, M., Weismantel, R.: Integer polynomial optimization in fixed dimension. Math. Oper. Res. **31**(1), 147–153 (2006). https://doi.org/10.1287/moor.1050.0169
9. Grötschel, M., Lovász, L., Schrijver, A.: Geometric Algorithms and Combinatorial Optimization. Algorithms and Combinatorics, vol. 2, 2nd edn. Springer, Berlin (1993). https://doi.org/10.1007/978-3-642-78240-4. corrected ed
10. Heinz, S.: Complexity of integer quasiconvex polynomial optimization. J. Complex. **21**(4), 543–556 (2005)
11. Heinz, S.: Quasiconvex functions can be approximated by quasiconvex polynomials. ESAIM Control Optim. Calc. Var. **14**(4), 795–801 (2008). https://doi.org/10.1051/cocv:2008010
12. Hemmecke, R., Onn, S., Weismantel, R.: A polynomial oracle-time algorithm for convex integer minimization. Math. Program. **126**(1), 97–117 (2011). https://doi.org/10.1007/s10107-009-0276-7
13. Hildebrand, R., Köppe, M.: A new Lenstra-type algorithm for quasiconvex polynomial integer minimization with complexity $2^{O(n \log n)}$. Discret. Optim. **10**(1), 69–84 (2013). https://doi.org/10.1016/j.disopt.2012.11.003
14. Khachiyan, L., Porkolab, L.: Integer optimization on convex semialgebraic sets. Discret. Comput. Geom. **23**(2), 207–224 (2000). https://doi.org/10.1007/PL00009496

15. Lenstra, H.: Integer programming with a fixed number of variables. Math. Oper. Res. **8**(4), 538–548 (1983). https://doi.org/10.1287/moor.8.4.538
16. Nemirovski, A., Yudin, D.: Evaluation of the information complexity of mathematical programming problems. Ekonomika i Matematicheskie Metody **13**(2), 3–45 (1976). (in Russian)
17. Nemirovsky, A., Yudin, D.: Problem Complexity and Method Efficiency in Optimization. Wiley, New York (1983)
18. Oertel, T.: Integer convex minimization in low dimensions. Thes. doct. phylosophy. Eidgenössische Technische Hochschule, Zürich (2014)
19. Oertel, T., Wagner, C., Weismantel, R.: Convex integer minimization in fixed dimension. https://arxiv.org/pdf/1203.4175.pdf (2012)
20. Oertel, T., Wagner, C., Weismantel, R.: Integer convex minimization by mixed integer linear optimization. Oper. Res. Lett. **42**(6), 424–428 (2014). https://doi.org/10.1016/j.orl.2014.07.005
21. Veselov, S., Gribanov, D., Zolotykh, N., Malishev, D., Chirkov, A.: Minimizing a symmetric quasiconvex function on a two-dimensional lattice. J. Appl. Ind. Math. **12**(3), 587–594 (2018). https://doi.org/10.1134/S199047891803016X

Dynamic Sparsification for Quadratic Assignment Problems

Maximilian John([✉]) and Andreas Karrenbauer

Max Planck Institute for Informatics, Saarbrücken, Germany
{maximilian.john,andreas.karrenbauer}@mpi-inf.mpg.de

Abstract. We present a framework for optimizing sparse quadratic assignment problems. We propose an iterative algorithm that dynamically generates the quadratic part of the assignment problem and, thus, solves a sparsified linearization of the original problem in every iteration. This procedure results in a hierarchy of lower bounds and, in addition, provides heuristic primal solutions in every iteration. This framework was motivated by the task of the French government to design the French keyboard standard, which included solving sparse quadratic assignment problems with over 100 special characters; a size where many commonly used approaches fail. The design of a new standard often involves conflicting opinions of multiple stakeholders in a committee. Hence, there is no agreement on a single well-defined objective function that can be used for an extensive one-shot optimization. Instead, the process is highly interactive and demands rapid prototyping, e.g., quick primal solutions, on-the-fly evaluation of manual changes, and prompt assessments of solution quality. Particularly concerning the latter aspect, our algorithm is able to provide high-quality lower bounds for these problems in several minutes.

Keywords: Quadratic assignment · Integer programming · Linearization · Keyboard optimization

1 Introduction

Assignment problems aim at finding the cheapest one-to-one correspondence between n items and locations. Already in 1946, Birkhoff [4] showed that the optimal assignment can be found in $\mathcal{O}(n^3)$ time if the objective function is linear. However, linear objective functions cannot capture pairwise dependencies between variables. Koopmans and Beckmann [16] investigated a variant with quadratic terms in the cost function. This quadratic optimization problem includes several practical applications, such as the keyboard layout problem [6], the facility location problem [19], the traveling salesman problem [5], and many others.

M. Khachay et al. (Eds.): MOTOR 2019, LNCS 11548, pp. 232–246, 2019.
https://doi.org/10.1007/978-3-030-22629-9_17

As expected, great modeling power comes with increased hardness; there are problems of only $n = 30$ items that cannot be solved to optimality in reasonable time. From a complexity theoretical point of view, Queyranne [23] showed that the quadratic assignment problem (QAP) is NP-hard to approximate within any constant factor, even if the quadratic cost can be factorized to a symmetric block diagonal matrix and a distance matrix describing a line metric.

To cope with the hardness, researchers have proposed many ideas over the last decades. Many of them are based on linearizations, e.g., the classical results by Gilmore [11] and Lawler [17], and by Kaufman and Broeckx [15]. These linearizations can be considered light, meaning that their space requirements are linear in the input size and the corresponding relaxations can be solved quickly, e.g., in a branch-&-bound framework. Moreover, the latter approach seems to be amenable for primal heuristics of state-of-the-art MIP solvers to compute good incumbents. However, their lower bounds deteriorate quickly with increasing input size, which negatively impacts the performance of branch-&-bound. More recently, new improved linearizations have been developed by Xia and Yuan [25] and Zhang [26] who combine the ideas of the light-weight approaches mentioned above.

On the contrary, the formulations by Frieze and Yadegar [10] and by Adams and Johnson [1] compute very strong lower bounds at the expense of $\mathcal{O}(n^4)$ additional variables. Both approaches yield equivalent formulations for the QAP and are commonly referred to as RLT1 formulations, a more general concept proposed by Sherali and Adams [24]. Huber and Riedl showed [13] that the Adams-Johnson formulation dominates the one of Xia and Yuan. For many instances of practical size and especially in our scenario, RLT1 and other similar more powerful approaches could not produce any result within the given resources, which is expected due to the sheer size of the problem.

Recently, semidefinite programming relaxations for QAPs [14,22,27] have become more popular. Peng et al. [20] showed that these approaches can indeed often produce good lower bounds for the QAP.

1.1 Keyboard Optimization as Assignment Problems

Already in the 70s, Pollatschek [21] as well as Burkard and Offermann [6] proposed to optimize keyboard layouts as a quadratic assignment problem. They consider the assignment problem with mixed linear and quadratic terms in the objective function. Formally, let us assume that the n items and locations are numbered from 1 to n. In the following, we refer to the set of items as $[n] := \{1, \ldots, n\}$. Furthermore, let $x_{ik} \in \{0, 1\}$ denote the decision of whether or not to assign item i to location k (or in our case: assign character i to key slot k). With c_{ik} being the linear assignment cost and $q_{ijk\ell}$ the quadratic assignment cost, we obtain the following quadratic program.

$$
\begin{aligned}
\min \sum_{i,k=1}^{n} c_{ik} x_{ik} + & \sum_{i,j,k,\ell=1}^{n} q_{ijk\ell} x_{ik} x_{j\ell} & \\
\text{subject to} \sum_{i=1}^{n} x_{ik} &= 1 & \forall k \in [n] \\
\sum_{k=1}^{n} x_{ik} &= 1 & \forall i \in [n] \\
x_{ik} &\in \{0,1\} & \forall i,k \in [n]
\end{aligned}
\tag{1}
$$

The term $q_{ijk\ell}$ describes the cost of simultaneously assigning items i and j to the locations k and ℓ, respectively. In keyboard problems, this term usually factors into $q_{ijk\ell} = p_{ij} \cdot d_{k\ell}$, where p_{ij} denotes the empirical probability of typing letter j after letter i and $d_{k\ell}$ is the time between pressing the key slots k and ℓ.

Typically, integer linear programs are relaxed by dropping the integrality constraints of the variables. In this case, however, the assignment polytope is well-understood, we can quickly solve linear assignment problems with over a million variables [18]. The hardness of QAPs comes, therefore, not from the polytope, but from the quadratic terms of the objective function. It is possible to exploit the structure of the objective so that special cases become more tractable, in some cases polynomial approximation algorithms have been developed [3]. In this work, we define a relaxation of the QAP, too. While also keeping the integrality constraints untouched, we exploit the sparse nature of the quadratic objective function and modify the quadratic terms to obtain upper and lower bounds for the original problem.

1.2 Designing the French Keyboard Standard

In 2015, the French Ministry of culture discussed the concerns about not having an official French keyboard standard [7]. The commonly used *AZERTY* layout did not provide typing frequent special characters like À, œ, etc. One year later, AFNOR, the French national organization for standardization, was issued with the task to design the new standard [8], which should support all missing special characters that are used in the French language. Two major options were discussed: optimizing the whole keyboard from scratch or keeping the most frequent characters (*A*–*Z*) fixed and only optimizing the addition of over 100 *special characters* to maintain familiarity and facilitate learnability. We participated in the endeavor for the latter.

The process of defining the standard consisted of several rounds of gathering data (details can be found in [9]), modeling the problem as a QAP, finding a (near-)optimal keyboard, which was then proposed to an official committee who expressed further wishes for the objective, modified the weighting of its components, and added or removed certain characters. Eventually, the objective function stabilized as a conic combination of four different measures: **performance** – special characters, which are often used in combination with fixed characters, should be close together in order to minimize the time to type; **ergonomics** – frequently used special characters should be quite central on the keyboard to avoid unhealthy stretches of the fingers; **intuitiveness** – special characters should be placed close to similar fixed characters and other special characters to simplify finding them on the keyboard; **familiarity** – frequent special characters should be placed close to their position in the original AZERTY keyboard if this character was already present there. Note that every interaction between special characters and fixed characters can be modeled as linear expressions because we are not allowed to change the position of these fixed characters, so their impact to the objective function is constant. Hence, 3 out of 4 of these measures describe a linear objective function. The quadratic part of the objective function models similarity of two special characters, which is part of the intuitiveness measurement. Since there is only a restricted number of similar special characters, e.g., é and è or % and ‰, this explains the sparsity of the quadratic objective function.

After a first consensus had been found, a public inquiry organized by AFNOR provided feedback on the proposed design. Not only after the public inquiry, but after every iteration of this feedback loop, the underlying model was updated and new solutions were heuristically computed; which our algorithm then showed to be near-optimal. Why is this last step important? In contrast to one-shot optimization with a single well-defined objective, the committee, which consists of multiple stakeholders with different interests and opinions, discusses several solutions, evaluates the impact of manual changes on different parts of the objective, and alters the optimization model to find a compromise. Deciding model changes based on very sub-optimal solutions is pointless since the observed solution might not properly represent the current model. Therefore, it is not only important to find near-optimal solutions, but also to have a sharp picture of the their quality, for allowing a well-founded discussion and decision process. Additionally, it is important that such solutions to updated models and corresponding bounds can be computed as fast as possible, ideally even in real-time.

Fig. 1. The new French keyboard standard (NF Z71-300). Special characters (blue) and diacritic marks (red) were added to the old AZERTY layout. More information about the new standard can be found on https://norme-azerty.fr (Color figure online)

Finally, the expert committee agreed on a layout for the new French keyboard standard [2], which is depicted in Fig. 1 and was launched on 2 April 2019[1].

[1] https://normalisation.afnor.org/actualites/faq-clavier-francais/ – retr. 2019-04-03.

1.3 Our Contribution

The goal of our framework is to utilize the power of the RLT1 approach while avoiding the computational overhead. We present an algorithm that dynamically generates the quadratic terms of the QAP, which leads to a hierarchy of lower bounds and heuristic primal solutions at the same time. In contrast to a classic column-generation approach, our algorithm guarantees a sequence of non-decreasing lower bounds in every step instead of non-increasing upper bounds. This iterative framework produces a $(1+\varepsilon)$-approximation[2] for the QAP for any $\varepsilon \geq 0$. We show the success of our framework during the design process of the new French keyboard standard. The lower bounds computed by our algorithm showed very small optimality gaps within several minutes for sparse QAPs with over 100 items and 130 locations. All examples shown in this paper are real-world instances created during this standardization process. Hence, our tool is usable to provide almost real-time feedback with very limited resources, for example, on a laptop.

2 Algorithm

Linear relaxations for quadratic programs have been extensively studied over the last decades and are still a good starting point for many new ideas. However, the disadvantage of standalone linear relaxations is either high space complexity or an inefficient bound generation. We want to overcome this issue for QAPs with sparse quadratic objectives.

Let $\mathcal{S} \subseteq [n]^4$ be a set of indices. We define the following subproblem of (1).

$$\min \quad \sum_{i,k=1}^{n} c_{ik} x_{ik} + \sum_{(i,j,k,\ell)\in\mathcal{S}} q_{ijk\ell} y_{ijk\ell} \tag{2a}$$

$$\text{subject to} \quad \sum_{i=1}^{n} x_{ik} = 1 \qquad \forall k \in [n]$$

$$\sum_{k=1}^{n} x_{ik} = 1 \qquad \forall i \in [n]$$

$$\sum_{j:(i,j,k,\ell)\in\mathcal{S}} y_{ijk\ell} \leq x_{ik} \qquad \forall i,k,\ell \in [n] \tag{2b}$$

$$\sum_{\ell:(i,j,k,\ell)\in\mathcal{S}} y_{ijk\ell} \leq x_{ik} \qquad \forall i,j,k \in [n] \tag{2c}$$

$$\sum_{i:(i,j,k,\ell)\in\mathcal{S}} y_{ijk\ell} \leq x_{j\ell} \qquad \forall j,k,\ell \in [n] \tag{2d}$$

[2] We give no polynomial time guarantee. The existence of a PTAS would imply P = NP.

$$\sum_{k:(i,j,k,\ell)\in\mathcal{S}} y_{ijk\ell} \le x_{j\ell} \qquad \forall i,j,\ell \in [n] \qquad (2e)$$

$$x_{ik} + x_{j\ell} \le 1 + y_{ijk\ell} \qquad \forall (i,j,k,\ell) \in \mathcal{S} \qquad (2f)$$

$$y_{ijk\ell} \in [0,1] \qquad \forall (i,j,k,\ell) \in \mathcal{S}$$

$$x_{ik} \in \{0,1\} \qquad \forall i,k \in [n]$$

Note that it is feasible to add symmetry constraints for the y-variables of the form $y_{ijk\ell} = y_{ji\ell k}$ inspired by the Adams-Johnson formulation because they simulate the commutative multiplication of x_{ik} and $x_{j\ell}$, however, we could not observe any performance gain, and thus, omit them.

We first show that the proposed formulation is exact in the boundary case $\mathcal{S} = [n]^4$.

Lemma 1. *Let $\mathcal{S} = [n]^4$ and let $z^{(1)} = x^{(1)}$, $z^{(2)} = (x^{(2)}, y^{(2)})$ be optimal solutions of* (1) *and (2), respectively.*

Then $cost(z^{(1)}) = cost(z^{(2)})$.

Proof. Let $(i,j,k,\ell) \in [n]^4$ and consider the linear inequalities (2b)–(2f). If one of $x^{(2)}_{ik}$ and $x^{(2)}_{j\ell}$ is 0, then at least one of the inequalities (2b) to (2e) forces $y^{(2)}_{ijk\ell}$ to 0. On the other hand, if both $x^{(2)}_{ik} = x^{(2)}_{j\ell} = 1$, constraint (2f) sets $y^{(2)}_{ijk\ell}$ to 1. Therefore, and because $x^{(2)}$ is a binary variable, we can interpret $y^{(2)}_{ijk\ell}$ as the product $x^{(2)}_{ik} \cdot x^{(2)}_{j\ell}$.

Since $\mathcal{S} = [n]^4$, this observation holds for all variables and the formulations (1) and (2) coincide. □

Despite this result, we emphasize that using $\mathcal{S} = [n]^4$ leads to an intractable problem size for most practical input instances, e.g., more than 100 000 000 variables in our application. Even for sparse problems, reducing $\mathcal{S}$ to all the indices with nonzero contribution to the quadratic objective term may not suffice as, e.g., still about 2 000 000 variables remain in our case. To overcome this issue, we select an increasing sequence of subsets $\mathcal{S}$, with each subset being significantly smaller than $[n]^4$. Lemma 2 explains why it is beneficial to do so.

Lemma 2. *Let $\mathcal{S} \subset [n]^4$ and z^* be the optimal solution of (2). Then $cost(z^*)$ is a lower bound for* (1).

Proof. Let $(P, c, q), (P', c', q')$ be the polytopes and objective functions of (2) defined over $\mathcal{S}$ and $[n]^4$, respectively. Clearly, the set of constraints of P form a subset of the constraints of P'. Hence, every feasible solution in P' is also feasible in P, i.e., $P' \subseteq P$.

Setting $q_{ijk\ell} = 0$ for all $(i,j,k,\ell) \notin \mathcal{S}$, we can write (2a) as

$$\sum_{i,k=1}^{n} c_{ik} x_{ik} + \sum_{(i,j,k,\ell)\in[n]^4} q_{ijk\ell} y_{ijk\ell}.$$

Since all terms in the objective functions are assumed to be non-negative, it holds for every $(i,j,k,\ell) \in [n]^4$ that $q_{ijk\ell} \le q'_{ijk\ell}$, which concludes the proof. □

This lemma shows that dynamically increasing $\mathcal{S}$ yields a hierarchy of integer linear programs with increasing bounds for the original QAP. The proposed iterative algorithm later in this section is a natural consequence of Lemma 2. It remains to show how to initialize and update the set $\mathcal{S}$. We remark here that it is advisable to solve the integer linear programs close to optimality instead of considering their linear programming relaxations. Although dropping the integrality constraints drastically reduces the computation time with growing $\mathcal{S}$, the resulting lower bounds have shown to be significantly worse than the ones obtained by solving the integral versions for the same amount of time.

We now present two variants of the algorithm, which differ only in the procedure on how to grow $\mathcal{S}$.

Variant 1: Conservative Growth. First, we choose an arbitrary $\varepsilon \geq 0$. We will show later that the algorithm then produces a $(1+\varepsilon)$-approximation of the optimal assignment. Note, however, that our algorithm allows to choose $\varepsilon = 0$, then computing an optimal solution. Assume that for a given index set $\mathcal{S}$, we computed an optimal binary solution (x^*, y^*) with objective value V. We build the candidate set

$$C := \{(i,j,k,\ell) \notin \mathcal{S} : x^*_{ik} = x^*_{j\ell} = 1 \text{ and } q_{ijk\ell} > 0\} \tag{3}$$

and sort C in an ascending order with respect to $q_{ijk\ell}$. Note that $|C| \leq n^2$. Formally, we define the function $\pi : [|C|] \to [n]^4$ such that for every $i < j \in \{1, \dots, |C|\}$, it holds $q_{\pi(i)} \leq q_{\pi(j)}$. Let s be the index that satisfies the following equation.

$$s = \max\left\{ t \in \{0, \dots, |C|\} : \sum_{\alpha=1}^{t} q_{\pi(\alpha)} \leq \varepsilon \cdot V \right\} \tag{4}$$

Intuitively, we skip the s smallest positive cost values that sum up to a certain threshold and add the rest of the indices to our active set $\mathcal{S}$.

The complete algorithm is presented in Algorithm 1. Theorem 1 shows that the update step eventually yields a $(1+\varepsilon)$-approximation.

Input : number of items/locations n, linear cost c, quadratic cost q, precision parameter ε

Result: Optimal assignment or upper/lower bound if aborted

1 $\mathcal{S} \leftarrow \emptyset$;
2 **do**
3     $(x^*, y^*) \leftarrow$ opt. sol. of (2) with $\mathcal{S}$;
4     $V \leftarrow$ evaluate x^* at (1);
5   C, π, s as in equations (3)-(4);
6     $\mathcal{S} \leftarrow \mathcal{S} \cup \{\pi(i)\}_{i=s+1}^{|C|}$;
7 **while** $\mathcal{S}$ *changed in line 6*;
8 **return** x^*;

Algorithm 1. The complete algorithm (conservative version)

Theorem 1. *Let $\varepsilon \geq 0$. If line 6 of Algorithm 1 does not add any index to $\mathcal{S}$, then the x-part of the current solution (x^*, y^*) is $(1+\varepsilon)$-optimal for problem* (1).

Proof. Since $C \cap \mathcal{S} = \emptyset$ by definition, the only reason why $\mathcal{S}$ did not change is that $s = |C|$. In particular, this means that

$$\sum_{\alpha \in C} q_\alpha \leq \varepsilon \cdot cost(x^*, y^*)$$

Let $\tilde{x}$ be the optimal solution of (1). We evaluate (x^*, y^*) on the complete objective function of the QAP and interpret $y^*_{ijk\ell} = x^*_{ik} x^*_{k\ell}$, which is a valid assumption already shown in the proof of Lemma 1. Then, we obtain an upper bound for $\tilde{x}$.

$$\begin{aligned}
OPT &= \sum_{i,k=1}^{n} c_{ik}\tilde{x}_{ik} + \sum_{i,j,k,\ell=1}^{n} q_{ijk\ell}\tilde{x}_{ik}\tilde{x}_{j\ell} \\
&\leq \sum_{i,k=1}^{n} c_{ik}x^*_{ik} + \sum_{(i,j,k,\ell)\in\mathcal{S}} q_{ijk\ell}x^*_{ik}x^*_{j\ell} + \sum_{(i,j,k,\ell)\notin\mathcal{S}} q_{ijk\ell}x^*_{ik}x^*_{j\ell} \\
&= \sum_{i,k=1}^{n} c_{ik}x^*_{ik} + \sum_{(i,j,k,\ell)\in\mathcal{S}} q_{ijk\ell}x^*_{ik}x^*_{j\ell} + \sum_{(i,j,k,\ell)\in C} q_{ijk\ell}x^*_{ik}x^*_{j\ell} \\
&\leq cost(x^*, y^*) + \varepsilon cost(x^*, y^*) = (1+\varepsilon)cost(x^*, y^*)
\end{aligned}$$

□

Variant 2: Progressive Growth. We change the definition of the candidate set C in Eq. (3) to

$$C' := \left\{(i,j,k,\ell) \notin \mathcal{S} : x^*_{ik} = 1 \vee x^*_{j\ell} = 1 \text{ and } q_{ijk\ell} > 0\right\}. \tag{5}$$

This means we consider a tuple as a candidate if at least one of the corresponding x-variables were set to 1 in the previous optimal solution (instead of requiring both variables to be 1). The rest of the algorithm remains the same. In this second variant, $|C'| \leq n^3$, i.e., we potentially add more terms to the model. This can improve the evolution of lower bounds because we consider a more substantial portion of the model more quickly. As a trade-off, we potentially add more irrelevant terms than the conservative variant and, additionally, we could quickly arrive at a model of a size that exceeds the resources of the computer used to run the algorithm. Note that Theorem 1 also holds for this variant of the algorithm, the proof is analogue to the proof shown above with the extra information that $C \subset C'$.

Variant 3: Hybrid Strategy. To achieve a balance between the fast evolution of lower bounds in variant 2 and the moderate growth of model size in variant 1, we propose the hybrid strategy that kick-starts with the progressive variant 2 and switches to the conservative variant 1 before the model size grows too large. The evaluation shows that this strategy is indeed superior to both standalone variants. For all instances, there is a critical point where the amount of generated quadratic terms would grow so large that an MIP solver cannot compute the integer optimal solution within a reasonable time. Therefore, we switch to the conservative variant at this critical point, which grows the model more slowly while still steadily improving the lower bound.

3 Evaluation

We applied our algorithm within several stages of the French keyboard standardization process. The instances consist of over 100 special characters and 130 keys (in order to achieve the classic QAP formulation, one can generate dummy characters symbolizing that a key is left empty), and the objective function consists of a conic combination of a sparse quadratic and dense linear cost terms. The quadratic term can be factorized into a sparse matrix F, which describes the association score (similarity) of two different special characters, and a dense matrix D, describing the distances between two key slots. The weight of the quadratic part ranges between 30% and 50%. Additionally, some instances fix few characters like punctuation symbols to fixed slots or require that the capital versions of special characters are placed on the shifted slot of the same letter (e.g., È is placed on the shifted slot of è) whereas other instances also allow them to be on the Alt-Shift or Alt version of this slot.

We evaluated the instances on a single Intel(R) Xeon(R) CPU E5-2680 v3 @ 2.50 GHz processor core with 16 GB of RAM. We compare our algorithm against the formulation of Xia and Yuan [25] as a state-of-the-art lightweight linearization for the QAP. As already mentioned before, stronger formulations like RLT1 could not compute any lower bound within the given resources, often because the model size already exceeded the available RAM. We use Gurobi version 8.1 [12] as the underlying solver for both approaches.

We first discuss the impact of the variant choice on one example instance. More specifically, we test the hybrid strategy and the effect of the switch from progressive to conservative at different iterations τ. Figure 2 shows the evolution of lower bounds for $\tau = 1, \ldots, 10$. Since naturally the bound evolves faster during the first seconds and minutes, the graph shows a more detailed view on the evolution within this first period. Setting $\tau = 1$ leads to using the conservative variant from the beginning while setting $\tau = 10$ implies that the strategy switch does not occur within the given time window of 12 h because the model of the last iteration is already too large to be solved efficiently. We observe that as long as the model size is moderately low, the progressive variant achieves better results at every time stamp. However, after roughly 45 min, the model size for this variant already grows notably large so that the next iterations takes quite a long time. After yet another size increase, the ILP solver could not compute an optimal solution within the remaining 10 h. Note that depending on the available resources (time limit and hardware) as well as the particular instance (dimension and sparsity), the critical point at which a switch from the progressive to the conservative variant is valuable varies. Since all our instances are of similar size and sparsity, the critical point for this evaluation is at the 9th iteration.

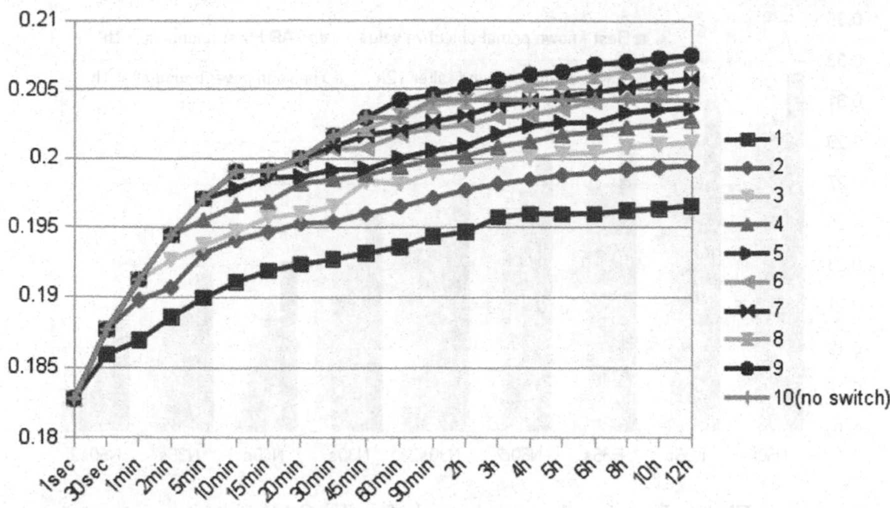

Fig. 2. The evolution of the lower bound when switching variants for instance N50s. The numbers in the legend describe the iteration at which the switch was triggered.

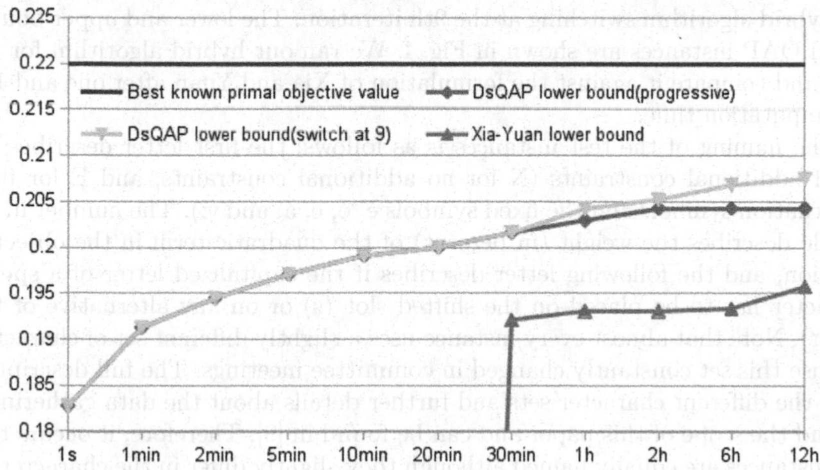

Fig. 3. The evolution of the lower bounds within 12 h of computation time for instance N50s.

Figure 3 compares the evolution of the lower bounds of our algorithm using only variant 2, switching after 9 iterations, and the formulation of Xia and Yuan within a total time period of 12 h for the same example instance. It is important to note that the setup time for the Xia-Yuan formulation is around 25 min for every instance because over 10 000 linear assignment problems are solved beforehand. Therefore, the first bound for the original QAP is only produced after 25 min.

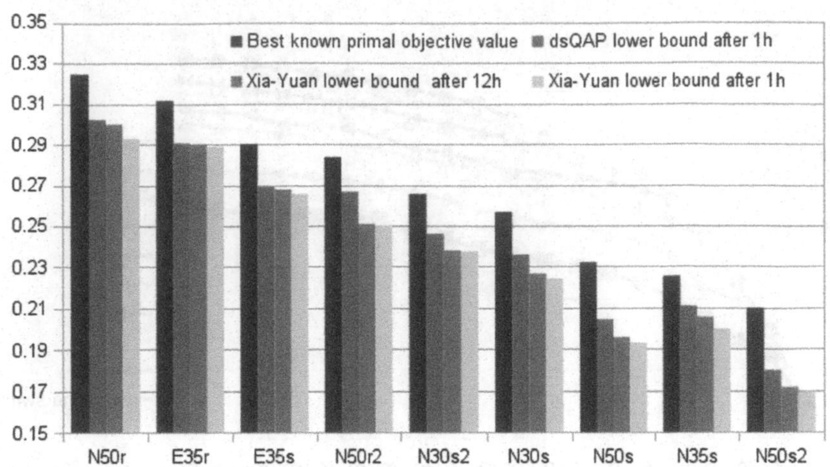

Fig. 4. Lower and upper bounds for the QAP instances

To avoid visual clutter in the following figures, we only depict the results of the hybrid algorithm switching at the 9th iteration. The lower and upper bounds for all QAP instances are shown in Fig. 4. We ran our hybrid algorithm for one hour and compare it against the formulation of Xia and Yuan after one and 12 h of computation time.

The naming of the test instances is as follows: the first letter describes the set of additional constraints (N for no additional constraints, and E for fixed punctuation symbols and the fixed symbols è, é, ê, à, and €). The number in the middle describes the weight (in percent) of the quadratic term in the objective function, and the following letter describes if the capitalized letter of a special character has to be placed on the shifted slot (s) or on any alternative of this slot (r). Note that almost every instance uses a slightly different set of characters because this set constantly changed in committee meetings. The full description of all the different character sets and further details about the data gathering is beyond the scope of this paper and can be found in [9]. Therefore, it occurs that two instances are equally named although they slightly differ in the character set used. In this case, one of the instance names ends with 2 for better differentiation.

We can see that within one hour, we outperform the formulation of Xia and Yuan for every instance independent of its time limit being one hour or 12 h. Although we slightly improved the lower bounds of all instances, this is not the true benefit of our framework. What we really want to emphasize here is how fast we achieve high-quality lower bounds, which is especially important in the practical application of our algorithm. In this highly interactive environment with countless model updates and changes, receiving valuable feedback of an optimization method after only several minutes can greatly improve the dynamics of an expert committee that discusses different proposals and has to decide the next steps towards a final keyboard standard.

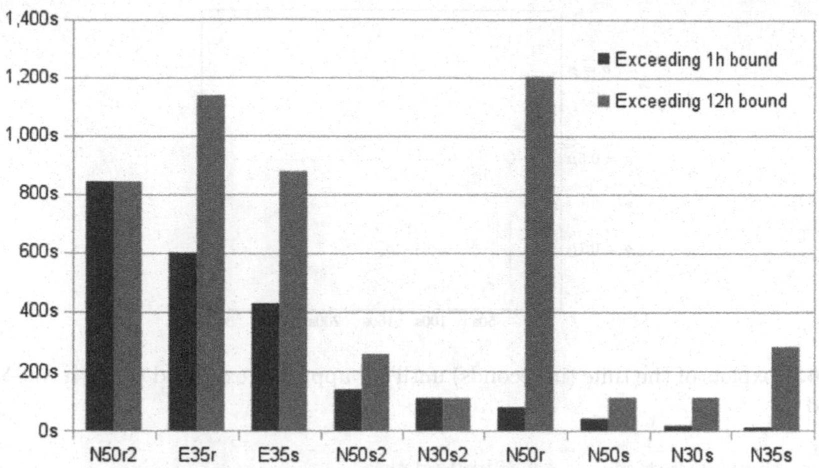

Fig. 5. The time we need to exceed the bounds of Xia and Yuan after 1 h and 12 h (in seconds)

We measure the time our algorithm needs to exceed the lower bounds that the Xia-Yuan formulation produces after 1 h and 12 h, respectively. Figure 5 shows that we achieve this goal within several minutes for all instances. In the worst case, it takes 20 min to exceed the 12 h bound of Xia-Yuan. Hence, for every instance, we achieve superior lower bounds within the setup time of 25 min that is needed for the creation of the Xia-Yuan linearization.

3.1 Robustness Analysis

To analyze the robustness of our approach, we vary the nonzero values of the quadratic cost matrix with additive noise generated by a normal distribution with 0 mean and standard deviation σ.

Recall that the quadratic matrix Q is the Kronecker product of the dense matrix D containing the distances between the keys and the sparse matrix F encoding the similarity between the special characters. We only add noise to the entries in F while keeping its entries non-negative. More specifically, consider $f_{ij} > 0$ and $\delta_{ij} \sim N(0, \sigma)$, then we set $f'_{ij} = f_{ij} + \delta_{ij}$ if $f'_{ij} > 0$, otherwise we recompute δ_{ij}. Let μ be the average value of all nonzero entries in the association matrix A, then we set σ to 10%, 50%, and 100% of μ. For this evaluation, we use the instance N35s as a base instance and generate 20 randomly variated instances for each of the three variance values.

Figure 6 shows the boxplots of the time (in seconds) our approach needed to exceed the bound that the Xia-Yuan formulation achieves after 12 h. In every of the 60 instances in total, we exceeded said bound after at most five minutes. Note that the Xia-Yuan formulation has a setup time for around 25 min for instances of this size. This means we can consistently produce high quality bounds during the setup time of the competing approach.

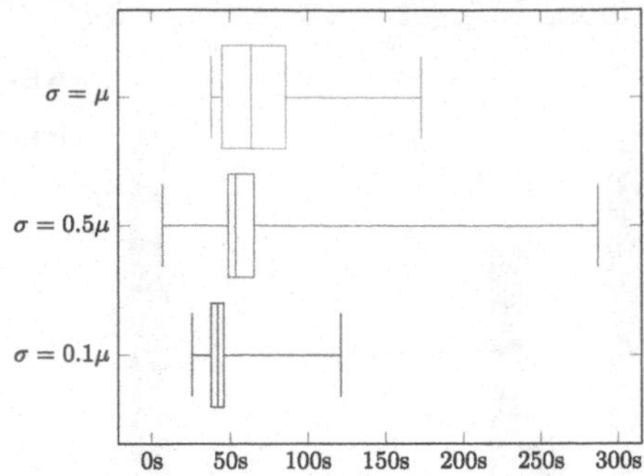

Fig. 6. Boxplots of the time (in seconds) until our approach exceeded the 12 h Xia-Yuan bound

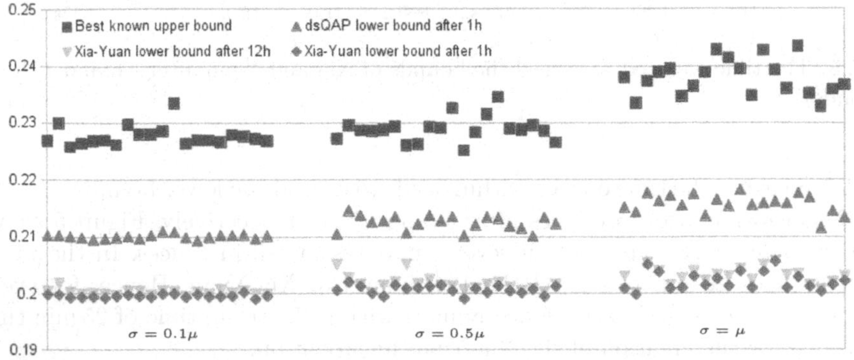

Fig. 7. Bounds for 60 randomly variated instances (20 each) with variance σ

Moreover, Fig. 7 shows the lower and upper bounds for the 20 runs each with $\sigma \in \{0.1\mu, 0.5\mu, \mu\}$, respectively. We observe that the results of these randomized instances are very consistent with the results of the original evaluation, independent of the variance.

4 Conclusion

We presented a lightweight framework for sparse quadratic assignment problems that combines powerful linearization techniques and ideas from column-generation. It is lightweight in a sense that it can generate good bounds for sparse QAPs of huge size (over 100 items) on a normal laptop. Our algorithm was used in the process of defining the new French keyboard standard. The evaluation, which is based on real data gathered during this standardization process, showed that we can compete with state-of-the-art linearization techniques. We showed that we can produce high quality lower bounds within several minutes, which serves the purpose of almost real-time feedback in such a dynamic interactive optimization process.

References

1. Adams, W., Johnson, T.: Improved linear programming-based lower bounds for the quadratic assignment problem. DIMACS 512 Ser. Discret. Math. Theor. Comput. Sci. **16**, 43–77 (1994). https://doi.org/10.1090/dimacs/016/02
2. AFNOR: Interfaces utilisateurs - Dispositions de clavier bureautique français, NF Z71–300 Avril 2019
3. Arkin, E.M., Hassin, R., Sviridenko, M.: Approximating the maximum quadratic assignment problem. Inf. Process. Lett. **77**(1), 13–16 (2001). https://doi.org/10.1016/S0020-0190(00)00151-4
4. Birkhoff, D.: Tres observaciones sobre el algebra lineal. Universidad Nacional de Tucuman Revista Serie A **5**, 147–151 (1946)
5. Burkard, R.E., Çela, E., Pardalos, P.M., Pitsoulis, L.S.: The Quadratic Assignment Problem, pp. 1713–1809. Springer, Boston (1998). https://doi.org/10.1007/978-1-4613-0303-9_27
6. Burkard, R., Offermann, J.: Entwurf von Schreibmaschinentastaturen mittels quadratischer Zuordnungsprobleme. Zeitschrift für Oper. Res. **21**, 121–132 (1977)
7. DGLFLF: Rapport au Parlement sur l'emploi de la langue française. Government Report (2015). http://www.culture.gouv.fr/Thematiques/Langue-francaise-et-langues-de-France/La-DGLFLF/Nos-priorites/Rapport-au-Parlement-sur-l-emploi-de-la-langue-francaise-2015. From the Délégation générale à la langue française et aux langues de France of the Ministère de la Culture et de la Communication (in French)
8. DGLFLF: Vers une norme française pour les claviers informatiques. Government Publication (2016). http://www.culture.gouv.fr/Thematiques/Langue-francaise-et-langues-de-France/Politiques-de-la-langue/Langues-et-numerique/Les-technologies-de-la-langue-et-la-normalisation/Vers-une-norme-francaise-pour-les-claviers-informatiques. From the Délégation générale à la langue française et aux langues de France of the Ministère de la Culture et de la Communication (in French)
9. Feit, A.M.: Assignment Problems for Optimizing Text Input. G5 artikkeliväitöskirja (2018). http://urn.fi/URN:ISBN:978-952-60-8016-1
10. Frieze, A., Yadegar, J.: On the quadratic assignment problem. Discrete Appl. Math. **5**(1), 89–98 (1983). https://doi.org/10.1016/0166-218X(83)90018-5
11. Gilmore, P.C.: Optimal and suboptimal algorithms for the quadratic assignment problem. SIAM J. Appl. Math. **10**, 305–313 (1962)
12. Gurobi Optimization, L.: Gurobi Optimizer Version 8.1 (2019). http://www.gurobi.com
13. Huber, C., Riedl, W.: The Quadratic Assignment Problem: the Linearization of Xia and Yuan is Weaker than the Linearization of Adams and Johnson and a Family of Cuts to Narrow the Gap, preprint on webpage at https://arxiv.org/abs/1710.02472
14. John, M., Karrenbauer, A.: A Novel SDP Relaxation for the Quadratic Assignment Problem Using Cut Pseudo Bases, pp. 414–425. Springer, Cham (2016). https://doi.org/10.1007/978-3-319-45587-7_36
15. Kaufman, L., Broeckx, F.: An algorithm for the quadratic assignment problem using Benders' decomposition. Eur. J. Oper. Res. **2**(3), 207–211 (1978). https://doi.org/10.1016/0377-2217(78)90095-4

16. Koopmans, T., Beckmann, M.J.: Assignment Problems and the Location of Economic Activities. Cowles Foundation Discussion Papers 4, Cowles Foundation for Research in Economics, Yale University (1955). http://EconPapers.repec.org/RePEc:cwl:cwldpp:4
17. Lawler, E.L.: The quadratic assignment problem. Manag. Sci. **9**(4), 586–599 (1963). https://doi.org/10.1287/mnsc.9.4.586
18. Lee, Y., Orlin, J.B.: On Very Large Scale Assignment Problems, pp. 206–244. Springer, Boston (1994). https://doi.org/10.1007/978-1-4613-3632-7_12
19. Nugent, C., Vollman, T., Ruml, J.: An experimental comparison of techniques for the assignment of facilities to locations. Oper. Res. **16**(1), 150–173 (1968). https://doi.org/10.1287/opre.16.1.150
20. Peng, J., Mittelmann, H., Li, X.: A new relaxation framework for quadratic assignment problems based on matrix splitting. Math. Program. Comput. **2**(1), 59–77 (2010). https://doi.org/10.1007/s12532-010-0012-6
21. Pollatschek, M., Gershoni, N., Radday, Y.: Optimization of the typewriter keyboard by simulation. Angewandte Mathematik **10** (1976)
22. Povh, J., Rendl, F.: Copositive and Semidefinite relaxations of the quadratic assignment problem. Discret. Optim. **6**(3), 231–241 (2009). https://doi.org/10.1016/j.disopt.2009.01.002
23. Queyranne, M.: Performance ratio of polynomial heuristics for triangle inequality quadratic assignment problems. Oper. Res. Lett. **4**(5), 231–234 (1986). https://doi.org/10.1016/0167-6377(86)90007-6
24. Sherali, H.D., Adams, W.P.: A hierarchy of relaxations and convex hull characterizations for mixed-integer zero-one programming problems. Discret. Appl. Math. **52**(1), 83–106 (1994). https://doi.org/10.1016/0166-218X(92)00190-W
25. Xia, Y., Yuan, Y.X.: A new linearization method for quadratic assignment problems. Optim. Methods Softw. **21**(5), 805–818 (2006). https://doi.org/10.1080/10556780500273077
26. Zhang, H., Beltran-Royo, C., Ma, L.: Solving the quadratic assignment problem by means of general purpose mixed integer linear programming solvers. Ann. OR **207**, 261–278 (2013)
27. Zhao, Q., Karisch, S.E., Rendl, F., Wolkowicz, H.: Semidefinite programming relaxations for the quadratic assignment problem. J. Comb. Optim. **2**(1), 71–109 (1998). https://doi.org/10.1023/A:1009795911987

On Vertex Adjacencies in the Polytope of Pyramidal Tours with Step-Backs

Andrei Nikolaev(✉)

P.G. Demidov Yaroslavl State University, Yaroslavl, Russia
andrei.v.nikolaev@gmail.com

Abstract. We consider the traveling salesperson problem in a directed graph. The pyramidal tours with step-backs are a special class of Hamiltonian tours for which the traveling salesperson problem is solved by dynamic programming in polynomial time. The polytope of pyramidal tours with step-backs PSB(n) is defined as the convex hull of the characteristic vectors of all possible pyramidal tours with step-backs in a complete directed graph. The skeleton of PSB(n) is the graph whose vertex set is the vertex set of PSB(n) and the edge set is the set of geometric edges or one-dimensional faces of PSB(n). The main result of the paper is a necessary and sufficient condition for vertex adjacencies in the skeleton of the polytope PSB(n) that can be verified in polynomial time.

Keywords: Traveling salesperson problem · Directed graph · Pyramidal tour with step-backs · Polytope · 1-skeleton · Vertex adjacency

1 Introduction

We consider a classic asymmetric traveling salesperson problem: for a given complete weighted digraph $D_n = (V, E)$ it is required to find a Hamiltonian tour of minimum weight. We denote by HT_n the set of all Hamiltonian tours in D_n. With each Hamiltonian tour $x \in HT_n$ we associate a characteristic vector $x^v \in \mathbb{R}^E$ by the following rule:

$$x_e^v = \begin{cases} 1, & \text{if an edge } e \in E \text{ is contained in the tour } x, \\ 0, & \text{otherwise.} \end{cases}$$

The polytope

$$\mathrm{ATSP}(n) = \mathrm{conv}\{x^v \mid y \in HT_n\}$$

is called *the asymmetric traveling salesperson polytope.*

The *skeleton* of a polytope P (also called 1-*skeleton*) is the graph whose vertex set is the vertex set of P (the characteristic vectors x^v for the traveling salesperson problem) and edge set is the set of geometric edges or one-dimensional faces of P. Many papers are devoted to the study of 1-skeletons associated with

M. Khachay et al. (Eds.): MOTOR 2019, LNCS 11548, pp. 247–263, 2019.
https://doi.org/10.1007/978-3-030-22629-9_18

combinatorial problems. On the one hand, the vertex adjacencies in 1-skeleton are of great interest for the development of algorithms to solve problems based on local search technique (when we choose the next solution as the best one among adjacent solutions). For example, various algorithms for perfect matching, set covering, independent set, a ranking of objects, problems with fuzzy measures, and many others are based on this idea [1,3,10,11,13]. On the other hand, some characteristics of 1-skeleton of the problem, such as the diameter and the clique number, estimate the time complexity for different computation models and classes of algorithms [4,5,9,15].

However, for such combinatorial problems as a knapsack, a set partition and set covering, an integer programming, a leaf-constrained and degree-constrained minimum spanning tree, a connected k-factor and some others already the question whether two vertices in 1-skeleton are adjacent is an NP-complete problem [8,17,22]. Historically, the first result of this type was obtained by Papadimitriou for the traveling salesperson polytope.

Theorem 1 (Papadimitriou, [19]). *The question whether two vertices of the polytope* ATSP(n) *are nonadjacent is NP-complete.*

In this regard, the study of 1-skeleton of the traveling salesperson problem has shifted to the study of individual faces of the polytope [20,21], the polytopes of related problems [2], as well as the polytopes of special cases of the traveling salesperson problem. In particular, for the polytope of the pyramidal tours, it was established that the verification of the vertex adjacency in 1-skeleton can be performed in linear time [6,7].

In this paper, we consider a 1-skeleton of a wider class of the pyramidal tours with step-backs.

2 Pyramidal Tours with Step-Backs

We suppose that the cities are labeled from 1 to n. Let τ be a Hamiltonian tour. We denote the successor of i-th city as $\tau(i)$. For any natural k, we denote the k-th successor of i as $\tau^k(i)$, the k-th predecessor of i as $\tau^{-k}(i)$.

The city i satisfying $\tau^{-1}(i) < i$ and $\tau(i) < i$ is called a *peak.*

A *pyramidal tour* is a Hamiltonian tour with only one peak n.

A *step-back peak* (Fig. 1) is the city i, such that either

$$\tau^{-1} < i,\ \tau(i) = i - 1 \text{ and } \tau^2(i) > i,$$

or

$$\tau^{-2} > i,\ \tau^{-1}(i) = i - 1 \text{ and } \tau(i) < i.$$

A *proper peak* is a peak i which is not a step-back peak. A *pyramidal tour with step-backs* is a Hamiltonian tour with exactly one proper peak n.

Traveling salesperson problem on pyramidal tours is one of the most studied polynomial special cases of the problem [14]. A more general class of pyramidal tours with step-backs was introduced in [12]. These tours are of interest, since, on

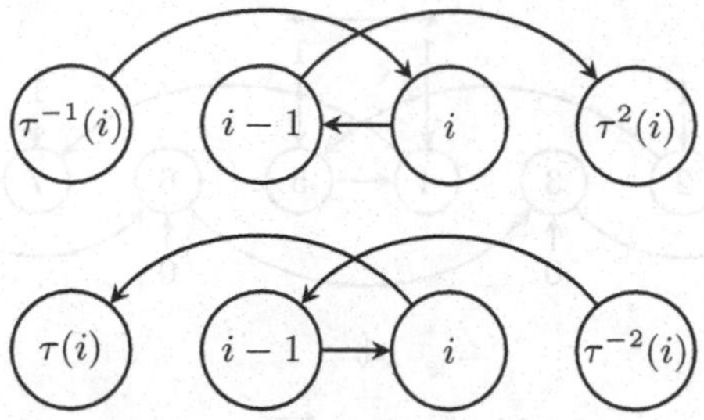

Fig. 1. A step-back in ascending and descending order

the one hand, the minimum cost pyramidal tour with step-backs can be found in $O(n^2)$ time by dynamic programming, and, on the other hand, there are known restrictions on the distance matrix that guarantee the existence of an optimal tour that is pyramidal with step-backs [12].

A generalization of pyramidal tours with step-backs is the class of quasi-pyramidal tours for which the traveling salesperson problem is fixed-parameter tractable [16,18].

We denote by $PSBT_n$ the set of all pyramidal tours with step-backs in the complete digraph $D_n = (V, E)$. With each pyramidal tour with step-backs $x \in PSBT_n$ we associate a characteristic vector $x^v \in \mathbb{R}^E$ by the following rule:

$$x_e^v = \begin{cases} 1, & \text{if an edge } e \in E \text{ is contained in the tour } x, \\ 0, & \text{otherwise.} \end{cases}$$

The polytope

$$\mathrm{PSB}(n) = \mathrm{conv}\{x^v \mid x \in PSBT_n\}$$

is called the *polytope of pyramidal tours with step-backs.*

Besides we use a special encoding to represent the pyramidal tours with step-backs. With each tour $x \in PSBT_n$ we associate an encoding vector $x^{0,1,sb}$ of length $n-2$, each coordinate corresponds to a city from 2 to $n-1$, by the following rule:

$$x_i^{0,1,sb} = \begin{cases} 1, & \text{if } i \text{ is visited by } x \text{ in ascending order,} \\ \overleftarrow{1\,1}, & \text{if } i \text{ is a step-back peak in ascending order,} \\ 0, & \text{if } i \text{ is visited by } x \text{ in descending order,} \\ \overrightarrow{0\,0}, & \text{if } i \text{ is a step-back peak in descending order.} \end{cases}$$

Note that a step-back peak i also involves the previous coordinate $i-1$. An example of a pyramidal tour with step-backs and the corresponding encoding vector $x^{0,1,sb}$ is shown in Fig. 2.

We denote by $x_{[i,j]}^{0,1,sb}$ a fragment of encoding on coordinates from i to j. The superscript indicates what we consider in the encoding: descending order (0), ascending order (1), or step-backs (sb). For example, $x_{[i,j]}^{1,sb}$ means a fragment of

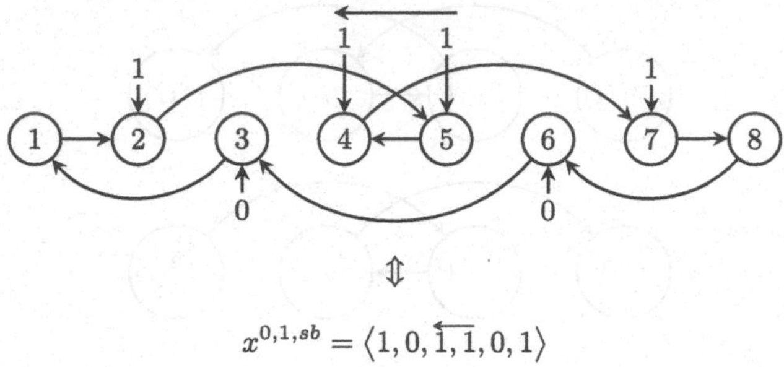

$$x^{0,1,sb} = \langle 1, 0, \overleftarrow{1, 1}, 0, 1 \rangle$$

Fig. 2. An example of a tour and the corresponding encoding

the encoding only in ascending order taking into account step-backs; $x^{0,1}_{[i,j]}$ – a fragment of the encoding disregarding step-backs, etc.

3 Auxiliary Statements

We denote by $x \cup y$ a multigraph that contains all edges of both tours x and y.

Lemma 1 (Sufficient condition for nonadjacency). *Given two tours x and y, if the multigraph $x \cup y$ includes a pair of other complementary pyramidal tours with step-backs, then the corresponding vertices x^v and y^v of the polytope* PSB(n) *are not adjacent.*

Proof. Let two complementary pyramidal tours with step-backs z and t be composed from the edges of x and y, then for the corresponding vertices the following equality takes place:

$$x^v + y^v = z^v + t^v. \tag{1}$$

We divide the equality (1) by 2 and obtain that the segment $[x^v, y^v]$ intersects with the segment $[z^v, t^v]$. Therefore, the vertices x^v and y^v of the polytope PSB(n) are not adjacent.

Lemma 2 (Necessary condition for nonadjacency). *If the vertices x^v and y^v of the polytope* PSB(n) *are not adjacent, then the multigraph $x \cup y$ includes at least two pyramidal tours with step-backs other than x and y.*

Proof. Let the vertices x^v and y^v of the polytope PSB(n) be not adjacent, then the segment $[x^v, y^v]$ intersects with the convex hull of some of the remaining vertices of the polytope PSB(n):

$$\begin{aligned} \alpha x^v + (1-\alpha) y^v &= \beta_1 z_1^v + \ldots + \beta_m z_m^v, \\ \forall i: \ \beta_i > 0 \text{ and } & \beta_1 + \ldots + \beta_m = 1. \end{aligned} \tag{2}$$

Note that $m \geq 2$, since the segment connecting two vertices of a convex polytope cannot intersect a third vertex. If at least one tour z_i includes an edge e that does not belong to the multigraph $x \cup y$, then the equality (2) is violated in the coordinate corresponding to the edge e.

Lemma 3. *Let x and y be two pyramidal tours with step-backs. Suppose that there are two edges e_x of x and e_y of y that no pyramidal tour with step-backs can include both edges e_x and e_y at the same time. Let the corresponding vertices x^v and y^v of the polytope* PSB(n) *be not adjacent. Then the convex combination of the remaining vertices of* PSB(n), *that coincides with the convex combination of x^v and y^v, cannot include with a nonzero coefficient any vertex corresponding to the tour without at least one edge of the pair e_x and e_y.*

Proof. Since the vertices x^v and y^v are not adjacent, their convex hull intersects with the convex hull of the remaining vertices of the polytope PSB(n):

$$\begin{gathered}\alpha x^v + (1-\alpha) y^v = \sum \beta_i z_i^v + \sum \beta_{j,x} z_{j,x}^v + \sum \beta_{k,y} z_{k,y}^v, \\ \sum \beta_i + \sum \beta_{j,x} + \sum \beta_{k,y} = 1,\end{gathered} \tag{3}$$

where $z_{j,x}$ are all pyramidal tours with step-backs, containing the edge e_x, and $z_{j,y}$ are all pyramidal tours with step-backs, containing the edge e_y. The remaining tours z_i do not contain edges e_x and e_y, and no pyramidal tour with step-backs can include both edges e_x and e_y at the same time. The equality (3) in the coordinates, corresponding to e_x and e_y, takes the form of a system

$$\begin{cases} \alpha = \sum \beta_{j,x}, \\ 1-\alpha = \sum \beta_{k,y}. \end{cases}$$

Therefore, $\sum \beta_i = 0$.

4 Necessary and Sufficient Condition for Adjacency

We consider 12 blocks of the following form (a wavy line means that the corresponding coordinate can either contain a step-back or not):

$$U_{11} = \left\langle \begin{matrix} 1 \\ 1 \end{matrix} \right\rangle, \; U_{00} = \left\langle \begin{matrix} 0 \\ 0 \end{matrix} \right\rangle, \; U_{1111} = \left\langle \begin{matrix} \overleftarrow{1\;\;1} \\ \overleftarrow{1\;\;1} \end{matrix} \right\rangle, \; U_{0000} = \left\langle \begin{matrix} \overrightarrow{0\;\;0} \\ \overrightarrow{0\;\;0} \end{matrix} \right\rangle,$$

$$L_{1110} = \left\langle \begin{matrix} \overleftarrow{1\;\;1} \\ 1\;\;\tilde{0} \end{matrix} \right\rangle, \; L_{1011} = \left\langle \begin{matrix} 1\;\;\tilde{0} \\ \overleftarrow{1\;\;1} \end{matrix} \right\rangle, \; L_{0001} = \left\langle \begin{matrix} \overrightarrow{0\;\;0} \\ 0\;\;\tilde{1} \end{matrix} \right\rangle, \; L_{0100} = \left\langle \begin{matrix} 0\;\;\tilde{1} \\ \overrightarrow{0\;\;0} \end{matrix} \right\rangle,$$

$$R_{1101} = \left\langle \begin{matrix} \overleftarrow{1\;\;1} \\ \tilde{0}\;\;1 \end{matrix} \right\rangle, \; R_{0111} = \left\langle \begin{matrix} \tilde{0}\;\;1 \\ \overleftarrow{1\;\;1} \end{matrix} \right\rangle, \; R_{0010} = \left\langle \begin{matrix} \overrightarrow{0\;\;0} \\ \tilde{1}\;\;0 \end{matrix} \right\rangle, \; R_{1000} = \left\langle \begin{matrix} \tilde{1}\;\;0 \\ \overrightarrow{0\;\;0} \end{matrix} \right\rangle.$$

Theorem 2. *Vertices x^v and y^v of the polytope* PSB(n) *are not adjacent if and only if the following conditions are satisfied.*

- *There exists a city i (called a left block) such that the tours x and y on the coordinate i (coordinates i and $i+1$ for double blocks) have the form of U, L, or $i = 1$.*

- *There exists a city j (called a right block) such that the tours x and y on the coordinate j (coordinates $j-1$ and j for double blocks) have the form of U, R, or $j = n$.*
  We denote by i_a the first city after the left block: $i_a = i+1$ for single blocks and $i_a = i+2$ for double blocks. We denote by j_b the last city before the right block: $j_b = i-1$ for single blocks and $j_b = j-2$ for double blocks.
  Two blocks cut the encoding of the tours into three parts: the left (less than i_a), the central (from i_a to j_b) and the right (larger than j_b).
- *In the central part, the coordinates of $x^{0,1}$ and $y^{0,1}$ completely coincide: $x^{0,1}_{[i_a,j_b]} = y^{0,1}_{[i_a,j_b]}$.*
  We say that two tours
  - *differ in the left part if $x^{0,1,sb}_{[1,i_a-1]} \neq y^{0,1,sb}_{[1,i_a-1]}$,*
  - *differ in the right part if $x^{0,1,sb}_{[j_b+1,n]} \neq y^{0,1,sb}_{[j_b+1,n]}$,*
  - *differ in the central part in ascending order if $x^{1,sb}_{[i_a,j_b]} \neq y^{1,sb}_{[i_a,j_b]}$,*
  - *differ in the central part in descending order if $x^{0,sb}_{[i_a,j_b]} \neq y^{0,sb}_{[i_a,j_b]}$.*

  The remaining conditions are divided into four cases depending on the values of $x^{0,1}_i$ and $x^{0,1}_j$.
  1. *If $x^{0,1}_i = x^{0,1}_j = 1$, then the tours differ*
     - *in the central part in ascending order;*
     - *in the left part, or in the central part in descending order, or in the right part.*
  2. *If $x^{0,1}_i = x^{0,1}_j = 0$, then the tours differ*
     - *in the central part in descending order;*
     - *in the left part, or in the central part in ascending order, or in the right part.*
  3. *If $x^{0,1}_i = 1, x^{0,1}_j = 0$, then the tours differ*
     - *in the central part in ascending order or in the right part;*
     - *in the central part in descending order or in the left part.*
  4. *If $x^{0,1}_i = 0, x^{0,1}_j = 1$, then the tours differ*
     - *in the central part in descending order or in the right part;*
     - *in the central part in ascending order or in the left part.*

  Cities 1 and n can be considered in the encoding as visited in ascending or descending order, if required.

Proof. Necessity. Let the vertices x^v and y^v of PSB(n) be not adjacent, then by Lemma 2 there exists a pyramidal tour with step-backs $z \subset x \cup y$, different from x and y, such that the vertex z^v is in a convex combination

$$\alpha x^v + (1-\alpha) y^v = \beta z^v + \sum \beta_i z_i$$

with a nonzero coefficient.

We choose the city i with the smallest number such that z enters i along an edge of the tour x, and leaves along an edge of the tour y. We choose the city j with the smallest number such that z enters j along an edge of the tour y, and

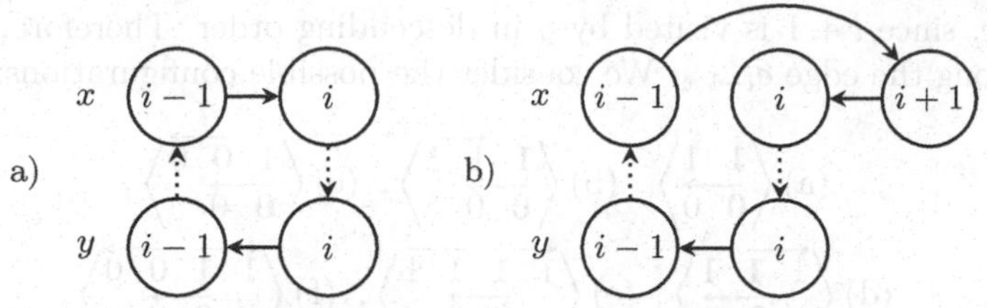

Fig. 3. Transition $1 \to 0$ (cases 1 (a) and (b))

leaves along an edge of the tour x. By construction, the tour z contains edges of both x and y, therefore such cities exist.

Part 1. Let us prove that the city i (city j) cannot be visited by the tours x and y in opposite orders. Without loss of generality, we consider the case when the city i is visited by z and x in ascending order, and by y in descending order. The remaining cases are treated similarly since they are completely symmetric.

Let us consider an edge $e_{i,y}$ of the tour y that leaves i and is included in the tour z. It can lead either to a city with a smaller number or to a city with a larger number.

1. Let the edge $e_{i,y}$ lead to a city with a smaller number. However, the edge $e_{i,y}$ is a part of z in ascending order. The only possible option is that $e_{i,y}$ is an edge of the form $(i-1) \leftarrow (i)$ that will be a step-back in ascending order of the tour z. In the multigraph $x \cup y$ there are two edges leaving $i-1$: $e_{i-1,x}$ of x and $e_{i-1,y}$ of y. The edge $e_{i-1,y}$ leads to a city with a smaller number, since $i-1$ is visited by y in descending order. Thus, it cannot be a part of z, otherwise, z is not a pyramidal tour with step-backs. Therefore, z can go only along the edge $e_{i-1,x}$ to a city with a larger number. Since the cities $i-1$ and i are visited by x in ascending order, only two configurations are possible (the city i is in bold):
$$\text{(a)} \left\langle \begin{matrix} 1 & \mathbf{1} \\ 0 & \mathbf{0} \end{matrix} \right\rangle, \ \text{(b)} \left\langle \begin{matrix} 1 & \overleftarrow{\mathbf{1} \ 1} \\ 0 & \mathbf{0} \end{matrix} \right\rangle .$$
In both of them, the tour z goes to a city that has already been passed before: (a) i, (b) $i+1$ (Fig. 3).
Hereinafter, unless stated otherwise, the following notation is used in the figures: solid edges – edges of z, dashed – edges of $(x \cup y) \backslash z$, dotted – transitions of z between edges of x and y.
2. Let the edge $e_{i,y}$ lead to a city with a larger number. However, $e_{i,y}$ is a part of y in descending order. Consequently, this is an edge of the form $(i) \to (i+1)$ that was a step-bask of y. The next edge of z has to go to a city with a larger number since we cannot return to i. Two edges $e_{i+1,y}$ of y and $e_{i+1,x}$ of x leave the city $i+1$. The edge $e_{i+1,y}$ is directed to a city with a smaller

number, since $i+1$ is visited by y in descending order. Therefore, z can go only along the edge $e_{i+1,x}$. We consider the possible configurations:

$$\text{(a)} \left\langle \begin{array}{l} \mathbf{1}\;\;1 \\ \overrightarrow{\mathbf{0}\;\;0} \end{array} \right\rangle, \text{(b)} \left\langle \begin{array}{l} \mathbf{1}\;\;\overleftarrow{1\;\;1} \\ \overrightarrow{\mathbf{0}\;\;0} \end{array} \right\rangle, \text{(c)} \left\langle \begin{array}{l} \mathbf{1}\;\;\overrightarrow{0\;\;0} \\ \overrightarrow{\mathbf{0}\;\;0} \end{array} \right\rangle,$$

$$\text{(d)} \left\langle \begin{array}{l} \overleftarrow{1\;\;\mathbf{1}}\;\;1 \\ \quad\;\overrightarrow{\mathbf{0}\;\;0} \end{array} \right\rangle, \text{(e)} \left\langle \begin{array}{l} \overleftarrow{1\;\;\mathbf{1}}\;\;\overleftarrow{1\;\;1} \\ \quad\;\overrightarrow{\mathbf{0}\;\;0} \end{array} \right\rangle, \text{(f)} \left\langle \begin{array}{l} \overleftarrow{1\;\;\mathbf{1}}\;\;\overrightarrow{0\;\;0} \\ \quad\;\overrightarrow{\mathbf{0}\;\;0} \end{array} \right\rangle.$$

(a) Transitions in cities i and $i+1$ between tours x and y do not make sense, since the edge $(i) \rightarrow (i+1)$ is included in both tours (Fig. 4).

(b) We consider the edges $(i) \rightarrow (i+2)$ of x and $(i) \leftarrow (i+r)$ of y (Fig. 4). None of them is included in z. The edge $(i) \rightarrow (i+2)$ cannot be a part of descending order, the edge $(i) \leftarrow (i+r)$ cannot be a part of ascending order. Therefore, no pyramidal tour with step-backs can contain both edges at the same time. By Lemma 3, the vertex z^v cannot be included in a convex combination that coincides with a convex combination of x^v and y^v with a nonzero coefficient. We got a contradiction.

(c) We consider the edges $(i-r) \leftarrow (i+2)$ of x and $(i-s) \leftarrow (i+1)$ of y (Fig. 5). None of them is included in z since the cities $i+1$ and $i+2$ are visited in ascending order. However, only these two edges lead from cities with numbers greater than i to cities with numbers less than i. The tour z cannot return to the city 1. We got a contradiction.

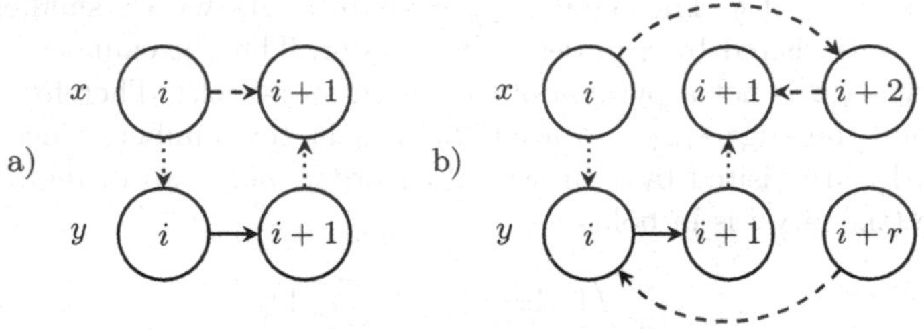

Fig. 4. Transition $1 \rightarrow 0$ (cases 2 (a) and (b))

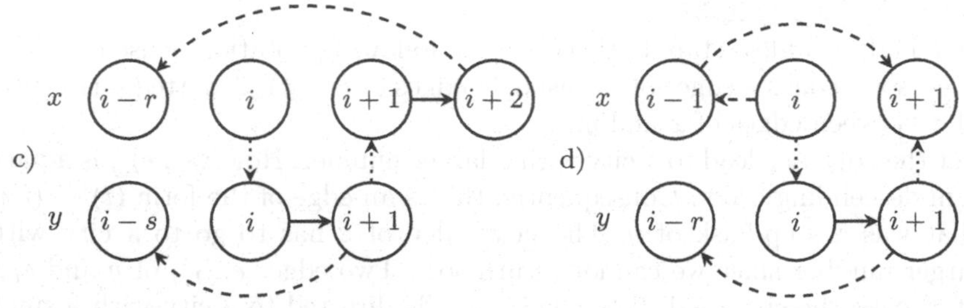

Fig. 5. Transition $1 \rightarrow 0$ (cases 2 (c) and (d))

(d) We consider the edges $(i-1) \rightarrow (i+1)$ of x and $(i-r) \leftarrow (i+1)$ of y (Fig. 5). As in case (b), these two edges are not included in z and no pyramidal tour with step-backs can contain both these edges at the same time. By Lemma 3, we got a contradiction.

(e) The configuration has the form shown in Fig. 6 (the edges of z are solid). We consider the edge $e_{i-1,y}$ of y that leaves the city $i-1$.
   - Let the edge $e_{i-1,y}$ be directed to a city with a larger number. Suppose that there exists a pyramidal tour with step-backs t that contains both the edges $(i+1) \leftarrow (i+2)$ and $(i) \leftarrow (i+s)$ that are not included in z (Fig. 7, the edges of t are solid). There are two edges from i: $(i) \rightarrow (i+1)$ and $(i-1) \leftarrow (i)$. The edge $(i) \rightarrow (i+1)$ cannot be part of t, since in this case two edges $(i) \rightarrow (i+1)$ and $(i+1) \leftarrow (i+2)$ of t enter the city $i+1$. Therefore, the edge $(i-1) \leftarrow (i)$ is a part of t, and the cities $i-1$ and i are visited by t in descending order. However, both edges leaving the city $i-1$ are directed to the cities with numbers at least $i+1$. Thus, no pyramidal tour with step-backs t can contain both edges $(i+1) \leftarrow (i+2)$ and $(i) \leftarrow (i+s)$ that are not included in z. By Lemma 3, we got a contradiction.
   - Let the edge $e_{i-1,y}$ be directed to a city with a smaller number. Suppose that there exists a pyramidal tour with step-backs t that contains both the edges $(i-1) \leftarrow (i)$ and $(i-r) \leftarrow (i+1)$ that are not included in z (Fig. 8, the edges of t are solid). There are two edges directed to $i+1$: $(i) \rightarrow (i+1)$ and $(i+1) \leftarrow (i+2)$. The edge $(i) \rightarrow (i+1)$ cannot be part of t, since in this case two edges $(i) \rightarrow (i+1)$ and $(i-1) \leftarrow (i)$ of t leave the city i. Therefore, the edge $(i+1) \leftarrow (i+2)$ is a part of t and the cities $i+1$ and $i+2$ are visited by t in descending order. In this case, the tour t has the edge $(i-p) \leftarrow (i-1)$ of two edges that leave the city $i-1$. Consequently, the cities i and $i-1$ are also visited in descending order. However, no pyramidal tour with step-backs can go along the edges $(i+1) \leftarrow (i+2)$ and $(i-r) \leftarrow (i+1)$ and visit the city i in descending order. Thus, no pyramidal tour with step-backs t can contain both the edges $(i-1) \leftarrow (i)$ and $(i-r) \leftarrow (i+1)$ that are not included in z. By Lemma 3, we got a contradiction.

(f) Similar to the case (c) there are only three edges: $(i-1) \leftarrow (i)$, $(i-r) \leftarrow (i+2)$, and $(i-s) \leftarrow (i+1)$ that lead to the cities with numbers less than i (Fig. 9), and none of them are included in z. The tour z cannot return to the city 1. We got a contradiction.

Thus, if the tour z enters the city i along an edge of x and leaves along an edge of y, or vice versa, then i is visited by x and y in the same order. Since from the city 1 there are only edges leading to the cities that are visited in ascending order by x and y, the tour z in ascending order includes only edges of ascending orders of x and y, in descending order – only edges of descending orders.

Part 2. Let us prove that transitions between the edges of x and y can only be performed by blocks U, L, R from the statement of the theorem. Without loss of generality, we assume that the city i is visited by x and y in ascending order

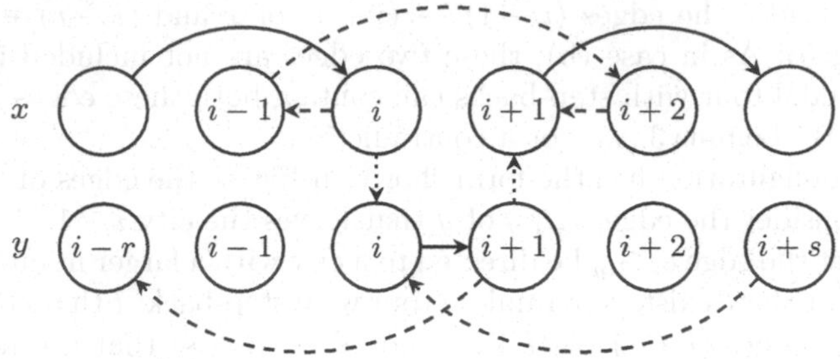

Fig. 6. Transition 1 → 0 (case 2 (e))

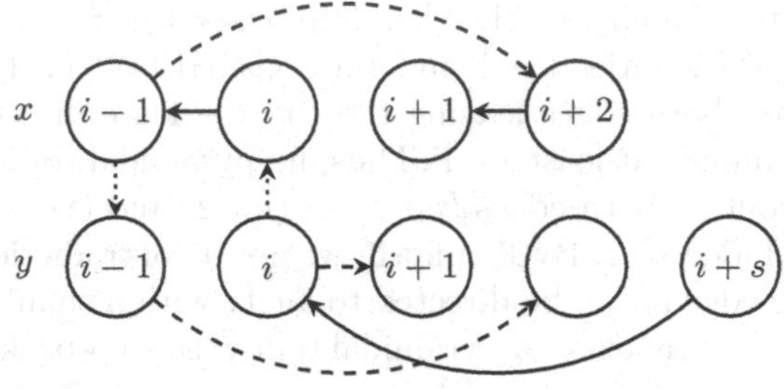

Fig. 7. Case 2 (e), $e_{i-1,y}$ is directed to a city with a larger number

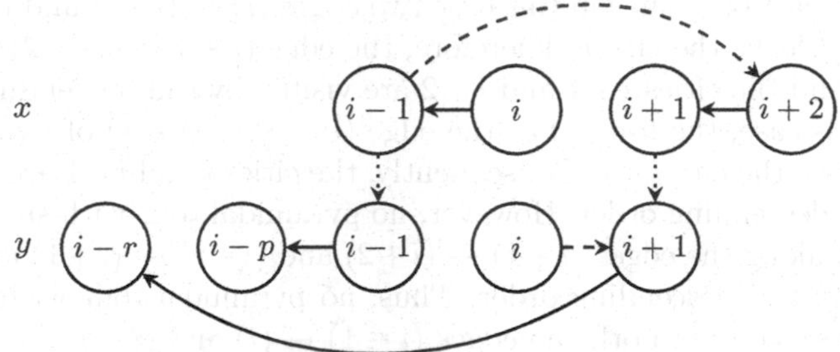

Fig. 8. Case 2 (e), $e_{i-1,y}$ is directed to a city with a smaller number

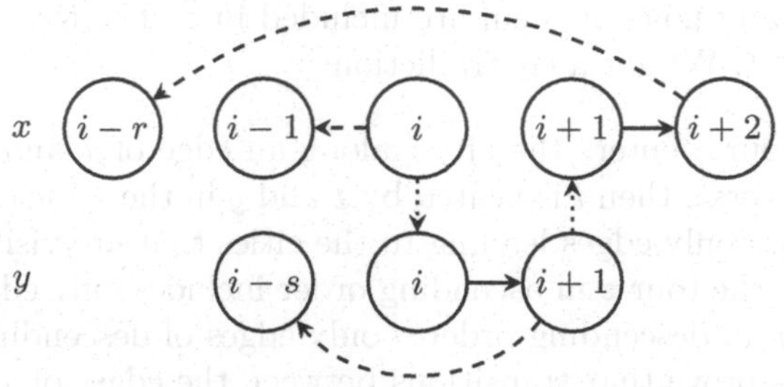

Fig. 9. Transition 1 → 0 (case 2 (f))

($x_i^{0,1} = y_i^{0,1} = 1$). The case $x_i^{0,1} = y_i^{0,1} = 0$ is treated similarly. We consider the possible configurations:

$$\text{(a)}\left\langle\begin{array}{c}1\\1\end{array}\right\rangle,\ \text{(b)}\left\langle\begin{array}{cc}1 & \\ \multicolumn{2}{c}{\overleftarrow{1\;\;1}}\end{array}\right\rangle,\ \text{(c)}\left\langle\begin{array}{cc} & 1\\ \multicolumn{2}{c}{\overleftarrow{1\;\;1}}\end{array}\right\rangle,\ \text{(d)}\left\langle\begin{array}{cc}\multicolumn{2}{c}{\overleftarrow{1\;\;1}}\\ & 1\end{array}\right\rangle,\ \text{(e)}\left\langle\begin{array}{cc}\multicolumn{2}{c}{\overleftarrow{1\;\;1}}\\ 1 & \end{array}\right\rangle,$$

$$\text{(f)}\left\langle\begin{array}{cc}\multicolumn{2}{c}{\overleftarrow{1\;\;1}}\\ \multicolumn{2}{c}{\overleftarrow{1\;\;1}}\end{array}\right\rangle,\ \text{(g)}\left\langle\begin{array}{ccc} & \multicolumn{2}{c}{\overleftarrow{1\;\;1}}\\ \multicolumn{2}{c}{\overleftarrow{1\;\;1}} & \end{array}\right\rangle,\ \text{(h)}\left\langle\begin{array}{ccc}\multicolumn{2}{c}{\overleftarrow{1\;\;1}} & \\ & \multicolumn{2}{c}{\overleftarrow{1\;\;1}}\end{array}\right\rangle.$$

(a) The transition has the form of a block U_{11}.

(b) Suppose that the city $i+1$ is visited by x in ascending order ($x_{i+1}^{0,1} = 1$):

$$\text{(b1)}\left\langle\begin{array}{cc}1 & 1\\ \multicolumn{2}{c}{\overleftarrow{1\;\;1}}\end{array}\right\rangle,\ \text{(b2)}\left\langle\begin{array}{ccc}1 & \multicolumn{2}{c}{\overleftarrow{1\;\;1}}\\ \multicolumn{2}{c}{\overleftarrow{1\;\;1}} & \end{array}\right\rangle.$$

In the case (b1) none of the edges entering the city $i+1$ was included in the tour z, therefore, z cannot be a Hamiltonian tour.
In the case (b2), the tour z can visit the city $i+1$ only along the edge $(i+1) \leftarrow (i+2)$ of x. At the same time, z can enter the city $i+2$ only along an edge of the tour y in ascending order:

$$\text{(b21)}\left\langle\begin{array}{ccc}1 & \multicolumn{2}{c}{\overleftarrow{1\;\;1}}\\ \multicolumn{2}{c}{\overleftarrow{1\;\;1}} & 1\end{array}\right\rangle,\ \text{(b22)}\left\langle\begin{array}{cccc}1 & \multicolumn{2}{c}{\overleftarrow{1\;\;1}} & \\ \multicolumn{2}{c}{\overleftarrow{1\;\;1}} & \multicolumn{2}{c}{\overleftarrow{1\;\;1}}\end{array}\right\rangle.$$

In the case (b21), transition between the tours x and y does not make sense, since the edge $(i) \rightarrow (i+2)$ is contained in both tours (Fig. 10). The case (b22) contains a transition of the configuration (g), the impossibility of which will be considered separately.
Thus, the city $i+1$ can be visited by x only in descending order ($x_{i+1}^{0,1} = 0$). The transition has the form of a block:

$$L_{1011} = \left\langle\begin{array}{cc}1 & \tilde{0}\\ \multicolumn{2}{c}{\overleftarrow{1\;\;1}}\end{array}\right\rangle.$$

(c, d, e) Similar to configuration (b), the transitions have the form of the blocks R_{0111}, R_{1101}, and L_{1110}.

(f) The transition has the form of a block U_{1111}.

(g) The tour z enters the city i by the edge $(i) \leftarrow (i+1)$ of x and leaves by the edge $(i-1) \leftarrow (i)$ of y. A pyramidal tour with step-backs cannot contain both these edges in ascending order since this is a double step-back (Fig. 10).

(h) The edges $(i-1) \leftarrow (i)$ and $(i) \leftarrow (i+1)$ are not included in z and no pyramidal tour with step-backs can contain both these edges at the same time (Fig. 11). By Lemma 3, we got a contradiction.

Thus, the transition between the edges of x and y is possible only at blocks U, L, R from the statement of the theorem. Besides, it can be done at cities 1 and n which can be considered as universal blocks.

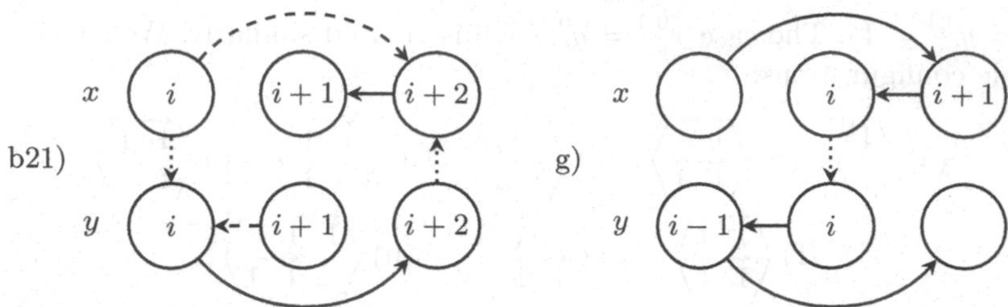

Fig. 10. Transition 1 → 1 (cases (b21) and (g))

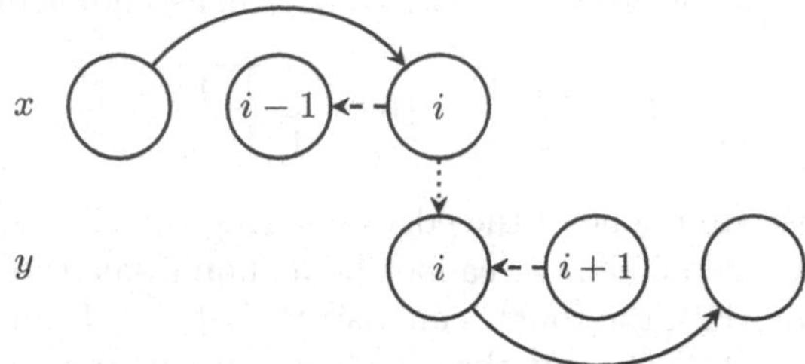

Fig. 11. Transition 1 → 1 (case (h))

Part 3. Let us prove that the remaining conditions of the theorem are satisfied. By construction, i is the city with the smallest number, such that z enters i by an edge of x and leaves by an edge of y, j is the city with the smallest number, such that z enters j by an edge of y and leaves by an edge of x.

The coordinates $x_i^{0,1}$ and $x_j^{0,1}$ can take one of the four combinations of the values 0 and 1 that are described in the statement of the theorem. Without loss of generality we assume that $i < j$ and $x_i^{0,1} = 1$, other cases are treated similarly.

Since i and j are the cities with the smallest numbers where the tour z makes transitions between edges of x and y, the traversal diagram has the form shown in Fig. 12. In particular, the tour z visits $i - 1$ by the edges of x.

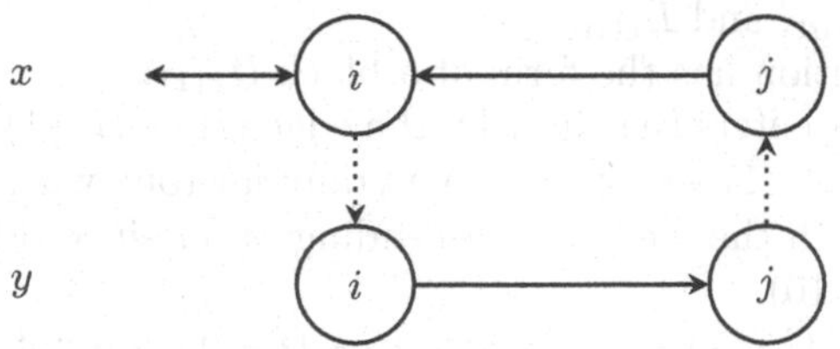

Fig. 12. The traversal diagram of the tour z if $i < j$

First, we prove that transition at i has the form of a left block U or L. Suppose the contrary, then the block has one of two possible forms:

$$R_{1101} = \left\langle \begin{matrix} \overleftarrow{1\ 1} \\ \tilde{0}\ 1 \end{matrix} \right\rangle,\ R_{0111} = \left\langle \begin{matrix} \tilde{0}\ 1 \\ \overleftarrow{1\ 1} \end{matrix} \right\rangle.$$

In the case of R_{1101}, the tour z skips the city $i-1$ in ascending order, since the edge $(i-1) \leftarrow (i)$ of x is not a part of z, and then again skips $i-1$ in descending order. In the case of R_{0111}, the tour z visits the city $i-1$ in ascending order, since this time $(i-1) \leftarrow (i)$ of y is a part of z, and then again visits $i-1$ in descending order. In both cases, the tour is not Hamiltonian. Similarly, we can prove that the transition at j cannot have the form of a left block L.

We denote by i_a the first city after the left block: $i_a = i+1$ for single blocks and $i_a = i+2$ for double blocks. We denote by j_b the last city before the right block: $j_b = i-1$ for single blocks and $j_b = j-2$ for double blocks.

Therefore, by construction, in the central part between the blocks, the tour z in ascending order goes along the edges of y: $z^{1,sb}_{[i_a,j_b]} = y^{1,sb}_{[i_a,j_b]}$, in descending order – along the edges of x: $z^{0,sb}_{[i_a,j_b]} = x^{0,sb}_{[i_a,j_b]}$. While on the left side z moves along the edges of x in both directions: $z^{0,1,sb}_{[1,i_a-1]} = x^{0,1,sb}_{[1,i_a-1]}$ (Fig. 12).

We combine the conditions for the central part:

$$\begin{cases} z^{1,sb}_{[i_a,j_b]} = y^{1,sb}_{[i_a,j_b]}, \\ z^{0,sb}_{[i_a,j_b]} = x^{0,sb}_{[i_a,j_b]} \end{cases} \Rightarrow \quad x^{0,1}_{[i_a,j_b]} = y^{0,1}_{[i_a,j_b]}.$$

Indeed, if for at least one city in the central part the coordinates of $x^{0,1}_{[i_a,j_b]}$ and $y^{0,1}_{[i_a,j_b]}$ do not match, then the tour z will either skip this city or visit it twice.

1. If $x^{0,1}_j = 1$, then both cities i and j are visited by x, y, z in ascending order. We verify the remaining conditions of the theorem.
   - If the first condition is not satisfied:
   $$x^{1,sb}_{[i_a,j_b]} = y^{1,sb}_{[i_a,j_b]} = z^{1,sb}_{[i_a,j_b]},$$
   then the transitions at the cities i and j do not make sense, since all the edges of y that are part of z as a result are also contained in the tour x.
   - If the second condition is not satisfied:
   $$\begin{cases} x^{0,sb}_{[i_a,j_b]} = y^{0,sb}_{[i_a,j_b]} = z^{0,sb}_{[i_a,j_b]}, \\ x^{0,1,sb}_{[1,i_a-1]} = y^{0,1,sb}_{[1,i_a-1]} = z^{0,1,sb}_{[1,i_a-1]}, \\ x^{0,1,sb}_{[j_b+1,n]} = y^{0,1,sb}_{[j_b+1,n]} = z^{0,1,sb}_{[j_b+1,n]}, \end{cases}$$
   then the tour z completely coincides with the tour y.
2. If $x^{0,1}_j = 0$, then the city i is visited by x, y, z in ascending order, the city j – in descending order. We verify the remaining conditions of the theorem.

- If the first condition is not satisfied:

$$\begin{cases} x^{1,sb}_{[i_a,j_b]} = y^{1,sb}_{[i_a,j_b]} = z^{1,sb}_{[i_a,j_b]}, \\ x^{0,1,sb}_{[j_b+1,n]} = y^{0,1,sb}_{[j_b+1,n]} = z^{0,1,sb}_{[j_b+1,n]}, \end{cases}$$

then the tour z completely coincides with the tour x.
- If the second condition is not satisfied:

$$\begin{cases} x^{0,sb}_{[i_a,j_b]} = y^{0,sb}_{[i_a,j_b]} = z^{0,sb}_{[i_a,j_b]}, \\ x^{0,1,sb}_{[1,i_a-1]} = y^{0,1,sb}_{[1,i_a-1]} = z^{0,1,sb}_{[1,i_a-1]}, \end{cases}$$

then the transitions at the cities i and j do not make sense, since all the edges of x that are part of z as a result are also contained in the tour y.

Thus, if the vertices x^v and y^v of the polytope PSB(n) are not adjacent, then the conditions of the theorem are satisfied.

Sufficiency. Suppose that sufficient conditions of the theorem are satisfied. We consider the pyramidal tour with step-backs z, constructed as described in Table 1, and the pyramidal tour with step-backs t, constructed as $t = (x \cup y) \backslash z$. The multigraph $x \cup y$ includes a pair of complementary pyramidal tours with step-backs z and t, different from x and y. Thus, by Lemma 1 the vertices x^v and y^v of the polytope PSB(n) are not adjacent. Examples of the first and third sufficient conditions are shown in Fig. 13.

Table 1. Construction of the tour z

1. If $x^{0,1}_i = x^{0,1}_j = 1$, then	2. If $x^{0,1}_i = x^{0,1}_j = 0$, then
$z^{0,1,sb}_k = \begin{cases} x^{0,1,sb}_k, & \text{if } k < i_a, \\ y^{1,sb}_k, & \text{if } i_a \leq k \leq j_b, \\ x^{0,sb}_k, & \text{if } i_a \leq k \leq j_b, \\ x^{0,1,sb}_k, & \text{if } k > j_b. \end{cases}$	$z^{0,1,sb}_k = \begin{cases} x^{0,1,sb}_k, & \text{if } k < i_a, \\ x^{1,sb}_k, & \text{if } i_a \leq k \leq j_b, \\ y^{0,sb}_k, & \text{if } i_a \leq k \leq j_b, \\ x^{0,1,sb}_k, & \text{if } k > j_b. \end{cases}$
3. If $x^{0,1}_i = 1$, $x^{0,1}_j = 0$, then	4. If $x^{0,1}_i = 0$, $x^{0,1}_j = 1$, then
$z^{0,1,sb}_k = \begin{cases} x^{0,1,sb}_k, & \text{if } k < i_a, \\ y^{1,sb}_k, & \text{if } i_a \leq k \leq j_b, \\ x^{0,sb}_k, & \text{if } i_a \leq k \leq j_b, \\ y^{0,1,sb}_k, & \text{if } k > j_b. \end{cases}$	$z^{0,1,sb}_k = \begin{cases} x^{0,1,sb}_k, & \text{if } k < i_a, \\ x^{1,sb}_k, & \text{if } i_a \leq k \leq j_b, \\ y^{0,sb}_k, & \text{if } i_a \leq k \leq j_b, \\ y^{0,1,sb}_k, & \text{if } k > j_b. \end{cases}$

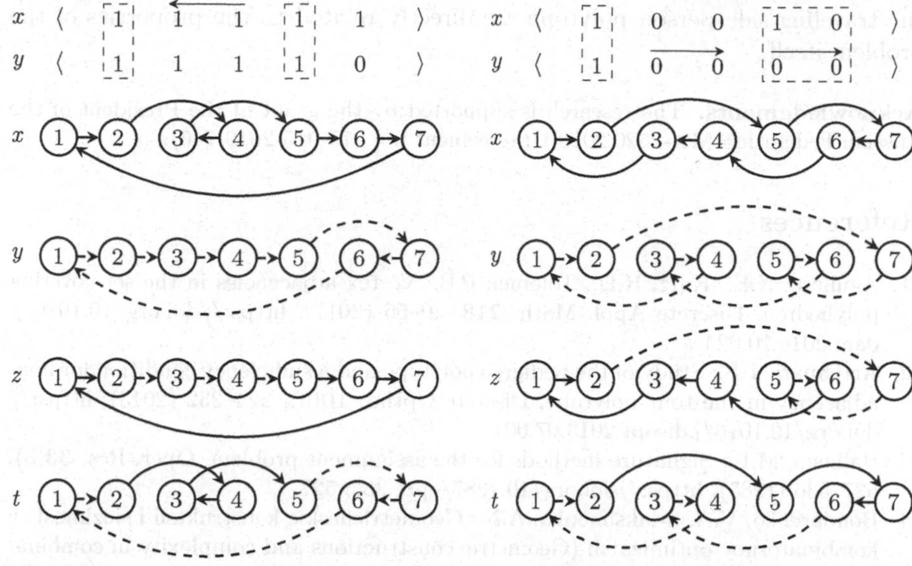

Fig. 13. Examples of first and third sufficient conditions

Theorem 3. *The question whether two vertices of the polytope* PSB(n) *are adjacent can be verified in polynomial time.*

Proof. We consider two pyramidal tours with step-backs x and y, and the corresponding vertices x^v and y^v of the polytope PSB(n).

In the encodings $x^{0,1,sb}$ and $y^{0,1,sb}$ there are $O(n)$ possible positions for the left block. Similarly, there are $O(n)$ possible positions for the right block. For each pair of blocks, the verification of the remaining conditions will require a single pass along the vectors $x^{0,1,sb}$ and $y^{0,1,sb}$ that can be performed in time $O(n)$. Thus, the vertex adjacency test by an exhaustive search of all possible cases of the Theorem 2 will require at most $O(n^3)$ operations.

In fact, the test can be performed in linear time $O(n)$. A single pass through the encodings $x^{0,1,sb}$ and $y^{0,1,sb}$ is enough to consistently find the left block, then the right block, then check the remaining conditions.

5 Conclusion

The general formulation of the traveling salesperson problem and the verification of vertex adjacency in 1-skeleton of the traveling salesperson polytope are NP-complete [19]. At the same time, the traveling salesperson problem for pyramidal tours and pyramidal tours with step-backs is solvable by dynamic programming in polynomial time [12,14]. We have established that the vertex adjacency in 1-skeleton of the polytope of pyramidal tours [7] and pyramidal tours with step-backs can be verified in polynomial time. Thus, the properties of 1-skeleton of the traveling salesperson polytope are directly related to the properties of the problem itself.

Acknowledgments. The research is supported by the grant of the President of the Russian Federation MK-2620.2018.1 (agreement no. 075-015-2019-746).

References

1. Aguilera, N.E., Katz, R.D., Tolomei, P.B.: Vertex adjacencies in the set covering polyhedron. Discrete Appl. Math. **218**, 40–56 (2017). https://doi.org/10.1016/j.dam.2016.10.024
2. Arthanari, T.S.: Study of the pedigree polytope and a sufficiency condition for non-adjacency in the tour polytope. Discrete Optim. **10**(3), 224–232 (2013). https://doi.org/10.1016/j.disopt.2013.07.001
3. Balinski, M.L.: Signature methods for the assignment problem. Oper. Res. **33**(3), 527–536 (1985). https://doi.org/10.1287/opre.33.3.527
4. Bondarenko, V.A., Maksimenko, A.N.: Geometricheskie konstruktsii i slozhnost' v kombinatornoy optimizatsii (Geometric constructions and complexity in combinatorial optimization). LKI, Moscow (2008). [in Russian]
5. Bondarenko, V.A., Nikolaev, A.V.: On graphs of the cone decompositions for the min-cut and max-cut problems. Int. J. Math. Sci. **2016** (2016). Article ID 7863650, 6 p. https://doi.org/10.1155/2016/7863650
6. Bondarenko, V.A., Nikolaev, A.V.: Some properties of the skeleton of the pyramidal tours polytope. Electron. Notes Discrete Math. **61**, 131–137 (2017). https://doi.org/10.1016/j.endm.2017.06.030
7. Bondarenko, V.A., Nikolaev, A.V.: On the skeleton of the polytope of pyramidal tours. J. Appl. Ind. Math. **12**, 9–18 (2018). https://doi.org/10.1134/S1990478918010027
8. Bondarenko, V.A., Nikolaev, A.V., Shovgenov, D.A.: 1-skeletons of the spanning tree problems with additional constraints. Autom. Control Comput. Sci. **51**(7), 682–688 (2017). https://doi.org/10.3103/s0146411617070033
9. Bondarenko, V.A., Nikolaev, A.V., Shovgenov, D.A.: Polyhedral characteristics of balanced and unbalanced bipartite subgraph problems. Autom. Control Comput. Sci. **51**(7), 576–585 (2017). https://doi.org/10.3103/s0146411617070276
10. Chegireddy, C.R., Hamacher, H.W.: Algorithms for finding K-best perfect matchings. Discrete Appl. Math. **18**, 155–165 (1987). https://doi.org/10.1016/0166-218X(87)90017-5
11. Combarro, E.F., Miranda, P.: Adjacency on the order polytope with applications to the theory of fuzzy measures. Fuzzy Sets Syst. **161**, 619–641 (2010). https://doi.org/10.1016/j.fss.2009.05.004
12. Enomoto, H., Oda, Y., Ota, K.: Pyramidal tours with step-backs and the asymmetric traveling salesman problem. Discrete Appl. Math. **87**, 57–65 (1998). https://doi.org/10.1016/S0166-218X(98)00048-1
13. Gabow, H.N.: Two algorithms for generating weighted spanning trees in order. SIAM J. Comput. **6**, 139–150 (1977). https://doi.org/10.1137/0206011
14. Gilmore, P.C., Lawler, E.L., Shmoys, D.B.: Well-solved special cases. In: Lawler, E., Lenstra, J.K., Rinnooy Kan, A., Shmoys, D. (eds.) The Traveling Salesman Problem: A Guided Tour of Combinatorial Optimization, pp. 87–143. Wiley, Chichester (1985)

15. Grötschel, M., Padberg, M.: Polyhedral theory. In: Lawler, E., Lenstra, J.K., Rinnooy Kan, A., Shmoys, D. (eds.) The Traveling Salesman Problem: A Guided Tour of Combinatorial Optimization, pp. 251–305. Wiley, Chichester (1985)
16. Khachay, M., Neznakhina, K.: Generalized pyramidal tours for the generalized traveling salesman problem. In: Gao, X., Du, H., Han, M. (eds.) COCOA 2017. LNCS, vol. 10627, pp. 265–277. Springer, Cham (2017). https://doi.org/10.1007/978-3-319-71150-8_23
17. Matsui, T.: NP-completeness of non-adjacency relations on some 0–1 polytopes. In: Proceedings of ISORA 1995. Lecture Notes in Operations Research, vol. 1, pp. 249–258 (1995)
18. Oda, Y.: An asymmetric analogue of van der Veen conditions and the traveling salesman problem. Discrete Appl. Math. **109**, 279–292 (2001). https://doi.org/10.1016/S0166-218X(00)00273-0
19. Papadimitriou, C.H.: The adjacency relation on the traveling salesman polytope is NP-Complete. Math. Program. **14**, 312–324 (1978). https://doi.org/10.1007/BF01588973
20. Sierksma, G.: The skeleton of the symmetric traveling salesman polytope. Discrete Appl. Math. **43**, 63–74 (1993). https://doi.org/10.1016/0166-218X(93)90169-O
21. Sierksma, G., Teunter, R.H., Tijssen, G.A.: Faces of diameter two on the Hamiltonian cycle polytype. Oper. Res. Lett. **18**(2), 59–64 (1995). https://doi.org/10.1016/0167-6377(95)00035-6
22. Simanchev, R.Yu.: On the vertex adjacency in a polytope of connected k-factors. Trudy Inst. Mat. i Mekh. UrO RAN **24**(2), 235–242 (2018). https://doi.org/10.21538/0134-4889-2018-24-2-235-242

Routing Open Shop with Two Nodes, Unit Processing Times and Equal Number of Jobs and Machines

Mikhail Golovachev[2] and Artem V. Pyatkin[1,2(✉)]

[1] Sobolev Institute of Mathematics, Koptyug Ave., 4, 630090 Novosibirsk, Russia
artem@math.nsc.ru

[2] Novosibirsk State University, Pirogova Str., 2, 630090 Novosibirsk, Russia
mik-golovachev2@mail.ru

Abstract. In the Routing Open Shop problem n jobs are located in the nodes of an edge-weighted graph $G = (V, E)$ and m machines must process all jobs in such a way that each machine processes only one job at a time and each job is processed by only one machine at a time. The goal is to minimize the makespan, i. e. the time when the last machine comes back to the initial node called a depot (at the beginning all machines are in the depot). This problem is NP-hard even when the graph contains only two nodes. In this paper we consider the case of $G = K_2$ when all processing times and travel times are unit. We pose the conjecture that the problem is polynomially solvable in this case, i. e. that the makespan depends only on the number of machines and the loads of the nodes and can be calculated in time $O(\log mn)$. We provide some bounds on the makespan for the case of $m = n$ depending on the loads distribution.

Keywords: Routing Open Shop · Unit processing times · Complexity · Scheduling · Polynomial time · Makespan bounds

1 Introduction

The Open Shop is one of the classical scheduling problems. There are given a set J of n jobs, a set M of m machines and a matrix of processing times p_{ij} for each machine M_i and job J_j. The task is to find a schedule with the minimum makespan to process each job on each machine in an arbitrary order so that each machine processes only one job at a time and each job is processed by only one machine at a time. Here all "switching" (used for changing jobs) times for all machines are assumed to be zero. In the Routing Open Shop problem the jobs are located in the nodes (vertices) of an edge-weighted graph $G = (V, E)$ where the weight of an edge is equal to a travel time needed for a machine to move

The research was supported by the program of fundamental scientific researches of the SB RAS, project 0314-2019-0014 and by the Russian Foundation for Basic Research, project 17-01-00170.

M. Khachay et al. (Eds.): MOTOR 2019, LNCS 11548, pp. 264–276, 2019.
https://doi.org/10.1007/978-3-030-22629-9_19

from one node to another (if two jobs are in the same node then the switching times are still zero). All machines at the beginning are in the same node called the depot. The makespan is equal to the time when the last machine comes back to the depot after processing all jobs. Note that the Routing Open Shop problem generalizes two well-known NP-hard problems, namely, Open Shop and metric Traveling Salesman Problem.

The Open Shop problem was first considered in [8]. It was proved there that it is polynomially solvable for $m = 2$ and NP-hard for $m \geq 3$. Moreover, for an arbitrary number of machines m there is no c-approximation algorithm for $c < 5/4$ unless P=NP [14]. If all processing times are unit then this problem is equivalent to an edge coloring of a bipartite graph which is polynomially solvable (the fastest algorithm can be found in [7]). The Routing Open Shop was introduced in [1] and proved to be NP-hard even for $m = 2$ and $G = K_2$ in [2]. Kononov [9] suggested an FPTAS for the latter case of this problem, while the best possible 6/5-approximation with respect to the standard lower bound algorithm for it was presented in [1]. If preemption is allowed then the Routing Open Shop problem is NP-hard for $G = K_2$ and arbitrary m, but polynomially solvable for $m = 2$ [12] (note that the preemptive Open Shop is polynomially solvable [8] for an arbitrary m). In the case of unit processing times Routing Open Shop was proved to be fixed parameter tractable in [3,4] parametrized by $m + |V|$. However, in the case of unit processing time and arbitrary m the problem complexity remains unknown even for $G = K_2$.

In this paper we consider the Routing Open Shop with $G = K_2$, unit processing times and unit travel times. Main results of the paper are obtained for the case of $m = n$. Note that in spite of the well-known symmetry in the traditional Open Shop problem between the sets of jobs and machines, the existence of the depot makes Routing Open Shop substantially different from the scheduling with job transportations [6,11,13], so the results of this paper, most probably, cannot be applied there.

2 Preliminaries

For a string x denote by $x^{[i]}$ its cyclic shift by i positions to the right. For instance, if $x = (1, 3, 2, 6, 5, 4)$ then $x^{[2]} = (5, 4, 1, 3, 2, 6)$.

Denote by a the number of jobs in the depot and by b the number of jobs in the second node. Clearly, $a + b = n$. We may assume that the jobs $J_1, \ldots, J_a$ are in the depot and the jobs $J_{a+1}, \ldots, J_n$ are not.

It is convenient to present a schedule as a table $m \times n$ filled in by positive integers, where the value k in a cell (i, j) means that the machine M_i processes the job J_j in the interval $[k - 1, k]$. Denote by t_{ij} a number in a cell (i, j). Then the table defines a correct schedule with the makespan K if the following conditions are satisfied:

1. $t_{ij} \in \{1, \ldots, K\}$ if $j \in \{1, \ldots, a\}$ and $t_{ij} \in \{2, \ldots, K - 1\}$ if $j \in \{a + 1, \ldots, n\}$;

2. $t_{ij_1} \neq t_{ij_2}$ and $t_{i_1j} \neq t_{i_2j}$ whenever $i_1 \neq i_2$ and $j_1 \neq j_2$;
3. For every $i \in \{1,\ldots,m\}, j_1 \in \{1,\ldots,a\}, j_2 \in \{a+1,\ldots,n\}$ the inequality $|t_{ij_1} - t_{ij_2}| \geq 2$ holds.

We refer to the columns $1,\ldots,a$ as a *left part* of the table and to the columns $a+1,\ldots,n$ as a *right part* of the table.

Denote by K^* the optimal makespan. The first and second conditions imply that $K^* \geq m+2$, while the second and third conditions result in $K^* \geq n+2$. So, we have the following lower bound:

$$K^* \geq \max\{m,n\} + 2. \tag{1}$$

We call any table satisfying the second condition a *Latin rectangle* (a Latin square, if $n = m$). For any integer $j \geq i$ denote by $L(i,j)$ the Latin square with a side of size $j-i+1$ filled in by the numbers $i,\ldots,j$ and by $L'(i,j)$ and $L''(i,j)$ the Latin rectangles obtained from $L(i,j)$ by deleting the last string or the last column, respectively. The easiest way to construct $L(i,j)$ is taking the string $x_0 = (i, i+1, \ldots, j)$ and putting $x_k = x_0^{[k]}$ for $k = 1,\ldots,j-i$. Note that the idea of using Latin rectangles for unit-time Open Shop scheduling first appeared in [5] (see [10, Chapter 8] for further details).

The following upper bound was proved in [3,4]:

$$K^* \leq \max\{m,n\} + 4. \tag{2}$$

The corresponding schedule can be obtained by the following procedure (the detailed proof can be found in [3,4]):

Step 1. If $m > n$ then add $m-n$ columns to the left part of the table (fictive jobs in the depot) and increase a by $m-n$. Put the first string $x_1 = (1,\ldots,n)$ and $x_i = x_1^{[i-1]}$ for $i = 2,\ldots,m$.
Step 2. Color each cell (i,j) green, if $i \leq j \leq a$, yellow, if $j \geq \max\{i, a+1\}$, orange, if $j \leq \min\{i-1, a\}$ and red, if $a < j < i$.
Step 3. Increase the numbers in yellow, orange and red cells by 1,2, and 3, respectively.

The obtained table defines a correct schedule with the makespan at most $\max\{m,n\}+4$. Moreover, if $m \leq a+1$ then there are no red cells, and we obtain an optimal solution with $K^* = n+2$.

It follows from the bounds (1) and (2) that the optimal makespan $K^* = \max\{m,n\} + s$ where $s \in \{2,3,4\}$; so, for solving the problem these three cases should be characterized. Further in the paper we consider the only the case of $m = n$. Note that the general case cannot be reduced to this one, since adding one more job or machine can increase the optimal makespan (even if the lower bound (1) remains the same). For instance, in Table 1a and b there are two examples with the minimum possible makespan 9 ($m = 7, n = 6$ and $m = 6, n = 7$ respectively), but adding a job (in any node) to the example (a) or a machine to (b) results in an instance with the makespan 10, as follows from Theorem 1 and Corollary 1 below. Nevertheless, solving the case $m = n$ looks important since it could provide ideas useful for the general case.

Table 1. Examples of correct schedules

a)

1	2	3	4	6	8
9	1	2	3	7	5
8	9	1	2	4	6
7	8	9	1	5	3
2	3	4	5	8	7
6	7	8	9	2	4
5	6	7	8	3	2

b)

1	2	3	4	5	7	8
9	1	2	3	4	6	7
8	9	1	2	3	5	6
7	8	9	1	2	4	5
6	7	8	9	1	3	4
5	6	7	8	9	2	3

c)

1	2	*3*
3	1	*2*
2	**3**	*1*

$\longrightarrow$

1	2	3	4	5	6	8	9	10
3	1	2	6	4	5	10	8	9
2	3	1	5	6	4	9	10	8
9	10	11	1	2	3	5	6	7
11	9	10	3	1	2	7	5	6
10	11	9	2	3	1	6	7	5
6	7	8	9	10	11	2	3	4
8	6	7	11	9	10	4	2	3
7	8	6	10	11	9	3	4	2

3 Main Results

The subject of study in this section is the following

Problem 1. Consider the Routing Open Shop with $G = K_2$, $m = n$ machines and jobs, where a jobs are in the depot and $b = n - a$ jobs are in the second node. All processing times and travel times between the nodes are unit. Find out whether the optimal makespan $K^* = n + 2$ or $K^* = n + 3$ or $K^* = n + 4$.

The first theorem provides the criterion when the lower bound (1) is achievable.

Theorem 1. *In Problem* 1 *the optimal makespan* $K^* = n + 2$ *if and only if* $a \equiv 0 \pmod{b}$.

Proof. *Sufficiency.* Let $k = a/b$. Consider a Latin square $L(1, k+1)$ obtained by k cyclic shifts of the string $(1, \ldots, k+1)$. Color all cells below the main diagonal orange, all cells in the last column yellow and all remaining cells green. Let a cell contain a number t. Then substitute it by $L((t-1)b+1, tb)$ if the cell is green, by $L((t-1)b+2, tb+1)$ if the cell is yellow, and by $L((t-1)b+3, tb+2)$ if the cell is orange. Apply the same operation for all cells. This procedure for $a = 6$ and $b = 3$ is illustrated in Table 1c where *italic* and **bold** numbers correspond to yellow and orange cells respectively. It is easy to check that in each string the number in the yellow cell is larger than numbers in green cells and smaller than numbers in orange cells; also, for every column in the left part of the table every number in a green cell is less than any number in an orange cell. These two facts clearly result in the correctness of the obtained schedule of makespan $K^* = n + 2$.

Necessity. If $K^* = n + 2$ then all numbers from $\{2, \ldots, n+1\}$ are met in each column of the right part of the table. In particular, there are b strings containing 2 there. Since all machines work without waits (except for the travels between the nodes), all number from $\{3, \ldots, b+1\}$ are also met in these strings in the right part of the table, i. e. the intersection of these strings and columns induces in the table a Latin square $L(2, b+1)$. Similarly, the strings containing $b + 2$

in the right part of the table induce there $L(b+2, 2b+1)$, etc. However, it is possible only if $n \equiv 0 \pmod b$, and thus $a \equiv 0 \pmod b$. □

Corollary 1. *If $a \equiv b-1 \pmod b$ or $a \equiv 0 \pmod{b+1}$ then $K^* = n+3$.*

Proof. Add a new machine and a new job to the first or the second node respectively and apply Theorem 1. □

Another completely solved case is $a < b$.

Theorem 2. *Let $a < b$. Then the optimal makespan $K^* = n+3$ if $b-a=1$ or $a=1, b=3$ or $a=2, b=4$ and $K^* = n+4$ in all other cases.*

Proof. Assume $K^* = n+3$, i. e. each machine has at most one wait. Since each machine must process b jobs in the second node and come back, it must arrive there at the time at least $a+2$ and cannot leave the second node before the time $b+2$. Therefore, in the left part of the table all numbers are either at most $a+1$ or at least $b+3$, i. e. there are at most $2a+2$ of them. Since all numbers in every column must be distinct, we have $2a+2 \geq a+b$ and hence $b-a \leq 2$. Suppose $b = a+2$. Since the left part of the table contains $a(a+b) = a(2a+2)$ cells, each of $2a+2$ available numbers is used exactly a times. In particular, this holds for the numbers $a+1$ and $b+3$. So, there are a machines that arrive to the second node at time $a+2$ (already having one wait in the depot) and a machines that leave the second node at time $b+2 = a+4$ (and thus they must work without waits there). But then at the moment $a+3$ at least $2a$ machines must process jobs in the second node, implying $2a \leq b = a+2$, i. e. $a \leq 2$.

To finish the proof note that the case $a = b-1$ follows from Corollary 1 and the correct schedules in cases $a=1, b=3$ and $a=2, b=4$ are presented in Table 2a and b respectively. □

Table 2. Schedules in Theorem 2 and Proposition 1

a)

1	3	5	6
2	6	4	5
7	2	3	4
6	4	2	3

b)

1	2	6	4	8	7
3	1	7	6	5	8
2	3	8	7	6	5
8	9	2	3	4	6
7	8	5	2	3	4
9	7	4	5	2	3

c)

$L'(1, a)$	$L''(a+2, 2a)$
$L'(a+2, 2a+1)$	$L''(2, a)$

So, later on we consider only the case $a > b$.

Proposition 1. *If $a - b \in \{1, 2\}$ then $K^* = n+3$.*

Proof. The case $a = b+1$ follows from Corollary 1 and the schedule for the case $a = b+2$ is presented in Table 2c (note that $n = 2(a-1)$). □

However, if $a - b = k \geq 3$ then the optimal makespan can be $n + 4$ provided that b is large enough.

Theorem 3. *If $a = b+k$ where $k \geq 3$ and $b \geq 2k+1$ then the optimal makespan $K^* = n+4$.*

Proof. Assume the opposite, $K^* = n + 3$. Then only numbers $2, \ldots, n+2$ can meet in the right part of the table. Let each number i meets there x_i times. Clearly, $x_i \in \{0, \ldots, b\}$ and $\sum_{i=2}^{n+2} x_i = nb$, and hence

$$d := \sum_{i=2}^{n+2} (b - x_i) = b. \tag{3}$$

Note also that if i is the minimum (maximum) number in a string in the right part of the table then the maximum (minimum) number there is at most $b + i$ (at least $i - b$), for otherwise the corresponding machine would have more than one wait. Let α_1 and α_2 (β_1 and β_2) be the number of strings with the minimum number 2 and 3 respectively (maximum number $n+2$ and $n+1$ respectively) in the right part of the table, and denote by γ the number of other strings. Clearly, $n = \alpha_1+\alpha_2+\beta_1+\beta_2+\gamma, x_2 = \alpha_1, x_3 \leq \alpha_1+\alpha_2, x_{n+2} = \beta_1$, and $x_{n+1} \leq \beta_1+\beta_2$. Note that $n - b + 1 = b + k + 1 > b + 3$ since $k > 2$. Therefore, $x_{b+3} \leq \alpha_2 + \gamma$ and $x_{n-b+1} \leq \beta_2 + \gamma$. Put $I = \{2, 3, b + 3, n - b + 1, n + 1, n + 2\}$. We have $d \geq \sum_{i \in I}(b - x_i) \geq 6b - 2n = 2b - 2k > b$ since $b > 2k$, contradicting (3). □

Corollary 2. *If $b + 3 \leq a \leq (3b - 1)/2$ (thus, $b \geq 7$) then $K^* = n + 4$.*

On the other hand, if a is much more than b then $K^* = n + 3$, as follows from Corollary 3 below.

Theorem 4. *If $a = kb + l$ where $k \geq l \geq 1$ then the optimal makespan $K^* = n+3$.*

Proof. Clearly, $n = kb + b + l = l(b + 1) + (k - l + 1)b$. If $k = l$ then the result follows from Corollary 1, so assume $k - l \geq 1$. Partition n strings of the table into l blocks of size $b + 1$ and $(k - l + 1)$ blocks of size b. The right part of the table consists of the following Latin rectangles and squares (listed from the highest to the lowest):

$$L''(n-b+2, n+2), L''(n-2b+1, n-b+1), \ldots, L''((k-l+1)b+3, (k-l+2)b+3),$$

$$L((k - l)b + 2, (k - l + 1)b + 1), L((k - l - 1)b + 2, (k - l)b + 1), \ldots, L(2, b + 1).$$

Denote by x_i the first string of ith block ($i = 0, 1, \ldots, k$) in the left part of the table (i. e., each x_i has length a). Then the left part of ith block contains the strings $x_i, x_i^{[1]}, \ldots, x_i^{[b]}$ if $i \in \{0, \ldots, l-1\}$ and $x_i, x_i^{[1]}, \ldots, x_i^{[b-1]}$ if $i \in \{l, \ldots, k\}$. So, it is sufficient to specify the first string of each block. They are obtained by

Table 3. The block structure in Theorem 4

1	2	3	4	5	6	7	8	9	10	11	12	13	14	15	16	17	18	19	21	22	23	24
19	1	2	3	4	5	6	7	8	9	10	11	12	13	14	15	16	17	18	25	21	22	23
18	19	1	2	3	4	5	6	7	8	9	10	11	12	13	14	15	16	17	24	25	21	22
17	18	19	1	2	3	4	5	6	7	8	9	10	11	12	13	14	15	16	23	24	25	21
16	17	18	19	1	2	3	4	5	6	7	8	9	10	11	12	13	14	15	22	23	24	25
25	26	22	23	24	1	2	3	4	5	6	7	8	9	10	11	12	13	14	16	17	18	19
14	25	26	22	23	24	1	2	3	4	5	6	7	8	9	10	11	12	13	20	16	17	18
13	14	25	26	22	23	24	1	2	3	4	5	6	7	8	9	10	11	12	19	20	16	17
12	13	14	25	26	22	23	24	1	2	3	4	5	6	7	8	9	10	11	18	19	20	16
11	12	13	14	25	26	22	23	24	1	2	3	4	5	6	7	8	9	10	17	18	19	20
20	21	17	18	19	25	26	22	23	24	1	2	3	4	5	6	7	8	9	11	12	13	14
9	20	21	17	18	19	25	26	22	23	24	1	2	3	4	5	6	7	8	15	11	12	13
8	9	20	21	17	18	19	25	26	22	23	24	1	2	3	4	5	6	7	14	15	11	12
7	8	9	20	21	17	18	19	25	26	22	23	24	1	2	3	4	5	6	13	14	15	11
6	7	8	9	20	21	17	18	19	25	26	22	23	24	1	2	3	4	5	12	13	14	15
15	16	12	13	14	20	21	17	18	19	25	26	22	23	24	1	2	3	4	6	7	8	9
4	15	16	12	13	14	20	21	17	18	19	25	26	22	23	24	1	2	3	9	6	7	8
3	4	15	16	12	13	14	20	21	17	18	19	25	26	22	23	24	1	2	8	9	6	7
2	3	4	15	16	12	13	14	20	21	17	18	19	25	26	22	23	24	1	7	8	9	6
10	11	7	8	15	16	12	13	14	20	21	17	18	19	25	26	22	23	24	2	3	4	5
24	10	11	7	8	15	16	12	13	14	20	21	17	18	19	25	26	22	23	5	2	3	4
23	24	10	11	7	8	15	16	12	13	14	20	21	17	18	19	25	26	22	4	5	2	3
22	23	24	10	11	7	8	15	16	12	13	14	20	21	17	18	19	25	26	3	4	5	2

concatenation of some partial strings defined below. The string $z_j = (1, \ldots, j)$ has length j. Put

$$w_j = (n - j(b+1) + 4, \ldots, n - (j-1)(b+1) + 3)^{[2]} =$$

$$= (n-(j-1)(b+1)+2, n-(j-1)(b+1)+3, n-j(b+1)+4, \ldots, n-(j-1)(b+1)+1)$$

for $j = 1, \ldots, l$. Each w_j has length $b+1$. Put also

$$y_{k-l} = ((k-l+1)b+2, (k-l+1)b+3, (k-l)b+3, \ldots, (k-l+1)b)$$

and

$$y_j = (jb+3, \ldots, (j+1)b+2)^{[2]} = ((j+1)b+1, (j+1)b+2, jb+3, \ldots, (j+1)b)$$

for $j = 1, \ldots, k-l-1$. All these strings are of length b.

Now define

$$x_0 = z_{kb+l}, x_k = y_1 \ldots y_{k-l} w_l \ldots w_1$$

$$x_j = w_j \ldots w_1 z_{(k-j)b+l-j} \text{ for } j = 1, \ldots, l,$$

$$x_{k-j} = y_{j+1} \ldots y_{k-l} w_l \ldots w_1 z_{jb} \text{ for } j = 1, \ldots, k-l-1.$$

An example of the schedule structure for $a = 19, b = 4$ (i. e. $k = 4, l = 3$) is presented in Table 3.

Let us check that this table defines a correct schedule. Since the maximum number in the left and right parts of the table are respectively $n+3$ (in w_1) and $n+2$, the condition 1 is true. It is easy to see that the intervals of the numbers in the Latin rectangles and squares do not intersect, implying that the condition 2 holds for columns in the right part of the table.

To check the conditions 2 and 3 for strings note that in the left part of the lowest block k the numbers from the window $\{1,\ldots,b+2\}$ are absent and the numbers from the set $\{2,\ldots,b+1\}$ are used in the right part of this block. Each next block shifts both the window and the set by b until the block l is reached. In the block j such that $0 \le j \le l-1$ the window is $\{a-j(b+1)+1,\ldots,a-j(b+1)+b+3\}$ and the set is $\{n-(j+1)b-j+2,\ldots,n-jb-j+2\}$. So, the condition 3 holds for every string. Besides, it is easy to verify that no number meets twice in any x_j, implying the condition 2 for strings.

In order to verify the condition 2 for the columns in the left part of the table we will determine the positions of each number $i \in \{1,\ldots,n+3\}$ in the left part of the table.

First let $i = \alpha b + \beta$ where $0 \le \alpha \le (k-l)$ and $1 \le \beta \le b$. In this case i meets in z_j for $j \ge i$ and in the table it covers the diagonal from the cell $(1,i)$ to the cell $(a-i+1,a)$. Also, if $\beta \ne 1$ it covers the diagonal from $(a-i+2,1)$ to $(a-i+\beta,\beta-1)$. Since $\beta-1 < i$ even if $\alpha = 0$ all these columns with i are distinct. However, if $i \ge b+3$ then i also meets in some y_j. Namely, if $1 \le \alpha \le k-l, 3 \le \beta \le b$ then i meets in y_α and covers a diagonal from $(n-i+\beta+1,\beta)$ to $(n,i-1)$). If $2 \le \alpha \le k-l, \beta \in \{1,2\}$ then i meets in $y_{\alpha-1}$ and covers a diagonal from $(n-i+b+\beta+1,\beta)$ to $(n-b,i-1)$).

The number $i = (k-l+1)b+1$ meets only in z_j for $j \ge i$ and in the table it covers the diagonals from $(1,i)$ to $(a-i+1,a)$ and from $(a-i+2,1)$ to $(a-i+b+1,b)$.

Now let $i = (k-l)b+\alpha(b-1)+\beta$ where $1 \le \alpha \le l-1, 1 \le \beta \le b+1$. Again, as a part of corresponding z_j, i covers the diagonal from $(1,i)$ to $(a-i+1,a)$ and also the diagonal from $(a-i+2,1)$ to $(a-i+\beta,\beta-1)$ unless $\beta = 1$. If $\alpha = 1$ and $\beta \in \{1,2\}$ then i meets in y_{k-l} and covers a diagonal from $(n-(k-l)b+1,\beta)$ to $(n,i-b-2)$. If $1 \le \alpha \le l, 3 \le \beta \le b+1$ then i is a part of $w_{l-\alpha+1}$ and covers a diagonal from $(n-i+\beta+2,\beta)$ to $(n,i-2)$. If $2 \le \alpha \le l+1, \beta \in \{1,2\}$ then i meets in $w_{l-\alpha+2}$ and covers a diagonal from $(n-i+b+\beta+3,\beta)$ to $(n,i-b+3)$.

Finally, let $i = kb+l+\beta = a_\beta$ for $1 \le \beta \le b+3$. If $\beta \in \{1,2\}$ then i meets only in w_2 and covers the diagonal from $(2b+3,\beta)$ to $(n,i-b-3)$. Otherwise, i is only a part of w_1. If $3 \le \beta \le b+1$ then i covers the diagonal from $(b+2,\beta)$ to $(n-2-\beta,a)$ and $(n+1-\beta,1)$ to $(n,\beta-2)$. If $\beta \in \{b+2,b+3\}$ then i covers the diagonal from $(b+2,\beta-b+1)$ to $(n,a+\beta-b-3)$.

Now it is easy to verify that for every i all columns containing i are different and thus, the schedule is correct. □

The lower bound on k in Theorem 4 can be slightly decreased, but the schedule construction is a bit different in this case.

Theorem 5. *If* $a = (k-1)b+k$ *then the optimal makespan* $K^* = n+3$.

Table 4. The cases $a = 8, b = 5$ and $a = 9, b = 6$

1	2	3	4	5	6	7	8	15	11	12	13	14
8	1	2	3	4	5	6	7	11	12	13	14	15
7	8	1	2	3	4	5	6	12	14	15	10	11
6	7	8	1	2	3	4	5	14	15	10	12	13
5	6	7	8	1	2	3	4	13	10	11	15	12
4	5	6	15	16	1	2	3	10	13	8	11	9
3	12	13	14	15	16	1	2	6	8	5	9	7
2	3	12	13	14	15	16	1	8	5	9	6	10
9	10	11	12	13	14	15	16	3	4	6	7	2
16	9	10	11	12	13	14	15	4	6	7	2	3
15	16	9	10	11	12	13	14	5	7	2	3	4
14	15	16	9	10	11	12	13	7	2	3	4	5
13	14	15	16	9	10	11	12	2	3	4	5	6

1	2	3	4	5	6	7	8	9	11	13	14	15	16	17
9	1	2	3	4	5	6	7	8	17	11	13	14	15	16
8	9	1	2	3	4	5	6	7	16	17	11	13	14	15
7	8	9	1	2	3	4	5	6	15	16	17	12	13	14
6	7	8	9	1	2	3	4	5	14	15	16	17	12	13
5	6	7	8	9	1	2	3	4	13	14	15	16	17	12
4	5	14	15	16	17	1	2	3	8	12	10	7	11	9
3	4	5	14	15	16	17	1	2	10	8	12	9	7	11
2	3	4	5	14	15	16	17	1	12	10	8	11	9	7
18	10	11	12	13	14	15	16	17	7	2	3	4	5	6
17	18	10	11	12	13	14	15	16	6	7	2	3	4	5
16	17	18	10	11	12	13	14	15	5	6	7	2	3	4
15	16	17	18	10	11	12	13	14	4	5	6	8	2	3
14	15	16	17	18	10	11	12	13	3	4	5	6	8	2
13	14	15	16	17	18	10	11	12	2	3	4	5	6	8

Proof. In this case $n = a + b = k(b + 1)$, i. e. the set of strings of the table can be partitioned into k blocks of $b + 1$ strings. The right part of the table consists of the following Latin rectangles (listed from the lowest to the highest):

$$L''(2, b+2), L''(b+4, 2b+4), L''(2b+5, 3b+5), \ldots, L''((k-1)b+k+2, kb+k+2).$$

Denote by x_i the first string of ith block ($i = 1, \ldots, k$) in the left part of the table. Then for every $i = 1, \ldots, k$ the left part of ith block contains strings $x_i, x_i^{[1]}, \ldots, x_i^{[b]}$, and again, it is sufficient to specify only the first string of each block. They are obtained by concatenation of some partial strings defined below. The string $z_j = (1, \ldots, j)$ has length j. Put

$$w_j = ((k-j)b + k - j + 4, \ldots, (k-j+1)b + k - j + 4)^{[2]}, \text{ for } j = 1, \ldots, k-2$$

and $w = (b + 4, \ldots, 2b + 5)^{[2]}$. Note that w is of length $b + 2$ while each w_j has length $b + 1$. Put

$$x_1 = z_a, \quad x_k = ww_{k-2} \ldots w_1$$

and

$$x_j = w_j \ldots w_1 z_{a-j(b+1)} \text{ for } j = 2, \ldots, k-1.$$

Let us verify the correctness of the schedule. Condition 1 clearly holds as well as condition 2 for columns in the right part of the table. It is easy to see that ith block for $i = 1, \ldots, k-1$ misses the numbers $(k-i)b + k + 2 - i \ldots, (k-i+1)b + k + 4 - i$ in the left part of the table, while the numbers $(k-i)b + k + 3 - i \ldots, (k-i+1)b + k + 3 - i$ are used in the right part of the table. The kth block uses the numbers $b+4, \ldots, n+3$ in the left part of the table and the numbers $2, \ldots, b+2$ in the right part of the table. Since each number is used at most once in each string, the conditions 2 and 3 hold for strings.

Table 5. The case $a = 10, b = 6$

1	2	3	4	5	6	7	8	9	10	18	12	13	15	16	17
10	1	2	3	4	5	6	7	8	9	17	18	12	13	15	16
9	10	1	2	3	4	5	6	7	8	16	17	18	12	14	15
8	9	10	1	2	3	4	5	6	7	15	16	17	18	12	14
7	8	9	11	1	2	3	4	5	6	14	15	16	17	18	13
6	7	8	9	11	1	2	3	4	5	13	14	15	16	17	18
5	6	7	17	16	19	1	2	3	4	12	13	14	9	10	11
4	5	6	7	17	16	19	1	2	3	11	10	9	14	13	12
3	4	16	15	18	17	14	13	1	2	8	7	6	11	9	10
2	3	4	16	15	18	17	14	13	1	10	9	11	6	7	8
14	13	12	10	9	15	16	17	18	19	2	3	4	5	6	7
19	14	13	12	10	9	15	16	17	18	7	2	3	4	5	6
18	19	14	13	12	10	9	15	16	17	6	8	2	3	4	5
17	18	19	14	13	12	10	9	15	16	5	6	8	2	3	4
16	17	18	19	14	13	12	10	9	15	4	5	7	8	2	3
15	16	17	18	19	14	13	12	10	9	3	4	5	7	8	2

In order to verify the condition 2 for the columns in the left part of the table we will determine the positions of each number $i \in \{1, \ldots, n+3\}$ in the left part of the table.

Clearly, the number 1 covers in the table the diagonal from the cell $(1, 1)$ to the cell $(a-1, a-2)$. Each $i = \alpha(b+1)+\beta$ where $0 \le \alpha \le k-2$ and $2 \le \beta \le b+2$ meets in z_j for all $j \ge i$ and covers the diagonal from $(1, i)$ to $(a - i + 1, a)$ and if $\beta \ne 2$ it also covers the diagonal from $(a-i+2, 1)$ to $(a - \alpha(b+1) - 1, \beta+2)$.

Each number $i = b + 3 + \beta$ where $1 \le \beta \le b + 2$ belongs to w and thus it covers a diagonal from $(a, \beta+2)$ to $(n, \beta+b+2)$ if $\beta \le b$ and the diagonal from $(a, \beta - b)$ to (n, β) if $\beta \in \{b+1, b+2\}$.

Now let $i = \alpha(b+1)+\beta$ where $2 \le \alpha \le k-2$ and $4 \le \beta \le b+4$; then i meets in $w_{k-\alpha}$. If $\beta \in \{b+3, b+4\}$ then i covers the diagonals from $(n-\alpha(b+1)+1, \beta-b-2)$ to $(a - 1, i - 2b - 4)$ and from $(a, i - 2b - 2)$ to $(n, i - b - 2)$. If $4 \le \beta \le b+2$ then i covers the diagonals from $(n - \alpha(b+1) + 1, \beta - 1)$ to $(a-1, i-b-3)$ and from $(a, i - b - 1)$ to $(n, i - 1)$.

Finally, let $i = a + \beta$ where $3 \le \beta \le b + 3$. All these numbers meet only in w_1. If $3 \le \beta \le b + 1$ then i covers three diagonals, namely, from $(b + 2, \beta)$ to $(a - 1, i - b - 3)$, from $(a, i - b - 1)$ to $(n - \beta + 1, a)$ and from $(n - \beta + 2, 1)$ to $(n, \beta-1)$. The number $n+2$ covers the diagonals from $(b+2, 1)$ to $(a-1, a-b-2)$ and from $(a, a - b)$ to (n, a), while the number $n + 3$ covers the diagonals from $(b+2, 2)$ to $(a - 1, a - b - 1)$ and from $(a, a - b + 1)$ to $(n - 1, a)$ and also meets in the cell $(n, 1)$.

Now it is easy to verify that for every i all columns containing i are different and thus, the schedule is correct. □

Table 6. The case $a = 16, b = 6$

1	2	3	4	5	6	7	8	9	10	11	12	13	14	15	16	18	19	20	21	22	23
16	1	2	3	4	5	6	7	8	9	10	11	12	13	14	15	24	18	19	20	21	22
15	16	1	2	3	4	5	6	7	8	9	10	11	12	13	14	23	24	18	19	20	21
14	15	16	1	2	3	4	5	6	7	8	9	10	11	12	13	22	23	24	18	19	20
13	14	15	16	1	2	3	4	5	6	7	8	9	10	11	12	21	22	23	24	18	19
12	13	14	15	16	1	2	3	4	5	6	7	8	9	10	11	20	21	22	23	24	18
11	12	13	14	15	16	1	2	3	4	5	6	7	8	9	10	19	20	21	22	23	24
19	20	21	22	23	24	25	1	2	3	4	5	6	7	8	9	11	12	13	14	15	16
9	19	20	21	22	23	24	25	1	2	3	4	5	6	7	8	16	11	12	13	14	15
8	9	19	20	21	22	23	24	25	1	2	3	4	5	6	7	15	16	11	12	13	14
7	8	18	19	20	21	22	23	24	25	1	2	3	4	5	6	14	15	16	10	12	13
6	7	8	18	19	20	21	22	23	24	25	1	2	3	4	5	13	14	15	16	10	12
5	6	7	8	18	19	20	21	22	23	24	25	1	2	3	4	12	13	14	15	16	10
4	5	17	23	14	15	16	20	21	22	18	24	25	1	2	3	8	10	12	7	11	9
3	4	5	17	24	14	15	16	20	21	22	23	18	25	1	2	9	8	10	12	7	11
2	3	4	5	17	25	14	15	16	20	21	22	23	24	18	1	10	9	8	11	12	7
10	11	12	13	9	17	18	14	15	16	20	21	22	23	24	25	7	2	3	4	5	6
25	10	11	12	13	9	17	18	14	15	16	20	21	22	23	24	6	7	2	3	4	5
24	25	10	11	12	13	9	17	18	14	15	16	20	21	22	23	5	6	7	2	3	4
23	24	25	10	11	12	13	19	17	18	14	15	16	20	21	22	4	5	6	8	2	3
22	18	24	25	10	11	12	13	19	17	23	14	15	16	20	21	3	4	5	6	8	2
21	22	23	24	25	10	11	12	13	19	17	18	14	15	16	20	2	3	4	5	6	8

Corollary 3. *If* $a \geq b^2 - 3b$ *then* $K^* \leq n + 3$.

Proof. Let $a \equiv l \pmod{b}$. Then $a = kb + l$ where $k \geq b - 3$ and $l < b$. If $l = 0$ or $l = b - 1$ then the statement follows respectively from Theorem 1 or Corollary 1. Otherwise, it follows from Theorem 4 or Theorem 5. □

At last, we show that the bound $b \geq 7$ from Corollary 2 on the minimum b for which the makespan can reach $n + 4$ cannot be improved.

Proposition 2. *If* $b \leq 6$ *then* $K^* \leq n + 3$.

Proof. Let $a = kb + l$ where $l < b$. The cases $l = 0, 1, b - 1$ follow respectively from Theorems 1, 4, and Corollary 1. The case $l = 2$ is resolved either by Theorem 4 or by Proposition 1. This covers all possibilities for $b \leq 4$. The cases $l = 3, k \geq 2$ and $l = 4, k \geq 3$ follow from Theorems 4 and 5. Finally, the schedules for the remaining four cases $b = 5, a = 8$ and $b = 6, a \in \{9, 10, 16\}$ are given in Tables 4, 5 and 6. □

4 Conclusions

In this paper we partially characterized the makespan in Problem 1 for some values of a and b. Our conjecture is that in all remaining open cases the makespan is $n+3$. If this were true then Problem 1 would be polynomially solvable. Indeed, its input size is $O(\log n)$ (it is sufficient to specify just a and b), and the conditions of Theorems 1, 2, 3 can be checked in time $O(\log n)$. Note that the schedule in the form of table has size $O(n^2)$, i. e. in the table representation it cannot be bounded by a polynomial of the input size. However, since the constructed schedules have quite regular structure, it is possible to specify functions computable in $O(\log n)$ time which output completion time of an operation for every machine-job pair.

Note that in the general case ($m \neq n$) the situation is more unclear since even the criterion from Theorem 1 does not work, as shown in Table 1a and b. However, we conjecture that the problem is polynomially solvable in the general case.

References

1. Averbakh, I., Berman, O., Chernykh, I.: A 6/5-approximation algorithm for the two-machine routing open shop problem on a 2-node network. Eur. J. Oper. Res. **166**, 3–24 (2005). https://doi.org/10.1016/j.ejor.2003.06.050
2. Averbakh, I., Berman, O., Chernykh, I.: The routing open-shop problem on a network: complexity and approximation. Eur. J. Oper. Res. **173**, 531–539 (2006). https://doi.org/10.1016/j.ejor.2005.01.034
3. van Bevern, R., Pyatkin, A.V.: Completing partial schedules for open shop with unit processing times and routing. In: Kulikov, A.S., Woeginger, G.J. (eds.) CSR 2016. LNCS, vol. 9691, pp. 73–87. Springer, Cham (2016). https://doi.org/10.1007/978-3-319-34171-2_6
4. van Bevern, R., Pyatkin, A.V., Sevastyanov, S.V.: An algorithm with parameterized complexity of constructing the optimal schedule for the routing open shop problem with unit execution times. Siberian Electron. Math. Rep. **16**, 42–84 (2019). https://doi.org/10.33048/semi.2019.16.003
5. Bräsel, H., Kluge, D., Werner, F.: A polynomial algorithm for the $[n/m/0, t_{ij} = 1, tree/cmax]$ open shop problem. Eur. J. Oper. Res. **72**, 125–134 (1994). https://doi.org/10.1016/0377-2217(94)90335-2
6. Brucker, P., Knust, S., Cheng, T.C.E., Shakhlevich, N.V.: Complexity results for flow-shop and open-shop scheduling problems with transportation delays. Ann. Oper. Res. **129**, 81–106 (2004). https://doi.org/10.1023/b:anor.0000030683.64615.c8
7. Cole, R., Ost, K., Schirra, S.: Edge-coloring bipartite multigraphs in $O(E \log D)$ time. Combinatorica **21**, 5–12 (2001). https://doi.org/10.1007/s004930170002
8. Gonzalez, T., Sahni, S.: Open shop scheduling to minimize finish time. J. ACM **23**, 665–679 (1976). https://doi.org/10.1145/321978.321985
9. Kononov, A.V.: On the routing open shop problem with two machines on a two-vertex network. J. Appl. Ind. Math. **6**, 318–331 (2012). https://doi.org/10.1134/s1990478912030064
10. Leung, J.Y. (ed.): Handbook of Scheduling - Algorithms, Models, and Performance Analysis. Chapman and Hall/CRC (2004). http://www.crcnetbase.com/isbn/978-1-58488-397-5
11. Lushchakova, I., Soper, A., Strusevich, V.: Transporting jobs through a two-machine open shop. Naval Res. Logist. **56**, 1–18 (2009). https://doi.org/10.1002/nav.20323

12. Pyatkin, A.V., Chernykh, I.D.: The open shop problem with routing at a two-node network and allowed preemption. J. Appl. Indust. Math. **6**, 346–354 (2012). https://doi.org/10.1134/s199047891203009x
13. Strusevich, V.: A heuristic for the two-machine open-shop scheduling problem with transportation times. Discrete Appl. Math. **93**(2), 287–304 (1999). https://doi.org/10.1016/S0166-218X(99)00115-8
14. Williamson, D.P., et al.: Short shop schedules. Oper. Res. **45**, 288–294 (1997). https://doi.org/10.1287/opre.45.2.288

Combinatorial Optimization

On $(1+\varepsilon)$-approximate Data Reduction for the Rural Postman Problem

René van Bevern[1(✉)], Till Fluschnik[2], and Oxana Yu. Tsidulko[1,3]

[1] Department of Mechanics and Mathematics, Novosibirsk State University, Novosibirsk, Russian Federation
rvb@nsu.ru, tsidulko@math.nsc.ru

[2] Algorithmics and Computational Complexity, Fakultät IV, TU Berlin, Berlin, Germany
till.fluschnik@tu-berlin.de

[3] Sobolev Institute of Mathematics of the Siberian Branch of the Russian Academy of Sciences, Novosibirsk, Russian Federation

Abstract. Given a graph $G = (V, E)$ with edge weights and a subset $R \subseteq E$ of required edges, the NP-hard Rural Postman Problem (RPP) is to find a closed walk of minimum total weight containing all edges of R. The number b of vertices incident to an odd number of edges of R and the number c of connected components formed by the edges in R are both bounded from above by the number of edges that has to be traversed additionally to the required ones. We show how to reduce any RPP instance I to an RPP instance I' with $2b + O(c/\varepsilon)$ vertices in $O(n^3)$ time so that any α-approximate solution for I' gives an $\alpha(1+\varepsilon)$-approximate solution for I, for any $\alpha \geq 1$ and $\varepsilon > 0$. That is, we provide a polynomial-size approximate kernelization scheme (PSAKS). We make first steps towards a PSAKS with respect to the parameter c.

Keywords: Eulerian extension · Lossy kernelization · Parameterized complexity

1 Introduction

In the framework of lossy kernelization [15,29], we study trade-offs between the provable effect of data reduction and the provably achievable solution quality for the following classical vehicle routing problem [31].

Problem 1.1 (Rural Postman Problem, RPP).

Input: A graph $G = (V, E)$ with n vertices, edge weights $\omega : E \to \mathbb{N} \cup \{0\}$, and a multiset R of *required edges* of G.

Task: Find a closed walk W^* in G containing each edge of R and minimizing the total weight $\omega(W^*)$ of the edges on W^*.

M. Khachay et al. (Eds.): MOTOR 2019, LNCS 11548, pp. 279–294, 2019.
https://doi.org/10.1007/978-3-030-22629-9_20

We call any closed walk containing all edges of R an *RPP tour*. RPP has direct applications in snow plowing, street sweeping, meter reading [7,13], vehicle depot location [18], drilling, and plotting [17,20]. The undirected version occurs especially in rural areas, where service vehicles can operate in both directions even on one-way roads [13]. Moreover, RPP is a special case of the Capacitated Arc Routing Problem (CARP) [19] and used in "route first, cluster second" algorithms for CARP [1,6,34], which are notably the only ones with proven approximation guarantees [4,24,35]. Improved approximations for RPP automatically lead to better approximations for CARP.

Unfortunately, containing the metric Traveling Salesman Problem as a special case, RPP is APX-hard [25]. While there is a folklore polynomial-time 3/2-approximation, we aim for $(1+\varepsilon)$-approximations for all $\varepsilon > 0$. Since finding such approximations typically requires exponential time, we present data reduction rules for this task. Their effectiveness depends on the desired approximation factor.

1.1 Our Contributions and Outline of This Paper

In Sect. 2, we introduce basic notation. In Sect. 3, we prove basic structural properties of optimal RPP solutions. In Sect. 4, we show our main theorem:

Theorem 1.2. *For any $\varepsilon > 0$, any RPP instance (G, R, ω) can be reduced to an RPP instance (G', R', ω') in $O(n^3 + |R|)$ time such that*

(i) the number of vertices in G' is $2b + O(c/\varepsilon)$,
(ii) the number of required edges is $|R'| \leq 4b + O(c/\varepsilon)$,
(iii) the maximum edge weight with respect to ω' is $O((b + c)/\varepsilon)$,
(iv) any α-approximate solution for I' for some $\alpha \geq 1$ can be transformed into an $\alpha(1 + \varepsilon)$-approximate solution for I in polynomial time,

where b is the number of vertices of G incident to an odd number of edges in R and c is the number of connected components formed by the edges in R.

Notably, the α-approximate solution for I' in Theorem 1.2 may be obtained by any means, for example exact algorithms or heuristics. Thus, Theorem 1.2 can be used to speed up expensive heuristics without much loss in the solution quality. In terms of the recently introduced concept of lossy kernelization [29], Theorem 1.2 yields a *polynomial-size approximate kernelization scheme (PSAKS)*.

Remark 1.3. We can prove that Theorem 1.2 cannot be generalized to $\varepsilon = 0$ unless the polynomial-time hierarchy collapses.[1] In fact, we can prove the stronger result that RPP is WK[1]-complete [23] even when parameterized by a larger parameter—the number and cost $d = \omega(W^*) - \omega(R) + |W^*| - |R|$ of edges traversed additionally to the required ones. That is, *exactly* solving RPP presumably cannot be polynomial-time reduced to solving instances of size polynomial in d (and thus, also not to solving instances of size polynomial in $b + c \leq 3d/2$).

[1] All omitted proofs can be found in the full version of this paper, available on arXiv: https://arxiv.org/abs/1812.10131.

1.2 Related Work

Classical Complexity. Being a generalization of the metric TSP, RPP is APX-hard [25]. There is a folklore polynomial-time 3/2-approximation (we refer to arc routing surveys [5,13] for a detailed algorithmic description).

Parameterized Complexity. Dorn et al. [9] showed an $O(4^d \cdot n^3)$-time algorithm for the directed RPP, where $d = |W^*| - |R|$ is the minimum number of deadheading arcs in an optimal solution W^*. It can be easily adapted to the undirected RPP. Sorge et al. [32] showed an $O(4^{c \log b^2} \operatorname{poly}(n))$-time algorithm for the directed RPP, where c is the number of (weakly) connected components induced by the required arcs in R and $b = \sum_{v \in V} |\operatorname{indeg}(v) - \operatorname{outdeg}(v)|$. It is not obvious whether this algorithm can be adapted to the undirected RPP maintaining its running time. Gutin et al. [22] showed a randomized algorithm that solves the directed and undirected RPP in $f(c)\operatorname{poly}(n)$ time if edge weights are bounded polynomially in n. The existence of a deterministic algorithm with this running time is open [5,22,33].

Exact Kernelization. RPP can easily be reduced to a problem kernel with $2|R|$ vertices [5]. In contrast, Sorge et al. [32] showed that, unless the polynomial-time hierarchy collapses, the directed RPP has no problem kernel of size polynomial in the number of deadheading arcs. This can be strengthened to WK[1]-hardness, also for the directed RPP (see Remark 1.3).

Lossy Kernelization. Recently the concept of approximate kernelization has gained increased interest [15,29]. In this context, Eiben et al. [11] called for finding connectivity-constrained problems that do not have polynomial-size kernels but α-approximate polynomial-size kernels. We exhibit that RPP is such a problem (see Theorem 1.2 and Remark 1.3). Among the so far few known lossy kernels [11,12,27–29], our Theorem 1.2 stands out since it shows a time *and* size efficient PSAKS, which is a property previously observed only in results of Krithika et al. [27].

2 Preliminaries

Sets and Multisets. By $\mathbb{N}$ we denote the set of natural numbers including zero. For two multisets A and B, $A \uplus B$ is the multiset obtained by adding the multiplicities of elements in A and B. By $A \setminus B$ we denote the multiset obtained by subtracting the multiplicities of elements in B from the multiplicities of elements in A. Finally, given some weight function $\omega\colon A \to \mathbb{N}$, the *weight* of a multiset A is $\omega(A) := \sum_{e \in A} \nu(e)\omega(e)$, where $\nu(e)$ is the multiplicity of e in A.

Graphs. We generally consider *multigraphs* $G = (V, E)$ with a set $V(G) := V$ of *vertices*, a multiset $E(G) := E$ over $\{\{u, v\} \mid u, v \in V\}$ of *(undirected) edges*, and *edge weights* $\omega\colon E \to \mathbb{N}$. Graphs are allowed to have loops and parallel edges. For a multiset R of edges, we denote by $V(R)$ the set of their incident vertices.

Paths and Cycles. A *walk from v_0 to v_ℓ in G* is a sequence $w = (v_0, e_1, v_1, e_2, v_2, \ldots, e_\ell, v_\ell)$ such that e_i is an edge with end points v_{i-1} and v_i for each $i \in \{1, \ldots, \ell\}$. If $v_0 = v_\ell$, then we call w a *closed walk*. If all vertices on w are pairwise distinct, then w is a *path*. If only its first and last vertex coincide, then w is a *cycle*. By $E(w)$ we denote the multiset of edges on w. The *length* of walk w is its number $|w| := \ell = |E(w)|$ of edges. The *weight* of walk w is $\omega(w) := \sum_{i=1}^{\ell} \omega(e_\ell)$. An *Euler tour for G* is a closed walk that traverses each edge of G exactly as often as it is present in G. A graph is *Eulerian* if it allows for an Euler tour.

Connectivity and Blocks. Two vertices u, v of G are *connected* if there is a path from u to v in G. A *connected component* of G is a maximal subgraph of G in which the vertices are mutually connected. A vertex v of G is a *cut vertex* if removing v and its incident edges increases the number of connected components of G. A *biconnected component* or *block* of G is a maximal subgraph without cut vertices.

Edge- and Vertex-Induced Subgraphs. For a subset $U \subseteq V$ of vertices, *the subgraph $G[U]$ of $G = (V, E)$ induced by U* consists of the vertices of U and all edges of G between them (respecting multiplicities). For a multiset R of edges of G, $G\langle R\rangle := (V(R), R)$ is the graph *induced by the edges in R*. For a walk w, we also denote $G\langle w\rangle := G\langle E(w)\rangle$. Note that $G\langle R\rangle$ and $G\langle w\rangle$ do not contain isolated vertices yet might contain edges with a higher multiplicity than G and, therefore, are not necessarily sub(multi)graphs of G.

Lossy Kernelization. Kernelization is a notion of provably effective data reduction [21,26] from parameterized complexity theory [8]. Since RPP does not have polynomial-size kernels (see Remark 1.3) and is hard to approximate at the same time, we consider *approximate* kernelization [29]:

Definition 2.1 (polynomial-size approximate kernelization scheme). *A* polynomial-size approximate kernelization scheme (PSAKS) *for an optimization problem L with parameter k consists of two algorithms: for each constant $\varepsilon > 0$,*

(i) the first algorithm reduces an instance I of L to an instance I' of size poly(k) *in polynomial time,*
(ii) the second algorithm turns any α-approximate solution for I' into an $\alpha \cdot (1 + \varepsilon)$-approximate solution for I in polynomial time.

We will use the following lemma to shrink edge weights. It is a generalization of an idea implicitly used for weight reduction in a proof of Lokshtanov et al. [29, Theorem 4.2] and shrinks weights faster and more significantly than a theorem of Frank and Tardos [16] that is frequently used in the exact kernelization of weighted problems [2,14,30].

Lemma 2.2 (lossy weight reduction). *Let $\mathcal{F} \subseteq \mathbb{Q}^n_{\geq 0}$ and $\omega \in \mathbb{Q}^n_{\geq 0}$ such that*
- *$\|\omega\|_\infty \leq \beta$ for some $\beta \in \mathbb{Q}$ and*
- *$\|x\|_1 \leq N$ for some $N \in \mathbb{N}$ and all $x \in \mathcal{F}$.*

Then, for any $\varepsilon > 0$, in linear time, we can compute $\bar{\omega} \in \mathbb{N}^n$ such that

(i) $\|\bar{\omega}\|_\infty \leq N/\varepsilon$ and
(ii) for any $x \in \mathcal{F}$ with $\bar{\omega}^\top x \leq \alpha \cdot \bar{\omega}^\top \bar{x}^*$, one has $\omega^\top x \leq \alpha \cdot \omega^\top x^* + \varepsilon\beta$,

where $\alpha \in \mathbb{Q}$, $x^* \in \arg\min\{\omega^\top x \mid x \in \mathcal{F}\}$, and $\bar{x}^* \in \arg\min\{\bar{\omega}^\top x \mid x \in \mathcal{F}\}$.

3 Solution Structure

In this section, we prove fundamental properties of optimal solutions to RPP. To make these hold, we first establish the triangle inequality.

Proposition 3.1 ([3]). *In $O(n^3)$ time, an RPP instance (G, R, ω) can be turned into an RPP instance (G', R, ω') such that*

- *G' is a complete graph on the vertex set of G,*
- *ω' satisfies the triangle inequality, and*
- *any α-approximate RPP tour for (G', R, ω') can be turned into an α-approximate RPP tour for (G, R, ω) in polynomial time.*

Remark 3.2. Since Proposition 3.1 does not change R, it affects neither the number of connected components nor the number of odd-degree vertices of $G\langle R\rangle = G'\langle R\rangle$. Thus, it is sufficient to prove Theorem 1.2 for RPP with triangle inequality.

Now, consider any RPP tour W for an RPP instance (G, R, ω). Then $G\langle W\rangle$ is an Eulerian supergraph of $G\langle R\rangle$ with total edge weight $\omega(W)$. Moreover, any Eulerian supergraph $G\langle W'\rangle$ of $G\langle R\rangle$ yields an RPP tour for (G, R, ω) of total weight $\omega(W')$. Thus, *RPP tours* one-to-one correspond to *Eulerian extensions* [33]:

Definition 3.3 (Eulerian extension, edge-minimizing). *An* Eulerian extension *(EE) for an RPP instance (G, R, ω) is a multiset S of edges such that $G\langle R \uplus S\rangle$ is Eulerian.*

We say that an Eulerian extension S is edge-minimizing *if there is no Eulerian extension S' with $|S'| < |S|$ and $\omega(S') \leq \omega(S)$.*

We exploit that a graph without isolated vertices is Eulerian if and only if it is connected and *balanced*:

Definition 3.4 (balanced). *A vertex is* balanced *if it has even degree. A graph is* balanced *if each of its vertices is balanced.*

Thus, solving RPP reduces to finding a minimum-weight set S of edges such that $G\langle R \uplus S\rangle$ is connected and balanced. Since an Euler tour in the Eulerian graph $G\langle R \uplus S\rangle$ is computable in linear time using Hierholzer's algorithm, we can easily recover an RPP tour from an Eulerian extension.

Proposition 3.5. *Let (G, R, ω) be an RPP instance.*

(i) From any RPP tour W for (G, R, ω), one can compute an Eulerian extension S of cost $\omega(W) = \omega(R) + \omega(S)$ in time linear in $|W|$.
(ii) From any Eulerian extension S for (G, R, ω), one can compute an RPP tour W of cost $\omega(W) = \omega(R) + \omega(S)$ in time linear in $|R| + |S|$.

When assuming the triangle inequality, any RPP tour can be shortcut so as not to contain vertices that are not incident to required edges. Thus:

Observation 3.6. Any edge-minimizing Eulerian extension S for an RPP instance (G, R, ω) satisfies $V(S) \subseteq V(R)$.

Moreover, since an edge-minimizing Eulerian extension uses balanced vertices only to make connections between components, we can prove:

Lemma 3.7. *Let (G, R, ω) be an RPP instance and c be the number of connected components of $G\langle R\rangle$. At most $2c - 2$ balanced vertices in $G\langle R\rangle$ are incident to edges of an edge-minimizing Eulerian extension and this bound is tight.*

Using the triangle inequality, any RPP tour using a vertex more than once can be shortcut, yielding the following lemma:

Lemma 3.8. *An edge-minimizing Eulerian extension contains exactly one edge incident to each unbalanced vertex of $G\langle R\rangle$ and either no or two edges incident to each balanced vertex of $G\langle R\rangle$.*

We now establish some inequalities used in the analysis of our algorithm.

Definition 3.9. *In the context of an RPP instance (G, R, ω), we denote by*

- R – *the set of required arcs,*
- c – *the number of connected components in $G\langle R\rangle$,*
- b – *the number of imbalanced vertices in $G\langle R\rangle$,*
- W^* – *a minimum-weight RPP tour with a minimum number of edges,*
- D – *a minimum-weight edge-minimizing Eulerian extension for (G, R, ω),*
- T – *a minimum-weight set of edges such that $G\langle R \uplus T\rangle$ is connected, of minimum cardinality,*
- M – *a minimum-weight set of edges such that $G\langle R \uplus M\rangle$ is balanced, of minimum cardinality.*

Notably, when assuming the triangle inequality, M is simply a minimum-weight perfect matching on the b imbalanced vertices in $G\langle R\rangle$ [10].

Lemma 3.10. *The following relations hold:*

$$\omega(W^*) = \omega(R) + \omega(D), \quad (3.1) \qquad |W^*| = |R| + |D|, \quad (3.5)$$
$$\omega(M) \leq \omega(D), \quad (3.2) \qquad 2b = |M| \leq |D|, \quad (3.6)$$
$$\omega(T) \leq \omega(D), \quad (3.3) \qquad c - 1 = |T| \leq |D|, \quad (3.7)$$
$$\omega(D) \leq \omega(M) + 2\omega(T), \quad (3.4) \qquad |D| \leq |M| + 2|T|, \quad (3.8)$$

where $|S| \leq |M| + 2|T|$ *holds for* any *edge-minimizing Eulerian extension* S.

4 Approximate Kernelization Schemes for the Rural Postman Problem

In this section, we prove Theorem 1.2. To this end, in Sect. 4.1, we present three data reduction rules. In Sect. 4.2, we then show how to apply these rules to obtain a polynomial-size approximate kernelization scheme (PSAKS) of size $2b+O(c/\varepsilon)$, proving Theorem 1.2. Finally, in Sect. 4.3, we discuss some problems that one faces when trying to improve it to a PSAKS of size $O(c)$.

4.1 Data Reduction Rules

Since, by Observation 3.6, no edge-minimizing Eulerian extension uses vertices outside of $V(R)$, the following is immediate.

Reduction Rule 4.1. Let (G, R, ω) be an RPP instance with triangle inequality. Delete all vertices that are not incident to edges in R.

Proposition 4.2. *Reduction Rule 4.1 turns an RPP instance* (G, R, ω) *into an RPP instance* (G', R, ω) *such that*

- *any edge-minimizing Eulerian extension for* (G, R, c) *is one for* (G', R, c) *and*
- *any Eulerian extension for* (G', R, c) *is one for* (G, R, c).

The next data reduction rule shrinks the set of required edges. This will be crucial since other data reduction rules only reduce the number of vertices, yet may leave the multiset of required edges between them unbounded.

Reduction Rule 4.3. Let (G, R, ω) be an instance of RPP and C be a cycle in $G\langle R\rangle$ such that $G\langle R \setminus C\rangle$ has the same number of connected components as $G\langle R\rangle$, then delete the edges of C from R.

Lemma 4.4. *Using Reduction Rule 4.3, one can in* $O(|R|)$ *time compute a set* $R' \subseteq R$ *of required edges with the following properties.*

(i) Any Eulerian extension for (G, R', ω) *is one for* (G, R, ω) *and vice versa.*
(ii) The number of edges in each connected component of $G\langle R'\rangle$ *with* k *vertices is at most* $\max\{1, 2k-2\}$.

We finally present a data reduction rule that removes balanced vertices. To this end, the following lemma in particular shows that removing a balanced vertex with all its incident edges changes the balance of an even number of vertices. This allows us to restore their original balance by adding a matching to the set of required edges, not increasing the total weight of required edges. This will be crucial to prove that our reduction rules maintain approximation factors.

Lemma 4.5. *Let $\Gamma = (V, E)$ be a multigraph, $\omega\colon \{\{u, v\} \mid u, v \in V\} \to \mathbb{N}$ satisfy the triangle inequality, and F be an even-cardinality submultiset of edges incident to a common vertex $v \in V$. Then*

(i) The set $U \subseteq V \setminus \{v\}$ of vertices incident to an odd number of edges of F has even cardinality.
(ii) For any matching M_v in the complete graph on U, $\omega(M_v) \leq \omega(F)$ and $|M_v| \leq |F|$.

We now use Lemma 4.5 to define an operation that allows us to remove a balanced vertex from $G\langle R\rangle$. It is illustrated in Fig. 1.

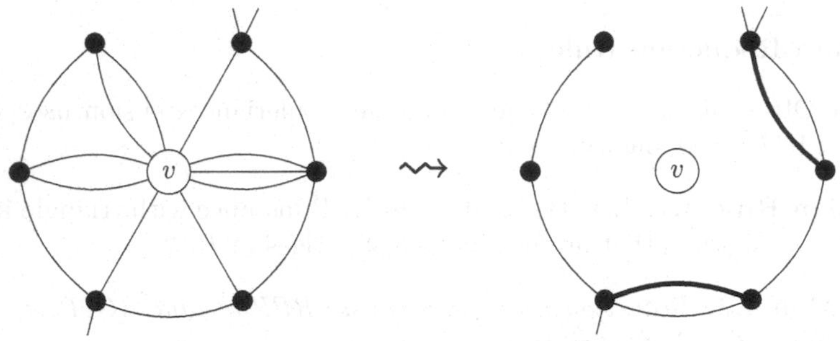

Fig. 1. Illustration of Definition 4.6(a). Only required edges are shown. Thick edges on the right are the added matching M_v.

Definition 4.6 (vertex extraction). *Let (G, R, ω) be an RPP instance with ω satisfying the triangle inequality, v be a vertex that*

- *is balanced in a connected component of $G\langle R\rangle$ with at least three vertices and*
- *not a cut vertex of $G\langle R\rangle$ or contained in exactly two blocks of $G\langle R\rangle$,*

and let $R_v \subseteq R$ be the required edges incident to v. The result of extracting *v is a set R' constructed as follows:*

(a) If v is not a cut vertex of $G\langle R\rangle$, then $R' = (R \setminus R_v) \uplus M_v$, where M_v is any perfect matching on the set of vertices incident to an odd number of edges of R_v.

(b) If v is a cut vertex of $G\langle R\rangle$ contained in exactly two blocks A and B of $G\langle R\rangle$, then $R' = (R \setminus R_v) \uplus M_v \uplus \{\{a, b\}\}$, where a is a neighbor of v in A, b is a neighbor of v in B, and M_v is any perfect matching on the set of vertices incident to an odd number of edges of $R_v \setminus \{\{a, v\}, \{b, v\}\}$.

Lemma 4.7. *Let (G, R, ω) be an RPP instance and R' be the result of extracting a balanced vertex v of $G\langle R\rangle$. Then the following properties hold.*

(i) $V(R') = V(R) \setminus \{v\}$.
(ii) $\omega(R') \leq \omega(R)$ and $|R'| \leq |R|$.
(iii) Each vertex of $G\langle R'\rangle$ is balanced if and only if it is balanced in $G\langle R\rangle$.
(iv) Two vertices of $G\langle R'\rangle$ are connected if and only if they are so in $G\langle R\rangle$.
(v) Any multiset S of edges with $V(S) \subseteq V(R')$ is an Eulerian extension for (G, R', ω) if and only if it is one for (G, R, ω).

We can now turn Definition 4.6 into a data reduction rule. Its parameter $\gamma \in \mathbb{Q}$ allows a trade-off between aggressivity and introduced error.

Reduction Rule 4.8. Let (G, R, ω) be an RPP instance with $G = (V, E)$, ω satisfying the triangle inequality, and $\gamma \in \mathbb{Q}$. Let C_i be the vertices in connected component $i \in \{1, \ldots, c\}$ of $G\langle R\rangle$ and $B_i \subseteq C_i$ be an inclusion-maximal set of vertices such that, for each $u, v \in B_i$ with $u \neq v$, one has $\omega(\{u, v\}) > \gamma$. Finally, let

$$B := \bigcup_{i=1}^{c} B_i.$$

Now, initially let $R' := R$ and, as long as $G\langle R'\rangle$ contains a vertex $v \in V \setminus B$ that can be extracted using Definition 4.6, replace R' by the result of extracting v.

Lemma 4.9. *Let (G, R, ω) be an RPP instance with ω satisfying the triangle inequality. Then, Reduction Rule 4.8 in $O(n^3)$ time yields a multiset R' of edges such that*

(i) $\omega(R') \leq \omega(R)$ and $V(R') \subseteq V(R)$.
(ii) Any multiset S of edges with $V(S) \subseteq V(R')$ is an Eulerian extension for (G, R', ω) if and only if it is one for (G, R, ω).
(iii) Any edge-minimizing Eulerian extension S for (G, R, ω) can be turned into an Eulerian extension S' for (G, R', ω) such that $\omega(S') \leq \omega(S) + 2\gamma \cdot (2c - 2)$.
(iv) $G\langle R'\rangle$ contains at most $2b + 2c + 4\omega(R)/\gamma$ vertices.

Proof. (i) and (ii) follow from Lemma 4.7 since R' is the result of a sequence of vertex extractions.

(iii) We turn S into an Eulerian extension S' with $V(S') \subseteq V(R')$ and then apply (ii). First, since S is edge-minimizing and ω satisfies the triangle inequality, by Observation 3.6, $V(S) \subseteq V(R)$. By Reduction Rule 4.8, the vertices in $X := V(R) \setminus V(R')$ are not in B and, thus, for each $v \in X \cap C_i$, we find a vertex $v' \in B_i$ such that $\omega(\{v, v'\}) \leq \gamma$. Note that $v' \in V(R')$. Since each vertex in X is balanced in $G\langle R\rangle$, by Lemma 3.8, each vertex $v \in X \cap V(S)$ is incident to exactly *two* edges $\{v, u\}$ and $\{v, w\}$ of S (possibly, $u = w$). Since $\{v, v'\} \subseteq C_i$, $S' := (S \setminus \{\{v, u\}, \{v, w\}\}) \uplus \{v', u\} \uplus \{v', w\}$ is also an Eulerian extension for (G, R, ω). Moreover, $\omega(S') \leq \omega(S) + 2\gamma$. Doing this replacement for each $v \in X \cap V(S)$, we finally obtain an Eulerian extension S' for (G, R, ω) with $V(S') \subseteq V(R')$ and $\omega(S') \leq \omega(S) + 2\gamma \cdot |X \cap V(S)|$. Since each vertex in X is balanced in $G\langle R\rangle$, by Lemma 3.7, $|X \cap V(S)| \leq 2c - 2$. Finally, by (ii), S' is an Eulerian extension for (G, R', ω).

(iv) The vertices of $G\langle R'\rangle$ can be partitioned into $X \uplus Y \uplus Z$, where X are imbalanced in $G\langle R'\rangle$, Y are balanced and in B, and Z are balanced but not in B.

By Lemma 4.7(iii), the vertices in X are imbalanced in $G\langle R\rangle$ also. Thus,

$$|X| \leq b. \tag{4.1}$$

We next analyze $|Y|$. For $i \in \{1, \ldots, c\}$, let $R_i \subseteq R$ be the edges between vertices in C_i, T_i^* be a tree of the smallest weight in $G\langle R_i\rangle$ connecting all vertices in B_i, T_i be a minimum-weight spanning tree in $G[B_i]$, and H_i be a minimum-weight Hamiltonian cycle in $G[B_i]$. Doubling all edges of T_i^* yields a closed walk in $G\langle R_i\rangle$ containing the vertices in B_i. Using the triangle inequality of ω, it can be shortcut to a Hamiltonian cycle in $G[B_i]$. Thus, $\omega(T_i) \leq \omega(H_i) \leq 2\omega(T_i^*)$.[2] We thus get

$$(|B_i| - 1)\gamma = \sum_{e \in T_i} \gamma < \sum_{e \in T_i} \omega(e) = \omega(T_i) \leq 2\omega(T_i^*) \leq 2\omega(R_i) \quad \text{and thus}$$

$$|Y| \leq |B| = \sum_{i=1}^{c} |B_i| < \sum_{i=1}^{c} \left(\frac{2\omega(R_i)}{\gamma} + 1 \right) = 2\omega(R)/\gamma + c. \tag{4.2}$$

Finally, we analyze $|Z|$. Definition 4.6 is not applicable to any vertex $v \in Z$, since it would have been removed by Reduction Rule 4.8. Thus, v is a cut vertex contained in at least three blocks of $G\langle R'\rangle$ or the connected component of $G\langle R'\rangle$ containing v consists of only two vertices. To analyze $|Z|$, for each $i \in \{1, \ldots, c\}$, consider $X_i := X \cap C_i$, $Z_i := Z \cap C_i$, the set $R_i' \subseteq R'$ of edges between vertices in C_i, and the *block-cut tree* T_i of $G\langle R_i'\rangle$: the vertices of T_i are the cut vertices and the blocks of $G\langle R_i'\rangle$ and there is an edge between a cut vertex v and a block A of $G\langle R_i'\rangle$ in T_i if v is contained in A. Then either $|Z_i| \leq 2$ or the vertices in Z_i have degree at least three in T_i. Therefore, T_i has at most $|X_i| + |Y_i|$ leaves. Since a tree with ℓ leaves has at most $\ell - 1$ vertices of degree three, $|Z_i| \leq \max\{2, |X_i| + |Y_i| - 1\}$. Thus,

$$|Z| = \sum_{i=1}^{c} |Z_i| \leq |X| + \sum_{i=1}^{c} |Y_i| = |X| + |Y|. \tag{4.3}$$

Combining (4.1), (4.2), (4.3), and that $|V(R')| = |X| + |Y| + |Z|$, (iv) follows. □

[2] That is, T_i is the folklore 2-approximation of a Steiner tree with terminals B_i in $G\langle R_i\rangle$.

4.2 A Polynomial-Size Approximate Kernelization Scheme for the Parameter $b+c$ (proof of Theorem 1.2)

This section proves Theorem 1.2. We describe how to transform a given RPP instance I and $\varepsilon > 0$ into an RPP instance I' such that any α-approximate solution for I' can be transformed into an $\alpha(1+\varepsilon)$-approximate solution for I. Due to Proposition 3.1, we assume that $I = (G, R, \omega)$ has been preprocessed in $O(n^3)$ time so as to satisfy the triangle inequality.

Shrinking the Graph. Choose $\varepsilon_1 + \varepsilon_2 = \varepsilon$. Apply Reduction Rule 4.8 with

$$\gamma = \frac{\varepsilon_1 \cdot \omega(R)}{4c-4}, \tag{4.4}$$

which, by Lemma 4.9, in $O(n^3)$ time gives an instance (G, R_1, ω) with

$$|V(R_1)| \leq 2b + 2c + \frac{16c-16}{\varepsilon_1}. \tag{4.5}$$

To (G, R_1, ω) we apply Reduction Rule 4.3, which, by Lemma 4.4, in $O(|R_1|)$ time gives an instance (G, R_2, ω) with

$$R_2 \subseteq R_1 \qquad \text{and} \qquad |R_2| \leq 4b + 4c + \frac{32c-32}{\varepsilon_1}. \tag{4.6}$$

Finally, applying Reduction Rule 4.1 to (G, R_2, ω) in linear time yields an instance (G_2, R_2, ω) such that

$$|V(G_2)| \leq |V(R_2)| \leq |V(R_1)|. \tag{4.7}$$

Shrinking Edge Weights. Since $G\langle R \uplus T\rangle$ is connected, due to the triangle inequality of ω, each edge $e = \{u, v\}$ of G, and thus of its subgraph G_2, satisfies $\omega(e) \leq \omega(R) + \omega(T)$. Moreover, by Lemma 3.10, any edge-minimizing Eulerian extension for (G_2, R_2, ω) has at most $|M| + 2|T| = b/2 + 2c - 2$ edges. Thus, we can apply Lemma 2.2 with $\beta = \omega(R) + \omega(T)$ and $N = |R_2| + b/2 + 2c - 2$ to (G_2, R_2, ω) to get an instance (G_2, R_2, ω_2) such that for all edges e,

$$\omega(e) \leq \frac{|R_2| + b/2 + 2c - 2}{\varepsilon_2}. \tag{4.8}$$

In Lemma 2.2, set $\mathcal{F}$ just contains all vectors x that encode RPP tours W induced by edge-minimizing Eulerian extensions for (G_2, R_2, ω) (it has a 1 for each edge of G_2 in W and a 0 for each edge of G_2 not in W). We finally return (G_2, R_2, ω_2), whose construction takes $O(n^3 + |R|)$ time, as required by Theorem 1.2.

Kernel Size Analysis. The returned instance satisfies Theorem 1.2(i) due to (4.5) and (4.7), (ii) due to (4.6), and (iii) due to (4.8).

Approximation Factor Analysis. It remains to prove Theorem 1.2(iv), that is, that we can lift an α-approximate solution for (G_2, R_2, ω_2) to an $\alpha(1+\varepsilon)$-approximate solution for (G, R, ω).

An optimal RPP tour for (G, R, ω) has cost $\omega(W^*) = \omega(R) + \omega(D)$ by (3.1), where D is a minimum-cost Eulerian extension. By Lemma 4.9(iii) and (4.4), there is an Eulerian extension D' for (G, R_1, ω) with

$$\omega(D') \leq \omega(D) + 2\gamma(2c-2) = \omega(D) + \varepsilon_1 \cdot \omega(R). \tag{4.9}$$

By Lemma 4.4, D' is an Eulerian extension for (G, R_2, ω) and, by Proposition 4.2, for (G_2, R_2, ω). Then D' is also an Eulerian extension for (G_2, R_2, ω_2). Thus, an optimal RPP tour for (G_2, R_2, ω_2) has cost at most $\omega_2(R_2) + \omega_2(D')$. By Proposition 3.5, an α-approximate solution for (G_2, R_2, ω_2), can be turned into an Eulerian extension S such that

$$\omega_2(R_2) + \omega_2(S) \leq \alpha(\omega_2(R_2) + \omega_2(D')). \tag{4.10}$$

By Proposition 4.2, S is an Eulerian extension for (G, R_2, ω). By Lemma 4.4, S is an Eulerian extension for (G, R_1, ω), and by Lemma 4.9, it is one for (G, R, ω), since $V(S) \subseteq V(G_2) = V(R_2) \subseteq V(R_1) \subseteq V(R)$. Thus, by Proposition 3.5, S can be turned into an RPP tour of cost $\omega(R) + \omega(S)$ for (G, R, ω). We analyze this cost. By (4.10) and Lemma 2.2 with $\beta = \omega(R) + \omega(T)$,

$$\omega(R_2) + \omega(S) \leq \alpha(\omega(R_2) + \omega(D')) + \varepsilon_2(\omega(R) + \omega(T)).$$

Using $\omega(R_2) \leq \omega(R_1) \leq \omega(R)$ from Lemmas 4.4 and 4.9, and $\alpha \geq 1$, we get

$$\begin{aligned}
\omega(R) + \omega(S) &\leq \alpha(\omega(R) + \omega(D')) + \varepsilon_2(\omega(R) + \omega(T)) \\
&\leq \alpha(\omega(R) + \omega(D')) + \varepsilon_2(\omega(R) + \omega(D)) && \text{using (3.7)} \\
&\leq \alpha(\omega(R) + \omega(D) + \varepsilon_1\omega(R)) + \varepsilon_2(\omega(R) + \omega(D)) && \text{using (4.9)} \\
&\leq \alpha(1 + \varepsilon_1 + \varepsilon_2)(\omega(R) + \omega(D)) = \alpha(1+\varepsilon)\omega(W^*) && \text{using (3.1)}.
\end{aligned}$$

Thus, we got an $\alpha(1+\varepsilon)$-approximation for (G, R, c). □

4.3 Towards a Polynomial-Size Approximate Kernelization Scheme for the Parameter c

In the previous section we have shown a polynomial-size approximate kernelization scheme (PSAKS) for RPP parameterized by $b + c$. An obvious question is whether there is a PSAKS for the parameter c. Unfortunately, we leave this question open, yet make some first steps and discuss the difficulties in resolving this question.

To get a PSAKS for c, one has to reduce the number of imbalanced vertices in $G\langle R\rangle$. One idea is adding to R cheap edges of a minimum-weight perfect matching M on imbalanced vertices, since this is optimal if it happens to connect $G\langle R\rangle$.

Reduction Rule 4.10. Let (G, R, ω) be an RPP instance with triangle inequality and $\delta \in \mathbb{Q}$. Add to R a subset $M^* \subseteq M$ of edges with $\sum_{e \in M^*} \omega(e) \leq \delta$.

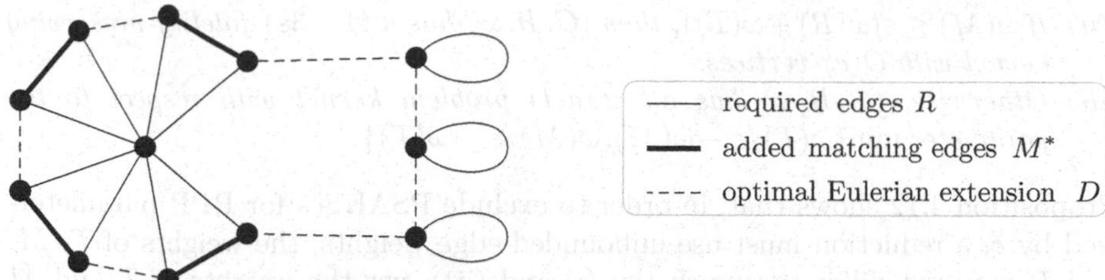

Fig. 2. Example showing that the bound given in Observation 4.11(iii) is tight: adding the edges in M^* to R breaks the only optimal Eulerian extension D (dashed). To fix it, one either has to double all edges of D or add all edges of M^* to D. Note that the star can be arbitrarily enlarged.

Observation 4.11. Let $R' = R \uplus M^*$ be obtained by applying Reduction Rule 4.10 to R.

(i) There are at most $2(|M| - |M^*|)$ imbalanced vertices in $G\langle R'\rangle$.
(ii) For any Eulerian extension S' for (G, R', ω), $S = S' \uplus M^*$ is an Eulerian extension for (G, R, ω) and $\omega(R) + \omega(S) = \omega(R') + \omega(S')$.
(iii) For any Eulerian extension S for (G, R, ω), $S' = S \uplus M^*$ is an Eulerian extension for (G, R', ω) with $\omega(S') \leq \omega(S) + \delta$.

We expect that Reduction Rule 4.10 will indeed have some impact in practice when choosing $\delta = \varepsilon(\omega(R) + \omega(M))$, for example. Yet to show a PSAKS, it is unsuitable:

1. To reduce the number of imbalanced vertices in $G\langle R\rangle$ to some constant, we have to add all but a constant number of edges of M to R, yet, by Observation 4.11(iii), each added edge potentially contributes to the error and thus would merely retain a 2-approximation. Unfortunately, Fig. 2 shows that the bound given by Observation 4.11(iii) is tight.
2. Reduction Rule 4.10 increases the total weight of required edges. This makes it unusable for a PSAKS, since, in the resulting instance, a solution might be $(1+\varepsilon)$-approximate merely due to the fact that the lower bound $\omega(R)$ on the solution is sufficiently large (we will see this below).

Given the difficulties of showing a PSAKS for c, it is tempting to disprove its existence. However, the existing tools for excluding PSAKSes [29] also exclude polynomial-size kernels from which only *optimal* solutions can be lifted to $(1+\varepsilon)$-approximate solutions for the input instance. In terms of Fellows et al. [15], these are so-called $(1+\varepsilon)$-*fidelity-preserving* kernels. However, we can easily build a $(1+\varepsilon)$-fidelity-preserving kernel with size polynomial in $\omega(T)$, that is, of size polynomial in c in case that the edge weights are bounded by poly(c):

Proposition 4.12. *Let (G, R, ω) be an instance of RPP with triangle inequality.*

(i) If $\omega(T) \leq \varepsilon(\omega(R) + \omega(M))$, then one can find a $(1+2\varepsilon)$-approximate RPP tour for (G, R, ω) in polynomial time.

(ii) *If* $\omega(M) \leq \varepsilon(\omega(R) + \omega(T))$, *then* (G, R, ω) *has a* $(1 + 3\varepsilon)$*-fidelity-preserving kernel with* $O(c)$ *vertices.*
(iii) *Otherwise,* (G, R, ω) *has an (exact) problem kernel with respect to the parameter* $\min\{\omega(T)/\varepsilon - \omega(M), \omega(M)/\varepsilon - \omega(T)\}$.

Proposition 4.12 shows that, in order to exclude PSAKSes for RPP parameterized by c, a reduction must use unbounded edge weights, the weights of T, M, and R may not differ too much (by (i) and (ii)), yet the weights of T and M must not be too close either (by (iii)). Given these restrictions, we conjecture:

Conjecture 4.13. RPP has a PSAKS with respect to the parameter c.

5 Conclusion

Our main algorithmic contribution is a polynomial-size approximate kernelization scheme for the Rural Postman Problem parameterized by $b + c$, where b is the number of vertices incident to an odd number of required edges and c is the number of connected components formed by the required edges. In future work, we plan to implement the algorithm and to evaluate it on real-world data.

Notably, the approach taken by Reduction Rule 4.8, namely reducing all vertices that do not belong to some inclusion-maximal set B, does not generalize well to asymmetric distances, so that the main open question besides resolving Conjecture 4.13 is whether the scheme for the parameter $b + c$ presented in this work can be generalized to the *directed* Rural Postman Problem. We point out that, using known ideas [4], one can reduce any instance I of the directed or undirected RPP to an instance I' with c vertices in $O(n^3 \log n)$ time such that any α-approximation for I' yields an $(\alpha + 1)$-approximation for I.

Acknowledgments. René van Bevern and Oxana Yu. Tsidulko are supported by the Russian Foundation for Basic Research, project 18-501-12031 NNIO_a, and by the Ministry of Science and Higher Education of the Russian Federation under the 5-100 Excellence Programme. Till Fluschnik is supported by the German Research Foundation, project TORE (NI 369/18).

References

1. Belenguer, J.M., Benavent, E., Lacomme, P., Prins, C.: Lower and upper bounds for the mixed capacitated arc routing problem. Comput. Oper. Res. **33**(12), 3363–3383 (2006)
2. van Bevern, R., Fluschnik, T., Tsidulko, O.Yu.: Parameterized algorithms and data reduction for safe convoy routing. In: Proceedings of 18th ATMOS, OASIcs, vol. 65, pp. 10:1–10:19. Schloss Dagstuhl-Leibniz-Zentrum für Informatik (2018)
3. van Bevern, R., Hartung, S., Nichterlein, A., Sorge, M.: Constant-factor approximations for capacitated arc routing without triangle inequality. Oper. Res. Lett. **42**(4), 290–292 (2014)

4. van Bevern, R., Komusiewicz, C., Sorge, M.: A parameterized approximation algorithm for the mixed and windy capacitated arc routing problem: theory and experiments. Networks **70**(3), 262–278 (2017)
5. van Bevern, R., Niedermeier, R., Sorge, M., Weller, M.: Complexity of arc routing problems. In: Arc Routing: Problems, Methods, and Applications, MOS-SIAM Series on Optimization, vol. 20. SIAM (2014)
6. Brandão, J., Eglese, R.: A deterministic tabu search algorithm for the capacitated arc routing problem. Comput. Oper. Res. **35**(4), 1112–1126 (2008)
7. Corberán, Á., Laporte, G. (eds.): Arc Routing: Problems, Methods, and Applications. SIAM, Philadelphia (2014)
8. Cygan, M., et al.: Parameterized Algorithms. Springer, Cham (2015). https://doi.org/10.1007/978-3-319-21275-3
9. Dorn, F., Moser, H., Niedermeier, R., Weller, M.: Efficient algorithms for Eulerian extension and Rural Postman. SIAM J. Discrete Math. **27**(1), 75–94 (2013)
10. Edmonds, J., Johnson, E.L.: Matching, Euler tours and the Chinese postman. Math. Program. **5**, 88–124 (1973)
11. Eiben, E., Hermelin, D., Ramanujan, M.S.: Lossy kernels for hitting subgraphs. In: Proceedings of 42nd MFCS, LIPIcs, vol. 83, pp. 67:1–67:14. Schloss Dagstuhl-Leibniz-Zentrum für Informatik, Dagstuhl (2017)
12. Eiben, E., Kumar, M., Mouawad, A.E., Panolan, F., Siebertz, S.: Lossy kernels for connected dominating set on sparse graphs. In: Proceedings of 35th STACS, LIPIcs, vol. 96, pp. 29:1–29:15. Schloss Dagstuhl-Leibniz-Zentrum für Informatik (2018)
13. Eiselt, H.A., Gendreau, M., Laporte, G.: Arc routing problems, Part II: The Rural Postman Problem. Oper. Res. **43**(3), 399–414 (1995)
14. Etscheid, M., Kratsch, S., Mnich, M., Röglin, H.: Polynomial kernels for weighted problems. J. Comput. Syst. Sci. **84**, 1–10 (2017)
15. Fellows, M.R., Kulik, A., Rosamond, F.A., Shachnai, H.: Parameterized approximation via fidelity preserving transformations. J. Comput. Syst. Sci. **93**, 30–40 (2018)
16. Frank, A., Tardos, É.: An application of simultaneous diophantine approximation in combinatorial optimization. Combinatorica **7**(1), 49–65 (1987)
17. Ghiani, G., Improta, G.: The laser-plotter beam routing problem. J. Oper. Res. Soc. **52**(8), 945–951 (2001)
18. Ghiani, G., Laporte, G.: Eulerian location problems. Networks **34**(4), 291–302 (1999)
19. Golden, B.L., Wong, R.T.: Capacitated arc routing problems. Networks **11**(3), 305–315 (1981)
20. Grötschel, M., Jünger, M., Reinelt, G.: Optimal control of plotting and drilling machines: a case study. Z. Oper. Res. **35**(1), 61–84 (1991)
21. Guo, J., Niedermeier, R.: Invitation to data reduction and problem kernelization. ACM SIGACT News **38**(1), 31–45 (2007)
22. Gutin, G., Wahlström, M., Yeo, A.: Rural Postman parameterized by the number of components of required edges. J. Comput. Syst. Sci. **83**(1), 121–131 (2017)
23. Hermelin, D., Kratsch, S., Sołtys, K., Wahlström, M., Wu, X.: A completeness theory for polynomial (Turing) kernelization. Algorithmica **71**(3), 702–730 (2015)
24. Jansen, K.: Bounds for the general capacitated routing problem. Networks **23**(3), 165–173 (1993)
25. Karpinski, M., Lampis, M., Schmied, R.: New inapproximability bounds for TSP. J. Comput. Syst. Sci. **81**(8), 1665–1677 (2015)

26. Kratsch, S.: Recent developments in kernelization: A survey. Bull. EATCS **113** (2014)
27. Krithika, R., Majumdar, D., Raman, V.: Revisiting connected vertex cover: FPT algorithms and lossy kernels. Theor. Comput. Syst. **62**(8), 1690–1714 (2018)
28. Krithika, R., Misra, P., Rai, A., Tale, P.: Lossy kernels for graph contraction problems. In: Proceedings 36th FSTTCS, LIPIcs, vol. 65, pp. 23:1–23:14. Schloss Dagstuhl-Leibniz-Zentrum für Informatik, Dagstuhl (2016)
29. Lokshtanov, D., Panolan, F., Ramanujan, M.S., Saurabh, S.: Lossy kernelization. In: Proceedings 49th STOC, pp. 224–237. ACM (2017)
30. Marx, D., Végh, L.A.: Fixed-parameter algorithms for minimum-cost edge-connectivity augmentation. ACM Trans. Algorithms **11**(4), 27:1–27:24 (2015)
31. Orloff, C.S.: A fundamental problem in vehicle routing. Networks **4**(1), 35–64 (1974)
32. Sorge, M., van Bevern, R., Niedermeier, R., Weller, M.: From few components to an Eulerian graph by adding arcs. In: Kolman, P., Kratochvíl, J. (eds.) WG 2011. LNCS, vol. 6986, pp. 307–318. Springer, Heidelberg (2011). https://doi.org/10.1007/978-3-642-25870-1_28
33. Sorge, M., van Bevern, R., Niedermeier, R., Weller, M.: A new view on Rural Postman based on Eulerian extension and matching. J. Discrete Algorithms **16**, 12–33 (2012)
34. Ulusoy, G.: The fleet size and mix problem for capacitated arc routing. Eur. J. Oper. Res. **22**(3), 329–337 (1985)
35. Wøhlk, S.: An approximation algorithm for the capacitated arc routing problem. Open Oper. Res. J. **2**, 8–12 (2008)

A 2-Approximation Algorithm for the Graph 2-Clustering Problem

Victor Il'ev[1,2(✉)], Svetlana Il'eva[1], and Alexander Morshinin[2]

[1] Dostoevsky Omsk State University, Omsk, Russia
iljev@mail.ru
[2] Sobolev Institute of Mathematics SB RAS, Omsk, Russia
morshinin.alexander@gmail.com

Abstract. We study a version of the graph 2-clustering problem. In this version, for a given undirected graph, one has to find a nearest 2-cluster graph, i.e., the graph on the same vertex set with exactly 2 nonempty connected components each of which is a complete graph. The distance between two graphs is the number of noncoinciding edges.

The problem under consideration is NP-hard. In 2004, Bansal, Blum, and Chawla presented a simple polynomial time 3-approximation algorithm for the similar correlation clustering problem in which the number of clusters doesn't exceed 2. In 2008, Coleman, Saunderson, and Wirth presented a 2-approximation algorithm for this problem applying local search to every feasible solution obtained by the 3-approximation algorithm of Bansal, Blum, and Chawla.

Unfortunately, the method of proving the performance guarantee of the Coleman, Saunderson, and Wirth's algorithm is not suitable for the graph 2-clustering. Coleman, Saunderson, and Wirth used switching technique that allows to reduce clustering any graph to the equivalent problem whose optimal solution is the complete graph, i.e., the cluster graph consisting of the single cluster.

In the graph 2-clustering problem any optimal solution has to consist of exactly 2 clusters, so we need another approximation algorithm and another method of proving a bound on its worst-case behaviour. We present a polynomial time 2-approximation algorithm for the 2-clustering problem on general graphs. In contrast to the proof of Coleman, Saunderson, and Wirth, our proof of the performance guarantee of this algorithm doesn't use switchings.

Keywords: Graph clustering · Approximation algorithm · Performance guarantee

1 Introduction

We study a version of the graph clustering problem equivalent to the well-known 2-correlation clustering. In this version, for a given undirected graph, one has to

M. Khachay et al. (Eds.): MOTOR 2019, LNCS 11548, pp. 295–308, 2019.
https://doi.org/10.1007/978-3-030-22629-9_21

find a nearest 2-cluster graph, i.e., the graph on the same vertex set with exactly 2 nonempty connected components each of which is a complete graph.

We consider only *simple* graphs, i.e., undirected graphs without loops and multiple edges. A graph is called a *cluster graph* if each of its connected components is a complete graph.

Let V be a finite set. Denote by $\mathcal{M}(V)$ the set of all cluster graphs on the vertex set V; let $\mathcal{M}_k(V)$ be the set of all cluster graphs on V consisting of exactly k nonempty connected components, and let $\mathcal{M}_{\leq k}(V)$ be the set of all cluster graphs on V consisting of at most k connected components, $2 \leq k \leq |V|$.

If $G_1 = (V, E_1)$ and $G_2 = (V, E_2)$ are graphs on the same vertex set V, then the *distance* $\rho(G_1, G_2)$ between them is defined as follows

$$\rho(G_1, G_2) = |E_1 \Delta E_2| = |E_1 \setminus E_2| + |E_2 \setminus E_1|,$$

i.e., $\rho(G_1, G_2)$ is the number of noncoinciding edges in G_1 and G_2.

The following variants of the graph clustering problem with bounded number of clusters were studied in the literature under different names: **Graph Approximation Problem** [1,5], **k-Correlation Clustering** [2,3], **MinDisAgree**[k] [4], **k-Cluster Editing** [6], etc.

$\mathbf{GC_{\leq k}}$. Given a graph $G = (V, E)$ and an integer k, $2 \leq k \leq |V|$, find a graph $M^* \in \mathcal{M}_{\leq k}(V)$ such that

$$\rho(G, M^*) = \min_{M \in \mathcal{M}_{\leq k}(V)} \rho(G, M).$$

$\mathbf{GC_k}$. Given a graph $G = (V, E)$ and an integer k, $2 \leq k \leq |V|$, find a graph $M^* \in \mathcal{M}_k(V)$ such that

$$\rho(G, M^*) = \min_{M \in \mathcal{M}_k(V)} \rho(G, M).$$

In 2004, Shamir, Sharan, and Tsur [6] showed that problem $\mathbf{GC_k}$ is NP-hard for any fixed $k \geq 2$. In 2006, Giotis and Guruswami [4] published a more simple proof of the same result. At the same time, Ageev, Il'ev, Kononov, and Talevnin [1] independently proved that problems $\mathbf{GC_2}$ and $\mathbf{GC_{\leq 2}}$ are NP-hard even on 3-regular graphs and deduced from this that both the problems $\mathbf{GC_k}$ and $\mathbf{GC_{\leq k}}$ on general graphs are NP-hard for any fixed $k \geq 2$.

In 2004, Bansal, Blum, and Chawla [2] presented a simple polynomial time 3-approximation algorithm for problem $\mathbf{GC_{\leq 2}}$. In 2006, Giotis and Guruswami [4] presented a randomized *PTAS* for problem **MinDisAgree**[k] equivalent to $\mathbf{GC_{\leq k}}$ (for any fixed $k \geq 2$). In 2008, Coleman, Saunderson, and Wirth [3] pointed out that complexity of *PTAS* from [4] makes it practically useless and presented a 2-approximation algorithm for problem $\mathbf{GC_{\leq 2}}$ applying local search to every feasible solution obtained by the 3-approximation algorithm from [2].

Unfortunately, the method of proving the performance guarantee of the Coleman, Saunderson, and Wirth's algorithm is not suitable for the graph 2-clustering problem $\mathbf{GC_2}$. Coleman, Saunderson, and Wirth used switching technique that allows to reduce clustering any graph to the equivalent problem whose optimal solution is the complete graph, i.e., the cluster graph consisting of the single cluster.

In problem $\mathbf{GC_2}$ any optimal solution has to consist of exactly 2 nonempty clusters, so we need another approximation algorithm and another method of proving a bound on its worst-case behaviour. We present a modified 2-approximation algorithm for problem $\mathbf{GC_2}$. In contrast to the proof of Coleman, Saunderson, and Wirth, our proof of the performance guarantee of this algorithm doesn't use switchings.

2 Problem $\mathbf{GC_2}$

Consider the special case of problem $\mathbf{GC_k}$ with $k = 2$.

$\mathbf{GC_2}$. Given a graph $G = (V, E)$, find a graph $M^* \in \mathcal{M}_2(V)$ such that

$$\rho(G, M^*) = \min_{M \in \mathcal{M}_2(V)} \rho(G, M).$$

Let us introduce the following notation.

Let $N_G(v)$ be the set of vertices adjacent to v in the graph $G = (V, E)$, and $\overline{N}_G(v) = V \setminus (N_G(v) \cup \{v\})$.

Let $G_1 = (V, E_1)$ and $G_2 = (V, E_2)$ be graphs on the same vertex set V, $n = |V|$. Denote by $D(G_1, G_2)$ the graph on the vertex set V with the edge set $E_1 \Delta E_2$. Note that $\rho(G_1, G_2)$ is equal to the number of edges in the graph $D(G_1, G_2)$.

For nonempty sets $V_1, V_2 \subseteq V$ such that $V_1 \cap V_2 = \emptyset$ and $V_1 \cup V_2 = V$ we denote by $M(V_1, V_2)$ the cluster graph in $\mathcal{M}_2(V)$ with connected components induced by V_1, V_2. The sets V_1 and V_2 are called *clusters*.

The following lemma is the straight corollary of the "handshaking lemma".

Lemma 1. *Let $d_{\min}$ be the minimum vertex degree in the graph $D(G_1, G_2)$. Then the following inequality holds:*

$$\rho(G_1, G_2) \geq \frac{nd_{\min}}{2}.$$

Observe also the following useful property of the graph $D(G, M^*)$, where M^* is an optimal solution to problem $\mathbf{GC_2}$ on an arbitrary graph G.

Lemma 2. *Let $M^* = M(X^*, Y^*) \in \mathcal{M}_2(V)$ be an optimal solution to problem $\mathbf{GC_2}$ on an arbitrary n-vertex graph $G = (V, E)$, where $|X^*| \geq 2, |Y^*| \geq 2$. Then for each vertex $v \in V$ the following inequality holds:*

$$d_D(v) \leq \frac{n}{2},$$

where $D = D(G, M^*)$.

Proof. Suppose the opposite. Let $w \in V$ be a vertex such that $d_D(w) > \frac{n}{2}$, i.e., $d_D(w) = \frac{n}{2} + c$, where $c > 0, \frac{n}{2} + c \in \mathbb{N}$, and $\frac{n}{2} + c \leq n - 1$. Consider the graph $\widetilde{M} \in \mathcal{M}_2(V)$ obtained from M^* by moving the vertex w to another cluster. It is obvious that $\widetilde{M}$ is a feasible solution to problem $\mathbf{GC_2}$ on the graph G (since $|X^*| \geq 2, |Y^*| \geq 2$). Consider the graph $\widetilde{D} = D(G, \widetilde{M})$. It is easy to see that the degree of the vertex w in the graph $\widetilde{D}$ is equal to the non-degree of the vertex w in the graph D:

$$d_{\widetilde{D}}(w) = n - 1 - \frac{n}{2} - c = \frac{n}{2} - 1 - c.$$

Obviously, the graphs D and $\widetilde{D}$ differ only in the edges of the form $wu, u \in V$, other edges in these graphs are the same. Therefore,

$$\begin{aligned}\rho(G, \widetilde{M}) - \rho(G, M^*) &= |N_{\widetilde{D}}(w)| - |N_D(w)| = d_{\widetilde{D}}(w) - d_D(w)\\ &= \frac{n}{2} - 1 - c - \frac{n}{2} - c = -(1 + 2c) < 0.\end{aligned}$$

Hence the graph M^* is not optimal, contradicting the condition of the lemma.
Lemma 2 is proved.

Let $G = (V, E)$ be an arbitrary graph. For any vertex $v \in V$ and a set $A \subseteq V$ we denote by A_v^+ the number of vertices $u \in A$ such that $(v, u) \in E$, and by A_v^- the number of vertices $u \in A \setminus \{v\}$ such that $(v, u) \notin E$.

We rewrite Lemma 2 in a more convenient form.

Lemma 3. *Let $G = (V, E)$ be an arbitrary graph, $|V| = n$, and $M^* = M(X^*, Y^*)$ be an optimal solution to problem $\mathbf{GC_2}$ on the graph G, where $|X^*| \geq 2, |Y^*| \geq 2$. Then, for each vertex $v \in V$, the following inequalities hold:*

1. *if $v \in X^*$, then $(X^*)_v^- + (Y^*)_v^+ \leq \frac{n}{2}$;*
2. *if $v \in Y^*$, then $(X^*)_v^+ + (Y^*)_v^- \leq \frac{n}{2}$.*

We will use the following approximation algorithm for problem $\mathbf{GC_2}$. In contrast to the 3-approximation algorithm of Bansal, Blum, and Chawla [2] for problem $\mathbf{GC_{\leq 2}}$, our algorithm looks over only 2-cluster graphs, i.e., feasible solutions to problem $\mathbf{GC_2}$.

Algorithm $\mathbf{A_1}$.
Input: graph $G = (V, E)$.
Step 1. For each ordered pair of vertices $(v, w) \in V \times V$, $v \neq w$, define the cluster graph $M_{v,w} = M(X, Y) \in \mathcal{M}_2(V)$, where $X = \{v\} \cup \big(N_G(v) \setminus \{w\}\big)$, $Y = V \setminus X$.
Step 2. Among all $M_{v,w}$ choose the nearest to G cluster graph M_1:

$$\rho(G, M_1) = \min_{\substack{(v, w) \in V \times V, \\ v \neq w}} \rho(G, M_{v,w}).$$

Output: cluster graph $M_1 = M(X,Y) \in \mathcal{M}_2(V)$.

In fact, algorithm $\mathbf{A_1}$ differs from the algorithm of Bansal, Blum, and Chawla [2] by the only additional operation which excludes each vertex $w \neq v$ from the set X.

The following bound on worst-case behaviour of algorithm $\mathbf{A_1}$ takes place.

Theorem 1. *For any graph $G = (V, E)$ the following inequality holds:*

$$\rho(G, M_1) \leq 3\rho(G, M^*),$$

where $M_1 \in \mathcal{M}_2(V)$ is the solution returned by algorithm $\mathbf{A_1}$ and M^* is an optimal solution to problem $\mathbf{GC_2}$ on the graph G.

Proof. Let $M^* = M(X^*, Y^*)$ and let v be a vertex of the minimum degree in the graph $D = D(G, M^*)$. Without loss of generality, we can assume that $v \in X^*$. Obviously, there is a vertex w such that $w \in Y^*$.

By the definition of the graph D we have

$$X^* = \{v\} \cup (N_G(v) \setminus N_D(v)) \cup (\overline{N}_G(v) \cap N_D(v)). \tag{1}$$

Consider the graph $M_{v,w} = M(X,Y) \in \mathcal{M}_2(V)$, where $X = \{v\} \cup (N_G(v) \setminus \{w\})$, $Y = V \setminus X$. Clearly,

$$N_G(v) \setminus \{w\} = \Big((N_G(v) \setminus N_D(v)) \cup (N_G(v) \cap N_D(v))\Big) \setminus \{w\}.$$

Further, we show that the graph $M_{v,w}$ can be obtained from the graph M^* by moving at most $d_{\min}$ vertices to another cluster, where $d_{\min} = d_D(v)$ is the minimum degree of vertices in the graph D. The following two cases are possible.

Case 1. Vertices v and w aren't adjacent in G, i.e., $w \in \overline{N}_G(v) \cap \overline{N}_D(v)$. Then $N_G(v) \setminus \{w\} = N_G(v)$. Calculate the cardinality of the set $X^* \Delta X$. By the definition of the graph $M_{v,w}$,

$$\begin{aligned} X &= \{v\} \cup (N_G(v) \setminus \{w\}) = \{v\} \cup N_G(v) \\ &= \{v\} \cup (N_G(v) \setminus N_D(v)) \cup (N_G(v) \cap N_D(v)). \end{aligned}$$

So, using (1), we have

$$X^* \Delta X = (X^* \setminus X) \cup (X \setminus X^*) = (\overline{N}_G(v) \cap N_D(v)) \cup (N_G(v) \cap N_D(v)) = N_D(v).$$

Thus, $|X^* \Delta X| = |N_D(v)| = d_D(v) = d_{\min}$, so the graph $M_{v,w}$ can be obtained from the graph M^* by moving $d_{\min}$ vertices of the set $N_D(v)$ to another cluster.

Case 2. Vertices v and w are adjacent in G, i.e., $w \in N_G(v) \cap N_D(v)$. Then $d_{\min} \geq 1$. Calculate the cardinality of the set $X^* \Delta X$. By the definition of the graph $M_{v,w}$,

$$\begin{aligned} X &= \{v\} \cup (N_G(v) \setminus \{w\}) = (\{v\} \cup N_G(v)) \setminus \{w\} \\ &= (\{v\} \cup (N_G(v) \setminus N_D(v)) \cup (N_G(v) \cap N_D(v))) \setminus \{w\}. \end{aligned}$$

So, using (1) and inclusion $w \in N_G(v) \cap N_D(v)$, we obtain

$$X^* \Delta X = \left((\overline{N}_G(v) \cap N_D(v)) \cup (N_G(v) \cap N_D(v)) \right) \setminus \{w\} = N_D(v) \setminus \{w\}.$$

So $|X^* \Delta X| = |N_D(v)| - 1 = d_{\min} - 1$. Hence the graph $M_{v,w}$ can be obtained from the graph M^* by moving $d_{\min} - 1$ vertices of the set $N_D(v) \setminus \{w\}$ to another cluster.

Thus, it is shown that the graph $M_{v,w}$ can be obtained from the graph M^* by moving at most $d_{\min}$ vertices to another cluster. Note that moving $d_{\min}$ vertices may increase the objective function by at most $nd_{\min}$. So, by Lemma 1, we have

$$\rho(G, M_{v,w}) \le \rho(G, M^*) + nd_{\min} \le \rho(G, M^*) + 2\rho(G, M^*) = 3\rho(G, M^*).$$

The graph $M_{v,w}$ is constructed among all graphs at step 1 of algorithm $\mathbf{A_1}$.

Theorem 1 is proved.

Remark 1. Among all graphs constructed by algorithm $\mathbf{A_1}$ at step 1, there is the cluster graph $M_{v,w} = M(X, Y)$ such that

(a) $d_D(v) = \min\limits_{u \in V} d_D(u)$;
(b) $M_{v,w}$ is obtained from M^* by moving at most $d_{\min}$ vertices to another cluster;
(c) $v \in X \cap X^*, w \in Y \cap Y^*$.

The proof follows from the proof of Theorem 1.

3 A 2-Approximation Algorithm for GC_2

Consider the following local search procedure.

Procedure LS(M, X, Y, x, y).
Input: cluster graph $M = M(X, Y) \in \mathcal{M}_2(V)$, $x \in X, y \in Y$.
Iteration 0. Set $X_0 = X, Y_0 = Y$.
Iteration $k (k \ge 1)$.

Step 1. For each vertex $u \in V \setminus \{x, y\}$ calculate the following quantity $\delta_k(u)$ (possible variation of the value of the objective function in case of moving the vertex u to another cluster. If $\delta_k(u) > 0$, then this quantity is said to be the *local improvement for the vertex u at iteration k*):

$$\delta_k(u) = \begin{cases} (X_{k-1})_u^- - (X_{k-1})_u^+ + (Y_{k-1})_u^+ - (Y_{k-1})_u^- & \text{for } u \in X_{k-1} \setminus \{x\}, \\ (Y_{k-1})_u^- - (Y_{k-1})_u^+ + (X_{k-1})_u^+ - (X_{k-1})_u^- & \text{for } u \in Y_{k-1} \setminus \{y\}. \end{cases}$$

Step 2. Choose the vertex $u_k \in V \setminus \{x, y\}$ such that

$$\delta_k(u_k) = \max_{u \in V \setminus \{x,y\}} \delta_k(u).$$

Step 3. If $\delta_k(u_k) \le 0$, then set $X' = X_{k-1}$ and $Y' = Y_{k-1}$. **STOP.** Return $M' = M(X', Y')$.

Step 4. If $u_k \in X_{k-1}$, then set $X_k = X_{k-1} \setminus \{u_k\}$, $Y_k = Y_{k-1} \cup \{u_k\}$. Else $u_k \in Y_{k-1}$, then set $X_k = X_{k-1} \cup \{u_k\}$, $Y_k = Y_{k-1} \setminus \{u_k\}$. **Go to iteration $k+1$.**

Output: cluster graph $M' = M(X', Y') \in \mathcal{M}_2(V)$.

Remark 2. The cluster graph M' returned by procedure **LS** always belongs to the set $\mathcal{M}_2(V)$.

This is obvious because the vertices $x \in X$ and $y \in Y$ always lie in different clusters.

Consider the following approximation algorithm for problem $\mathbf{GC_2}$ that can be viewed as extension of algorithm $\mathbf{A_1}$ when local search procedure is applied to every feasible solution obtained by algorithm $\mathbf{A_1}$.

Algorithm $\mathbf{A_2}$.

Input: graph $G = (V, E)$.

Step 1. For each ordered pair of vertices $(v, w) \in V \times V, v \neq w$, do the following:

Step 1.1. Define a cluster graph $M_{v,w} = M(X, Y) \in \mathcal{M}_2(V)$, where $X = \{v\} \cup (N_G(v) \setminus \{w\}), Y = V \setminus X$.

Step 1.2. Run local search procedure $\mathbf{LS}(M_{v,w}, X, Y, v, w)$. Denote the resulting graph by $M'_{v,w}$.

Step 2. Among all locally-optimal solutions $M'_{v,w}$ choose the nearest to G cluster graph M':

$$\rho(G, M') = \min_{\substack{(v,w) \in V \times V, \\ v \neq w}} \rho(G, M'_{v,w}).$$

Step 3. For each $u \in V$ define the cluster graph $M''_u = M(X'', Y'') \in \mathcal{M}_2(V)$, where $X'' = V \setminus \{u\}, Y'' = \{u\}$.

Step 4. Among all graphs M''_u choose the nearest to G cluster graph M'':

$$\rho(G, M'') = \min_{u \in V} \rho(G, M''_u).$$

Step 5. If $\rho(G, M') \leq \rho(G, M'')$, then set $M_2 = M'$, else set $M_2 = M''$.

Output: cluster graph $M_2 = M(X, Y) \in \mathcal{M}_2(V)$.

The running time of algorithm $\mathbf{A_2}$ is $O(n^6)$. This is greater than the running time of the algorithm of Coleman, Saunderson, and Wirth [3] for problem $\mathbf{GC_{\leq 2}}$ because of more complicated step 1.1.

Let $G = (V, E)$ be an arbitrary graph, let $M^* = M(X^*, Y^*)$ be an optimal solution to problem $\mathbf{GC_2}$ on the graph G, and let $M = M(X, Y)$ be an arbitrary feasible solution to problem $\mathbf{GC_2}$ on the graph G. We will compare the distance $\rho(G, M^*)$ and the distance $\rho(G, M)$ in the following way.

Lemma 4. $\rho(G,M) - \rho(G,M^*)$

$$= \sum_{u\in X\cap Y^*} \left((X\cap X^*)_u^- - (X\cap X^*)_u^+ + (Y\cap Y^*)_u^+ - (Y\cap Y^*)_u^-\right) + \sum_{u\in Y\cap X^*} \left((Y\cap Y^*)_u^- - (Y\cap Y^*)_u^+ + (X\cap X^*)_u^+ - (X\cap X^*)_u^-\right).$$

Proof. Calculate the distance between graphs G and M^*. Taking into account that $X^* = (X\cap X^*)\cup(Y\cap X^*)$ and $Y^* = (X\cap Y^*)\cup(Y\cap Y^*)$, we have

$$\begin{aligned}\rho(G,M^*) = &\frac{1}{2}\sum_{u\in X\cap X^*}(X\cap X^*)_u^- + \frac{1}{2}\sum_{u\in Y\cap X^*}(Y\cap X^*)_u^- \\ &+ \frac{1}{2}\sum_{u\in X\cap Y^*}(X\cap Y^*)_u^- + \frac{1}{2}\sum_{u\in Y\cap Y^*}(Y\cap Y^*)_u^- \\ &+ \sum_{u\in X\cap Y^*}(X\cap X^*)_u^+ + \sum_{u\in Y\cap X^*}(Y\cap Y^*)_u^+ + \sum_{u\in Y\cap X^*}(X\cap X^*)_u^- \\ &+ \sum_{u\in Y\cap X^*}(X\cap Y^*)_u^+ + \sum_{u\in Y\cap Y^*}(X\cap X^*)_u^+ + \sum_{u\in X\cap Y^*}(Y\cap Y^*)_u^-.\end{aligned}$$

Similarly, calculate the distance between graphs G and M.

$$\begin{aligned}\rho(G,M) = &\frac{1}{2}\sum_{u\in X\cap X^*}(X\cap X^*)_u^- + \frac{1}{2}\sum_{u\in Y\cap X^*}(Y\cap X^*)_u^- \\ &+ \frac{1}{2}\sum_{u\in X\cap Y^*}(X\cap Y^*)_u^- + \frac{1}{2}\sum_{u\in Y\cap Y^*}(Y\cap Y^*)_u^- \\ &+ \sum_{u\in X\cap Y^*}(X\cap X^*)_u^- + \sum_{u\in Y\cap X^*}(Y\cap Y^*)_u^- + \sum_{u\in Y\cap X^*}(X\cap X^*)_u^+ \\ &+ \sum_{u\in Y\cap X^*}(X\cap Y^*)_u^+ + \sum_{u\in Y\cap Y^*}(X\cap X^*)_u^+ + \sum_{u\in X\cap Y^*}(Y\cap Y^*)_u^+.\end{aligned}$$

Thus, $\rho(G,M) - \rho(G,M^*) =$

$$\sum_{u\in X\cap Y^*} \left((X\cap X^*)_u^- - (X\cap X^*)_u^+ + (Y\cap Y^*)_u^+ - (Y\cap Y^*)_u^-\right) + \sum_{u\in Y\cap X^*} \left((Y\cap Y^*)_u^- - (Y\cap Y^*)_u^+ + (X\cap X^*)_u^+ - (X\cap X^*)_u^-\right).$$

Lemma 4 is proved.

The main result of this paper is the following bound on the worst-case behaviour of algorithm $\mathbf{A_2}$.

Theorem 2. *For any graph $G = (V,E)$ the following inequality holds:*

$$\rho(G, M_2) \leq 2\rho(G, M^*),$$

where $M_2 \in \mathcal{M}_2(V)$ is the solution returned by algorithm $\mathbf{A_2}$ and M^* is an optimal solution to problem $\mathbf{GC_2}$ on the graph G.

Proof. Let $M^* = M(X^*, Y^*)$. We may assume that $|X^*| \geq 2$ and $|Y^*| \geq 2$ since otherwise $M_2 = M^*$ due to step 3 of algorithm $\mathbf{A_2}$.

Let v be a vertex of the minimum degree in the graph $D = D(G, M^*)$. Since step 1 of algorithm $\mathbf{A_1}$ and step 1.1 of algorithm $\mathbf{A_2}$ are the same, then by Remark 1 among all graphs constructed by algorithm $\mathbf{A_2}$ at step 1.1 there exists the graph $M_{v,w} = M(X, Y)$ such that

(a) $d_D(v) = \min_{u \in V} d_D(u)$;

(b) $M_{v,w}$ is obtained from M^* by moving at most $d_{\min}$ vertices to another cluster;

(c) $v \in X \cap X^*, w \in Y \cap Y^*$.

Consider the performance of procedure $\mathbf{LS}(M_{v,w}, X, Y, v, w)$ on the graph $M_{v,w} = M(X, Y)$.

It is easy to see that

$$|X \cap Y^*| \cup |Y \cap X^*| \leq d_{\min}.$$

Local search procedure $\mathbf{LS}$ starts with $X_0 = X$ and $Y_0 = Y$. At every iteration k either $\mathbf{LS}$ moves some vertex $u_k \in V \setminus \{v, w\}$ to another cluster, or no vertex is moved and $\mathbf{LS}$ finishes.

Consider in detail iteration $t+1$ such that

- at every iteration $k \in \{1, ..., t\}$ procedure $\mathbf{LS}$ selects some vertex $u_k \in (X \cap Y^*) \cup (Y \cap X^*)$;
- at iteration $t+1$ either procedure $\mathbf{LS}$ selects some vertex $u_{t+1} \in \big((X \cap X^*) \cup (Y \cap Y^*)\big) \setminus \{v, w\}$, or iteration $t+1$ is the last iteration of $\mathbf{LS}$.

Let us introduce the following quantities:

$$\alpha_{t+1}(u) = \begin{cases} (X_t \cap X^*)^-_u - (X_t \cap X^*)^+_u + (Y_t \cap Y^*)^+_u - (Y_t \cap Y^*)^-_u & \text{for } u \in X_t \cap Y^*, \\ (Y_t \cap Y^*)^-_u - (Y_t \cap Y^*)^+_u + (X_t \cap X^*)^+_u - (X_t \cap X^*)^-_u & \text{for } u \in Y_t \cap X^*. \end{cases}$$

$$\beta_{t+1}(u) = \begin{cases} (X_t \cap Y^*)^-_u - (X_t \cap Y^*)^+_u + (Y_t \cap X^*)^+_u - (Y_t \cap X^*)^-_u & \text{for } u \in X_t \cap Y^*, \\ (Y_t \cap X^*)^-_u - (Y_t \cap X^*)^+_u + (X_t \cap Y^*)^+_u - (X_t \cap Y^*)^-_u & \text{for } u \in Y_t \cap X^*. \end{cases}$$

It is not difficult to see that for each vertex $u \in (X_t \cap Y^*) \cup (Y_t \cap X^*)$ there holds

$$\delta_{t+1}(u) = \alpha_{t+1}(u) + \beta_{t+1}(u). \tag{2}$$

Indeed, if $u \in X_t \cap Y^*$, then the local improvement $\delta_{t+1}(u)$ is equal to

$$\begin{aligned} \delta_{t+1}(u) &= (X_t)^-_u - (X_t)^+_u + (Y_t)^+_u - (Y_t)^-_u \\ &= (X_t \cap X^*)^-_u - (X_t \cap X^*)^+_u + (Y_t \cap Y^*)^+_u - (Y_t \cap Y^*)^-_u \\ &+ (X_t \cap Y^*)^-_u - (X_t \cap Y^*)^+_u + (Y_t \cap X^*)^+_u - (Y_t \cap X^*)^-_u \\ &= \alpha_{t+1}(u) + \beta_{t+1}(u). \end{aligned}$$

For vertices $u \in Y_t \cap X^*$ equality (2) can be proved similarly.

Consider the cluster graph $M_t = M(X_t, Y_t)$. Then by Lemma 4

$$\rho(G, M_t) - \rho(G, M^*) = \sum_{u \in X_t \cap Y^*} \alpha_{t+1}(u) + \sum_{u \in Y_t \cap X^*} \alpha_{t+1}(u).$$

Since at all iterations preceding iteration $t+1$ only vertices of the set $(X \cap Y^*) \cup (Y \cap X^*)$ were moved, then

$$|X_t \cap Y^*| + |Y_t \cap X^*| = r \leq d_{\min}. \tag{3}$$

Hence

$$\rho(G, M_t) - \rho(G, M^*) \leq r \max\{\alpha_{t+1}(u) : u \in (X_t \cap Y^*) \cup (Y_t \cap X^*)\}. \tag{4}$$

Now let us prove that at iteration $t+1$ the following statement is true:

$$\forall u \in (X_t \cap Y^*) \cup (Y_t \cap X^*) \quad \alpha_{t+1}(u) \leq \frac{n}{2}. \tag{5}$$

We do this in two stages.

I. First, for each vertex $u \in V \setminus \{v, w\}$, we estimate the local improvement $\delta_{t+1}(u)$, i.e., decreasing of the value of the objective function in case of moving u to another cluster.

(1) Prove that for all $u \in (X_t \cap X^*) \cup (Y_t \cap Y^*) \setminus \{v, w\}$ the following inequality holds:

$$\delta_{t+1}(u) \leq 2\big(|Y_t \cap X^*| + |X_t \cap Y^*|\big) + 1. \tag{6}$$

(a) Let $u \in (X_t \cap X^*) \setminus \{v\}$. Observe that

$$(Y_t \cap X^*)_u^+ + (Y_t \cap X^*)_u^- + (X_t \cap Y^*)_u^+ + (X_t \cap Y^*)_u^- = |Y_t \cap X^*| + |X_t \cap Y^*|, \tag{7}$$

$$(X_t \cap X^*)_u^+ + (X_t \cap X^*)_u^- + (Y_t \cap Y^*)_u^+ + (Y_t \cap Y^*)_u^- = n - 1 - |Y_t \cap X^*| - |X_t \cap Y^*|. \tag{8}$$

By Lemma 3 we obtain

$$(X_t \cap X^*)_u^- + (Y_t \cap Y^*)_u^+ \leq (X^*)_u^- + (Y^*)_u^+ \leq \frac{n}{2}. \tag{9}$$

The local improvement $\delta_{t+1}(u)$ for the vertex $u \in (X_t \cap X^*) \setminus \{v\}$ is equal to

$$\begin{aligned}\delta_{t+1}(u) &= (X_t)_u^- - (X_t)_u^+ + (Y_t)_u^+ - (Y_t)_u^- \\ &= (X_t \cap X^*)_u^- - (X_t \cap X^*)_u^+ + (Y_t \cap Y^*)_u^+ - (Y_t \cap Y^*)_u^- \\ &\quad + (X_t \cap Y^*)_u^- - (X_t \cap Y^*)_u^+ + (Y_t \cap X^*)_u^+ - (Y_t \cap X^*)_u^-.\end{aligned}$$

Add and subtract $(X_t \cap X^*)_u^- + (Y_t \cap Y^*)_u^+$. Then

$$\begin{aligned}\delta_{t+1}(u) &= 2\big((X_t \cap X^*)_u^- + (Y_t \cap Y^*)_u^+\big) - (X_t \cap X^*)_u^- - (Y_t \cap Y^*)_u^+ \\ &- (X_t \cap X^*)_u^+ - (Y_t \cap Y^*)_u^- + (X_t \cap Y^*)_u^- - (X_t \cap Y^*)_u^+ + (Y_t \cap X^*)_u^+ - (Y_t \cap X^*)_u^-.\end{aligned}$$

So, using (9) and (8), we have

$$\begin{aligned}\delta_{t+1}(u) \le 2\tfrac{n}{2} - \big((X_t\cap X^*)^-_u + (Y_t\cap Y^*)^+_u + (X_t\cap X^*)^+_u + (Y_t\cap Y^*)^-_u\big)\\ + (X_t\cap Y^*)^-_u - (X_t\cap Y^*)^+_u + (Y_t\cap X^*)^+_u - (Y_t\cap X^*)^-_u\\ = n - n + 1 + |Y_t\cap X^*| + |X_t\cap Y^*|\\ + (X_t\cap Y^*)^-_u - (X_t\cap Y^*)^+_u + (Y_t\cap X^*)^+_u - (Y_t\cap X^*)^-_u.\end{aligned}$$

Since all terms are non-negative, then

$$\begin{aligned}(X_t\cap Y^*)^-_u - (X_t\cap Y^*)^+_u + (Y_t\cap X^*)^+_u - (Y_t\cap X^*)^-_u\\ \le (X_t\cap Y^*)^-_u + (X_t\cap Y^*)^+_u + (Y_t\cap X^*)^+_u + (Y_t\cap X^*)^-_u.\end{aligned}$$

So, using (7), we have

$$\begin{aligned}\delta_{t+1}(u) \le |Y_t\cap X^*| + |X_t\cap Y^*| + 1\\ + (X_t\cap Y^*)^-_u - (X_t\cap Y^*)^+_u + (Y_t\cap X^*)^+_u - (Y_t\cap X^*)^-_u\\ \le |Y_t\cap X^*| + |X_t\cap Y^*| + 1 + |Y_t\cap X^*| + |X_t\cap Y^*|\\ = 2\big(|Y_t\cap X^*| + |X_t\cap Y^*|\big) + 1.\end{aligned}$$

Thus, for each vertex $u \in (X_t\cap X^*)\setminus\{v\}$ inequality (6) is proved.

(b) If $u \in (Y_t\cap Y^*)\setminus\{w\}$, then one can prove inequality (6) by symmetric replacement X_t, X^* with Y_t, Y^* respectively.

(2) Prove that for all $u \in (Y_t\cap X^*)\cup(X_t\cap Y^*)$

$$\delta_{t+1}(u) \le 2\big(|Y_t\cap X^*| + |X_t\cap Y^*|\big) + 1. \tag{10}$$

(a) If at iteration $t+1$ of the local search procedure **LS** a vertex $u_{t+1} \in \big((X_t\cap X^*)\cup(Y_t\cap Y^*)\big)\setminus\{v,w\}$ is moved, then using (6) we obtain

$$\forall u \in (Y_t\cap X^*)\cup(X_t\cap Y^*)\quad \delta_{t+1}(u) \le \delta_{t+1}(u_{t+1}) \le 2\big(|Y_t\cap X^*| + |X_t\cap Y^*|\big) + 1.$$

(b) If iteration $t+1$ of procedure **LS** is the last, then the following inequalities hold:

$$\forall u \in (Y_t\cap X^*)\cup(X_t\cap Y^*)\quad \delta_{t+1}(u) \le 0.$$

In this case inequalities (10) are obvious.

II. Now prove inequalities (5).

(1) First, prove that

$$\forall u \in Y_t\cap X^*\quad \alpha_{t+1}(u) \le \frac{n}{2}.$$

Suppose the opposite, i.e., there is a vertex $p \in Y_t\cap X^*$ such that $\alpha_{t+1}(p) > \frac{n}{2}$. By (2), $\delta_{t+1}(p) = \alpha_{t+1}(p) + \beta_{t+1}(p)$, whence

$$\beta_{t+1}(p) = \delta_{t+1}(p) - \alpha_{t+1}(p) < \delta_{t+1}(p) - \frac{n}{2}.$$

By (10), $\delta_{t+1}(p) \le 2\big(|Y_t \cap X^*| + |X_t \cap Y^*|\big) + 1$, therefore

$$\beta_{t+1}(p) < 2\big(|Y_t \cap X^*| + |X_t \cap Y^*|\big) + 1 - \frac{n}{2}. \qquad (11)$$

Since $d_{\min} = \min_{u \in V} d_D(u)$, then

$$d_D(p) = (Y_t \cap X^*)_p^- + (X_t \cap X^*)_p^- + (X_t \cap Y^*)_p^+ + (Y_t \cap Y^*)_p^+ \ge d_{\min}.$$

So, using (3), we obtain

$$(Y_t \cap X^*)_p^- + (X_t \cap X^*)_p^- + (X_t \cap Y^*)_p^+ + (Y_t \cap Y^*)_p^+ \ge |Y_t \cap X^*| + |X_t \cap Y^*|. \quad (12)$$

Since $p \in Y_t \cap X^*$, then

$$(Y_t \cap X^*)_p^+ + (Y_t \cap X^*)_p^- + (X_t \cap Y^*)_p^+ + (X_t \cap Y^*)_p^- = |Y_t \cap X^*| + |X_t \cap Y^*| - 1, \quad (13)$$

$$(X_t \cap X^*)_p^+ + (X_t \cap X^*)_p^- + (Y_t \cap Y^*)_p^+ + (Y_t \cap Y^*)_p^- = n - |Y_t \cap X^*| - |X_t \cap Y^*|. \quad (14)$$

By the definition,

$$\beta_{t+1}(p) = (Y_t \cap X^*)_p^- - (Y_t \cap X^*)_p^+ + (X_t \cap Y^*)_p^+ - (X_t \cap Y^*)_p^-.$$

It follows from (13) that

$$-(Y_t \cap X^*)_p^+ - (X_t \cap Y^*)_p^- = (Y_t \cap X^*)_p^- + X_t \cap Y^*)_p^+ - |Y_t \cap X^*| - |X_t \cap Y^*| + 1,$$

hence

$$\begin{aligned} \beta_{t+1}(p) &= (Y_t \cap X^*)_p^- + (X_t \cap Y^*)_p^+ + (Y_t \cap X^*)_p^- + (X_t \cap Y^*)_p^+ \\ &\quad - |Y_t \cap X^*| - |X_t \cap Y^*| + 1 \\ &= 2\big((Y_t \cap X^*)_p^- + (X_t \cap Y^*)_p^+\big) - |Y_t \cap X^*| - |X_t \cap Y^*| + 1. \end{aligned}$$

By (12),

$$(Y_t \cap X^*)_p^- + (X_t \cap Y^*)_p^+ \ge |Y_t \cap X^*| + |X_t \cap Y^*| - (X_t \cap X^*)_p^- - (Y_t \cap Y^*)_p^+,$$

so

$$\begin{aligned} \beta_{t+1}(p) &\ge 2\big(|Y_t \cap X^*| + |X_t \cap Y^*| - (X_t \cap X^*)_p^- - (Y_t \cap Y^*)_p^+\big) \\ &\quad - |Y_t \cap X^*| - |X_t \cap Y^*| + 1 \\ &= |Y_t \cap X^*| + |X_t \cap Y^*| + 1 - 2(X_t \cap X^*)_p^- - 2(Y_t \cap Y^*)_p^+. \end{aligned}$$

Add and subtract $(X_t \cap X^*)_p^+$ and $(Y_t \cap Y^*)_p^-$. Then

$$\begin{aligned} \beta_{t+1}(p) &\ge |Y_t \cap X^*| + |X_t \cap Y^*| + 1 \\ &\quad + (Y_t \cap Y^*)_p^- - (Y_t \cap Y^*)_p^+ + (X_t \cap X^*)_p^+ - (X_t \cap X^*)_p^- \\ &\quad - \big((Y_t \cap Y^*)_p^- + (Y_t \cap Y^*)_p^+ + (X_t \cap X^*)_p^+ + (X_t \cap X^*)_p^-\big). \end{aligned}$$

So, using (14), we have

$$\begin{aligned}\beta_{t+1}(p) \geq |Y_t \cap X^*| + |X_t \cap Y^*| + 1\\ + (Y_t \cap Y^*)_p^- - (Y_t \cap Y^*)_p^+ + (X_t \cap X^*)_p^+ - (X_t \cap X^*)_p^- - n\\ + |Y_t \cap X^*| + |X_t \cap Y^*| = 2\big(|Y_t \cap X^*| + |X_t \cap Y^*|\big) + 1 - n\\ + (Y_t \cap Y^*)_p^- - (Y_t \cap Y^*)_p^+ + (X_t \cap X^*)_p^+ - (X_t \cap X^*)_p^- .\end{aligned}$$

Since $p \in Y_t \cap X^*$, then

$$\alpha_{t+1}(p) = (Y_t \cap Y^*)_p^- - (Y_t \cap Y^*)_p^+ + (X_t \cap X^*)_p^+ - (X_t \cap X^*)_p^-,$$

whence

$$\begin{aligned}\beta_{t+1}(p) &\geq 2\big(|Y_t \cap X^*| + |X_t \cap Y^*|\big) + 1 - n + \alpha_{t+1}(p)\\ &> 2\big(|Y_t \cap X^*| + |X_t \cap Y^*|\big) + 1 - \frac{n}{2}.\end{aligned}$$

This contradicts to inequality (11). Therefore, for each $u \in Y_t \cap X^*$ inequality (5) holds (due to arbitrariness of the vertex p).

(2) For all $u \in X_t \cap Y^*$ the inequality

$$\alpha_{t+1}(u) \leq \frac{n}{2}$$

can be proved similarly by symmetric replacement X_t, X^* with Y_t, Y^* respectively.

Thus, inequalities (5) hold.

Using (3), (4), (5), and Lemma 1, we obtain

$$\begin{aligned}\rho(G, M'_{v,w}) - \rho(G, M^*) &\leq r \max\{\alpha_{t+1}(u) : u \in (X_t \cap Y^*) \cup (Y_t \cap X^*)\}\\ &\leq r\frac{n}{2} \leq d_{\min}\frac{n}{2} \leq \rho(G, M^*).\end{aligned}$$

Thus,

$$\rho(G, M'_{v,w}) \leq 2\rho(G, M^*).$$

The graph $M'_{v,w}$ is constructed among all graphs at step 1.2 of algorithm $\mathbf{A_2}$.

Theorem 2 is proved.

References

1. Ageev, A.A., Il'ev, V.P., Kononov, A.V., Talevnin, A.S.: Computational complexity of the graph approximation problem. Diskretnyi Analiz i Issledovanie Operatsii. Ser. 1 **13**(1), 3–11 (2006). (in Russian). English transl. in: J. Appl. Ind. Math. **1**(1), 1–8 (2007)
2. Bansal, N., Blum, A., Chawla, S.: Correlation clustering. Mach. Learn. **56**, 89–113 (2004)

3. Coleman, T., Saunderson, J., Wirth, A.: A local-search 2-approximation for 2-correlation-clustering. In: Halperin, D., Mehlhorn, K. (eds.) ESA 2008. LNCS, vol. 5193, pp. 308–319. Springer, Heidelberg (2008). https://doi.org/10.1007/978-3-540-87744-8_26
4. Giotis, I., Guruswami, V.: Correlation clustering with a fixed number of clusters. Theory Comput. **2**(1), 249–266 (2006)
5. Il'ev, V.P., Fridman, G.Š.: On the problem of approximation by graphs with fixed number of components. Doklady AN SSSR. **264**(3), 533–538 (1982) (in Russian). English transl. in: Soviet Math. Dokl. **25**(3), 666–670 (1982)
6. Shamir, R., Sharan, R., Tsur, D.: Cluster graph modification problems. Discrete Appl. Math. **144**(1–2), 173–182 (2004)

Approximation Scheme for the Capacitated Vehicle Routing Problem with Time Windows and Non-uniform Demand

Michael Khachay[1,2,3](✉) and Yuri Ogorodnikov[1,2]

[1] Krasovsky Institute of Mathematics and Mechanics, Ekaterinburg, Russia
{mkhachay,yogorodnikov}@imm.uran.ru

[2] Ural Federal University, Ekaterinburg, Russia

[3] Omsk State Technical University, Omsk, Russia

Abstract. The Capacitated Vehicle Routing Problem with Time Windows (CVRPTW) is the well-known combinatorial optimization problem having numerous valuable applications in operations research. Unlike the classic CVRP (without time windows constraints), approximation algorithms with theoretical guarantees for the CVRPTW are still developed much less, even for the Euclidean plane. In this paper, perhaps for the first time, we propose an approximation scheme for the planar CVRPTW with non-uniform splittable demand combining the well-known instance decomposition framework by A. Adamaszek et al. and Quasi-Polynomial Time Approximation Scheme (QPTAS) by L. Song et al. Actually, for any $\varepsilon \in (0,1)$ the scheme proposed finds a $(1+\varepsilon)$-approximate solution of the problem in polynomial time provided the capacity q and the number p of time windows does not exceed $2^{\log^{\delta} n}$ for some $\delta = O(\varepsilon)$. For any fixed p and q the scheme is Efficient Polynomial Time Approximation Scheme (EPTAS) with subquadratic time complexity.

Keywords: Capacitated vehicle routing problem · Time windows · Splittable demand · Polynomial time approximation scheme

1 Introduction

The Capacitated Vehicle Routing Problem (CVRP) is the famous combinatorial optimization problem, which was introduced by Dantzig and Ramser in their seminal paper [9] and has a wide range of relevant applications in practice (see, e.g. [30]). In the simplest setting of the problem, we are given by a finite set of customers having the same unit demand and a fleet of identical capacitated vehicles located initially at a single depot. The goal is to construct a collection of vehicle routes minimizing the total transportation cost and servicing all the customers.

The Capacitated Vehicle Routing Problem with Time Windows (CVRPTW) [20,30] is an extension of the CVRP, where service of each customer should start

M. Khachay et al. (Eds.): MOTOR 2019, LNCS 11548, pp. 309–327, 2019.
https://doi.org/10.1007/978-3-030-22629-9_22

at a specified time interval, called a time window. CVRP with hard windows is widely applicable in natural gas distribution [7], dial-ride company planning [11], continent-scale distribution of building materials [23], low-carbon economy [26], and other practical transportation problems [25].

The problem is well-investigated by specialists in the field of exact methods, heuristics, and meta-heuristics. Recently, a significant progress was achieved in solving practically important instances of the CVRPTW by local-search heuristics [13], Tabu-search [29], genetic [31], memetic [6,21], ant colony algorithms [22], and their combinations (see, e.g. [8]).

Nevertheless, approximation results for this problem in the class of algorithms with theoretical guarantees are still extremely rare. To the best of our knowledge, they are exhausted by the Quasi-Polynomial Time Approximation Scheme (QPTAS) proposed in [28] and extended recently to the case of multiple depots [27] and our recent Efficient Polynomial-Time Approximation Schemes (EPTAS) for the CVRPTW with any fixed capacity and number of time windows. In addition, all known results relate to the special setting of the problem, where all customers have the same unit demand.

Our Contribution. In this paper, perhaps for the first time for the CVRPTW with non-uniform demand, we propose an approximation scheme with theoretically proved time complexity bounds. Our scheme extends the decomposition framework introduced by Adamaszek et al. in [1] for efficient approximation of the simplest unit-demand CVRP on the Euclidean plane to more general case of the problem to take into account additional time windows constraints and a non-uniform splittable customer demand.

The rest of this paper is structured as follows. In Sect. 2, we give a short overview of known approximation results for the CVRPTW in the class of algorithms with theoretical bounds. In Sect. 3, we recall the mathematical setting of the CVRPTW with non-uniform demand. Section 4 presents the mail idea of the proposed scheme. In subsequent sections, we discuss this scheme in detail. We start in Sect. 5 with basic known results needed for the subsequent constructions. Then, in Sect. 6 we present our approximation scheme with a proof of its accuracy bound. Time complexity bounds are proved in the Sect. 7. Finally, in Sect. 8, we summarize the results obtained and list some questions that still remain open.

2 Related Work

Being an extension of the well-known strongly NP-hard Traveling Salesman Problem (TSP) [30], the Capacitated Vehicle Routing Problem is also strongly NP-hard even in the Euclidean plane [24] provided the capacity q belongs to the input. The metric CVRP remains intractable and APX-hard even for any fixed $q \geq 3$ and for the two-valued $\{1, 2\}$-metric.

For the Euclidean CVRP, the first approximation results date to the seminal paper by Haimovich and Rinnooy Kan [12], where the first PTAS for the CVRP on the plane and capacity $q = o(\log\log n)$ and first constant-factor algorithms for an arbitrary metric were introduced. Then, in [3] an improved scheme, whose running time retains polynomial for the wider range $q = O(\log n/\log\log n)$, was proposed.

The ideas proposed by Arora in his celebrated paper [2] were used by Das and Mathieu to design their Quasi-Polynomial Time Approximation Scheme (QPTAS) [10] for the general case of the planar Euclidean CVRP. Their QPTAS finds a $(1+\varepsilon)$-approximate solution of this problem (for the case, when q is a part of the instance) in time $n^{(\log n)^{O(1/\varepsilon)}}$. Using this QPTAS as a black-box, Adamaszek, Czumaj, and Lingas [1] showed that $(1+\varepsilon)$-approximate solution of the planar CVRP can be found in polynomial time, if $q \le 2^{\log^{\delta} n}$ for some $\delta = \delta(\varepsilon)$. Some aforementioned results were extended to the case of Euclidean spaces of an arbitrary finite dimension [14,18,19] and several special graphs [4,5].

Unlike CVRP, approximability of the Euclidean CVRPTW is much less investigated. To the best of our knowledge, the family of known approximation algorithms for this problem is exhausted by a Quasi-Polynomial Time Approximation Scheme (QPTAS) developed in [27,28] for the general case of the problem and approximation schemes for the case of $\max\{p, q\} = o(\log\log n)$ and $p^3q^4 = O(\log n)$, where p is the number of time windows, proposed in [16] and [15,17], respectively.

All aforementioned results for the CVRPTW relate to the simplest setting of the problem, when all customers have the same unit demand. In this paper, we try to bridge this gap and to propose an approximation scheme for the case of the CVRPTW with a non-uniform splittable demand.

3 Problem Statement

We consider the Euclidean Capacitated Vehicle Routing Problem with Time Windows and non-uniform Splittable customer Demand (CVRPTW-SD). For the sake of simplicity, we restrict ourselves to the case of the Euclidean plane, pairwise disjoint time windows, and a single depot. An instance of the CVRPTW-SD is defined by

- a set $X = \{x_1, \ldots, x_n\}$ of customer locations (*customers*) on the Euclidean plane and a dedicated location y also known as *depot*, such that, for any locations $v_1, v_2 \in X \cup \{y\}$, transportation cost associated with the direct move from v_1 to v_2 coincides with $\|v_1 - v_2\|_2$
- a natural-valued function d specifying *demand* $d(x)$ of any customer $x \in X$ that should be serviced by one or more vehicle *routes*
- an unbounded fleet of *vehicles* having the same integer *capacity* q and located initially in the depot y
- a linearly ordered set $\mathcal{T} = \{T_1, \ldots, T_p\}$ of the consecutive *time windows*, such that the demand $d(x)$ of any customer x should be fulfilled within the given time window $T(x) \in \mathcal{T}$; we assume that, for any $1 \le j < p$, the time window T_j *precedes* T_{j+1} and use the notation $T_j \prec T_{j+1}$.

The goal is to satisfy the demand of each customer minimizing the total transportation cost with respect to the capacity and time windows constraints.

Mathematically, an instance of the CVRPTW-SD is given by a complete node- and edge-weighted graph $G = (X \cup y, E, d, w)$, natural number q, and a partition

$$X_1 \cup \ldots \cup X_p = X, \quad \text{where } X_j = \{x_i \in X : T(x_i) \in \mathcal{T}\}, \; (j \in \{1, \ldots, p\}). \quad (1)$$

To any customer node $x_i \in X$, the weighting function d assigns[1] the positive integer demand $d_i = d(x_i)$, while the function w defines the transportation cost $w(v_1, v_2) = \|v_1 - v_2\|_2$ for any edge $e = \{v_1, v_2\} \in E$.

A *feasible route* is an ordered pair $\mathcal{R}_j = (R_j, D_j)$, where $R_j = y, x_{i_1}, \ldots, x_{i_s}, y$ is a simple cycle in the graph G and the n-tuple $D_j = (d_{1j}, \ldots, d_{nj})$ satisfying time windows

$$T(x_{i_l}) \preceq T(x_{i_{l+1}}), \quad (1 \le l < s)$$

and capacity

$$1 \le d_{i_l j} \le d_{i_l}, \quad (1 \le l \le s)$$
$$d_{ij} = 0, \quad i \notin \{i_1, \ldots, i_s\}$$
$$\sum_{i=1}^{n} d_{ij} \le q$$

constraints, where d_{ij} is a part of the i-th customer demand covered by the route R_j. To any feasible route $\mathcal{R}$, we assign the transportation cost

$$w(\mathcal{R}) = w(y, x_{i_1}) + w(x_{i_1}, x_{i_2}) + \ldots + w(x_{i_s}, y).$$

The goal is to find, for some $m \ge 1$, a minimum cost multi-cover $\mathcal{U} = (\mathcal{R}_1, \ldots, \mathcal{R}_m)$ of the graph G, satisfying the total customer demand, i.e.

$$\sum_{j=1}^{m} d_{ij} = d_i, \quad (1 \le i \le n).$$

In the sequel, we propose a novel approximation scheme for this problem, which is an Efficient Sub-Quadratic Approximation Scheme for any fixed capacity q and the number p of time windows retaining the polynomial running time, when $\max\{p, q\} \le 2^{\log^{\delta} n}$ for some $\delta = \delta(\varepsilon)$.

4 Main Idea

Our scheme extends the approach proposed by Adamaszek et al. in [1] to the more general case of the Capacitated Vehicle Routing Problem augmented by non-uniform splittable customer demand and time windows constraints. In this section, we give a short overview of the scheme, which consists of the following stages.

[1] Without loss of generality, we can can assume that $d(y) = 0$, for the depot y.

Preprocessing. To any customer x_i, we assign the distance $r_i = w(y, x_i)$ from the depot y and relabel the customers in non-increasing order of these distances, i.e., $r_1 \geq \ldots \geq r_n$. Then, given an $\varepsilon > 0$, we set a tolerance threshold

$$\rho = \frac{r_1\varepsilon}{N}, \text{ where } N = \sum_{i=1}^{n} \left\lceil \frac{d_i}{q} \right\rceil, \tag{2}$$

and exclude all the customers x_i, for which $r_i \leq \rho$.

Rounding. We reduce the given instance of the CVRPTW-SD to an appropriate instance of the special kind, which we call *rounded.* To proceed with such a reduction, we draw a number of circles centered at the depot y and separating them into equal sectors by rays spreading from this depot and introduce an accuracy dependent grid consisting of *locations*, which are the intersections between circles and rays. We divide each location to p *slots* by the number of given time windows and move any customer x_i to the corresponding slot of the closest location. Finally, we show that any $(1+\varepsilon)$-approximate solution of the rounded instance obtained can be efficiently transformed to a $(1+O(\varepsilon))$-approximate solution of the initial one. Therefore, in the sequel, we assume that the given instance is rounded.

Decomposition. At this stage, we decompose the given rounded CVRPTW-SD instance into a number of independent subinstances of two kind, we call them *white* and *gray*, which can be solved in parallel. We show that, to obtain an $(1+O(\varepsilon))$-approximate solution of the initial instance, it is sufficient to construct $(1+\varepsilon)$-approximate solution of any white subinstance and approximate any gray subinstance by an appropriate adaptation of the well-known Iterative Tour Partition (ITP) heuristic [12]. Then, following [1], we show that any subinstance (white or gray) can be efficiently reduced to an equivalent one, whose total demand does not exceed a polynomial of the capacity q, the number of time windows p and $1/\varepsilon$.

Blackboxing. Finally, we complete our approximation scheme by employing the QPTAS proposed in [28] and our extension of the ITP heuristic [16] to find approximate solutions of all the reduced white and gray subinstances, respectively.

5 Preliminaries

We start with some necessary definitions and facts. All of them remain valid not only for the planar setting of the CVRPTW considered in this paper but also for the general metric CVRPTW defined by an arbitrary non-negative edge-weighting function w satisfying the triangle inequality.

Lemma 1. *For any instance of the CVRPTW-SD, such that* $r_1 \geq \ldots \geq r_n$, $r_i = w(y, x_i)$, *the following inequality*

$$\mathrm{OPT} \geq \max\left\{\mathrm{TSP}^*(X \cup \{y\}), 2r_1, \frac{2}{q}\sum_{i=1}^{n} d_i r_i\right\} \tag{3}$$

is valid, where $\mathrm{TSP}^*(X \cup \{y\})$ *is the optimum value of the TSP instance defined by the graph* $G = G(X \cup \{y\}, E, w)$.

Proof. Since the inequalities $\mathrm{OPT} \geq \mathrm{TSP}^*(X \cup \{y\}) \geq 2r_1$ are a straightforward consequence of the triangle inequality, we prove the bound $\mathrm{OPT} \geq \frac{2}{q}\sum_{i=1}^{n} d_i r_i$.

Let $\mathcal{U} = \{\mathcal{R}_1, \ldots, \mathcal{R}_m\}$ be an arbitrary optimum solution of the given CVRP-TW-SD instance. For each $j \in [m] = \{1, \ldots, m\}$, introduce the non-empty subset $X(\mathcal{R}_j) = \{x_i \in X : d_{ij} > 0\}$ of customers visited by the route $\mathcal{R}_j$. Since, for any $x_i \in X(\mathcal{R}_j)$, $2r_i = w(y, x_i) + w(x_i, y) \leq w(\mathcal{R}_j)$, by the triangle inequality, the following equation

$$\frac{d_{ij}}{\sum_{l=1}^{n} d_{lj}} w(\mathcal{R}_j) \geq \frac{2d_{ij}}{\sum_{l=1}^{n} d_{lj}} r_i$$

is valid for each customer $x_i \in X$. Therefore,

$$w(\mathcal{R}_j) = \frac{\sum_{i=1}^{n} d_{ij}}{\sum_{l=1}^{n} d_{lj}} w(\mathcal{R}_j) \geq \frac{2\sum_{i=1}^{n} d_{ij} r_i}{\sum_{l=1}^{n} d_{lj}} \geq \frac{2}{q}\sum_{i=1}^{n} d_{ij}.r_i,$$

since $q \geq \sum_{l=1}^{n} d_l j$, and

$$w(\mathcal{U}) = \sum_{j=1}^{m} w(\mathcal{R}_j) \geq \frac{2}{q}\sum_{j=1}^{m}\sum_{i=1}^{n} d_{ij} r_i = \frac{2}{q}\sum_{i=1}^{n} r_i \sum_{j=1}^{m} d_{ij} = \frac{2}{q}\sum_{i=1}^{n} d_i r_i.$$

Lemma is proved.

The well-known Iterative Tour Partition (ITP) heuristic introduced in [12] for the metric Capacitated Vehicle Routing Problem (CVRP) with unit demand can be defined as follows. Consider an instance of the metric CVRP defined by the complete edge-weighted graph $G = G(X \cup \{y\}, E, w)$ and capacity q. Suppose, we are given by an arbitrary Hamiltonian cycle H in the subgraph $G\langle X \rangle$ induced by the customer subset X. Starting at some customer x, cut the cycle H onto $l = \lceil n/q \rceil$ chains, where $n = |X|$, such that each chain, except maybe the last one, visits q customers exactly. For any chain obtained, connect its endpoints with the depot y directly constructing the set $S(x)$ of l routes. Proceed with the similar procedure taking each other customer $x \in X$ as a staring point and output the route set

$$S_{\mathrm{ITP}} = \arg\min\{w(S(x)) : x \in X\}$$

of the minimum cost. The following lemma [12] gives an upper bound for the cost of the obtained solution.

Lemma 2.

$$w(S_{\text{ITP}}) \leq 2\left\lceil\frac{n}{q}\right\rceil\frac{\sum_{i=1}^{n} r_i}{n} + (1-1/q)w(H) \leq \left(1+\frac{q}{n}\right)\cdot\frac{2}{q}\sum_{i=1}^{n} r_i + (1-1/q)w(H).$$

In [17], we extend ITP heuristic to the case of the metric CVRPTW with uniform demand. For the sake of completeness, we present this technique in this paper in Algorithm 1, which can be easily adapted to the case of the metric CVRPTW with non-uniform splittable demand.

Algorithm 1. The ITP heuristic for the metric CVRPTW

Input: an instance of the metric CVRPTW defined by a complete graph $G(X \cup \{y\}, E, w)$, capacity q, and partition $X_1 \cup \ldots \cup X_p = X$

Parameter: β-approximation algorithm $\mathcal{A}_\beta$ for the metric TSP

Output: an approximate solution S_{ITP} of the given CVRPTW instance

1: using $\mathcal{A}_\rho$ obtain a ρ-approximate metric TSP solution H for the subgraph $G\langle X\rangle$
2: by shortcutting, split the cycle H into smaller cycles $H_1, \ldots, H_p$, s.t. H_j spans customers from X_j
3: **for each** cycle H_j **do**
4:     **for each** $x \in X_j$ **do**
5:         starting from the node x, split the cycle H_j into $l_j = \lceil |X_j|/q\rceil$ chains, s.t. each of them, except maybe one, spans q vertices
6:         connecting endpoints of each chain with the depot y directly, construct a set $S(x)$ of l_j routes
7:     **end for**
8:     put $S_j = \arg\min\{w(S(x)) \colon x \in X_j\}$
9: **end for**
10: output the solution $S_{\text{ITP}} = S_1 \cup \ldots \cup S_p$.

Indeed, to obtain an ITP-based approximate solution in this case, we represent each customer x_i with demand d_i by the family of d_i its unit-demand copies and reduce the initial instance to the obtained instance of the metric CVRPTW with unit demand defined by the auxiliary graph on $D = 1 + \sum_{i=1}^{n} d_i$ nodes. For the weight $w(S_{\text{ITP}})$ of the resulting solution, we obtain the following bound.

Lemma 3.

$$w(S_{ITP}) \leq 2\cdot\left(\frac{2}{q}\sum_{i=1}^{n} d_i r_i\right) + pw(H) \leq 2\cdot\left(\frac{2}{q}\sum_{i=1}^{n} d_i r_i\right) + p\beta\cdot\text{TSP}^*(X).$$

Proof. Indeed, applying Lemma 2 to each customer subset X_j, we obtain

$$w(S_j) \leq \left(1+\frac{q}{D_j}\right)\cdot\frac{2}{q}\sum_{x_i\in X_j} d_i r_i + (1-1/q)w(H_j) \leq 2\cdot\left(\frac{2}{q}\sum_{x_i\in X_j} d_i r_i\right) + w(H),$$

where $D_j = \sum_{x_i\in X_j} d_i$. Since $w(S_{\text{ITP}}) = \sum_{j=1}^{p} w(S_j)$ and $W(H) \leq \beta\text{TSP}^*(X)$, Lemma is proved.

Finally, we present the following fact taken from [1], which helps us to reduce the instance in question to the equivalent one with much less total demand and to prove a polynomial time complexity bound of the scheme proposed. Hereinafter, we call a feasible route $\mathcal{R}$ *non-trivial*, if it visits at least two distinct customers, i.e. $|X(\mathcal{R})| > 1$. Otherwise, the route is called *trivial*.

Lemma 4. *For any instance of the CVRPTW-SD, there exists an optimum solution* $\mathcal{U} = \{\mathcal{R}_1, \ldots, \mathcal{R}_m\}$*, such that, among its* m *routes, at most* $|X|$ *are non-trivial.*

Actually, Lemma 4 was proven in [1] for a more restricted case, i.e. the unit-demand CVRP free of the time windows constraints. But this result can be easily extended to the CVRPTW with splittable non-uniform demand. For the sake of brevity, we skip the proof this claim, postponing it to the forthcoming paper.

6 Approximation Scheme

It this section, we describe our approximation scheme following the overview presented in Sect. 4 and prove its correctness.

Suppose, we are given by $\varepsilon \in (0, 1)$ and an instance of the Euclidean CVRPTW-SD on the plane defined by a complete node- and edge-weighted graph $G = (X \cup \{y\}), E, d, w)$, capacity $q \in \mathbb{N}$, and partition $X_1 \cup \ldots \cup X_p = X$ induced by an ordered set $\mathcal{T} = \{T_1, \ldots, T_p\}$ of consecutive disjoint time windows (see Sect. 3 for details). In this section, we show how to construct an $(1 + \varepsilon)$-approximate solution of this instance.

6.1 Instance Preprocessing

Discuss the details of an approximation scheme proposed by us. Firstly, reordering the customers X by decreasing their distances $r(x)$ to the depot y. Then, we can notify that some customers can be ignoring with respect to the fixed ε. We start with assigning to each customer x_i the distance $r_i = w(x_i, y)$ from the depot y and reordering them by descending the distances $r_1 \geq \ldots \geq r_n$. Then, we show that, during construction an $(1+\varepsilon)$-approximate solution we can ignore the customers, which are located sufficiently close to the depot in accordance to formula (2).

Lemma 5. *Demand of all customers, for which* $r_i \leq \rho$*, can be serviced by routes of at most* $\varepsilon \cdot \mathrm{OPT}$ *total cost.*

Proof. Indeed, for any customer x_i, its demand d_i can be serviced by $\lceil d_i/q \rceil$ trivial routes, each of them has the cost $2r_i$. Therefore, for the total cost C_ρ, we have

$$C_\rho \leq \sum_{i=1}^{n} 2r_i \left\lceil \frac{d_i}{q} \right\rceil \leq 2\rho \cdot N \leq 2N \frac{\varepsilon r_1}{N} \leq 2\varepsilon r_1 \leq \varepsilon \cdot \mathrm{OPT},$$

where the last inequality follows from Lemma 1. Lemma is proved.

In the sequel, without loss of generality we assume that the equation $\rho \le r_i \le r_1$ holds for any customer $x_i \in X$.

6.2 Rounding

In this section, we reduce the given instance to a special one, which we call *rounded*. To proceed with this reduction, we introduce the accuracy dependent grid induced by the circles centered at the depot y of radii

$$\rho_i = \rho\left(1 + \frac{\varepsilon}{q}\right)^i, \quad 0 \le i \le \lceil \log_{1+\frac{\varepsilon}{q}} N/\varepsilon \rceil \tag{4}$$

and rays spreading from y dividing each disk into $s = \lceil 2\pi q/\varepsilon \rceil$ equal circular sectors with central angle $2\pi/s$. We call *locations* the obtained intersection points between rays and circles. To any location, we assign p *slots*, by the number of different time windows. Then, we move each customer $x_i \in X_j$ to the j-th slot of the nearest location such that, each slot accumulates the total demand of all customers that are moved to it. Since the number of circles and rays are

$$\log_{1+\frac{\varepsilon}{q}} \frac{N}{\varepsilon} = \Theta\left(\frac{q}{\varepsilon} \log \frac{N}{\varepsilon}\right) \text{ and } \Theta\left(\frac{q}{\varepsilon}\right),$$

respectively, the total number of slots is $\Theta\left(p \cdot \left(\frac{q}{\varepsilon}\right)^2 \log \frac{N}{\varepsilon}\right)$.

Thus, we reduce the initial instance to the special instance of the Euclidean CVRPTW-SD (we call it *rounded*), whose customers are slots assigned to grid locations.

Lemma 6. *The proposed reduction changes the cost of any solution by at most* $\varepsilon \cdot \mathrm{OPT}$.

Proof. Indeed, consider an arbitrary customer x with demand $d(x)$ located at a distance $r(x)$ from the depot y, between two neighboring circles of radii ρ_i and ρ_{i+1} (Fig. 1). It is easy to verify that the distance between x and the nearest location l has the following upper bound

$$\|x - l\|_2 \le p_1 + p_2 \le r(x)\alpha/2 + (\rho_{i+1} - \rho_i)/2.$$

Therefore,

$$\|x - l\|_2 \le r(x)\frac{\varepsilon}{q},$$

since $\alpha \le \varepsilon/q$, $\rho_{i+1} = \rho_i(1 + \varepsilon/q)$, and $r(x) \ge r(x)$, by construction. Since an arbitrary feasible solution visits each customer x_i by at most d_i routes, the total change of its cost induced by moving all the customers to slots at the closest locations does not exceed

$$\varepsilon \cdot \frac{2}{q} \sum_{i=1}^{n} d_i r_i \le \varepsilon \cdot \mathrm{OPT},$$

by Lemma 1. Lemma is proved.

Thanks to Lemma 6, in the rest of this paper, we can assume without loss of generality that each CVRPTW-SD instance considered is rounded.

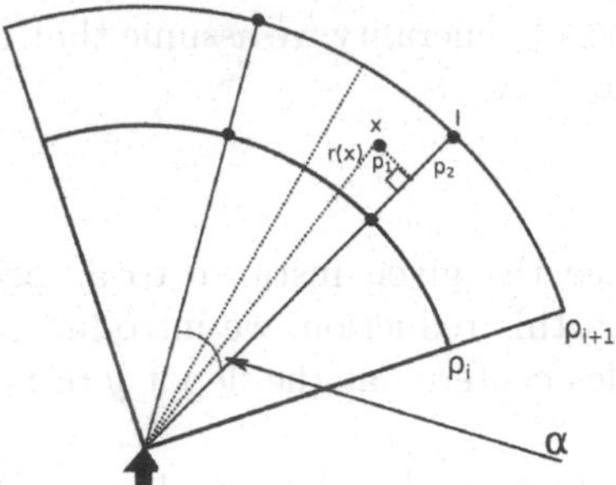

Fig. 1. Moving the customer x to the slot at the nearest location l

6.3 Instance Decomposition

In this section, we show that any rounded instance of the CVRPTW-SD can be decomposed into an appropriate collection of subinstances, which can be solved in parallel, such that $(1+\varepsilon)$-approximate solution of the initial instance can be combined from the approximate solutions of the subinstances obtained.

We start this decomposition with partitioning the enclosing disk (of radius r_1) to *rings*, such that each ring (except maybe the most inner one) consists of $k = \lceil \log_{1+\frac{\varepsilon}{q}} \frac{5}{\varepsilon} \rceil$ consecutive circles. Then, each regular ring K has an inner radius $r_{in} = \rho(1+\varepsilon/q)^i$ for some $0 \le i \le \lceil \log_{1+\frac{\varepsilon}{q}} N/\varepsilon \rceil$ and the outer one $r_{out} = r_{in}(1+\varepsilon/q)^k$. By $W(K)$ we denote a width of the ring K. Since

$$W(K) = r_{in}\left(\left(1+\frac{\varepsilon}{q}\right)^{\lceil \log_{1+\frac{\varepsilon}{q}} \frac{5}{\varepsilon} \rceil} - 1\right) \ge r_{in}\left(\left(1+\frac{\varepsilon}{q}\right)^{\log_{1+\frac{\varepsilon}{q}} \frac{5}{\varepsilon}} - 1\right)$$
$$= r_{in}\left(\frac{5}{\varepsilon} - 1\right) \ge r_{in}\left(\frac{5}{\varepsilon} - \frac{1}{\varepsilon}\right) = 2r_{in}\frac{2}{\varepsilon},$$

we obtain the following upper bound

$$2r_{in} \le \frac{\varepsilon}{2} \cdot W(K) \tag{5}$$

for the length of the inner radius of any ring K in terms of its width $W(K)$, which is important for the subsequent constructions.

At the second step, for a positive integer $a = \lceil (20p\beta + 4)/\varepsilon \rceil$ and some number $b \in \{0, \ldots, a-1\}$, whose choice will be explained later, we color all the rings obtained in white and gray, starting from the outer one, such that the ring K_i is painted gray, if $i \equiv b \pmod a$. Here β is an approximation factor of the algorithm used for solving the auxiliary TSP instances and the choice of b will be explained later, in Lemma 10.

In the sequel, we show that such a coloring leads to a successful decomposition of the initial rounded CVRPTW-SD instance. Let us discuss it in detail. First, we prove Lemma 7 that holds for much more general white-gray ring colorings.

Indeed, by $\mathfrak{F}_1, \ldots, \mathfrak{F}_\alpha$ and $\mathrm{OPT}(\mathfrak{F}_i)$ denote the maximal (by inclusion) families of consecutive white rings and the optimum value of the CVRPTW-SD subinstance induced by slots located in rings of the family $\mathfrak{F}_i$, respectively.

Lemma 7. *For any white-gray coloring of rings obtained by the following rules: (i) any monochromatic pair of the adjacent rings is white; (ii) the outer ring is white as well, the following equation*

$$\sum_{i=1}^{\alpha} \mathrm{OPT}(\mathfrak{F}_i) \leq \left(1 + \frac{\varepsilon}{2}\right) \mathrm{OPT}$$

is valid.

Proof. Indeed, let $\mathcal{U} = \{\mathcal{R}_1, \ldots \mathcal{R}_m\}$ be an arbitrary optimum solution of the initial rounded instance of the CVRPTW-SD. By the following recurrent procedure, transform any route $\mathcal{R} \in \mathcal{U}$ to an appropriate collection of routes, such that each new route visits the slots located in a single family of white rings exclusively. For the given route $\mathcal{R}$, consider the outermost white ring family visited by this route, say $\mathfrak{F}_1$, and the adjacent gray ring K (Fig. 2). Including $2 \cdot l$ inner radii r_{in} and l chords of the ring K, split the route $\mathcal{R}$ into subroutes $\mathcal{R}_g(1), \ldots, \mathcal{R}_g(l)$, each of them visits no slots outside $\mathfrak{F}_1$ and a single subroute $\mathcal{R}_b$ located in the interior of the ring K. Thanks to Eq. (5) and the triangle inequality, such a transformation results in the increase of the transportation cost by at most

$$4 \cdot r_{in} \cdot l \leq 2l \cdot \varepsilon/2 \cdot W(K) \leq \varepsilon/2 \cdot w(\mathcal{R} \cap K),$$

where $w(\mathcal{R} \cap K)$ denotes the partial cost of the route $\mathcal{R}$ related to its intersection with the ring K. Continuing this transformation procedure recursively

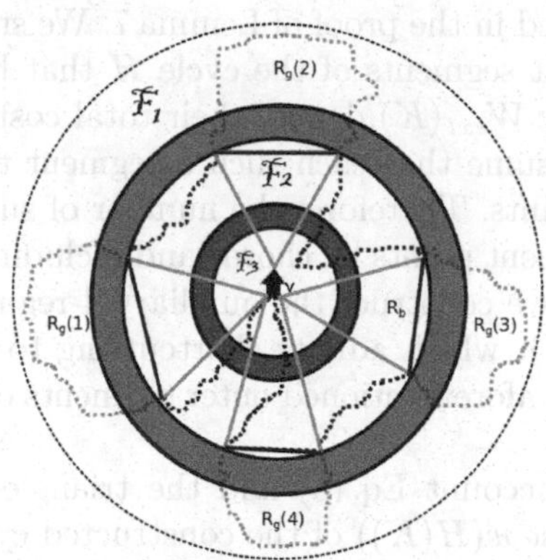

Fig. 2. Splitting of the route $\mathcal{R}$ into $\mathcal{R}_g(1), \ldots, \mathcal{R}_g(l)$ and $\mathcal{R}_b$ subroutes

(proceeding with the subroute $\mathcal{R}_b$ and so on), we obtain that the total cost increasing caused by such a transformation for the route $\mathcal{R}$ does not exceed

$$\frac{\varepsilon}{2} \cdot \sum_{j=1}^{\alpha} w(\mathcal{R} \cap K_j),$$

where the summation is performed over all gray rings $K_1 \ldots, K_\alpha$. Therefore, the total cost of the obtained routes is at most

$$w(\mathcal{U}) + \frac{\varepsilon}{2} \sum_{i=1}^{m} \sum_{j=1}^{\alpha} w(\mathcal{R}_i \cap K_j) \leq (1 + \varepsilon/2) w(\mathcal{U}).$$

Lemma follows from the obvious observation that, for any family $\mathfrak{F}_i$, the optimum value $\mathrm{OPT}(\mathfrak{F}_i)$ does not exceed the total cost of the subroutes produced by the above recursive procedure that visit this family.

For any gray ring K, by $\mathrm{TSP}^*(K)$ we denote the optimum value of the Euclidean TSP for the slots located in this ring. Evidently, each $\mathrm{TSP}^*(K)$ does not exceeds the optimum value of the TSP instance induced by all slots and the depot, we denote this value by TSP^*. The following lemma gives much more accurate bound.

Lemma 8. *Let $K_1, \ldots, K_\alpha$ be gray rings. Then,*

$$\sum_{i=1}^{\alpha} \mathrm{TSP}^*(K_i) \leq (1 + \pi\varepsilon)\, \mathrm{TSP}^*.$$

Proof. Let H be an arbitrary minimum cost Hamiltonian cycle passing through all the slots and the depot, such that $w(H) = \mathrm{TSP}^*$ (Fig. 3a). To obtain the desired bound, we employ the following recursive tour splitting procedure similar to the procedure provided in the proof of Lemma 7. We start with the outermost gray ring K and cut out segments of the cycle H that belong to this ring and its exterior (Fig. 3b). By $W_{ext}(K)$ denote their total cost. Further, without loss of generality, we can assume that each such a segment touches the inner circle of the ring K in two points. Therefore, the number of such points is even.

Connecting the adjacent points by chords and including the perfect matching as it is done in Fig. 4a, we construct the auxiliary 4-regular multi-graph having the Eulerian cycle $E(K)$, which admits shortcutting to the Hamiltonian cycle $H(K)$ containing all the aforementioned outer segments of the cycle H (Fig. 4b).

Again, taking into account Eq. (5) and the triangle inequality, obtain the upper bound for the cost $w(H(K))$ of the constructed cycle $H(K)$

$$\begin{aligned} w(H(K)) &\leq w(E(K)) \leq W_{ext}(K) + 4\pi \cdot r_{in} \\ &\leq W_{ext}(K) + \pi\varepsilon \cdot W(K) \leq W_{ext}(K) + \pi\varepsilon \cdot w(H \cap K), \end{aligned}$$

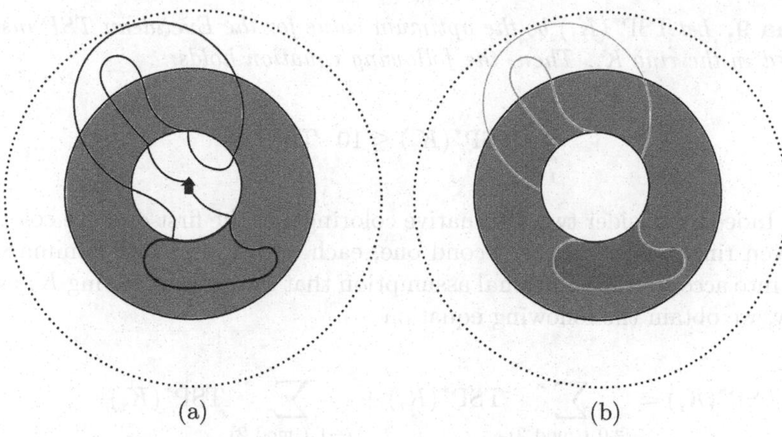

Fig. 3. (a) the initial cycle (b) cutting out the outer segments

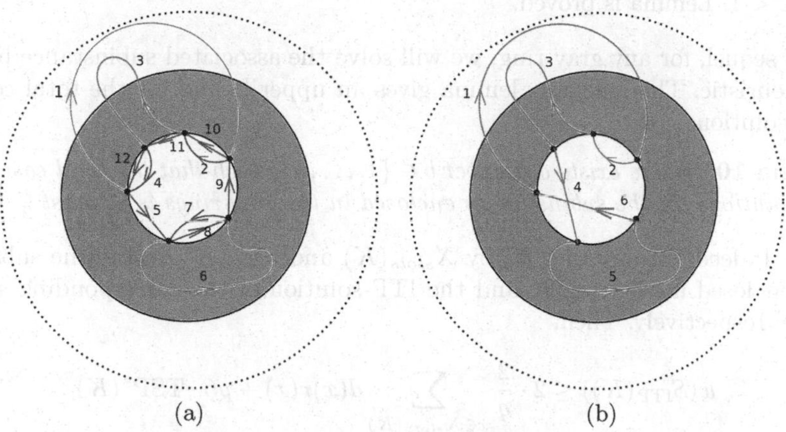

Fig. 4. (a) constructing the Eulerian cycle (b) shortcutting to the Hamiltonian cycle

where $w(H \cap K)$ denotes the cost of the segment of the cycle H that belongs to the ring K. Proceeding with this procedure recursively and summing over all the gray rings, we obtain the final bound

$$\sum_{i=1}^{\alpha} \mathrm{TSP}^*(K_i) \leq \sum_{i=1}^{\alpha} w(H(K_i)) \leq (1+\pi\varepsilon)\, w(H) = (1+\pi\varepsilon)\,\mathrm{TSP}^*.$$

Lemma is proved.

Further, applying Lemma 8 twice to the alternating coloring, where each family $\mathfrak{F}_i$ consists of a single ring, we estimate the total cost of the optimum Hamiltonan cycles for all rings obtained at the first step of instance decomposition.

Lemma 9. *Let* $\mathrm{TSP}^*(K_i)$ *be the optimum value for the Euclidean TSP instance enclosed in the ring* K_i*. Then, the following equation holds:*

$$\sum_{i=1}^{k} \mathrm{TSP}^*(K_i) \leq 10 \cdot \mathrm{TSP}^*.$$

Proof. Indeed. Consider two alternative colorings. In the first one, we color gray each even ring, whilst, in the second one, each odd. Employing Lemma 8, and taking into account the additional assumption that the outermost ring K_1 cannot be gray, we obtain the following equation

$$\sum_{i=1}^{k} \mathrm{TSP}^*(K_i) = \sum_{i\equiv 0 \pmod 2} \mathrm{TSP}^*(K_i) + \sum_{i\equiv 1 \pmod 2} \mathrm{TSP}^*(K_i)$$
$$\leq 2\,(1+\pi\varepsilon)\,\mathrm{TSP}^* + \mathrm{TSP}^*(K_1) \leq 2\,(1+\pi\varepsilon)\,\mathrm{TSP}^* + \mathrm{TSP}^* \leq 10 \cdot \mathrm{TSP}^*,$$

since $\varepsilon < 1$. Lemma is proved.

In the sequel, for any gray ring, we will solve the associated subinstance by the ITP heuristic. The following lemma gives an upper bound for the total cost of these solutions.

Lemma 10. *There exists a number* $b \in \{1, \ldots, a\}$*, such that the total cost of all ITP solutions for the subinstances enclosed in the gray rings is at most* $\frac{\varepsilon}{2} \cdot \mathrm{OPT}$.

Proof. Indeed, for any ring K, by $X_{slots}(K)$ and $S_{\mathrm{ITP}}(K)$ denote the subset of slots enclosed in the ring K and the ITP-solution of the corresponding subinstance, respectively. Then,

$$w(S_{\mathrm{ITP}}(K)) \leq 2 \cdot \frac{2}{q} \sum_{x\in X_{slots}(K)} d(x)r(x) + p\beta \cdot \mathrm{TSP}^*(K)$$

Therefore, by Lemmas 3, 9, and 1,

$$\sum_{b=0}^{a-1} \sum_{i\equiv b \pmod a} w(S_{\mathrm{ITP}}(K_i)) \leq 2 \cdot \frac{2}{q} \sum_{x\in X_{slots}} d(x)r(x) + p\beta \cdot \sum_{i=1}^{k} \mathrm{TSP}^*(K_i)$$
$$\leq 2 \cdot \mathrm{OPT} + 10p\beta \cdot \mathrm{TSP}^* \leq (2 + 10p\beta)\mathrm{OPT}.$$

Hence, there exists b, such that

$$\sum_{i\equiv b \pmod a} w(S_{\mathrm{ITP}}(K_i)) \leq \frac{2+10p\beta}{a}\mathrm{OPT} \leq \frac{\varepsilon}{2}\mathrm{OPT},$$

by construction. Lemma is proved.

To complete the decomposition of the given instance, we perform white-gray coloring of the rings driven by the parameters a and b and obtain subinstances defined by white families $\mathfrak{F}_i$ and gray rings K_j. Hereinafter, we call them *white* and *gray*, respectively. Then, by Lemma 4, we reduce each subinstance with σ slots and an arbitrarily large demand to the equivalent one, whose total demand does not exceed $\sigma^2 q$. Finally, we find a $(1+\varepsilon/2)$-approximate solution and an ITP-solution of any reduced white and gray subinstance, respectively. Our first result follows straightforwardly from Lemmas 7 and 10.

Theorem 1. *For any $\varepsilon \in (0,1)$, the proposed decomposition provides $(1+\varepsilon)$-approximate solution for the initial rounded CVRPTW-SD instance.*

6.4 Approximate Algorithms for Subinstances

As we mentioned in Sect. 4, to find an approximate solution for an arbitrary white subinstance we apply as a black box the quasi-polynomial approximation scheme (QPTAS) proposed by Song et al. in [28], whilst, each gray subinstance we approximate with our recent modification [16,17] of the well-known ITP heuristic.

7 Time Complexity Bounds

As shown in Sect. 6, the proposed scheme consists of the following stages: pre-processing, rounding, instance decomposition and the main stage dealing with the approximation of white and gray subinstances.

It can be easily verified that the first three stages can be carried out in time $O(n \log n)$, where n is the number of distinct customers. To estimate time complexity of the final stage, recall that the Song's QPTAS and our modification of the ITP are developed for the case of CVRPTW with unit demand. Therefore, in our case, their complexity bounds should be represented in terms of the total customer demand D defining the instance in question, i.e. $O\left(D^{\log^{O(1/\varepsilon)} D}\right)$ and $O(D^3)$, respectively (see [16,28]).

Thanks to Lemma 4, for any subinstance (white or gray) obtained during the decomposition of the given rounded instance, its total demand $D = \sigma^2 q$, where σ is the number of slots engaged, which in turn is determined by the number of circles included into an appropriate family of white rings or to the gray ring. By construction, each ring contains

$$O\left(\log_{1+\frac{\varepsilon}{q}} \frac{1}{\varepsilon}\right) = O\left(\frac{q}{\varepsilon} \cdot \log \frac{1}{\varepsilon}\right)$$

circles, each of them consists of $O(pq/\varepsilon)$ slots. Therefore, for any gray subinstance,

$$\sigma = \sigma_g = O\left(\frac{pq^2}{\varepsilon^2} \cdot \log \frac{1}{\varepsilon}\right),$$

while any white instance is determined by

$$\sigma = \sigma_w = a \cdot \sigma_g = O\left(\frac{(pq)^2}{\varepsilon^3} \cdot \log \frac{1}{\varepsilon}\right)$$

slots. Further, the number I of white (or gray) subinstances is

$$I = O\left(\frac{\log \frac{N}{\varepsilon}}{a \log \frac{1}{\varepsilon}}\right) = O\left(\frac{\varepsilon \log \frac{N}{\varepsilon}}{p \log \frac{1}{\varepsilon}}\right),$$

where N is defined by Eq. (2). Thus, we proved our second result.

Theorem 2. *Time complexity of the proposed scheme is*

$$O\left(I \cdot \mathcal{K}(p, q, \varepsilon) + n \log n\right), \tag{6}$$

where

$$\mathcal{K}(p, q, \varepsilon) = (\sigma_w^2 q)^{(\log(\sigma_w^2 q))^{O(1/\varepsilon)}} + (\sigma_g^2 q)^3. \tag{7}$$

Notice that the proposed scheme admits near to linear parallel speedup, since all the subinstances obtained at the stage of instance decomposition can be solved independently.

Corollary 1. *For any fixed $\varepsilon \in (0, 1)$, the running time of the proposed scheme does not exceed $O(n \log N)$, if $p = \Omega(1)$, $q = \Omega(1)$, and*

$$\max\{p, q\} \le 2^{\log^\delta n} \tag{8}$$

for some $\delta = O(\varepsilon)$.

Proof. Fix an arbitrary $\varepsilon \in (0, 1)$ and obtain an upper bound for (6). Indeed, $I = O(\log N)$, $\sigma_w^2 q = C p^4 q^5$ for some constant $C > 0$, and the first term in (7) dominates the second one. Then, Eq. (8) implies that, for $n \gg 1$

$$\log \sigma_w^2 q = \log C + 4 \log p + 5 \log q \le 10 \log^\delta n \le \log^{2\delta} n.$$

Therefore,

$$\begin{aligned}\mathcal{K}(p, q, \varepsilon) = 2^{(\log(Cp^4 q^5))^{O(1/\varepsilon)}} &\le 2^{(\log^{2\delta} n)^{O(1/\varepsilon)}} \\ &= 2^{(\log n)^{C_1 \delta/\varepsilon}} = \left(2^{\log n}\right)^{\frac{\log^{C_1\delta/\varepsilon} n}{\log n}} \le 2^{\log n} = n\end{aligned}$$

any time, when $C_1 \delta/\varepsilon \le 1$, where C_1 is some positive constant, and $n > 1$. Hence, for the fixed ε,

$$I \cdot \mathcal{K}(p, q, \varepsilon) + n \log n = O(n \log N).$$

Corollary is proved.

Corollary 2. *For any fixed p and q the proposed scheme is EPTAS with time complexity $O\left(\left(\frac{1}{\varepsilon^8}\right)^{\left(\log \frac{1}{\varepsilon}\right)^{O(1/\varepsilon)}} \cdot \log N + n \log n\right)$.*

8 Conclusion

In this paper, perhaps for the first time for the Euclidean Capacitated Vehicle Routing Problem with Time Windows and non-uniform splittable demand, a polynomial time approximation scheme is proposed. The scheme is based on the instance decomposition framework developed in [1] and uses the QPTAS from [28] and our modification [16] of the Iterative Tour Partition as a black box. For any fixed $\varepsilon \in (0,1)$ and the total customer demand D, the scheme finds a $(1+\varepsilon)$-approximate solution of the problem in time $O(n \log D)$ any time provided that $\max\{p,q\} \leq 2^{\log^{\delta} n}$ for some $\delta = O(\varepsilon)$. Furthermore, for any fixed capacity q and the number p of time windows, the proposed scheme is an EPTAS, which significantly outperforms by the time complexity bound the previous best result [17] for the CVRPTW on the Euclidean plane.

For future work we left the open questions related to a possible extension of the proposed scheme to an arbitrary finite dimension Euclidean space and to the case of non-splittable demand.

References

1. Adamaszek, A., Czumaj, A., Lingas, A.: PTAS for k-tour cover problem on the plane rof moderately large values of k. Int. J. Found. Comput. Sci. **21**(06), 893–904 (2010). https://doi.org/10.1142/S0129054110007623
2. Arora, S.: Polynomial Time Approximation Schemes for Euclidean Traveling Salesman and other geometric problems. J. ACM **45**, 753–782 (1998)
3. Asano, T., Katoh, N., Tamaki, H., Tokuyama, T.: Covering points in the plane by k-tours: towards a polynomial time approximation scheme for general k. In: Proceedings of the Twenty-Ninth Annual ACM Symposium on Theory of Computing, STOC 1997, pp. 275–283. ACM, New York (1997). https://doi.org/10.1145/258533.258602
4. Becker, A., Klein, P.N., Saulpic, D.: A quasi-polynomial-time approximation scheme for vehicle routing on planar and bounded-genus graphs. In: Pruhs, K., Sohler, C. (eds.) 25th Annual European Symposium on Algorithms, ESA 2017, Vienna, Austria, 4–6 September 2017. LIPIcs, vol. 87, pp. 12:1–12:15. Schloss Dagstuhl - Leibniz-Zentrum fuer Informatik (2017). https://doi.org/10.4230/LIPIcs.ESA.2017.12. http://www.dagstuhl.de/dagpub/978-3-95977-049-1
5. Becker, A., Klein, P.N., Saulpic, D.: Polynomial-time approximation schemes for k-center, k-median, and capacitated vehicle routing in bounded highway dimension. In: Azar, Y., Bast, H., Herman, G. (eds.) 26th Annual European Symposium on Algorithms, ESA 2018, Helsinki, Finland, 20–22 August 2018. LIPIcs, vol. 112, pp. 8:1–8:15. Schloss Dagstuhl - Leibniz-Zentrum fuer Informatik (2018). https://doi.org/10.4230/LIPIcs.ESA.2018.8. http://www.dagstuhl.de/dagpub/978-3-95977-081-1
6. Blocho, M., Czech, Z.: A parallel memetic algorithm for the vehicle routing problem with time windows. In: 2013 Eighth International Conference on P2P, Parallel, Grid, Cloud and Internet Computing, pp. 144–151 (2013). https://doi.org/10.1109/3PGCIC.2013.28

7. Cassettari, L., Demartini, M., Mosca, R., Revetria, R., Tonelli, F.: A multi-stage algorithm for a capacitated vehicle routing problem with time constraints. Algorithms **11**(5) (2018). https://doi.org/10.3390/a11050069. http://www.mdpi.com/1999-4893/11/5/69
8. Chen, X., Kong, Y., Dang, L., Hou, Y., Ye, X.: Exact and metaheuristic approaches for a bi-objective school bus scheduling problem. PLOS ONE **10**(7), 1–20 (2015). https://doi.org/10.1371/journal.pone.0132600
9. Dantzig, G., Ramser, J.: The truck dispatching problem. Manag. Sci. **6**, 80–91 (1959)
10. Das, A., Mathieu, C.: A quasipolynomial time approximation scheme for Euclidean capacitated vehicle routing. Algorithmica **73**, 115–142 (2015). https://doi.org/10.1007/s00453-014-9906-4
11. Gschwind, T., Irnich, S.: Effective handling of dynamic time windows and its application to solving the dial-a-ride problem. Transp. Sci. **49**(2), 335–354 (2015)
12. Haimovich, M., Rinnooy Kan, A.H.G.: Bounds and heuristics for capacitated routing problems. Math. Oper. Res. **10**(4), 527–542 (1985). https://doi.org/10.1287/moor.10.4.527
13. Hashimoto, H., Yagiura, M.: A path relinking approach with an adaptive mechanism to control parameters for the vehicle routing problem with time windows. In: van Hemert, J., Cotta, C. (eds.) EvoCOP 2008. LNCS, vol. 4972, pp. 254–265. Springer, Heidelberg (2008). https://doi.org/10.1007/978-3-540-78604-7_22
14. Khachai, M.Y., Dubinin, R.D.: Approximability of the vehicle routing problem in finite-dimensional Euclidean spaces. Proc. Steklov Inst. Math. **297**(1), 117–128 (2017). https://doi.org/10.1134/S0081543817050133
15. Khachai, M., Ogorodnikov, Y.: Polynomial time approximation scheme for the capacitated vehicle routing problem with time windows. Trudy instituta matematiki i mekhaniki UrO RAN **24**(3), 233–246 (2018). https://doi.org/10.21538/0134-4889-2018-24-3-233-246
16. Khachay, M., Ogorodnikov, Y.: Efficient PTAS for the Euclidean CVRP with time windows. In: van der Aalst, W.M.P., et al. (eds.) AIST 2018. LNCS, vol. 11179, pp. 318–328. Springer, Cham (2018). https://doi.org/10.1007/978-3-030-11027-7_30
17. Khachay, M., Ogorodnikov, Y.: Improved polynomial time approximation scheme for capacitated vehicle routing problem with time windows. In: Evtushenko, Y., Jaćimović, M., Khachay, M., Kochetov, Y., Malkova, V., Posypkin, M. (eds.) OPTIMA 2018. CCIS, vol. 974, pp. 155–169. Springer, Cham (2019). https://doi.org/10.1007/978-3-030-10934-9_12
18. Khachay, M., Dubinin, R.: PTAS for the Euclidean capacitated vehicle routing problem in R^d. In: Kochetov, Y., Khachay, M., Beresnev, V., Nurminski, E., Pardalos, P. (eds.) DOOR 2016. LNCS, vol. 9869, pp. 193–205. Springer, Cham (2016). https://doi.org/10.1007/978-3-319-44914-2_16
19. Khachay, M., Zaytseva, H.: Polynomial time approximation scheme for single-depot Euclidean capacitated vehicle routing problem. In: Lu, Z., Kim, D., Wu, W., Li, W., Du, D.-Z. (eds.) COCOA 2015. LNCS, vol. 9486, pp. 178–190. Springer, Cham (2015). https://doi.org/10.1007/978-3-319-26626-8_14
20. Kumar, S., Panneerselvam, R.: A survey on the vehicle routing problem and its variants. Intell. Inf. Manag. **4**, 66–74 (2012). https://doi.org/10.4236/iim.2012.43010
21. Nalepa, J., Blocho, M.: Adaptive memetic algorithm for minimizing distance in the vehicle routing problem with time windows. Soft Comput. **20**(6), 2309–2327 (2016). https://doi.org/10.1007/s00500-015-1642-4

22. Necula, R., Breaban, M., Raschip, M.: Tackling dynamic vehicle routing problem with time windows by means of ant colony system. In: 2017 IEEE Congress on Evolutionary Computation (CEC), pp. 2480–2487 (2017). https://doi.org/10.1109/CEC.2017.7969606
23. Pace, S., Turky, A., Moser, I., Aleti, A.: Distributing fibre boards: a practical application of the heterogeneous fleet vehicle routing problem with time windows and three-dimensional loading constraints. Procedia Comput. Sci. **51**, 2257–2266 (2015). https://doi.org/10.1016/j.procs.2015.05.382. International Conference on Computational Science, ICCS 2015
24. Papadimitriou, C.: Euclidean TSP is NP-complete. Theor. Comput. Sci. **4**, 237–244 (1977)
25. Savelsbergh, M., van Woensel, T.: 50th anniversary invited article - city logistics: challenges and opportunities. Transp. Sci. **50**(2), 579–590 (2016). https://doi.org/10.1287/trsc.2016.0675
26. Shen, L., Tao, F., Wang, S.: Multi-depot open vehicle routing problem with time windows based on carbon trading. Int. J. Environ. Res. Public Health **15**(9), 2025 (2018). https://doi.org/10.3390/ijerph15092025
27. Song, L., Huang, H.: The Euclidean vehicle routing problem with multiple depots and time windows. In: Gao, X., Du, H., Han, M. (eds.) COCOA 2017. LNCS, vol. 10628, pp. 449–456. Springer, Cham (2017). https://doi.org/10.1007/978-3-319-71147-8_31
28. Song, L., Huang, H., Du, H.: Approximation schemes for Euclidean vehicle routing problems with time windows. J. Comb. Optim. **32**(4), 1217–1231 (2016). https://doi.org/10.1007/s10878-015-9931-5
29. Ting, C.K., Liao, X.L., Huang, Y.H., Liaw, R.T.: Multi-vehicle selective pickup and delivery using metaheuristic algorithms. Inf. Sci. **406–407**, 146–169 (2017). https://doi.org/10.1016/j.ins.2017.04.001. http://www.sciencedirect.com/science/article/pii/S0020025517306436
30. Toth, P., Vigo, D.: Vehicle Routing: Problems, Methods, and Applications. MOS-SIAM Series on Optimization, 2nd edn. SIAM, Philadelphia (2014)
31. Vidal, T., Crainic, T.G., Gendreau, M., Prins, C.: A hybrid genetic algorithm with adaptive diversity management for a large class of vehicle routing problems with time-windows. Comput. Oper. Res. **40**(1), 475–489 (2013). https://doi.org/10.1016/j.cor.2012.07.018

Local Search Approach for the Medianoid Problem with Multi-purpose Shopping Trips

Sergey Khapugin[1] and Andrey Melnikov[1,2(✉)]

[1] Novosibirsk State University, Novosibirsk 630090, Russia
s.khapugin@g.nsu.ru
[2] Sobolev Institute of Mathematics, Novosibirsk 630090, Russia
melnikov@math.nsc.ru

Abstract. We consider a modification to the classic medianoid problem, where facilities of different types are present on the market. A newcomer firm opens facilities providing a specific type of products and competes with existing facilities of that type. Each customer requires multiple products of different types and chooses the shortest route visiting facilities providing the needed types of products. A local search approach to maximize the market share of the newcomer firm is proposed, utilizing upper and lower bounds for the customers' trip lengths to avoid time-consuming computations.

Keywords: Competitive location · Multi-purpose trips · Local search

1 Introduction

Location of commercial facilities is a long-term strategic decision that must take into account as much information regarding the market as it is possible. One of the most critical aspects is competition between suppliers of similar products or services. The models of competitive location were firstly introduced by Hotelling [8], and now form a broad class of optimization problems considering such aspects as design, pricing, capacity management, shipment, and others [1,4,9,11].

In the present paper, we consider a generalization of the classic competitive location model called the medianoid problem. It consists of finding locations of r new facilities providing that the total weight of customers that prefer the new facilities to the existing competitor's ones is maximized. The model was introduced by S. L. Hakimi in the paper [6], where it is shown to be NP–hard. We focus on the discrete case, where both the set of customers and the set of potential locations for new facilities are finite.

Assumptions about the customers' behavior are one of the essential ingredients of competitive models since they determine procedures to estimate the

The research is supported by Russian Science Foundation (project 17-11-01021).

M. Khachay et al. (Eds.): MOTOR 2019, LNCS 11548, pp. 328–341, 2019.
https://doi.org/10.1007/978-3-030-22629-9_23

major objectives such as market share and income. These assumptions are often called *behavior rules.* In the literature, the most widely–used rules are binary and proportional ones assuming that the customer's demand is captured by a single facility or is proportionally distributed among several facilities, respectively. Such measurable factors like travel distance, quality of the facility, and others are used to determine the distribution of the customer's budget.

The present work aims to focus on the influence of facilities that do not compete with the firm but may be in demand for customers. A concentration of these facilities allows the customers to make multi-purpose trips when several different types of products or services can be obtained together during a single tour. The paper of Marianov et al. [14] considers multi-purpose shopping trips in the context of non-essential products. The experiments there show that the total demand depends on the relative location of facilities of different types since spending the trip budget on getting several products provides higher utility to a customer. The authors conclude that taking into account multi-purpose trips is beneficial to the decision maker. In the present paper, we deal with essential products and investigate how the binary customer behavior rule, assuming that each customer patronizes the closest facility, is affected by multi-purpose shopping trips.

To find a quality location of the firm's facilities in a situation when the customers optimize their shopping trips and get service from the facilities visited by the shortest route, we developed the randomized local search procedure. During the computations, the procedure stores estimations of the routes' lengths to avoid solving the customers' problems directly. The information about routes' lengths gathered during the workflow is used to compute an upper bound for the firm's market share and estimate the quality of the solution obtained. The procedure computing the upper bound is to solve a bi-criteria mixed-integer problem with lexicographic order of the objective functions. It finds an optimal location of the firm's facilities when the information about routes' lengths is sufficient.

The paper is organized as follows. In Sect. 2, we introduce all the necessary notations and formulate the medianoid problem with multi-purpose shopping trips in terms of bi-level mathematical programming. All the ingredients of the methods developed are given in Sect. 3. Section 4 presents the results of numerical experiments with artificially generated instances, and Sect. 5 concludes the paper.

2 Mathematical Model

Let us introduce the necessary notations to formalize the problem to find an optimal location of the firm's facilities. We suppose that the firm produces some type of products or just a *product* for short. This type of products is further indexed with t_0. The firm enters the market by opening r facilities in some of a finite set of candidate sites I. Denote the set of locations where the firm decided to open its facilities with I_F. The same product is supplied by existing facilities located in sites I_L and belonging to competitors. Without loss of generality, we assume that $I_L \cap I = \emptyset$.

Denote the set of customers interested in the product t_0 with J. Let $\mathcal{T}$ be the set of all products other than t_0 that are of interest for the customers from J. Further, we refer to these products as *additional*. Suppose that, for each product $t \in \mathcal{T}$, we are given with the set I_t of facilities providing that product.

Suppose that, for each customer $j \in J$, we are given with a subset $T_j \subseteq \mathcal{T}$ of all the additional products that the customer needs. Denote the initial location of the customer $j \in J$ by s_j. The customer then decides how to get all the needed products while minimizing the distance traveled. More specifically, the customer $j \in J$ starts from s_j, visits a single facility in I_t for each $t \in T_j$, one of the facilities in I_L or I_F, and returns to s_j, while minimizing the route's length. To find the shortest route, the customer must solve a generalized traveling salesman problem (GTSP). We say that the customer is captured by the firm if their route visits one of the facilities opened by the firm.

From a certain perspective, an initial location of the customer $j \in J$ can be considered as a facility providing an exclusive product t_j that is in demand for this customer only. For consistency, denote $I_{t_0} = I \cup I_L$, as the set of all the sites that can potentially provide the product t_0, and $I_{t_j} = \{s_j\}$. With this in mind, let $\overline{T}_j = T_j \cup \{t_0\} \cup \{t_j\}$ be the set of all the products the customer needs, including their exclusive product.

Let S be the set of all customer locations, $S = \{s_j | j \in J\}$; let $\mathcal{I}$ be the set of all the facilities that provide additional products, $\mathcal{I} = \bigcup_{t \in \mathcal{T}} I_t$. We assume that we are given a full weighted directed graph G with a set of vertices $V = S \cup \mathcal{I} \cup I \cup I_L$, and we will consider the weight of an edge w_{ij} in this graph to be the distance between two locations. Note that the edge weights have to be non-negative, but do not need to be symmetric or follow the triangle inequality.

For each $T \subseteq \mathcal{T} \cup \{t_0\} \cup S$, define $I_T = \bigcup_{t \in T} I_t$ as the set of all the locations that provide either of the products in T. When the customer j solves their GTSP, they only travel within a certain subset of V, which is denoted by $V_j = I_{\overline{T}_j}$.

To formulate the mathematical model of the problem, we use the following variables. Let $z_i, i \in I$ be the decision variables, with $z_i = 1$ if the firm decides to open a facility in the location $i \in I$, and 0 otherwise. In other words, $z_i = 1 \iff i \in I_F, z_i = 0$ otherwise.

Let z_{ij}, $i \in I$, $j \in J$, be an indicator variable that equals one if and only if the customer j is served by the facility i opened by the firm, and zero otherwise. If that is the case, the firm earns a profit equal to b_j.

Finally, let $\gamma^j_{gh}, j \in J; g, h \in V_j$ be a $(0, 1)$–variable that indicates whether or not the customer j travels along the edge gh during the shortest tour.

With this in mind, the mathematical model can be formulated as the following bi-level program with multiple lower-level problems of the customers.

$$\max_{(z_i),(z_{ij})} \sum_{i \in I} \sum_{j \in J} b_j z_{ij}, \tag{1}$$

s.t.:

$$\sum_{i \in I} z_i = r, \tag{2}$$

$$z_{ij} = \sum_{g \in V_j} \gamma^j_{gi}, i \in I, j \in J, \tag{3}$$

$$z_i, z_{ij} \in \{0, 1\}, i \in I, j \in J, \tag{4}$$

where, for each $j \in J$, (γ^j_{gh}) solve the routing problem of the customer j:

$$\min_{(\gamma^j_{gh})} \sum_{g \in V_j} \sum_{h \in V_j} w_{gh} \gamma^j_{gh} \tag{5}$$

s.t.:

$$\sum_{g \in V_j} \gamma^j_{gh} = \sum_{g \in V_j} \gamma^j_{hg}, h \in V \tag{6}$$

$$\sum_{g \in I_t} \sum_{h \in I_t} \gamma^j_{gh} = 0, t \in \overline{T}_j, \tag{7}$$

$$\sum_{g \in I_T} \sum_{h \in V_j \setminus I_T} \gamma^j_{gh} \geq 1, T \subset \overline{T}_j, \tag{8}$$

$$\sum_{g \in V_j} \gamma^j_{gi} \leq z_i, i \in I, \tag{9}$$

$$\gamma^j_{gh} \in \{0, 1\}, g, h \in V_j \tag{10}$$

The objective function (1) represents the total firm's profit obtained from the customers captured. Due to assumptions of the model, the firm opens r facilities, which is guaranteed by the constraint (2). Next, the constraint (3) defines the connection between z_{ij} and γ^j_{gh}, ensuring that the customer j is considered captured by the facility i if the customer's shortest tour visits i.

Note that each of the lower-level problems (5)–(10) is a GTSP problem in the sub-graph of G with the subsets to visit being $I_t, t \in T_j$. The constraints (6) ensure flow conservation in and out of a given vertex, (7) forbids travel within the same subset, while (8) forbids the tour to be split into unconnected cycles by ensuring that for each possible combination of subsets there exists a single edge that leads to the outside of these subsets. Finally, the constraint (9) forbids the tour to pass through facilities in I that have not been opened by the firm.

Third, note that, given values for z_i, γ^j_{gh} are calculated by solving the problem (5)–(9). We are to highlight that the customer may have several different optimal tours that can lead to different values of the objective function (1). In our experiments, we deal with a pessimistic formulation of the problem (1)–(10) assuming the a customer chooses a tour avoiding to visit firm's facilities, if they could. It is equivalent to the assumption that a customer is captured by the firm's facility only when the route through this facility is shorter than the ones visiting the competitors' facilities by some $\varepsilon > 0$. Once the values of (z_i) are fixed, (3) is enough to compute the best values for z_{ij}, and the corresponding value of the objective function. Thus, the problem under study can be considered as a problem to maximize a pseudo–Boolean function $f(z)$ depending on the vector $z = (z_i)$ of location variables' values. This function maps the vector z to the value of the objective function (1) on the corresponding solution of the problem (1)–(10).

3 Algorithm

Being an NP–hard problem even in a case of single product type, the medianoid problem with multi–purpose trips is a reasonable target for metaheuristic algorithms that explore the search space using local information about the landscape of the objective function. The representation of the problem in the form of maximizing the pseudo-Boolean function $f(\cdot)$ allows working in a convenient space of $(0, 1)$-vectors with r non-zero components. At the same time, given z, computing the objective function $f(z)$ requires to solve multiple NP–hard customers' problems. Thus, an intensive evaluation of the objective function significantly decreases the capabilities of the algorithm.

The GTSP is a classic combinatorial optimization problem arising in many practical applications [12], and the literature on algorithmic approaches to this problem is vast [7,17]. In our computations, we do not apply a specialized method to solve arising GTSPs but delegate them to the Gurobi solver for mixed-integer programming problems (MIP) [5]. On the other side, the main efforts are focused on developing tools that allow calling the solver less frequently and decrease the computational cost of exploring the solution space. The quality of the solution obtained by the algorithm can be estimated when compared with the upper bound provided by the procedure introduced in Subsect. 3.3.

3.1 Local Search

To find a quality solution of the problem to maximize the function $f(\cdot)$, we apply the randomized local search scheme introduced in [15] to solve large instances of bi-level competitive location model. For each $z \in \{0, 1\}^{|I|}$ we define its swap neighborhood

$$Swap(z) = \{y \in \{0, 1\}^{|I|} | H(z, y) = 2, H(x, 0) = H(y, 0)\},$$

where $H(\cdot, \cdot)$ is the Hamming distance between two vectors. The neighborhood $Swap(z)$ contains all the $(0, 1)$–vectors that are obtained from z by moving a single non-zero component to another position.

The swap neighborhood contains $r(|I| - r)$ elements and evaluating all of them is a costly procedure. The algorithm utilizes a randomization technique to increase the exploring rate and prevent sticking in a local optimum.

A randomized neighborhood of the vector z is a randomly chosen subset $RSwap(z) \subseteq Swap(z)$. In our implementation, the probability of the element to be taken in $RSwap(z)$ is the same for all the elements. The number of solutions in $RSwap(z)$ is derived from two parameters: q_1 and q_2. The parameters define the relative quality of the best solution in $RSwap(z)$ and the probability that this quality is achieved. For instance, $q_1 = q_2 = 0.5$ means that the best element of $RSwap(z)$ produces a not worse value of the objective function than 50% of the set $Swap(z)$ with a probability not less than 0.5.

The local search starts at a random feasible solution z^1 and performs similar iterations. The iteration s of the search starts from the solution z^s and consists of making N steps. Each step moves the search to the best neighbor of the current solution with respect to the neighborhood $RSwap(\cdot)$. The iteration $s+1$ starts from the solution z^{s+1} being the best solution found on iteration s. Notice that the solutions z^{s+1} and z^s can be equal. The process is repeated, until the time limit T is reached. The best found solution is returned as the local search result.

To tune the parameters N, q_1 and q_2, the irace package was used [13]. It implements a modified F-race algorithm finding parameters' values that outperform other settings on a given set of instances. After tuning the parameters, the setting $q_1 = 0.8, q_2 = 0.85, N = |I|$ was found as the most efficient.

3.2 Objective Function Evaluation and Tour Length Estimates

Since explicit calculating the optimal routes is a time-consuming procedure, it is desirable to avoid it as much as possible. The approach used relies on storing bounds for tour lengths. Let us discuss it in detail.

For each pair $i \in I_{t_0}, j \in J$, define the ij-tour, which is the shortest tour that passes through s_j, i, and through one facility from I_t for each $t \in T_j$. For the ij-tour, let us introduce three values: l_{ij}, L_{ij} and u_{ij}, which are, respectively: the lower bound for the tour's length, the exact value of the tour's length and the upper bound for it. Over the course of the local search, values l_{ij} and u_{ij} are stored and modified in such a way that the inequality $l_{ij} \leq L_{ij} \leq u_{ij}$ always holds.

Consider some feasible solution $z \in \{0,1\}^{|I|}$ such that $\sum_{i \in I} z_i = r$. Then, having $I_F = \{i \in I | z_i = 1\}$, let us discuss the procedure to evaluate $f(z)$. Consider the problem (5)–(10), where the customer's options for acquiring the product t_0 are limited to a non-empty subset of facilities $C \subseteq I_F \cup I_L$. On the base of the Gurobi MIP solver, we implement a procedure $\mathbf{GTSP}(j, C, B, b)$, where B and b, such that $B > b \geq 0$, are numerical input parameters. The procedure may finish with the following three results:

1. An optimal tour passes through $i^* \in C$ and has the length $L_{i^*j} < B$.
2. It is proven that no ij-tour with $i \in C$ exists with $L_{ij} < B$.
3. A tour passing through some $i_1 \in C$ is shorter than b, but no proof of optimality.

Both the second and the third exit conditions stop the branch-and-bound process prematurely, potentially saving time. The second exit condition is reached when the solver makes sure that an optimal tour has a length greater than B. The third exit condition is reached when a feasible solution better than a quality threshold b is found. Since proving optimality takes more time than yielding good solutions, this option leads to time savings when finding an optimal solution is not necessary.

From here, the procedure to check, if a customer is captured by the firm or not, becomes one of building C and choosing B and b. In the first case, the procedure returns 1 and 0 otherwise.

Algorithm 1, answering the question if the customer is captured by the firm or not, starts from finding the current best candidate $i_0 \in I_L \cup I_F$ and then

Algorithm 1. Detect if the customer is captured

$i_0 \leftarrow \arg\min_{i \in I_F \cup I_L} u_{ij}$
$C \leftarrow \{i \in I_F \cup I_L | l_{ij} \leq u_{i_0 j}\}$
if $C \subseteq I_F$ **then**
  return 1
else if $C \subseteq I_L$ **then**
  return 0
end if
if $u_{i_0 j} = l_{i_0 j}$ **then**
  $C \leftarrow C \setminus \{i_0\}$ // *the exact length of the* $i_0 j$*-tour is known*
end if
$B := u_{i_0 j}$
if $i_0 \in I_L$ **and** $C \subseteq I_F$ **then**
  $b := l_{i_0 j}$
else
  $b := 0$
end if
Call **GTSP**(j, C, B, b)
if (An optimal ij-tour found through $i^* \in C$, with length $L < B$) **then**
  $l_{i^* j} \leftarrow L$
  $u_{i^* j} \leftarrow L$
  for all $i \in C \setminus \{i^*\}$ **do**
    $l_{ij} \leftarrow \max\{l_{ij}, L\}$ // *since* i^* *is optimal, other* ij*-tours are not shorter*
  end for
  return $|\{i^*\} \cap I_F|$
else if (It is proven that no ij-tour with $i \in C$ exists with $L < B$) **then**
  for all $i \in C$ **do**
    $l_{ij} \leftarrow \max\{l_{ij}, B\}$
  end for
  return $|\{i_0\} \cap I_F|$
else if (An ij-tour found through $i_1 \in C$ with length $L < b$, but no proof of optimality) **then**
  $u_{i_1 j} \leftarrow L$
  return 1
end if

ensures that those facilities that are proven to have a worse tour than the one through i_0 are not considered. After that, it is checked if solving GTSP can be avoided entirely, that is, if all the candidates belong to one owner. If the exact length of the $i_0 j$-tour is known, then i_0 is removed from the candidate list. Then, the quality threshold B for the GTSP solver is set up to make sure that only solutions with ij-tours shorter than the $i_0 j$-tour are considered.

The lower quality threshold is set up afterward. It is $b = l_{i_0 j}$ only if $i_0 \in I_L$ and all the other candidates are from I_F. After that, the optimizer is launched and the results are interpreted. If an optimal $i^* j$-tour is found, then it is shorter than all other tours, and by our assumption we conclude that $l_{ij} = \min\{l_{ij}, L\}$. If it is proven that no ij-tours shorter than B exist, then the values l_{ij} are updated accordingly. Finally, if some good enough i_1j-tour is found, but no proof is given, then the value u_{i_1j} is updated.

The only thing that remains uncovered yet for a complete description of the local search algorithm is how the initial values for l_{ij} and u_{ij} are set up. There are numerous heuristics for the GTSP which the local search can benefit from [10], but solving the GTPS is out of scope of this study, thus, we applied a very straightforward approach to estimate the lengths of the tours for the beginning. Namely, the upper bounds are set up as lengths of randomly generated feasible GTSP solutions. The lower bounds are set up as the distance to travel from s_j to i and back: $l_{ij} = \omega_{is_j} + \omega_{s_ji}$, which is a feasible lower bound when the triangle inequality holds. When it is not the case, other trivial estimates M and 0, respectively, are used, where M is a sufficiently large value. Even the second variant yields a significant improvement over computing the lengths explicitly.

3.3 Finalizing the Algorithm and Upper Bound Procedure

During the computational process, the local search algorithm continuously updates the values l_{ij}, u_{ij}. These values are used to construct a lexicographical bi-objective MIP [3] called *estimating problem*. Since the local search avoids computing optimal customers' tours, the information about lengths of these tours may not be full after the termination of the algorithm. Further, we say that the customer $j \in J$ is *potentially captured* by the firm's facility $i \in I$ if $l_{ij} < u_j^L$, i. e. the information about tours of this customer, gathered during the local search process, is insufficient to claim that the j prefers some facility from I_L to i. It is clear, that, given the location of the firm's facilities, the total weight of all potentially captured customers is not less than the actual value of the firm's market share. Thus, the primary objective function of the estimating problem being the total weight of potentially captured customers provides an upper bound for the function $f(\cdot)$. The secondary objective function shows the total weight of the potentially captured customers that can not be proven to be captured by the firm using the information about the lengths of customers' tours. While minimizing this function, we aim to reduce the gap between the upper bound value and the actual market share.

Consider a set of values l_{ij}, u_{ij} computed during the local search process, and a solution z^{LS}, which is the best solution found. For all $j \in J$, let us denote: $l_j^L = \min_{i \in I_L} l_{ij}$, $u_j^L = \min_{i \in I_L} u_{ij}$, the most optimistic and the most pessimistic estimates of the shortest ij-tour for all $i \in I_L$, respectively. Introduce two matrices: c_{ij} and $\widetilde{c}_{ij}$, $i \in I, j \in J$ defined as follows:

$$c_{ij} = \begin{cases} 1 \text{ if } l_{ij} < u_j^L \\ 0 \text{ , otherwise} \end{cases} \qquad \widetilde{c}_{ij} = \begin{cases} 1 \text{ if } u_{ij} < l_j^L \\ 0 \text{ otherwise} \end{cases}$$

As such, $c_{ij} = 1$ if j is potentially captured by i, whereas $\widetilde{c}_{ij} = 1$ if i captures j for sure. Note that $c_{ij} \geq \widetilde{c}_{ij}$

To formulate the MIP, we use additional $(0, 1)$-variables (v_j) and $(\widetilde{v}_j)$, $j \in J$, where v_j indicates if some firm's facility potentially captures j. At the same time,

$\widetilde{v}_j$ equals one if some firm's facility surely captures j, and zero otherwise. With the introduced notations, the bi-objective estimating problem can be written as follows:

$$UB = \max_{(z_i),(v_j),(\widetilde{v}_j)} \sum_{j \in J} b_j v_j, \tag{11}$$

$$Gap = \min_{(z_i),(v_j),(\widetilde{v}_j)} \sum_{j \in J} b_j (v_j - \widetilde{v}_j), \tag{12}$$

s.t.

$$v_j \le \sum_{i \in I} c_{ij} z_i, j \in J \tag{13}$$

$$\widetilde{v}_j \le \sum_{i \in I} \widetilde{c}_{ij} z_i, j \in J \tag{14}$$

$$\sum_{i \in I} z_i = r \tag{15}$$

$$v_j, \widetilde{v}_j, z_i \in \{0,1\}, \quad i \in I, j \in J. \tag{16}$$

The first objective function (11) represents the total demand that can be potentially captured by the firm. The second objective function shows the magnitude of deviation of the actual value of market share from the value UB. The constraints (13) and (14) have a form of covering constraints with matrices (c_{ij}) and $(\widetilde{c}_{ij})$ indicating if the element is covered or not. Note that integrality constraints for variables (v_j) and $(\widetilde{v}_j)$ can be relaxed with the optimal solution guaranteed to be integer.

Consider the lexicographically optimal solution $((z_i), (v_j), (\widetilde{v}_j))$ of the problem (11)–(16) and the corresponding values UB and Gap of its objective functions. First, note that the value UB is an upper bound for the function $f(\cdot)$. Moreover, if $Gap = 0$, then $z^* = (z_i)$ maximizies the function $f(\cdot)$. The overall algorithm consists of running the local search procedure with time limit stopping criteria and then solving the problem (11)–(16) providing values UB and Gap.

4 Numerical Experiments

To analyze the impact of the algorithm's ingredients and the model's behavior, we performed numerical experiments with artificial data. From the application side, our interest is to compare the model's predictions with the classic $(r|X_p)$-medianoid problem. The numerical data was taken from the benchmark "Discrete location problems" [2], compiling instances of the $(r|p)$-centroid problem and results of experiments with these instances. The experiments are performed by a MacBook Pro with an Intel Core i5 2.9 GHz dual-core processor and 8 GB of RAM under control of the Windows 8.1 operating system. The algorithm was written in C# and utilized Gurobi 7.5.2 running with default settings.

The $(r|p)$-centroid problem is a classic bi-level location problem considering two players deciding where to open their facilities. The first player, called the

Leader, opens their facilities first. The second player, called the Follower, opens their facilities after the Leader but has the benefit of knowing the Leader's solution. Customers choose facilities based on how close they are to the facility. A more detailed description of this problem can be found, for instance, in the work [16].

In the instances considered, the customers are points on the Euclidean plane. The sites to open facilities coincide with the locations of the customers. This means that $S = V(G)$, $I \cup I_L = S$. The instances from [2] are provided with optimal Leader's solutions if the authors found them, or the best Leader's solutions obtained so far. For each instance, we fix the best Leader's solution and consider the set of facilities opened by the Leader, I_L, as the existing competitors' ones in our model. The rest of the possible locations are given to the firm to open its facilities in: $I = S \setminus I_L$. It is assumed that the firm opens the same number of facilities as the competitors: $r = |I_L|$.

Next, the additional facilities are generated. For each $t \in \mathcal{T}$ the set I_t is built as an optimal solution of the p-median problem with $p = k$ and randomly generated weights (b_j^t), $j \in J$ of the customers from the uniform distribution on the integer range $\{1, 2, \dots 25\}$. In other words, the set I_t is chosen to contain k elements of the set S such that the sum over the set of customers J of distances from a customer j to the nearest element of S weighed by b_j^t is minimized. It is set $|I_{t_1}| = |I_{t_2}|, t_1, t_2 \in \mathcal{T}$ meaning that the number of facilities providing additional products is the same for all product types. In the test instances generated, the customers share the same shopping lists having $T_j = \mathcal{T}$ for all $j \in J$, whereas the number of additional product types equals two, $|\mathcal{T}| = 2$. This value was chosen because solving the GTSP explicitly turned out to be time-consuming on our hardware for $|\mathcal{T}| \geq 3$.

The base $(r|p)$-centroid instances are unweighted ones ($b_j \equiv 1$) with codes 111, 211, 311, and weighted instances coded by 111(W), 211(W), 311(W), 411(W) and 511(W), where (b_j) are uniformly distributed on the integer range $\{1, \dots, 200\}$. We consider values $r = 5, 10$ for unweighted instances and $r = 5, 10, 15, 20$ for the weighted ones.

A comparison was made between a benchmark solution, z^0, which is the optimal Follower solution in the original $(r|p)$-centroid problem and the best solution found by the algorithm. This solution is chosen from the best solution found during the local search, z^{LS}, and the solution found by solving the bi-objective problem described in Subsect. 3.3, z^{BO}.

The local search was given a time limit of two minutes. The bi-objective problem was solved by the optimizer in under a second in all of the tested instances. Let $\overline{z}$ denote the best solution from z^{LS} and z^{BO}.

Table 1 provides the following values:

r is the number of facilities open by the firm;

$|I_t|$ is the size of the set of additional facilities for each type;

Imp is the improvement for the best solution found by the algorithm over the benchmark solution:

$$Imp = (F(\overline{z}) - F(z^0))/F(z^0) \cdot 100\%;$$

Table 1. Summarized results of the numerical experiments

r	$\|I_t\|$	Improvement, %	SP of best solution	GTSPs solved	Solutions explored
5	5	76.1%	3	19015	1060
5	10	54.9%	2.5	6584	458
5	20	21%	4.5	1385	174
10	5	82.3%	5.2	15674	660
10	10	51.5%	6.83	7344	382
10	20	20.5%	7.5	1622	86
15	5	32.7%	5.4	12985	422
15	10	52.1%	7.6	6713	225
15	20	27%	10.4	1607	62
20	5	35%	4.4	10299	265
20	10	34.8%	8.6	6130	165
20	20	22.9%	13.2	1585	42

It was noticed that high-quality solutions tend to share locations with the facilities providing additional products. SP is a numerical expression of this tendency computed as the number of times the facilities from I_F share location with some additional facility, i.e.,

$$SP = \sum_{i \in I_F} \sum_{t \in T} |I_t \cap \{i\}|;$$

"GTSPs solved" is the number calls to the MIP solver to solve a GTSP;
"Solutions explored" is the number of evaluations of the objective function.

The values for Imp, SP, "GTSPs solved" and "Solutions explored" are averaged out across all the instances tested.

To summarize the results, first notice that the higher r and $|I_t|$, the fewer solutions are explored and the fewer GTSPs are solved. It is expected, considering that GTSPs become more difficult to solve as the number of nodes in a subset to visit grows. Consequently, one would expect that the quality of solutions would decrease as $|I_t|$ increases. This trend is most apparent for $r = 5, 10$, the Improvement measure decreases by 20—30% each $|I_t|$ step.

Also worthy to note that the SP of the best solution is on average significantly higher than the expected value for random solutions, which can be roughly estimated as $(2|I_t|/100) \cdot r$.

Table 2 provides a more detailed comparison between the medianoid models with single-purpose and multi-purpose shopping trips. For the numerical data based on the first three weighted and unweighted $(r|p)$-centroid instances and different values of r and $|I_t|$, we list the following data.

f_s is the market share computed for the solution z^0 in the single-purpose model with no additional products;

Table 2. Detailed results on some instances

Instance #	f_s	f_m	f_{LS}	f_{BO}	GTSPs solved	Solutions explored
$r = 5, \|I_t\| = 5$						
111	53	22	22	22	20955	1276
211	52	38	53	53	19850	484
311	55	28	40	47	20671	1072
$r = 5, \|I_t\| = 10$						
111	53	54	52	64	6495	544
211	52	35	47	62	6744	616
311	55	53	75	79	6331	562
$r = 5, \|I_t\| = 15$						
111	53	53	54	57	1305	148
211	52	35	47	62	1387	148
311	55	46	41	59	1483	132
$r = 10, \|I_t\| = 5$						
111	50	34	74	74	21694	1336
211	51	65	70	70	15417	712
311	52	43	78	78	15026	580
$r = 10, \|I_t\| = 10$						
111	50	25	45	45	8810	460
211	51	38	53	53	6970	208
311	52	25	45	45	7529	436
$r = 10, \|I_t\| = 20$						
111	50	49	67	65	1700	124
211	51	35	33	47	1569	76
311	52	55	55	66	1504	112
$r = 15, \|I_t\| = 5$						
111(W)	47.1	41.7	43.7	43.7	10607	232
211(W)	48.9	57.1	64.8	64.8	14278	424
311(W)	48.7	35.9	57.8	57.8	19600	664
$r = 15, \|I_t\| = 10$						
111(W)	47.1	51.3	67.9	67.9	5141	208
211(W)	48.9	45.7	54.7	54.7	8060	268
311(W)	48.7	23.4	46.0	46.0	6701	184
$r = 15, \|I_t\| = 20$						
111(W)	47.1	31.0	33.7	29.0	1432	40
211(W)	48.9	41.7	54.6	51.9	1869	100
311(W)	48.7	23.2	36.3	30.5	1655	52
$r = 20, \|I_t\| = 5$						
111(W)	48.1	47.2	64.4	64.4	10673	436
211(W)	48.4	32.1	38.7	38.7	12558	220
311(W)	47.7	26.3	47.3	47.3	10635	220
$r = 20, \|I_t\| = 10$						
111(W)	48.1	24.6	39.0	39.0	5579	136
211(W)	48.4	33.8	52.8	52.8	6795	256
311(W)	47.7	29.5	34.4	34.0	5911	112
$r = 20, \|I_t\| = 20$						
111(W)	48.1	37.2	42.9	41.0	1410	52
211(W)	48.4	16.1	24.3	17.7	1444	28
311(W)	47.7	31.4	41.5	31.0	1774	40

f_m is the market share computed for z^0 when multi-purpose trips are taken into account;
f_{LS} is the market share computed with multi-purpose trips for z^{LS};
f_{BO} is the market share computed with multi-purpose trips for z^{BO};

This table shows that optimal solutions of the model with single-purpose shopping could perform poorly in a situation when the customers actually make multi-purpose trips. At the same time, taking into account the locations of facilities providing additional products allows finding locations of a new-comer firm's facilities, which are more attractive for the customers and capture a greater market share. When speaking about the accuracy of the upper bound computed by the bi-objective estimating problem, it is insufficient to measure the quality of the solutions obtained for the instances considered. The reason is that the local search algorithm does not have enough time to gather enough information about the trip lengths, which ends up causing some facilities to be assumed to capture all or most of the customers optimistically. It prevents the upper bound for the objective function from being a non-trivial estimation.

Overall, the algorithm performs well on the instances considered, reliably and quickly producing solutions of significantly higher quality than the benchmark.

5 Conclusion and Future Work

In this paper, we considered the medianoid problem with multi-purpose shopping trips. It aims to maximize a market share of a new-comer firm opening its facilities in a competitive environment. The model takes into account the presence of facilities that do not compete with the firm but that are of interest for the customers. The customers are assumed to solve a generalized traveling salesman problem (GTSP) with the aim to minimize the length of their shopping tours visiting facilities of the types needed. Thus, the model generalizes the classic medianoid problem, where customers get service in the nearest facility.

To find a quality solution of the model, we developed a local search procedure storing estimations for lengths of the shortest customers' tours to avoid solving GTSPs explicitly. The gathered information about lengths of the tours is used to construct an estimating problem in a form of bi-objective MIP providing upper bound for the firm's market share.

Numerical experiments show that taking into account the information about facilities, which are of interest for the customers, may bring significant benefits for the entering firm. The necessity to solve customers' routing problems makes exploring the search space more difficult for metaheuristics due to costly procedure to estimate the objective function. At the same, the suggested technique to store the additional information about tours' lengths has demonstrated its efficiency in speeding-up the computations and increasing exploring capabilities of the method in a situation of time limitation.

Our future work is focused on improving the quality of tour lengths' estimations used by the method. Plenty of algorithmic approaches on the GTSP presented in the literature can be applied for these purposes. Efficient strategies to balance between exploring the search space of the firm's location and collecting the information about optimal customers' routes are subject for future research as well. Finally, studying the model to find a Stackelberg equilibrium in a situation where the competitors make sequential decisions about location of their facilities is planned.

References

1. Ashtiani, M.: Competitive location: a state-of-art review. Int. J. Ind. Eng. Comput. **7**(1), 1–18 (2016). https://doi.org/10.5267/j.ijiec.2015.8.002
2. Benchmarks library "Discrete location problems". http://math.nsc.ru/AP/benchmarks/Competitive/p_med_comp_tests_eucl_eng.html. Accessed 01 Apr 2019
3. Ehrgott, M.: Multicriteria Optimization. Springer, Heidelberg (2006)
4. Eiselt, H., Laporte, G.: Sequential location problems. Eur. J. Oper. Res. **96**(2), 217–231 (1997). https://doi.org/10.1016/S0377-2217(96)00216-0
5. Gurobi Optimization, I.: Gurobi optimizer reference manual (2019). http://www.gurobi.com. Accessed 01 Apr 2019
6. Hakimi, S.: On locating new facilities in a competitive environment. Eur. J. Oper. Res. **12**, 29–35 (1983)
7. Helsgaun, K.: Solving the equality generalized traveling salesman problem using the Lin-Kernighan-Helsgaun algorithm. Math. Program. Comput. **7**(3), 269–287 (2015). https://doi.org/10.1007/s12532-015-0080-8
8. Hotelling, H.: Stability in competition. Econ. J. **39**, 41–57 (1929)
9. Karakitsiou, A.: Modeling Discrete Competitive Facility Location. SO. Springer, Cham (2015). https://doi.org/10.1007/978-3-319-21341-5
10. Khachai, M., Neznakhina, E.: Approximation schemes for the generalized traveling salesman problem. Proc. Steklov Inst. Math. **299**, 97–105 (2017). https://doi.org/10.1134/S0081543817090127
11. Kress, D., Pesch, E.: Sequential competitive location on networks. Eur. J. Oper. Res. **217**, 483–499 (2012). https://doi.org/10.1016/j.ejor.2011.06.036
12. Laporte, G., Asef-Vaziri, A., Sriskandarajah, C.: Some applications of the generalized travelling salesman problem. J. Oper. Res. Soc. **47**(12), 1461–1467 (1996). https://doi.org/10.1057/jors.1996.190
13. López-Ibáñez, M., Dubois-Lacoste, J., Pérez Cáceres, L., Stützle, T., Birattari, M.: The irace package: iterated racing for automatic algorithm configuration. Oper. Res. Perspect. **3**, 43–58 (2016). https://doi.org/10.1016/j.orp.2016.09.002
14. Marianov, V., Eiselt, H., Lüer-Villagra, A.: Effects of multipurpose shopping trips on retail store location in a duopoly. Eur. J. Oper. Res. **269**(2), 782–792 (2018). https://doi.org/10.1016/j.ejor.2018.02.024
15. Mel'nikov, A.: Randomized local search for the discrete competitive facility location problem. Autom. Remote Control **75**, 700–714 (2014). https://doi.org/10.1134/S0005117914040109
16. Roboredo, M.C., Pessoa, A.A.: A branch-and-cut algorithm for the discrete $(r|p)$-centroid problem. Eur. J. Oper. Res. **224**(1), 101–109 (2013). https://doi.org/10.1016/j.ejor.2012.07.042
17. Smith, S.L., Imeson, F.: GLNS: an effective large neighborhood search heuristic for the generalized traveling salesman problem. Comput. Oper. Res. **87**, 1–19 (2017). https://doi.org/10.1016/j.cor.2017.05.010

Flow Shop with Job–Dependent Buffer Requirements—a Polynomial–Time Algorithm and Efficient Heuristics

Alexander Kononov[1], Julia Memar[2(✉)], and Yakov Zinder[2]

[1] Sobolev Institute of Mathematics, Siberian Branch of the Russian Academy of Sciences, Novosibirsk, Russia
alvenko@math.nsc.ru

[2] University of Technology Sydney, PO Box 123, Broadway, NSW 2007, Australia
{julia.memar,yakov.zinder}@uts.edu.au

Abstract. The paper is concerned with the two-machine flow shop, where each job needs storage space (a buffer requirement) during the entire time of its processing. The buffer requirement is determined by the duration of job's first operation. The goal is to minimise the time needed for the completion of all jobs. This scheduling problem is NP-hard in the strong sense even for very restricted cases such as the case with a given order of jobs processing on one of the machines. The paper contributes to the efforts of establishing the borderline between the NP-hard and polynomial-time solvable cases by proving that there exists a polynomial-time algorithm which constructs an optimal schedule if the duration of each operation does not exceed one-fifth of the buffer capacity. The presented polynomial-time algorithm is used as a basis for a heuristic for the general case. This heuristic is complemented by a Lagrangian relaxation based heuristic and a bin-packing based constructive heuristic. The heuristics are tested by computational experiments.

Keywords: Flow shop · Buffer · Makespan · Polynomial-time algorithm · Lagrangian relaxation · Heuristic

1 Introduction

In this paper, we consider a two-machine flow shop problem with a limited buffer. Each job seizes the portion of the buffer from the start of its processing on the first-stage machine and releases this portion only after its completion on the second-stage machine. It is assumed that the buffer requirement of each job is equal to the job's processing time on the first stage of the flow shop. The capacity of the buffer cannot be exceeded at any time. The objective is to minimize the maximum completion time of all jobs.

A. Kononov—Research of the first author is partially supported by RFBR grant 17-07-00513.

M. Khachay et al. (Eds.): MOTOR 2019, LNCS 11548, pp. 342–357, 2019.
https://doi.org/10.1007/978-3-030-22629-9_24

The flow shop problem with a buffer has been extensively studied in the literature. Most of the papers, however, consider flow shops with an intermediate buffer between stages, where the buffer capacity is limited by a number of jobs [1,2,4,17]. The models with the buffer requirement, which varies from job to job, and the buffer is occupied by a job for its entire processing, has been studied only recently, though such models better reflect the real-world applications [20]. The scheduling problem, considered in this paper, arises in supply chains when the change of vehicles involves unloading and loading, using certain storage space [6]; in multimedia systems where files for presentations are downloaded from remote storage and stored in a limited memory [15,16].

The considered scheduling problem is NP-hard in the strong sense [14]. The problem remains NP-hard in the strong sense even under the restriction that, on one of the machines, the jobs are to be processed in a given sequence [8]. Furthermore, as has been proven in [7], there are instances for which the set of all optimal schedules does not contain a permutation schedule, that is, a schedule in which both machines process the jobs in the same order. Even the decision problem, requiring an answer to the question of whether or not the given instance is one of the instances that do not have an optimal schedule that is a permutation one, is an NP-complete. This paper contributes to the efforts of establishing the borderline between the NP-hard and polynomial-time solvable cases by proving that there exists a polynomial-time algorithm which constructs an optimal schedule if the duration of each operation does not exceed one-fifth of the buffer capacity.

In what follows, the presented polynomial-time algorithm is also used as a basis for a heuristic for the general case. This heuristic is complemented by a Lagrangian relaxation based heuristic and a bin-packing based constructive heuristic. The heuristics are tested by computational experiments. These algorithms contribute to the existing publications aimed at the development of optimisation procedures for the general case [12,13,15].

The paper is organised as follows. Section 2 provides the problem's description. Section 3 presents a polynomial-time algorithm and a proof that this algorithm constructs an optimal schedule. Section 4 describes three heuristics, a heuristic which is based on the polynomial-time algorithm in Sect. 3 (this heuristic will be referred to as barrier heuristic), a Lagrangian relaxation based heuristic, and a bin-packing based heuristic. A lower bound on the optimal value of makespan is introduced in Sect. 5. The results of computational experiments are provided in Sect. 6. Section 7 concludes the paper.

2 Description of the Problem

We are given a set of jobs $N = \{1, ..., n\}$ and two machines M_1 and M_2. Each job has two operations. The first operation of job i must be processed on the machine M_1 for a given amount of time a_i and the second operation of job i must be processed on the machine M_2 for b_i time units, the job's second operation can commence only after the first operation has been completed.

Once job processing has been started, it cannot be interrupted. Each machine can process at most one job at a time, and each job can be processed by at most one machine at a time; the processing of jobs commences at time $t = 0$. Each job i seizes $\omega(i)$ units of buffer space when its processing has started on the first machine; this portion of the buffer is released only when the job has been completed on the second machine. Similar to [12,13,15,16] it is assumed that $\omega(i) = a_i$ for each job i. At any point in time t, the total buffer requirement of all jobs that started their processing before or at t and have a completion time of their second operation greater than t cannot exceed Ω - the buffer capacity. A schedule σ specifies for each $j \in N$ the points in time $S_j^1(\sigma)$ and $S_j^2(\sigma)$, when job j starts processing, and $C_j^1(\sigma)$ and $C_j^2(\sigma)$, when job j completes processing on machine M_1 and M_2, correspondingly. Thus we have $S_j^1(\sigma) + a_j = C_j^1(\sigma)$ and $S_j^2(\sigma) + b_j = C_j^2(\sigma)$. The goal is to minimise maximum completion time $C_{max}(\sigma) = \max_{j \in N} C_j^2(\sigma)$. Following [12] we call this problem the PP-problem.

3 Polynomial–Time Algorithm

The PP-problem is strongly NP-hard [16]. However, the problem is easily solvable if the buffer size is large enough, for example, when all jobs can be simultaneously placed in the buffer. In this case, the problem is equivalent to the two-machine flow shop problem, and it can be solved in $O(n \log n)$ time by Johnson's algorithm [11]. Thus, the computational complexity of the problem depends on the relationship between the size of jobs and the size of the buffer.

Let I be an instance of the PP-problem such that

$$\max_{i \in N}\{a_i, b_i\} \leq \frac{\Omega}{5}. \tag{1}$$

Assume for a moment that there is no buffer restriction. Then a permutation schedule σ^J constructed by Johnson's rule [11] is optimal. Johnson's rule can be formulated as follows:

- partition N into two sets: $L_1 = \{i \in N : a_i < b_i\}$ and $L_2 = \{i \in N : a_i \geq b_i\}$;
- first schedule the jobs from L_1 in non-decreasing order of a_i, and then schedule the jobs from L_2 in non-increasing order of b_i.

Further assume that the jobs are numbered according to the sequence constructed by Johnson's rule, then $C_{max}(\sigma^J)$ can be expressed as following:

$$C_{max}(\sigma^J) = \max_k \left(\sum_{i=1}^{k} a_i + \sum_{i=k}^{n} b_i \right). \tag{2}$$

Let k' be the number at which the maximum is reached in (2). Denote by $Idle_1$ and $Idle_2$ the total idle time in the interval $[0, C_{max}(\sigma^J)]$ on machines M_1 and M_2, respectively. Then we have

$$Idle_1 = C_{max}(\sigma^J) - \sum_{i=1}^{n} a_i = \sum_{i=k'}^{n} b_i - \sum_{i=k'+1}^{n} a_i \leq \sum_{i=k'}^{n} b_i \quad (3)$$

and

$$Idle_2 = C_{max}(\sigma^J) - \sum_{i=1}^{n} b_i = \sum_{i=1}^{k'} a_i - \sum_{i=1}^{k'-1} b_i \leq \sum_{i=1}^{k'} a_i. \quad (4)$$

Let $n_A = \lceil \frac{5Idle_2}{\Omega} \rceil$ and $n_B = \lceil \frac{5Idle_1}{\Omega} \rceil$. We introduce set X of n_A jobs with $a_i = 0$, $b_i = \frac{Idle_2}{n_A}$ for $i = 1, \dots n_A$ and set Y of n_B jobs with $a_i = \frac{Idle_1}{n_B}$, $b_i = 0$ for $i = 1, \dots n_B$. Observe that $a_i \leq \frac{\Omega}{5}$ for $i \in Y$ and $b_i \leq \frac{\Omega}{5}$ for $i \in X$. Moreover, (3) and (4) imply that $n_A + n_B \leq n + 1$.

Consider the modified instance I' of the problem with the set of jobs $N' = N \cup X \cup Y$. We note that

$$\sum_{i \in N'} a_i = \sum_{i \in N'} b_i. \quad (5)$$

If the buffer capacity is unlimited we will show that the optimal makespan $OPT(I')$ of I' is equal to $C_{max}(\sigma^J)$. Since the schedule σ^J is feasible, an operation of each job on M_1 precedes an operation of the same job on M_2. Keeping the order of operations, we shift all operations on M_1 to the left and all operations on M_2 to the right without changing the makespan. Thus, in the new schedule the machine M_1 is idle during the period of time from $C_{max}(\sigma^J) - Idle_1$ to $C_{max}(\sigma^J)$ and the machine M_2 is idle during the time period from 0 to $Idle_2$. We schedule first all jobs in X in an arbitrary order, then all jobs in N in the same order as in the schedule σ^J and finally all jobs in Y in an arbitrary order. In this case, the jobs of the set X are completed at time 0 on M_1 and occupy the interval $[0, Idle_2]$ on M_2 and the jobs of the set Y are processed in the interval $[C_{max}(\sigma^J) - Idle_1, C_{max}(\sigma^J)]$ on M_1 and after that are executed at time $C_{max}(\sigma^J)$ on M_2. Denote the obtained schedule by σ'. Then $C_{max}(\sigma') = C_{max}(\sigma^J)$ and both machines work without idle times in the interval $[0, C_{max}(\sigma^J)]$. Moreover, the permutation of jobs, induced by σ', does not contradict Johnson's rule. Denote this permutation by π^J.

Now we consider the instance I' of the PP-problem with the set of jobs N' and the buffer with capacity Ω. Let $L_1' = X \cup L_1$ and $L_2' = L_2 \cup Y$ and set $H = \frac{2\Omega}{5}$. Let π_1 be the permutation of jobs from L_1' in a non-decreasing order of a_i, and π_2 be the permutation of jobs from L_2' in a non-increasing order of b_i. Denote by $l_{k,i}(\pi)$ the total processing time of first k jobs in a permutation π on machine M_i, i.e. $l_{k,1}(\pi) = \sum_{j=1}^{k} a_{\pi(j)}$ and $l_{k,2}(\pi) = \sum_{j=1}^{k} b_{\pi(j)}$. Let $R_k(\pi) = l_{k,2}(\pi) - l_{k,1}(\pi)$ and $n' = n + n_A + n_B$. In what follows, $\pi = \emptyset$ indicates that sequence π is not specified. Starting with $\pi = \emptyset$, Algorithm 1 below constructs a schedule by sequentially determining the order in which the jobs are to be processed.

Algorithm 1.

1: Set $i = 1, i_1 = 1, i_2 = 1, \pi = \emptyset, R_0(\emptyset) = 0$.
2: **while** $i \leq n'$ **do**
3:    **if** $R_{i-1}(\pi) < H$ **and** $i_1 \leq |L'_1|$ **then**
4:       set $\pi(i) = \pi_1(i_1)$, $i = i + 1$, $i_1 = i_1 + 1$;
5:    **else**
6:       set $\pi(i) = \pi_2(i_2)$, $i = i + 1$, $i_2 = i_2 + 1$;
7:    **end if**
8:    set $S^1_{\pi(i)}(\sigma) = l_{i-1,1}(\pi)$ and $S^2_{\pi(i)}(\sigma) = l_{i-1,2}(\pi)$;
9: **end while**
10: **return** schedule σ.

Lemma 1. *Algorithm 1 finds an optimal schedule for instance I'.*

Proof. First we will show that Algorithm 1 works correctly and builds a feasible schedule. Observe that the following inequalities hold:

$$R_i(\pi) > 0, \text{ for } 1 \leq i \leq n' - 1; \tag{6}$$

$$R_{i-1}(\pi) < R_i(\pi), \text{ if } \pi(i) \text{ is in } L'_1; \tag{7}$$

$$R_{i-1}(\pi) \geq R_i(\pi), \text{ if } \pi(i) \text{ is in } L'_2. \tag{8}$$

It is easy to check that the operator **if** works correctly. Indeed, if $R_{i-1}(\pi) \geq H$ or $i_1 = |L'_1|$, then (5) and (6) imply that there are still unassigned jobs in L'_2 and $i_2 \leq |L'_2|$.

We will prove that the following conditions are satisfied:

(a) the schedule σ has no overlapping jobs on the same machine;
(b) the schedule σ has no overlapping operations of the same job;
(c) the schedule σ does not violate the buffer constraint.

(a): Since $l_{0,i}(\pi) = 0$ for $i = 1, 2$, the job $\pi(1)$ is $\pi_1(1)$ and it starts at time 0 on both machines. According to the step 8 of the algorithm, each next job $\pi(i)$ starts its operation on either stage at the time of completion of the previous job $\pi(i-1)$. Hence, the schedule σ has no overlapping jobs on the same machine.

(b): We note that if $R_i(\pi) < H$ for all i, then the permutation π is Johnson's permutation and the schedule σ coincides with the schedule σ'. The feasibility σ' implies that the schedule σ has no overlapping operations of the same job.

Let $\pi(k)$ be the first job in π such that $R_k(\pi) \geq H$. Let $\pi(h)$ be the last job from L'_1 in the permutation π, i.e., $\pi(h) = \pi_1(|L'_1|)$. Then $R_j \geq R_{j+1}$ for all $j \geq h$, and, hence, $k \leq h$. If $k = h$, then $\pi(1), \ldots, \pi(h) \in L'_1$ and $\pi(h+1), \ldots, \pi(n') \in L'_2$. So, the schedule σ coincides with the schedule σ'. If $k < h$ we partition the permutation π into three subsequences: $(\pi(1), \ldots, \pi(k))$, $(\pi(k+1), \ldots, \pi(h))$, and $(\pi(h+1), \ldots, \pi(n'))$. By virtue of (7), the first k jobs in the π are from the set L'_1. Hence, σ and σ' are the same for the first k jobs. Consequently, for $j = 1, \ldots, k$, the operations of job $\pi(j)$ do not overlap.

Since $R_k(\pi) \geq H$ and $k < h$, for each $k+1 \leq j \leq h$, the algorithm assigns a job from L'_2 when $R_{j-1}(\pi) \geq H$, and from L'_1, otherwise. Since $H = 2\Omega/5$ and the processing time of each operation does not exceed $\Omega/5$, we have

$$\frac{\Omega}{5} \leq R_j(\pi) \leq \frac{3\Omega}{5}, \tag{9}$$

for any $k \leq j \leq h$. Hence, for a job j, $k+1 \leq j \leq h$, we obtain

$$C^1_{\pi(j)}(\sigma) = S^1_{\pi(j)}(\sigma) + a_{\pi(j)} = l_{j-1,1}(\pi) + a_{\pi(j)}$$
$$= l_{j-1,2}(\pi) + a_{\pi(j)} - R_{j-1}(\pi) \leq S^2_{\pi(j)}(\sigma).$$

Thus, the operations of job $\pi(j)$ do not overlap for $j = k+1, \ldots, h$.

Finally, we observe, that set of h first jobs are the same for π and π^J. Moreover, the machines M_1 and M_2 work without idle time in both schedules σ and σ'. Thus, we have $C^1_{\pi(h)}(\sigma) = C^1_{\pi^J(h)}(\sigma')$ and $C^2_{\pi(h)}(\sigma) = C^2_{\pi^J(h)}(\sigma')$. Moreover, we have $\pi(j) = \pi^J(j)$ for all $j > h$. Hence, σ and σ' are the same for the last $n' - h$ jobs and feasibility of σ' implies that operations of job $\pi(j)$ do not overlap for $j = h+1, \ldots, n'$.

(c): For any $j \in N'$ consider the buffer consumption at its starting time S^1_j. Let $k(j)$ be the job with the smallest completion time such that $C^2_{k(j)} \geq S^1_j$. The buffer consumption at S^1_j does not exceed the buffer requirements of job $k(j)$ and all jobs within interval $[C^1_{k(j)}, C^2_{k(j)}]$ and the job j. By virtue of (1) and (9) this buffer load does not exceed

$$\frac{\Omega}{5} + R_{k(j)}(\pi) + \frac{\Omega}{5} \leq \frac{\Omega}{5} + \frac{3\Omega}{5} + \frac{\Omega}{5} = \Omega,$$

hence the buffer capacity is observed every time a job starts its processing.

Since both machines proceed jobs without idle time, the makespan of σ coincides with the load of the machine and σ is an optimal schedule. □

It remains to note that after removing all jobs from set $X \cup Y$ from schedule σ we obtain a feasible schedule for the original instance I of the problem. Thus, we get the following result.

Theorem 1. *There exists an $O(n \log n)$ algorithm that constructs an optimal schedule for any instance of PP-problem which satisfies (1).*

Observe, that the Theorem 1 shows that the condition (1) implies that

- The problem (under this condition) is polynomially solvable;
- The optimal makespan coincides with $LB^{Johnson}$.

4 Heuristics

In this section we describe three different approaches to construct a schedule:

- barrier heuristic;
- Lagrangian relaxation decomposition based heuristic;
- bin-packing based heuristic.

In all three heuristics we will use the "Wait" Algorithm which is described in [8] and can be summarised as follows:

- the "Wait" Algorithm follows the given permutation on each stage by placing jobs one by one, assigning a current job for each stage first, before proceeding to the next job; the second operation of a job is placed only after the first operation has completed;
- if there is no sufficient space in the buffer for the current job, the "Wait" Algorithm waits till one or more jobs have completed on the second stage to allow space for the current job.

4.1 Barrier Heuristic

The polynomial-time algorithm described in Sect. 3 constructs an optimal schedule for a particular case of the considered scheduling problem. Hence there is a reasonable expectation that the permutation obtained with the help of this algorithm would allow to construct good quality schedules for arbitrary instances of the problem. The barrier heuristic can be summarised as follows:

Algorithm 2. Barrier heuristic

1: Ignoring the buffer constraint and applying Johnson's rule construct a permutation schedule σ^J;
2: Set $a_{\max} = \max_{i \in N} a_i$;
3: Set $Idle_1 = C_{\max}(\sigma^J) - \Sigma_{i \in N} a_i$, and $Idle_2 = C_{\max}(\sigma^J) - \Sigma_{i \in N} b_i$;
4: Set $n_x = \left\lceil \frac{Idle_2}{a_{\max}} \right\rceil$, $x = \frac{Idle_2}{n_x}$, $n_y = \left\lceil \frac{Idle_1}{a_{\max}} \right\rceil$, and $y = \frac{Idle_1}{n_y}$;
5: Create a new instance I' adding to the set N n_x "dummy" jobs i with $a_i = 0$, $b_i = x$ and n_y "dummy" jobs j with $a_j = y$, $b_j = 0$;
6: Set a value for the barrier H and construct a schedule for the instance I' by Algorithm 1 for this value of H;
7: Let π be a permutation of jobs obtained by Algorithm 1, apply the "Wait" Algorithm for the instance I' and the permutation π ignoring the "dummy" jobs.

4.2 Lagrangian Relaxation Based Heuristic

Lagrangian relaxation is an efficient method for solving problems of combinatorial optimization [5,8–10,18]. Lagrangian relaxation is obtained by relaxing some of the constraints of the integer formulation of the problem. This relaxation allows to decompose the dual problem into subproblems. The Lagrangian heuristic is an iterative procedure, which at each iteration solves the subproblems by a recursive algorithm for the current set of Lagrangian multipliers. At each iteration of the Lagrangian heuristic, job's starting times, obtained during the decomposition stage, provide the order of jobs on M_1 and M_2. To construct a feasible schedule and update the best upper bound we will be using the permutation formed by the starting times on M_1 and the "Wait" Algorithm. The Lagrangian multipliers are updated with standard gradient method [5]. After all iterations, the schedule with the smallest value of the objective function is chosen as the result of the Lagrangian heuristic. The integer formulation, Lagrangian relaxation and decomposition, and the recursive procedure below is the adaptation of the model, discussed in [8], for the objective function of the maximum completion time.

Integer Formulation

It is easy to see that for any $i \in N$ its completion time $C_i \leq \sum_{i \in N}(a_i + b_i)$. However, smaller planning horizon T improves convergence of an algorithm. To obtain the tighter T, we run "Wait" Algorithm with the permutation defined by non-increasing order of $a_i + b_i$, and set T to the resulting value of the makespan. Define x_{it}^m, $i \in N$, $0 \leq t < T$, $m \in \{1, 2\}$, as

$$x_{it}^m = \begin{cases} 1, & \text{if } S_i^m = t; \\ 0, & \text{otherwise.} \end{cases} \tag{10}$$

Denote by $C_{max} = \max_{i \in N} \sum_{t=1}^{T-1} t x_{it}^2 + b_i$. The considered scheduling problem can be formulated as:

$$\min\ C_{max} \tag{11}$$

subject to

$$\sum_{t=0}^{T-1} x_{it}^m = 1, \quad \text{for } 1 \leq i \leq n \text{ and } m \in \{1, 2\} \tag{12}$$

$$\sum_{i=1}^{n} \sum_{\tau=\max\{0, t-p_i^m+1\}}^{t} x_{i\tau}^m \leq 1, \quad \text{for } 0 \leq t < T \text{ and } m \in \{1, 2\} \tag{13}$$

$$\sum_{t=1}^{T-1} t x_{it}^2 - \sum_{t=1}^{T-1} t x_{it}^1 \geq a_i, \quad \text{for } 1 \leq i \leq n \tag{14}$$

$$\sum_{i=1}^{n} \omega(i) \left(\sum_{\tau=0}^{t} x_{i\tau}^1 - \sum_{\tau=0}^{t-b_i} x_{i\tau}^2 \right) \leq \Omega, \quad \text{for } 0 \leq t < T \tag{15}$$

$$\sum_{t=1}^{T-1} t x_{it}^2 + b_i \leq C_{max}, \quad \text{for } 1 \leq i \leq n \tag{16}$$

$$x_{it}^m \in \{0, 1\}, \quad \text{for } 1 \leq i \leq n,\ 0 \leq t < T, \text{ and } m \in \{1, 2\};\ C_{max} \geq 0 \tag{17}$$

Lagrangian Relaxation and Decomposition

To obtain Lagrangian relaxation, for Lagrangian multipliers $v_{tm} \geq 0$ and $u_t \geq 0$ we dualize the constraints (13), (15). To dualize (16) we use technique described in [19]. For multipliers $\lambda_i \geq 0$ with at least one $\lambda_j > 0$ we aggregate (16):

$$\sum_{i=1}^{n} \lambda_i \left(\sum_{t=1}^{T-1} t x_{it}^2 + b_i \right) \leq \sum_{i=1}^{n} \lambda_i C_{max}$$

or

$$\sum_{i=1}^{n} \frac{\lambda_i}{\sum_{j=1}^{n} \lambda_j} \left(\sum_{t=1}^{T-1} t x_{it}^2 + b_i \right) \leq C_{max}. \tag{18}$$

Denote by $q_i = \frac{\lambda_i}{\sum_{j=1}^{n} \lambda_j}$, $i \in N$; hence we obtain the Lagrangian relaxation:

$$\min C_{max} + \sum_{i=1}^{n} q_i \left(\sum_{t=1}^{T-1} t x_{it}^2 + b_i \right) - C_{max} + \sum_{t=0}^{T-1} \sum_{m=1}^{2} v_{tm} \left(\sum_{i=1}^{n} \sum_{\tau=\max\{0,t-p_i^m+1\}}^{t} x_{i\tau}^m - 1 \right)$$
$$+ \sum_{t=0}^{T-1} u_t \left[\sum_{i=1}^{n} \omega(i) \left(\sum_{\tau=0}^{t} x_{i\tau}^1 - \sum_{\tau=0}^{t-b_i} x_{i\tau}^2 \right) - \Omega \right]$$

subject to (12), (14) and (17).

Let (v, u, q) be the sets of the all Lagrangian multipliers, and $L(v, u, q)$ be the optimal value of the Lagrangian relaxation above. For each $i \in N$ denote by $L_i(v, u, q)$ the optimal value of the following integer linear program:

$$\min q_i \sum_{t=1}^{T-1} t x_{it}^2 + \sum_{t=0}^{T-1} \sum_{m=1}^{2} v_{tm} \sum_{\tau=\max\{0,t-p_i^m+1\}}^{t} x_{i\tau}^m + \omega(i) \sum_{t=0}^{T-1} u_t \left(\sum_{\tau=0}^{t} x_{i\tau}^1 - \sum_{\tau=0}^{t-b_i} x_{i\tau}^2 \right) \tag{19}$$

subject to

$$\sum_{t=0}^{T-1} x_{it}^m = 1, \quad \text{for } m \in \{1, 2\} \tag{20}$$

$$\sum_{t=1}^{T-1} t x_{it}^2 - \sum_{t=1}^{T-1} t x_{it}^1 \geq a_i \tag{21}$$

$$x_{it}^m \in \{0, 1\}, \quad \text{for } 0 \leq t < T, \text{ and } m \in \{1, 2\} \tag{22}$$

Therefore, for the chosen set of Lagrangian multipliers (v, u, q), $L(v, u, q)$ could be computed as the sum of all $L_i(v, u, q)$ and a linear combination of parameters:

$$L(v, u, q) = \sum_{i=1}^{n} L_i(v, u, q) + \sum_{i=1}^{n} q_i b_i - \sum_{t=0}^{T-1} \sum_{m=1}^{2} v_{tm} - \Omega \sum_{t=0}^{T-1} u_t \tag{23}$$

Consequently, for the given set (v, u, q), the $L(v, u, q)$ can be found by solving n separate integer problems (19)–(22). For each problem we will use the reclusive procedure described in [8], with some minor current changes. For convenience of the reader we summarise the procedure below.

Recursive Procedure
If job i starts at time s on M_1 and at time r on M_2, then by virtue of (20), only $x^1_{is} = 1$ and $x^2_{ir} = 1$, and the value of objective function (19) is

$$q_i r + \sum_{t=s}^{s+a_i-1} v_{t1} + \sum_{t=r}^{r+b_i-1} v_{t2} + \omega(i) \sum_{t=s}^{r+b_i-1} u_t.$$

Define the function $f(r)$ for $a_i \le r \le T - b_i$ as

$$f(r) = \min_{0 \le s \le r-a_i} \left(\sum_{t=s}^{s+a_i-1} v_{t1} + \omega(i) \sum_{t=s}^{r+b_i-1} u_t \right). \tag{24}$$

Observe that the initial value of $f(r) = f(a_i) = \sum_{t=0}^{a_i-1} v_{t1} + \omega(i) \sum_{t=0}^{b_i-1} u_t$, and hence for $r > a_i$ the following recursive relation holds:

$$f(r) = \min \left[f(r-1) + \omega(i) u_{r+b_i-1}, \sum_{t=r-a_i}^{r-1} v_{t1} + \omega(i) \sum_{t=r-a_i}^{r+b_i-1} u_t \right] \tag{25}$$

Hence the value of $L_i(v, u, q)$ can be found as

$$L_i(v, u, q) = \min_{a_i \le r \le T-b_i} \left(f(r) + q_i r + \sum_{t=r}^{r+b_i-1} v_{t2} \right)$$

4.3 Bin-Packing Heuristic

As the name suggests, the bin-packing heuristic utilises the idea of bin-packing [3], and the heuristic can be summarised as follows:

Algorithm 3. Bin-packing heuristic

1: Sort jobs from N in non-increasing order of $\omega(i)$;
2: Partition all jobs into "bins" of size Ω by going through the list of jobs, and assigning the current job on the list to the first bin the job "fits in": that is if the total first operation requirement of all jobs in the bin does not exceed Ω as well as the total second operation requirement of all jobs in the bin does not exceed Ω. If the job does not fit to any of the existing bins, create a new bin;
3: Create a permutation of jobs by sorting the bins in non-decreasing order of total buffer requirement of jobs in a bin and sorting the jobs in each bin according to Johnson's rule;
4: Use the permutation and "Wait" Algorithm to construct a schedule.

---

5 Lower Bound

The proposed lower bound method is taking into account the buffer capacity Ω and can be computed as follows. Assume that all jobs $i \in N$ are numbered in the non-increasing order of a_i. Let $LargeJobs = \{1, 2, ..., k\}$ be the set of the jobs i with buffer requirement $a_i > \frac{\Omega}{2}$. Obviously, that no two jobs from this set can be in the buffer together. Hence in any schedule the time, required to process a subset $B_l = \{1, 2, ..., l\} \subseteq LargeJobs$, $1 \leq l \leq k$, is at least $\Sigma_{i=1}^{l}(a_i + b_i)$. In addition, for the subset B_l there might be a subset of smaller jobs $S_l = \{i > l : a_i + a_l > \Omega\}$ such that none of the smaller jobs from S_l can occupy the buffer together with any job from B_l. Note that $N = S_0$ when $LargeJobs = \emptyset$. Let $C(S_l)^J$ be the maximum completion time in a permutation schedule where the jobs from S_l are scheduled according to Johnson's rule and there is no buffer restriction. The minimum time required to process sets B_l and S_l is at least $\Sigma_{i=1}^{l}(a_i + b_i) + C(S_l)^J$. Hence the following lower bound denoted as LB^{buffer}:

$$LB^{buffer} = \max_{1 \leq l \leq k} \{\Sigma_{i=1}^{l}(a_i + b_i) + C(S_l)^J\} \tag{26}$$

Denote by $LB^{Johnson} = C(N)^J$. Hence the lower bound LB can be found as

$$\max\{LB^{buffer}, LB^{Johnson}\} \tag{27}$$

It is easy to see, that if there are no large jobs, then $LB^{buffer} = LB^{Johnson}$.

6 Computational Experiments

The computational experiments were run for the barrier, Lagrangian and bin-packing heuristics and aimed to compare their performance against the lower bound (27). The computational experiments were conducted by the second author on a personal computer with Intel Core $i5$ processor CPU@1.70 Ghz, with Ubuntu 14.04 LTS, and base memory 4096 MB. The algorithms were implemented in the C programming language.

The test instances were generated randomly with a_i for a job i chosen from the interval $[1, 20]$, and b_i chosen from the interval $[1, 50]$. There were 50 instances in each tested set. In what follows, we will describe an instance as $n - \Omega k$, where n is the number of jobs, and Ωk is the size of the buffer. The experiments were run for sets of instances with 25, 50 and 100 jobs and for buffer sizes $\Omega 1 = a_{max}$, $\Omega 3 = 3a_{max}$ and $\Omega 5 = 5a_{max}$, where a_{max} is the maximum buffer requirement among all jobs of an instance. Subgradient algorithm in the Lagrangian heuristic was run for 300 iterations; the value of the barrier in the barrier heuristic was set to $H = 2a_{max}$; there was a 30 min time limit per instance for all heuristics.

The results of the computational experiments are represented by the box-plot charts on Figs. 1, 2, 3, which were obtained as follows. For each instance, the duality gap DG was calculated as $DG = \frac{UB - LB}{LB}$, where UB is the makespan value provided by either barrier, Lagrangian or bin-packing heuristic, and LB

is calculated according to (27). For the instances with the small buffer $\Omega 1$ the Lagrangian heuristic provided solutions with smaller duality gap than both the barrier and bin-packing heuristic did. Remarkably, once the buffer size is larger (sizes $\Omega 3$ and $\Omega 5$), all three heuristics provide solutions with the duality gap within $0\% - 3\%$ for the absolute majority of the instances. Moreover, the barrier heuristic solved to optimality all instances with the buffer $\Omega 5$ as it provided the same value of the objective function as the corresponding value of the lower bound. Here we note that the condition (1) does not hold for the $\Omega 5$ instances, as for all tested instances $a_{max} < b_{max}$, where a_{max} and b_{max} are the maximum processing time on the first and second stage, correspondingly, among all jobs of an instance.

Both barrier and bin-packing heuristics have spent within 0.00004–0.0008 s per 25–100 jobs instances, while the Lagrangian heuristic's CPU time per instance varied between 6–28 s per 25 jobs instance, between 1–3.5 min per 50 jobs instance and between 6–20 min per 100 jobs instance. In comparison, when fifteen $25 - \Omega 5$ instances were tested by running a straightforward integer program (CPLEX), it took 30 min per instance for to deliver optimal solutions for only a third of the instances, however CPLEX determined that the provided solution is optimal for less than a half of these instances. For another fifteen $25 - \Omega 3$ instances, CPLEX found an optimal solution for only 20% of instances, however it did not recognised any of these values as optimal within the given time.

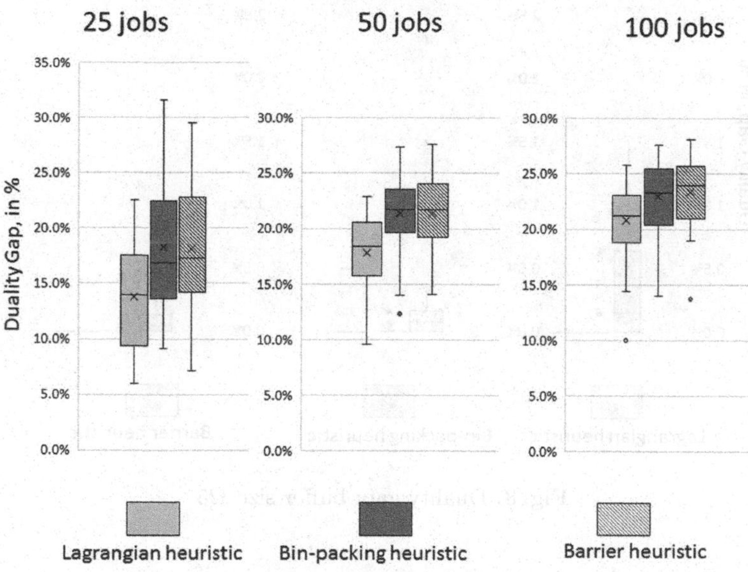

Fig. 1. Duality gap: buffer size $\Omega 1$

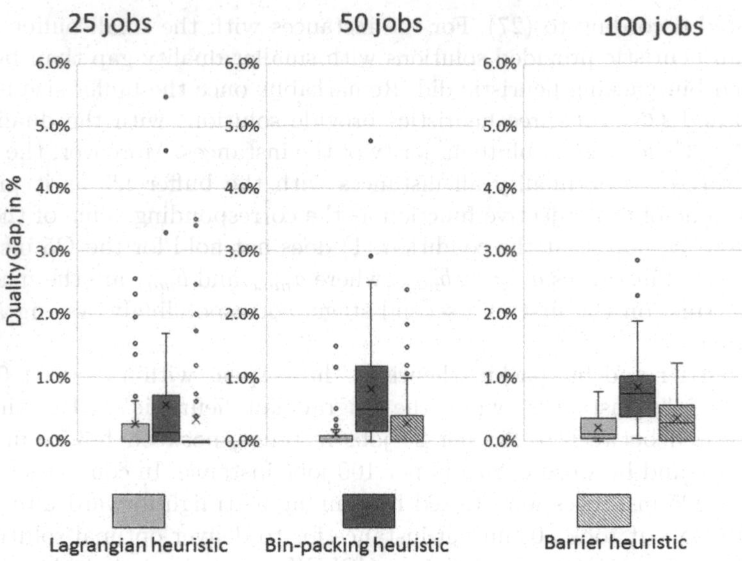

Fig. 2. Duality gap: buffer size $\Omega 3$

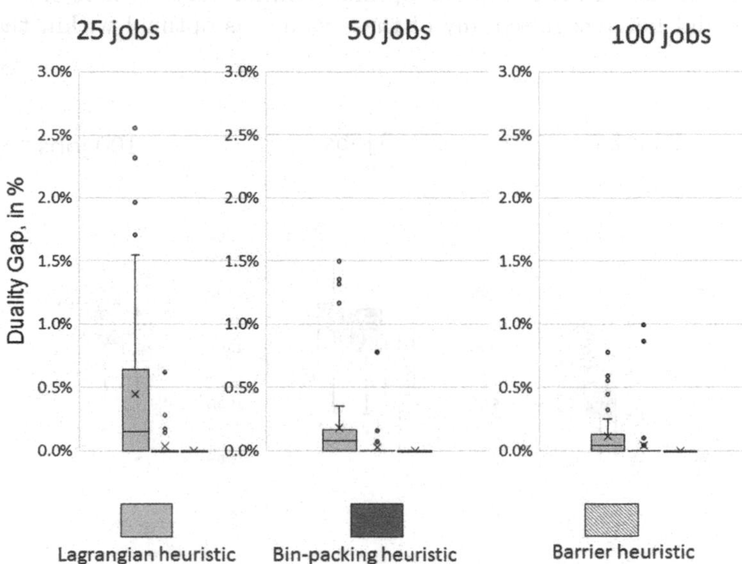

Fig. 3. Duality gap: buffer size $\Omega 5$

The box-plot chart on Fig. 4 compares the values of LB^{buffer} and $LB^{Johnson}$ for instances with the smaller buffer $\Omega 1$ and for each instance shows by how much, in %, LB^{buffer} is tighter (has larger value) than $LB^{Johnson}$; for each

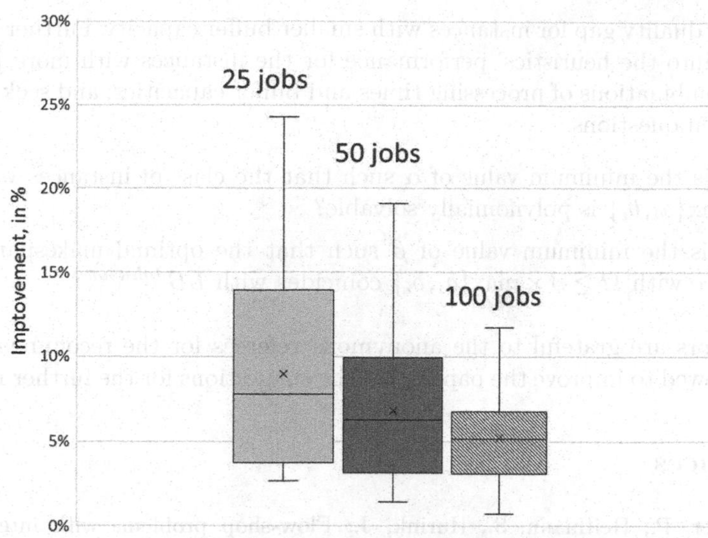

Fig. 4. Lower bound improvement, in %

instance the improvement is calculated as $\frac{LB^{buffer} - LB^{Johnson}}{LB^{Johnson}}$. The LB^{buffer} improves the lower bound by 1%–24% across all instances.

In summary, the computational experiments demonstrated that the barrier and bin-packing heuristics are fast algorithms which generate near optimal/optimal schedules for the instances with larger buffer size. The Lagrangian heuristics provide tighter results for the instances with smaller buffer. The proposed method to calculate lower bound considerably tightens the duality gap for instances with smaller buffer.

7 Conclusion

In this paper we discussed the two-machine flow shop with a limited buffer and the objective function of the maximum completion time. The buffer requirement equals to the processing time of a job on the first stage and it varies from job to job; the job occupies the buffer for its entire processing, and the buffer capacity cannot be violated at any point of time. We established the borderline between the NP-hard and polynomial-time solvable cases of the considered problem by proving that there exists a polynomial-time algorithm which constructs an optimal schedule if the duration of each operation does not exceed one-fifth of the buffer capacity. We also introduced three heuristics: barrier, Lagrangian and bin-packing heuristics and an efficient method to calculate lower bound. The computational experiments demonstrated, that the barrier and bin-packing heuristics construct nearly optimal/optimal schedules for the instances with larger buffer capacity, and the Lagrangian heuristic produced tighter solutions for the instances with smaller buffer capacity. The proposed lower bound allowed to tighten duality gap for instances with smaller buffer capacity. Further research will look into the heuristics' performance for the instances with more jobs and various combinations of processing times and buffer capacities, and seek answers to the open questions:

- What is the minimum value of α such that the class of instances with $\Omega \geq \alpha \times \max_{i \in N}\{a_i, b_i\}$ is polynomially solvable?
- What is the minimum value of β such that the optimal makespan of any instance with $\Omega \geq \beta \times \max_{i \in N}\{a_i, b_i\}$ coincides with $LB^{Johnson}$?

The authors are grateful to the anonymous referees for the recommendations, which allowed to improve the paper, and the suggestions for the further research.

References

1. Brucker, P., Heitmann, S., Hurink, J.: Flow-shop problems with intermediate buffers. OR Spectr. **25**(4), 549–574 (2003)
2. Brucker, P., Knust, S.: Complex Scheduling. Springer, Heidelberg (2012). https://doi.org/10.1007/978-3-642-23929-8
3. Coffman Jr., E.G., Garey, M.R., Johnson, D.S.: An application of bin-packing to multiprocessor scheduling. SIAM J. Comput. **7**(1), 1–17 (1978)
4. Emmons, H., Vairaktarakis, G.: Flow Shop Scheduling. Springer, Boston (2013). https://doi.org/10.1007/978-1-4614-5152-5
5. Fisher, M.L.: The lagrangian relaxation method for solving integer programming problems. Manag. Sci. **50**, 1861–1871 (2004)
6. Fung, J., Singh, G., Zinder, Y.: Capacity planning in supply chains of mineral resources. Inf. Sci. **316**, 397–418 (2015)
7. Fung, J., Zinder, Y.: Permutation schedules for a two-machine flow shop with storage. Oper. Res. Lett. **44**(2), 153–157 (2016)
8. Gu, H., Kononov, A., Memar, J., Zinder, Y.: Efficient lagrangian heuristics for the two-stage flow shop with job dependent buffer requirements. J. Discrete Algorithms **52–53**, 143–155 (2018)
9. Gu, H., Memar, J., Zinder, Y.: Scheduling batch processing in flexible flowshop with job dependent buffer requirements: lagrangian relaxation approach. In: Rahman, M.S., Sung, W.-K., Uehara, R. (eds.) WALCOM 2018. LNCS, vol. 10755, pp. 119–131. Springer, Cham (2018). https://doi.org/10.1007/978-3-319-75172-6_11
10. Irohara, T.: Lagrangian relaxation algorithms for hybrid flow-shop scheduling problems with limited buffers. Int. J. Biomed. Soft Comput. Hum. Sci. **15**(1), 21–28 (2010)
11. Johnson, S.M.: Optimal two-and three-stage production schedules with setup times included. Nav. Res. Logist. Q. **1**(1), 61–68 (1954)
12. Kononov, A., Hong, J.S., Kononova, P., Lin, F.C.: Quantity-based buffer-constrained two-machine flowshop problem: active and passive prefetch models for multimedia applications. J. Sched. **15**(4), 487–497 (2012)
13. Kononova, P., Kochetov, Y.A.: The variable neighborhood search for the two machine flow shop problem with a passive prefetch. J. Appl. Ind. Math. **7**(1), 54–67 (2013)

14. Lin, F.C., Hong, J.S., Lin, B.M.: A two-machine flowshop problem with processing time-dependent buffer constraints-an application in multimedia presentations. Comput. Oper. Res. **36**(4), 1158–1175 (2009)
15. Lin, F.C., Hong, J.S., Lin, B.M.: Sequence optimization for media objects with due date constraints in multimedia presentations from digital libraries. Inf. Syst. **38**(1), 82–96 (2013)
16. Lin, F.C., Lai, C.Y., Hong, J.S.: Minimize presentation lag by sequencing media objects for auto-assembled presentations from digital libraries. Data Knowl. Eng. **66**(3), 382–401 (2008)
17. Pinedo, M.L.: Scheduling: Theory, Algorithms, and Systems. Springer, New York (2012). https://doi.org/10.1007/978-1-4614-2361-4
18. Tang, L.X., Xuan, H.: Lagrangian relaxation algorithms for real-time hybrid flowshop scheduling with finite intermediate buffers. J. Oper. Res. Soc. **57**(3), 316–324 (2006)
19. van de Velde, S.L.: Machine scheduling and lagrangian relaxation (1991)
20. Witt, A., Voß, S.: Simple heuristics for scheduling with limited intermediate storage. Comput. Oper. Res. **34**(8), 2293–2309 (2007)

Pareto-Based Hybrid Algorithms for the Bicriteria Asymmetric Travelling Salesman Problem

Yulia V. Kovalenko[1(✉)] and Aleksey O. Zakharov[2]

[1] Sobolev Institute of Mathematics, Novosibirsk, Russia
julia.kovalenko.ya@yandex.ru
[2] Saint Petersburg State University, St. Petersburg, Russia
a.zakharov@spbu.ru

Abstract. We consider the bicriteria asymmetric travelling salesman problem (bi-ATSP): Given a complete directed graph where each arc is associated to a couple of positive weights, the aim is to find the Pareto set, consisting of all non-dominated Hamiltonian circuits. We propose new hybrid algorithms for the bi-ATSP using the adjacency-based representation of solutions and the operators that use the Pareto relation. Our algorithms are based on local search and evolutionary methods. The local search combines principles of the well-known Pareto Local Search procedures and Variable Neighborhood Search approach, realizing the search in width and depth. A genetic algorithm with NSGA-II scheme is applied to improve and extend a set of Pareto local optima by means of evolutionary processes. The experimental evaluation shows applicability of the algorithms to various structures of the bi-ATSP instances generated randomly and constructed from benchmark asymmetric instances with single objective.

Keywords: The Pareto set · Genetic algorithm · Local search · Computational experiment

1 Introduction

The travelling salesman problem (TSP) is one of the most popular problems in combinatorial optimization [3]. Given a complete graph where each arc (or edge) is associated with a positive weight, we search for a circuit (or cycle) visiting every vertex of the graph exactly once and minimizing the total weight. In this paper, we consider the bicriteria asymmetric TSP (bi-ATSP) which is a special case of the multicriteria asymmetric TSP [8], where an arc is associated to a couple of weights. Note that in the asymmetric problem (ATSP) weights of an arc depend on direction in contrast to the symmetric case (STSP), and we need to consider this feature when algorithms are developed.

Optimal solution to a multicriteria optimization problem (MOP) is usually supposed to be the Pareto set [8,22]. The bi-ATSP is NP-hard and intractable

M. Khachay et al. (Eds.): MOTOR 2019, LNCS 11548, pp. 358–373, 2019.
https://doi.org/10.1007/978-3-030-22629-9_25

(see, e.g. [4]). Moreover, in [2], the non-approximability bounds were obtained for the multicriteria ATSP with weights 1 and 2. The results are based on the non-existence of a small size approximating set. Therefore, metaheuristics, in particular local search heuristics (LSs) and hybrid methods, are appropriate to approximate the Pareto set of the bi-ATSP.

Local search methods are successfully applied for solving a wide variety of NP-hard single-objective and multi-objective combinatorial optimization problems. There are two main approaches to construct a LS for multi-objective instances. The first one explores different aggregations of the objective functions, and uses a good local search algorithm to find a local optimum for single-objective version of the problem [20]. The second approach is the Pareto local search (PLS), which explores neighborhoods of a set of non-dominated solutions and determines acceptance criterion for new solutions based on the Pareto relation [19]. Hybrid algorithms have been also developed (see e.g. [16]). For example, non-dominated solutions are generated by solving a number of aggregated subproblems in the first phase and PLS is then adopted for obtaining better approximation in the second phase. A variant of LS heuristics is Variable Neighborhood Search (VNS) [11], which systematically changes the neighborhood within a randomized local search algorithm. VNS demonstrated competitive results on a wide variety of intractable problems [11].

Moreover, LSs are often used as subroutines in metaheuristics (evolutionary algorithms, ant and bee colony optimization methods and others) in order to improve solutions obtained in searching operators. One of the effective algorithms of this class is Genetic Local Search (GLS) (see, e.g., [12,15,23]).

NSGA [28] is a generational multi-objective evolutionary algorithm (MOEA) based on Pareto-dominance. It sorts a population into different non-domination levels and uses the well-known sharing function approach, which has been found to maintain sustainable diversity in a population with appropriate setting of its associated parameters. In NSGA-II [7], the sharing function approach is replaced with a crowded-comparison approach, which does not require any user-defined parameter for maintaining diversity among population members. NSGA-II has one of the best results in the literature on MOEAs for the MOPs with two or three objectives. In [6], a fast implementation of a steady-state version of the NSGA-II is proposed for two dimensions.

Various meta-heuristics and heuristics have been developed for the multicriteria STSP, such as LSs, MOEAs, multi-objective ant colony optimization methods, memetic algorithms and others (see, e.g., [1,10,13,15–17,19,20,23]). Symmetry of weights plays an important role in the operators and processes of the abovementioned algorithms. At the same time, we have not found in the literature any multi-objective metaheuristic proposed specifically to the multicriteria ATSP and experimentally tested on instances with non-symmetric weights of arcs. So, in this paper we propose new VNS with PLS scheme and GLS with NSGA-II scheme for the bi-ATSP using adjacency-based representation of solutions. Computational experiment is carried out on instances, generated randomly or constructed from the well-known benchmark ATSP instances with single objective. The results of the experiment indicate the viability and effectiveness of our algorithms. Performance of the algorithms is estimated by hypervolume, two set coverage metric, spread measure, and ε-indicator [26,32].

In [10,23], NSGA-II was adopted to the multicriteria symmetric travelling salesman problem, and the experimental evaluation was performed on symmetric instances from TSPLIB library [25]. Psychas et al. [23] developed and experimentally studied NSGA-II on the multicriteria symmetric travelling salesman problem, in which a solution is encoded as a floating point in interval $(0, 1]$. In [17], NSGA-II was successfully integrated with SPEA2, MOEA/D and 2-opt improving for symmetric bi-objective TSP. Various LSs were proposed and demonstrated competitive results on bi-STSP instances (see, e.g., [16,19,20]).

Previously, in [30,31], we proposed the MOEA based on NSGA-II scheme, which applies a problem-specific heuristic to generate the initial population and uses reproduction operators based on the preservation of the adjacencies found in the parents taking into account the Pareto-dominance. In comparison to the MOEA from [30] the current GLS applies new local search heuristics to generate the initial population and improve offspring after crossover operations. In addition, the GLS realizes a more effective non-dominated sorting [14] and a new rule for adaptation of reproduction operators. Here, more careful experimental analysis and evaluation of our GLS are carried out.

2 Problem Statement

An instance of the bicriteria asymmetric travelling salesman problem [3] (bi-ATSP) is given by a complete graph $G = (V, E)$, where $V = \{v_1, \dots, v_n\}$ is the set of vertices and E is the set of arcs. Each arc $e \in E$ is characterized by couple of weights $w(e) = (w_1(e), w_2(e))$, which can represent travel distance, travel time, expenses, number of flight changes, etc. A feasible solution to an instance of the bi-ATSP is a Hamiltonian circuit (tour), i.e., a circuit through the entire set of vertices. We denote by $\mathcal{C}$ all possible $(n-1)!$ tours of graph G. The weight of a tour C is a two-dimensional vector $W(C) = (W_1(C), W_2(C))$, where $W_j(C) = \sum_{e \in C} w_j(e)$ is the sum of arc weights in the tour, $j = 1, 2$.

We say that one solution (tour) C^* dominates another solution C if the inequality $W(C^*) \leq W(C)$ holds. The relation $W(C^*) \leq W(C)$ means that $W(C^*) \neq W(C)$ and $W_i(C^*) \leqslant W_i(C)$ for $i = 1,\ 2$. This relation $\leq$ is also called *Pareto relation.* A set of all non-dominated solutions is called *the set of Pareto-optimal solutions* [8,22] $P_W(\mathcal{C}) = \{C \in \mathcal{C} \mid \nexists C^* \in \mathcal{C} : W(C^*) \leq W(C)\}$. If we denote $\mathcal{W} = W(\mathcal{C})$, then *the Pareto set* is defined as $P(\mathcal{W}) = \{y \in \mathcal{W} \mid \nexists y^* \in \mathcal{W} : y^* \leq y\}$. We assume that the Pareto set is specified except for a collection of equivalence classes, generated by equivalence relation $C' \sim C''$ iff $W(C') = W(C'')$.

The aim of solving the bi-ATSP is to find the set of Pareto-optimal tours, which gives the Pareto set.

3 Local Search Procedures

In the single objective case, a local search algorithm starts from an initial feasible solution. It moves iteratively from one solution to a better neighboring solution and terminates at a local optimum. In the multiobjective case, another approach is usually used. The non-dominated solutions that were previously produced are compared to a solution from the neighborhood during the local search. The number of steps of the algorithm and the time complexity of one step depend essentially on the neighborhood.

In general, k-opt neighborhood for the bi-ATSP is defined as the set of tours that can be obtained from a given tour by replacing k arcs, where k is not less than 3 and odd for the asymmetric setting. As shown previously, such neighborhood structure is more appropriate for TSP instances (see, e.g., [11,16,19,20]).

Let C be a tour, $\mathcal{N}$ be a neighborhood structure, and $\mathcal{N}(C)$ denote the set of solutions in the neighborhood of C. We say that solution C is a Pareto local optimum with respect to $\mathcal{N}$ if and only if there is no C' in $\mathcal{N}(C)$ such that $W(C') \leq W(C)$.

3.1 Improving One Solution

Here we propose an algorithm that allows to locally improve an arbitrary feasible tour. Our Local Search Heuristic named LS_{one} is a typical local search heuristic that explores a subset of 3-opt neighborhood, and uses the well-known "first improving move" strategy.

We try to improve the current tour by changing three of its arcs (see Fig. 1). To this end, we consider arcs of the current tour as candidates for arc (v_{i_1}, v_{i_2}) to be deleted in the order defined by the Pareto relation. The last level consists of all non-dominated arcs. Arcs of the previous level are the non-dominated arcs, when arcs of the last level are discounted, and so on. Arcs of the same level are ordered at random. In addition, we test only the $\alpha\%$ first arcs as candidates for (v_{i_1}, v_{i_2}).

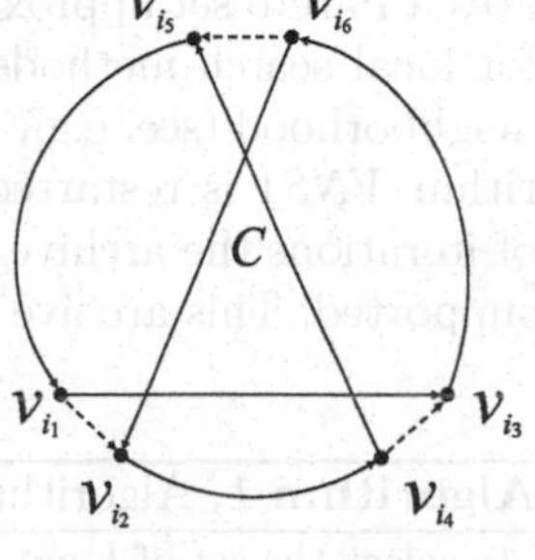

Fig. 1. 3-change.

Later, in our search, the possibilities of choosing v_{i_3} (arc (v_{i_1}, v_{i_3}) is added) is considered in the following sequence. For each vertex v we store a list of the remaining vertices in the sequence defined by the Pareto relation such that the non-dominated arcs have the highest priority. Considering candidates for v_{i_3}, we start at the beginning of v_{i_1}'s list and proceed down the list during the search. Moreover, only the $\beta\%$ first vertices are stored in the sorted list assigned to each vertex, this allows to reduce the running time and memory usage.

Finally, among all vertices belonging to the closed circuit C created by (v_{i_1}, v_{i_3}), we choose a vertex v_{i_5} that would produce a favorable 3-change in terms of the Pareto dominance. Local Search Heuristic stops if no favorable

3-change is possible, otherwise it proceeds to the next step with a new tour obtained.

The presented algorithm attempts to greedy improve the given solution returning only one tour and can be considered as a search in depth.

3.2 Variable Neighborhood Search

One of the well-known approaches to solve different optimization problems is Variable Neighborhood Search (VNS) [11]. VNS is a metaheuristic, which systematically changes the neighborhood within a randomized local search algorithm.

Initially, VNS was proposed for single-objective optimization problems. To the best of our knowledge, there are two versions of VNS for single objective ATSP. In the first VNS the well-known HyperOpt (the problem is splitted into small subproblems, which are solved by an exact algorithm) and k-opt local search approaches are combined [5]. The second one uses insertion and swap neighborhood structures [21]. We extend the single-objective VNS to the multi-objective one for the considered problem.

We realize the following two schemes of VNS attempting to find a Pareto local optimum set for the bi-ATSP (the Pareto local optimum solutions) with respect to 3-opt neighborhood structure, searching in width.

The first VNS algorithm called *VNS*1 works such that only one solution is constructed at each iteration (see Algorithm 1).

Preliminary computational experiment showed that one run of *VNS*1 generates a Pareto set approximation of small size. Note that this property is natural for local search methods, where each time only one solution is selected from a neighborhood (see, e.g., [20]). So, we propose the following restarting rule. Algorithm *VNS*1 is restarted every time as soon as during the given number N_{rst} of iterations the archive is not updated. In addition, an external elite archive is supported. This archive stores non-dominated solutions over all restarts.

Algorithm 1. Algorithm *VNS*1

1: select the set of k-opt neighborhood structures $\mathcal{N}_k$, $k = 3, 5, \dots, \gamma \leq n$
2: generate an initial set of non-dominated tours named as archive (randomly construct N_{VNS} number of tours, apply local optimization by LS_{one} to each of them and choose the non-dominated ones)
3: pick a random non-visited tour C from the archive and set $k := 3$; if such solution does not exist then terminate
4: *Shaking*: generate a tour C' at random from $\mathcal{N}_k(C)$
5: *Local improving*: apply local search procedure LS_{one} to C' as initial solution; let C'' be the obtained tour
6: *Moving*: if C'' is not dominated by any solution from the archive, then we add it to the archive, remove from the archive all solutions that are dominated by C'', mark C as visited tour and go to step 3; otherwise set $k := k + 2$ and go to step 4 (if $k > \gamma$, then mark C as visited tour and go to step 3)

The second VNS algorithm called *VNS*2 is PLS with changes of the neighborhood. In PLS, a list of potentially efficient solutions is updated. A neighborhood of the chosen solution is explored, and if the neighbor is not dominated by a solution of the list, the neighbor is added to this list. The algorithm stops when there is no more possibility to find new non-dominated neighbors. In addition, our algorithm *VNS*2 changes the neighborhood during the search process and works in accordance with Algorithm 2.

Algorithm 2. Algorithm *VNS*2

1: select the set of k-opt neighborhood structures $\mathcal{N}_k$, $k = 3, 5, \ldots, \gamma \leq n$
2: generate an initial set of non-dominated tours named as archive (randomly construct N_{VNS} number of tours, apply local optimization by LS_{one} to each of them and choose the non-dominated ones)
3: pick a random non-visited tour C from the archive and set $k := 3$; if all solutions in the archive are visited then terminate
4: *Shaking*: generate a tour C' at random from $\mathcal{N}_k(C)$
5: *Local descent*: explore 3-opt neighborhood of C'; whenever a non-dominated solution $\bar{C}$ with respect to the archive is found in the 3-opt neighborhood of C', we add it to the archive and remove from the archive all solutions that are dominated by $\bar{C}$
6: *Moving*: if there is a neighbor of C' that dominates a solution from the archive, then we mark C as visited tour and go to step 3; otherwise set $k := k + 2$ and go to step 4 (if $k > \gamma$, then mark C as visited tour and go to step 3)

Algorithm *VNS*2 performs the search in 3-opt neighborhood of a tour as in algorithm LS_{one}.

In [19], the structure of the Pareto local optimum sets was analyzed on randomly generated symmetric instances and problems constructed from symmetric series KRO of TSPLIB library [25]. It was experimentally shown that a Pareto local optimum set is partitioned into small number of clusters, where the minimum number of different edges between a solution of a cluster and at least one solution of the same cluster is at most 3. So, we expect that analogous property may take place for asymmetric instances, and exploring 3-opt neighborhood will have success.

4 Genetic Local Search

The genetic algorithm (GA) is a random search method that models a process of evolution of a population of *individuals* [24], being sample solutions to the considered optimization problem. In genetic local search, local optimization is improved by means of evolutionary processes of GA.

4.1 NSGA-II Scheme

We develop a GLS based on Non-dominated Sorting Genetic Algorithm II (NSGA-II) [7]. Our algorithm *GLS* is initiated by generating the initial population, where N_{GLS} random solutions are locally optimized by LS_{one}. The population size N_{GLS} remains constant during execution of *GLS*. Each population is sorted based on the non-domination relation (the Pareto relation), and ranks for solutions are assigned. All non-dominated solutions of the population are set rank 1, solutions have rank $k + 1$, if they are non-dominated when solutions of ranks $1, \ldots, k$ are discounted. Sorting is computed in $O(N_{GLS}\log(N_{GLS}))$ time by means of the more effective algorithm [14] than the one proposed in the original version of NSGA-II [7]. To get an estimate of the density of solutions surrounding a solution C in a non-dominated level of the population, two nearest solutions on each side of this solution are identified for each of the objectives. The estimation of solution C is called *crowding distance* and it is computed as a normalized perimeter of the cuboid formed in the criterion space by the nearest neighbors. The crowding distances of individuals in all non-dominated levels are computed in $O(N_{GLS}\log(N_{GLS}))$ time (see e.g. [7]).

At each iteration of *GLS* we select pairs of parent solutions from the current population P_{t-1} using *s-tournament selection* [24]. Then we mutate parents, and create offspring, applying a crossover (recombination) to each pair of parents. Offspring compose population Q_{t-1}.

The next population P_t is composed from the best N_{GLS} solutions of the current population P_{t-1} and the offspring population Q_{t-1}. Population $Q_{t-1} \cup P_{t-1}$ is sorted based on the non-domination relation, and the crowding distances of individuals are calculated. The best N_{GLS} solutions are selected using the rank and the crowding distance as follows. Between two solutions with differing non-domination ranks, we prefer the solution with the lower rank. If both solutions belong to the same level, then we prefer the solution with the bigger crowding distance.

Algorithm *GLS* aims at constructing an approximation with size limit in contrast to the PLS. This may be important when the Pareto set is required to be approximated by a relatively small number of points of good quality.

4.2 Recombination and Mutation

The experimental results of [9,29] for the TSP indicate that reproduction operators with the adjacency-based representation of solutions have an advantage over operators, which emphasize the order of the vertices in parent solutions. We suppose that a feasible solution to the bi-ATSP is encoded as a list of arcs.

Our recombination combines Directed Edge Crossover (DEC) operator and local search improving. DEC is transmitting and respectful, i.e. offspring is constructed only from arcs presented in parents, moreover, arcs shared by both parents are copied into the offspring. Our version of DEC operator tries to construct an offspring of good quality, taking into account the Pareto relation (see details in [30]).

The recombination is organized as follows. We select two parents, apply DEC operator, then obtain a tour, and explore its 3-opt neighborhood like in LS_{one}. All non-dominated solutions of the neighborhood, also compared to its parents, form offspring. The recombination is applied to parents pairs until the total number of offspring is less than N_{GLS}.

The presented approach allows us to avoid creating a clone of parents and to maintain a diverse set of good solutions in the population.

The mutation is applied to each parent with probability p_{mut}, which is a tunable parameter of the GA. We use a mutation operator proposed in [9] for the one-criteria ATSP. It performs a random jump within 3-opt neighborhood, trying to improve a parent solution in terms of one of the criteria. Each time one of two objectives is used in mutation with equal probability.

5 Computational Experiment

This section presents the results of computational experiment on the bi-ATSP instances. Our algorithms (*GLS*, *VNS*1, *VNS*2) were programmed in C++ and tested on a computer with Intel Xeon E5420 2.5 GHz processor, 16 Gb RAM. On the basis of preliminary computational experiment we set the tournament size $s = 10$, the mutation probability $p_{mut} = 0.1$. All vertices are taken into account when 3-opt neighborhood is explored in LS_{one} for constructing initial population in *GLS* and initial archive in *VNS*1 and *VNS*2, i.e. $\alpha = \beta = 100\%$. However, the percentages of tested vertices $\alpha = \beta = 50\%$ in searching process of the proposed algorithms. Also in algorithm *VNS*1 we set N_{rst} to the number of elements of the archive at step 2 in Algorithm 1.

We estimate the performance of our algorithms and compare them. The experiment is arranged in the following form. Algorithm *VNS*2 is terminated when all tours from the archive are visited, but it continues no more than 5000 iterations. Another two algorithms are given CPU resource T approximately equal to the average run time of *VNS*2 over 30 trials. Thus, all compared algorithms have similar average run time. Note that there exists MOOLIBRARY library [18], which contains test instances of some discrete multicriteria problems. However the multicriteria TSP is not presented in this library, so we generate the bi-ATSP test instances randomly and construct them from benchmark ATSP instances with one objective, as well.

The most popular metrics *hypervolume* (HV), *two set coverage metric* (C), *ε-indicator* (I_ε), and *spread measure* (Δ) are used to compare the performance of algorithms on medium-size instances [26,32]. These metrics provide the analysis of experimental results from different sides: accuracy, diversity, cardinality, giving many aspect estimates. Later on, we say that vector $a \in \mathcal{W}$ weakly dominates vector $b \in \mathcal{W}$ iff $a_1 \leqslant b_1$ and $a_2 \leqslant b_2$ [22]. As for approximation sets, weak dominance means that any vector from "worse" set is weakly dominated by a vector from "better" set [32]. Obviously, any set weakly dominates itself.

We construct the following three series with $n = 50$: S10, S20, S10,20. Each series consists of five problems with integer weights $w_1(\cdot)$ and $w_2(\cdot)$ of arcs randomly generated from intervals [1,10] and [1,10] for S10, [1,20] and [1,20] for S20, [1,10] and [1,20] for S10,20.

We also tested two series composed from benchmark single-criterion instances. Series SftvRand consists of seven ATSP instances from TSPLIB library [25]: ftv33, ftv35, ftv38, ftv44, ftv47, ftv55, ftv64 are used for the first criterion. Arc weights for the second criterion are generated randomly from interval $[1, w_1^{max}]$, where w_1^{max} is the maximum arc weight on the first criterion. Here each instance is named by adding the letter R after the original name. Series SND contains five paired combinations of instances ND4944 and ND61442 (ND120, n = 120); ND82040 and ND82041m (ND122, n = 122), ND61443 and ND82042 (ND128, n = 128), ND102641m and ND122641a (ND152, n = 152), ND81744 and ND82043m (ND154, n = 154) presented in [27]. These asymmetric instances are constructed from one single-vehicle routing problem.

The size N_{VNS} in *VNS*1 and *VNS*2 was set to 100 for series S10, S20, S10,20 and SftvRand, and was set to 200 for series SND. For algorithm *GLS* the size N_{GLS} of initial population is equaled to the doubled average size of approximations obtained by *VNS*2 for series S10, S20, S10,20 and SftvRand, and N_{GLS} equals the average size of approximations obtained by *VNS*2 for series SND.

Tables 1, 2, 3, 4, 5 and Figs. 2, 3, 4 show the obtained results. In Table 1 we denote the following average data obtained by the corresponding algorithm over 30 runs: the number of elements in the Pareto set approximation as K, standard deviation of elements number as σ_K, Euclidean distance between consecutive points in approximation set as δ, and its standard deviation as σ_δ. In Tables 2, 3, 4 and 5, the coverage metric and ε-indicator are presented, and each entry is averaged over all pairwise comparisons of runs of corresponding two algorithms. We denote approximation sets obtained by algorithms by the name of algorithms itself in notation of binary metrics. Also we note that standard deviation σ_δ is known as spread measure.

Initially, we compare *VNS*2 and *GLS* on series S10, S20, S10,20 and SftvRand. Average hypervolume for *GLS* is at most 1% less than average hypervolume for *VNS*2. Algorithm *VNS*2 generates the number of points only in 1.1 times greater than *GLS* on average. Both algorithms demonstrate similar spread of solutions in approximations (see values of metrics δ and σ_δ). According to [32] since values of ε-indicator between approximations constructed by *VNS*2 and *GLS* tend to 1 and also differ by at most 3.75% (for series S10, S20, S10,20) and at most 0.63% (for series SftvRand), approximation sets are close to each other. Moreover, all values are greater than 1, and we could not claim that one approximation weakly dominates another (see Table 4). So, the algorithms generate close to each other sets with identical "distribution" along some function.

However, an approximation of *VNS*2 covers at least 55% points found by *GLS*, but an approximation of *GLS* covers at most 50% points found by *VNS*2 as shown by coverage metric (for two non-dominated sets we calculate the fraction of points in each set that are weakly dominated by at least one solution in the

Table 1. Number of points in approximation and spread measure

Instance	*VNS*2		*GLS*		*VNS*1	
	$K(\sigma_K)$	$\delta(\sigma_\delta)$	$K(\sigma_K)$	$\delta(\sigma_\delta)$	$K(\sigma_K)$	$\delta(\sigma_\delta)$
S10	60 (2.8)	5.2 (6.2)	58 (3.1)	5.2 (6)	45 (3.2)	7.2 (7.2)
S10,20	84 (3.5)	6.1 (8.4)	79 (3.6)	6.3 (8.4)	58 (3.9)	9.4 (10.8)
S20	110 (5.3)	7.2 (8.3)	100 (5.6)	7.8 (8.8)	68 (4.7)	11.8 (12.4)
ftv33R	141 (8.4)	44.6 (64.3)	136 (9.5)	43.7 (54)	88(6.4)	72.1 (105.7)
ftv35R	189 (8.3)	42 (37.3)	176 (6.5)	45.2 (36.9)	102 (8.5)	77.8 (64.9)
ftv38R	175 (11.6)	41.6 (39.1)	161 (9.8)	45.1 (37.9)	100 (6.9)	72.1 (67.6)
ftv44R	201 (14.4)	44.6 (45.6)	185 (9.7)	47.2 (44)	97 (8)	101.2 (119.7)
ftv47R	253 (20.6)	41.3 (44.3)	219 (14.4)	48 (48.7)	114 (8.4)	93.4 (97.4)
ftv55R	341 (21.5)	36.2 (43.7)	307 (20.1)	39 (42.8)	135 (7.4)	89 (87.2)
ftv64R	389 (22.1)	37.5 (41.2)	323 (16.1)	44.5 (44.8)	150 (11.3)	98.9 (95.8)
ND120	618 (53.3)	78.9 (313.1)	464 (11.1)	79.8 (273.9)	121 (7.3)	479 (1070.6)
ND122	551 (36.5)	154.6 (757)	415 (12.6)	147.4 (684.5)	138 (8.5)	670 (1528.9)
ND128	784 (69.4)	70.8 (254.8)	539 (27.9)	84.9 (304.4)	151 (5.8)	375 (783)
ND152	774 (43.3)	70 (252)	584 (7.9)	81 (295.2)	136 (12.3)	424.6 (705.5)
ND154	686 (65.9)	55.13 (119.1)	510 (13)	59.45 (191.8)	142 (19)	287.80 (565.8)

other set). Note that different nature of algorithms *VNS*2 and *GLS* (shaking in *VNS*2 and evolutionary processes in *GLS*) allows us to construct different non-dominated solutions.

Algorithms *VNS*2 and *GLS* demonstrate more successful results than algorithm *VNS*1 on all considered instances in terms of all presented performance metrics. We believe that this is due to the fact, that only one solution is generated on each iteration of *VNS*1. Outperformance becomes more significant on series SND, where data are not random. Moreover, the results for series SND clearly indicate the superior performance of *GLS*, which approximation covers at least 65% points found by *VNS*2 and at least 98% points found by *VNS*1. While the fraction of a set obtained by *GLS* which is covered by a set of *VNS*1 or *VNS*2 composes no more than 26%. The comparisons based on hypervolume lead to the same conclusions (see Fig. 4).

Also values of ε-indicator on series SND show that approximation sets obtained by *GLS* weakly dominate and do not equal approximations obtained by *VNS*1 in four out of five instances due to conditions $I_\varepsilon(GLS, VNS1) \leqslant 1$ and $I_\varepsilon(VNS1, GLS) > 1$ are valid (see Table 5) [32].

The statistical analysis of experimental data was carried out using the Wilcoxon signed-rank test at a 0.05 significance level. We test for each instance the difference between values of hypervolume reached in 30 runs of algorithms *VNS*2 and *GLS*. As a result, we obtain that

1. on series S10, S20, S10,20 algorithm *VNS*2 demonstrates better values of hypervolume than algorithm *GLS* (in all 15 cases the difference between hypervolumes is statistically significant);

2. on series SftvRand the difference between hypervolumes is statistically significant only on instance ftv44R, where *VNS*2 outperforms *GLS*;
3. on series SND algorithm *GLS* shows significantly better values of hypervolume than *VNS*2.

Table 2. Two set coverage metric (series S10, S20, S10,20, SftvRand), %

C-metric	S10	S20	S10,20	ftv33R	ftv35R	ftv38R	ftv44R	ftv47R	ftv55R	ftv64R
$C(VNS2, GLS)$	74.9	79.5	72.1	63.2	60.7	55.9	73.1	63.4	59.2	57.1
$C(VNS2, VNS1)$	78.2	88.2	83.3	70.8	90.6	80.7	89.7	87.4	91.7	93.8
$C(GLS, VNS2)$	29.5	24.1	31.3	50.1	58.2	44.5	33.2	38.7	43.1	44.5
$C(GLS, VNS1)$	53.9	56.2	59.6	65.7	86.1	72.9	76.1	76.1	86.3	86.2
$C(VNS1, VNS2)$	25.6	11.8	17.5	23.6	8.2	14.7	5.5	8.4	4.9	4.1
$C(VNS1, GLS)$	47.7	40.1	38.9	27.1	10.3	19.3	14.4	14.5	8.8	9.7

Table 3. Two set coverage metric (series SND), %

C-metric	ND120	ND122	ND128	ND152	ND154
$C(VNS2, GLS)$	0.39	25.58	0.98	5.76	0.05
$C(VNS2, VNS1)$	86.26	98.71	76.14	85.72	84.14
$C(GLS, VNS2)$	98.78	65.06	97	91.28	99.87
$C(GLS, VNS1)$	100	98.47	100	100	100
$C(VNS1, VNS2)$	7.42	0.47	15.16	6.92	10.01
$C(VNS1, GLS)$	0	0.54	0	0	0

Table 4. ε-Indicator (series S10, S20, S10,20, SftvRand)

I_ε	S10	S20	S10,20	ftv33R	ftv35R	ftv38R	ftv44R	ftv47R	ftv55R	ftv64R
$I_\varepsilon(VNS2, GLS)$	1.03	1.03	1.038	1.033	1.022	1.031	1.045	1.033	1.032	1.034
$I_\varepsilon(VNS2, VNS1)$	1.02	1.02	1.025	1.019	1.012	1.017	1.013	1.013	1.013	1.011
$I_\varepsilon(GLS, VNS2)$	1.05	1.07	1.059	1.04	1.022	1.03	1.044	1.036	1.031	1.039
$I_\varepsilon(GLS, VNS1)$	1.05	1.06	1.05	1.02	1.016	1.021	1.021	1.019	1.018	1.018
$I_\varepsilon(VNS1, VNS2)$	1.06	1.09	1.078	1.079	1.052	1.066	1.101	1.083	1.073	1.106
$I_\varepsilon(VNS1, GLS)$	1.05	1.08	1.075	1.078	1.054	1.068	1.105	1.084	1.071	1.099

Table 5. ε-Indicator (series SND)

I_ε	ND120	ND122	ND128	ND152	ND154
$I_\varepsilon(VNS2, GLS)$	1.00043	1.00012	1.00046	1.00034	1.00043
$I_\varepsilon(VNS2, VNS1)$	1.00011	1.00002	1.00017	1.00010	1.00009
$I_\varepsilon(GLS, VNS2)$	1.00002	1.00011	1.00005	1.00009	0.99996
$I_\varepsilon(GLS, VNS1)$	0.99988	1.00002	0.99983	0.99995	0.99979
$I_\varepsilon(VNS1, VNS2)$	1.00117	1.00142	1.00093	1.00156	1.00098
$I_\varepsilon(VNS1, GLS)$	1.00145	1.00144	1.00117	1.00169	1.00122

We also tested a version of GLS named *GLS*1, where an offspring is locally improved by LS_{one}, and, therefore, only one tour is created in recombination. The experiment clearly showed an advantage of *GLS* over *GLS*1. The algorithm *GLS*1 on average gave only 60% points of Pareto set approximation obtained by *GLS* within the same CPU time limit.

Generally speaking, genetic local search and variable neighborhood search are based on the same basic principles "search locally" and "explore variety", meaning searching non-dominated solutions in a neighborhood and exploration of various regions. In our algorithms, in each pair of *GLS*1, *VNS*1 and *GLS*, *VNS*2 principle "search locally" is the same, and difference occurs in "explore variety" principle. *GLS*s use evolutionary process, and *VNS*s apply shaking procedure. This leads to different subsets of non-dominated solutions, which do not completely cover each other.

Aforementioned procedures of "explore variety" work in different ways, and this gives different behavior of the proposed algorithms on various problem structures, as we have seen changes in performance on random and non-random generated series. So, we could not establish the leader, which certainly outperforms other algorithms. Summing up the obtained experimental results we conclude that *VNS*2 and *GLS* may be used successfully in combination, for example:

- alternative using of algorithms by means of some learning technics;
- apply shaking on *GLS* iterations;
- partition points obtained in *VNS*2 into pairs and apply the recombination to extend the approximation set.

Let us notice that PLSs showed comparable or better results than GLSs in terms of coverage metric and R measure [26] on symmetric instances constructed from series KRO of TSPLIB library (see details in [19,20]). However, running times reported for PLSs appear to be higher than for GLSs. Both approaches are based on exploring k-opt neighborhood.

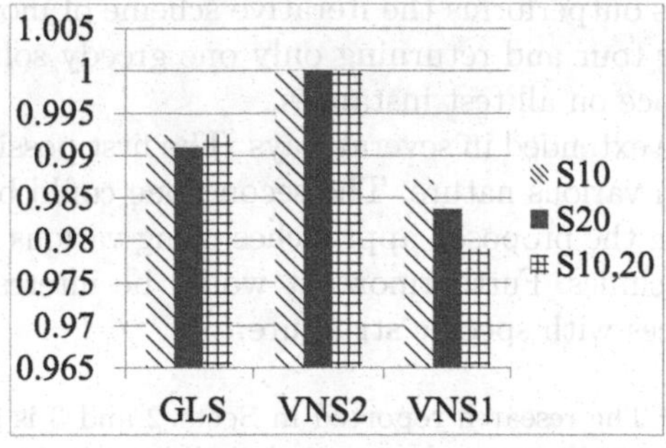

Fig. 2. Relative hypervolume on series S10, S20, S10,20

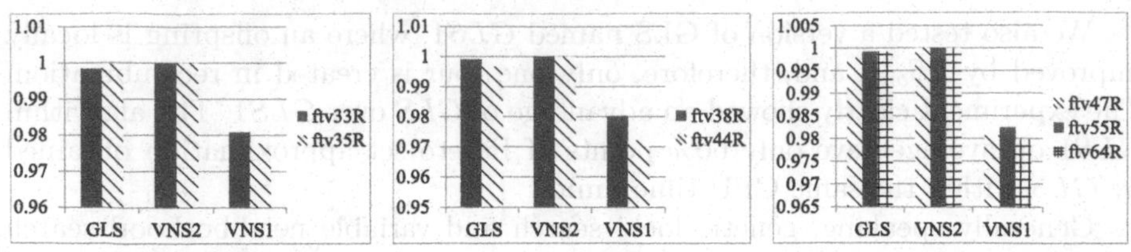

Fig. 3. Relative hypervolume on series SftvRand

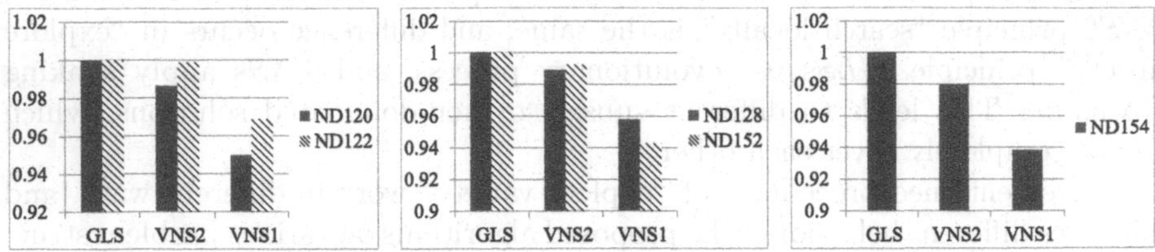

Fig. 4. Relative hypervolume on series SND

6 Conclusion

New hybrid algorithms aimed at finding the approximation of the Pareto set are proposed for the bi-ATSP. Starting from Pareto local optima, our algorithms realize shaking procedures or reproduction operations with subsequent heuristic local search.

An experimental evaluation on randomly generated and benchmark instances shows that the behavior and outcome of an algorithm depend on problem structure. Variable neighborhood search based on PLS approach and genetic local search with NSGA-II scheme are competitive with each other.

Results of the experiment also indicate that at local search improving stage, the strategy of exploring a neighborhood of the given tour and returning all non-dominated neighbors outperforms the iterative scheme of moving from a tour to a better neighboring tour and returning only one greedy solution as the result. Superiority takes place on all test instances.

Our study can be extended in several ways. The first possibility is to consider more objectives with various nature. The second one could be devoted to investigation of combining the proposed approaches using various neighborhoods and machine learning technics. Furthermore, it would be interesting to investigate the bi-ATSP instances with specific structure.

Acknowledgement. The research reported in Sects. 2 and 3 is supported by RFBR grant 19-47-540005, and the research reported in Sect. 4 is supported by the Ministry of Science and Higher Education of the Russian Federation, government program of Sobolev Institute of Mathematics SB RAS, project N 0250-2019-0001 (Yu. Kovalenko). The research reported in Sect. 5 is supported by RFBR grant 17-07-00371 (A. Zakharov).

References

1. Angel, E., Bampis, E., Gourves, L.: A dynasearch neighborhood for the bicriteria traveling salesman problem. In: Gandibleux, X., Sevaux, M., Sorensen, K., T'kindt, V. (eds.) Metaheuristics for Multiobjective Optimisation. LNCS, vol. 535, pp. 153–176. Springer, Heidelberg (2004). https://doi.org/10.1007/978-3-642-17144-4_6
2. Angel, E., Bampis, E., Gourvès, L., Monnot, J.: (Non)-approximability for the multi-criteria *TSP*(1,2). In: Liśkiewicz, M., Reischuk, R. (eds.) FCT 2005. LNCS, vol. 3623, pp. 329–340. Springer, Heidelberg (2005). https://doi.org/10.1007/11537311_29
3. Ausiello, G., Crescenzi, P., Gambosi, G., Kann, V., Marchetti-Spaccamela, A., Protasi, M.: Complexity and Approximation. Springer, Heidelberg (1999). https://doi.org/10.1007/978-3-642-58412-1
4. Bökler, F.: The multiobjective shortest path problem is NP-Hard, or is it? In: Trautmann, H., et al. (eds.) EMO 2017. LNCS, vol. 10173, pp. 77–87. Springer, Cham (2017). https://doi.org/10.1007/978-3-319-54157-0_6
5. Burke, E.K., Cowling, P.I., Keuthen, R.: Effective local and guided variable neighbourhood search methods for the asymmetric travelling salesman problem. In: Boers, E.J.W. (ed.) EvoWorkshops 2001. LNCS, vol. 2037, pp. 203–212. Springer, Heidelberg (2001). https://doi.org/10.1007/3-540-45365-2_21
6. Buzdalov, M., Yakupov, I., Stankevich, A.: Fast implementation of the steady-state NSGA-II algorithm for two dimensions based on incremental non-dominated sorting. In: GECCO-15, pp. 647–654 (2015). https://doi.org/10.1145/2739480.2754728
7. Deb, K., Pratap, A., Agarwal, S., Meyarivan, T.: A fast and elitist multi-objective genetic algorithm: NSGA-II. IEEE Trans. Evol. Comput. **6**(2), 182–197 (2002). https://doi.org/10.1109/4235.996017
8. Ehrgott, M.: Multicriteria Optimization. Springer, Heidelberg (2005). https://doi.org/10.1007/3-540-27659-9
9. Eremeev, A.V., Kovalenko, Y.V.: Genetic algorithm with optimal recombination for the asymmetric travelling salesman problem. In: Lirkov, I., Margenov, S. (eds.) LSSC 2017. LNCS, vol. 10665, pp. 341–349. Springer, Cham (2018). https://doi.org/10.1007/978-3-319-73441-5_36
10. Garcia-Martinez, C., Cordon, O., Herrera, F.: A taxonomy and an empirical analysis of multiple objective ant colony optimization algorithms for the bi-criteria TSP. Eur. J. Oper. Res. **180**, 116–148 (2007). https://doi.org/10.1016/j.ejor.2006.03.041
11. Hansen, P., Mladenović, N., Todosijevic, R., Hanafi, S.: Variable neighborhood search: basics and variants. EURO J. Comput. Optim. **5**(3), 423–454 (2017). https://doi.org/10.1007/s13675-016-0075-x
12. Jaszkiewicz, A.: Genetic local search for multi-objective combinatorial optimization. Eur. J. Oper. Res. **137**(1), 50–71 (2002). https://doi.org/10.1016/S0377-2217(01)00104-7
13. Jaszkiewicz, A., Zielniewicz, P.: Pareto memetic algorithm with path relinking for bi-objective traveling salesperson problem. Eur. J. Oper. Res. **193**, 885–890 (2009). https://doi.org/10.1016/j.ejor.2007.10.054
14. Jensen, M.T.: Reducing the run-time complexity of multiobjective EAs: the NSGA-II and other algorithms. IEEE Trans. Evol. Comput. **7**(5), 503–515 (2003). https://doi.org/10.1109/TEVC.2003.817234

15. Kumar, R., Singh, P.K.: Pareto evolutionary algorithm hybridized with local search for biobjective TSP. In: Abraham, A., Grosan, C., Ishibuchi, H. (eds.) Hybrid Evolutionary Algorithms. SCI, vol. 14, pp. 361–398. Springer, Heidelberg (2007). https://doi.org/10.1007/978-3-540-73297-6_14
16. Lust, T., Teghem, J.: The multiobjective traveling salesman problem: a survey and a new approach. In: Coello Coello, C.A., Dhaenens, C., Jourdan, L. (eds.) Advances in Multi-Objective Nature Inspired Computing. SCI, vol. 272, pp. 119–141. Springer, Berlin (2010). https://doi.org/10.1007/978-3-642-11218-8_6
17. Moraes, D., Sanches, D., Rocha, J., Garbelini, J., Castoldi, M.: A novel multi-objective evolutionary algorithm based on subpopulations for the bi-objective traveling salesman problem. Soft Comput. 1–12 (2018). https://doi.org/10.1007/s00500-018-3269-8
18. Multiobjective optimization library. http://home.ku.edu.tr/~moolibrary/. Accessed 09 Feb 2019
19. Paquete, L., Chiarandini, M., Stützle, T.: Pareto local optimum sets in the biobjective traveling salesman problem: an experimental study. In: Gandibleux, X., Sevaux, M., Sörensen, K., T'kindt, V. (eds.) Metaheuristics for Multiobjective Optimisation. LNEMS, vol. 535, pp. 177–199. Springer, Heidelberg (2004). https://doi.org/10.1007/978-3-642-17144-4_7
20. Paquete, L., Stützle, T.: A two-phase local search for the biobjective traveling salesman problem. In: Fonseca, C.M., Fleming, P.J., Zitzler, E., Thiele, L., Deb, K. (eds.) EMO 2003. LNCS, vol. 2632, pp. 479–493. Springer, Heidelberg (2003). https://doi.org/10.1007/3-540-36970-8_34
21. Piriyaniti, I., Pongchairerks, P.: Variable neighbourhood search algorithms for asymmetric travelling salesman problems. Int. J. Oper. Res. **18**(2), 157–170 (2013). https://doi.org/10.1504/IJOR.2013.056104
22. Podinovskiy, V.V., Noghin, V.D.: Pareto-optimal'nye resheniya mnogokriterial'nyh zadach (Pareto-optimal solutions of multicriteria problems). Fizmatlit, Moscow (2007, in Russian)
23. Psychas, I.D., Delimpasi, E., Marinakis, Y.: Hybrid evolutionary algorithms for the multiobjective traveling salesman problem. Expert. Syst. Appl. **42**(22), 8956–8970 (2015). https://doi.org/10.1016/j.eswa.2015.07.051
24. Reeves, C.R.: Genetic algorithms for the operations researcher. INFORMS J. Comput. **9**(3), 231–250 (1997)
25. Reinelt, G.: TSPLIB - a traveling salesman problem library. ORSA J. Comput. **3**(4), 376–384 (1991). https://doi.org/10.1287/ijoc.3.4.376
26. Riquelme, N., Von Lucken, C., Baran, B.: Performance metrics in multi-objective optimization. In: 2015 Latin American Computing Conference (CLEI), pp. 1–11. IEEE (2015). https://doi.org/10.1109/CLEI.2015.7360024
27. Soler, D.N., Martinez, E., Mico, J.: A transformation for the mixed general routing problem with turn penalties. J. Oper. Res. Soc. **59**(4), 540–547 (2008). https://doi.org/10.1057/palgrave.jors.2602385
28. Srinivas, N., Deb, K.: Multiobjective optimization using nondominated sorting in genetic algorithms. Evol. Comput. **2**(3), 221–248 (1994). https://doi.org/10.1162/evco.1994.2.3.221
29. Whitley, D., Starkweather, T., McDaniel, S., Mathias, K.: A comparison of genetic sequencing operators. In: Proceedings of the Fourth International Conference on Genetic Algorithms, pp. 69–76. Morgan Kaufmann, New York (1991)

30. Zakharov, A., Kovalenko, Y.: Construction and reduction of the pareto set in asymmetric travelling salesman problem with two criteria. Vestnik of Saint Petersburg University. Appl. Math. Comput. Sci. Control. Process. **14**(4), 378–392 (2018). https://doi.org/10.21638/11702/spbu10.2018.410
31. Zakharov, A.O., Kovalenko, Y.V.: Reduction of the pareto set in bicriteria asymmetric traveling salesman problem. In: Eremeev, A., Khachay, M., Kochetov, Y., Pardalos, P. (eds.) OPTA 2018. CCIS, vol. 871, pp. 93–105. Springer, Cham (2018). https://doi.org/10.1007/978-3-319-93800-4_8
32. Zitzler, E., Thiele, L., Laumanns, M., Fonseca, C.M., da Fonseca, V.G.: Performance assessment of multiobjective optimizers: an analysis and review. IEEE Trans. Evol. Comput. **7**(2), 117–132 (2003). https://doi.org/10.1109/TEVC.2003.810758

Simulated Annealing Approach to Verify Vertex Adjacencies in the Traveling Salesperson Polytope

Anna Kozlova and Andrei Nikolaev(✉)

P. G. Demidov Yaroslavl State University, Yaroslavl, Russia
fyz95@mail.ru, andrei.v.nikolaev@gmail.com

Abstract. We consider 1-skeletons of the symmetric and asymmetric traveling salesperson polytopes whose vertices are all possible Hamiltonian tours in the complete directed or undirected graph, and the edges are geometric edges or one-dimensional faces of the polytope. It is known that the question whether two vertices of the symmetric or asymmetric traveling salesperson polytopes are nonadjacent is NP-complete. A sufficient condition for nonadjacency can be formulated as a combinatorial problem: if from the edges of two Hamiltonian tours we can construct two complementary Hamiltonian tours, then the corresponding vertices of the traveling salesperson polytope are not adjacent. We consider a heuristic simulated annealing approach to solve this problem. It is based on finding a vertex-disjoint cycle cover and a perfect matching. The algorithm has a one-sided error: the answer "not adjacent" is always correct, and was tested on random and pyramidal Hamiltonian tours.

Keywords: Traveling salesperson problem · Hamiltonian tour · Traveling salesperson polytope · 1-skeleton · Vertex adjacency · Simulated annealing · Vertex-disjoint cycle cover · Perfect matching

1 Introduction

We consider a classical traveling salesperson problem on a complete directed or undirected graph.

Symmetric traveling salesperson problem. Given a complete weighted graph $K_n = (V, E)$, it is required to find a Hamiltonian cycle of minimum weight.

Asymmetric traveling salesperson problem. Given a complete weighted digraph $D_n = (V, A)$, it is required to find a Hamiltonian tour of minimum weight.

We denote by HC_n the set of all Hamiltonian cycles in K_n and by HT_n the set of all Hamiltonian tours in D_n. With each Hamiltonian cycle $x \in HC_n$ we associate a characteristic vector $x^v \in \mathbb{R}^E$ by the following rule:

$$x_e^v = \begin{cases} 1, & \text{if the cycle } x \text{ contains an edge } e \in E, \\ 0, & \text{otherwise.} \end{cases}$$

M. Khachay et al. (Eds.): MOTOR 2019, LNCS 11548, pp. 374–389, 2019.
https://doi.org/10.1007/978-3-030-22629-9_26

With each Hamiltonian tour $y \in HT_n$ we associate a characteristic vector $y^v \in \mathbb{R}^A$ by the following rule:

$$y_a^v = \begin{cases} 1, & \text{if the tour } y \text{ contains an edge } a \in A, \\ 0, & \text{otherwise.} \end{cases}$$

The polytope

$$\mathrm{TSP}(n) = \mathrm{conv}\{x^v \mid x \in HC_n\}$$

is called *the symmetric traveling salesperson polytope*, and the polytope

$$\mathrm{ATSP}(n) = \mathrm{conv}\{y^v \mid y \in HT_n\}$$

is called *the asymmetric traveling salesperson polytope.*

The 1-skeleton of a polytope P is the graph whose vertex set is the vertex set of P (characteristic vectors x^v for the traveling salesperson problem) and edge set is the set of geometric edges or one-dimensional faces of P. Many papers are devoted to the study of 1-skeletons associated with combinatorial problems. On the one hand, the vertex adjacency in 1-skeleton is of great interest for the development of algorithms to solve problems based on local search technique (when we choose the next solution as the best one among adjacent solutions). For example, various algorithms for perfect matching, set covering, independent set, a ranking of objects, problems with fuzzy measures, and many others are based on this idea [2,4,11,13,17]. On the other hand, some characteristics of 1-skeletons, such as the diameter and the clique number, estimate the time complexity for different computation models and classes of algorithms [6–8,19].

Unfortunately, the classical result by Papadimitriou states that the construction of 1-skeleton of the traveling salesperson polytope is NP-complete for both directed and undirected graphs.

Theorem 1 (Papadimitriou, [26]). *The question whether two vertices of the polytopes* TSP(n) *or* ATSP(n) *are nonadjacent is NP-complete.*

As a result, there are a large number of papers on the diameter and the clique number of 1-skeleton of TSP(n) and ATSP(n) [6,28,29], but little progress with adjacency relation. We can only note the polynomial time algorithms to test vertex adjacencies in the pedigree polytope [3] and the polytope of pyramidal tours [9,10] which are directly related to the traveling salesperson problem.

However, the vertex adjacency test for TSP(n) and ATSP(n) is an interesting problem itself. It can be solved with a geometric approach by constructing and analyzing the facet description of the polytope using convex hull algorithms. Although, this seems not very promising because both polytopes have a superexponential number of vertices and facets [16,19]. In particular, the largest known symmetric traveling salesperson polytope for a problem with 10 cities has a conjectured complete description with 51 043 900 866 facets [12].

Another approach is combinatorial. In [27] the sufficient condition for vertex adjacency in the traveling salesperson polytope was reformulated in a combinatorial form.

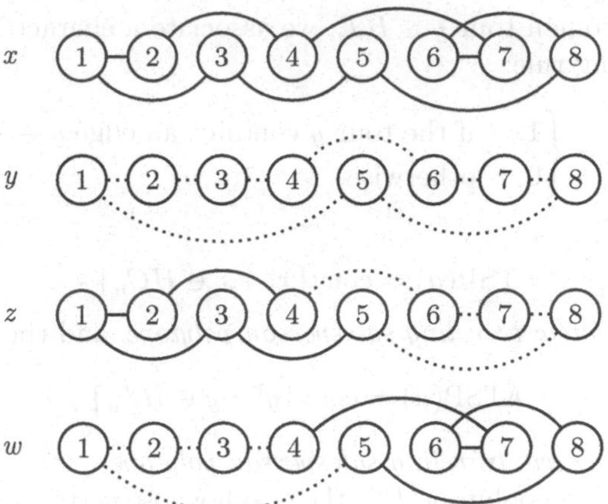

Fig. 1. Two complementary tours z and w are constructed from the edges of x and y

Lemma 1 (Sufficient condition for nonadjacency). *If from the edges of two Hamiltonian tours x and y it is possible to construct two complementary Hamiltonian tours z and w, then the corresponding vertices x^v and y^v of the polytope* TSP(n) *(or* ATSP(n)*) are not adjacent.*

From the geometric point of view, the Lemma 1 means that the segment connecting two vertices x^v and y^v intersects with the segment connecting two other vertices z^v and w^v of the polytope TSP(n) (or ATSP(n) correspondingly), thus it cannot be an edge in 1-skeleton. An example of a satisfied sufficient condition is shown in Fig. 1.

Let us formulate the sufficient condition for vertex nonadjacency of the traveling salesperson polytope in the form of a combinatorial problem.

Instance. Let x and y be two Hamiltonian tours.

Question. Does the multigraph $x \cup y$ include a pair of Hamiltonian tours z and w different from x and y such that

$$z \cup w = x \cup y \text{ and } z \cap w = \emptyset?$$

We denote by $x \cup y$ a multigraph that contains all edges of both tours x and y (Fig. 2).

In this formulation, the problem is close to the 2-peripatetic salesperson problem in which it is required to find two Hamiltonian tours of minimum weight without common edges. The 2-peripatetic salesperson is NP-complete even for 4-regular graphs [14]. Much attention was paid to the development of approximation algorithms for this problem (see, for example, [1,5,18]).

However, the combinatorial form of the sufficient condition for nonadjacency has a number of differences from the 2-peripatetic salesperson problem:

- this is a decision problem, not an optimization one;

- the graph is a 4-regular graph (or digraph) of a special form constructed as a union of two Hamiltonian tours;
- it is required to find two Hamiltonian tours different from x and y.

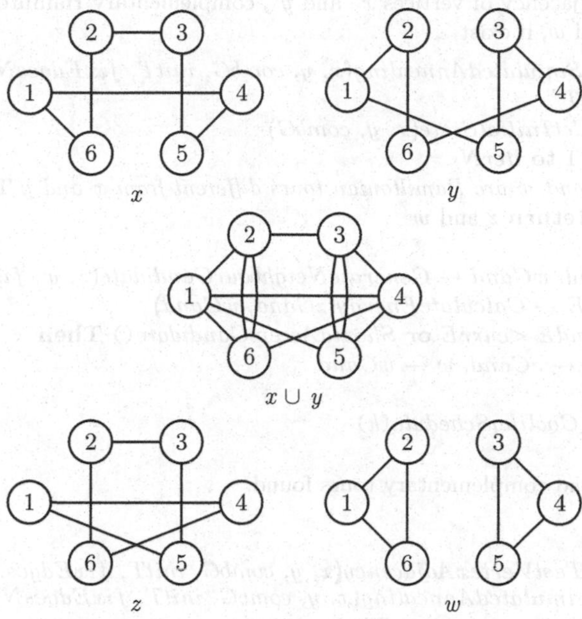

Fig. 2. An example of $w = (z \cup y) \backslash z$ that is not a Hamiltonian tour

Note also that if from the edges of two Hamiltonian tours x and y it is possible to construct another Hamiltonian tour z, then all the remaining edges $(x \cup y) \backslash z$ almost certainly does not form a Hamiltonian tour (Fig. 2). Thus, instead of algorithms for a single Hamiltonian tour in the multigraph $x \cup y$, in this paper we consider a heuristic simulated annealing approach to test vertex adjacency in the symmetric and asymmetric traveling salesperson polytopes based on finding a vertex-disjoint cycle cover and a perfect matching. We have chosen the simulated annealing heuristic, since it is easy to implement, and traditionally it shows good results for the traveling salesperson problem [21].

2 Simulated Annealing

The simulated annealing borrows the concept from annealing in metallurgy where a metal material is repeatedly heated, kneaded, and cooled to enlarge the size of its crystals to eliminate defects [21].

We consider a general scheme of the Algorithm 1. It is required to minimize the energy function specified for the current system state. The algorithm starts from a certain initial state: at each step, a neighbor candidate state is generated

Algorithm 1. Simulated Annealing Algorithm

```
Input  : Hamiltonian tours x and y (or 2/4-regular graph combG),
         initial temperature initT, number of iterations iterN, size of a queue
         of fixed edges fixEdgesN
Output: adjacency of vertices x^v and y^v, complementary Hamiltonian tours z
         and w, if exist

Procedure SimulatedAnnealing(x, y, combG, initT, fixEdgesN)
    T ← initT
    z, w ← GetInitialState(x, y, combG)
    For k ← 1 to iterN
        If z and w are Hamiltonian tours different from x and y Then
            Return z and w
        End
        zCand, wCand ← GenerateNeighbourCandidate(z, w, fixEdgesN)
        candE ← CalculateEnergy(zCand, wCand)
        If candE <currE or ShouldAcceptCandidate() Then
            z ← zCand, w ← wCand
        End
        T ← CoolingSchedule(k)
    End
    Return no complementary tours found;
End

Procedure TestVertexAdjacency(x, y, combG, initT, fixEdgesN)
    z, w ← SimulatedAnnealing(x, y, combG, initT, fixEdgesN)
    If z and w are not empty Then
        Return vertices x^v and y^v are not adjacent
    Else
        Return vertices x^v and y^v are probably adjacent
    End
End
```

which energy is compared to the energy of the previous state. If the energy decreases, the system transits to the new state, otherwise, it may transit with a certain probability (to prevent falling into the local minimum).

The algorithm receives input data in one of the following formats:

1. Two Hamiltonian tours $x = [a_1, \ldots, a_N]$ and $y = [b_1, \ldots, b_N]$, given as the permutations of vertices in a complete graph (or digraph) K_N;
2. 2/4-regular graph (2—for directed and 4—for undirected graphs) of size N, i.e. the union of two Hamiltonian tours, given as the adjacency list.

Other input parameters: the initial value of temperature *initT*, the maximum number of iterations *iterN*, and the size of a queue of fixed edges *fixEdgesN*. The algorithm stops when the solution is found or when the number of iterations exceeds the value of the parameter *iterN*. As an output, the algorithm returns two complementary Hamiltonian tours z and w constructed from the edges of x and y. By the sufficient condition (Lemma 1), the corresponding vertices x^v

and y^v of the traveling salesperson polytope are not adjacent. If the algorithm cannot find the complementary tours, then it returns that the corresponding vertices are probably adjacent. Thus, the algorithm has a one-sided error: the answer "not adjacent" is always correct, while the answer "probably adjacent" leaves the possibility that the vertices actually are not adjacent.

3 Generation of the Initial State

To generate the initial system state and neighbor candidate states, we construct a vertex-disjoint cycle cover of the multigraph $x \cup y$. *A vertex-disjoint cycle cover* of a graph G is a set of cycles with no vertices in common which are subgraphs of G and contain all vertices of G.

If x and y are undirected Hamiltonian cycles, then all vertices in the multigraph $x \cup y$ have degrees equal to 4. Let z be a vertex-disjoint cycle cover of $x \cup y$, then all the remaining edges form a graph $w = (x \cup y) \backslash z$ with all vertex degrees being equal to 2. Thus, w is also a vertex-disjoint cycle cover of $x \cup y$.

If x and y are directed Hamiltonian tours, then all vertices in the multigraph $x \cup y$ have both indegrees and outdegrees equal to 2. Let z be a vertex-disjoint cycle cover of $x \cup y$, then in the digraph $w = (x \cup y) \backslash z$ all vertices have both indegrees and outdegrees equal to 1. Thus, w is also a vertex-disjoint cycle cover of $x \cup y$.

Finding a vertex-disjoint cycle cover of both the directed and undirected graph can be performed in polynomial time by a reduction to perfect matching [30]. Let us recall that *a perfect matching* is a set of pairwise nonadjacent edges which matches all vertices of the graph. The procedures for directed and undirected graphs are somewhat different. We consider them separately.

Let x and y be undirected Hamiltonian cycles.

Step 1. From the multigraph $x \cup y = G = (V, E)$, we construct a new graph $G' = (V', E')$. With each vertex $v \in V$ we associate a gadget G_v that is a complete bipartite subgraph $K_{4,2}$ (note that the degree of v equals 4) as it is shown in Fig. 3:
- there are 4 vertices in the outer part (v_a, v_b, v_c and v_d) that correspond to 4 edges incident to v in G (edges A, B, C, D); these vertices are connected with other gadgets;
- there are 2 vertices in the inner part (v_1 and v_2) that are connected only with the vertices of the outer part.

Step 2. A perfect matching in G' corresponds to a vertex-disjoint cycle cover in the original graph G. Indeed, a perfect matching has to cover both inner vertices v_1 and v_2. Therefore, it includes exactly one edge of $\{(v_1, v_a), (v_1, v_b), (v_1, v_c), (v_1, v_d)\}$ and exactly one edge of $\{(v_2, v_a), (v_2, v_b), (v_2, v_c), (v_2, v_d)\}$. Both of these edges cover exactly two vertices of $\{v_a, v_b, v_c, v_d\}$. The other two vertices have to be covered by the edges that correspond to the edges of G (Fig. 4). We include these edges into z, then the degree of each vertex v in the graph z equals 2, and thus, z is a vertex-disjoint cycle cover of the multigraph $x \cup y$.

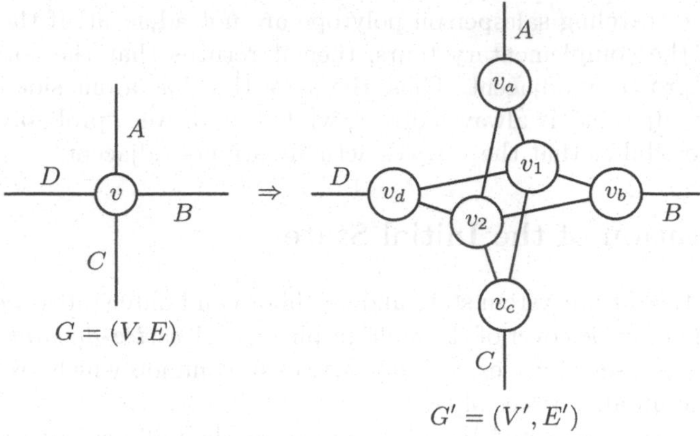

Fig. 3. Construction of the graph G' for the symmetric problem

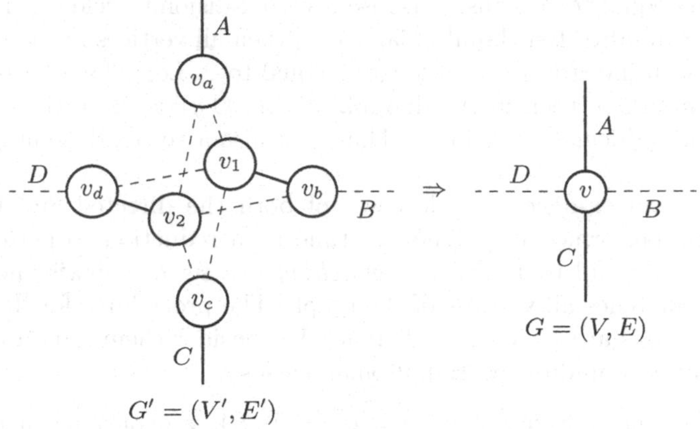

Fig. 4. A perfect matching in G' and a vertex-disjoint cycle cover in G

A perfect matching in a general undirected graph can be found by Edmond's algorithm [15] in $O(V^2E)$ time or using Micali-Vazirani matching algorithm [24] in $O(\sqrt{V}E)$ time. We have chosen Edmond's algorithm as a more simple one to implement. Note that replacing it with a more efficient Micali-Vazirani algorithm does not require changing the rest of the algorithm.

Let x and y be directed Hamiltonian tours.

Step 1. From the directed multigraph $x \cup y = D = (V, A)$, we construct a bipartite graph $D' = (L, R, E)$. With each vertex $v \in V$ we associate a pair of vertices $v_L \in L$ and $v_R \in R$, and with each edge $(u, v) \in A$ we associate a new edge (u_L, v_R) in the bipartite graph D' (Fig. 5).

Step 2. A perfect matching in the bipartite graph D' corresponds to a vertex-disjoint directed cycle cover in the original graph D. Indeed, every vertex of D is a head of exactly one edge and a tail of exactly one edge of a perfect matching in D' (Fig. 6).

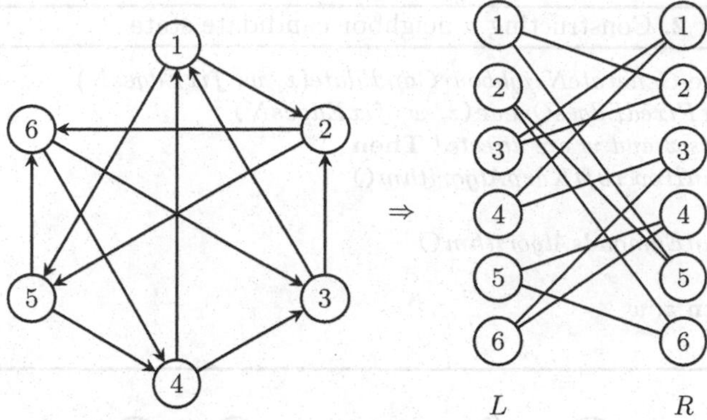

Fig. 5. Construction of the bipartite graph D' for the asymmetric problem

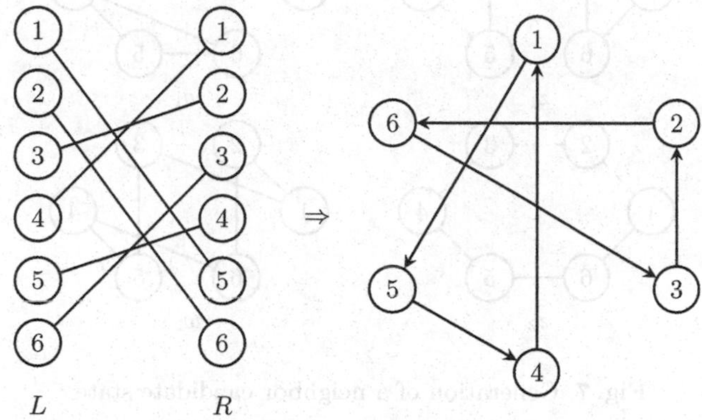

Fig. 6. A perfect matching in D' corresponds to a vertex-disjoint cycle cover of D

A perfect matching in a bipartite graph can be found by Hopcroft-Karp algorithm [20] in $O(\sqrt{V}E)$ time.

4 Generation of a Neighbor Candidate State

A process of constructing a neighbor candidate state is shown in Procedure 2.

The algorithm receives as input the current state as the vertex-disjoint cycle covers z and w, and the parameter $fixEdgesN$ that set the size of a queue of edges that are fixed in the graph z and the corresponding perfect matching. When this limit is exceeded, the first edge of the queue is deleted.

In order to find a neighbor candidate state we chose an edge of w with endpoints in two different connected components of z and add it to the queue of fixed edges (Fig. 7, fixed edges of z are dashed, an edge of w that is added to the queue is dashed-dotted). Such edge always exists due to the connectivity of the multigraph $x \cup y$. If z contains exactly one connected component, then

Algorithm 2. Constructing a neighbor candidate state

```
Procedure GenerateNeighbourCandidate(z, w, fixEdgesN)
|  UpdateFixedEdgesQueue(z, w, fixEdgesN)
|  If tours z and w are directed Then
|  |  RunHopcroftKarpAlgorithm()
|  Else
|  |  RunEdmondsAlgorithm()
|  End
|  Return z, w
End
```

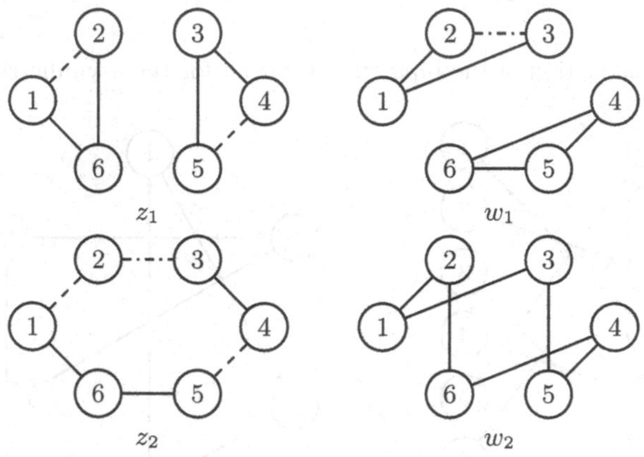

Fig. 7. Generation of a neighbor candidate state

the graphs z and w can be swapped. The idea of this procedure is to reduce the number of connected components in z and w. The neighbor candidate state is constructed by the perfect matching algorithms with fixed edges forming the initial matching.

5 Cooling Schedule

If a neighbor candidate state has two complementary Hamiltonian tours different from x and y (or any two complementary Hamiltonian tours, if the input is a 2/4-regular graph), then the algorithm successfully stops and returns a solution.

Otherwise, the energy function is calculated for a neighbor candidate state. We have chosen the following function (Procedure 3): (number of connected components in z) + (number of connected components in w).

At each step of the algorithm, the candidate states correspond to two vertex-disjoint cycle covers z and w. Therefore, if the total number of connected components in z and w equals 2, then z and w are Hamiltonian tours.

Algorithm 3. Energy function and cooling schedule

```
Procedure CalculateEnergy(z, w)
| Return CountComponents(z) + CountComponents(w)
End

Procedure CoolingSchedule(k)
| Return initT / k
End
```

If the energy function has decreased compared to the previous state, then we accept a transition to the neighbor candidate state.

If the energy function has not decreased, then we make a transition with probability

$$P = e^{-\frac{candE - currE}{T}},$$

where T is the current temperature, $currE$ is the current energy value, and $candE$ is the energy of the considered neighbor candidate state. Such a transition is necessary to avoid the problem of falling into the local minimum.

The current temperature gradually decreases from $initT$ to 0, and its function depends on the initial temperature $initT$ and the index of the current iteration k (Procedure 3).

Note that after a small modification the algorithm can be used to solve the Hamiltonian decomposition problem [23].

HAMILTONIAN DECOMPOSITION PROBLEM. Given a $2k$-regular graph G, is it possible to find k edge-disjoint Hamiltonian cycles in G?

Every candidate state will correspond to k vertex-disjoint cycle covers of G, and the energy function will be equal to the total number of connected components.

6 Experiments

The algorithm to verify vertex adjacencies in the polytopes TSP(n) and ATSP(n) was implemented as a console application with different input parameters. Some of them are described below:

- **--N**—number of vertices in the input graph/tour;
- **--times**—number of times to run the algorithm;
- **--iterN**—number of iterations in the simulated annealing algorithm;
- **--stateCandidate=random/match**—how to generate a neighbor candidate state:

random: random exchange of edges between tours
match: constructing a vertex-disjoint cycle cover and a perfect matching;

- **--exEdgesN**—number of edges to randomly exchange between tours (used only for --stateCandidate=random);

--ansN—multistart: number of repeatedly runs of the algorithm (used only for --stateCandidate=random);

--fixEdgesN—the size of a queue of edges that can be fixed in the initial matching (used only for --stateCandidate=match).

We tested the algorithm on random directed and undirected Hamiltonian tours, and also on directed and undirected pyramidal tours.

A Hamiltonian tour

$$\tau = (1, i_1, i_2, \ldots, i_r, n, j_1, j_2, \ldots, j_{n-r-2})$$

is called *pyramidal* if

$$i_1 < i_2 < \ldots < i_r \text{ and } j_1 > j_2 > \ldots > j_{n-r-2}.$$

We have chosen the pyramidal tours for experiments since for them the vertex adjacencies in the corresponding polytopes can be easily verified. In particular, the following problem: given two pyramidal tours x and y, is it possible to construct two complementary pyramidal tours z and w from the edges of x and y, can be solved in linear time [9,10]. Thus, we run the algorithm on pyramidal tours, for which it is known that the sufficient condition of Lemma 1 is satisfied. This allows us to estimate the error percentage when the algorithm could not find complementary tours that are guaranteed to exist.

The results of the tests for undirected pyramidal tours are presented in Table 1 (Edmond's algorithm is used). The algorithm was run with the number of iterations $iterN = 8000$, the initial temperature $initT = 1000$ and the number of fixed edges $fixEdgesN = \lfloor N/3 \rfloor$. In the previous version of the program [22] a different method to generate neighbor candidate state was implemented—the exchange of random edges between two subgraphs z and w. Its results are also shown in the table for comparison. For the exchange of random edges the following input parameters were used: the number of iterations $iterN = 50000$, the initial temperature $initT = 1000$, the number of multistart attempts $ansN = 5$ and the number of edges to exchange $exEdgesN = 3$.

The following parameters were calculated for both algorithms:

- TNF_{avg} is the average executing time in milliseconds when the complimentary Hamiltonian tours were not found;
- TF_{avg} is the average time in milliseconds when the algorithm managed to find complimentary tours;
- T_{avg} is the total average time in milliseconds for all tests;
- *Acc* is the accuracy parameter that indicates how often the algorithm could find the right solution (since the complementary tours are guaranteed to exist).

A dash in the table means that for all conducted experiments the case specified in the column never happened. For instance, for the size $N = 8$ the algorithm was able to find complimentary tours for all input tours, that is why TF_{avg} and T_{avg} columns have the same values.

Table 1. Results for undirected pyramidal tours with number of tests $times = 50$

N	Exchange of random edges				Cycle cover and perfect matching			
	Tours are not found, TNF_{avg}	Tours are found, TF_{avg}	Average time, T_{avg}	Accuracy, %, Acc,	Tours are not found, TNF_{avg}	Tours are found, TF_{avg}	Average time, T_{avg}	Accuracy, %, Acc
8	–	9,84	9,84	100	–	5,57	5,57	100
16	5066,57	382,79	851,17	90	–	22,08	22,08	100
24	6549,75	1403,32	5005,82	30	9096,65	33,34	214,61	98
32	7832,24	1330,37	7312,09	8	–	224,93	224,93	100
40	10035,64	2351,53	9728,28	4	18656,05	610,26	971,18	98
48	13455,39	–	13455,39	0	27695,76	978,18	3649,94	90
64	16243,99	108,64	15921,28	2	48377,16	988,05	12361,43	76
96	-				98409,50	12293,68	53629,27	52
128					158485,17	21982,49	120264,40	28
192					334841,38	26165,54	297800,28	12
256					569912,07	10740,39	513994,91	10

Note that compared to the exchange of random edges, both the accuracy of the algorithm and the size of solved problems have increased. The accuracy can also be adjusted by increasing the number of iterations or changing the maximum number of fixed edges in the queue.

The results of the tests for directed pyramidal tours and the Hopcroft-Karp algorithm are presented in Table 2. The given computed parameters are the same as in Table 1. The input parameters are similar to the case with undirected tours. Here, for all test graphs with the number of vertices less than 100, the algorithm works with the accuracy of 100%.

We analyzed the cases where the algorithm could not find complementary cycles, although they exist. The algorithm works best if the multigraph $x \cup y$ has a lot of Hamiltonian cycles, and it faces difficulties if $x \cup y$ has a unique pair of complementary cycles z and w.

Finally, Table 3 shows the test results for random directed and undirected Hamiltonian tours. Since we can not estimate the accuracy of the algorithm in this case, the different parameter is presented – the percentage of experiments where complimentary tours were found and the vertices are not adjacent.

From the table, it can be concluded that for random tours the algorithm works even faster than for pyramidal tours. Note that for undirected graphs the algorithm finds complementary tours more often than for directed graphs. This is due to the fact that the 1-skeleton of the asymmetric traveling salesperson polytope is generally much denser than the 1-skeleton of the symmetric polytope. For example, the diameter of the 1-skeleton of ATSP(n) is 2 [25], while the best known upper bound for the diameter of the 1-skeleton of TSP(n) is 4 [28]. Besides, for undirected cycles, the algorithm was able to find a solution for almost all cases. We can conclude that for the symmetric traveling salesperson polytope TSP(n) two random vertices are not adjacent with a very high probability.

Table 2. Results for directed pyramidal tours (Hopcroft-Karp algorithm), with the number of tests $times = 50$ and number of fixed edges $fixEdgesN = [N/3]$

N	Tours are not found, TNF_{avg}	Tours are found, TF_{avg}	Average time, T_{avg}	Accuracy, %, Acc
8	–	2,27	2,27	100
16	–	5,03	5,03	100
24	–	19,75	19,75	100
32	–	19,14	19,14	100
40	–	40,89	40,89	100
48	–	95,38	95,38	100
64	–	689,20	689,20	100
96	–	330,21	330,21	100
128	129532,21	4514,36	9515,07	96
192	242480,51	15783,70	70190,93	76
256	407471,21	13373,64	178894,62	58

Table 3. Results for random Hamiltonian tours with the number of tests $times = 50$

N	Undirected tours				Directed tours			
	Tours are not found, TNF_{avg}	Tours are found, TF_{avg}	Average time, T_{avg}	Percentage not adjacent, %	Tours are not found, TNF_{avg}	Tours are found, TF_{avg}	Average time, T_{avg}	Percentage not adjacent, %
8	1604,66	15,02	491,91	30	3118,99	4,05	2682,90	14
16	–	13,75	13,756	100	6799,17	4,71	4624,95	32
24	–	28,52	28,52	100	10594,29	9,38	8265,61	22
32	–	46,64	46,64	100	14764,29	39,19	10052,26	32
40	–	75,27	75,27	100	19184,31	19,33	15734,62	18
48	–	87,86	87,86	100	25214,48	142,08	19197,11	24
64	–	235,37	235,37	100	38886,56	238,95	29611,13	24
96	–	481,37	481,37	100	74654,25	1150,20	61423,52	18
128	–	827,42	827,42	100	121790,32	1851,29	95403,73	22
192	–	4064,84	4064,84	100	252321,67	8979,84	213386,98	16
256	–	8875,01	8875,01	100	442527,79	62699,56	389351,84	14

The largest instance that was solved by the algorithm had random Hamiltonian tours on 5 000 vertices and required 2 797 275 ms. However, due to the long waiting time for several tests, we limited the presented experiments to tours of size under 256 vertices.

7 Conclusion

The construction and study of 1-skeletons of the polytopes associated with intractable problems is of interest for the development and analysis of combinatorial algorithms. However, for such problems as the traveling salesperson even determining whether two vertices are adjacent or not is an NP-complete problem. This paper proposes an original heuristic approach based on simulated annealing to verify vertex adjacencies in 1-skeleton of the traveling salesperson polytope. The algorithm has a one-sided error: the answer "not adjacent" is always correct, while the answer "probably adjacent" leaves the possibility that the vertices actually are not adjacent. The algorithm showed good practical results during the experiments.

Acknowledgments. The research is supported by the grant of the President of the Russian Federation MK-2620.2018.1 (agreement no. 075-015-2019-746).

References

1. Ageev, A.A., Pyatkin, A.V.: A 2-approximation algorithm for the metric 2-peripatetic salesman problem. In: Kaklamanis, C., Skutella, M. (eds.) WAOA 2007. LNCS, vol. 4927, pp. 103–115. Springer, Heidelberg (2008). https://doi.org/10.1007/978-3-540-77918-6_9
2. Aguilera, N.E., Katz, R.D., Tolomei, P.B.: Vertex adjacencies in the set covering polyhedron. Discrete Appl. Math. **218**, 40–56 (2017). https://doi.org/10.1016/j.dam.2016.10.024
3. Arthanari, T.S.: On pedigree polytopes and Hamiltonian cycles. Discrete Math. **306**(14), 1474–1492 (2006). https://doi.org/10.1016/j.disc.2005.11.030
4. Balinski, M.L.: Signature methods for the assignment problem. Oper. Res. **33**(3), 527–536 (1985). https://doi.org/10.1287/opre.33.3.527
5. Baburin, A.E., Della Croce, F., Gimadi, E.K., Glazkov, Y.V., Paschos, V.Th.: Approximation algorithms for the 2-peripatetic salesman problem with edge weights 1 and 2. Discrete Appl. Math. **157**, 1988–1992 (2009). https://doi.org/10.1016/j.dam.2008.06.025
6. Bondarenko, V.A.: Nonpolynomial lower bounds for the complexity of the traveling salesman problem in a class of algorithms. Autom. Rem. Contr. **44**, 1137–1142 (1983)
7. Bondarenko, V.A., Maksimenko, A.N.: Geometricheskie konstruktsii i slozhnost' v kombinatornoy optimizatsii (Geometric constructions and complexity in combinatorial optimization), LKI, Moscow (2008). (in Russian)
8. Bondarenko, V.A., Nikolaev, A.V.: On graphs of the cone decompositions for the min-cut and max-cut problems. Int. J. Math. Sci. **2016** (2016). Article ID 7863650, 6 p. https://doi.org/10.1155/2016/7863650
9. Bondarenko, V.A., Nikolaev, A.V.: Some properties of the skeleton of the pyramidal tours polytope. Electron. Notes Discrete Math. **61**, 131–137 (2017). https://doi.org/10.1016/j.endm.2017.06.030
10. Bondarenko, V.A., Nikolaev, A.V.: On the skeleton of the polytope of pyramidal tours. J. Appl. Ind. Math. **12**, 9–18 (2018). https://doi.org/10.1134/S1990478918010027

11. Chegireddy, C.R., Hamacher, H.W.: Algorithms for finding K-best perfect matchings. Discrete Appl. Math. **18**, 155–165 (1987). https://doi.org/10.1016/0166-218X(87)90017-5
12. Christof, T., Reinelt, G.: Decomposition and parallelization techniques for enumerating the facets of combinatorial polytopes. Int. J. Comput. Geom. Appl. **11**, 423–437 (2001). https://doi.org/10.1142/S0218195901000560
13. Combarro, E.F., Miranda, P.: Adjacency on the order polytope with applications to the theory of fuzzy measures. Fuzzy Set. Syst. **161**, 619–641 (2010). https://doi.org/10.1016/j.fss.2009.05.004
14. De Kort, J.B.J.M.: Bounds for the symmetric 2-peripatetic salesman problem. Optim. **23**, 357–367 (1992). https://doi.org/10.1080/02331939208843770
15. Edmonds, J.: Paths, trees, and flowers. Can. J. Math. **17**, 449–467 (1965). https://doi.org/10.4153/CJM-1965-045-4
16. Fiorini, S., Massar, S., Pokutta, S., Tiwary, H.R., De Wolf, R.: Exponential lower bounds for polytopes in combinatorial optimization. J. ACM **62** (2015). Article No. 17. https://doi.org/10.1145/2716307
17. Gabow, H.N.: Two algorithms for generating weighted spanning trees in order. SIAM J. Comput. **6**, 139–150 (1977). https://doi.org/10.1137/0206011
18. Glebov, A.N., Zambalaeva, D.Z.: A polynomial algorithm with approximation ratio 7/9 for the maximum two peripatetic salesmen problem. J. Appl. Ind. Math. **6**, 69–89 (2012). https://doi.org/10.1134/S1990478912010085
19. Grötschel, M., Padberg, M.: Polyhedral theory. In: Lawler, E., Lenstra, J.K., Rinnooy Kan, A., Shmoys, D. (eds.) The Traveling Salesman Problem: A Guided Tour of Combinatorial Optimization, pp. 251–305. Wiley, Chichester (1985)
20. Hopcroft, J.E., Karp, R.M.: An $n^{5/2}$ algorithm for maximum matchings in bipartite graphs. SIAM J. Comp. **2**(4), 225–231 (1973). https://doi.org/10.1137/0202019
21. Kirkpatrick, S., Gelatt, C.D., Vecchi, M.P.: Optimization by simulated annealing. Science **220**(4598), 671–680 (1983). https://doi.org/10.1126/science.220.4598.671
22. Kozlova, A.P., Nikolaev, A.V.: Proverka smezhnosti vershin mnogogrannika zadachi kommivoyazhyora (Verification of vertex adjacency in the traveling salesperson polytope). Zametki po informatike i matematike (Notes on Computer Science and Mathematics) **10**, 51–58 (2018). (in Russian)
23. Kühn, D., Osthus, D.: Hamilton decompositions of regular expanders: a proof of Kelly's conjecture for large tournaments. Adv. Math. **237**, 62–146 (2013). https://doi.org/10.1016/j.aim.2013.01.005
24. Micali, S., Vazirani, V.V.: An $O(\sqrt{|V|}\cdot|E|)$ algorithm for finding maximum matching in general graphs. In: Proceedings of the 21st IEEE Symposium on Foundations of Computer Science, pp. 17–27 (1980). https://doi.org/10.1109/SFCS.1980.12
25. Padberg, M.W., Rao, M.R.: The travelling salesman problem and a class of polyhedra of diameter two. Math. Program. **7**(1), 32–45 (1974). https://doi.org/10.1007/BF01585502
26. Papadimitriou, C.H.: The adjacency relation on the traveling salesman polytope is NP-Complete. Math. Program. **14**, 312–324 (1978). https://doi.org/10.1007/BF01588973
27. Rao, M.R.: Adjacency of the traveling salesman tours and 0-1 vertices. SIAM J. Appl. Math. **30**, 191–198 (1976). https://doi.org/10.1137/0130021

28. Rispoli, F.J., Cosares, S.: A bound of 4 for the diameter of the symmetric traveling salesman polytope. SIAM J. Discrete Math. **11**(3), 373–380 (1998). https://doi.org/10.1137/S0895480196312462
29. Sierksma, G.: The skeleton of the symmetric traveling salesman polytope. Discrete Appl. Math. **43**, 63–74 (1993). https://doi.org/10.1016/0166-218X(93)90169-O
30. Tutte, W.T.: A short proof of the factor theorem for finite graphs. Can. J. Math. **6**, 347–352 (1954). https://doi.org/10.4153/CJM-1954-033-3

Less Is More: Tabu Search for Bipartite Quadratic Programming Problem

Dragan Urošević[1(✉)], Yiad Ibrahim Yousef Alghoul[2(✉)], Zhazira Amirgaliyeva[4,5(✉)], and Nenad Mladenović[2,3(✉)]

[1] Mathematical Institute SASA, Belgrade, Serbia
draganu@turing.mi.sanu.ac.rs
[2] Emirates College of Technologies, Abu Dhabi, UAE
iyad.alghoul@ect.ac.ac
[3] Ural Federal University, Ekaterinburg, Russia
nenadmladenovic12@gmail.com
[4] Al-Farabi Kazakh National University, Almaty, Kazakhstan
zh.amirgaliyeva@gmail.com
[5] Institute of Information and Computational Technologies, Almaty, Kazakhstan

Abstract. Having defined a complete bipartite graph G, with weights associated with both vertices and edges, the Bipartite Quadratic Programming problem (BQP) consists in selecting a subgraph that maximizes the sum of the weights associated with the chosen vertices and the edges that connect them. Applications of the BQP arise in mining discrete patterns from binary data, approximation of matrices by rank-one binary matrices, computation of the cut-norm of a matrix, etc. In addition, BQP is also known in the literature under different names such as maximum weighted induced subgraph, maximum weight bi-clique and maximum cut on bipartite graphs. Since the problem is NP-hard, many heuristic methods have been proposed in the literature to solve it. In this paper, we apply the recent Less is more approach, whose basic idea is to design a heuristic as simple as possible, i.e., a method that uses a minimum number of ingredients but provides solutions of better quality than the current state-of-the-art. To reach that goal, we propose a simple hybrid heuristic based on Tabu search, that uses two neighborhood structures and relatively simple rule for implementation of short-term memory operation. In addition, a simple rule for calculation of tabu list length is introduced. Computational results were compared favorably with the current state-of-the-art heuristics. Despite its simplicity, our heuristic was able to find 6 new best known solutions on very well studied test instances.

Keywords: Discrete optimization · Graphs · Bipartite quadratic programming · Tabu search · Variable neighborhood search

M. Khachay et al. (Eds.): MOTOR 2019, LNCS 11548, pp. 390–401, 2019.
https://doi.org/10.1007/978-3-030-22629-9_27

1 Introduction

1.1 BPQ Problem

The bipartite unconstrained 0–1 quadratic programming problem (BQP) is defined on a complete bipartite graph $G = (V, E)$. Set of vertices V consists of two subsets. The first subset represents the vertices in the left-hand side and it is denoted by I, while the second subset represents the vertices in the right-hand side and it is denoted by J. The set of edges connecting these two subsets is denoted by E. The weight is associated with each vertex, as well as with each edge. The weight of a vertex $i \in I$ is denoted by a_i, the weight of vertex $j \in J$ is denoted by b_j, while the weight of an edge (i, j) with $i \in I$ and $j \in J$ is denoted by c_{ij}. Having such a defined complete bipartite graph G, the BQP consists of selecting a subgraph that maximizes the sum of the weights associated with the chosen vertices and the edges that connect them. More formally, the problem may be defined as the following 0–1 quadratic programming problem:

$$\max \sum_{i \in I} a_i x_i + \sum_{j \in J} b_j y_j + \sum_{i \in I, j \in J} c_{ij} x_i y_j \tag{1}$$

$$x_i, y_j \in \{0, 1\} \; i \in I, j \in J, \tag{2}$$

where binary variables x_i and y_j are used to indicate if a certain node is selected or not. More precisely, a variable x_i takes value 1 if and only if a node $i \in I$ is selected. Analogously, a variable y_j takes value 1 if and only if a node $j \in J$ is selected.

1.2 Applications and Previous Work

Applications of the BQP arise in mining discrete patterns from binary data, approximation of matrices by rank-one binary matrices, computation of the cut-norm of a matrix, etc. In addition, BQP is also known in the literature under different names such as maximum weight induced subgraph [12], maximum weight biclique [2] and maximum cut on bipartite graphs [1]. Also, matrix factorization [6] is a similar problem that can be reduced to BQP.

Mainly, the BQP has been studied from the theoretical perspective of evaluation of the complexity of the problem and proposition of exact methods for some cases. Recently, the BQP attracted researchers' attention to develop heuristic approaches that enable them to tackle large scale instances that turned out to be elusive for exact methods. Among them, the first work from the heuristic perspective is by Karapetyan and Punnen [9]. In this work, the authors propose 24 heuristics that may be classified as fast-heuristics, slow-heuristics and row-merge heuristics. After that Duarte et al. [4] proposed two solution construction procedures, two mechanisms to perform neighborhood exploration, which are together with a perturbation procedure exploited within ILS framework. Glover et al. [7] proposed heuristic algorithms based on tabu search and Very Large Scale Neighborhood search, as well as a hybrid algorithm combining them.

Recently, Karapetyan et al. [10] proposed new metaheuristic scheme that they named Conditional Markov Chain Search (CMCS). This scheme is conceived for automated generation of a multi-component metaheuristic and was successfully applied to the BQP. The components that Karapetyan et al. [10] considered may be split into two categories: hill climbers, components used to possibly improve a solution, and mutations, components used to diversify search. In total, the authors designed and compared 5 heuristics stemming from this scheme. One of those heuristics turned out to belong to the family of variable neighborhood search (VNS) heuristics. This heuristic uses three neighborhood structures in the hill climbing phase and a shaking procedure based on flipping values of 16 randomly chosen variables that correspond to the nodes in the set I. According to the computational results, designed in such way VNS heuristic turn out to be dominated by the best one stemming from CMCS scheme, which also showed superiority in comparison with previously proposed heuristics.

1.3 Contribution and Outline

Analyzing numerous previous heuristic approaches, we found that they are becoming more and more complex. We decided to apply the recent Less is more approach [3,11,13], whose basic idea is to design a heuristic as simple as possible, i.e., a heuristic that uses a minimum number of components but provides solutions of better quality than the current state-of-the-art. In this paper, we propose Tabu Search based heuristics with only two neighborhoods and a relatively simple rule for tabu list implementation (especially, a relatively simple rule for calculation of a tabu list length).

This paper is organized as follows. In Sect. 2, a description of the proposed Tabu Search is given. In Sect. 3 results of detailed experiments are reported. Concluding remarks are given in Sect. 4.

2 Tabu Search Based Heuristic for the BQP

In this section, we give a detailed description of the main ingredients of our heuristic (solution representation and neighborhood structures explored) as well as how these elements are integrated into our heuristic.

2.1 Solution Representation

The solution of the BQP in our heuristic is presented as (m', n', x, y) where m' and n' denote number of elements selected from the sets I and J, respectively, while x and y are two arrays used to store elements chosen from the sets I and J, respectively. More precisely, let us denote by m and n the cardinalities of the sets I and J, then the array x has the form $x_1, x_2, ..., x_{m'-1}, x_{m'}, x_{m'+1}, ..., x_m$ meaning that elements $x_1, x_2, ..., x_{m'}$ are included into a solution (i.e., subset $I' \subset I$) while elements $x_{m'+1}, x_{m'+2}, ..., x_m$ are not included into a solution. Analogously, we have the array y of the length n (i.e, $y_1, y_2, ..., y_{n'-1}, y_{n'}, y_{n'+1}, ..., y_n$)

meaning that elements $y_1, y_2, ..., y_{n'}$ are included into a solution (i.e., subset $J' \subset J$) while elements $y_{n'+1}, y_{n'+2}, ..., y_n$ are not included into a solution.

For such represented solution we also maintain two auxiliary arrays sxc and syc with the lengths m and n, respectively. The elements of these arrays are defined as

$$sxc_i = \sum_{j\in J'} c_{ij} = \sum_{j=1}^{n'} c_{iy_j},$$

$$syc_j = \sum_{i\in I'} c_{ij} = \sum_{i=1}^{m'} c_{x_ij}.$$

In other words, each element sxc_i (resp. syc_j) represents the total weight of the edges connecting vertex i (resp. j) with vertices in the subset J' (resp. I'). Using these two arrays the objective function value for the current solution (m', n', x, y) can be calculated (expressed) in the following way:

$$f(x,y) = \sum_{i=1}^{m'} a_{x_i} + \sum_{j=1}^{n'} b_{y_j} + \sum_{i=1}^{m'} sxc_{x_i} = \sum_{i=1}^{m'} a_{x_i} + \sum_{j=1}^{n'} b_{y_j} + \sum_{j=1}^{n'} syc_{y_j}.$$

As it will be shown in the subsequent sections, the arrays sxc (syc) enable us to speed-up local search procedures. Note that the complexity of calculating arrays sxc and syc is $O(mn)$.

Note also that for the fixed subset $I' \subset I$, optimal subset $J' \subset J$ giving the maximum value of the objective function (not absolutely maximal, but maximal for fixed subset $I' \subset I$) in linear time. In other words, the complexity of computing optimal subset J' is $O(n)$. Namely, it can be shown that for the fixed subset $I' \subset I$, the subset $J' \subset J$ that gives the maximal value of the objective function is equal to:

$$J'(I') = \{j \in J | b_j + syc_j > 0\}.$$

Obviously, the objective function can be rewritten in the following form:

$$f(I', J') = f(x,y) = \sum_{i=1}^{m'} a_{x_i} + \sum_{j=1}^{n'} (b_{y_j} + syc_{y_j})$$

so, by inserting an element j having property that $b_j + syc_j < 0$, the objective value decreases.

Similarly, for a fixed subset $J' \subset J$, an optimal subset $J' \subset I$ giving maximum value of the objective function value can be calculated in linear time in the following way:

$$I'(J') = \{i \in I | a_i + sxc_i > 0\}.$$

So, a solution can be represented by a pair (m', x), where m' is the number of elements selected into the subset $I' \subset I$, while x is an array (permutation of set $\{1, 2, ..., m\}$) having property that its first m' elements are included into the subset I', while the rest of the array x are labels of elements not included in the subset I'.

Similarly, a solution can be represented by a pair (n', y), where n' is the number of elements selected into the subset $J' \subset J$, while y is an array (permutation of set $\{1, 2, ..., n\}$) having property that the first n' elements are included into the subset J' while the rest of the array y are labels of elements not included into the subset J'.

In the rest of the text, we suppose that the solution is represented by a pair (m', x). i.e. by representation of the subset $I' \subset I$.

2.2 Neighborhoods

Drop Neighborhood. Drop (remove) neighborhood N_{remI} consists of all solutions obtained by removing one element from the current subset (solution) $I' \subset I$.

In order to explore this neighborhood, for each $k \in I'$, we calculate the subset $J'(I \backslash \{k\})$ optimal for the subset $I' \backslash \{k\}$, and also calculate the objective function value for the pair of subsets $I' \backslash \{k\}$ and $J'(I' \backslash \{k\})$. The calculated change of the objective function value (i.e. new value of objective function) has the complexity $O(n)$. Namely, the objective function value for a solution obtained after removing element k from I' is equal to

$$f_n(I' \setminus \{k\}) = \sum \{a_i | i \in I' \setminus \{k\}\} + \sum \{b_j + syc_j - c_{kj} | b_j + syc_j - c_{kj} > 0\},$$

or

$$f_n(I' \setminus \{k\}) = \sum \{a_i | i \in I'\} - a_k + \sum \{b_j + syc_j - c_{kj} | b_j + syc_j - c_{kj} > 0, j = 1, 2, ..., n\}.$$

Calculation of objective function value after removing one vertex from subset $I' \subset I$ is presented in Algorithm 1. So, exploration of complete Remove neighborhood has complexity $O(mn)$ (of course, by using an auxiliary data structure syc).

Algorithm 1. Calculation of the objective function value after removing x_k $(1 \le k \le m')$ from the current solution x. Value sa denotes the sum of weights of nodes from the subset $I' \subset I$.

Function `remIov`(m', x, syc, sa, k);
$ov \leftarrow sa - a_{x_k}$;
for $j \leftarrow 1$ **to** n **do**
    if $b_j + syc_j - c_{x_k j} > 0$ **then**
        $ov \leftarrow ov + b_j + syc_j - c_{x_k j}$
    end
end
return ov;

Add Neighborhood. Add (Insert) neighborhood N_{insI} consists of all solutions obtained by inserting into the solution one element that is currently not in the subset (solution) $I' \subset I$.

In order to explore this neighborhood, for each $k \notin I'$ we calculate a subset $J'(I' \cup \{k\})$ optimal for the subset $I' \cup \{k\}$, and also calculate the objective function value for the pair of subsets $I' \cup \{k\}$ and $J'(I' \cup \{k\})$. The calculated change of the objective function value (i.e. new value of objective function) has the complexity $O(n)$. Namely, objective function value for solution obtained after inserting an element k into the subset I' is equal to

$$f_n(I' \cup \{k\}) = \sum\{a_i | i \in I' \cup \{k\}\} + \sum\{b_j + syc_j + c_{kj} | b_j + syc_j + c_{kj} > 0,\ j = 1, 2, ..., n\},$$

or

$$f_n(I' \cup \{k\}) = \sum\{a_i | i \in I'\} - a_k + \sum\{b_j + syc_j - c_{kj} | b_j + syc_j - c_{kj} > 0,\ j = 1, 2, ..., n\}.$$

Calculation of an objective function value after inserting one vertex into the subset $I' \subset I$ is presented in Algorithm 2. So, exploration of complete Insert neighborhood has complexity $O(mn)$ (of course, by using an auxiliary data structure syc).

Algorithm 2. Calculation of the objective function value after inserting x_k $(m' + 1 \leq k \leq m)$ into the current solution x. With sa is denoted as the sum of weights of nodes from subset $I' \subset I$.

Function insIov(m', x, syc, sa, k);
$ov \leftarrow sa + a_{x_k}$;
for $j \leftarrow 1$ **to** n **do**
    if $b_j + syc_j + c_{x_k j} > 0$ **then**
        $ov \leftarrow ov + b_j + syc_j + c_{x_k j}$
    end
end
return ov;

2.3 Tabu Search for the BQP

Algorithm 3 represents Tabu search for BQP. Tabu list contains vertices which change status in recent iterations and because of that are "forbidden" for using (removing from solution or inserting into solution) in a current iteration. Length of tabu list (i.e. number of recent iterations) is not fixed. Length of tabu list changes according to the procedure previously proposed by Galinier *et al.* [5]. Complete Tabu search algorithm consists of sequence of iterations. Iterations are numbered by positive integer numbers starting from 1. Length of a tabu list changes in iterations b_i $(i = 0, 1, 2, ...)$, in the following way: length of a tabu list in iterations from interval $[b_i, b_{i+1}]$ is equal $a_i + rand(2)$, where

$$a_i = \left\lfloor \frac{c_{i \bmod 15}}{8} T_{\max} \right\rfloor.$$

In this formula:

- c is an array with 15 elements as follows: 1, 2, 1, 4, 1, 2, 8, 1, 2, 1, 4, 1, 2, 1;
- $T_{\max}$ is a maximal allowed length of tabu list; and
- $rand(n)$ represents a random number between 0 and n.

Sequence b_i is defined as following:

$$b_0 = 1 \quad b_{i+1} = b_i + 5 \times a_i.$$

In other words, a length of tabu list changes periodically (shortest period is 15) and can take one of the following four values (with "small noise" not greater than 2):

$$\left\lfloor 1 \times \frac{T_{\max}}{8} \right\rfloor, \left\lfloor 2 \times \frac{T_{\max}}{8} \right\rfloor, \left\lfloor 4 \times \frac{T_{\max}}{8} \right\rfloor, \left\lfloor 8 \times \frac{T_{\max}}{8} \right\rfloor.$$

Number 5 in the formula for calculating b_{i+1} is also proposed by [5]. Value $T_{\max}$ is selected after detailed experimentation and set by the following equation:

$$T_{\max} = \min(m'_{best}, m - m'_{best})/2$$

where m'_{best} is number of elements in subset I'_{best} representing the best solution obtained during the execution of tabu search.

3 Computational Results

For comparison, we use test instances generated by [9]. All the details on how the test instances are generated can be found in Sect. 4 of [9]. Test instances can be classified into five types: random graphs, max biclique graphs, max induced subgraph, maxcut, and matrix factorization. For each graph type, 17 test instances of different sizes are generated giving in total 85 test instances.

Test instances are divided into three groups according to their sizes:

- 7 small instances ($m \in \{20, 25, 30, 35, 40, 45, 50\}$ and $n = 50$),
- medium instances ($m \in \{200, 400, 600, 800, 1000\}$ and $n = 1000$),
- large instances ($m \in \{1000, 2000, 3000, 4000, 5000\}$ and $n = 5000$).

Our method is implemented on programming language C++. All experiments are performed on a machine with an Intel(R) Core(TM) i5-3470 with CPU 3.20GHz and 16GB RAM. We decided to perform experiments only on medium and large instances (small instances are relatively "easy" for solving). All instances from medium size and large size are executed ten times. Stopping condition was execution time in seconds. For medium size instances, execution time was set to 1000 second, while execution time for large size instances was set to 10000 s.

Algorithm 3. Tabu search for BQP

```
Function BQPTS();
(m', x, syc) ← InitialSolution;
f_curr ← ObjValue(m', x, syc);
f_best ← f_curr;
sa ← SumA(m', x);
while stopping condition is not met do
    f_maxRNT ← −∞; f_maxINT ← −∞;
    f_maxRT ← −∞; f_maxIT ← −∞;
    for i ← 1 to m' do
        f_tmp ← remIov(m', x, syc, sa, i);
        if x_i ∈ Tabu then
            if f_tmp > f_maxRT then
                f_maxRT ← f_tmp; i_maxRT ← i;
            end
        else
            if f_tmp > f_maxRNT then
                f_maxRNT ← f_tmp; i_maxRNT ← i;
            end
        end
    end
    for i ← m' + 1 to m do
        f_tmp ← insIov(m', x, syc, sa, i);
        if x_i ∈ Tabu then
            if f_tmp > f_maxIT then
                f_maxIT ← f_tmp; i_maxIT ← i;
            end
        else
            if f_tmp > f_maxINT then
                f_maxINT ← f_tmp; i_maxINT ← i;
            end
        end
    end
    if max(f_maxIT, f_maxRT) > max(f_maxINT, f_maxRNT) and
      f_curr + max(f_maxIT, f_maxRT) > f_best then
        if f_maxIT > f_maxRT then
            insI(m', x, syc, sa, i_maxIT); f_curr ← f_maxIT;
        else
            remI(m', x, syc, sa, i_maxRT); f_curr ← f_maxRT;
        end
    else
        if f_maxINT > f_maxRNT then
            insI(m', x, syc, sa, i_maxINT); f_curr ← f_maxINT;
        else
            remI(m', x, syc, sa, i_maxRNT); f_curr ← f_maxRNT;
        end
    end
    if f_curr > f_best then
        f_best ← f_curr;
    end
end
```

In Table 1 results obtained on medium size instances are presented. In the first column of Table 1 the names of instances are given. Column 2 contains the best known values for the corresponding instances. Columns 3, 4, 5 and 6 contain percentage deviation of results obtained by executing CMCS [10] from the best known results with time limit 1 s, 10 s, 100 s and 1000 s, respectively. Column 7 contains a percentage deviation of results obtained by TS [4] with time limit 1000 s. Columns 8, 9 and 10 contain best, average and worst results obtained by ten times execution of our Tabu Search (as already said in the previous paragraph, a time limit was set to 1000 s). Column 11 contains time when the solution reported as best has been found for the first time. Column 12 contains a percentage deviation of the best solution obtained by Tabu Search from best known approach (negative value indicates that our Tabu Search reaches the solution better than the previous best known method).

In Table 2 results obtained on large size instances are presented. The format of this table is the same as the format of Table 1.

Table 1. Comparison of different methods on moderate size instances

		CMCS [10]				TS [4]	TS				
Instance	Best known	1s	10s	100s	1000 s	1000 s	Best	Average	Worst	Time	Dev
Rand200x1000	612947	0.00	0.00	0.00	0.00	0.00	612947	612947	612947	2.77	0.00
Rand400x1000	951950	0.05	0.00	0.00	0.00	0.00	951950	951950	951950	0.60	0.00
Rand600x1000	1345748	0.00	0.00	0.00	0.00	0.00	1345748	1345748	1345748	3.86	0.00
Rand800x1000	1604925	0.09	0.00	0.00	0.00	0.01	1604925	1604925	1604925	14.49	0.00
Rand1000x1000	1830236	0.04	0.04	0.02	0.00	0.07	1830236	1830236	1830236	29.70	0.00
Biclique200x1000	2150201	0.00	0.00	0.00	0.00	0.00	2150201	2150201	2150201	144.67	0.00
Biclique400x1000	4051884	0.27	0.09	0.00	0.00	0.00	4051884	4051884	4051884	1.18	0.00
Biclique600x1000	5501111	0.59	1.48	0.47	0.47	0.65	5501111	5500173	5497187	420.39	0.00
Biclique800x1000	6703926	0.68	0.56	0.04	0.04	0.79	6703926	6703926	6703926	80.15	0.00
Biclique1000x1000	8680142	0.10	0.35	0.35	0.11	0.91	8680142	8680142	8680142	1.36	0.00
MaxInduced200x1000	513081	0.00	0.00	0.00	0.00	0.00	513081	513081	513081	0.33	0.00
MaxInduced400x1000	777028	0.01	0.00	0.00	0.00	0.00	777028	777028	777028	2.08	0.00
MaxInduced600x1000	973711	0.00	0.00	0.00	0.00	0.00	973711	973711	973711	2.21	0.00
MaxInduced800x1000	1205533	0.01	0.00	0.00	0.00	0.07	1205533	1205533	1205533	20.77	0.00
MaxInduced1000x1000	1415622	0.03	0.03	0.03	0.01	0.06	1415622	1415622	1415622	193.69	0.00
BMaxCut200x1000	617700	1.59	0.06	0.00	0.00	0.14	617700	617700	617700	7.90	0.00
BMaxCut400x1000	951726	1.34	0.40	0.00	0.00	1.13	951726	951726	951726	48.12	0.00
BMaxCut600x1000	1239982	1.83	0.53	0.53	0.37	2.00	1239982	1239214.25	1236322	365.41	0.00
BMaxCut800x1000	1545820	1.74	1.05	0.08	0.08	1.66	1545820	1544237.5	1540902	437.96	0.00
BMaxCut1000x1000	1816688	1.83	0.46	0.23	0.23	2.47	1814056	1811155.75	1803470	507.42	0.14
MatrixFactor200x1000	6283	0.18	0.00	0.00	0.00	0.00	6283	6283	6283	0.18	0.00
MatrixFactor400x1000	9862	0.00	0.00	0.00	0.00	0.00	9862	9862	9862	0.50	0.00
MatrixFactor600x1000	12902	0.05	0.00	0.00	0.00	0.03	12902	12902	12902	1.50	0.00
MatrixFactor800x1000	15466	0.49	0.00	0.00	0.00	0.19	15466	15466	15466	6.51	0.00
MatrixFactor1000x1000	18813	0.08	0.03	0.00	0.00	0.11	18813	18813	18813	93.17	0.00
Average	1782131.48	0.44	0.20	0.07	0.05	0.41	1782026.20	1781778.66	1781102.68	95.48	0.01

Table 2. Comparison of different methods on large size instances

Instance	Best known	CMCS [10]				TS [4]	TS				
		10s	100s	1000 s	10000 s	10000 s	Best	Average	Worst	Time	Dev
Rand1000X5000	7183221	0.04	0.04	0.01	0.01	0.01	7183221	7183221.0	7183221	44.51	0.00
Rand2000X5000	11098093	0.18	0.07	0.07	0.02	0.09	11098715	11097632.1	11096692	5914.27	-0.01
Rand3000X5000	14435941	0.16	0.12	0.11	0.07	0.22	14434759	14433443.8	14432775	2408.98	0.01
Rand4000X5000	18069396	0.14	0.07	0.01	0.01	0.19	18067473	18063949.0	18062468	4453.69	0.01
Rand5000X5000	20999474	0.26	0.11	0.08	0.07	0.25	20992020	20988157.8	20984020	3141.69	0.04
Biclique1000X5000	38495688	0.22	0.08	0.02	0.00	0.02	38435916	38397458.8	38370869	643.70	0.16
Biclique2000X5000	64731072	1.67	0.52	0.19	0.28	0.94	64756649	64731248.5	64710827	201.56	-0.04
Biclique3000X5000	98204538	1.68	0.43	0.01	0.04	1.5	98204538	98204538.0	98204538	75.81	0.00
Biclique4000X5000	128500727	0.38	0.38	0.22	0.00	2.19	128500727	128498216.3	128480641	876.47	0.00
Biclique5000X5000	163628686	0.38	0.00	0.00	0.00	1.01	163628686	163628686.0	163628686	199.52	0.00
MaxInduced1000X5000	5465051	0.01	0.01	0.00	0.00	0.02	5465051	5465051.0	5465051	290.50	0.00
MaxInduced2000X5000	8266136	0.1	0.01	0.01	0.00	0.12	8266136	8265941.3	8265602	4759.89	0.00
MaxInduced3000X5000	11090573	0.15	0.08	0.04	0.03	0.18	11089348	11088467.3	11087531	3320.41	0.01
MaxInduced4000X5000	13496469	0.29	0.20	0.06	0.05	0.36	13496041	13493433.5	13490305	3432.96	0.00
MaxInduced5000X5000	16021337	0.19	0.14	0.08	0.08	0.29	16021100	16019793.0	16018828	5454.68	0.00
BMaxCut1000X5000	6644232	2.98	2.69	2.17	1.20	1.70	6664946	6656808.5	6651058	5514.34	-0.31
BMaxCut2000X5000	10352878	5.39	3.75	3.39	1.80	2.58	10359416	10345767.3	10321518	5826.62	-0.06
BMaxCut3000X5000	13988920	3.49	2.69	1.99	1.81	3.45	14024740	13984576.3	13933552	7914.67	-0.26
BMaxCut4000X5000	17090794	4.36	3.34	3.31	2.31	4.28	17056434	17019868.3	16953064	8307.92	0.20
BMaxCut5000X5000	20134370	3.15	2.49	2.34	1.79	3.90	20233688	20111129.5	20032680	8983.47	-0.49
MatrixFactor1000X5000	71485	0.11	0.07	0.00	0.00	0.02	71485	71485.0	71485	239.36	0.00
MatrixFactor2000X5000	108039	0.19	0.16	0.06	0.04	0.09	108044	108034.1	108025	5115.26	0.00
MatrixFactor3000X5000	144255	0.17	0.16	0.14	0.11	0.26	144257	144213.8	144180	3843.12	0.00
MatrixFactor4000X5000	179493	0.26	0.13	0.10	0.10	0.29	179438	179413.5	179382	5681.04	0.03
MatrixFactor5000X5000	211088	0.21	0.16	0.13	0.04	0.33	210998	210961.0	210889	4420.08	0.04
Average		1.05	0.72	0.58	0.39	0.97	27547753.04	27535659.77	27523515.48	3642.58	-0.03

From these tables we can conclude the following:

- regarding the medium size instances, our Tabu search reach the best known solution for 23 out of 25 instances (CMCS reaches best known for 18 instances, while TS [4] reaches best known for 10 instances);
- for 21 instances out of 25 instances, our Tabu search reaches the best known solution in all ten executions;
- average percentage deviation of the best solution obtained by our Tabu search is 0.01% (while the average percentage deviation for CMCS is 0.05% and the average percentage deviation for TS [4] is 0.41%);
- regarding the large instances, for 8 out of 25 instances the new best known solutions have been found;
- average percentage deviation of the best solution obtained by our Tabu search is -0.03% (while the average percentage deviation for CMCS is 0.39% and the average percentage deviation for TS [4] is 0.97%).

4 Concluding Remarks

In this paper, we present a new simple Tabu search heuristic for solving Quadratic Bipartite Programming Problem (BQP). In designing a heuristic, we follow the recent Less is more approach (LIMA) [3,11,13], whose basic idea is in forcing simplicity and user-friendliness, i.e., use of the minimum number of ingredients, but in a way to outperform the current state-of-the-art heuristics. Our method uses two neighborhood structures, and thus may be seen as a hybrid between TS and Variable neighborhood search (VNS) [8]. Proposed heuristic uses only the short-term memory mechanism and only one parameter, i.e., the tabu list length. Moreover, the frequency-based memory and long-term memory (diversification) are not used at all.

Despite its simplicity, our Tabu search based hybrid method outperforms all existing methods for solving the BQP problem. When compared with a few most successful heuristics from the literature on commonly used test instances, average results reported by our method had the smallest deviation from the best known values. Moreover, we reported 6 new best known solutions on large test instances.

Future research may include the following: (i) proposition of new neighborhood structures; (ii) changing the concept of tabu list by modifying the rule for calculation of the tabu list length, (iii) suggestion of a nested VNS strategy within TS, without increasing the number of parameters, etc.

Acknowledgements. The research has been supported in part by Research Grants 174010 and III 044006 of the Serbian Ministry of Education, Science and Technological Development. The research is also partially covered by the framework of the Grant Number BR05236839 "Development of information technologies and systems for stimulation of personality's sustainable development as one of the bases of development of digital Kazakhstan". The research is partly supported by the Ministry of Education and Science, Republic of Kazakhstan (Institute of Information and Computer Technologies), project no. AP05133090.

References

1. Alon, N., Naor, A.: Approximating the cut-norm via grothendieck's inequality. SIAM J. Comput. **35**(4), 787–803 (2006)
2. Ambühl, C., Mastrolilli, M., Svensson, O.: Inapproximability results for maximum edge biclique, minimum linear arrangement, and sparsest cut. SIAM J. Comput. **40**(2), 567–596 (2011)
3. Costa, L.R., Aloise, D., Mladenovic, N.: Less is more: basic variable neighborhood search heuristic for balanced minimum sum-of-squares clustering. Inf. Sci. **415**, 247–253 (2017). https://doi.org/10.1016/j.ins.2017.06.019. http://www.sciencedirect.com/science/article/pii/S0020025517307934
4. Duarte, A., Laguna, M., Martí, R., Sánchez-Oro, J.: Optimization procedures for the bipartite unconstrained 0–1 quadratic programming problem. Comput. Oper. Res. **51**, 123–129 (2014)
5. Galinier, P., Boujbel, Z., Fernandes, M.C.: An efficient memetic algorithm for the graph partitioning problem. Ann. Oper. Res. **19**(1), 1–22 (2011)
6. Gillis, N., Glineur, F.: Low-rank matrix approximation with weights or missing data is np-hard. SIAM J. Matrix Anal. Appl. **32**(4), 1149–1165 (2011)
7. Glover, F., Ye, T., Punnen, A.P., Kochenberger, G.: Integrating tabu search and vlsn search to develop enhanced algorithms: a case study using bipartite boolean quadratic programs. Eur. J. Oper. Res. **241**(3), 697–707 (2015)
8. Hansen, P., Mladenovic, N., Todosijevic, R., Hanafi, S.: Variable neighborhood search: basics and variants. EURO J. Comput. Optim. **5**(3), 423–454 (2017)
9. Karapetyan, D., Punnen, A.P.: Heuristic algorithms for the bipartite unconstrained 0–1 quadratic programming problem. arXiv preprint arXiv:1210.3684 (2012)
10. Karapetyan, D., Punnen, A.P., Parkes, A.J.: Markov chain methods for the bipartite boolean quadratic programming problem. Eur. J. Oper. Res. **260**(2), 494–506 (2017)
11. Mladenovic, N., Todosijevic, R., Urosevic, D.: Less is more: basic variable neighborhood search for minimum differential dispersion problem. Inf. Sci. **326**, 160–171 (2016)
12. Punnen, A.P., Sripratak, P., Karapetyan, D.: The bipartite unconstrained 0–1 quadratic programming problem: polynomially solvable cases. Discrete Applied Mathematics **193**(Supplement C), 1–10 (2015)
13. Silva, K., Aloise, D., de Souza, S.X., Mladenovic, N.: Less is more: simplified Nelder-Mead method for large unconstrained optimization. Yugoslav J. Oper. Res. **28**(2), 153–169 (2018). http://yujor.fon.bg.ac.rs/index.php/yujor/article/view/609

Black-Box Optimization in an Extended Search Space for SAT Solving

Oleg Zaikin(✉) and Stepan Kochemazov

ISDCT SB RAS, Irkutsk, Russia
zaikin.icc@gmail.com, veinamond@gmail.com

Abstract. The Divide-and-Conquer approach is often used to solve hard instances of the Boolean satisfiability problem (SAT). In particular, it implies splitting an original SAT instance into a series of simpler subproblems. If this split satisfies certain conditions, then it is possible to use a stochastic pseudo-Boolean black-box function to estimate the time required for solving an original SAT instance with the chosen decomposition. One can use black-box optimization methods to minimize the function over the space of all possible decompositions. In the present study, we make use of peculiar features which stem from the NP-completeness of the Boolean satisfiability problem to improve this general approach. In particular, we show that the search space over which the black-box function is minimized can be extended by adding solver parameters and the SAT encoding parameters into it. In the computational experiments, we use the SMAC algorithm to optimize such black-box functions for hard SAT instances encoding the problems of cryptanalysis of several stream ciphers. The results show that the proposed approach outperforms the competition.

Keywords: Black-box optimization · Discrete optimization · Monte Carlo method · SAT · Cryptanalysis

1 Introduction

In multidisciplinary research, it is often necessary to optimize a function for which there is no reliable information on its derivative. Moreover, in many cases a function is not even defined analytically. These features make it impossible to tackle such optimization problems using traditional methods. This fact led to the development of derivative-free and black-box optimization methods [2] that seldom require more than being able to compute the value of a function in a point of the search space, thus treating the function as a black box.

One particular class of optimization problems which can be dealt with only using black-box optimization methods contains many problems related to the algorithms for solving some NP-complete problem. Such an algorithm usually implements exhaustive search, heavily modified by various heuristics. Because of its heuristic nature and the NP-completeness of the underlying problem,

M. Khachay et al. (Eds.): MOTOR 2019, LNCS 11548, pp. 402–417, 2019.
https://doi.org/10.1007/978-3-030-22629-9_28

the only arguments to be made regarding the behaviour of such algorithm on any input rely on treating it like a black box.

In the present paper, we study the applications of black-box optimization methods to the Boolean satisfiability problem (SAT) [4]. In particular, we consider them in the context of the divide-and-conquer SAT solving approach. As it follows from the name, it implies that an original SAT instance is split into disjoint simpler subproblems that can be solved independently. If the original problem has a small number of solutions (or none) and also if all subproblems constructed during "division" have more or less the same complexity, then it is possible to use the Monte-Carlo method [27] to estimate how long it will take to solve the original problem. For this purpose a small sample of subproblems is processed and the resulting time is scaled to the total number of subproblems. In [32] it was proposed, that the division method that consists in choosing a subset of variables in a SAT instance and varying all possible assignments of their values, satisfies the aforementioned criteria. It means that for a given SAT instance and a fixed algorithm for solving SAT the set of variables for splitting formula into subproblems can be viewed as an input to a stochastic discrete black-box function computing the estimation of the runtime required for solving the original problem. Thus, one can attempt to optimize this function over the space of all possible subsets of variables in order to find the one that yields minimal runtime estimation.

On the one hand, the NP-completeness of SAT forces one to use heuristic algorithms for its solving and gives no guarantees that a particular SAT instance will be solved in a reasonable time. On the other hand, in the context of black-box optimization it may open new horizons. Indeed, the heuristic parameters of SAT-solving algorithms can naturally be used to extend the search space for the black-box function considered. Another prominent feature of SAT that distinguishes it from the usual cases is that any original problem can be translated into SAT form by a multitude of ways. Due to the heuristic nature of SAT solving algorithms, they all are worth being considered. Thus, the encoding parameters can also be used to extend the search space in question. We investigate this direction of research and show that taking these parameters into consideration makes it possible to obtain much better runtime estimations. In our experiments, we use the Sequential Model-based optimization Algorithm Configuration [18]. As the test cases, we consider the SAT instances encoding the cryptanalysis of the Alternating step and Bivium keystream generators.

The paper has the following structure. In the next section, the Divide-and-Conquer approach for solving SAT is described. In Sect. 3 the problem of finding a good Divide-and-Conquer based decomposition is considered as a stochastic black-box optimization problem. In particular, in this section a new stochastic discrete black-box objective function is proposed, which operates in an extended search space. In Sect. 4, the certain extended search space of the proposed type is described. In Sect. 5, the computational study of the proposed objective function in the extended search space is presented. Finally, the related work is discussed and conclusions are drawn.

2 Background and Notation

The Boolean satisfiability problem (SAT) is without exaggeration one of the most well-studied combinatorial problems. In the general formulation as a decision problem it consists in answering the question whether an arbitrary Boolean formula is satisfiable, i.e. if there exists such an assignment of all Boolean variables in this formula that it takes the value of True over them. In practice, it is typically considered for the Boolean formulas in Conjunctive Normal Form (CNF). To put it more formally, assume that C is a Boolean formula in CNF over a set of n Boolean variables X, $x_i \in \{0, 1\}$, $x_i \in X$, $i = 1, \ldots, n$. If the formula is satisfiable – it means that there exists such an assignment $\alpha \in \{0, 1\}^n$, that once each variable x_i is set to value α_i, the formula is evaluated as True. If there is no such assignment, then the formula is called unsatisfiable.

Since SAT is an NP-complete problem, it means that a lot of hard practical problems can actually be reduced to it. Their hardness does not magically disappear during this reduction, but makes it possible to apply to these problems the well-developed apparatus for solving SAT. In the recent two decades, there have been achieved a tremendous progress in this area. One of the directions in SAT solving is aimed at parallel SAT solving. It is of particular importance for hard SAT instances. There are two classes of parallel SAT solving techniques [12]: Portfolio (competition-based) and Divide-and-Conquer (cooperation-based). According to the Portfolio approach [15], many different sequential SAT solvers solve the same original SAT instance simultaneously, until any of them finds a satisfying assignment, or proves that the given formula is unsatisfiable. In the Divide-and-Conquer approach, the original instance is decomposed into a family of simpler subproblems that are solved separately by sequential solvers. One of the relative advantages of the Divide-and-Conquer approach consists in the fact that since it deals with simplified subproblems, it is more convenient from the general point of view, for example it can easily be scaled to any amount of computational resources by splitting more or less, the subproblems can be processed independently, it does not require one to wait for large amounts of time like months or years, etc.

There exist several Divide-and-Conquer decomposition techniques, e.g., scattering [20] or Cube-and-Conquer [17]. In the present study, we are interested mostly in such variants of Divide-and-Conquer solving, that make it reasonable to assume that all the subproblems will be solved in more or less the same time. In particular, we employ the following decomposition. As before, assume that C denotes a Boolean formula in CNF and X is a set of its Boolean variables. Assume that a subset S of variables from X is given, $|S| = k, k < n$. A simplified formula produced by assigning values $\alpha = (\alpha_1, \ldots, \alpha_k)$ to variables from S is denoted as $C[\alpha/S]$. A set of simplified formulas produced by instantiating all possible different assignments of values to variables from S in C is called a *decomposition* of an original formula C and is denoted as $D_S[C] = \{C[\alpha/S], \alpha \in \{0, 1\}^{|S|}\}$. Note, that $|D_S[C]| = 2^k$. Since all subproblems in the decomposition differ only in values of several variables it is reasonable to assume that all subproblems represent similarly weakened variants of the original one. The SAT instances forming

D_S can be processed by independent SAT solvers in parallel. Note, that if the original formula is unsatisfiable, then the unsatisfiability of all simplified formulas must be proven. Otherwise once the satisfying assignment is found for any subproblem, the processing of all the remaining subproblems can be interrupted.

Note that finding a good set S to decompose a problem is actually a very hard task. For example, if a SAT instance encodes a cryptanalysis instance then finding such set is (with some reservations) equal to a construction of a guess-and-determine attack on the underlying cipher [3]. For some SAT instances the choice of S may be prompted by the structure of the original problem (see, e.g., [33,40]). For others, it might not matter much which variables are put into S because they are equally inter-dependant on each other. In [10,32,35] there was proposed a general approach to finding S which essentially involved using a special black-box objective function for this purpose. It will be discussed in detail in the following section.

3 Optimization of Black-Box Functions for Divide-and-Conquer SAT Solving

Assume that a SAT instance C over a set of Boolean variables $X, |X| = n$, a SAT solver A and an integer $N, N \ll 2^n$ are given. As an input the function F takes a set $S \subseteq X$. The value of the function is calculated using the Monte Carlo method [27] as follows. First, a random sample $R = \{C[\beta_1/S], \ldots, C[\beta_N/S]\}$ of size N is constructed by choosing randomly $\{\beta_1, \ldots, \beta_N\}$ from $\{0,1\}^{|S|}$. Then the solver A is launched on each subproblem from R. The notation $T_A(C[\beta_i/S])$ stands for the runtime of A on the subproblem $C[\beta_i/S]$. Finally, the value of F is the estimation of the time it would take to solve all subproblems from $D_S[C]$ by the solver A.

$$F_{N,C,A}(S) = 2^{|S|} \times \frac{1}{N} \times \sum_{i=1}^{N} T_A(C[\beta_i/S]). \quad (1)$$

Function (1) is a *stochastic* function, because the Monte Carlo method is used for its calculation. The function's input S is a subset of $X, |X| = n$. Any possible input S can be represented as a Boolean vector of size n, where the $i - th$ element is 1 if $i \in S$, and 0, otherwise. It means that the function (1) maps B^n onto $\mathbb{R}$, so (1) is a *pseudo-Boolean* function. It is also a *black-box* function, because its analytic form is unknown. This function is *costly*, because the experiment involving solving N SAT instances to construct the Monte Carlo estimation requires a lot of computational resources for large N. Thus, (1) is a stochastic costly pseudo-Boolean black-box function.

Now let us assume that P_A is the set of discrete parameters for algorithm A and P_C is the set of discrete parameters of the encoding used to transform an original instance into SAT instance C. Without the loss of generality, assume that p_A and p_C denote assignments of parameters from P_A and P_C, respectively. $C(p_C)$ stands for a SAT instance constructed with parameter values specified by p_C, $A(p_A)$ is the solver A with parameters values p_A. Then the corresponding stochastic costly discrete black-box function over the extended set of parameters will look as follows:

$$G_N(p_C, p_A, S) = 2^{|S|} \times \frac{1}{N} \times \sum_{i=1}^{N} T_{A(p_A)}(C(p_C)[\beta_i/S]). \quad (2)$$

Note, that pseudo-Boolean function (1) can be considered as a special case of discrete function (2) with constant p_A and p_C. Usually in the role of p_A for minimizing (1) the default parameters of A are used.

Also, it is possible to consider one more special case of (2), when it is intended to find good values of P_A and P_C with constant S. The corresponding stochastic costly discrete black-box function is presented below.

$$H_{N,S}(p_C, p_A) = 2^{|S|} \times \frac{1}{N} \times \sum_{i=1}^{N} T_{A(p_A)}(C(p_C)[\beta_i/S]). \quad (3)$$

In the computational experiments (see Sect. 5), the following optimization strategies were employed.

1. Standard: minimize $F_{N,C,A}$ over possible $S \in 2^n$;
2. Extended: minimize G_N over possible $\{S, p_C, p_A\}$, $S \in 2^n$, $p_C \in P_C$, $p_A \in P_A$;
3. Combined: minimize $F_{N,C,A}$ over possible $S \in 2^n$, then use the found S_{best} as a constant to minimize $H_{N,S}$ over possible $\{p_C, p_A\}$, $p_C \in P_C$, $p_A \in P_A$.

In the next section, certain values of parameters P_A and P_C, used to optimize (2) and (3), are described. In Sect. 5, the proposed optimization strategies are compared on hard optimization problems.

4 Description of the Extended Search Space

As it was noted above, the extension of the parameter space which makes it possible to introduce minimization problems for functions (2) and (3) is possible because at their core lies the solving of SAT instances using some SAT solver. Since SAT is an NP-complete problem and thus is solved mainly via heuristic algorithms, there are a lot of possibilities for additional parameters. The particular cases of encoding of an original problem into SAT and tuning SAT solvers' parameters are detailed below.

4.1 SAT Encodings Parameters

The peculiar feature of NP problems consists in the fact that there can be several equivalent formulations of the same problem instance. Here, there are only reasonable limits on their number since essentially, one can introduce nonsensical changes that formally result in a different SAT instance while changing nothing essential.

Let us consider an example. Assume that we have the following Boolean formula:

$$x_1 \oplus x_2 \oplus x_3 \oplus x_4 \tag{4}$$

and we are interested in finding if it is satisfiable or not. In order to apply modern SAT solvers to it, we need to transform it into Conjunctive Normal Form. CNF is a conjunction of *clauses*, where a *clause* is a disjunction of literals, and *literals* are formulas of the kind $\in \{x, \neg x\}$ (x is a Boolean variable). It can be done directly by constructing a truth table for (4) and writing CNF directly based on it. The result will look like this:

$$\begin{array}{rl} & (\neg x_1 \vee \neg x_2 \vee \neg x_3 \vee \neg x_4) \wedge (x_1 \vee x_2 \vee x_3 \vee x_4) \\ \wedge & (x_1 \vee x_2 \vee \neg x_3 \vee \neg x_4) \wedge (x_1 \vee \neg x_2 \vee x_3 \vee \neg x_4) \\ \wedge & (x_1 \vee \neg x_2 \vee \neg x_3 \vee x_4) \wedge (\neg x_1 \vee x_2 \vee x_3 \vee \neg x_4) \\ \wedge & (\neg x_1 \vee x_2 \vee \neg x_3 \vee x_4) \wedge (\neg x_1 \vee \neg x_2 \vee x_3 \vee x_4) \end{array} \tag{5}$$

Meanwhile, it is possible to introduce two new auxiliary variables x_5 and x_6 and transform (4) into the equivalent Boolean formula that looks as follows:

$$(x_5 \equiv x_1 \oplus x_2) \wedge (x_6 \equiv x_3 \oplus x_4) \wedge (x_5 \oplus x_6)$$

And the resulting CNF:

$$\begin{array}{l} (x_5 \vee x_1 \vee \neg x_2) \wedge (x_5 \vee \neg x_1 \vee x_2) \wedge (\neg x_5 \vee x_1 \vee x_2) \\ \wedge (\neg x_5 \vee \neg x_1 \vee \neg x_2) \wedge (x_6 \vee x_3 \vee \neg x_4) \wedge (x_6 \vee \neg x_3 \vee x_4) \\ \wedge (\neg x_6 \vee x_3 \vee x_4) \wedge (\neg x_6 \vee \neg x_3 \vee \neg x_4) \wedge (x_5 \vee \neg x_6) \wedge (\neg x_5 \vee x_6) \end{array} \tag{6}$$

Thus, essentially, CNFs (5) and (6) encode the very same original Boolean formula (4). Moreover, it is easy to show, that from any satisfying assignment of one of them it is easy to effectively construct one for another [36]. What is the difference? The CNF (5) has no auxiliary variables and contains less clauses than (6) (8 vs 10). However, the average clause size in (5) is bigger and the number of distinct literals is also bigger (32 literals vs 28). Generally speaking, the only way to answer the most important question: which CNF is better for some SAT solver—is to launch the solver in question on both and wait until both are solved.

The empirical evaluation shows that rational considerations apply to SAT encodings only to a certain extent. Sometimes, the encoding that has lower number of variables, clauses and literals just works worse than the competition. Thus, it is important to be able to construct and test different SAT encodings for hard problems. Note, that the corresponding area concerned with devising new methods for encoding different predicates to SAT is developing quite actively in the recent years [31].

In our experiments we consider the problems of cryptanalysis of the following cryptographic keystream generators: Bivium [7] and the alternating step generator (ASG) [14]. ASG is considered in two variants: ASG-72 and ASG-96, which are described in [40]. Thus, 3 cryptanalysis problems were considered, all in the form of the so-called *known plaintext attack*. Informally, it means that we need to find the a *secret key* of a generator given the keystream sequence it produced from this secret key. Note, that both these generators have already been studied in the context of the SAT-based cryptanalysis [10,32,35,40], but the previous studies did not consider an extended parameter space for the objective function.

The different encodings for the considered problems were constructed in much the same way as it was shown for (5) and (6), however, since the algorithms and the corresponding Boolean formulas are much more complex compared to (4), the differences are more subtle. To construct the SAT encodings for all considered problems we employed the TRANSALG software system [30]. In case of the Bivium keystream generator, there were constructed two distinct encodings, that differ in the way the auxiliary variables are introduced and handled. For the cryptanalysis of ASG we considered three variants of SAT encodings that differ in the size of a keystream fragment: 76, 80, 84 for ASG-72 and 112, 116, 120 for ASG-96. It was shown in [40] that they yield quite different results.

It is clear, that the different encodings for the same problem do not use the same variables, thus making it hard to switch between them while preserving the set S used to decompose the problem, because, for example, the variables $\{x_1, x_5, x_{42}\}$ for one SAT instance may correspond to completely different entities in another SAT instance. However, they have at least two sets of variables with one-to-one correspondence: the input variables, which encode the secret key of analyzed keystream generator, and output variables that contain the values assigned to keystream bits produced by a generator. Since the values of output variables are set (remind, that we want to find the secret key for a given keystream fragment), it means that only input variables remain in common. Thankfully, due to specifics of propositional encoding techniques, they form the so-called *Strong Backdoor set* [39], meaning that once all values of input variables are set, it is possible to determine if the corresponding SAT instance is satisfiable in polynomial time, thus in a way making them the most valuable. The feature of the TRANSALG tool is that it makes the input variables appear before all other, e.g. they are always represented by a set $X^{in} = \{x_1, \ldots, x_k\}$, $k > 0$. Due to the reasons outlined above, we can assume that X^{in} is common for all encodings of each particular problem. Thus in all experiments the set S that is used to decompose a SAT instance can only be a subset of set X^{in}.

4.2 SAT Solver Parameters

The most promising results in the SAT-based cryptanalysis were achieved by SAT solvers based on the Conflict-Driven Clause Learning (CDCL) concept [24]. CDCL performs a depth-first search of the space of partial truth assignments. This search is augmented in several ways. The most important is *Clause learning*. which consists in learning new clauses when the search reaches a conflict state. If a conflict cannot be resolved by backtracking then the formula is unsatisfiable. If all the variables are assigned and no conflict is detected then the formula is satisfiable.

In the role of A (see Sect. 3), a CDCL solver MAPLELCMDISTCHRONOBT [28] was chosen, because it won the Main track at the SAT competition 2018. In Table 1, we show the nine parameters of this solver which were chosen for tuning.

Values of these parameters affect the basic CDCL algorithm, as well as various heuristics added to boost it. The first 7 parameters originally have real value, while the last 2 parameters are integer (and all their possible values are presented in the table). The discretization of the first 7 parameters was performed in order to obtain a set of discrete variables for the search space. Thus, an arbitrary p_A is a 9-element vector with values from Table 1. Based on the performed discretization, there are $5^7 \times 3^2 = 703\,125$ possible assignments of p_A in total.

4.3 Search Spaces Sizes

According to Subsect. 4.1, there is one additional variable which refers to SAT encoding type. In the cases of ASG-72 and ASG-96, this variable has 3 values, while for Bivium it has two values. According to Subsect. 4.2, there are 9 solver's variables with 703 125 values in total.

In Table 2, the size of the search space for each considered pair (optimization strategy, problem) is shown.

Table 1. The parameters (P_A) of the solver MAPLELCMDISTCHRONOBT.

Name	Possible values	Default value
step-size-dec	0.000001; 0.00001; 0.0001; 0.001; 0.01	0.000001
min-step-size	0.01; 0.03; 0.06; 0.1; 0.3	0.06
cla-decay	0.999; 0.95; 0.9; 0.8; 0.5	0.999
var-decay	0.9; 0.8; 0.7; 0.6; 0.5	0.8
step-size	0.1; 0.2; 0.3; 0.4; 0.5	0.4
chrono	−1, 10, 100, 1000, 10000	100
confl-to-chrono	1000, 2000, 4000, 8000, 16000	4000
ccmin-mode	0, 1, 2	2
phase-saving	0, 1, 2	2

Table 2. Search spaces sizes for each pair (strategy, problem).

	Standard	Extended	Combined
ASG-72	2^{72}	$2^{72} \times 2\,109\,375$	2^{72}, then 2 109 375
ASG-96	2^{96}	$2^{96} \times 2\,109\,375$	2^{96}, then 2 109 375
Bivium	2^{177}	$2^{177} \times 1\,406\,250$	2^{177}, then 1 406 250

5 Computational Experiments

To solve three black-box optimization problems, described in Sect. 4, three optimization strategies from Sect. 3 were used: Standard, Extended and Combined. Let us remind, that the Standard strategy have been already used in previous studies, while to the best of our knowledge the Extended and Combined strategies are proposed in this paper.

To minimize the objective functions (1), (2), (3), the SMAC (sequential model-based algorithm configuration) [18] tool was chosen. SMAC is an implementation of the Sequential Model-based optimization (SMBO) framework [21]. According to SMBO, a regression model is constructed, that predicts values of an objective function and as a result recommends in which points an objective function should be calculated. SMAC's prediction model is based on the *random forest* machine learning algorithm [6]. SMAC can be used for optimizing costly discrete black-box functions. The additional reason why SMAC was chosen is that it has been widely used for tuning SAT solving algorithms (see [11,19]). Note, that in two optimization strategies out of three (Combined and Extended) it is in fact required to solve a similar problem – to tune SAT solver's parameters.

The objective functions were implemented on the basis of the ALIAS tool [22]. This tool was configured to work with the MapleLCMDistChronoBT sequential SAT solver [28] which was briefly described in Subsect. 4.2.

All the experiments were conducted on the "Academician V.M. Matrosov" computing cluster [8]. Each computational node of this cluster is equipped with 2 × 18-core Intel Xeon E5-2695 CPUs and 128 Gb RAM. In all experiments SMAC operated on 1 computing node. While ALIAS calculates the Monte Carlo estimation in the multi-threaded mode, at each moment of time SMAC operates with exactly one point from a search space.

In all experiments N was equal to 1000, so the SAT solver needed to solve 1000 simplified versions of an original instance to obtain a Monte-Carlo-based objective function value at a point. Since SAT solvers are essentially heuristic algorithms, they can work for very large amounts of time during the calculation of the objective function value at some point. That is why the time limit on the SAT solver runtime was introduced. If for any subproblem from a random sample the runtime of SAT solver exceeded the imposed time limit, the processing of the sample was interrupted with the objective function value set to $1e100$. Two values of time limit were tried: 5 s and 10 s. It turned out, that this parameter is quite important.

Since all the objective functions are stochastic (see Sect. 3), SMAC was launched 3 times on each configuration (problem, optimization strategy, time limit) on one cluster's computational node (i.e. on 36 CPU cores) to alleviate the effect of randomness. As it is recommended by SMAC's developers, each launch was for one day. Thus, $3 \times 3 \times 2 \times 3 = 54$ 1-day launches were performed in total. Note, that in the case of the Combined strategy, the function (1) was first minimized for 12 h, then the point with the best found estimation was used as a starting point for minimizing function (3) for another 12 h. The obtained results are shown in Table 3. The best results for each of three optimization problems are marked with bold.

Table 3. Objective functions values found by different strategies for ASG-72, ASG-96 and Bivium. The best values are marked with bold.

Strategy	Solver limit 10 s			Solver limit 5 s		
	Launch 1	Launch 2	Launch 3	Launch 1	Launch 2	Launch 3
ASG-72						
Standard	3.42e+7	1.49e+7	2.2e+7	7.47e+6	**2.26e+6**	7.62e+6
Extended	3.6e+7	3.14e+7	9.19e+6	4.99e+7	2.23e+7	1.53e+7
Combined	3.8e+6	3.6e+6	1.98e+7	1.42e+7	1.34e+7	4.01e+6
ASG-96						
Standard	1.73e+11	3.19e+10	4.82e+11	3.07e+11	4.29e+11	1.41e+12
Extended	**7.26e+9**	1.52e+11	2.38e+11	2.6e+11	7.8e+11	5.95e+11
Combined	1.33e+10	3.23e+11	7.99e+10	1.07e+11	6.36e+11	4.85e+11
Bivium						
Standard	3.27e+19	2.95e+19	3.49e+19	2.4e+19	9.18e+18	1.1e+19
Extended	1.14e+19	9.25e+18	7.99e+18	7.72e+18	**4.27e+18**	8.88e+18
Combined	4.3e+19	1.59e+19	5.74e+19	1.82e+19	4.46e+19	1.55e+19

Results from Table 3 are also shown in Figs. 1, 2 and 3. The best results for each pair (problem, solver time limit) are marked with a horizontal gold dotted line.

In Table 4 the number of calculations of the objective functions performed by SMAC is shown. For the Combined strategy, the numbers for both stages are presented.

Note, that while 24 h on 36 CPU cores were given for SMAC in all launches (that is similar to 864 h on 1 CPU core), the number of calculations did not exceed 5000 in each launch. To perform one calculation of an objective function, it took about 10–40 m (on 1 CPU core) on average. Thus, all three objective functions are extremely costly.

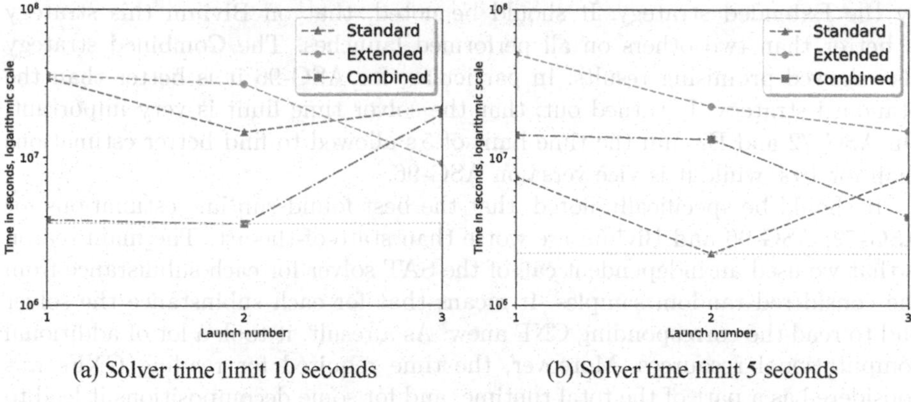

(a) Solver time limit 10 seconds (b) Solver time limit 5 seconds

Fig. 1. Results of 3 launches of each strategy on ASG-72.

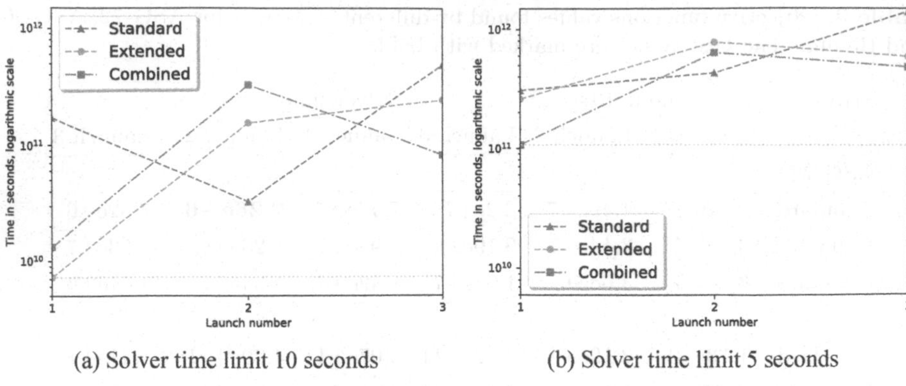

(a) Solver time limit 10 seconds (b) Solver time limit 5 seconds

Fig. 2. Results of 3 launches of each strategy on ASG-96.

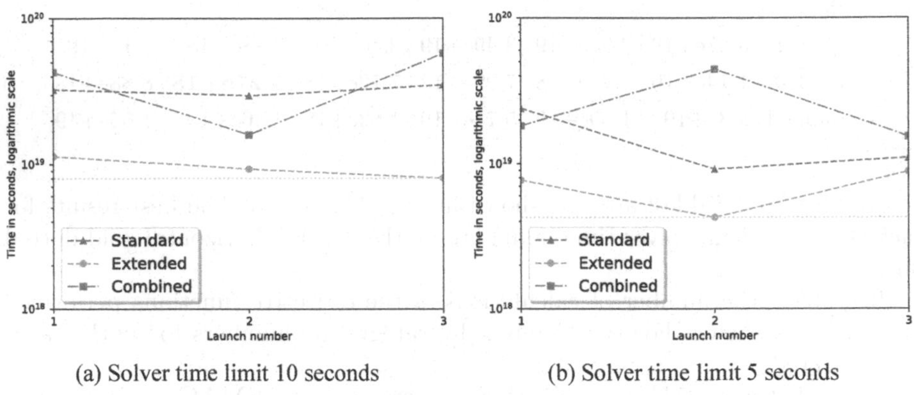

(a) Solver time limit 10 seconds (b) Solver time limit 5 seconds

Fig. 3. Results of 3 launches of each strategy on Bivium.

According to the results, on ASG-72 the Standard strategy found the best runtime estimation. As for ASG-96 and Bivium, the best results were achieved by the Extended strategy. It should be noted, that on Bivium this strategy is better than two others on all performed launches. The Combined strategy also showed promising results. In particular, for ASG-96 it is better than the Standard strategy. It turned out, that the solver time limit is very important. On ASG-72 and Bivium the time limit of 5 s allowed to find better estimations than for 10 s, while it is vice versa on ASG-96.

It should be specifically noted, that the best found runtime estimations for ASG-72, ASG-96 and Bivium are worse than state-of-the-art. The main reason is that we used an independent call of the SAT solver for each subinstance from the considered random samples. It means that for each subinstance the solver had to read the corresponding CNF anew. As a result, it took a lot of additional computational resources. Moreover, the time required for reading CNFs was considered as a part of the total runtime, and for some decompositions it lead to

Table 4. The number of calculations of the objective functions performed by SMAC. For the Combined strategy, the numbers of two stages are divided by "+".

Strategy	Solver limit 10 s			Solver limit 5 s		
	Launch 1	Launch 2	Launch 3	Launch 1	Launch 2	Launch 3
ASG-72						
Standard	4147	4199	4183	1343	1612	1635
Extended	2544	2513	1549	2236	1765	2241
Combined	969 + 780	728 + 555	729 + 557	969 + 1630	728 + 3461	729 + 794
ASG-96						
Standard	2342	2331	2351	2502	2539	2534
Extended	2111	3120	3121	2777	2813	2877
Combined	671 + 460	1102 + 815	1101 + 880	1185 + 1786	1256 + 1831	1237 + 1775
Bivium						
Standard	2173	2229	2234	2902	2903	2884
Extended	2186	2190	2189	4155	4183	4204
Combined	1125 + 829	1127 + 818	1128 + 815	1124 + 3424	1116 + 2523	941 + 2666

overly pessimistic estimations. However, the goal of this study was to compare several search spaces in the similar conditions, and this goal was achieved. In the nearest future, we are planning to improve the proposed approach by handling the mentioned drawback.

6 Discussion

The present study is a step forward in solving hard SAT instances by black-box optimization methods. The related works are discussed below.

The local search metaheuristics have been widely used for cryptanalysis. In [23] the cryptanalysis of several polyalphabetic ciphers was performed using the hill climbing and simulated annealing algorithms. In [5], an algebraic cryptanalysis of the Trivium stream cipher was considered as an optimization problem, which was also studied by hill climbing and simulated annealing. In [37], Mixed Integer Linear Programming was used to find guessing strategies for algebraic cryptanalysis of the block cipher EPCBC.

In [25], a stochastic local search algorithm was used to solve hard cryptanalysis problems in the SAT form. In [32] the SAT-based cryptanalysis of the stream ciphers A5/1, Bivium and Grain was considered as an optimization problem, which in turn was solved by a tabu search algorithm. The objective function minimized in that paper is essentially the objective function (1), described in Sect. 3 in detail. In [34], a completely different objective function combined with the similar optimization algorithm were used to analyze the Magma and AES-128 block ciphers and the Trivium stream cipher. Implicitly, the objective function (1) was also employed to analyze the block cipher GOST and several keystream generators (Crypto-1, Hitag2, A5/1, Bivium) [9,10,26,33,35], but in these papers the sets for decomposing an original SAT instance were constructed manually.

To the best of our knowledge one of the first examples of the study of different SAT encodings for cryptanalysis instances are the papers [29,40]. In [40], the alternating step generator and two of its modifications were analyzed using SAT. An objective function based on a modified version of (1) was minimized by a hill climbing algorithm. The master thesis [29] is devoted to the analysis of SAT-based cryptanalysis of the SHA-1 hash function.

Generally speaking, each objective function, considered in this study, can be minimized by any discrete black-box optimization algorithm that suits for costly functions: evolutionary [1] algorithm; genetic algorithm [38]; variable neighborhood search [16]; tabu search [13].

The SMAC tool has been widely used for tuning SAT solving algorithms (see, e.g., [11]). It was used as one of the automated algorithm configuration tools on the Configurable SAT Solver Challenge [19] in 2014, where the solvers were compared by the performance achieved after a fully automated configuration step. SMAC was applied to the problem of finding good decompositions for SAT solving in [22,41].

As far as the authors are aware, the extended search space, that includes not only decomposition set, but also SAT solver and SAT encoding parameters, for optimizing the black-box function that estimates the runtime of divide-and-conquer SAT solving was considered for the first time in the present paper.

7 Conclusions and Future Work

In the present paper, an improvement on the stochastic discrete black-box optimization for SAT solving was made. Two new optimization strategies were proposed, which are based on the usage of an extended search space. Three hard discrete black-box optimization problems were analyzed by these strategies and also by the standard one. For two out of three problems, the proposed strategies showed better results.

In the computational experiments, only SMAC algorithm was used to minimize all three considered objective functions. The main reason is that these functions are extremely costly, and we could afford a meticulous computational study of only one such algorithm. However, in the nearest future, we are planning to try several other black-box optimization algorithms, namely, local search, tabu search, evolutionary algorithm, genetic algorithm.

Acknowledgements. Authors thank Dr. Alexander Semenov for valuable preliminary discussions regarding the SAT algorithms parameter tuning in the context of Divide-and-Conquer solving. We would also like to thank anonymous reviewers for their valuable comments that made it possible to improve the quality of the present paper. The research was partially supported by Council for Grants of the President of the Russian Federation (grant no. MK-4155.2018.9) and by Russian Foundation for Basic Research (grant no. 19-07-00746-a).

References

1. Ashlock, D.: Evolutionary Computation for Modeling and Optimization, 1st edn. Springer, New York (2010)
2. Audet, C., Hare, W.: Derivative-Free and Blackbox Optimization. Springer Series in Operations Research and Financial Engineering. Springer, Cham (2017). https://doi.org/10.1007/978-3-319-68913-5
3. Bard, G.V.: Algebraic Cryptanalysis, 1st edn. Springer, Boston (2009). https://doi.org/10.1007/978-0-387-88757-9
4. Biere, A., Heule, M., van Maaren, H., Walsh, T. (eds.): Handbook of Satisfiability, Frontiers in Artificial Intelligence and Applications, vol. 185. IOS Press, Amsterdam (2009)
5. Borghoff, J., Knudsen, L.R., Matusiewicz, K.: Hill climbing algorithms and Trivium. In: Biryukov, A., Gong, G., Stinson, D.R. (eds.) SAC 2010. LNCS, vol. 6544, pp. 57–73. Springer, Heidelberg (2011). https://doi.org/10.1007/978-3-642-19574-7_4
6. Breiman, L.: Random forests. Mach. Learn. **45**(1), 5–32 (2001)
7. De Cannière, C., Preneel, B.: Trivium. In: Robshaw, M., Billet, O. (eds.) New Stream Cipher Designs - The eSTREAM. LNCS, vol. 4986, pp. 244–266. Springer, Heidelberg (2008). https://doi.org/10.1007/978-3-540-68351-3_18
8. Irkutsk supercomputer center of SB RAS. http://hpc.icc.ru
9. Courtois, N.T., Gawinecki, J.A., Song, G.: Contradiction immunity and guess-then-determine attacks on GOST. Tatra Mt. Math. Publ. **53**(1), 2–13 (2012)
10. Eibach, T., Pilz, E., Völkel, G.: Attacking bivium using SAT solvers. In: Kleine Büning, H., Zhao, X. (eds.) SAT 2008. LNCS, vol. 4996, pp. 63–76. Springer, Heidelberg (2008). https://doi.org/10.1007/978-3-540-79719-7_7
11. Falkner, S., Lindauer, M., Hutter, F.: SpySMAC: automated configuration and performance analysis of SAT solvers. In: Heule, M., Weaver, S. (eds.) SAT 2015. LNCS, vol. 9340, pp. 215–222. Springer, Cham (2015). https://doi.org/10.1007/978-3-319-24318-4_16
12. Le Frioux, L., Baarir, S., Sopena, J., Kordon, F.: PaInleSS: a framework for parallel SAT solving. In: Gaspers, S., Walsh, T. (eds.) SAT 2017. LNCS, vol. 10491, pp. 233–250. Springer, Cham (2017). https://doi.org/10.1007/978-3-319-66263-3_15
13. Glover, F.: Future paths for integer programming and links to artificial intelligence. Comput. OR **13**(5), 533–549 (1986)
14. Günther, C.G.: Alternating step generators controlled by De Bruijn sequences. In: Chaum, D., Price, W.L. (eds.) EUROCRYPT 1987. LNCS, vol. 304, pp. 5–14. Springer, Heidelberg (1988). https://doi.org/10.1007/3-540-39118-5_2
15. Hamadi, Y., Jabbour, S., Sais, L.: Manysat: a parallel SAT solver. JSAT **6**(4), 245–262 (2009)
16. Hansen, P., Mladenović, N.: Variable neighborhood search: principles and applications. Eur. J. Oper. Res. **130**(3), 449–467 (2001). https://doi.org/10.1016/S0377-2217(00)00100-4
17. Heule, M.J.H., Kullmann, O., Biere, A.: Cube-and-conquer for satisfiability. In: Hamadi, Y., Sais, L. (eds.) Handbook of Parallel Constraint Reasoning, pp. 31–59. Springer, Cham (2018). https://doi.org/10.1007/978-3-319-63516-3_2
18. Hutter, F., Hoos, H.H., Leyton-Brown, K.: Sequential model-based optimization for general algorithm configuration. In: Coello, C.A.C. (ed.) LION 2011. LNCS, vol. 6683, pp. 507–523. Springer, Heidelberg (2011). https://doi.org/10.1007/978-3-642-25566-3_40

19. Hutter, F., Lindauer, M., Balint, A., Bayless, S., Hoos, H., Leyton-Brown, K.: The configurable SAT solver challenge (CSSC). Artif. Intell. **243**, 1–5 (2015). https://doi.org/10.1016/j.artint.2016.09.006
20. Hyvärinen, A.E.J., Junttila, T., Niemelä, I.: A distribution method for solving SAT in grids. In: Biere, A., Gomes, C.P. (eds.) SAT 2006. LNCS, vol. 4121, pp. 430–435. Springer, Heidelberg (2006). https://doi.org/10.1007/11814948_39
21. Jones, D.R., Schonlau, M., Welch, W.J.: Efficient global optimization of expensive black-box functions. J. Glob. Optim. **13**(4), 455–492 (1998)
22. Kochemazov, S., Zaikin, O.: ALIAS: a modular tool for finding backdoors for SAT. In: Beyersdorff, O., Wintersteiger, C.M. (eds.) SAT 2018. LNCS, vol. 10929, pp. 419–427. Springer, Cham (2018). https://doi.org/10.1007/978-3-319-94144-8_25
23. Lasry, G.: A Methodology for the Cryptanalysis of Classical Ciphers with Search Metaheuristics. Ph.D. thesis, University of Kassel, Germany (2018). http://www.upress.uni-kassel.de/katalog/abstract.php?978-3-7376-0458-1
24. Marques-Silva, J.P., Lynce, I., Malik, S.: Conflict-driven clause learning SAT solvers. In: Biere et al. [4], pp. 131–153
25. Massacci, F.: Using Walk-SAT and Rel-SAT for cryptographic key search. In: IJCAI 1999, pp. 290–295 (1999)
26. Mcdonald, C., Charnes, C., Pieprzyk, J.: Attacking Bivium with MiniSat. Technical Report 2007/040, ECRYPT Stream Cipher Project (2007)
27. Metropolis, N., Ulam, S.: The Monte Carlo method. J. Amer. statistical assoc. **44**(247), 335–341 (1949)
28. Nadel, A., Ryvchin, V.: Chronological backtracking. In: Beyersdorff, O., Wintersteiger, C.M. (eds.) SAT 2018. LNCS, vol. 10929, pp. 111–121. Springer, Cham (2018). https://doi.org/10.1007/978-3-319-94144-8_7
29. Nossum, V.: SAT-based preimage attacks on SHA-1. Ph.D. thesis, University of OSLO, Department of Informatics (2012)
30. Otpuschennikov, I., Semenov, A., Gribanova, I., Zaikin, O., Kochemazov, S.: Encoding cryptographic functions to SAT using TRANSALG system. In: ECAI 2016. FAAI, vol. 285, pp. 1594–1595. IOS Press (2016)
31. Prestwich, S.D.: CNF encodings. In: Biere et al. [4], pp. 75–97
32. Semenov, A., Zaikin, O.: Algorithm for finding partitionings of hard variants of boolean satisfiability problem with application to inversion of some cryptographic functions. SpringerPlus **5**(1), 1–16 (2016)
33. Semenov, A., Zaikin, O., Bespalov, D., Posypkin, M.: Parallel logical cryptanalysis of the generator A5/1 in BNB-grid system. In: Malyshkin, V. (ed.) PaCT 2011. LNCS, vol. 6873, pp. 473–483. Springer, Heidelberg (2011). https://doi.org/10.1007/978-3-642-23178-0_43
34. Semenov, A., Zaikin, O., Otpuschennikov, I., Kochemazov, S., Ignatiev, A.: On cryptographic attacks using backdoors for SAT. In: AAAI 2018, pp. 6641–6648 (2018)
35. Soos, M., Nohl, K., Castelluccia, C.: Extending SAT solvers to cryptographic problems. In: Kullmann, O. (ed.) SAT 2009. LNCS, vol. 5584, pp. 244–257. Springer, Heidelberg (2009). https://doi.org/10.1007/978-3-642-02777-2_24
36. Tseitin, G.S.: On the complexity of derivation in propositional calculus. In: Siekmann, J.H., Wrightson, G. (eds.) Automation of Reasoning. Symbolic Computation (Artificial Intelligence), pp. 466–483. Springer, Heidelberg (1983). https://doi.org/10.1007/978-3-642-81955-1_28

37. Walter, M., Bulygin, S., Buchmann, J.: Optimizing guessing strategies for algebraic cryptanalysis with applications to EPCBC. In: Kutyłowski, M., Yung, M. (eds.) Inscrypt 2012. LNCS, vol. 7763, pp. 175–197. Springer, Heidelberg (2013). https://doi.org/10.1007/978-3-642-38519-3_12
38. Whitley, D.: A genetic algorithm tutorial. Stat. Comput. **4**(2), 65–85 (1994). https://doi.org/10.1007/BF00175354
39. Williams, R., Gomes, C.P., Selman, B.: Backdoors to typical case complexity. In: IJCAI 2003, pp. 1173–1178 (2003)
40. Zaikin, O., Kochemazov, S.: An improved SAT-based guess-and-determine attack on the alternating step generator. In: Nguyen, P., Zhou, J. (eds.) ISC 2017. LNCS, vol. 10599, pp. 21–38. Springer, Cham (2017). https://doi.org/10.1007/978-3-319-69659-1_2
41. Zaikin, O., Kochemazov, S.: Pseudo-boolean black-box optimization methods in the context of divide-and-conquer approach to solving hard SAT instances. In: OPTIMA 2018 (Supplementary Volume), pp. 76–87. DEStech Publications, Inc. (2018)

Optimal Control and Approximation

A Control Problem for Parabolic Systems with Incomplete Information

Boris I. Ananyev[1,2(✉)]

[1] N.N. Krasovskii Institute of Mathematics and Mechanics UB of RAS, Kovalevskaya Street 16, Yekaterinburg 620990, Russia
abi@imm.uran.ru

[2] Ural Federal University, Mira 19, Yekaterinburg, Russia

Abstract. In this paper, abstract parabolic control systems in Hilbert space are considered. The state of the system is unknown, but there is an equation of measurement in discrete times. The initial state and disturbances are restricted by joint integral constraints. According to measurements, the information set is introduced that contains the true state of the system. This set includes all the states of the system that are compatible with the measurements. The preliminary aim of control consists in minimization of the terminal criterion depending of the information set. We suggest some statements of the problem based on the separation of control and observation processes. The optimal instants of transition from estimation to control are looked for as well. The approach is applied to distributed systems with partial derivatives and to systems with the deviation of time of retarded and neutral types. The approximation scheme are suggested and examples are considered.

Keywords: Control · Evolutionary systems · Information sets · Incomplete information

1 Introduction and Preliminaries

First of all we indicate that problems of control under incomplete information were investigated in many books and papers [3–8]. The authors use either the stochastic approach [7] or the minimax deterministic one going back to [3] and developed in subsequent works. We keep to the deterministic problem formulation in [3,4]. Similar formulations were used and modified in [9–11]. In this work, we continue and complement [12,13] trying to generalize some results from [14,15] on the case of infinite-dimensional systems. The algorithm of solution is developed and special cases are considered for parabolic and hyperbolic partial differential systems. Examples are examined. We consider also finite dimensional and numerical approximations for the problem.

1.1 Weak Solutions of Evolutionary Systems

Let V, H be two real Hilbert spaces with norms $\|\cdot\|$ and $|\cdot|$ respectively. Suppose that $V \subset H$, V is dense imbedded in H and separable, $|v| \leq \gamma\|v\|$

M. Khachay et al. (Eds.): MOTOR 2019, LNCS 11548, pp. 421–433, 2019.
https://doi.org/10.1007/978-3-030-22629-9_29

for every $v \in V$. The last inequality means that the imbedding V into H is continuous and the dual space V^* contains $H^* = H$. The spaces H and H^* are identified. Let further $a(u, v)$ be a continuous, bilinear and coercive form on V, such that $a(v, v) \geq \alpha\|v\|^2$, $\forall v \in V$.

Let a function $f : [0, T] \to H$ be measurable and $\int_0^T \|f(t)\|^2 dt < \infty$. For every point $z_0 \in H$ there exists a unique continuous in H function $z(t) \in V$, $t > 0$, such that

$$d\langle z(t), v\rangle/dt + a(z(t), v) = \langle f(t), v\rangle, \quad \forall v \in V, \quad z(0) = z_0. \tag{1}$$

Here $z(t)$ is implicitly supposed to be weakly absolutely continuous (see [1]).

The form $a(u, v)$ defines a linear continuous operator $u \to Au \in V^*$ according to the equality $a(u, v) = \langle Au, v\rangle$. Define by $D(A)$ the set of all elements $h \in V$, for which $Ah \in H \subset V^*$. The operator $-A$ on H is an infinitesimal closed generator for some strongly continuous semigroup $S(t) : H \to H$ (see [1,2]). Besides the solution of (1) has a form

$$z(t) = S(t)z_0 + \int_0^t S(t-s)f(s)ds, \tag{2}$$

where the integral is understood in Bochner's sense [2]. Remark that the solution of (1) may be considered as a generalized solution of Cauchy problem

$$\dot{z} + Az = f(t), \quad z(0) = z_0 \in H. \tag{3}$$

The generalized solution of (3) exists, is unique and may be represented by (2). The solution $z(t)$ is weakly differentiable in H, i.e. the weak limit $\lim_{\delta\to 0}(z(t+\delta) - z(t))/\delta = dz(t)/dt$ there exists a.e. on $[0, T]$ in weak topology of H.

2 The System and Measurements

Consider a controlled system of the form

$$\dot{z} + Az = Bu(t) + C\xi(t), \quad z \in H. \tag{4}$$

Suppose that the operator A is defined by continuous bilinear form $a(u, v)$ given on a separable Hilbert space $V \subset H$; B and C are continuous linear operators from Hilbert spaces H_1 and H_2 to the H, respectively. Let $L_2(0, T, H_i)$ be the Hilbert space of weakly measurable functions $f(t) \in H_i$ such that $\int_0^T \|f(t)\|^2 dt \leq \infty$. According to Subsect. 1.1, an each pair of functions $u(\cdot) \in L_2(0, T; H_1)$ and $\xi(\cdot) \in L_2(0, T; H_2)$ along with an initial state $z_0 \in H$ defines a unique weak solution $z(t; z_0, u, \xi)$ of (4). This solution satisfies the equation

$$d\langle z(t), v\rangle/dt + a(z(t), v) = \langle Bu(t) + C\xi(t), v\rangle, \quad \forall v \in V, \quad z(0) = z_0,$$

and may be represented as

$$z(t) = S(t)z_0 + \int_0^t S(t-s)(Bu(s) + C\xi(s))ds. \tag{5}$$

In what follows the state $z(t)$ of (4) or (5) is unknown. The available information about it may be described as follows. Given a uniform partition $0 = t_0 < t_1 < \cdots < t_N = T$ of $[0,T]$, $t_i - t_{i-1} = T/N = \delta$, at the instants t_i a vector $y_i = Gz(t_{i-1}) + w_i$ is observed, where $G : H \to R^m$ is a finite-dimensional linear operator. Unknown disturbances $\xi(\cdot)$, the initial state z_0, and vectors w_i are restricted by the joint constraint

$$\|z_0\|^2_{P_0} + \int_0^T \|\xi(t)\|^2_Q dt + \sum_{i=1}^N \|w_i\|^2_R \le 1. \tag{6}$$

Here and further we use the notation $\|u\|^2_F = \langle u, Fu\rangle$ for a self-adjoint positive and coercive operator F; $\langle \cdot\,,\cdot\rangle$ is an inner product in the corresponding space. The operators P_0, Q, and the matrix R are supposed to be similar to F. Besides, we have a constraint on the control $u(\cdot)$:

$$\int_0^T \|u(t)\|^2_F dt \le 1. \tag{7}$$

2.1 Transformation to a Discrete-Time System

System (5) with measurements and controls may be represented in a discrete-time form

$$\begin{aligned} z_i &= Sz_{i-1} + \eta_i + \xi_i, \quad \text{where} \quad S = S(\delta), \quad z_i = z(t_i), \\ \eta_i &= \int_{t_{i-1}}^{t_i} S(t_i - s)Bu(s)ds, \quad \xi_i = \int_{t_{i-1}}^{t_i} S(t_i - s)C\xi(s)ds, \\ y_i &= Gz_{i-1} + w_i, \quad i \in 1:N. \end{aligned} \tag{8}$$

Let us derive constraints on parameters in (8). Denote by ξ_i^N the set of elements $\{\xi_i, \dots, \xi_N\}$. The symbol w_i^N has the same meaning. If $i = 1$, we write $\xi^N = \xi_1^N$. Find first the support function (see, for example, [6]) of all the parameters $\{z_0, \xi^N, w^N\}$ according to constraints (6). Let $\chi_A(s)$ be a characteristic function. We have

$$\begin{aligned} &\max_{z_0,\xi(\cdot),w^N} \left\{ \langle k, z_0\rangle + \sum_{i=1}^N \Big(\langle l_i, \xi_i\rangle + \langle m_i, w_i\rangle\Big)\right\} \\ &= \max_{z_0,\xi(\cdot),w^N} \left\{ \langle k, z_0\rangle + \int_0^T \left\langle \sum_{i=1}^N \chi_{[t_{i-1},t_i]}(s) C^* S^*(t_i - s) l_i, \xi(s) \right\rangle ds + \sum_{i=1}^N \langle m_i, w_i\rangle \right\} \\ &= \sqrt{\langle k, P_0^{-1}k\rangle + \sum_{i=1}^N \Big(\langle l_i, \mathbf{C} l_i\rangle + \langle m_i, R^{-1} m_i\rangle\Big)}, \end{aligned}$$

where the self-adjoint positive operator $\mathbf{C}$ is defined as

$$\mathbf{C}l = \int_0^\delta S(\delta - s)CQ^{-1}C^*S^*(\delta - s)lds$$
$$= \int_{t_{i-1}}^{t_i} S(t_i - s)CQ^{-1}C^*S^*(t_i - s)lds.$$

that does not depend on i. Doing the same with the control, we obtain

$$\max_{u(\cdot)}\left\{\sum_{i=1}^N \langle l_i, \eta_i\rangle\right\} = \max_{u(\cdot)}\left\{\int_0^T \left\langle \sum_{i=1}^N \chi_{[t_{i-1},t_i]}(s)B^*S^*(t_i - s)l_i, u(s)\right\rangle ds\right\}$$
$$= \sqrt{\sum_{i=1}^N \left(\langle l_i, \mathbf{B}l_i\rangle\right)},$$

where the self-adjoint positive operator $\mathbf{B}$ is defined as

$$\mathbf{B}l = \int_0^\delta S(\delta - s)BF^{-1}B^*S^*(\delta - s)lds. \tag{9}$$

Now defining $\mathbb{B} = \mathbf{B}^{1/2}$ and $\mathbb{C} = \mathbf{C}^{1/2}$ we come to the conclusion.

Lemma 1. *The discrete-time system* (8) *with constraints* (6), (7) *is equivalent to the system*

$$z_i = Sz_{i-1} + \mathbb{B}u_i + \mathbb{C}v_i, \quad \textit{with constraints} \tag{10}$$
$$\sum_{i=1}^N \|u_i\|^2 \le 1, \quad \|z_0\|^2_{P_0} + \sum_{i=1}^N \left(\|v_i\|^2 + \|w_i\|^2_R\right) \le 1,$$
$$y_i = G_i z_{i-1} + w_i, \quad i \in 1 : N.$$

Proof. It follows from the fact that the support functions of the sets $\{\mathbb{B}u^N\}$ and $\{z_0, \mathbb{C}v^N, w^N\}$ coincide with functions found above. □

Note that the states z_i of system (10) are not the approximations of $z(t_i)$. We have the equality $z_i = z(t_i)$ under some parameters in the systems.

3 Estimation for Discrete-Time Evolutionary Systems

For system (10) the *information set* $\mathcal{Z}_j(y, u)$ (see [4]) is defined as follows.

Definition 1. The set $\mathcal{Z}_j(y, u) \subset H$ is said to be *informational* if it consists of all vectors z_j for which there exist elements z_0, v_i, w_i, such that Eq. (10) are fulfilled for all $i \in 1 : j$, constraints in (10) hold, and measurements $y_i = Gz_{i-1} + w_i$ are valid for all $i \in 1 : j$.

Introduce the linear operator $\mathbb{S}(z, v) = Sz + \mathbb{C}v$. The representation of $\mathcal{Z}_i(y, u)$ is given by

Theorem 1. *The information set is the ellipsoid* $\mathcal{Z}_i(y,u) = \{z : \|z-\hat{z}_i\|^2_{P_i} + h_i \le 1\}$ *with parameters given by the formulas*

$$P_i^{-1} = SJ_i^{-1}S^* + \mathbf{C}, \quad J_i = P_{i-1} + G^*RG, \qquad (11)$$
$$\hat{z}_i = \mathbb{B}u_i + S\check{z}_i, \quad \check{z}_i = \hat{z}_{i-1} + J_i^{-1}G^*R(y_i - G\hat{z}_{i-1}),$$
$$\hat{z}_0 = 0, \quad h_i = h_{i-1} + \|y_i - G\hat{z}_{i-1}\|^2_{\mathcal{G}_i},$$
$$h_0 = 0, \quad \mathcal{G}_i^{-1} = GP_{i-1}^{-1}G^* + R^{-1}.$$

The sum $\|z-\hat{z}_i\|^2_{P_i} + h_i$ *is a minimum of relation* $\|z_0\|^2_{P_0} + \sum_{j=1}^i (\|v_j\|^2 + \|w_j\|^2_R)$ *under the assumption that parameters* z_0, v_j, w_j *submit the boundary condition* $z_i = z$ *due to Eq.* (10).

Proof. Theorem 1 may be proved by induction. Let $u_i = 0$ and $F_i(z,v) = \|v\|^2 + \|y_i - Gz\|^2_R$. Introduce some axillary sets and functions:

$$\mathcal{V}_i(y) = \{(z,v) \in H \times H : V_{i-1}(z) + F_i(z,v) \le 1\},$$
$$\mathcal{Z}_i(y) = \mathbb{S}\mathcal{V}_i(y), \quad V_0(z) = \|z\|^2_{P_0}, \quad i \in 1:N,$$
$$V_i(z_i) = \begin{cases} \min\limits_{(z,v)\in\mathcal{V}_i(y)} \{V_{i-1}(z) + F_i(z,v) : z_i = \mathbb{S}(z,v)\}, & z_i \in \mathcal{Z}_i(y), \\ 2, & z_i \notin \mathcal{Z}_i(y). \end{cases} \qquad (12)$$

The set $\mathcal{V}_i(y)$ is said to be *compartible with signal* at the instant i, the set $\mathcal{Z}_i(y)$ is *informational* at the instant i. So, the sets $\mathcal{Z}_i(y)$ are images of $\mathcal{V}_i(y)$ according to (10). Let the signal y^N be realized under the elements z_0^*, v_i^*, w_i^*, $i \in 1:N$. Then the constraints in (10) are fulfilled with these elements. We assert that sets $\mathcal{V}_i(y)$ and $\mathcal{Z}_i(y)$ are not empty for all $i \in 1:N$. The function $V_i(z_i)$ is equal to the minimum of functional $\tilde{F}_i(z_0, v^i, y) = \|z_0\|^2_{P_0} + \sum_{j=1}^i F_j(z_{j-1}, v_j)$ over all the elements z_0, v^i, satisfying to (10) and the boundary condition $z_i = \mathbb{S}(z_{i-1}, v_i)$. The informational sets $\mathcal{Z}_i(y)$ are expressed by the inequality $\mathcal{Z}_i(y) = \{z \in H : V_i(z) \le 1\}$. Note that the functional $\tilde{F}_i(z_0^*, v^{i*}, y) \le 1$ for all $i \in 1:N$. Therefore, the pair $(z_{i-1}^*, v_i^*) \in \mathcal{V}_i(y)$ and the element $z_i^* \in \mathcal{Z}_i(y)$ $\forall i$. The sets in (12) are not empty. The relation $\mathcal{Z}_i(y) = \{z \in H : V_i(z) \le 1\}$ is obvious for $i = 1$. Indeed, we have

$$\mathcal{V}_1(y) = \{(z,v) : \|z\|^2_{P_0} + \|v\|^2 + \|y_1 - Gz\|^2_R = \|z - \check{z}_1\|^2_{J_1} + \|v\|^2 + h_1 \le 1\},$$
$$\mathcal{Z}_1(y) = \mathbb{S}\mathcal{V}_1(y) = \{z : \|z - \hat{z}_1\|^2_{P_1} + h_1 = V_1(z) \le 1\}.$$

Here we use the known inverse operator formula $R - RG(P + G^*RG)^{-1}G^*R = (R^{-1} + GP^{-1}G^*)^{-1}$. Let the relation $\mathcal{Z}_{i-1}(y) = \{x \in H : V_{i-1}(x) \le 1\}$ be valid and formulas (11), (12), $i \ge 2$, be fulfilled for $i-1$. Now, from (12) it follows that the inclusion $z_i \in \mathcal{Z}_i(y)$ results in the existence of pair $(z_{i-1}, v_i) \in \mathcal{V}_i(y)$, for which $z_i = \mathbb{S}(z_{i-1}, v_i)$. Therefore, $V_i(z_i) \le 1$. Conversely, if the last inequality is valid, then by definition there exists a pair such that $z_i = \mathbb{S}(z_{i-1}, v_i) \in \mathbb{S}\mathcal{V}_i(y) = \mathcal{Z}_i(y)$. Moving back in indexes, we obtain that the inclusion $z \in \mathcal{Z}_i(y)$ is equivalent to the existence of the set (z_0, v^i), for which $\tilde{F}_i(z_0, v^i, y) \le 1$ and

$z = \mathbb{S}(z_{i-1}, v_i)$ under Eq. (10). So, we get $\min_{z_0,v^i} \tilde{F}_i(z_0, v^i, y) = V_i(z)$ under the boundary condition $z = \mathbb{S}(z_{i-1}, v_i)$. Suppose that $V_{i-1}(z) = \|z-\hat{z}_{i-1}\|^2_{P_{i-1}}+h_{i-1}$, $i \geq 2$. Then

$$\begin{aligned}\mathcal{V}_i(y) &= \big\{(z,v) : \|z-\hat{z}_{i-1}\|^2_{P_{i-1}} + h_{i-1} + \|v\|^2 + \|y_i - Gz\|^2_R \\ &\qquad = \|z-\check{z}_i\|^2_{J_i} + \|v\|^2 + h_i \leq 1\big\}, \\ \mathcal{Z}_i(y) &= \mathbb{S}\mathcal{V}_i(y) = \big\{z : \|z-\hat{z}_i\|^2_{P_i} + h_i = V_i(z) \leq 1\big\}.\end{aligned}$$

We see that values $y_i - G\hat{z}_{i-1}$ and h_i do not depend on controls u_i. Therefore, the values $\mathbb{B}u_i$ are added additively only for the second equality in (11). □

4 Problem Formulation and General Solution

We are going to formulate a problem in which processes of estimation and control are separate in time. At first the estimation is provided under given control and we get the information set $\mathcal{Z}_i(y,u)$. After that the minimax off-line procedure is realized. Our main control problem consists in finding of the instant i of finishing observation and passing to the new control on the rest of time.

4.1 Minimax Off-Line Control

From now on we introduce the other *compatible set* $\mathbf{V}_i(y,u)$ of uncertain parameters consisting of all pairs (z_i, v^N_{i+1}) that are compatible with the signal y^i. The projection $\text{proj}_H \mathbf{V}_i(y,u)$ of the compatible set on H coincides with the information set $\mathcal{Z}_i(y,u)$. This new compatible set is defined by the formula

$$\mathbf{V}_i(y,u) = \left\{ (z, v^N_{i+1}) : \|z-\hat{z}_i\|^2_{P_i} + \sum_{j=i+1}^{N} \|v_j\|^2 \leq 1 - h_i \right\},$$

where parameters are given in (11). Let $\tilde{u} = u^N_{i+1}$ be some controls and $\mathcal{Z}_N(\tilde{u} \mid \mathbf{V}_i(y,u))$ be the attainability domain of first equation in (10) with respect to $\mathbf{V}_i(y,u)$ under given further controls $\tilde{u}$. Consider some functional $\Phi(\mathcal{Z})$ that defined on all bounded sets $\mathcal{Z} \subset H$. The primary objective of controls consists in minimization of the cost $\Phi(\mathcal{Z}_N(y,u))$ that depends on the information set. At the initial instant we choose optimal control $u^{N,0}$ that solves the problem $\Phi(\mathcal{Z}_N(u^N \mid \mathbf{V}_0)) \to \min_{u^N} = r_0$ and after that it is corrected. Here $\mathbf{V}_0 = \big\{(z, v^N) : \|z\|^2_{P_0} + \sum_{j=1}^N \|v_j\|^2 \leq 1\big\}$ and the measurements are not taken into account.

At any instant $i = 1, \ldots, N$ we solve the auxiliary control problem

$$\Phi(\mathcal{Z}_N(u \mid \mathbf{V}_i(y,u^0))) \to \min_{u \in \mathbf{U}_i(u^0)} = r_i(y,u^0), \tag{13}$$

where $u^0 = u^{N,0}$ is a control chosen at initial instant; $\mathbf{U}_i(u^0)$ is a set of controls after the instant i, i.e. $\mathbf{U}_i(u^0) = \big\{u^N_{i+1} : \sum_{j=i+1}^N \|u_j\|^2 \leq 1 - \sum_{j=1}^i \|u^0_j\|^2\big\}$. Suppose that there exists at least one optimal control $u^{N,i}_{i+1}$ in problem (13).

4.2 Finding of the Observation Stopping Time

Now we explain how to find the instant i of finishing observation and passing to the new optimal control $u_{i+1}^{N,i}$ of problem (13) on the rest of time. To do the choice we compare the value $r_i(y, u^0)$ with value of forecasting

$$r_i(s, y^i, u^s) = \max_{y_{i+1}^s \in Y_{s,i}(y^i, u^s)} r_s(y, u), \tag{14}$$

where $Y_{s,i}(y^i, u^s) = \{y_{i+1}^s\}$ is a set of all possible continuations of signal y^i up to the instant $s > i$. The value (14) is the worst result of control if the system is located in the position $\{y^i, u^i\}$ and up to the instant s the control u_{i+1}^s is used. We set $r_i(i, y^i, u^i) = r_i(y, u)$. Our problem can be repeated [14,15]. Introduce one more value $\underline{r}_i(y, u) = \min_{s \in i:N} r_i(s, y^i, u^s)$. Let us be already located in position $\{y^i, u^i\}$, where u^i is a part of control u^N previously found. In this case, we verify the condition $\underline{r}_i(y, u) < r_i(y, u)$, $(i \in 1 : N - 1)$. If this holds, then the control u_{i+1}^N does not change. Otherwise, we pass to the new control $u_{t+1}^{N,t}$, delivering the minimum in (13). So, the first instant i such that

$$\underline{r}_i(y, u) \geq r_i(y, u), \quad \text{where } i \in 1 : N - 1, \tag{15}$$

we call the *observation stopping time*. In this instant i the observation is stopped and we pass the optimal off-line control in problem (13).

Consider some particular cases. Let $u = u^{N,0}$. If $\underline{r}_1(y, u) \geq r_1(y, u)$, then the observation is stopped at first instant. From the other hand, suppose that relations (15) are not valid for all $i \in 1 : N - 1$ and $\sum_{i=1}^N \|u_i^0\|^2 < 1$. In this case, the observation continues all the time, but the resource of control is not exhausted at the last instant N. Therefore, we can solve the minimax problem $\Phi(\mathcal{Z}_N(y, u)) \to \min_{u_N}$, $\|u_N\|^2 \leq 1 - \sum_{i=1}^N \|u_i^0\|^2$, and regard optimal $\tilde{u}_N$ as an additional control action at the last instant.

4.3 An Algorithm of Repeated Correction

If we can continue observation after any stopping time, then the following algorithm of repeated correction can be proposed.

1. We find the value r_0 and optimal control $u^{N,0}$ before any observations.
2. At $i = 1$ we decide if this control has to be changed, i.e. if the value $\underline{r}_1(y, u^{N,0}) < r_1(y, u^{1,0})$ then the control $u^{N,0}$ should be kept. Otherwise, we pass to the new control $u_2^{N,1}$, delivering the minimum in (13).
3. In position $\{y^i, u^i\}$, where u^i is a part of control u^N previously found, we verify the condition (15), where $i \in 1 : N - 1$. If this holds, then we pass to the optimal control $u_{i+1}^{N,i}$, delivering the minimum in (13).
4. In any case, if at the last instant N the inequality $\sum_{i=1}^N \|u_i\|^2 < 1$ is obtained, we solve the minimax problem $\Phi(\mathcal{Z}_N(y, u)) \to \min_{u_N}$, $\|u_N\|^2 \leq 1 - \sum_{i=1}^N \|u_i\|^2$, and regard optimal $\tilde{u}_N$ as an additional control action at the last instant.

According to the algorithm, we obtain the sequence $\{\tau_1, \tau_2, \dots\}$ of instants where control has been changed. This sequence depends on the signal. In particular, the sequence may be empty when observations are bad for control, or it may coincide with the set $1 : N-1$, when, on the contrary, the observations give essential information. The values $r_i = r_{\tau_i}(y,u)$ form the nonincreasing sequence. Here the strong inequalities $r_i > r_{i+1}$ hold if $\tau_{i+1} - \tau_i \geq 2$. In the case $\tau_{i+1} - \tau_i = 1$ the strong inequality $r_i > r_{i+1}$ holds if and only if the signal $y_{\tau_{i+1}}$ is not the worst.

Instead of inequality (15) at every instant $i < N$, we may check the simpler condition $r_t(t+1, y^t, u^{t+1}) < r_t(y,u)$. If it is fulfilled, then the control u_{t+1}^N does not change. Otherwise, we pass to the new control $u_{t+1}^{N,t}$ in problem (13).

5 A Special Case of the Terminal Cost

Let the terminal functional has the form $\Phi(\mathcal{Z}) = \max_{z\in\mathcal{Z}} \|\Delta z\|$, where $\Delta : H \to R^k$ is a linear finite-dimensional operator and $\|\cdot\|$ is the Euclidean norm. In this case, we can obtain formulas (13)–(15) in more detail.

First of all we describe all the continuations of the signal.

Lemma 2. *A signal y_{i+1}^s is a continuation of the signal y^i iff there exists a sequence φ_{i+1}^s such that $\sum_{j=i+1}^{s} \|\varphi_i\|^2_{\mathcal{G}_i} \leq 1 - h_i$, and $\hat{z}_j = \mathbb{B}u_j + S(\hat{z}_{j-1} + J_j^{-1}G^*R\varphi_j)$, $y_j = G\hat{z}_{j-1} + \varphi_j$, for $j \in i+1 : s$.*

This lemma follows from Eq. (12). Below we use vectors $l \in R^k$ as column-vectors and the symbol l' is used for row-vector. Then we have the relation

$$r_i(y,u) = \max_{l'l\leq 1}\Bigg\{\gamma_i(l)\hat{z}_i - \Bigg(1 - \sum_{j=1}^{i}\|u_i\|^2\Bigg)^{1/2}\Bigg(\sum_{j=i+1}^{N}\gamma_j(l)\mathbf{B}\gamma_i^*(l)\Bigg)^{1/2}$$
$$+ (1-h_i)^{1/2}\left(\pi_0(i)(1-l'l) + l'\Delta P_{N,i}\Delta^* l\right)^{1/2}\Bigg\}, \tag{16}$$

where $\gamma_j(l) = \gamma_{j+1}(l)S$, $\gamma_N(l) = l'\Delta$; $P_{j,i} = SP_{j-1,i}S^* + \mathbf{C}$, $P_{i,i} = P_i^{-1}$; $\pi_0(i) = \max_{l'l\leq 1} l'\Delta P_{N,i}\Delta^* l$. Using Lemma 2, we obtain

$$r_i(s, y^i, u^i) = \max_{l'l\leq 1}\Bigg\{\gamma_i(l)\hat{z}_i + \sum_{j=i+1}^{s}\gamma_j(l)\mathbb{B}u_j - \Bigg(1 - \sum_{i=1}^{s}\|u_i\|^2\Bigg)^{1/2}$$
$$\cdot\Bigg(\sum_{j=s+1}^{N}\gamma_j(l)\mathbf{B}\gamma_j^*(l)\Bigg)^{1/2} + (1-h_i)^{1/2}(\pi_0(s)(1-l'l) + l'\Delta P_{N,i}\Delta^* l)^{1/2}\Bigg\}. \tag{17}$$

Formulas (16)–(17) are established similarly to [4,9]. In addition, let us note that optimal control is on the formula

$$u_j^0 = -\mathbb{B}\gamma_j^*(l^0)\Bigg(1 - \sum_{i=1}^{j}\|u_i\|^2\Bigg)^{1/2}\Bigg(\sum_{i=j+1}^{N}\gamma_i(l^0)\mathbf{B}\gamma_i^*(l^0)\Bigg)^{-1/2}, \quad j > i,$$

where l^0 is a maximizer in formula (16) which does not convert the corresponding sum into zero.

6 A Finite-Dimensional Approximation

Let us return to general relations in Sect. 1, where V is a separable Hilbert space and $a(u, v)$ is a bilinear form with properties:

$$a(v, v) \geq \alpha \|v\|^2, \quad a(u, v) \leq \beta \|u\| \|v\|. \tag{18}$$

Given finite-dimensional subspace $\mathcal{F} \subset V$, define Ritz's projector $\Pi : V \to \mathcal{F}$ as $a(v, u - \Pi u) = 0$, $\forall v \in \mathcal{F}$ (see [16]). The following estimate holds:

$$\|u - \Pi u\| \leq \beta d(u, \mathcal{F})/\alpha, \quad \text{where } d(u, \mathcal{F}) = \min_{v \in \mathcal{F}} \|u - v\|. \tag{19}$$

Consider an increasing sequence $\mathcal{F}^n$ of finite-dimensional subspaces $\mathcal{F}^n \subset \mathcal{F}^{n+1} \subset V$ such that the distance $d(u, \mathcal{F}^n) \to 0$ as $n \to \infty$ $\forall u \in V$. Such a sequence is called *complete*. The proof of following lemma may be found in [16] or somewhere.

Lemma 3. *Let $u : [0, T] \to V$ be a continuous function and $\mathcal{F}^n$ be a complete sequence of finite-dimensional subspaces. Then the real function $\|u(t) - \Pi^n u(t)\|$ tends to zero uniformly in $t \in [0, T]$, where $\Pi^n : V \to \mathcal{F}^n$ is the Ritz projector.*

Let H be another Hilbert space and let the space $V \subset H$ be densely imbedded in H as in Sect. 1. The linear operator A with a dense domain $D(A) \subset V$ has been defined as $a(u, v) = \langle Au, v \rangle_H$, $\forall v \in V$. The dual operator A^* is defined by the relation $a(u, v) = \langle u, A^* v \rangle_H$, $\forall u \in V$. The operator $-A^*$ is a infinitesimal generator for the semigroup $S^*(t)$ (see, for example, [17]). In addition, the function $\psi(t) = S^*(t)\psi$, where $\psi \in H$, is defined a weak solution of equation

$$d\langle v, \psi(t) \rangle_H / dt + a(v, \psi(t)) = 0 \quad \forall v \in V, \quad \psi(0) = \psi.$$

This equation is similar to (1). Let us remind that the inclusion $z_0 \in D(A)$ implies $z(t) = S(t) z_0 \in D(A)$ for all $t \geq 0$ and

$$dz(t)/dt + Az(t) = 0, \tag{20}$$

i.e. $z(t)$ is a strong solution of Eq. (20).

Suppose that the increasing sequence $\mathcal{F}^n \subset V$ of finite-dimensional subspaces is complete. Consider the problem

$$d\langle z^n(t), v^n \rangle / dt + a(z^n(t), v^n) = \langle f(t), v^n \rangle \ \forall v^n \in \mathcal{F}^n, \quad z^n(0) = z^n, \tag{21}$$

where one needs to find a function $z^n(t) \in \mathcal{F}^n$. The problem (21) is called the *Galerkin-type finite-dimensional approximation* of problem (1). We need the following

Theorem 2 ([18]). *Let $z^n \to z$ in the space H as $n \to \infty$. Then the solution $z^n(t)$ of problem* (21) *uniformly converges on $[0, T]$ to the solution $z(t)$ of problem* (1) *in the space H.*

Let $e_1, \ldots, e_n$ be a basis in the space $\mathcal{F}^n$. We set

$$z^n(t) = \sum_{j=1}^{n} q^j(t)e_j, \quad z^n = \sum_{j=1}^{n} q^j e_j.$$

A finite-dimensional approximation of problems in Sect. 4 with respect to the complete sequence $\mathcal{F}^n$ of subspaces is as follows. Problem (21) is equivalent to the solution of differential equations in matrix form:

$$M\dot{q} + Kq = \mathbf{f}(t), \quad q_0 = [q^1; \ldots; q^n], \quad \mathbf{f}(t) = [\langle f(t), e_1\rangle; \ldots; \langle f(t), e_n\rangle],$$

where M ($\det M \neq 0$) and K have elements $\langle e_i, e_j\rangle$ and $a(e_i, e_j)$ respectively. The solution of the system for our problems may be written similarly to (5):

$$q(t) = S^n(t)q_0 + \int_0^t S^n(t-s)(B^n\mathbf{u}(s) + C^n\mathbf{v}(s))ds, \tag{22}$$

where $S^n(t) = \exp(-M^{-1}Kt)$ is the transition matrix having $n \times n$-dimension, $\mathbf{u}(t)$ and $\mathbf{v}(t)$ are n-dimensional measurable functions. Matrices B^n and C^n have the similar structure and represent a multiplication of matrix M^{-1} and the square root of matrices with elements $\langle e_i, BF^{-1}B^*e_j\rangle$ and $\langle e_i, CQ^{-1}C^*e_j\rangle$ respectively. Constraints (6) and (7) are transformed to

$$\|q_0\|^2_{P_0^n} + \int_0^T \|\mathbf{v}(s)\|^2 ds + \sum_{i=1}^{N} \|w_i\|^2_R \leq \nu^n, \quad \int_0^T \|\mathbf{u}(s)\|^2 ds \leq \mu^n. \tag{23}$$

Measurement equation from (8) has the form

$$y_i = G^n q(t_i) + w_i, \quad G^n = [Ge_1, \ldots, Ge_n] \in R^{k\times n}. \tag{24}$$

Problems of Sects. 4 and 5 may be solved for relations (22)–(24) as described above.

Let us explain the appearance of numbers μ^n and ν^n in constraints (23). The matter is that the system (10) is infinite-dimensional and, therefore, the signal y^N of this system in some cases can not be realized in finite-dimensional approximation (22)–(24) if we set $\mu^n = 1$, $\nu^n = 1$. But under some $\mu^n > 1$, $\nu^n > 1$ the finite-dimensional formulas like (13)–(17) are valid. Moreover, we get

Theorem 3. *There exist sequences $\mu^n \downarrow 1$, $\nu^n \downarrow 1$ as $n \to \infty$ such that formulas like* (11)–(17) *for finite-dimensional approximation* (22)–(24) *hold and $r_i^n(y,u) \to r_i(y,u)$, $r_i^n(s, y^i, u^i) \to r_i(s, y^i, u^i)$ as $n \to \infty$ in relations* (16), (17).

In the general case, it is hard to obtain the estimates of velocity for convergence $\mu^n \downarrow 1$, $\nu^n \downarrow 1$ with respect to parameters α, β in (18), (19).

6.1 An Application to Heat Equation

Let the controlled system be described by the equations

$$z_t = z_{xx} + u(t)f(x), \quad x \in [0,l], \quad t \geq 0, \text{ with boundary conditions} \quad (25)$$
$$z_x(t,0) = z(t,0), \quad z_x(t,l) = -z(t,l).$$

Here $f(x)$ is a smooth function on $[0,l]$, $u(t)$ is a control. This system describe the heat process for the uniform bar. In our situation $H_1 = R$, $C = 0$, $V = H^1(0,l)$, $H = L_2(0,l)$ where $H^1(0,l)$ is the Sobolev space with parameter $k = 1$. The operator $B : R \to L_2(0,l)$ has the form $Bu = uf(x)$. Dual operator $B^* : L_2(0,l) \to R$ is written as $B^*\phi = \int_0^l f(x)\phi(x)dx$, $\phi \in L_2(0,l)$. The weak form of considered system is obtained by the multiplication of (25) by $\phi \in H^1(0,l)$ with subsequent integration on $[0,l]$ using boundary conditions. The form $a(\phi,\psi)$ may be written as

$$a(\phi,\psi) = \int_0^l \dot{\phi}(x)\dot{\psi}(x)dx + \phi(l)\psi(l) + \phi(0)\psi(0).$$

The coercivity follows from Friedrich's inequality. So, relation (1) for system (25) looks like

$$\partial \int_0^l z(t,x)\phi(x)dx/\partial t + a(z(t,\cdot),\phi(\cdot)) = u(t)\int_0^l f(x)\phi(x)dx$$

for all $\phi \in H^1(0,l)$, $z(0,x) = z_0(x)$.

Let us divide the segment $[0,l]$ by n subsegments of length l/n. Let x_i, $i \in 0:n$, be the points of partition. For the space $\mathcal{F}^n$ we consider piecewise-linear functions $e_i(x)$, for which $e_i(x_i) = 1$ and $e_i(x_j) = 0$ if $i \neq j$. The sequence of finite-dimensional subspaces $\mathcal{F}^n$ with basis $e_i(x)$, $i \in 0:n$, is complete. Therefore, we can perform the approximation. Suppose that measurement equations are of the form

$$y_i = \int_0^l b(x)z(t_{i-1},x)dx + w_i, \quad i \in 1:N, \text{ where } b(\cdot) \in L_2(0,l).$$

Consider the $(n+1)\times(n+1)$-matrices M with elements $M_{ij} = \int_0^l e_i(x)e_j(x)dx$ and K with elements $K_{ij} = a(e_i,e_j)$. The M is a three-diagonal symmetric matrix, where $M_{00} = M_{nn} = l/(3n)$ and other diagonal elements are equal to $2l/(3n)$. The secondary diagonal elements are equal to $l/(6n)$. The K is also a three-diagonal symmetric matrix, where $K_{00} = K_{nn} = n/l + 1$ and other diagonal elements are equal to $2n/l$. The secondary diagonal elements of K are equal to $-n/l$. If $f(x) \equiv 1$, then we obtain the finite-dimensional system

$$M\dot{q} + Kq = u(t)\mathbf{f}, \quad \text{where } \mathbf{f} = l[1;2;\ldots;2;1]/(2n) \in R^{n+1}. \quad (26)$$

Let $b(x) \equiv 1$. Then measurement Eq. (24) has the form

$$y_i = G^n q(t_{i-1}) + w_i \quad \text{where} \quad G^n = \mathbf{f}'.$$

Suppose that initial constraints (6), (7) may be written as

$$\int_0^l z^2(0,x)dx + \sum_{i=1}^{N} w_i^2 \le 1, \quad \int_0^T u^2(t)dt \le 1.$$

It follows from this that constraints (23) are:

$$\|q_0\|_M^2 + \sum_{i=1}^{N} w_i^2 \le \mu^n, \quad \int_0^T u^2(t)dt \le 1.$$

We need not to increase the constraints for $u(\cdot)$, but we do it for q_0 and w_i in order to include the sequence y^N in the scope. After that we need to convert the continuous system (26) to discrete one of the type (9), (10). Many solved examples of such a finite-dimensional problems where considered in [12,13,19].

7 Conclusion

We considered a control problem with incomplete information for abstract parabolic control systems in Hilbert space. Information about the system state are known in discrete instants. According to measurements, the information set was introduced that contained the true state of the system. This set included all the states of the system that were compatible with the measurements. For the terminal criterion depending of the information set, we suggested some statements of the problem based on the separation of control and observation processes. The optimal instants of transition from estimation to control were looked for as well. The approach was applied to distributed systems with partial derivatives. The approximation scheme was suggested and example with heat equation was considered. In this research some aspects demand more detailed study. For example, we need to obtain the estimates for values μ^n, ν^n, and convergence speed for parameters in Theorem 3. It is interesting to expand the approach to the case of continuous measurements.

References

1. Lions, J.-L.: Some Aspects of the Optimal Control of Distributed Parameter Systems. SIAM, Philadelphia (1972)
2. Yosida, K.: Functional Analysis. Springer, Berlin (1980). https://doi.org/10.1007/978-3-642-61859-8
3. Krasovski, N.N.: Game problem on motion correction. Appl. Math. Mech. **33**(3), 386–396 (1969)
4. Kurzhanski, A.B.: Control and Observation under Conditions of Uncertainty. Nauka, Moscow (1977). (in Russian)
5. Kurzhanski, A.B., Valyi, I.: Ellipsoidal Calculus for Estimation and Control. Birkhäuser, Boston (1997)

6. Kurzhanski, A.B., Varaiya, P.: Dynamics and Control of Trajectory Tubes, Theory and Computation. Systems & Control: Foundations & Applications, vol. 85. Birkhäuser, Basel (2014). https://doi.org/10.1007/978-3-319-10277-1
7. Liptser, R.S., Shiryayev, A.N.: Statistics of Random Processes, I General Theory, II Applications. Springer, New York (1978). https://doi.org/10.1007/978-3-662-13043-8
8. Schweppe, F.: Uncertain Dynamic Systems. Prentice-Hall, Englewood Cliffs (1973)
9. Anan'ev, B.I.: Minimax quadratic problem of motion correction. Prikl. Matem. i Mech. **41**(3), 436–445 (1977). (in Russian)
10. Ananev, B.I., Anikin, S.A.: Problem of reconstructing input signals under communication constraints. Autom. Remote Control **70**(7), 1153–1164 (2009)
11. Ananev, B.I.: Correction of motion under communication constraints. Autom. Remote Control **71**(3), 367–378 (2010)
12. Ananyev, B.I.: A control problem for evolutionary systems with incomplete information. In: IEEE Xplore Digital Library. Add. 09 July 2018: Stability and Oscillations of Nonlinear Control Systems (STAB): 14th International Conference (Pyatnitskiy's Conference), Moscow, 30 May-1 June, Proceedings (2018). https://doi.org/10.1109/STAB.2018.8408340
13. Ananyev, B.I., Gusev, M.I., Filippova, T.F.: Control and estimation of dynamical systems states with uncertainty. N.N. Krasovskii Institute of Mathematics and Mechanics UB of RAS. Siberian Branch of RAS, Novosibirsk (2018). (in Russian)
14. Ananyev, B.I., Gredasova, N.V.: Multistage motion correction of a linear-quadratic controlled system. Bull. UGTU-UPI **4**(56), 280–288 (2005)
15. Ananyev, B.I., Gredasova, N.V.: Multistage correction of quasi-linear systems under discrete observations. Proc. Inst. Math. Mech. UB RAS **13**(4), 3–13 (2007)
16. Strang, G., Fix, G.J.: An Analysis of the Finite Element Method. Prentice-Hall, Englewood Cliffs (1973)
17. Curtain, R.F., Pritchard, A.J.: Infinite Dimensional Linear Systems Theory. Springer, Berlin (1978)
18. Kerimov, A.K.: On the Galerkin's approximation of optimal control problems for distributed systems of parabolic type. J. Comp. Math. Math. Phys. **19**(4), 851–865 (1979). (in Russian)
19. Gredasova, N.V.: Problems of multiple correction for controlled systems. Dissertation. IMM UB of RAS, Yekaterinburg (2012). (in Russian)

Best Approximation of a Differentiation Operator on the Set of Smooth Functions with Exactly or Approximately Given Fourier Transform

Vitalii V. Arestov[1,2(✉)]

[1] Ural Federal University, Ekaterinburg 620000, Russia
vitalii.arestov@urfu.ru
[2] Krasovskii Institute of Mathematics and Mechanics,
Ural Branch of the Russian Academy of Sciences, Ekaterinburg 620990, Russia
http://work.imkn.urfu.ru/arestov

Abstract. Let Y^n, $n \geqslant 2$, be the set of continuous bounded functions on the numerical axis with the following two properties: (1) the Fourier transform of a function is a function of bounded variation on the axis (in particular, a summable function); (2) a function is $n-1$ times continuously differentiable, its derivative of order $n-1$ is locally absolutely continuous, and the nth order derivative is bounded, more exactly, belongs to the space L_∞. In the space Y^n, consider the class $\mathcal{Q}^n$ of functions, for which the L_∞-norm of the nth order derivative is bounded by a constant, for example, by 1. The following two approximation problems are discussed: the best approximation of the differentiation operator D^k of order k, $1 \leqslant k < n$, by bounded operators on the class $\mathcal{Q}^n$ and the optimal calculation of the differentiation operator D^k on functions from the class $\mathcal{Q}^n$ under the assumption that their Fourier transform is given with a known error in the space of functions of bounded variation, in particular, in the space L of functions summable on the axis. In interrelation with these two problems, we discuss the exact Kolmogorov type inequality in the space Y^n between the uniform norm of the kth order derivative of a function, the variation of the Fourier transform of the function, and the L_∞-norm of its derivative of order n.

Keywords: Functions with exactly or approximately given Fourier transform · Kolmogorov inequality · Optimal differentiation method

---

This work was supported by the Russian Foundation for Basic Research (project no. 18-01-00336) and by the Russian Academic Excellence Project (agreement no. 02.A03.21.0006 of August 27, 2013, between the Ministry of Education and Science of the Russian Federation and Ural Federal University).

M. Khachay et al. (Eds.): MOTOR 2019, LNCS 11548, pp. 434–448, 2019.
https://doi.org/10.1007/978-3-030-22629-9_30

Introduction

In the present paper, we discuss the following three related problems.

(1) Exact inequality between the uniform norm (more exactly, the norm of the space $C(-\infty,\infty)$) of the kth order derivative of a function, the variation of the Fourier transform of the function, and the L_∞-norm of its nth order derivative, $0 < k < n$. This inequality can be considered as a nonclassical variant of the Kolmogorov inequality.
(2) Stechkin's problem on the best uniform approximation of the kth order differentiation operator on the class of functions with bounded nth order derivative, $0 < k < n$, by bounded operators in the corresponding spaces.
(3) Optimal (i.e., with the smallest possible error) calculation in the space $C(-\infty,\infty)$ of the kth order derivative of a function with bounded nth order derivative, $0 < k < n$, by the Fourier transform of the function approximately given with a known error in measure (and, in particular, in the space $L(-\infty,\infty)$).

We study all three problems simultaneously, taking into account their interrelation. Section 2 is devoted to these questions.

In Sect. 1, we discuss three similar problems in the space $C(-\infty,\infty)$ of continuous bounded functions on the axis: the classical variant of the Kolmogorov inequality, Stechkin's problem on the best approximation of differentiation operators by bounded operators, and optimal calculation of derivatives of smooth functions given approximately in the uniform norm on the axis. Solutions of all these three problems are known; we present information about these problems and the methods for their study to the extent that need for the further. This topic can be found, for example, in the author's review paper [3] and in monographs [5,12].

1 Three Interrelated Problems in the Space of Continuous Bounded Functions on the Axis

All functional spaces considered in what follows are complex. The space $L_\infty = L_\infty(-\infty,\infty)$ consists of measurable essentially bounded functions on the axis and is equipped with the norm

$$\|f\|_{L_\infty} = \text{ess sup}\{|f(t)| : t \in (-\infty,\infty)\}.$$

The space L_∞ contains the space $C = C(-\infty,\infty)$ of continuous bounded functions on the axis; it is equipped with the norm

$$\|f\|_C = \sup\{|f(t)| : t \in (-\infty,\infty)\}.$$

The space $C(-\infty,\infty)$ contains the space $C_0 = C_0(-\infty,\infty)$ of functions vanishing at infinity.

1.1 Kolmogorov Inequality

Let $W^n = W^n_{\infty,\infty}$, $n \geqslant 1$, be the space of functions $f \in C$ that are $n-1$ times continuously differentiable on the axis and such that the derivative $f^{(n-1)}$ of order $n-1$ is locally absolutely continuous, and the nth order derivative $f^{(n)}$ belongs to the space L_∞.

The following inequality holds on the set W^n with finite constant (independent of the function f but essentially depending on k and n) for $0 < k < n$:

$$\|f^{(k)}\|_C \leqslant C_{n,k} \|f\|_C^{(n-k)/n} \|f^{(n)}\|_{L_\infty}^{k/n}, \quad f \in W^n. \tag{1.1}$$

This fact was proved [11] by Hardy and Littlewood in 1912.

Inequality (1.1) with exact constant was first obtained [10] by Hadamard for $n = 2$ and $k = 1$ in 1914. Bosse (alias Shilov, a student of A.N. Kolmogorov, 1937) obtained [6] the exact inequality (1.1) for $n = 3, 4$ and all $1 \leqslant k < n$ and for $n = 5$ and $k = 2$. In 1939, Kolmogorov found [13] an exact constant in inequality (1.1) for all $1 \leqslant k < n$; his method is the elegant Kolmogorov comparison theorem. Kolmogorov's result is one of the most striking and important in this subject area; in this connection, inequality (1.1) and similar inequalities with other norms on the axis and semi-axis are often called *Kolmogorov inequalities.*

The known Favard–Akhiezer–Krein function

$$f_n(t) = \frac{4}{\pi} \sum_{\ell=0}^{\infty} \frac{\sin((2\ell+1)t - n\pi/2)}{(2\ell+1)^{n+1}}, \quad t \in \mathbb{R}, \tag{1.2}$$

is extremal in inequality (1.1). Let us mention some properties of function (1.2) (see, for example, [14, Ch. 5, Sect. 5.4]):

$$f_n^{(m)} = f_{n-m}, \quad 1 \leqslant m \leqslant n;$$

$$f_0(t) = \frac{4}{\pi} \sum_{\ell=0}^{\infty} \frac{\sin((2\ell+1)t)}{(2\ell+1)} = \operatorname{sgn} \sin t, \quad t \in \mathbb{R}.$$

The uniform norm of function (1.2) is

$$M_n = \|f_n\|_C = \frac{4}{\pi} \sum_{\ell=0}^{\infty} \frac{(-1)^{\ell(n+1)}}{(2\ell+1)^{n+1}}. \tag{1.3}$$

Using the known values of the sums of the corresponding numerical series (see, for example, [15]), we, in particular, obtain

$$M_0 = 1, \quad M_1 = \frac{\pi}{2}, \quad M_2 = \frac{\pi^2}{8}, \quad M_3 = \frac{\pi^3}{24}, \quad M_4 = \frac{5\pi^4}{384}.$$

The extremality of the function f_n and its properties mentioned above imply that

$$C_{n,k} = M_{n-k} (M_n)^{-\frac{n-k}{n}}.$$

1.2 Approximation of the Differentiation Operator by Bounded Operators (Stechkin's Problem)

Denote by $\mathscr{B} = \mathscr{B}(C, C)$ the set of all bounded linear operators in the space C and by $\mathscr{B}(N)$, where $N > 0$, the set of operators $T \in \mathscr{B}$ such that $\|T\|_{C\to C} \leqslant N$.

In the space W^n, consider the class of elements

$$Q^n = Q^n_{\infty,\infty} = \{f \in W^n : \|f^{(n)}\|_{L_\infty} \leqslant 1\}.$$

Let $0 < k < n$ be integer. For an operator $T \in \mathscr{B}$, define

$$U(T) = \sup\{\|f^{(k)} - Tf\|_C : f \in Q^n\}. \tag{1.4}$$

The value (1.4) can be interpreted as the deviation (in the space C) of the operator T from the differentiation operator $D^k = d^k/dt^k$ on the class Q^n. For $N > 0$, the value

$$E_{n,k}(N) = \inf\{U(T) : T \in \mathscr{B}(N)\} \tag{1.5}$$

is the best approximation (in the space C) of the differentiation operator D^k on the class Q^n by the set of bounded linear operators $\mathscr{B}(N)$.

This is *Stechkin's problem*; it consists in calculating value (1.5) and finding an extremal operator at which the infimum in (1.5) is attained; we will also call it problem (1.5), and sometimes the problem $E_{n,k}(N)$.

Stechkin's result about the connection between problem (1.5) and inequality (1.1) is very useful in this topic. As a special case of Stechkin's more general result [17, inequality (6)], value (1.5) and the best constant in (1.1) are related by the inequality (see details in [3, Sects. 1, 4])

$$E_{n,k}(N) \geqslant k\left(\frac{C_{n,k}}{n}\right)^{\frac{n}{k}}\left(\frac{N}{n-k}\right)^{-\frac{n-k}{k}}, \quad N > 0. \tag{1.6}$$

In fact, we have the equality

$$E_{n,k}(N) = k\left(\frac{C_{n,k}}{n}\right)^{\frac{n}{k}}\left(\frac{N}{n-k}\right)^{-\frac{n-k}{k}}, \quad N > 0. \tag{1.7}$$

This fact was first obtained in [2] as a consequence of Domar's result [8]. Later on, it became clear that this fact is a special case of a more general Gabushin's result [9] about the best approximation of unbounded functionals by bounded ones.

The solution of problem (1.4) is known at present. First, note an auxiliary fact. Value (1.4) is homogeneous by N; more exactly,

$$E_{n,k}(N) = N^{-\gamma}E_{n,k}(1), \quad \gamma = \frac{n-k}{k}. \tag{1.8}$$

The origin of this formula is as follows. Following Stechkin, we assign to an operator $T \in \mathscr{B}(N)$ an operator T_h, $h > 0$, by the formula

$$(T_hf)(t) = h^{-k}(Tf_h)(th^{-1}), \tag{1.9}$$

in which $f_h(u) = f(hu)$. It is easy to verify that

$$\|T_h\|_{C\to C} = h^{-k}\|T\|_{C\to C}; \tag{1.10}$$

$$U(T_h) = h^{n-k}U(T). \tag{1.11}$$

From definition (1.9) and property (1.10), we conclude that, under the mapping $T \to T_h$, the set of operators $\mathscr{B}(N)$ is mapped one-to-one onto the set $\mathscr{B}(h^{-k}N)$.

As a consequence of (1.11), we have the following formula for all $h > 0$ and $N > 0$:

$$E_{n,k}(h^{-k}N) = h^{n-k}E_{n,k}(N), \tag{1.12}$$

which is more general in comparison with (1.8). Moreover, an operator $T \in \mathscr{B}(N)$ is extremal in the problem $E_{n,k}(N)$ if and only if the operator T_h is extremal in the problem $E_{n,k}(h^{-k}N)$.

It follows from the above argument that it is sufficient to solve problem (1.5) for a specific N; formulas (1.12) and (1.9) enable us to obtain its solution for all values of the parameter N.

First exact results in problem (1.5) were obtained by Stechkin. He proved [16,17] that, for $n = 2$ and $n = 3$ and $1 \leqslant k < n$, the following classical (finite-difference) operators $T^h_{n,k}$ are extremal:

$$(T^h_{2,1}f)(t) = (T^h_{3,1}f)(t) = \frac{f(t+h) - f(t-h)}{2h}, \quad N = h^{-1}, \tag{1.13}$$

$$(T^h_{3,2}f)(t) = \frac{f(t+h) - 2f(t) + f(t-h)}{h^2}, \quad N = \frac{4}{h^2}.$$

Arestov found [1] the solution of problem (1.5) for $n = 4$ and 5, and Buslaev solved [7] the problem for arbitrary $n \geqslant 6$. For $n \geqslant 4$, extremal operators are infinite with uniform nodes. More exactly, for example, the extremal operator for $k = 1$ has the form

$$T_{n,1}f(t) = h^{-1}\sum_{\ell=0}^{\infty} \alpha_\ell(f(t + (2\ell+1)h) - f(t - (2\ell+1)h));$$

the sequence $\{\alpha_\ell\}_{\ell\geqslant 0}$ is the sum of $[n/2] - 1$ geometric progressions. In the proof of these results, the lower bound (1.6) and the exact Kolmogorov inequality (1.1) were used.

1.3 Optimal Differentiation of Approximately Given Functions

Kolmogorov inequality (1.1) and Stechkin's problem (1.5) are related to one more important problem of optimal differentiation in the space W^n of functions given approximately with a known error in the uniform norm (see [3] and the references therein).

Suppose that we need to calculate the kth order derivative $f^{(k)}$ of a function f in the situation when: (1) the function f is given approximately with a known error $\delta > 0$; more exactly, instead of the function f, we know a function $g = f_\delta \in C$ with the property $\|f - f_\delta\|_C \leqslant \delta$; (2) we know the a priori information about the function f that $f \in W^n$ and the specific bound for the norm of the highest derivative holds: $\|f^{(n)}\|_{L_\infty} \leqslant A$. For definiteness, we assume that $A = 1$, i.e., $f \in Q^n$. The problem is to construct a method that, being applied to the function f_δ under the given information, recovers the kth order derivative of the function from the class Q^n optimally (in the best way).

The exact statement of the problem is as follows. Let $\mathscr{O} = \mathscr{O}(C, C)$ be the set of arbitrary (single-valued) mappings of the space C to itself. For a mapping $T \in \mathscr{O}$ and a parameter $\delta > 0$, define

$$\Delta(T) = \Delta^\delta_{n,k}(T) = \sup\{\|f^{(k)} - Tg\|_C \colon f \in Q^n,\ g \in C,\ \|g - f\|_C \leqslant \delta\}. \quad (1.14)$$

Quantity (1.14) can be interpreted as the error of recovery of the differentiation operator $D^k = d^k/dt^k$ on functions from the class Q^n given with the error δ by means of the method T. The smallest value

$$\rho_{n,k}(\delta) = \inf\{\Delta(T) \colon T \in \mathscr{O}\} \quad (1.15)$$

of quantity (1.14) over all mappings $T \in \mathscr{O}$ is the smallest recovery error. A mapping $T^\delta_{n,k} \in \mathscr{O}$ at which the infimum in (1.15) is attained is called an *optimal recovery method.*

Let us present the following well known considerations (see, for example, [3] and the references therein), which reflect the relation between Stechkin's problem (1.5), the (exact) inequality (1.1), and problem (1.15). For an operator $T \in \mathscr{B}(N)$ and functions $f \in Q^n$ and $g \in C$ with the property $\|f - g\| \leqslant \delta$, we have

$$\|f^{(k)} - Tg\|_C \leqslant \|f^{(k)} - Tf\|_C + \|T(f - g)\|_C \leqslant U(T) + N\delta.$$

This implies the bound

$$\rho_{n,k}(\delta) \leqslant E_{n,k}(N) + N\delta, \quad N > 0.$$

Substituting (1.7) into this inequality, we obtain

$$\rho_{n,k}(\delta) \leqslant k\left(\frac{C_{n,k}}{n}\right)^{\frac{n}{k}}\left(\frac{N}{n-k}\right)^{-\frac{n-k}{k}} + N\delta, \quad N > 0.$$

The right-hand side of this inequality, as a function of variable $N > 0$, takes the smallest value equal to $C_{n,k}\delta^{\frac{n-k}{n}}$ for

$$N = N(\delta) = \frac{n-k}{n} C_{n,k}\delta^{-\frac{k}{n}}. \quad (1.16)$$

Consequently, the inequality $\rho_{n,k}(\delta) \leqslant C_{n,k}\delta^{\frac{n-k}{n}}$ holds. On the other hand, the inverse inequality also holds; see details and the references in [3, Sect. 2]. Hence, we conclude that there holds the equality

$$\rho_{n,k}(\delta) = C_{n,k}\delta^{\frac{n-k}{n}}. \quad (1.17)$$

Moreover, if $T_{n,k}$ is an extremal operator in Stechkin's problem (1.5) for value (1.16) of the parameter N, then this operator is an optimal method in problem (1.15).

2 Three Problems in Spaces of Functions with Exactly or Approximately Given Fourier Transform

In what follows, we refer to some results of [4] using the notation of the present paper.

2.1 A Nonclassical Variant of the Kolmogorov Inequality

Let $L = L(-\infty, \infty)$ be the space of (complex-valued) measurable functions f summable on the numerical axis $\mathbb{R} = (-\infty, \infty)$; the space L is equipped with the norm

$$\|f\|_L = \int |f(t)|\, dt.$$

Hereinafter, we omit the integration set in integrals over the axis. For a function $f \in L$, its Fourier transform $\widehat{f}$ and inverse Fourier transform $\check{f}$ are defined by the formulas (see, for example, [18])

$$\widehat{f}\,(t) = \int e^{-2\pi t\eta i} f(\eta)\, d\eta, \quad \check{f}\,(t) = \int e^{2\pi t\eta i} f(\eta)\, d\eta.$$

In what follows, we apply the Fourier transform, the inverse Fourier transform, and some other classical operations in functional spaces to more general objects, which can be interpreted as generalized functions. Let $\mathscr{S}$ be the (topological vector) space of fast decreasing infinitely differentiable functions on $\mathbb{R}$, and let $\mathscr{S}'$ be the corresponding dual space of generalized functions (see, for example, [18]). We use the standard notation $\langle\theta, \phi\rangle$ for the value of a functional $\theta \in \mathscr{S}'$ at a function $\phi \in \mathscr{S}$. The space $\mathscr{S}'$ contains the set $\mathscr{L}$ of measurable functions f locally summable on the real axis and increasing at infinity not faster than a degree of $|t|$, more exactly, satisfying the condition $\int (1+|t|)^{\lambda} |f(t)| dt < \infty$ with some exponent $\lambda = \lambda(f) \in \mathbb{R}$. The formula

$$\langle f, \phi\rangle = \int f(t)\phi(t) dt, \quad \phi \in \mathscr{S},$$

puts in correspondence to a function $f \in \mathscr{L}$ a functional $f \in \mathscr{S}'$.

For a functional $\theta \in \mathscr{S}'$, the derivative of order $n \geqslant 1$ is the functional $\theta^{(n)} \in \mathscr{S}'$ defined by the relation $\langle\theta^{(n)}, \phi\rangle = (-1)^n \langle\theta, \phi^{(n)}\rangle$, $\phi \in \mathscr{S}$. The Fourier transform of a functional $\theta \in \mathscr{S}'$ is the functional $\widehat{\theta} \in \mathscr{S}'$ acting by the formula

$$\langle\widehat{\theta}, \phi\rangle = \langle\theta, \widehat{\phi}\rangle, \quad \phi \in \mathscr{S}.$$

Denote by V the space of (complex-valued) bounded Borel measures on $(-\infty,\infty)$. We will identify this space with the set of (complex-valued) functions z of bounded variation on $(-\infty,\infty)$ whose real and imaginary parts at the discontinuity points are between the limits on the right and on the left. The norm in the space V is the complete variation $\bigvee z$ of a measure (a function) $z \in V$.

Define the space $F = \check{V} = \{f \in C\colon \widehat{f} \in V\}$ of functions $f \in C$ whose Fourier transforms are bounded Borel measures (in general, complex-valued) on the axis; more exactly, the set of functions representable in the form

$$f(t) = \int e^{2\pi t\eta i}\, d\mu(\eta), \quad \text{where} \quad \mu = \mu_f = \widehat{f} \in V. \tag{2.1}$$

We will denote by $\|f\|_F$ the complete variation $\bigvee \mu$ of the measure μ in (2.1). The space $F = \check{V}$ is a Banach space with respect to this functional.

For $n \geqslant 1$, consider the space $Y^n = F \cap W^n_{\infty,\infty}$ of functions $f \in F$ that are $n-1$ times continuously differentiable on $(-\infty,\infty)$ and such that the derivatives $f^{(n-1)}$ of order $n-1$ are locally absolutely continuous on the axis, and $f^{(n)} \in L_\infty$. The embedding $Y^n \subset W^n_{\infty,\infty}$ is valid; moreover, if $f \in Y^n$, then $\|f\|_C \leqslant \bigvee \widehat{f}$. Therefore, the classical variant (1.1) of the Kolmogorov inequality implies the following inequality on the set Y^n:

$$\|f^{(k)}\|_C \leqslant K_{n,k} \left(\bigvee \widehat{f}\right)^{(n-k)/n} \|f^{(n)}\|^{k/n}_{L_\infty}, \quad f \in Y^n. \tag{2.2}$$

The best constants in this inequality and inequality (1.1) are related as follows:

$$K_{n,k} \leqslant C_{n,k}. \tag{2.3}$$

As we will see below, the latter inequality, depending on the values of the parameter n, can turn into an equality but can also be strict.

The following statement about function (1.2) will be used in what follows.

Lemma 1. *For all $n \geqslant 1$, the function f_n belongs to the space F and*

$$\|f_n\|_F = M_n^*, \quad M_n^* = \frac{4}{\pi} \sum_{\ell=0}^{\infty} \frac{1}{(2\ell+1)^{n+1}}. \tag{2.4}$$

Proof. Function (1.2) can be written in the exponential form

$$f_n(t) = \frac{2}{\pi i} \sum_{\ell=0}^{\infty} \frac{1}{(2\ell+1)^{n+1}} \left(e^{-in\pi/2} e^{i(2\ell+1)t} - e^{in\pi/2} e^{-i(2\ell+1)t}\right). \tag{2.5}$$

Using the Dirac δ-function, we define the measure $d\mu_n$ on the axis by the relation

$$d\mu_n(\eta) = \frac{2}{\pi i} \sum_{\ell=0}^{\infty} \frac{e^{-in\pi/2}}{(2\ell+1)^{n+1}} \delta(\eta - \eta_\ell) - \frac{2}{\pi i} \sum_{\ell=0}^{\infty} \frac{e^{in\pi/2}}{(2\ell+1)^{n+1}} \delta(\eta + \eta_\ell), \tag{2.6}$$

where $\eta_\ell = \dfrac{2\ell+1}{2\pi}$, $\ell \geqslant 0$. Representation (2.5) of the function f_n by means of measure (2.6) can be written in the form

$$f_n(t) = \int_{-\infty}^{\infty} e^{2\pi\eta t i} d\mu_n(\eta) = \widetilde{\mu_n}(t). \tag{2.7}$$

Representation (2.7) means that $f_n \in F$ and

$$\|f_n\|_F = \bigvee \mu_n = \frac{4}{\pi}\sum_{\ell=0}^{\infty} \frac{1}{(2\ell+1)^{n+1}}.$$

Lemma 1 is proved.

Lemma 2. *For all $n \geqslant 2$, the function f_n belongs to the space Y^n and provides the following bound for the best constant $K_{n,k}$ in inequality (2.2):*

$$K_{n,k} \geqslant M_{n-k}\left(M_n^*\right)^{-\frac{n-k}{n}}. \tag{2.8}$$

Proof. According to Lemma 1, the function f_n belongs to the space F and, hence, $f_n \in F \cap W^n = Y^n$. Substituting the function f_n into inequality (2.2), we obtain bound (2.8). Lemma 2 is proved.

Lemmas 1 and 2 imply the following statement.

Theorem 1. *For odd $n \geqslant 3$ and all $1 \leqslant k < n$, the best constants in inequalities (2.2) and (1.1) coincide; i.e., the following equality holds:*

$$K_{n,k} = C_{n,k}, \tag{2.9}$$

and the function f_n defined by formula (1.2) is extremal not only in inequality (1.1) but also in inequality (2.2).

Proof. For all $n \geqslant 2$, bounds (2.3) and (2.8) for $K_{n,k}$ are valid. According to formulas (1.3) and (2.4), for odd n, the uniform norm and the F-norm of the function f_n coincide: $M_n = M_n^*$. This implies all the assertions of Theorem 1.

2.2 The Best Approximation of the Differentiation Operator on the Set of Smooth Functions with Exactly Given Fourier Transform

In the space Y^n, consider the class $\mathcal{Q}^n = \{f \in Y^n \colon \|f^{(n)}\|_{L_\infty} \leqslant 1\}$. Consider the problem of the best uniform approximation of the kth order differentiation operator on class $\mathcal{Q}^n$ by the set $\mathscr{B}(F, C)$ of bounded linear operators from F to C:

$$\mathcal{E}_{n,k}(N) = \inf\{\mathcal{U}(T) \colon \|T\|_{\widehat{V}\to C} \leqslant N\} \tag{2.10}$$

$$\mathcal{U}(T) = \sup\{\|f^{(k)} - Tf\| \colon f \in \mathcal{Q}^n\}. \tag{2.11}$$

Using the same argument as in the proof of (1.7), we prove that value (2.10) and the best constant in (2.2) are related as follows:

$$\mathcal{E}_{n,k}(N) = k\left(\frac{K_{n,k}}{n}\right)^{\frac{n}{k}}\left(\frac{N}{n-k}\right)^{-\frac{n-k}{k}}, \quad N > 0. \tag{2.12}$$

For all values of the parameters, we have the inequality

$$\mathcal{E}_{n,k}(N) \leqslant E_{n,k}(N), \quad N > 0. \tag{2.13}$$

This fact follows from formulas (1.7) and (2.12) and inequality (2.3). In the proof of the following theorem, we will give another, "operator," justification of this inequality.

Theorem 2. *For odd* $n \geqslant 3$, $1 \leqslant k < n$, *and* $N > 0$, *the following equality holds:*

$$\mathcal{E}_{n,k}(N) = E_{n,k}(N). \tag{2.14}$$

Moreover, an operator extremal in problem (1.5) *is also extremal in problem* (2.10).

Proof. Equality (2.14), certainly, follows from formulas (1.7) and (2.12) and equality (2.9).

Let us prove the second assertion of the theorem. We have the embedding $\mathscr{B}(C,C) \subset \mathscr{B}(F,C)$ together with the corresponding inequality for the operator norms. Indeed, let $T \in \mathscr{B}(C,C)$. Then $\|Tf\|_C \leqslant \|T\|_{C\to C}\|f\|_C \leqslant \|T\|_{C\to C} \bigvee \widehat{f}$ for all functions $f \in F$. Consequently, if $T \in \mathscr{B}(C,C)$, then $T \in \mathscr{B}(F,C)$ and $\|T\|_{F\to C} \leqslant \|T\|_{C\to C}$. Further, for every operator $T \in \mathscr{B}(C,C)$, the values of deviations (2.11) and (1.4) are related by the inequality $\mathcal{U}(T) \leqslant U(T)$, because $\mathcal{Q}^n \subset Q^n$.

For all $n \geqslant 2$, we have the following relations for an operator T extremal in problem (1.5):

$$\mathcal{E}_{n,k}(N) \leqslant \mathcal{U}(T) \leqslant U(T) = E_{n,k}(N). \tag{2.15}$$

Hence, inequality (2.13) holds for all $n \geqslant 2$ again. If n is odd, then, by (2.14), it follows from (2.15) that the operator T is also extremal in problem (2.10). Theorem 2 is proved.

2.3 Optimal Differentiation of Functions from the Class $\mathcal{Q}^n$ when the Fourier Transform is Known with an Error

Let $\mathscr{O}(V,C)$ be the set of all mappings from the space V of functions of bounded variation to the space C of continuous bounded functions. For a mapping $\Upsilon \in \mathscr{O}(V,C)$ and a parameter $\delta > 0$, define

$$R(\Upsilon) = R^{\delta}_{n,k}(\Upsilon) = \sup\left\{\|f^{(k)} - \Upsilon g\|_C \colon f \in \mathcal{Q}^n,\ g \in V,\ \bigvee\left(g - \widehat{f}\right) \leqslant \delta\right\}. \tag{2.16}$$

The smallest value

$$\varrho_{n,k}(\delta) = \inf\{R(\Upsilon) \colon \Upsilon \in \mathscr{O}(V,C)\} \tag{2.17}$$

of quantity (2.16) over all mappings $\Upsilon \in \mathscr{O}(V,C)$ is the smallest error of recovery of the differentiation operator $D^k = d^k/dt^k$ on functions from the class $\mathcal{Q}^n$ whose Fourier transform is given with the error δ in variation. The problem is to calculate (or at least to estimate) value (2.17) and find a mapping $\Upsilon^{\delta}_{n,k} \in \mathscr{O}(V,C)$ at which the infimum in (2.17) is attained; it is called an *optimal recovery method.*

Theorem 3. *For odd $n \geqslant 3$, the equality*

$$\varrho_{n,k}(\delta) = C_{n,k}\delta^{\frac{n-k}{n}}$$

holds. Moreover, if $T_{n,k}$ is an extremal operator in Stechkin's problem (1.5) *for value* (1.16) *of the parameter N, then the operator defined on the space V by the formula*

$$\Upsilon^{\delta}_{n,k} g = T_{n,k}\widehat{g}, \quad g \in V, \tag{2.18}$$

is an optimal method in problem (2.17).

Proof. Using the same argument as in the proof of (1.17), we prove the following statement. For all $n \geqslant 2$, we have the equality

$$\varrho_{n,k}(\delta) = K_{n,k}\delta^{\frac{n-k}{n}}.$$

Moreover, if $T_{n,k}$ is an extremal operator in problem (2.10) for

$$N = \frac{n-k}{n} K_{n,k}\delta^{-\frac{k}{n}},$$

then the operator defined on the space V by formula (2.18) is an optimal method in problem (2.17).

By Theorems 1 and 2, this statement for odd n becomes Theorem 3. Theorem 3 is proved.

2.4 Even n

In Subsects. 2.1–2.3, it is shown that, for odd n, extremal problems (2.2), (2.10), and (2.17) in the space Y^n reduce to the corresponding problems in the space W^n, whose solutions are known and were described in Sect. 1. Theorems 1, 2 and 3 are, most likely, not valid for even n. They does not hold at least for $n = 2$ $(k = 1)$. The exact inequality (2.2) and the solution of problem (2.10) for $n = 2$ and $k = 1$ were obtained in [4]. Let us comment these two results.

In [4, Theorem 2], it was proved that the smallest possible constant in inequality (2.2) for $n = 2$ and $k = 1$ is

$$K_{2,1} = \frac{\pi}{2}\left(\frac{4}{\pi}\sum_{\ell=0}^{\infty}\frac{1}{(2\ell+1)^3}\right)^{-1/2}, \tag{2.19}$$

and the function f_2 is extremal. The following bounds are valid for constant (2.19):

$$\sqrt{\frac{\pi}{2}} < K_{2,1} < \sqrt{2}. \tag{2.20}$$

According to Hadamard's result [10], in this case, $C_{2,1} = \sqrt{2}$. The second inequality in (2.20) means that the strict inequality $K_{2,1} < C_{2,1}$ holds.

In [4, Theorem 5], the solution of problem (2.10) for $n = 2$ and $k = 1$ was obtained. The extremal operator $\Theta_{2,1}$ found in [4] has the form of convolution with a singular kernel; it is different from operator (1.13), which, according to Stechkin's result [17], is extremal in problem (1.5). Note that, in addition, $\Theta_{2,1} \notin \mathscr{B}(C, C)$.

2.5 The Problems in the Space of Functions with Summable Fourier Transform

Consider the space $F_0 = \check{L}$ of functions from $C_0(-\infty, \infty)$ whose Fourier transforms are summable functions. In other words, the space F_0 consists of functions representable in the form

$$f(t) = \int e^{2\pi t\eta i}\varphi(t)\,d\eta, \quad \text{where} \quad \varphi = \widehat{f} \in L.$$

We define the norm in the space F_0 by the formula $\|f\|_{F_0} = \|\widehat{f}\|_L$. With respect to this functional, the space F_0 is Banach. For $n \geqslant 2$, define $Y_0^n = F_0 \cap W^n$; this is the space of functions $f \in C_0(-\infty, \infty)$ that are $n-1$ times continuously differentiable on the axis and such that the derivative $f^{(n-1)}$ of order $n-1$ is locally absolutely continuous, and the nth order derivative $f^{(n)}$ belongs to the space L_∞.

In the author's opinion, analogs of inequality (2.2) and problems (2.10) and (2.17) in the space Y_0^n are of interest. One can expect that the corresponding pairs of problems on the spaces Y^n and Y_0^n are equivalent. Let us explain this assumption for inequality (2.2) and problem (2.17).

Denote by $K_{n,k}^0$ the smallest constant in the inequality

$$\|f^{(k)}\|_C \leqslant K_{n,k}^0 \|\widehat{f}\,\|_L^{(n-k)/n} \|f^{(n)}\|_{L_\infty}^{k/n}, \quad f \in Y_0^n. \tag{2.21}$$

Obviously, $K_{n,k}^0 \leqslant K_{n,k}$. Let us discuss the possible equality

$$K_{n,k}^0 = K_{n,k}. \tag{2.22}$$

For $n = 2$ and $k = 1$, equality (2.22) was proved in [4]. For this, in [4, Lemma 2], a special case of the following lemma was used.

Lemma 3. *For the function f_n with $n \geqslant 2$, there exists a family of functions $\{g_\alpha\} \subset Y_0^n$ depending on a parameter α, $0 < \alpha \leqslant \alpha_0 = 1/(2\pi)$, and possessing the following properties:*

(1) $\|\widehat{g_\alpha}\|_L = \bigvee \widehat{f_n}$, $0 < \alpha \leqslant \alpha_0$;
(2) $\lim\limits_{\alpha\to+0} \|g_\alpha^{(n)}\|_{L_\infty} = \|f_n^{(n)}\|_{L_\infty}$;
(3) $\lim\limits_{\alpha\to+0} \|g_\alpha^{(l)}\|_{C_0} = \|f_n^{(l)}\|_C$, $0 \leqslant l \leqslant n-1$;
(4) *for all* $A > 0$ *and* $0 \leqslant l \leqslant n-1$, *the following limit relation holds:*

$$\|f_n^{(l)} - g_\alpha^{(l)}\|_{C[-A,A]} \to 0, \quad \alpha \to +0.$$

The proof of this statement for $n \geqslant 3$ is similar to the proof of [4, Lemma 2].
Lemma 3 and Theorem 1 imply the following statement.

Corollary. *Property* (2.22) *holds for odd* $n \geqslant 3$ *and all* $1 \leqslant k \leqslant n-1$.

In conclusion, let us discuss an analog of problem (2.17) in the space Y_0^n. Let $\mathscr{O}(L,C)$ be the set of all (single-valued) mappings from the space L of summable functions to the space C of continuous bounded functions. For a mapping $\Upsilon \in \mathscr{O}(L,C)$ and a parameter $\delta > 0$, define

$$R^0(\Upsilon) = \sup\left\{\|f^{(k)} - \Upsilon g\|_C \colon f \in Y_0^n,\ \|f^{(n)}\|_{L_\infty} \leqslant 1,\ g \in L,\ \|g - \widehat{f}\|_L \leqslant \delta\right\}.$$

We are interested in the value

$$\varrho_{n,k}^0(\delta) = \inf\{R^0(\Upsilon) \colon \Upsilon \in \mathscr{O}(L,C)\} \tag{2.23}$$

of the smallest error of recovery of the differentiation operator $D^k = d^k/dt^k$ on functions of the space Y_0^n whose Fourier transform is given with the error δ in the L-norm.

General results related to problems of recovery (see, for example, [3, Theorem 2.1]) give the bound $\varrho_{n,k}^0(\delta) \geqslant K_{n,k}^0 \delta^{\frac{n-k}{n}}$ for value (2.23) in terms of the best constant in inequality (2.21). The corollary and Theorems 1 and 3 imply the statement similar to Theorem 3.

Theorem 4. *The following equality holds for odd* $n \geqslant 3$:

$$\varrho_{n,k}^0(\delta) = C_{n,k}\delta^{\frac{n-k}{n}}.$$

Moreover, if $T_{n,k}$ *is an extremal operator in Stechkin's problem* (1.5) *for value* (1.16) *of the parameter* N, *then the operator defined on the space* L *by the formula*

$$\Upsilon_{n,k}^\delta g = T_{n,k}\widehat{g}, \quad g \in L,$$

is an optimal method in problem (2.23).

3 Conclusion

By now, exact solutions were known for the following three extremal problems in the space W^n, $n \geqslant 2$, of functions $f \in C(-\infty,\infty)$ that are $n-1$ times continuously differentiable on the axis, their derivative $f^{(n-1)}$ of order $n-1$ is locally absolutely continuous, and the derivative $f^{(n)}$ of order n belongs to the space L_∞. A.N. Kolmogorov (1939) obtained exact inequality between the uniform norm of the derivative of order k, $1 \leqslant k \leqslant n-1$, of functions from W^n and the norms of the function and the nth order derivative. S.B. Stechkin (1967), V.V. Arestov (1967), and A.P.Buslaev (1981) solved Stechkin's problem on the best approximation of the differentiation operator of order k on the class $\{f \in W^n : \|f^{(n)}\|_{L_\infty} \leqslant 1\}$ by bounded linear operators in the space $C(-\infty,\infty)$. By the known scheme, this enabled obtaining a solution of the problem on optimal differentiation of functions from the class W^n given with a known error in the uniform norm. In the present paper, we discussed analogs of all these problems in the narrower space $Y^n = F \bigcap W^n$ of functions from W^n whose Fourier transform is a bounded Borel measure on the axis. Unexpectedly, it turned out that the situation with solutions of the problems is different for even and odd n. For odd n, solutions of the corresponding problems on the spaces W^n and Y^n coincide. This is not so for even n. The corresponding problems have different solutions at least for $n = 2$.

References

1. Arestov, V.V.: On the best approximation of differentiation operators. Math. Notes **1**(2), 100–103 (1967). https://doi.org/10.1007/BF01268057
2. Arestov, V.V.: On the best approximation of differentiation operators in the uniform norm. Cand. Diss., Sverdlovsk (1969). (in Russian)
3. Arestov, V.V.: Approximation of unbounded operators by bounded operators and related extremal problems. Russ. Math. Surv. **51**(6), 1093–1126 (1996). https://doi.org/10.1070/RM1996v051n06ABEH003001
4. Arestov, V.V.: Best uniform approximation of the differentiation operator by operators bounded in the space L_2. Trudy Inst. Mat. Mekhaniki URO RAN **24**(4), 34–56 (2018). https://doi.org/10.21538/0134-4889-2018-24-4-34-56. (in Russian)
5. Babenko, V.F., Korneichuk, N.P., Kofanov, V.A., Pichugov, S.A.: Neravenstva dlya proizvodnykh i ikh prilozheniya [Inequalities for Derivatives and Their Applications]. Naukova Dumka Publisher, Kiev (2003). (in Russian)
6. Bosse, Yu.G. (Shilov, G.E.): On inequalities between derivatives. In: Collection of Works of Student Scientific Societies of Moscow State University, vol. 1, pp. 68–72 (1937). (in Russian)
7. Buslaev, A.P.: Approximation of a differentiation operator. Math. Notes **29**(5), 372–378 (1981). https://doi.org/10.1007/BF01158361
8. Domar, Y.: An extremal problem related to Kolmogoroff's inequality for bounded functions. Arkiv för Mat. **7**(5), 433–441 (1968). https://doi.org/10.1007/BF02590991
9. Gabushin, V.N.: Best approximations of functionals on certain sets. Math. Notes **8**(5), 780–785 (1970). https://doi.org/10.1007/BF01146932
10. Hadamard, J.: Sur le module maximum d'une fonction et de ses dérivées. Soc. Math. France, Comptes rendus des Séances **41**, 68–72 (1914)
11. Hardy, G.H., Littlewood, J.E.: Contribution to the arithmetic theory of series. Proc. Lond. Math. Soc. **11**(2), 411–478 (1912)

12. Ivanov, V.K., Vasin, V.V., Tanana, V.P.: Theory of Linear Ill-Posed Problems and Its Applications. VSP, Utrecht (2002)
13. Kolmogorov, A.N.: On inequalities between upper bounds of consecutive derivatives of an arbitrary function defined on an infinite interval. In: Selected works. Mathematics and Mechanics, pp. 252–263. Nauka Publisher, Moscow (1985). (Moskov. Gos. Univ., Uchenye Zap. (Mat. 3) 30, 3–16 (1939)). (in Russian)
14. Korneichuk, N.P.: Extremal Problems of Approximation Theory. Nauka, Moscow (1976). (in Russian)
15. Gradshteyn, I.S., Ryzhik, I.M.: Table of Integrals, Series, and Products. Elsevier/Academic Press, London, Oxford (2007)
16. Stechkin, S.B.: Inequalities between norms of derivatives of arbitrary functions. Acta Sci. Math. **26**(3–4), 225–230 (1965). (in Russian)
17. Stechkin, S.B.: Best approximation of linear operators. Math. Notes **1**(2), 91–99 (1967). https://doi.org/10.1007/BF01268056
18. Stein, E.M., Weiss, G.: Introduction to Fourier Analysis on Euclidean Spaces. Princeton University Press, Princeton (1971)

Feedback Minimum Principle for Optimal Control Problems in Discrete-Time Systems and Its Applications

Vladimir Dykhta and Stepan Sorokin(✉)

Matrosov Institute for System Dynamics and Control Theory, Irkutsk, Russia
dykhta@gmail.com, sorsp@mail.ru

Abstract. The paper is devoted to a generalization of a necessary optimality condition in the form of the Feedback Minimum Principle for a nonconvex discrete-time free-endpoint control problem. The approach is based on an exact formula for the increment of the cost functional. This formula is completely defined through a solution of the adjoint system corresponding to a reference process. By minimizing that increment in control variable for a fixed adjoint state, we define a multivalued map, whose selections are feedback controls with the property of potential "improvement" of the reference process. As a result, we derive a necessary optimality condition (optimal process does not admit feedback controls of a "potential descent" in the cost functional). In the case when the well-known Discrete Maximum Principle holds, our condition can be further strengthened. Note that obtained optimality condition is quite constructive and may lead to an iterative algorithm for discrete-time optimal control problems. Finally, we present sufficient optimality conditions for problems, where Discrete Maximum Principle does not make sense.

Keywords: Exact formula of the cost functional increment · Feedback controls · Necessary optimality conditions · Feedback Minimum Principle · Maximum Principle · Method of feedback iterations

1 Introduction

The paper concerns necessary (and sufficient) global optimality conditions for the following discrete optimal control problem (problem (P)):

$$x(t+1) = f\big(t, x(t), u(t)\big), \quad x(0) = x_0, \tag{1}$$

$$u(t) \in U(t), \quad t \in T, \tag{2}$$

$$J(\sigma) = l\big(x(N)\big) \to \min .$$

Partially supported by the Russian Foundation for Basic Research, projects nos 17-01-00733, 18-31-20030, 18-31-00425.

M. Khachay et al. (Eds.): MOTOR 2019, LNCS 11548, pp. 449–460, 2019.
https://doi.org/10.1007/978-3-030-22629-9_31

Here, $x(t) \in R^n$, sets $U(t) \subset R^m$ are compact for all $t \in T := \{0, \dots, N-1\}$. By σ we denote collections of vectors $\{x(t), u(t)\} = \{x(0), \dots, x(N), u(0), \dots, u(N-1)\}$, i.e. admissible processes of problem (P) (pairs of trajectories and controls), D stands for the set of all admissible processes in problem (P), and $\bar{\sigma} = \{\bar{x}(t), \bar{u}(t)\} \in D$ is the reference (examined) process.

The functions $f(t, x, u)$ are assumed to be continuous with respect to (w.r.t.) (x, u) and continuously differentiable w.r.t. x for all $t \in T$, the cost function $l(x)$ is smooth.

First of all, we are interested in the necessary conditions for optimality of $\bar{\sigma}$, using feedback controls $\{v(t, x)\}$ with the property of descent w.r.t. the functional J. Such controls are constructed via special solution of discrete Hamilton-Jacobi type inequality for weakly decreasing functions $\varphi(t, x)$ [1]. This special solution (being support majorant for the cost function J at point $\bar{\sigma}$) is completely defined by the trajectory $\{\psi(t)\}$ adjoint to the process $\bar{\sigma}$.

Being applied to classical optimal control problems in differential systems, the discussed approach leads to a rather effective and constructive necessary optimality condition. This condition, called Feedback Minimum Principle (FMP) [2,3], essentially strengthens the Pontryagin Maximum Principle. In [1] an analogue of the feedback principle was obtained for discrete optimal control problems, linear in the state variable. The present work contains a generalization of the results [2,3] for nonlinear discrete problem (P).

To illustrate our necessary optimality conditions, we consider certain modifications of examples from [4,5], which were used as counter-examples for the Discrete Maximum Principle (DMP) [4–8]. In such modifications, necessary optimality conditions with feedback controls are more effective either if DMP is not applicable at all, or when it is not able to discard nonoptimal processes. In the second case, FMP does work and leads to an optimal process.

2 Construction of Feedback Descent Controls

For a discrete dynamic system, the property of weak decrease of a function $\varphi(t, x) : T \times R^n \to R$ means that for any initial position (t_*, x_*) there exists a trajectory $\{x(t)\}$, $t = t_*, \dots, N$, $x(t_*) = x_*$ (with a corresponding admissible control $\{u(t)\}$, $t = t_*, \dots, N-1$) such that $\varphi\big(t+1, x(t+1)\big) - \varphi\big(t, x(t)\big) \le 0$ for $t = t_*, \dots, N$. The following Hamilton-Jacobi type inequality guarantees the property of weak decrease:

$$\min_{u \in U(t)} \varphi\big(t+1, f(t, x, u)\big) - \varphi(t, x) \le 0 \quad \forall x \in R^n,\ t \in T. \tag{3}$$

Necessary optimality conditions, discussed below, use solutions of (3) under appropriate boundary conditions.

Let us describe the construction of a desired solution to inequality (3).

Introduce the Pontryagin function

$$H(t, x, \psi, u) = \langle \psi, f(t, x, u) \rangle,$$

the adjoint (for $\bar{\sigma}$) system

$$\psi(t) = H_x\big(t, \bar{x}(t), \psi(t), \bar{u}(t)\big), \quad \psi(N) = l_x\big(\bar{x}(N)\big) \tag{4}$$

(note that the terminal condition corresponds to the *minimum* condition of the Pontryagin function in DMP) and the function

$$\varphi^*(t, x) = \langle \psi(t) - l_x\big(\bar{x}(t)\big), x \rangle + l(x) \tag{5}$$

($\langle \cdot, \cdot \rangle$ stands for the scalar product). Due to the terminal condition in (4), we obtain

$$\varphi^*(N, x) = l(x). \tag{6}$$

It is easy to see that the following equality holds:

$$\sum_{t \in T} \Big[\varphi^*\Big(t+1, f\big(t, x(t), u(t)\big)\Big) - \varphi^*\big(t, x(t)\big)\Big] = \\ J(\sigma) - \varphi^*(0, x_0) \quad \forall \sigma \in D. \tag{7}$$

Introduce the function

$$K(t, x, u) = \varphi^*\big(t+1, f(t, x, u)\big) - \varphi^*(t, x). \tag{8}$$

Then, by equalities (5)–(7), one can obtain the exact formula for the increment of the cost functional J:

$$J(\sigma) - J(\bar{\sigma}) = \sum_{t \in T} \Big[K\big(t, x(t), u(t)\big) - K\big(t, \bar{x}(t), \bar{u}(t)\big)\Big] \quad \forall \sigma \in D. \tag{9}$$

Based on the previous formula (see also (7)) for any position (t, x) we define the set $U_*(t, x)$ of feedback controls, which may generate the deepest descent for functionals (9) and (7). Evidently, if $\bar{\sigma}$ is an optimal process, then descent controls do not exist for $\bar{\sigma}$.

The discussed idea leads to the following multivalued φ^*-extremal map:

$$U_*(t, x) = \operatorname*{Argmin}_{u \in U(t)} \Big[H\big(t, x, p(t+1), u\big) + l\big(f(t, x, u)\big)\Big], \quad t \in T, \tag{10}$$

where

$$p(t) = \psi(t) - l_x\big(\bar{x}(t)\big), \quad t = 0, \dots, N.$$

Any sequence of vectors $\{v(t, x)\}$, $t \in T$, satisfying the inclusion $v(t, x) \in U_*(t, x)$ on $T \times R^n$, generates a trajectory $\{x^v(t)\}$ of the discrete system

$$x(t+1) = f\Big(t, x(t), v\big(t, x(t)\big)\Big), \quad x(0) = x_0, \tag{11}$$

and the open-loop control $\big\{u^v(t) = v\big(t, x^v(t)\big)\big\}$. Denote by D_* the set of sequences $\nu = \big\{x^v(t), v(t, x)\big\}$, which may be obtained in this way. Let $J(\nu) = l\big(x^v(N)\big)$ $\forall \nu \in D_*$.

Theorem 1. *If process* $\bar{\sigma} = \{\bar{x}(t), \bar{u}(t)\}$ *is optimal for problem (P), then the following inequality holds:*

$$J(\bar{\sigma}) \leq J(\nu) \quad \forall \nu \in D_*.$$

In other words, there are no feedback descent controls at the point $\bar{\sigma}$ *which can be generated by the* φ^**-extremal map* $U_*(t,x)$.

To prove the Theorem, it is sufficient to note that any sequence $\nu \in D_*$ generates the pair $\sigma = \{x^v(t), u^v(t)\} \in D$ such that $J(\sigma) = J(\nu)$.

The presented idea does not demand multifunction $U_*(t,x)$ to be constructed by the extremal principle, at all. In fact, any map $V(t,x) \subset U(t)$ on $T \times R^n$ could be chosen instead of $U_*(t,x)$. Of course, such a casual mapping $V(t,x)$ is generically useless.

Let us show that φ^*-extremal multifunction (10) for feedback descent controls corresponds to a solution of the Hamilton-Jacobi inequality (3). The latter one is designed by some "calibration" of function φ^*.

Let $R(t)$ be a reachable set of system (1), (2) at t; obviously, $R(t)$ is a compact set in R^n $\forall t = 1,\ldots,N$. Given an open set $Q(t) \supseteq R(t)$ for all $t = 1,\ldots,N$, define (see also (8))

$$m(t) = \sup_{x \in Q(t)} \min_{u \in U(t)} K(t,x,u), \quad t \in T,$$

$$r(t) = r(t+1) - m(t), \quad r(N) = 0,$$

$$\widetilde{\varphi}(t,x) = \varphi^*(t,x) - r(t), \quad (t,x) \in T \times Q(t).$$

It is easy to check that function $\widetilde{\varphi}$ satisfies the condition of weak decrease (3) on $T \times Q(t)$, and the $\widetilde{\varphi}$-extremal multifunction for descent controls coincides with $U_*(t,x)$. This reasoning provides additional justification for using the set of feedback descent controls (10).

We also stress an important role of Theorem 1 for applications. If process $\bar{\sigma}$ does not satisfy this necessary optimality condition, then one has a process that improves $\bar{\sigma}$ (new process has a smaller value of the cost functional J).

Example 1. Consider a modification of Example 2 from [4, p. 431], which is used to show that DMP is not applicable to problems of optimal control for systems, obtained by a difference approximation of continuous ones, in general. Modification is due to a square term in the cost function. The example illustrates the applicability of Theorem 1 in contrast to DMP.

$$J = x^2(2) + y(2) \to \min;$$
$$x(t+1) = x(t) + \frac{1}{2}u(t), \quad x(0) = 0,$$
$$y(t+1) = y(t) + x^2(t) - u^2(t), \quad y(0) = 0,$$
$$|u(t)| \leq 1, \quad t = 0,1.$$

One can check that

$$\min J = \min_{|u(t)|\le 1,\ t=0,1} \Big\{ -\frac{1}{4}\Big[2u^2(1) + \big(u(1)-u(0)\big)^2\Big]\Big\} = -\frac{7}{4},$$

and the minimum is attained by $u_0^* = \pm 1$, $u_1^* = \mp 1$.

The Pontryagin function is

$$H = \psi(t+1)\Big[x(t) + \frac{1}{2}u(t)\Big] + y(t) + x^2(t) - u^2(t),$$

and the adjoint system writes

$$\psi(t) = \psi(t+1) + 2x(t), \quad \psi(2) = 2x(2).$$

Let us consider the process $\bar{\sigma}$ with $\bar{u} \equiv -1$,

$$\bar{x}(1) = -\frac{1}{2},\ \bar{x}(2) = -1,$$
$$\bar{y}(1) = -1,\ \bar{y}(2) = -\frac{7}{4},$$
$$\bar{\psi}(1) = -3,\ \bar{\psi}(2) = -2,$$

and $J(\bar{\sigma}) = -\frac{3}{4}$.

Let us test this process by the necessary optimality condition proposed Theorem 1. The selectors of $U_*(t,x)$ are described by the following conditions:

$$t=0: \quad -u_0 - \frac{3}{4}u_0^2 \to \min \quad \Rightarrow \quad U_*(0,x_0) = \{1\};$$

$$t=1: \quad -\frac{3}{4}u_1^2 + x_1u_1 \to \min \quad \Rightarrow \quad U_*(1,x_1) = \begin{cases} \{-1\}, & x_1 > 0, \\ \{1\}, & x_1 < 0, \\ \{-1,1\}, & x_1 = 0. \end{cases}$$

Any feedback control $v: v(t,x) \in U_*(t,x)$ generates process $\widetilde{\sigma}$ with $\widetilde{u}(0) = 1$, $\widetilde{u}(1) = -1$, $\widetilde{x}(1) = \frac{1}{2}$, $\widetilde{x}(2) = 0$, $\widetilde{y}(1) = 1$, $\widetilde{y}(2) = -\frac{7}{4}$, $J(\widetilde{\sigma}) = -\frac{7}{4}$. Obviously, $\widetilde{\sigma}$ brings a global solution. Notice that the optimal process $\widetilde{\sigma}$ does not satisfy DMP.

Example 2. This example is aimed to show that Theorem 1 is rather effective to discard nonoptimal DMP-extrema (control processes satisfying DMP).

Consider the following nonconvex problem:

$$J = y(2) - ax^2(2) \to \min;$$
$$x(t+1) = x(t) + (t-1)u(t), \quad x(0) = 0,$$
$$y(t+1) = y(t) + \big(u(t)-1\big)x(t), \quad y(0) = 0,$$
$$|u(t)| \le 1, \quad t = 0,1; \quad a > 0.$$

It is easy to check that

$$\min J = \min_{|u(t)|\le 1,\ t=0,1} \big\{ -\big(u(1)-1\big)u(0) - au^2(0)\big\} = -2 - a.$$

Let us specify some objects. The Pontryagin function:

$$H = \psi(t+1)\big(x(t) + (t-1)u(t)\big) + y(t) + \big(u(t) - 1\big)x(t),$$

and the adjoint system:

$$\psi(t) = \psi(t+1) + u(t) - 1, \quad \psi(2) = -2ax(2).$$

The H-minimum condition looks as follows: $\big[(t-1)\psi(t+1) + x(t)\big]u_t \to \min$; $|u_t| \leq 1$.

Consider the process $\bar{\sigma}$: $\bar{u} \equiv 1$, $\bar{x}(1) = \bar{x}(2) = -1$, $\bar{y} \equiv 0$, $\bar{\psi} \equiv 2a$, $J(\bar{\sigma}) = -a$. Notice that $\bar{\sigma}$ satisfies DMP.

Let us apply Theorem 1 to $\bar{\sigma}$. The φ^*-extremal map (10) is defined by the following optimization problems:

$$t = 0: \quad -au_0^2 \to \min \Rightarrow \quad U_*(0, x_0) = \{\pm 1\};$$

$$t = 1: \quad x_1 u_1 \to \min \Rightarrow \quad U_*(1, x_1) = \begin{cases} \{-1\}, & x_1 > 0, \\ \{1\}, & x_1 < 0, \\ [-1, 1], & x_1 = 0. \end{cases}$$

Choosing the feedback control $v(t, x)$:

$$v(0) = -1, \quad v(1, x) = \begin{cases} -1, & x \geq 0, \\ 1, & x < 0, \end{cases}$$

one can obtain the process $\widetilde{\sigma}$: $\widetilde{u} \equiv -1$, $\widetilde{x}(1) = \widetilde{x}(2) = 1$, $\widetilde{y}(1) = 0$, $\widetilde{y}(2) = -2$ with $J(\widetilde{\sigma}) = -2 - a < -a = J(\bar{\sigma})$.

Thus, Theorem 1 leads to the global extremum $\widetilde{\sigma}$, starting from $\bar{\sigma}$.

3 Feedback Minimum Principle

In continuous optimal control problems, FMP [2,3] states that an optimal trajectory of the considered problem is necessarily optimal for a certain auxiliary problem of dynamic optimization, called the accessory one. Moreover, the analogue of Theorem 1 was covered by FMP. Below we show that for discrete optimization problems the situation is significantly different compared to the continuous case. However, this is not surprising: for example, in continuous optimization problems, the Pontryagin Maximum Principle is a universal necessary condition, but in discrete problems, it is not always the case [4–8].

Denote by (P_*) the following discrete problem of closed-loop (feedback) control:

$$J(\nu) := l\big(x(N)\big) \to \min, \quad \nu \in D_*,$$

where pairs $\nu = \{x(t), v(t, x)\}$ satisfy system (11) and the inclusion $v(t, x) \in U_*(t, x)$ on $T \times R^n$.

In general, the pair $\{\bar{x}(t), \bar{u}(t)\}$ is not admissible in problem (P_*). The following *minimum condition* $M(\bar{\sigma})$ guarantees that $\bar{\sigma} \in D_*$:

$$\bar{u}(t) \in U_*(t, \bar{x}(t)) \quad \forall t \in T.$$

By (10), $M(\bar{\sigma})$ is equal to the condition

$$H(t, \bar{x}(t), p(t+1), \bar{u}(t)) + l\Big(f(t, \bar{x}(t), \bar{u}(t))\Big) =$$

$$\min_{u \in U(t)} \Big[H(t, \bar{x}(t), p(t+1), u) + l\Big(f(t, \bar{x}(t), u)\Big)\Big] \quad \forall t \in T.$$

Given a process $\bar{\sigma}$ satisfying condition $M(\bar{\sigma})$, introduce the feedback control $\bar{v}(t, x) \in U_*(t, x)$ in the following way:

$$\bar{v}(t, x) = \begin{cases} \bar{u}(t), & (t, x) \in orb\ \bar{x}(t), \\ \text{any } w(t, x) \in U_*(t, x), & (t, x) \notin orb\ \bar{x}(t), \end{cases} \tag{12}$$

where $orb\ \bar{x}(t) = \{\ (t, \bar{x}(t)) \mid t = 0, \ldots, N\}$ is the orbit of trajectory $\bar{x}(t)$.

It is easy to see that $\bar{\nu} = \{\bar{x}(t), \bar{v}(t, x)\} \in D_*$. Then by Theorem 1 one can derive FMP as follows:

Theorem 2. *Let process $\bar{\sigma} = \{\bar{x}(t), \bar{u}(t)\}$ be optimal for problem (P) and satisfy the minimum condition $M(\bar{\sigma})$. Then process $\bar{\nu} = \{\bar{x}(t), \bar{v}(t, x)\}$ with control (12) is optimal for problem (P_*).*

In the assumptions of this theorem, problem (P_*) appears to be *accessory* (for $\bar{\sigma}$) in the classical sense. It means that (P_*) is a variational type problem, designed to analyze the optimality of process $\bar{\sigma}$.

FMP, generalizing Theorem 1, is a very attractive theoretical result. However, it is difficult to solve the accessory problem in practice. Therefore, in applications, one normally applies Theorem 1 instead of Theorem 2 (using the "trial and error" method when choosing selectors of multifunction $U_*(t, x)$). In addition, the class of problems, for which FMP is valid, is restricted by condition $M(\bar{\sigma})$. Although this condition is often met, it is not necessary at all (see Example 2, where process $\bar{\sigma}$ with $\bar{u} \equiv 1$ is admissible in problem (P_*) but does not solve it).

Example 3. Consider a modification of Example 6.46 from [5, Vol. II, p. 249] which shows a failure of DMP for optimal processes. At the same time, FMP holds here for some values of parameters.

$$J = x^2(3) + y(3) \to \min;$$

$$x(t+1) = g\big(t, u(t)\big), \quad x(0) = 0,$$

$$y(t+1) = ax^2(t) + by(t) - \frac{a}{b} g^2\big(t, u(t)\big), \quad y(0) = 0,$$

$$u(t) \in U, \quad t = 0, 1, 2; \quad a > 0,\ b > 0,$$

the function $g(t, \cdot)$ is continuous, the set U is compact.

$$\min J = \min_{u(2) \in U} \frac{b-a}{b} g^2\big(2, u(2)\big).$$

It means that any admissible process $\bar{\sigma}$ with $\bar{u}(2) \in \operatorname{Argmin}_{u \in U} \frac{b-a}{b} g^2(2, u(2))$ is optimal.

The Pontryagin function is

$$H = \psi(t+1) g\big(t, u(t)\big) + \xi(t+1)\Big[a x^2(t) + b y(t) - \frac{a}{b} g^2\big(t, u(t)\big)\Big]$$

(here, $\xi(t)$ is the adjoint of y).

Let us consider any optimal process and denote it by $\bar{\sigma} = \big(\bar{x}(t), \bar{y}(t), \bar{u}(t)\big)$. The corresponding trajectory $(\bar{\psi}, \bar{\xi})$ of adjoint system (4) is

$$\begin{array}{lll} \bar{\psi}(1) = 2abg\big(0, \bar{u}(0)\big), & \bar{\psi}(2) = 2ag\big(1, \bar{u}(1)\big), & \bar{\psi}(3) = 2g\big(2, \bar{u}(2)\big), \\ \bar{\xi}(1) = b^2, & \bar{\xi}(2) = b, & \bar{\xi}(3) = 1. \end{array}$$

One can check that $\bar{\sigma}$ does not satisfy DMP. Moreover, for $t = 0, 1$, the Pontryagin function H reaches on $\bar{u}(t)$ its maximum (rather than minimum), and the H-minimum conditions are as follows:

$$t = 0: \quad ab\Big[2g\big(0, \bar{u}(0)\big) g\big(0, u_0\big) - g^2\big(0, u_0\big)\Big] \to \min, \quad u_0 \in U;$$

$$t = 1: \quad a\Big[2g\big(1, \bar{u}(1)\big) g\big(1, u_1\big) - g^2\big(1, u_1\big)\Big] \to \min, \quad u_1 \in U.$$

When $t = 2$, the "$H \to \min$" condition takes the form

$$2g\big(2, \bar{u}(2)\big) g\big(2, u_2\big) - \frac{a}{b} g^2\big(2, u_2\big) \to \min, \quad u_2 \in U.$$

Therefore, all optimal processes do not satisfy DMP.

Let us check FMP for optimal process $\bar{\sigma}$. The φ^*-extremal multifunction (10) leads to the following conditions:

$$t = 0: \quad (ab - 1)\Big[2g\big(0, \bar{u}(0)\big) g\big(0, u_0\big) - g^2\big(0, u_0\big)\Big] \to \min, \quad u_0 \in U;$$

$$t = 1: \quad (a - 1)\Big[2g\big(1, \bar{u}(1)\big) g\big(1, u_1\big) - g^2\big(1, u_1\big)\Big] \to \min, \quad u_1 \in U;$$

$$t = 2: \quad \frac{b-a}{b} g^2\big(2, u_2\big) \to \min, \quad u_2 \in U$$

(compare with the previous formulas). It means that condition $M(\bar{\sigma})$ is satisfied and $\bar{\sigma}$ is admissible for problem (P_*) only when $a \leq 1$ and $ab \leq 1$. By the way, the conditions of Theorem 1 for $\bar{\sigma}$ are evidently relaxed.

Example 4. Now we propose another case, where Theorems 1 and 2 accompany one another. Consider the following modification of Example 3 from [4, p. 432]:

$$\begin{gathered} J = a x^2(2) + y(2) \to \min; \\ x(t+1) = 2u(t), \quad x(0) = 0, \\ y(t+1) = y(t) + x^2(t) - u^2(t), \quad y(0) = 0, \\ |u(t)| \leq 1, \quad t = 0, 1; \quad a \in R. \end{gathered}$$

Obviously,

$$\min J = \min_{|u(t)|\le 1,\ t=0,1} \left\{3u_0^2 + (4a-1)u_1^2\right\}.$$

Therefore, the minimizing controls are the following:

- if $a > \frac{1}{4}$, then $u^* \equiv 0$;
- if $a = \frac{1}{4}$, then $u_0^* = 0,\ u_1^* \in [-1,1]$;
- if $a < \frac{1}{4}$, then $u_0^* = 0,\ u_1^* \in \{-1,1\}$.

Any optimal process $\bar{\sigma}$ does not satisfy DMP for $a > 0$, but it satisfies condition $M(\bar{\sigma})$ and Theorems 1 and 2 ($\forall a$). However, in the case $a < \frac{1}{4}$ the condition $M(\sigma)$ does not hold for the process $\sigma \equiv 0$. Nevertheless, by applying Theorem 1 the process σ could be discarded.

4 Comparison with Known Necessary Optimality Conditions

Theorems 1 and 2 offer certain necessary conditions for global optimality, and the scope of application of constructive Theorem 1 is unlimited. As is known [6], only necessary conditions for a weak minimum have similar universality—in the class of sufficiently small variations $|x(t) - \bar{x}(t)|$ and $|u(t) - \bar{u}(t)|$ for all t. Therefore, these local conditions of optimality are less effective than those obtained above (both theoretically and practically).

The DMP is a necessary condition for a strong minimum (variations $|u(t) - \bar{u}(t)|$ do not have to be small), and in this sense DMP is more attractive than the conditions for a weak minimum. However, this criterion is not universal—it is valid for problem (P) under certain convexity conditions on the set $f(t, x, U(t))$; the simplest of these conditions is the convexity of $f(t, x, U(t))$ $\forall x \in R^n$ and $t \in T$. Theorems 1 and 2 do not imply these assumptions; however, FMP contains the assumption $M(\bar{\sigma})$ on the reference process. Therefore, a direct comparison of FMP with DMP in their applicability is difficult. However, as the examples show, the combination of Theorems 1 and 2 exceeds DMP (for problems where DMP is applicable) in efficiency. Note also that in the case when l is linear, condition $M(\bar{\sigma})$ coincides with the extremal condition from DMP, but, along with $M(\bar{\sigma})$, FMP requires $\bar{\sigma}$ to be optimal for the accessory problem (P_*). This fact essentially strengthens the necessary condition.

The previous examples show that FMP is more applicable than DMP. The case of linear cost function can be found, e.g., in [1] (this example coincides with Example 2, excepting the linear cost function $J = y(2)$). Now, we present another eloquent case:

Example 5. This quadratic modification of Example 8 by [4, p. 433] presents the situation when all optimal processes do not satisfy DMP, while FMP does hold for each one.

$$J = x^2(2) + y(2) \to \min;$$
$$x(t+1) = u(t), \quad x(0) = 0,$$
$$y(t+1) = y(t) + x^2(t), \quad y(0) = 0,$$
$$u(t) \in \{-1, +1\}, \quad t = 0, 1.$$

Notice that any admissible process is optimal.

The Pontryagin function is

$$H = \psi(t+1)u(t) + x^2(t) + y(t),$$

and the adjoint system takes the form:

$$\psi(t) = 2x(t), \quad \psi(2) = 2x(2).$$

It is notable that any optimal process $\bar{\sigma}$ does not satisfy the DMP:

$$2\bar{x}(t+1)u_t \to \min \quad \Rightarrow \quad u_t^* = -\operatorname{sign} \bar{u}(t).$$

At the same time, FMP holds for all optimal processes: $M(\bar{\sigma})$ and FMP lead to the condition

$$u_t^2 \to \min; \quad u_t \in \{-1, +1\}.$$

5 Sufficient Optimality Conditions

We proceed with the natural inequality

$$\Delta J(\bar{\sigma}) = J(\sigma) - J(\bar{\sigma}) \geq 0 \quad \forall \sigma \in D,$$

where the increment $\Delta J(\bar{\sigma})$ is described by the exact formula (9) (see also (5) and (8)).

Let $R(t)$ denote a compact reachable set of discrete system (1), (2) at time t, and $E(t) \supseteq R(t)$ be its outer estimate by some compact set $E(t) \subset R^n$ (here, $t = 1, \ldots, N$). Introduce the function

$$\mu(t) = \min_{(x,u) \in E(t) \times U(t)} K(t, x, u). \tag{13}$$

Represent this formula in a more traditional form — introduce the following objects:

$$h(t, x, \psi) = \min_{u \in U(t)} \Big[H(t, x, \psi, u) + l\big(f(t, x, u)\big) \Big] \tag{14}$$

(an analogue of the lower Hamiltonian of problem (P)),

$$\mathcal{K}(t,x) = h\big(t,x,p(t+1)\big) - l(x) - \langle p(t), x\rangle \tag{15}$$

(the extended lower Hamiltonian). So, function $\mu(t)$ from (13) can be defined in the following way:

$$\mu(t) = \min_{x\in E(t)} \mathcal{K}(t,x), \quad t = 1,\ldots,N. \tag{16}$$

By the definition of function $\mu(t)$ and formula (9), one obtains the following sufficient optimality condition:

Theorem 3. *Let a process $\bar\sigma = \{\bar x(t), \bar u(t)\}$ satisfy the minimum condition:*

$$\mathcal{K}\big(t,\bar x(t)\big) = \mu(t), \quad t = 1,\ldots,N,$$

where functions $\mathcal{K}$ and μ are defined by equalities (13)–(16) on some compact sets $E(t) \supseteq R(t)$, $t = 1,\ldots,N$. Then $\bar\sigma$ is optimal for (P).

Theorem 3 gives a first-order sufficient optimality condition, since it uses only the first derivatives of the input data. However, no convexity assumptions are imposed.

Note that these conditions are well combined with the necessary conditions of Theorems 1 and 2, since they are formulated in the same constructions: FMP can be applied iteratively. Assumed that these iterations stop, the resulting process can be checked for optimality by Theorem 3.

6 Conclusion

In the paper nonlocal necessary and sufficient optimality conditions with feedback comparison controls are obtained for nonconvex discrete control problems. The main results are related to the necessary optimality conditions in the class of feedback descent controls (Theorems 1 and 2). These conditions are constructive, independent of DMP, and lead to an efficient iterative algorithm for improving the control (see, e.g., [9]).

This algorithm seems efficient for solving complex discrete control problems with terminal constraints on trajectory, using the methods of penalty functions, modified Lagrangians, etc. Indeed, in all these methods, the associated problems of unconditional optimization should be also solved globally.

Theoretically, it is of interest to generalize Theorems 1 and 2 to more complex problems with constraints, analyze the connection with the quasi-maximum condition [5,7] (the influence of the time quantization frequency on the optimality conditions), the effect of relaxation of the problem, etc. These questions are also important for the theory of optimal control in differential systems.

References

1. Sorokin, S.P.: Necessary feedback optimality conditions and nonstandard duality in problems of discrete system optimization. Autom. Remote Control **75**(9), 1556–1564 (2014)
2. Dykhta, V.A.: Variational necessary optimality conditions with feedback descent controls for optimal control problems. Doklady Math. **91**(3), 394–396 (2015)
3. Dykhta, V.A.: Positional strengthenings of the maximum principle and sufficient optimality conditions. Proc. Steklov Inst. Math. **293**(1), S43–S57 (2016)
4. Gabasov, R., Kirillova, F.M.: Qualitative Theory of Optimal Processes. Nauka, Moscow (1971). [in Russian]
5. Mordukhovich, B.S.: Variational Analysis and Generalized Differentiation I-II. Fundamental Principles of Mathematical Sciences, vol. 330-331. Springer, Heidelberg (2006)
6. Propoi, A.I.: Elements of the theory of optimal discrete processes. Nauka, Moscow (1973). [in Russian]
7. Mordukhovich, B.S.: Approximation Methods in Optimization and Control Problems. Nauka, Moscow (1988). [in Russian]
8. Boltyanskiy, V.G.: Optimal Control of Discrete Systems. Nauka, Moscow (1973). [in Russian]
9. Sorokin, S.P., Staritsyn, M.V.: Numeric algorithm for optimal impulsive control based on feedback maximum principle. Optim. Lett. (2018). https://doi.org/10.1007/s11590-018-1344-9

Estimates of the Minimal Eigenvalue of the Controllability Gramian for a System Containing a Small Parameter

Mikhail Gusev[1,2(✉)]

[1] N.N. Krasovskii Institute of Mathematics and Mechanics, 16 S.Kovalevskaya str., 620108 Ekaterinburg, Russia
[2] Ural Federal University, 19 Mira street, 620002 Ekaterinburg, Russia
gmi@imm.uran.ru

Abstract. We consider a linear time-invariant control system with right-hand side depending on a small parameter. Assuming that the system is controllable, we study the asymptotics of the minimal eigenvalue of a system's controllability Gramian and provide some bounds for the eigenvalue. These estimates are applied to the study of convexity properties of reachable sets for nonlinear control systems with integral constraints on control variables.

Keywords: Control system · Controllability Gramian · Small parameter · Reachable set · Integral constraints

1 Introduction

Consider the linear time-invariant control system

$$\dot{x}(t) = \varepsilon Ax(t) + Bu(t),\ t \in [0,1], \tag{1}$$

$x \in \mathbb{R}^n$, $u \in \mathbb{R}^r$, and $\varepsilon > 0$ is a small parameter. If the pair (A, B) is completely controllable then $(\varepsilon A, B)$ is also controllable for any $\varepsilon \neq 0$. In this case the minimal eigenvalue $\nu(W_\varepsilon)$ of the controllability Gramian W_ε of (1) is positive for every value $\varepsilon > 0$. In this paper we study the asymptotics of $\nu(W_\varepsilon)$ for small ε and apply it to propose sufficient conditions for the convexity of reachable sets for a nonlinear time-invariant control-affine system on a small time interval under quadratic integral constraints on control variables. The proof is based on the result of Polyak [15] on the convexity of reachable sets for a nonlinear control system with $\mathbb{L}_2$ norms of controls bounded from above by a sufficiently small number. The reachability properties of nonlinear systems with integral constraints and algorithms for the construction of reachable sets were investigated in the papers [10,15], [5,6]. The problems of control and estimation under integral constraints were studied in many papers (see, for example, [1,3,4,7,11]).

M. Khachay et al. (Eds.): MOTOR 2019, LNCS 11548, pp. 461–473, 2019.
https://doi.org/10.1007/978-3-030-22629-9_32

2 Preliminaries

Further we use the following notation. By $A^\top$ we denote the transpose of a real matrix A, I is an identity matrix, 0 stands for a zero vector or a zero matrix of appropriate dimension. For $x, y \in \mathbb{R}^n$ let $(x, y) = x^\top y$ denotes the inner product of two vectors, $x^\top = (x_1, \ldots, x_n)$, $\|x\| = (x, x)^{\frac{1}{2}}$ be the Euclidean norm, and $B(\bar{x}, r) = \{x \in \mathbb{R}^n : \|x - \bar{x}\| \leq r\}$ be a ball of radius $r > 0$ centered at $\bar{x}$. For a real $n \times n$ matrix A a spectral matrix norm induced by the Euclidean vector norm is denoted as $\|A\|$. The symbols $\mathbb{L}_1$, $\mathbb{L}_2$ and $\mathbb{C}$ stand for the spaces of summable, square summable and continuous functions respectively. The norms in these spaces are denoted as $\|\cdot\|_{\mathbb{L}_1}$, $\|\cdot\|_{\mathbb{L}_2}$, $\|\cdot\|_{\mathbb{C}}$.

Definition 1. *The symmetric matrix $W_\varepsilon(t)$ defined by the equality*

$$W_\varepsilon(t) = \int_0^t X_\varepsilon(t, \tau) B B^\top X_\varepsilon^\top(t, \tau) d\tau, \tag{2}$$

where $X_\varepsilon(t, \tau)$ is a fundamental Cauchy matrix of system (1) ($\dot{X}_\varepsilon(t, \tau) = A X_\varepsilon(t, \tau)$, $X(\tau, \tau) = I$) is called the controllability Gramian of the control system (1).

Differentiating equality (2) we get that $W_\varepsilon(t)$ is a solution of the linear differential equation

$$\dot{W}_\varepsilon = \varepsilon A W_\varepsilon + \varepsilon W_\varepsilon A^\top + B B^\top, \; W_\varepsilon(0) = 0. \tag{3}$$

Proposition 1. *The matrix $W_\varepsilon(t)$, $t > 0$ is positive definite for every $\varepsilon \neq 0$ if and only if the pair (A, B) is completely controllable.*

Proof. Really, complete controllability of (A, B) is equivalent to the equality

$$\mathrm{span}(B, AB, ..., A^{n-1}B) = \mathbb{R}^n,$$

where $\mathrm{span}(B, AB, ..., A^{n-1}B)$ denotes the linear span of the columns of the corresponding matrices. For $\varepsilon \neq 0$ we have

$$\mathrm{span}(B, AB, ..., A^{n-1}B) = \mathrm{span}(B, \varepsilon AB, ..., \varepsilon^{n-1} A^{n-1} B).$$

Thus $(\varepsilon A, B)$ is contrlollable iff (A, B) is controllable. Since controllability is equivalent to non singularity of the controllability Gramian, this implies the assertion.

Let us look for $W_\varepsilon(t)$ as a sum of a series in powers of ε

$$W_\varepsilon(t) = V_0(t) + \varepsilon V_1(t) + \varepsilon^2 V_2(t) + ..., \; V_k(0) = 0, \; k = 0, 1, ... \tag{4}$$

Differentiating (4) and equating multipliers in front of equal degrees of ε we get

$$\dot{V}_0(t) = BB^\top,\ \dot{V}_k(t) = AV_{k-1}(t) + V_{k-1}(t)A^\top,\ k = 1, 2, ... \tag{5}$$

After integration Eq. (5) we get

$$V_0(t) = tU_0,\ V_i(t) = \frac{t^{i+1}}{(i+1)!}AU_i,\ i = 1, 2, ...,$$

where

$$U_0 = BB^\top,\ U_i = AU_{i-1} + U_{i-1}A^\top,\ k = 1, 2, ... \tag{6}$$

Thus for $W_\varepsilon = W_\varepsilon(1)$ we have

$$W_\varepsilon = \sum_{k=0}^{\infty} \frac{\varepsilon^k}{(k+1)!} U_k. \tag{7}$$

By virtue of the estimate $\|U_k\| \le 2\|A\|\|U_{k-1}\| \le 2^k\|A\|^k\|U_0\|$ the series (7), (4) are majorized by the converging series

$$\sum_{k=0}^{\infty} \frac{(2\varepsilon\|A\|)^k}{(k+1)!}\|U_0\|.$$

As a result we arrive at the following statement.

Proposition 2. *The matrix $W_\varepsilon = W_\varepsilon(1)$ is represented by the sum of series (7), uniformly convergent on every bounded subset of $\mathbb{R}$.*

We deal with the estimates of the asymptotic behaviour of the minimal eigenvalue $\nu(W_\varepsilon)$ under $\varepsilon \to 0$. Note that all the matrices U_k in (7) are symmetric but not necessarily positive semi-definite. For U_0 we obviously have $\nu(U_0) \ge 0$. If $\nu(U_0) > 0$ then there exists $\alpha > 0$ such that $\nu(W_\varepsilon) \ge \alpha$ for sufficiently small ε. Further, we assume that $\nu(U_0) = 0$, hence $\nu(W_\varepsilon) \to 0$ as $\varepsilon \to 0$.

Definition 2. (See, for example, [12]**)** *The pair (A, B) is linearly equivalent to the pair (A_1, B_1) if there exists a nonsingular matrix S such that $A_1 = SAS^{-1}$, $B_1 = SB$.*

The linear equivalent pairs generate equations of the same control system in different systems of coordinates. The pair (A, B) is controllable iff (A_1, B_1) is controllable.

Lemma 1. *Let (A, B), (A_1, B_1) be linearly equivalent pairs and let W_ε, W_ε^1 be corresponding controllability Gramians. There exist $\alpha > 0$, $\beta > 0$ such that*

$$\alpha\nu(W_\varepsilon) \le \nu(W_\varepsilon^1) \le \beta\nu(W_\varepsilon)$$

for all ε.

Proof. Denote by U^1_k, $k = 0, 1, 2, ...$ matrices in the expansion (7) for W^1_ε. Then by induction we get

$$U^1_0 = B_1 B_1^\top = SBB^\top S^\top = SU_0 S^\top,\ U^1_k = A_1 U^1_{k-1} + U^1_{k-1} A_1^\top =$$

$$SAS^{-1} SU_{k-1} S^\top + SU_{k-1} S^\top (S^{-1})^\top A^\top S^\top = S(AU_{k-1} + U_{k-1} A^\top) S^\top = SU_k S^\top.$$

The last implies the equality $W^1_\varepsilon = SW_\varepsilon S^\top$ which means that W^1_ε and W_ε are congruent matrices. For symmetric congruent matrices the following is true (see, for example,[13, Theorem 4.5.9]): for any symmetric matrix D there exist the numbers θ_i, $\lambda_1(SS^\top) \le \theta_i \le \lambda_n(SS^\top)$, $i = 1, ..., n$ such that

$$\lambda_i(SDS^\top) = \theta_i \lambda_i(D).$$

Here λ_i denote eigenvalues of the matrices ordered by ascending. The last implies the assertion of the lemma.

3 Estimates for Minimal Eigenvalues of Controllability Gramian

Consider systems with a single control input. In this case A is an $n \times n$ matrix and B is a column n-vector.

Theorem 1. *Assume that system is completely controllable. If $n = 2$ then there exist $\alpha > 0, \beta > 0$ such that for all sufficiently small $\varepsilon > 0$ the following inequality holds*

$$\alpha\varepsilon^2 \le \nu(W_\varepsilon) \le \beta\varepsilon^2.$$

If $n \ge 3$ then there exists $\beta > 0$ such that for all sufficiently small $\varepsilon > 0$

$$0 < \nu(W_\varepsilon) \le \beta\varepsilon^{2n-2}. \tag{8}$$

Proof. Since the pair (A, B) is controllable, there exists a nonsingular matrix S such

$$A_1 = SAS^{-1} = \begin{pmatrix} 0 & 1 & 0 & ... & 0 \\ 0 & 0 & 1 & ... & 0 \\ ... & ... & ... & ... & ... \\ 0 & 0 & 0 & ... & 1 \\ a_1 & a_2 & a_3 & ... & a_n \end{pmatrix},\quad B_1 = SB = \begin{pmatrix} 0 \\ 0 \\ ... \\ 0 \\ 1 \end{pmatrix}. \tag{9}$$

Here $a_1, a_2, ..., a_n$ are the coefficients of the characteristic polynomial of the matrix A. Taking into account Lemma 1 we can assume without loss of generality that the pair (A, B) itself has the form (9).

For $m \ge 1$ denote

$$S^m(\varepsilon) = \sum_{k=0}^{m} \frac{\varepsilon^k}{(k+1)!} U_k,\ R^m(\varepsilon) = \sum_{k=m}^{\infty} \frac{\varepsilon^{(k-m)}}{(k+1)!} U_k,$$

then W_ε is represented by

$$W_\varepsilon = S^m(\varepsilon) + \varepsilon^{m+1} R^{m+1}(\varepsilon).$$

Consider $n = 2$, in this case

$$A = \begin{pmatrix} 0 & 1 \\ a_1 & a_2 \end{pmatrix}, \; B = \begin{pmatrix} 0 \\ 1 \end{pmatrix}. \tag{10}$$

$$U_0 = \begin{pmatrix} 0 & 0 \\ 0 & 1 \end{pmatrix}, \; U_1 = \begin{pmatrix} 0 & 1 \\ 1 & 2a_2 \end{pmatrix}, \; U_2 = \begin{pmatrix} 2 & 3a_2 \\ 3a_2 & 2a_1 + 4a_2^2 \end{pmatrix}.$$

Letting

$$W_\varepsilon = S^2(\varepsilon) + \varepsilon^3 R^3(\varepsilon),$$

we get

$$S^2(\varepsilon) = \begin{pmatrix} \varphi_1(\varepsilon) & \varphi_2(\varepsilon) \\ \varphi_2(\varepsilon) & \varphi_3(\varepsilon) \end{pmatrix}$$

where

$$\varphi_1(\varepsilon) = \frac{1}{3}\varepsilon^2, \; \varphi_2(\varepsilon) = \frac{a_2}{2}\varepsilon^2 + \frac{1}{2}\varepsilon, \; \varphi_3(\varepsilon) = \frac{a_1 + 2a_2^2}{3}\varepsilon^2 + a_2\varepsilon + 1.$$

Calculating the minimal eigenvalue of $S^2(\varepsilon)$ we get

$$\nu(S^2(\varepsilon)) = \frac{\varphi_1 + \varphi_2}{2}(1 - \sqrt{1 + \psi}),$$

where

$$\psi = -4\frac{\varphi_1\varphi_3 - \varphi_2^2}{(\varphi_1 + \varphi_3)^2}.$$

Since

$$\varphi_1\varphi_3 - \varphi_2^2 = \frac{\varepsilon^2}{3} - \frac{\varepsilon^2}{4} + o(\varepsilon^2) = \frac{\varepsilon^2}{12} + o(\varepsilon^2), \; (\varphi_1 + \varphi_3)^2 = O(1),$$

we have

$$\psi = -\frac{\varepsilon^2}{3} + o(\varepsilon^2), \; \nu(S^2(\varepsilon)) = \frac{\varepsilon^2}{12} + o(\varepsilon^2).$$

From the theorem on the perturbation of eigenvalues of a symmetric matrix [13] it follows that

$$|\nu(W_\varepsilon) - \nu(S^2(\varepsilon))| \le \|\varepsilon^3 R^3(\varepsilon)\| \le \varepsilon^3 \|R^3(0)\|/2$$

for sufficiently small positive ε. Hence,

$$\nu(W_\varepsilon) \ge \nu(S^2(\varepsilon)) - \varepsilon^3\|R^3(0)\|/2 \ge \frac{\varepsilon^2}{12} + o(\varepsilon^2),$$

this proves the fist part of the theorem.

For a square $n \times n$ matrix A denote by $d_i(A)$ diagonals parallel to the antidiagonal, counting from the right-bottom corner. Thus

$$d_1(A) = \{a_{nn}\},\ d_2(A) = \{a_{n(n-1)},\ a_{(n-1)n}\},$$

$$d_3(A) = \{a_{n(n-2)}, a_{(n-1)(n-1)}, a_{(n-2)n}\},$$

etc., $i = 1, 2, ..., 2n-1$. By induction we prove that all the elements of the matrix U_i, lying above the diagonal $d_i(U_{i-1})$, are equal to zero. Then the elements in the left top corners of the matrices U_i are equal to zero for all $i = 1, ..., 2n-3$. Hence, this matrices $S^i(\varepsilon)$ are singular, that implies inequalities $\nu(S^i(\varepsilon)) \leq 0$ for all $i = 1, ..., 2n-3$. From the equality

$$W_\varepsilon = S^{2n-3}(\varepsilon) + \varepsilon^{2n-2}R^{2n-2}(\varepsilon)$$

we get the estimate (8).

4 Convexity of Small Time Reachable Sets

Consider the control system

$$\dot{x}(t) = f_1(t, x(t)) + f_2(t, x(t))u(t),\ x(t_0) = x^0, \tag{11}$$

where $t_0 \leq t \leq \bar{t}_1$, $x \in \mathbb{R}^n$, $u \in \mathbb{R}^r$, the functions $f_1 : \mathbb{R}^{n+1} \to \mathbb{R}^n$, $f_2 : \mathbb{R}^{n+1} \to \mathbb{R}^{n\times r}$ are assumed to be continuous and continuously differentiable in x.

If f_1, f_2 satisfy the conditions:

$$\| f_1(t, x) \| \leq l_1(t)(1+ \| x \|),\ \| f_2(t, x) \|_{n\times r} \leq l_2(t), \tag{12}$$

where $l_1(\cdot) \in \mathbb{L}_1[t_0, \bar{t}_1]$, $l_2(\cdot) \in \mathbb{L}_2[t_0, \bar{t}_1]$, then for any $u(\cdot) \in \mathbb{L}_2[t_0, \bar{t}_1]$ there exists a unique absolutely continuous solution $x(t)$ of system (11) which is defined on the interval $[t_0, \bar{t}_1]$.

With $t_0 < t_1 \leq \bar{t}_1$ given consider the space of square integrable vector-functions on $[t_0, t_1]$ with an inner product defined as

$$(u(\cdot), v(\cdot)) = \int_{t_0}^{t_1} u^\top(t)v(t)dt.$$

For this space we also use the notation $\mathbb{L}_2 = \mathbb{L}_2[t_0, t_1]$.

All the trajectories of system (11) corresponding to controls from a Hilbert ball $B(0, \mu) = \{u(\cdot) \in \mathbb{L}_2[t_0, \bar{t}_1] : (u(\cdot), u(\cdot)) \leq \mu^2\}$ are lying in a compact set $D \subset \mathbb{R}^n$ (see, for example, [7]). Instead of assuming that inequalities (12) are satisfied we may further suppose that all the trajectories of system (11) are defined on the interval $[t_0, t_1]$ and belong to some compact set D.

Definition 3. *The set*

$$G(t_1) = \{x \in \mathbb{R}^n : \exists u(\cdot) \in \mathbb{L}_2[t_0, t_1] : (u(\cdot), u(\cdot)) \leq \mu^2,\ x = x(t_1, u(\cdot))\},$$

is called a reachable set of system (11) at a given time instant t_1.

Definition 4. *Let $u(t)$ be a control from $\mathbb{L}_2$, $x(t)$ be a corresponding trajectory. A linear control system*

$$\dot{\delta x} = A(t)\delta x + B(t)\delta v,\ \delta x(t_0) = 0, \tag{13}$$

where

$$A(t) = \frac{\partial f_1}{\partial x}(t, x(t)) + \frac{\partial}{\partial x}[f_2(t, x(t))u(t)],\ B(t) = f_2(t, x(t))$$

is said to be a linearization of (11) along the pair $(x(t), u(t))$.

Assumption 1. *The functions $f_1(t, x)$, $f_2(t, x)$ have continuous derivatives in x which satisfy the Lipschitz conditions: for all $t \in [t_0, \bar{t}_1]$, $x_1, x_2 \in D$*

$$\|\frac{\partial f_1}{\partial x}(t, x_1) - \frac{\partial f_1}{\partial x}(t, x_2)\| \leq l_3\|x_1 - x_2\|,$$

$$\|\frac{\partial f_2}{\partial x}(t, x_1) - \frac{\partial f_2}{\partial x}(t, x_2)\| \leq l_4\|x_1 - x_2\|,$$

where $l_i \geq 0$ for $i = 3, 4$.

For $u(\cdot)$, $u_i(\cdot) \in B(0, \mu) \subset \mathbb{L}_2[t_0, t_1]$, and corresponding trajectories $x(\cdot)$, $x_i(\cdot)$, $i = 1, 2$ denote as $A(t)$, $A_i(t)$, $B(t)$, $B_i(t)$ the matrices of the linearizations of system (13) along the pairs $(u(\cdot), x(\cdot))$ $(u_i(\cdot), x_i(\cdot))$. Let $X(t, s)$, $X_i(t, s)$, $i = 1, 2$ be fundamental matrices of the systems

$$\dot{x}(t) = A(t)x(t),\ \dot{x}(t) = A_i(t)x(t),\ i = 1, 2,\ t \in [t_0, t_1].$$

Lemma 2. *Suppose the Assumption 1 to be fulfilled. There exists a constant C such that*

$$\|X_1(t, s) - X_2(t, s)\| \leq C\|u_1(\cdot) - u_2(\cdot)\|_{\mathbb{L}_2},\ t, s \in [t_9, t_1],$$

for any $u_i(\cdot) \in B(0, \mu)$.

Proof. From the integral identities

$$x_i(t) = \int_{t_0}^{t} f_1(s, x_i(s))ds + \int_{t_0}^{t} f_2(s, x_i(s))u_i(s)ds,$$

we get

$$\begin{aligned} \|x_1(t) - x_2(t)\| &\leq \| \textstyle\int_{t_0}^{t} [f_1(s, x_1(s)) - f_1(s, x_2(s))]ds\| \\ + \| \textstyle\int_{t_0}^{t} [f_2(s, x_1(s)) - f_2(s, x_2(s))]u_1(s)ds\| &+ \| \textstyle\int_{t_0}^{t} f_2(s, x_2(s))(u_1(s) - u_2(s))ds\| \\ &\leq \textstyle\int_{t_0}^{t} (L_1 + L_2\|u_1(s)\|)\|x_1(s) - x_2(s)\|ds + k_1\|u_1(\cdot) - u_2(\cdot)\|_{\mathbb{L}_2}. \end{aligned}$$

Here L_1, L_2 are Lipschitz constants (with respect to x) for $f_1(s, x), f_2(s, x)$ on the set D, and

$$k_1 = ((t_1 - t_0) \max_{[t_0, t_1] \times D} \|f_2(t, x)\|)^{1/2}.$$

From the Grownwall inequality [16] we have

$$\|x_1(\cdot) - x_2(\cdot)\|_{\mathbb{C}} \leq K\|u_1(\cdot) - u_2(\cdot)\|_{\mathbb{L}_2}, \tag{14}$$

where

$$K = k_1 \exp\left(L_1(t_1 - t_0) + L_2\mu(t_1 - t_0)^{1/2}\right).$$

Denote $X(s) = X(t_1, s)$, the matrix $X(s)$ satisfies the equation

$$\dot{X}(s) = -A^\top(s)X(s),\ X(t_1) = I.$$

From the proof of Theorem 3 in [17] it follows that there exists $k_2 > 0$ such that

$$\|X(s)\| \leq k_2,\ s \in [t_0, t_1] \tag{15}$$

for all $u(\cdot) \in B(0, \mu)$.

Applying inequalities from Assumption 1 and using the scheme of the proof of inequality (14) we obtain the estimates

$$\int_{t_0}^{t_1} \| A_1(t) - A_2(t) \| \, dt \leq C_1\|u_1(\cdot) - u_2(\cdot)\|_{\mathbb{L}_2}, \tag{16}$$

$$\| B_1(\cdot) - B_2(\cdot) \|_{\mathbb{C}} \leq C_2\|u_1(\cdot) - u_2(\cdot)\|_{\mathbb{L}_2}, \tag{17}$$

where $C_1 > 0$, $C_2 > 0$ do not depend on $u_1(\cdot), u_2(\cdot)$.

Since

$$\frac{d}{dt}(X_1 - X_2) = -A_1^\top(t)(X_1 - X_2) + (A_2(t) - A_1(t))^\top X_2,$$

we get the following formula

$$X_1(t) - X_2(t) = \int_{t_1}^{t} Y(t, s)(A_2(s) - A_1(s))^\top X_2(s)ds.$$

Here $Y(t, \tau)$ is a fundamental matrix of the system

$$\dot{x} = -A_1^\top(t)x.$$

Inequality (15) imlplies that there exists $C_3 > 0$ such that

$$\|Y(t, s)\| \leq C_3,\ t, s \in [t_0, t_1] \tag{18}$$

for all $u_1(\cdot) \in B(0, \mu)$. From (16), (17), (18) we get

$$\|X_1(t) - X_2(t)\| \leq C_1C_3k_2\|u_1(\cdot) - u_2(\cdot)\|_{\mathbb{L}_2},\ t \in [t_0, t_1],$$

and hence

$$\|X_1(t, s) - X_2(t, s)\| \leq C_4\|u_1(\cdot) - u_2(\cdot)\|_{\mathbb{L}_2} \tag{19}$$

for some $C_4 > 0$ and all $t, s \in [t_0, t_1]$.

The reachable sets for nonlinear systems, as a rule, are not convex. This creates additional difficulties in the application of algorithms using optimization techniques in their construction [2,8,9,14].

From [15] it follows that if the linearization of system (11) along the trajectory $x(t,0)$, $x(t_0,0) = x^0$, corresponding to zero control, is controllable, then the reachable set $G(t_1)$ is convex for all sufficiently small $\mu > 0$. The paper [15] provides also an upper estimate of the value of μ which ensures the convexity of the reachable set. This estimate is as follows.

Define the map $F : \mathbb{L}_2 \to \mathbb{R}^n$ by the equality

$$F(u(\cdot)) = x(t_1),$$

here $x(t)$ is a trajectory of system (11), $cu(\cdot)$. The map has a continuous Fréchet derivative $F' : \mathbb{L}_2 \to \mathbb{R}^n$

$$F'(u(\cdot))\Delta u(\cdot) = \Delta x(t_1).$$

where $\Delta x(t)$ is a solution of the linearization along $(u(t), x(t))$ of system (11) with zero initial vector and the control $\Delta u(t)$.

From (4) we get that

$$F'(u(\cdot)) = X(t_1, s, u(\cdot))B(s, u(\cdot)),\ s \in [t_0, t_1],$$

where $X(t_1, s, u(\cdot))$ is a fundamental matrix of system (13), whose matrices $A(t) = A(t, u(\cdot))$, $B(t) = B(t, u(\cdot))$ depend on $u(\cdot)$. Using inequalities (17), (19) one may prove that $F'(u(\cdot))$ is Lipschitz continuous

$$\|F'(u_1(\cdot)) - F'(u_2(\cdot))\| \leq L\|u_1(\cdot) - u_2(\cdot)\|_{\mathbb{L}_2} \tag{20}$$

on $B(0, \mu)$.

The controllability of the linearization of system (11) along the trajectory $x(t,0)$, is equivalent to positivity of the minimal eigenvalue ν of the controllability Gramian $W = W(t_1)$ of this system. The estimate mentioned above in this case has the form

$$\mu \leq \frac{\sqrt{\nu}}{2L}. \tag{21}$$

Further we propose sufficient conditions for the reachable set $G(t_1)$ to be convex in the case when μ is fixed but the time interval $[t_0, t_1]$ is small, denote $t_1 - t_0 = \varepsilon$. Applying a change of variables $t = \varepsilon\tau + t_0$ and denoting $y(\tau) = x(\varepsilon\tau + t_0)$, $v(\tau) = \varepsilon u(\varepsilon\tau + t_0)$ we have

$$\dot{y}(\tau) = \tilde{f}_1(\tau, y(\tau)) + \tilde{f}_2(\tau, y)v(\tau),\ 0 \leq \tau \leq 1, y(0) = x^0, \tag{22}$$

where $\tilde{f}_1(\tau, y) = \varepsilon f_1(\varepsilon\tau + t_0, y)$, $\tilde{f}_2(\tau, y) = f_2(\varepsilon\tau + t_0, y)$, with constraints on the control $v(\cdot)$ given by the inequality

$$\int_0^1 v^\top(t)v(t)dt \leq (\mu\sqrt{\varepsilon})^2. \tag{23}$$

Let us assume that all the trajectories of system (11) corresponding to $u(\cdot) \in B(0,\mu) \subset \mathbb{L}_2[t_0,\bar{t}_1]$ belong to some compact set D. Let $t_1 \in (t_0,\bar{t}_1]$. Note that trajectories of (22), (23) are also lying in D for $\varepsilon \leq \bar{t}_1 - t_0$, and $y(\tau,0) = x(\varepsilon\tau + t_0, 0)$. The Lipschitz contstants of $\tilde{f}_1$, $\tilde{f}_2$, $\partial\tilde{f}_1/\partial y$, $\partial\tilde{f}_2/\partial y$ on the set D equal respectively to εL_1, L_2, εl_3, l_4. The inequality (14) being applied to system (22) implies

$$\|y_1(\cdot) - y_2(\cdot)\|_{\mathbb{C}} \leq \tilde{K}\|v_1(\cdot) - v_2(\cdot)\|_{\mathbb{L}_2}, \tag{24}$$

for some $\tilde{K}$, which does not depend on ε. Notation $\mathbb{C}$, $\mathbb{L}_2$ refers here to spaces of functions defined on a segment $[0,1]$. Similarly, from (16), (17) we get

$$\int_{t_0}^{t_1} \| \tilde{A}_1(t) - \tilde{A}_2(t) \| \, dt \leq (\tilde{C}_1\varepsilon + \tilde{C}_2)\|v_1(\cdot) - v_2(\cdot)\|_{\mathbb{L}_2}, \tag{25}$$

$$\| \tilde{B}_1(\cdot) - \tilde{B}_2(\cdot) \|_{\mathbb{C}} \leq \tilde{C}_3\|v_1(\cdot) - v_2(\cdot)\|_{\mathbb{L}_2}, \tag{26}$$

for some $\tilde{C}_i \geq 0$. The matrices $\tilde{A}_i(t)$, $\tilde{B}_i(t)$ denote here the matrices of the linearization of system (22) along the pairs $(y_i(\cdot), u_i(\cdot))$.

Note that in the case when $f_2(t,x)$ does not depend on x, a constant $\tilde{C}_2$ equals to zero.

The inequality (20) can be rewritten as follows

$$\|\tilde{F}'(v_1(\cdot)) - \tilde{F}'(v_2(\cdot))\| \leq L(\varepsilon)\|v_1(\cdot) - v_2(\cdot)\|_{\mathbb{L}_2} \tag{27}$$

where $\tilde{F}$ is an analog of the map F for the system (22) and a Lipschitz constant of $\tilde{F}'$ $L(\varepsilon) = \tilde{L}_1\varepsilon + \tilde{L}_2$ for some $\tilde{L}_i \geq 0$. As above $\tilde{L}_2 = 0$ if $f_2(t,x) = f_2(t)$. The inequality (21) takes the form here

$$4\mu^2\varepsilon L^2(\varepsilon) \leq \nu, \tag{28}$$

this inequality gives the sufficient conditions for the reachable set $\tilde{G}(1)$ of the system (22) under constraints (23) to be convex. Taking into account that $G(t_1) = \tilde{G}(1)$ we come to the following

Theorem 2. *Let $\nu(\varepsilon)$ be the minimal eigenvalue of the controllability Gramian of the linearization of system (22) along $x(t,0)$. Suppose that there exist $C > 0$, $\alpha > 0, \bar{\varepsilon} > 0$ such that for all $\varepsilon \leq \bar{\varepsilon}$*

$$\nu(\varepsilon) \geq C\varepsilon^{1-\alpha}$$

or

$$\nu(\varepsilon) \geq C\varepsilon^{3-\alpha}$$

in the case $f_2(t,x) = f_2(t)$. Then $G(t_1)$ is convex for all sufficiently small t_1.

5 Time-Invariant Systems on a Small Time Interval

Consider here the autonomous control system with a single input

$$\dot{x}(t) = f(x(t)) + Bu(t), \; x(0) = x^0, \; 0 \le t \le t_1, \tag{29}$$

where $0 \le t \le t_1$, $x \in \mathbb{R}^n$, $u \in \mathbb{R}$, $f : \mathbb{R}^n \to \mathbb{R}^n$ is a continuously differential mapping, B is an $n \times 1$ matrix (a column vector), x^0 is a fixed initial state, with control variables subjected to quadratic integral constraints $u(\cdot) \in B(0, \mu)$.

Suppose, as before, that there exists a compact set $D \subset \mathbb{R}^n$ containing all the trajectories of the system (29), and that $f(x)$ has a Lipschitz continuous derivative on this set.

Denote $A(t) = \frac{\partial f}{\partial x}(x(t,0))$ a matrix of the linearization of the system along $x(t,0)$. Suppose that $f(x^0) = 0$, in this case $x(t,0) \equiv 0$, hence,

$$A(t) = \frac{\partial f}{\partial x}(x(t,0)) = \frac{\partial f}{\partial x}(x^0) = A$$

is a constant matrix. Let W_ε be the controllability Gramian of the pair $(\varepsilon A, B)$ on the interval $[0,1]$ and $\nu(\varepsilon)$ be the minimal eigenvalue of W_ε. If the pair (A, B) is controllable then by Theorem 1 $\nu(W_\varepsilon) \ge \alpha\varepsilon^2$ if $n = 2$, and $\nu(W_\varepsilon) \le \beta\varepsilon^4$ if $n \ge 3$ for some $\alpha, \beta > 0$.

With this in mind from Theorem 2 we get the following:

Corollary 1. *If $n = 2$ and the linearization of the system (29) at the point x^0 is controllable then the reachable set $G(t_1)$ is convex for all sufficiently small t_1. For $n \ge 3$ the sufficient conditions of convexity of $G(t_1)$ are not satisfied.*

As an illustrative example consider the Duffing equation

$$\dot{x}_1 = x_2, \quad \dot{x}_2 = -x_1 - 10x_1^3 + u, \quad 0 \le t \le t_1 \tag{30}$$

which describes the motion of nonlinear stiff spring on impact of an external force u, under integral constraints

$$\int_0^{t_1} u^2(t)dt \le \mu^2,$$

and zero initial state $x_1(0) = 0$, $x_2(0) = 0$. Consider a Lyapunov-type function

$$V(x) = V(x_1, x_2) = \frac{5}{2}x_1^4 + \frac{1}{2}x_1^2 + \frac{1}{2}x_2^2.$$

Differentiating $V(x(t))$ along an arbitrary trajectory of the system (30) one get

$$\frac{dV}{dt}(x(t)) = x_2(t)u(t),$$

hence

$$V(x(t)) = \int_0^t x_2(\tau)u(\tau)d\tau \le \mu(\int_0^t x_2^2(\tau)d\tau)^{1/2} \le \mu(2\int_0^t V(\tau)d\tau)^{1/2}. \tag{31}$$

An analog of the Grownwall lemma (see, for example, [16]) being applied to differential inequality (31) yields $V(x(t)) \leq \mu^2 t$. Hence, all the trajectories of system (30) belongs to a compact set

$$D = \{x \in \mathbb{R}^2 : V(x) \leq \mu^2 t_1\}.$$

The linearization of (30) along $x(t) \equiv 0$ after time variable change

$$\dot{x}_1 = t_1 x_2, \quad \dot{x}_2 = -x_1 + u, \quad x(0) = (0, 0), \quad 0 \leq t \leq 1$$

is completely controllable. From Corollary 1 it follows that the reachable sets $G(t_1)$ in this example are convex for small t_1.

In the next figure the results of the numerical simulation are shown. These results are obtained using the proposed in [8] algorithm based on Pontryagin's maximum principle for boundary trajectories.

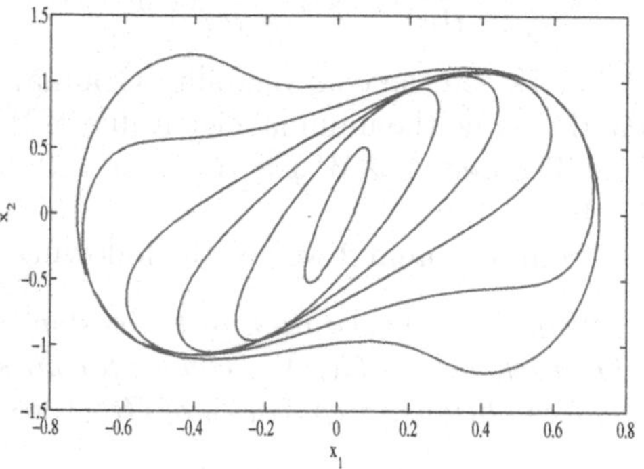

Fig. 1. The reachable sets for the Duffing system.

Figure 1 shows the plot of the reachable sets for $\mu^2 = 2$ at times $t_1 = 0.3, 0.5, 0.7, 0.9, 1.2, 1.5$ respectively. A larger set in the Figure corresponds to a larger value of t_1. This plot indicates that reachable sets are convex for small values of t_1 and lose their convexity as t_1 increase.

References

1. Anan'ev, B.I.: Motion correction of a statistically uncertain system under communication constraints. Autom. Remote Control **71**(3), 367–378 (2010)
2. Baier, R., Gerdts, M., Xausa, I.: Approximation of reachable sets using optimal control algorithms. Numer. Algebra Control Optim. **3**(3), 519–548 (2013)
3. Dar'in, A.N., Kurzhanskii, A.B.: Control under indeterminacy and double constraints. Differ. Equ. **39**(11), 1554–1567 (2003)
4. Filippova, T.F.: Estimates of reachable sets of impulsive control problems with special nonlinearity. In: AIP Conference Proceedings Application of Mathematics in Technical and Natural Sciences, 2016, vol. 1773, Article number 100004, pp. 1–10 (2016)

5. Guseinov, K.G., Ozer, O., Akyar, E., Ushakov, V.N.: The approximation of reachable sets of control systems with integral constraint on controls. Nonlinear Differ. Equ. Appl. **14**(1–2), 57–73 (2007)
6. Guseinov, Kh. G., Nazlipinar, A.S.: Attainable sets of the control system with limited resources.Trudy Inst. Mat. i Mekh. Uro RAN **16**(5), 261–268 (2010)
7. Gusev, M.: On reachability analysis of nonlinear systems with joint integral constraints. In: Lirkov, I., Margenov, S. (eds.) LSSC 2017. LNCS, vol. 10665, pp. 219–227. Springer, Cham (2018). https://doi.org/10.1007/978-3-319-73441-5_23
8. Gusev, M.I., Zykov, I.V.: On extremal properties of boundary points of reachable sets for a system with integrally constrained control. IFAC-PapersOnLine **50**(1), 4082–4087 (2017). https://doi.org/10.1016/j.ifacol.2017.08.792. 20th IFAC WORLD CONGRESS, 2017
9. Gusev, M.I.: Internal approximations of reachable sets of control systems with state constraints. Proc. Steklov Inst. Math. **287**(Suppl. 1), S77–S92 (2014)
10. Huseyin, N., Huseyin, A.: Compactness of the set of trajectories of the controllable system described by an affineintegral equation. Appl. Math. Comput. **219**, 8416–8424 (2013)
11. Kurzhanski, A.B., Varaiya, P.: Dynamic optimization for reachability problems. J. Optim. Theory Appl. **108**(2), 227–251 (2001)
12. Lee, E.B., Marcus, L.: Foundations of Optimal Control Theory. Willey, Hoboken (1967)
13. Horn, R.A., Jonson, C.R.: Matrix Analysis. Cambridge University Press, Cambridge (1986)
14. Patsko, V.S., Pyatko, S.G., Fedotov, A.: Three-dimensional reachability set for a nonlinear control system. J. Comput. Syst. Sci. Int. **42**(3), 320–328 (2003)
15. Polyak, B.T.: Convexity of the reachable set of nonlinear systems under l2 bounded controls. Dyn. Contin. Discrete Impuls. Syst. Ser. A Math. Anal. **11**, 255–267 (2004)
16. Walter, W.: Differential and Integral Inequalities. Springer, Berlin (1970). https://doi.org/10.1007/978-3-642-86405-6
17. Filippov, A.F.: Differential Equations with Discontinuous Righthand Sides. Kluwer Academic Press, Boston (1988)

Optimality Conditions and Numerical Algorithms for Hybrid Control Systems

Nadezhda Maltugueva(✉), Nikolay Pogodaev, and Olga Samsonyuk

Matrosov Institute for System Dynamics and Control Theory, Irkutsk 664033, Russia
{malt,n.pogodaev}@icc.ru, olga.samsonyuk@gmail.com

Abstract. For an optimal control problem with intermediate state constraints, we construct an iterative descent algorithm and prove a related necessary optimality condition. Finally, we show how these results can be applied to measure-driven multiprocesses.

Keywords: Necessary optimality condition · Optimal multiprocesses · Numerical method · Measure-driven multiprocesses

1 Introduction

In this paper we consider the following optimal control problem with intermediate state constraints:

$$\begin{cases} J = l\left(x(\theta_N)\right) \to \min, & \text{(cost)} \\ \dot{x} = f(t,x,u), \quad u(t) \in U, & \text{(dynamics)} \\ x(0) = x_0, & \text{(initial condition)} \\ x(\theta_i) \in A_i, \quad i = 1,\dots,N, & \text{(intermediate constraints)} \end{cases} \qquad (P)$$

where $U \subset \mathbb{R}^m$ is a compact set, $a = \theta_0 < \theta_1 < \cdots < \theta_N = b$ are given time moments on the segment $T = [a,b]$, and A_i, $i = 1,\dots N$, are given closed subsets of the phase space $\mathbb{R}^n$.

Let us remark that any control system with intermediate constrains can be naturally considered as a hybrid control system, or a multiprocess in the terminology of [5].

In what follows, we impose the usual regularity assumptions on f and l:

(H) $(t,x,u) \mapsto f(t,x,u)$ is continuous on $T \times \mathbb{R}^n \times U$, sublinear, i.e., $f(t,x,u) \le C(1+|x|)$ for all t, x, u, and continuously differentiable in x; $l\colon \mathbb{R}^n \to \mathbb{R}$ is continuously differentiable.

Note that, if (H) holds, the input-output map $u(\cdot) \mapsto x(\cdot)$ of the dynamical system is single-valued, so we may think of J as a function of u, i.e., $J = J[u]$.

Partially supported by the Russian Foundation for Basic Research, projects nos 18-31-20030, 18-31-00425, 18-01-00026, 17-01-00733.

M. Khachay et al. (Eds.): MOTOR 2019, LNCS 11548, pp. 474–488, 2019.
https://doi.org/10.1007/978-3-030-22629-9_33

Setting aside the methods involving integrating the Hamilton-Jacobi-Bellman equation, there are two main ways to tackle numerically an optimal control problem. The first approach consists in replacing it with a certain finite-dimensional optimization problem and then solving the latter (see, for instance, [4,21,22]). The second approach uses control variations in infinite dimensional spaces to construct a sequence of controls along which the cost of the problem monotonically decrease (see [9,12,20]). Roughly speaking, such algorithms can be thought of as gradient descent methods in infinite dimensional spaces; for this reason we will call them the *iterative descent algorithms.*

Methods of the first type usually involve time discretization and thus can be easily applied to the problem (P). On the other hand, methods of the second type deal, as a rule, only with unconstrained optimal control problems (in the sense that such problems contain no terminal or intermediate constraints). There were attempts to use them for solving hybrid problems similar to (P), but they led to highly complicated numerical algorithms [2,18]. In the present paper, we make another effort in this direction, but we base it on completely different ideas.

More precisely, we aim at constructing an iterative algorithm for (P) which, starting from an admissible u^0, generates a sequence of admissible controls u^k, $k \in \mathbb{N}$, with the descending property: $J[u^k] < J[u^{k-1}]$ for all $k \in \mathbb{N}$. To that end, we replace (P), for a given reference control $\bar{u}$, with an unconstrained optimal control problem $(P_{\bar{u}})$ with the property

$$J_{\bar{u}}[u] < J_{\bar{u}}[\bar{u}] \quad \Rightarrow \quad J[u] < J[\bar{u}].$$

Then, applying to $(P_{\bar{u}})$ any known descent method, we obtain a control u satisfying $J[u] < J[\bar{u}]$. As a byproduct we will prove a new necessary optimality condition, which, being rougher than the usual Hybrid maximum principle [1,6,8], has strong relations with the proposed algorithm. Finally, we show that optimal impulsive control problems with intermediate state constraints, interpreted as measure-driven optimal multiprocesses, can be reduced to the problem (P).

2 Preliminaries

We collect in this section several important definitions and lemmas that will be used throughout the paper.

2.1 Flows

Throughout this section, let $g\colon \mathbb{R}^+ \times \mathbb{R}^n \to \mathbb{R}^n$ be a time dependent vector field satisfying the usual regularity assumptions: $g = g(t, x)$ is measurable in t, continuously differentiable in x, and obeys the sublinear growth condition.

Definition 1. *The map $\Phi\colon \mathbb{R}^+ \times \mathbb{R}^+ \times \mathbb{R}^n \to \mathbb{R}^n$ defined by $\Phi^t_s(x) = y(t, s, x)$, where $y(\cdot, s, x)$ satisfies the Cauchy problem*

$$\begin{cases} \dot{y}(t) = g\left(t, y(t)\right), \\ y(s) = x, \end{cases} \tag{1}$$

*is called the **flow** of the time dependent vector field g.*

Recall some essential properties of the flow [3]:

(i) for any $s, t, \theta \in \mathbb{R}^+$, one has $\Phi_\theta^s \circ \Phi_t^\theta = \Phi_t^s$ and $\Phi_t^t = id$.
(ii) for any $s, t \in \mathbb{R}^+$, the map $x \mapsto \Phi_s^t(x)$ is a diffeomorphism and its derivative is given by $D\Phi_s^t(x) = M(t)$, where $M(\cdot)$ satisfies the linear matrix equation

$$\begin{cases} \dot{M}(t) = Dg\left(t, \Phi_s^t(x)\right) M(t), \\ M(s) = I. \end{cases} \tag{2}$$

Above id denotes the identity map, I the identity matrix, D the differentiation with respect to the spatial variable x.

2.2 Regular Sets

Given a closed set $S \subset \mathbb{R}^n$ and a point $x \in S$, let us denote by $N_S(x)$ and $N_S^L(x)$ the Bouligand and the limiting normal cones to S at x (see, e.g., [7] for their definitions).

Definition 2. *A closed set $S \subset \mathbb{R}^n$ is called regular at $x \in S$ if*

$$N_S(x) = N_S^L(x) = \operatorname{co} N_S^L(x).$$

For details we refer to [7, Section 10.3 and Theorem 11.36].

Example 1 (cf. [7, Corollary 10.44]). Let us take

$$S = \{x : \; g_i(x) \leq 0, \; i = 1, \ldots, k\},$$

where all g_i are continuously differentiable functions, and $I(x) = \{i : \; g_i(x) = 0\}$. Let $x \in S$ be such that $I(x) \neq \varnothing$. If the vectors $\{\nabla g_i(x), \; i \in I(x)\}$ are linearly independent then S is regular at x and

$$N_S(x) = \Big\{ \sum_{i \in I(x)} \lambda_i \nabla g_i(x) \; : \; \lambda_i \geq 0 \Big\}.$$

Definition 3. *Closed sets A_1, A_2 are said to be transversal at $x \in A_1 \cap A_2$ if $-N_{A_1}^L(x) \cap N_{A_2}^L(x) = \{0\}$.*

The following lemma helps to compute normal cones to the intersection of regular sets.

Lemma 1 (cf.[7, Theorem 11.39]). *Let A_1 and A_2 be regular and transversal at $x \in A_1 \cap A_2$. Then*

$$N_{A_1 \cap A_2}(x) = N_{A_1}(x) + N_{A_2}(x).$$

The next lemma describes how the normal cone to a set evolves when this set is transported by a vector field.

Lemma 2. *Let Φ be the flow of the time dependent vector field $(t,x) \mapsto g(t,x)$, $x(\cdot)$ be one of its integral curves, and $A \subset \mathbb{R}^n$ be a closed set such that $x(s) \in A$ for some $s \in \mathbb{R}^+$. Finally, let Ψ denote the flow of the time dependent vector field $(t,p) \mapsto -pDg(t,x(t))$. If A is regular at $x(s)$ then*

$$N_{\Phi_s^t(A)}\big(x(t)\big) = \Psi_s^t\big(N_A(x(s))\big) \quad \forall t \in \mathbb{R}^+.$$

Lemma 2 is a simple consequence of [7, Theorem 10.19].

3 Auxiliary Problem

Throughout this section, let $(\bar{x},\bar{u})$ be a fixed admissible process for (P), Φ and Ψ be the flows of the time dependent vector fields $(t,x) \mapsto f(t,x,\bar{u}(t))$ and $(t,p) \mapsto -pDf(t,\bar{x}(t),\bar{u}(t))$, respectively.

For any $i = 1,\dots N$, consider the following auxiliary problem:

$$\begin{cases} J_i = l \circ \Phi_{\theta_i}^{\theta_N}(x(\theta_i)) \to \min, & \text{(cost)} \\ \dot{x} = f(t,x,u), \quad u(t) \in U, & \text{(dynamics)} \\ x(\theta_{i-1}) = \bar{x}(\theta_{i-1}), & \text{(initial condition)} \\ x(\theta_i) \in \bigcap_{j=i}^N \Phi_{\theta_j}^{\theta_i}(A_j). & \text{(terminal condition)} \end{cases} \qquad (P_i)$$

The next lemma establishes a relation between the problems (P) and (P_i).

Lemma 3. *Let (x_i,u_i) be an admissible process in (P_i), $u\colon [0,T] \to U$ be defined by*

$$u(t) = \begin{cases} u_i(t), & t \in [\theta_{i-1},\theta_i], \\ \bar{u}(t), & \textit{otherwise}, \end{cases}$$

and $x(\cdot)$ be a solution of the Cauchy problem

$$\begin{cases} \dot{x}(t) = f\left(t,x(t),u(t)\right), \\ x(0) = x_0. \end{cases}$$

Then the pair (x,u) is an admissible process in (P). Moreover, for any $i = 1,\dots,N-1$, the inequality $J_i[u_i] < J_i[\bar{u}]$ implies $J[u] < J[\bar{u}]$.

Proof. Let us note that $x(t) = \bar{x}(t)$ for all $t \in [0,\theta_{i-1}]$. Moreover, from

$$x(\theta_i) \in \bigcap_{j=i}^N \Phi_{\theta_j}^{\theta_i}(A_j)$$

it follows that $x(\theta_j) \in A_j$, for all $j = i,\dots,N$. Thus, the process (x,u) is admissible in (P).

If $J_i[u_i] < J_i[\bar{u}]$ then, by definition, $l \circ \Phi_{\theta_i}^{\theta_N}(x(\theta_i)) < l \circ \Phi_{\theta_i}^{\theta_N}(\bar{x}(\theta_i))$. On the other hand, we have

$$l \circ \Phi_{\theta_i}^{\theta_N}(x(\theta_i)) = l(x(\theta_N)), \qquad l \circ \Phi_{\theta_i}^{\theta_N}(\bar{x}(\theta_i)) = l(\bar{x}(\theta_N)).$$

Hence $J[u] < J[\bar{u}]$, as desired.

This lemma immediately implies the following

Proposition 1. *If $\bar{u}$ is optimal in (P) then its restriction $\bar{u}|_{[\theta_{i-1},\theta_i]}$ on each time interval $[\theta_{i-1},\theta_i]$, $i = 1,\dots,N$, is optimal in (P_i).*

The above proposition, in turn, allows us to prove a necessary optimality condition for the original problem (P).

3.1 Necessary Optimality Condition

We state and prove the necessary optimality condition only in the generic case (under additional regularity assumptions). This greatly simplifies the proof and allows the reader to see the geometrical meaning of the result.

Regularity Assumption. Let $(\bar{x}, \bar{u})$ be an optimal pair in (P) and

$$E = \{x \,:\, l(x) \le l(\bar{x}(\theta_N))\}.$$

Then, for each $i = 1,\dots,N$,

(i) A_i is regular at $\bar{x}(\theta_i)$ and $Dl(\bar{x}(\theta_N)) \neq 0$;
(ii) the sets $\Phi_{\theta_N}^{\theta^i}(E)$, $\Phi_{\theta_j}^{\theta_i}(A_j)$, $j \ge i$, are pairwise transversal at $\bar{x}(\theta_i)$.

Note that (i) implies that each $\Phi_{\theta_j}^{\theta_i}(A_j)$, $j \ge i$, is regular at $\bar{x}(\theta_i)$.

Theorem 1 (necessary optimality condition). *Let $(\bar{x}, \bar{u})$ be an optimal process in (P) and the regularity assumption hold. Then there exists a family of arcs $p_i\colon [\theta_{i-1},\theta_i] \to \mathbb{R}^n$ satisfying, for a.e. $t \in [\theta_{i-1},\theta_i]$, the* ***adjoint equation***

$$\dot{p}_i(t) = -Df(t, \bar{x}(t), \bar{u}(t))p_i(t), \tag{3}$$

the ***maximum condition***

$$\langle p_i(t), f(\bar{x}(t), \bar{u}(t))\rangle = \max_{\omega \in U} \langle p_i(t), f(\bar{x}(t), \omega)\rangle, \tag{4}$$

the ***transversality condition***

$$-p_i(\theta_i) \in \Psi_{\theta_N}^{\theta_i}\left(N_E(\bar{x}(\theta_N))\right) + \sum_{j=i}^{N} \Psi_{\theta_j}^{\theta_i}\left(N_{A_j}(\bar{x}(\theta_j))\right), \tag{5}$$

and the ***nontriviality condition***

$$p_i(\theta_i) \neq 0. \tag{6}$$

Proof. Let $(\bar{x}, \bar{u})$ be optimal in (P). Then, by Proposition 1, its restriction on $[\theta_{i-1}, \theta_i]$ is optimal in (P_i) for each $i = 1, \ldots, N$.

Fix some i. According to the classical Pontryagin maximum principle [7, Theorem 22.2], if a pair $(\bar{x}, \bar{u})$ is optimal in (P_i) then there exist a scalar $\lambda_i \geq 0$ and an arc $p_i \colon [\theta_{i-1}, \theta_i] \to \mathbb{R}^n$ satisfying the nontriviality condition $(\lambda_i, p_i(\theta_i)) \neq 0$, the adjoint equation (3), the maximum condition (4), and the transversality condition

$$-p_i(\theta_i) \in N_{\bigcap_{j=i}^N \Phi_{\theta_j}^{\theta_i}(A_j)}(\bar{x}(\theta_i)) + \lambda_i D\left(l \circ \Phi_{\theta_N}^{\theta_i}\right)(\bar{x}(\theta_i)). \tag{7}$$

Note that

$$\Phi_{\theta_N}^{\theta_i}(E) = \left\{x \,:\, l \circ \Phi_{\theta_i}^{\theta_N}(x) \leq l \circ \Phi_{\theta_i}^{\theta_N}(\bar{x}(\theta_i))\right\}.$$

Taking into account the regularity assumption and Example 1, we conclude that

$$N_{\Phi_{\theta_N}^{\theta_i}(E)}(\bar{x}(\theta_i)) = \left\{\lambda D\left(l \circ \Phi_{\theta_N}^{\theta_i}\right)(\bar{x}(\theta_i)) \,:\, \lambda \geq 0\right\}.$$

Hence (7) can be equivalently expressed as

$$-p_i(\theta_i) \in N_{\bigcap_{j=i}^N \Phi_{\theta_j}^{\theta_i}(A_j)}(\bar{x}(\theta_i)) + N_{\Phi_{\theta_N}^{\theta_i}(E)}(\bar{x}(\theta_i)),$$

at the same time the nontriviality condition must bc substituted with the one given by (6).

Now, Lemma 1 implies that

$$N_{\bigcap_{j=i}^N \Phi_{\theta_j}^{\theta_i}(A_j)}(\bar{x}(\theta_i)) = \sum_{j=i}^N N_{\Phi_{\theta_j}^{\theta_i}(A_j)}(\bar{x}(\theta_i)),$$

while Lemma 2 together with the regularity assumption yields (5), as desired.

Remark 1. Let us describe the geometrical meaning of the transverslity condition (5). At the time moment θ_i it is constructed in the following way. We take all the targets ahead A_j, $j \geq i$, together with the favorable region E. Then we compute the normal cones to these sets at the points $\bar{x}(\theta_j)$, $j \geq i$, and $\bar{x}(\theta_N)$, respectively. Next we transfer these cones back to the time moment θ_i with flow Ψ of the adjoint system. Finally, we compute the sum of the translated cones to get the right-hand side of (5). See also Fig. 1.

Remark 2. Let us denote the right-hand side of (5) by C_i. It is easy to see that the sets C_i can be constructed recursively as follows:

$$\begin{cases} C_N = N_{A_N}\left(\bar{x}(\theta_N)\right) + N_E\left(\bar{x}(\theta_N)\right), \\ C_i = N_{A_i}\left(\bar{x}(\theta_i)\right) + \Psi_{\theta_{i+1}}^{\theta_i}\left(C_{i+1}\right), \; i < N. \end{cases}$$

Now, note that, for each $i = 0, \ldots, N-1$, the following implications hold:

$$-p_{i+1}(\theta_{i+1}) \in C_{i+1} \quad \Rightarrow \quad -p_{i+1}(\theta_i) \in \Psi^{\theta_i}_{\theta_{i+1}}(C_{i+1}) \quad \Rightarrow$$
$$-p_{i+1}(\theta_i) + N_{A_i}(\bar{x}(\theta_i)) \subset C_i.$$

If we replace, in the statement of Theorem 1, the transversality condition $-p_i(\theta_i) \in C_i$ with the more restrictive condition

$$-p_i(\theta_i) \in -p_{i+1}(\theta_i) + N_{A_i}(\bar{x}(\theta_i)),$$

we get exactly the jump condition of the Hybrid maximum principle [6,8]. This allows us to conclude that our necessary optimality condition is rougher than the Hybrid maximum principle (any extremal of the latter satisfies the assumptions of Theorem 1, but not vice versa).

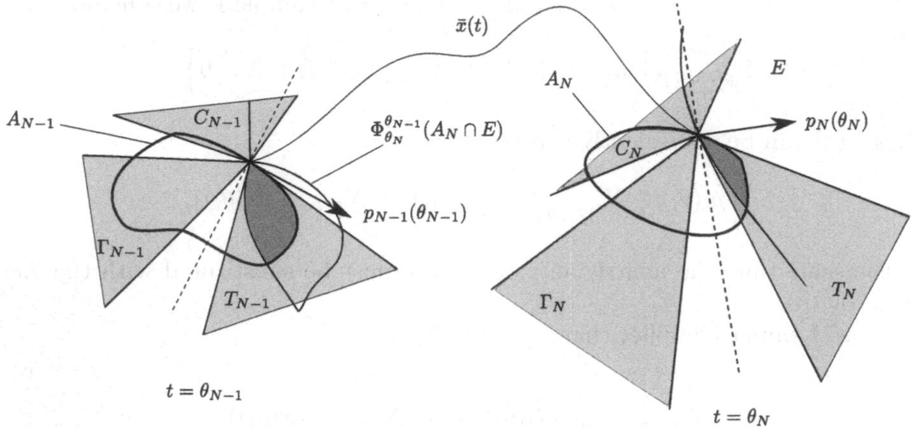

Fig. 1. The figure shows two snapshots of the phase space at the time moments $t = \theta_{N-1}$ and $t = \theta_N$. One the right snapshot, T_N is the tangent cone to $A_N \cap E$, i.e., the cone of "profitable directions", C_N is the corresponding normal cone, Γ_N is the cone of "feasible directions", i.e., those directions along which one can shift the terminal point $\bar{x}(\theta_N)$ by using needle variations of $\bar{u}$ on $[\theta_{N-1}, \theta_N]$. Since $\bar{u}$ is optimal, T_N and Γ_N are weakly separated by a hyperplane $p_N(\theta_N) \cdot v = 0$, hence $-p_N(\theta_N) \in C_N$. Similarly on the left snapshot, T_{N-1} is the tangent cone to the set $A_{N-1} \cap \Phi^{\theta_{N-1}}_{\theta_N}(A_N) \cap \Phi^{\theta_{N-1}}_{\theta_N}(E)$, C_{N-1} its normal cone, Γ_{N-1} is again the cone of "feasible directions". Since Γ_{N-1} and T_{N-1} must be weakly separated, one has $-p_{N-1}(\theta_{N-1}) \in C_{N-1}$. The regularity assumption allows us to express C_{N-1} and C_N in the form (5).

4 Numerical Method

4.1 General Scheme

Suppose that there is an iterative descent algorithm $\mathcal{A}_0$ that can be used to solve each auxiliary problem (P_i). In other words, starting from some admissible u^0, the algorithm $\mathcal{A}_0$ produces a sequence of admissible controls u^k, $k \in \mathbb{N}$, with the property $J_i[u^k] \leq J_i[u^{k-1}]$, for all $k \in \mathbb{N}$. Now, the following scheme provides a descent algorithm for (P).

Algorithm $\mathcal{A}$.

0. Let u^0 be an initial guess. Set $i := 1$, $k := 0$.
1. Make one iteration of $\mathcal{A}_0$ in the auxiliary problem (P_i) using u^k as the initial guess. This gives u_i^k satisfying $J_i[u_i^k] \leq J_i[u^k]$.
2. Set
$$u^{k+1}(t) := \begin{cases} u_i^k(t), & t \in [\theta_{i-1}, \theta_i], \\ u^k(t), & \text{otherwise}. \end{cases}$$
3. If $i < N$ then $i := i + 1$ else $i := 1$.
4. Return to step 1.

Indeed, according to Proposition 1, the algorithm produces a sequence u^k, $k \in \mathbb{N}$, satisfying $J[u^k] \leq J[u^{k-1}]$, for all $k \in \mathbb{N}$.

Definition 4. *We say that $\mathcal{A}_0$ is* ***strictly improving*** *for (P_i) if, for any initial guess u^0, the sequence u^k produced by $\mathcal{A}_0$ is such that*

(i) $J_i[u^k] = J_i[u^{k-1}]$ if u^{k-1} satisfies the Pontryagin maximum principle,
(ii) $J_i[u^k] < J_i[u^{k-1}]$ otherwise.

Examining the proof of Theorem 1, we conclude that if $\mathcal{A}_0$ is strictly improving for each (P_i), $i = 1, \ldots, N$, then $\mathcal{A}$ is strictly improving for (P) in the sense that, for any initial guess u^0, the sequence u^k produced by $\mathcal{A}$ is such that

(i) $J[u^k] = J[u^{k-1}]$ if u^{k-1} satisfies the necessary condition given by Theorem 1,
(ii) $J[u^k] < J[u^{k-1}]$ otherwise.

Roughly speaking, if $\mathcal{A}_0$ cannot improve the controls satisfying the Pontryagin maximum principle, then $\mathcal{A}$ cannot improve the controls satisfying the conditions of Theorem 1.

4.2 Implementation

As we have said before, most iterative descent algorithms can be applied only to problems with no terminal constraints. Hence, in practice, we use the penalty method to get rid of the terminal constraints in (P_i).

Suppose, for example, that all target sets A_i are uniformly prox-regular (see [17]). Then we can replace (P_i) with the following minimization problem

$$\begin{cases} l \circ \Phi_{\theta_i}^{\theta_N}(x(\theta_i)) + \sum\limits_{j=i}^{N} M_j\, d_{A_j}^2 \left(\Phi_{\theta_i}^{\theta_j}(x(\theta_i)) \right) \to \min, \\ \dot{x} = f(t, x, u), \quad u(t) \in U, \\ x(\theta_{i-1}) = \bar{x}(\theta_{i-1}), \end{cases} \qquad (\tilde{P}_i)$$

where $d_A(x)$ denotes the distance between a point x and a set A, M_j, $j = i, \ldots, N$, are sufficiently large positive constants.

Note that the uniform prox-regularity of A implies that $d^2_{A_j}$ is $C^{1,1}$ and $Dd^2_A(x) = 2(x - \mathrm{proj}_A(x))$ on a neighborhood of A. Hence, if $\Phi^{\theta_j}_{\theta_i}(x)$ is sufficiently close to A_j, we have

$$Dd^2_{A_j}\left(\Phi^{\theta_j}_{\theta_i}(x)\right) = 2\left(\Phi^{\theta_j}_{\theta_i}(x) - \mathrm{proj}_{A_j}\left(\Phi^{\theta_j}_{\theta_i}(x)\right)\right) \cdot D\Phi^{\theta_j}_{\theta_i}(x).$$

Recall that for computing $D\Phi^{\theta_j}_{\theta_i}(x)$ it suffices to solve a linear matrix differential equation of the form (2).

Thus, to compute the cost at a given point x, one must solve a nonlinear differential equation of the form (1). To compute the gradient of the cost, one must solve in addition a linear matrix differential equation of the form (2).

With these observations taken into account, any algorithm from [9,12,20] can be directly applied to $(\tilde{P}_i)$, and thus taken as $\mathcal{A}_0$.

4.3 Numerical Example

Here we apply the algorithm to solve the "generalized travelling salesman problem". Unlike the classical travelling salesman problem, we want to visit a number of targets $A_i \subset \mathbb{R}^2$ at time moments θ_i with the minimal fuel consumption.

The state of the salesman is described by a vector $x = (x_1, x_2, x_3, x_4) \in \mathbb{R}^4$, where (x_1, x_2) refers to the salesman's position and (x_3, x_4) to his or her velocity. We assume that the salesman's acceleration (u_1, u_2) can be controlled and the fuel consumption is proportional to the square of the acceleration.

Now, the problem can be formalized as follows:

$$\begin{cases} \int_0^{\theta_N} |u|^2\, dt \to \min, \\ \ddot{x} = u, \quad |u| \le a, \\ x(0) = x_0, \quad x(\theta_i) \in A_i, \end{cases}$$

where a denotes the maximal absolute value of the acceleration.

The usual trick, which consists in introducing new variables $y(t) = \dot{x}(t)$ and $z(t) = \int_0^t |u(s)|^2\, \mathrm{d}s$, allows us to rewrite the above problem in the form (P).

To deal with the auxiliary problems $(\tilde{P}_i)$ we use the method of needle linearization [20]. Let us remark that this method is strictly improving in the sense of Definition 4. Some results of the computations are presented in Fig. 2.

5 Applications for Measure-Driven Multiprocesses

5.1 An Optimal Impulsive Control Problem with Intermediate State Constraints

In this section, we address an optimal impulsive control problem and show how the results presented above can be applied for measure-driven multiprocesses.

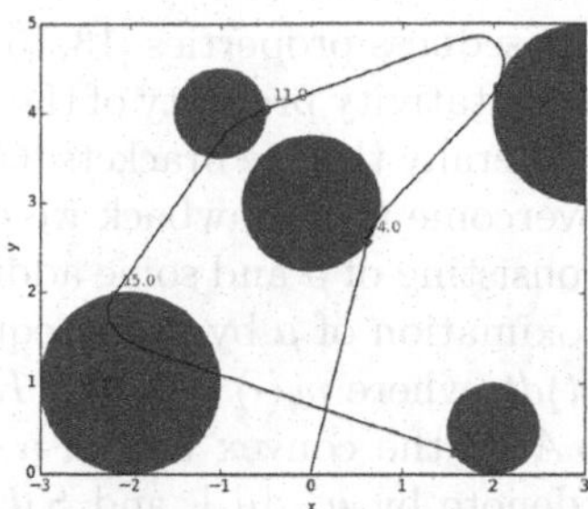

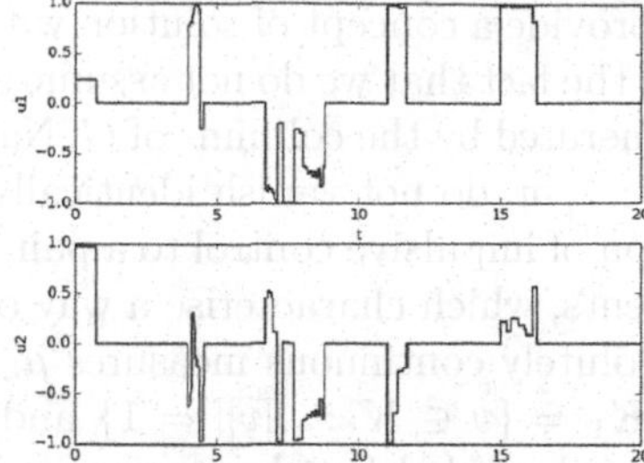

Fig. 2. The salesman must visit 5 targets $A_1, \ldots, A_5$ (the green balls on the plane) at the time moments 4, 7, 11, 15, 20, respectively. The control found by the algorithm is on the right figure, the path of the salesman is on the left. (Color figure online)

We consider the optimal impulsive control problem (P_{imp}):

$$J = l\big(x(b)\big) \to \min$$

stated on solutions of the measure-driven dynamic system

$$dx(t) = f\big(t, x(t), u(t)\big)\,dt + G\big(t, x(t)\big)\,\mu(dt), \quad x(a^-) = x_0, \tag{8}$$

$$x(\theta_i) \in A_i, \quad i = 1, \ldots, N, \tag{9}$$

$$u(t) \in U \text{ for } \mathcal{L}\text{-a.e. } t \in T, \quad \mu \in C^*(T, K). \tag{10}$$

Here, $T = [a, b] \subset \mathbb{R}$ is a fixed time interval, U is a given compact subset of $\mathbb{R}^r$, K is a closed convex cone from $\mathbb{R}^m$, A_i, $i = 1, \ldots, N$, are closed sets from $\mathbb{R}^n$, $\theta = \{\theta_1, \ldots, \theta_N\}$ is a given vector of time moments such that $a \le \theta_1 < \theta_2 < \cdots < \theta_N \le b$, $N < \infty$, and $x_0 \in \mathbb{R}^n$ is a given initial state. The symbol $\mathcal{L}$ stands for the Lebesgue measure on the real line. The dynamics (8)–(10) depends on two types of input signals: the "usual" control $u \in L^\infty(T, U)$, and the impulsive control μ, which is a vector measure. The functions $x(\cdot)$ define trajectories and are functions of bounded variation.

We posit the following assumptions:

(H_1) The function $l : \mathbb{R}^n \mapsto \mathbb{R}$ is continuous.

(H_2) The functions $f : T \times \mathbb{R}^n \times U \mapsto \mathbb{R}^n$, $G : T \times \mathbb{R}^n \mapsto \mathbb{R}^{n \times r}$ are continuous and locally Lipschitz continuous in x. Moreover, there exist constants $c_{1,2} > 0$ such that

$$\big\|f(t, x, u)\big\| \le c_1\,(1 + \|x\|), \quad \big\|G(t, x)\big\| \le c_2\,(1 + \|x\|)$$

for any $(t, x, u) \in T \times \mathbb{R}^n \times U$. Here, $\|\cdot\|$ denotes the vector norm defined by $\|x\| = \sum_{j=1}^{n} |x_j|$ or a consistent matrix norm of the proper dimension.

(H_3) The set $f(t, x, U) \doteq \big\{f(t, x, u) \mid u \in U\big\}$ is convex for all $(t, x) \in T \times \mathbb{R}^n$.

The solution concept for the control system (8)–(10) is given in Sect. 5.2. We note that any interpretation of (8)–(10) as a measure-driven differential equation cannot provide a concept of solution with well-posedness properties [13,15]. This is due to the fact that we do not assume any commutativity property of the vector fields generated by the columns of G. Namely, generally the Lie brackets $[G_i, G_j]$, $i, j = 1, \ldots, m$, do not vanish identically. To overcome this drawback we extend the notion of impulsive control to a pair $\pi(\mu)$ consisting of μ and some additional components, which characterise a way of approximation of μ by some sequences of $\mathcal{L}$-absolutely continuous measures $\mu_k = v_k(t)dt$, where $v_k(\cdot) \in L^\infty(T, K)$.

Let $K_1 \doteq \{v \in K \,:\, ||v|| = 1\}$ and let $co\, A$ be the convex hull of a set A. Given μ, a bounded Borel measure on T, we denote by $\mu_{\rm c}$, $|\mu_{\rm c}|$, and $S_{\rm d}(\mu)$ the continuous component in the Lebesgue decomposition of the measure μ, the total variation of $\mu_{\rm c}$, and the set on which the discrete component of μ is concentrated, i.e., $S_{\rm d}(\mu) \doteq \{s \in T \,:\, \mu(\{s\}) \neq 0\}$, respectively.

By an impulsive control π we mean a collection

$$\pi = \big(\mu, S, \{d_s, \omega_s(\cdot)\}_{s \in S}\big)$$

satisfying the following conditions:

(i) μ is a bounded K-valued Borel measure on T,

(ii) the set S is at most countable subset of the interval T, and $S_{\rm d}(\mu) \subseteq S$,

(iii) $d_s \in \mathbb{R}$, $d_s \geq \big\|\mu(\{s\})\big\|$ for all $s \in S$, and $\sum_{s \in S} d_s < \infty$;

(iv) the functions $\omega_s(\cdot) : [0, d_s] \to co\, K_1$, $s \in S$, are Borel measurable functions with the property

$$\int_0^{d_s} \omega_s(\tau) d\tau = \mu(\{s\}).$$

By $\mathcal{W}(T, K)$ we denote the set of all impulsive controls. Taken $\pi \in \mathcal{W}(T, K)$, one defines the function $V = V[\pi] : T \to \mathbb{R}$ by the relation:

$$V(t) = |\mu_{\rm c}|\big([a, t]\big) + \sum_{s \leq t,\, s \in S} d_s, \quad t \in (a, b], \quad V(a) = 0.$$

5.2 The Solution Concept and Comments

Let $BV(T, \mathbb{R}^n)$ be the space of $\mathbb{R}^n$-valued functions of bounded variation (BV-functions) and $BV^{\rm r}(T, \mathbb{R}^n)$ be the space of BV-functions which are right continuous on $(a, b]$.

We propose the solution concept for the control system (8)–(10) via the concept of graph completions for BV-functions.

Let $\tau_1 > 0$ be given. We say that $\eta : [0, \tau_1] \to [a, b]$ is a time reparametrization if $\eta(\cdot)$ is a nondecreasing Lipschitz continuous function such that $\eta(0) = a$, $\eta(\tau_1) = b$.

Given a time reparametrization $\eta(\cdot)$, we define the pseudoinverse function $\theta : [a, b] \to [0, \tau_1]$ by the rule:

$$\theta(t) = \inf\{\tau \in [0, \tau_1] \,:\, \eta(\tau) > t\}, \quad t \in (a, b], \quad \theta(a) = 0. \tag{11}$$

Given $\eta(\cdot)$ and its pseudoinverse $\theta(\cdot)$, let $S^\eta \doteq S_{\rm d}(\theta)$, $d_s^\eta \doteq \theta(s) - \theta(s-)$. Given $x \in BV^{\rm r}([a, b], \mathbb{R}^n)$, the time reparametrization $\eta(\cdot)$ is said to be consistent with $x(\cdot)$ if $S_{\rm d}(x) \subseteq S_{\rm d}(\theta)$.

Let $x \in BV^{\mathrm{r}}([a,b],\mathbb{R}^n)$, $\eta(\cdot)$ be a consistent time reparametrization, and $\theta(\cdot)$ be the pseudoinverse for $\eta(\cdot)$. Let $z^s : [0,d_s^\eta] \to \mathbb{R}^n$, $s \in S^\eta$ be the family of Lipschitz continuous functions such that $z^s(0) = x(s-)$, $z^s(d_s^\eta) = x(s)$, $s \in S^\eta$. Then, we say that $x_\eta \doteq \big(x(\cdot), \{z^s(\cdot)\}_{s\in S^\eta}\big)$ is a graph completion corresponding to $\eta(\cdot)$ for $x(\cdot)$.

Following [13,14], given controls (u,π), where $\pi = \big(\mu, S, \{d_s, \omega_s(\cdot)\}_{s\in S}\big) \in \mathcal{W}(T,K)$, $u \in L^\infty(T,\mathbb{R}^m)$, we define a function $x_{\mathrm{r}}(\cdot) \in BV^{\mathrm{r}}(T,\mathbb{R}^n)$ satisfying the following relations:

$$x_{\mathrm{r}}(t) = x_0 + \int_a^t f\big(\tau, x_{\mathrm{r}}(\tau), u(\tau)\big)\, d\tau + \int_a^t G\big(\tau, x_{\mathrm{r}}(\tau)\big)\, \mu_{\mathrm{c}}(d\tau)$$
$$+ \sum_{s\in S,\ s\le t} \big(z^s(d_s) - x_{\mathrm{r}}(s^-)\big), \quad t \in (a,b], \quad x_{\mathrm{r}}(a) = x_0, \tag{12}$$

$$\frac{dz^s(\tau)}{d\tau} = G\big(s, z^s(\tau)\big)\, \omega_s(\tau), \quad z^s(0) = x_{\mathrm{r}}(s^-),$$
$$\text{for } \mathcal{L}\text{-a.e. } \tau \in [0,d_s] \text{ and all } s \in S. \tag{13}$$

We note that the collection $\big(x_{\mathrm{r}}(\cdot), \{z^s(\cdot)\}_{s\in S}\big)$ is a graph completion for $x_{\mathrm{r}}(\cdot)$. Indeed, given π and $V = V[\pi]$, let $\theta_\mu(t) = t - a + V(t)$ for all $t \in T$. Denote $\tau_1 \doteq b - a + V(b)$ and consider the function $\eta_\mu : [0,\tau_1] \to [a,b]$ for which $\theta_\mu(\cdot)$ is the pseudoinverse defined by (11). Then, $x_{\eta_\mu} = \big(x_{\mathrm{r}}(\cdot), \{z^s(\cdot)\}_{s\in S}\big)$ and $\eta_\mu(\cdot)$ is a consistent time reparametrization for $x_{\mathrm{r}}(\cdot)$.

Next, we consider a set-valued function $X : T \to \mathrm{comp}(\mathbb{R}^n)$ defined as

(i) $X(t) = x_{\mathrm{r}}(t)$ for all $t \in T \setminus S$,
(ii) $X(s) = \big\{z^s(\tau) \,:\, \tau \in [0,d_s]\big\}$ for all $t = s \in S$.

Then, X is said to be the solution of the measure-driven equation (8) corresponding to the controls (u,π). We say that X satisfies the intermediate constraints (9) if

$$X(\theta_i) \cap A_i \neq \emptyset, \quad i = 1,\ldots,N. \tag{14}$$

In what follows, σ denotes a *feasible process* of problem (P_{imp}), i.e., a triple (X,u,π) satisfying the conditions (8), (14) together with

$$u \in L^\infty(T,U), \quad \pi \in \mathcal{W}(T,K), \tag{15}$$

and Σ stands for the set of all feasible processes.

We can interpret (P_{imp}) as a relaxation of a conventional optimal control problem. Indeed, let $\{\varepsilon_k\}$ be a sequence such that $\varepsilon_k \to 0$ as $k \to \infty$. We consider the following problem (P_{0,ε_k}):

$$J_{0,\varepsilon_k} = l\big(x(b)\big) \to \inf$$

subject to the dynamics

$$\dot{x}(t) = f\big(t, x(t), u(t)\big) + G\big(t, x(t)\big)\, v(t), \quad x(a) = x_0, \tag{16}$$
$$x(\theta_i) \in A_i + \varepsilon_k B, \quad i = 1,\ldots,N, \tag{17}$$
$$u(t) \in U,\ v(t) \in K \text{ for } \mathcal{L}\text{- a.e. } t \in T, \tag{18}$$

where $x \in AC(T, \mathbb{R}^n)$, $u \in L^\infty(T, \mathbb{R}^r)$, $v \in L^\infty(T, \mathbb{R}^m)$, $B \subset \mathbb{R}^n$ is the closed unit ball centered at zero. We say that $g = \big(x(\cdot), u(\cdot), v(\cdot)\big)$ is a feasible process for the problem (P_{0,ε_k}) if the components of g satisfy the relation (16)–(18). We denote by Σ_{ε_k} the collection of all feasible processes g.

In general, the problem (P_{0,ε_k}) does not have optimal solution with measurable controls $u(\cdot)$, $v(\cdot)$ and absolutely continuous trajectories $x(\cdot)$. This is due to the fact that the velocity set defined by the Eq. (16) is unbounded.

The following theorems clarify the sense in which the problem (P_{imp}) is considered as a relaxation of a conventional optimal control problem.

Theorem 2 (approximation of solutions). *Let* $\sigma = \big(X, u, \pi(\mu)\big) \in \Sigma$. *Then, there exist sequences* $\{\varepsilon_k\}$ *and* $\{g_k\}$ *such that:*
(i) $\varepsilon_k \to 0$ *as* $k \to \infty$,
(ii) $g_k = \big(x_k(\cdot), u_k(\cdot), v_k(\cdot)\big) \in \Sigma_{\varepsilon_k}$ *for all* $k \in \mathbb{N}$,
(iii) there exists a selection $x(\cdot)$ *of the set-valued function* X *such that* $x_k(t) \to x(t)$ *for all* $t \in T$.

Theorem 3 (existence of an optimal solution). *Let sequences* $\{\varepsilon_k\}$ *and* $\{x_k(\cdot), u_k(\cdot), v_k(\cdot)\}$ *be such that: (i)* $\varepsilon_k \to 0$ *as* $k \to \infty$,
(ii) for every $k \in \mathbb{N}$ *the process* $g_k = \big(x_k(\cdot), u_k(\cdot), v_k(\cdot)\big)$ *is feasible for* (P_{0,ε_k});
(iii) $\sup_k \int_a^b \|v_k(t)\| dt < +\infty$;
(iv) $l\big(\lim_{k\to\infty} x_k(b)\big) = \lim_{k\to\infty} \inf_{g \in \Sigma_{\varepsilon_k}} J_{\varepsilon_k}(g)$.
Then, there exists $\bar\sigma = \big(\bar X, \bar u, \bar\pi\big) \in \Sigma$, *where* $\bar\pi = \big(\bar\mu, \bar S, \{\bar d_s, \bar\omega_s(\cdot)\}_{s\in\bar S}\big)$ *and* $\bar X$ *is defined by the corresponding collection* $\big(\bar x_{\mathrm{r}}(\cdot), \{\bar z^s(\cdot)\}_{s\in\bar S}\big)$, *such that*

$$J(\bar\sigma) = \min_{\sigma\in\Sigma} J(\sigma)$$

and $x_k(t) \to \bar x_{\mathrm{r}}(t)$ *for all* $t \in T \setminus \bar S$.

The proofs of Theorems 2, 3 follow from the approximation results in [19].

5.3 Space-Time Representation for Measure-Driven Multiprocesses

Following [13,15,16,23], we consider a space-time representation of the measure-driven system (8)–(10). This representation reduces our impulsive model to a conventional variational problem with measurable compact-valued controls. The space-time problem (P_{a}) takes the form:

$$\hat J = l\big(y(\tau_b)\big) \to \min$$

subject to the relations:

$$\eta'(\tau) = \omega_0(\tau), \quad \eta(0) = a, \quad \eta(\tau_b) = b, \tag{19}$$

$$y'(\tau) = f\big(\eta(\tau), y(\tau), \nu(\tau)\big)\,\omega_0(\tau) + G\big(\eta(\tau), y(\tau)\big)\,\omega(\tau), \quad y(0) = x_0, \tag{20}$$

$$\big(\eta(\tau_i), y(\tau_i)\big) \in \{\theta_i\} \times A_i, \quad i = 1, \ldots, N, \tag{21}$$

$$\nu(\tau) \in U, \quad \big(\omega_0(\tau), \omega(\tau)\big) \in co\,\tilde K_1 \quad \text{for } \mathcal{L}\text{-a.e. } \tau \in [0, \tau_b]. \tag{22}$$

Here, $(\eta(\cdot), y(\cdot)) \in AC([0,\tau_b], \mathbb{R}^{n+1})$, $\nu(\cdot) \in L^\infty([0,\tau_b], \mathbb{R}^r)$, $(\omega_0(\cdot), \omega(\cdot)) \in L^\infty([0,\tau_b], \mathbb{R}^{m+1})$, $\rho = (\tau_1, \ldots, \tau_N, \tau_b)$ is a vector of non-fixed points of time such that $0 \le \tau_1 < \tau_2 < \cdots < \tau_N \le \tau_b$; $\tilde{K}_1 \doteq \{(\omega_0, \omega) \in [0,1] \times K : \ \omega_0 + \|\omega\| = 1\}$; prime indicates the derivative w.r.t τ.

Let ζ denote a feasible process of problem (P_{a}), i.e., a tuple $(\rho, \eta, y, \nu, \omega_0, \omega)$ satisfying the conditions (19)–(22), and Σ_{a} stands for the set of all feasible processes. Then, there is a one-to-one correspondence between the sets Σ and Σ_{a} [15]. Furthermore,

$$\min_{\sigma \in \Sigma} J(\sigma) = \min_{\zeta \in \Sigma_{\mathrm{a}}} \hat{J}(\zeta).$$

We note that applying the hybrid maximum principle [6] to the problem (P_{a}) allows us to obtain more general necessary optimality conditions for measure-driven multiprocesses than in [10,11], where special assumptions about optimal controls were posited.

By a standard way, the problem (P_{a}) is transformed to an optimal control problem with fixed intermediate time points. Thus, (P_{imp}) can be considered as a particular case of (P). Assuming additionally that the functions f, G are continuously differentiable in x, and l is differentiable, we can apply the necessary optimality condition presented by Theorem 1 and the numerical algorithm for our measure-driven multiprocesses.

References

1. Ashepkov, L.T.: Optimal Control of Discontinuous Systems. Nauka, Novosibirsk (1987)
2. Baturin, V.A., Maltugueva, N.S.: Second-order improvement method to solve problems of optimal control of the logic-dynamic systems. Autom. Remote Control **72**(4), 808–817 (2011)
3. Betounes, D.: Differential Equations: Theory and Applications, 2nd edn. Springer, New York (2010). https://doi.org/10.1007/978-1-4419-1163-6
4. Buss, M., Von Stryk, O., Bulirsch, R., Schmidt, G.: Towards hybrid optimal control. at - Automatisierungstechnik **48**(9), 448–459 (2000). https://doi.org/10.1524/auto.2000.48.9.448
5. Clarke, F.H., Vinter, R.B.: Applications of optimal multiprocesses. SIAM J. Control Optim. **27**(5), 1048–1071 (1989)
6. Clarke, F.H., Vinter, R.B.: Optimal multiprocesses. SIAM J. Control Optim. **27**(5), 1072–1091 (1989)
7. Clarke, F.H.: Functional Analysis, Calculus of Variations and Optimal Control, Graduate Texts in Mathematics, vol. 264. Springer, London (2013). https://doi.org/10.1007/978-1-4471-4820-3
8. Dmitruk, A.V., Kaganovich, A.M.: The hybrid maximum principle is a consequence of Pontryagin maximum principle. Syst. Control. Lett. **57**(11), 964–970 (2009)
9. Dykhta, V.A.: Variational necessary optimality conditions with feedback descent controls for optimal control problems. Dokl. Math. **91**(3), 394–396 (2015)
10. Dykhta, V.A., Samsonyuk, O.N.: Maximum principle for nonsmooth optimal control problems with multipoint state constraints. Russ. Math. (Izvestiya VUZ. Matematika) **45**(2), 16–29 (2001)

11. Dykhta, V.A., Samsonyuk, O.N.: A maximum principle for smooth optimal impulsive control problems with multipoint state constraints. Comput. Math. Math. Phys. **49**, 942–957 (2009)
12. Gurman, V.I., Baturin, V.A., Danilina, E.V., et al.: New methods for the improvement of control processes. (Novye metody uluchsheniya upravlyaemykh protsessov). Nauka, Novosibirsk. 184 p. R. 1.80 (TIB: FM 1097) (1987)
13. Miller, B.M.: The generalized solutions of nonlinear optimization problems with impulse control. SIAM J. Control Optim. **34**, 1420–1440 (1996)
14. Miller, B.M., Rubinovich, E.Y.: Impulsive Control in Continuous and Discrete-continuous Systems. Kluwer Academic/Plenum Publishers, New York (2003). https://doi.org/10.1007/978-1-4615-0095-7
15. Miller, B.M., Rubinovich, E.Y.: Discontinuous solutions in the optimal control problems and their representation by singular space-time transformations. Autom. Remote Control **74**, 1969–2006 (2013)
16. Motta, M., Rampazzo, F.: Space-time trajectories of nonlinear systems driven by ordinary and impulsive controls. Differ. Integr. Equ. **8**(2), 269–288 (1995)
17. Poliquin, R.A., Rockafellar, R.T., Thibault, L.: Local differentiability of distance functions. Trans. Am. Math. Soc. **352**(11), 5231–5249 (2000)
18. Rasina, I.V.: Iterative optimization algorithms for discrete-continuous processes. Autom. Remote Control **73**(10), 1591–1603 (2012). https://doi.org/10.1134/S0005117912100013
19. Samsonyuk, O.N.: Invariant sets for nonlinear impulsive control systems. Autom. Remote Control **76**(3), 405–418 (2015)
20. Srochko, V.A.: Iterative Methods for Solving Optimal Control Problems. Fizmatlit, Moscow (2000)
21. von Stryk, O., Bulirsch, R.: Direct and indirect methods for trajectory optimization. Ann. Oper. Res. **37**(1–4), 357–373 (1992)
22. Wei, S., Uthaichana, K., Žefran, M., Decarlo, R.A., Bengea, S.: Applications of numerical optimal control to nonlinear hybrid systems. Nonlinear Anal. Hybrid Syst. **1**(2), 264–279 (2007)
23. Zavalishchin, S.T., Sesekin, A.N.: Dynamic Impulse Systems, Mathematics and Its Applications, vol. 394. Kluwer Academic Publishers Group, Dordrecht (1997). https://doi.org/10.1007/978-94-015-8893-5

On Ellipsoidal Estimates for Reachable Sets of the Control System

Oxana G. Matviychuk[1,2](✉)

[1] Krasovskii Institute of Mathematics and Mechanics, Ural Branch of Russian Academy of Sciences, 16 Sofia Kovalevskaya str., 620990 Ekaterinburg, Russia
vog@imm.uran.ru

[2] Ural Federal University, 19 Mira str., 620002 Ekaterinburg, Russia
https://www.imm.uran.ru/

Abstract. The problem of the ellipsoidal estimation of the reachable set of the control system under uncertainties is considered. The matrix included in the differential equations of the system dynamics is uncertain and only bounds on admissible values of this matrix coefficients are known. It is assumed that the initial states of the system are unknown but belong to a given star-shaped symmetric nondegenerate polytope. This polytope may be a non-convex set. Under such conditions, the dynamical system is a nonlinear and reachable set loses convexity property. A Minkowski function is used in the investigation to describe the trajectory tubes and their set-valued estimates. The step by step algorithm for constructing external and internal ellipsoidal estimates of reachable sets for such bilinear control systems is proposed. Numerical experiments were performed. The results of these numerical experiments are included.

Keywords: Control system · Ellipsoidal calculus · Estimation · Reachable set

1 Introduction

The present paper deals with the problem of reachable sets estimation for bilinear control systems described by differential equations. The case of the set-membership description of uncertain parameters is considered here. The matrix included in the differential equations of the system dynamics is uncertain, but the bounds on admissible perturbations of the matrix are known. These systems can be used for simulation of various electrical, mechanical and other types of systems with unknown parameters bounded by certain limits [1,2]. As an example, we can indicate mechanical systems in which the stiffness or friction coefficients are given inaccurately. Electrical systems where the resistance, capacitance, inductance, or feedback coefficients are known with a certain accuracy can also be described by such systems.

The systems with uncertainty on initial data were considered in [3,8–10,17, 18,21]. The systems with convex initial sets were considered in many works.

M. Khachay et al. (Eds.): MOTOR 2019, LNCS 11548, pp. 489–500, 2019.
https://doi.org/10.1007/978-3-030-22629-9_34

However, in concrete applied problems the initial sets and reachable sets may be non-convex but have special properties. In the present paper it is assumed that the initial state of the system is bounded by a given star-shaped set [16,19]. In a common case this set may be a non-convex.

The most developed approaches for estimating reachable sets are the method of ellipsoidal calculus [3,4,6,9,17] and the method of polyhedral techniques [7]. In the present paper we continue the researches [11–14] and develop methods of ellipsoidal approximation. As the result of the study we present modified estimates of the reachable set of the system using a special structure of the initial set and unknown parameters. The external and internal ellipsoidal estimates of reachable sets for such bilinear control systems are considered here.

2 Problem Formulation

In this section, we introduce the main necessary notations used in the paper and give the basic formulation of the problem.

2.1 Basic Notations

Let $\operatorname{comp}\mathbb{R}^n$ be the set of all compact subsets of the $\mathbb{R}^n$ and $\operatorname{conv}\mathbb{R}^n$ be the set of all convex and compact subsets of $\mathbb{R}^n$. Here $\mathbb{R}^n$ is the n–dimensional vector space. Also let $\mathbb{R}^{n\times n}$ stands for the set of all real $n\times n$–matrices, $\widetilde{\mathbb{R}}^{n\times n}\subset\mathbb{R}^{n\times n}$ stands for the set of all symmetric positive definite matrices, and $x'y=(x,y)=\sum_{i=1}^n x_iy_i$ be the usual inner product of $x,y\in\mathbb{R}^n$ with prime as a transpose, $\|x\|=(x'x)^{1/2}$. Let $I\in\mathbb{R}^{n\times n}$ be the identity matrix, $\operatorname{tr}(A)$ be the trace of $n\times n$-matrix A (the sum of its diagonal elements), $\operatorname{diag} b=\operatorname{diag}\{b_i\}$ be the diagonal matrix A with $a_{ii}=b_i$ where b_i are components of the vector b. By the symbol $\overline{\operatorname{co}}\,A$ we denote closed convex hull of the set $A\subset\mathbb{R}^n$.

The Hausdorff distance between sets $A,B\in\mathbb{R}^n$ we denote by $h(A,B)$. Here $h(A,B)=\max\{h^+(A,B),\,h^-(A,B)\}$, with $h^+(A,B)$ and $h^-(A,B)$ being the Hausdorff semidistances between A and B, $h^+(A,B)=\sup\{d(x,B):\,x\in A\}$, $h^-(A,B)=h^+(B,A)$, $d(x,A)=\inf\{\|x-y\|:\,y\in A\}$.

By symbol

$$B(a,r)=\{x\in\mathbb{R}^n:\|x-a\|\le r\}$$

we denote the *ball* in $\mathbb{R}^n$ with radius $r>0$ and center $a\in\mathbb{R}^n$. By symbol

$$E(a,Q)=\{x\in\mathbb{R}^n:\,(Q^{-1}(x-a),(x-a))\le 1\}$$

denote the *ellipsoid* in $\mathbb{R}^n$ with symmetric positive definite $n\times n$-matrix Q and center $a\in\mathbb{R}^n$.

We suppose, that *parallelepiped* [7] $\mathcal{P}(p,P)$ in $\mathbb{R}^n$ is a set

$$\mathcal{P}(p,P)=\{x:x=p+\sum_{i=1}^n p^i\alpha_i,\ |\alpha_i|\le 1,\ i=\overline{1,n}\},\tag{1}$$

where $p \in \mathbb{R}^n$ is its center, and $P = \{p^1 \dots p^n\}$ is the orientation matrix ($\det P \neq 0$), p^i are the direction vectors. The unit cube is the parallelepiped with a unit orientation matrix $P = I$.

A set $Z \subseteq \mathbb{R}^n$ is called *star-shaped* (with center c) if $c + \lambda(Z - c) \subseteq Z$ for all $\lambda \in [0, 1]$. By symbol $\mathrm{St}(c, \mathbb{R}^n)$ we will denote the set of all star-shaped compact subsets $Z \subseteq \mathbb{R}^n$ with center c, $\mathrm{St}\,\mathbb{R}^n = \mathrm{St}(0, \mathbb{R}^n)$.

2.2 Problem Statement

Introduce the bilinear control system

$$\dot{x} = A(t)\,x + u(t), \quad x \in \mathbb{R}^n, \quad x_0 \in \mathcal{X}_0 \quad t \in [t_0, T], \tag{2}$$

where the matrix function $A(t) \in \mathbb{R}^{n \times n}$ is measurable, unknown and belongs to the set $\mathcal{A}$

$$A(t) \in \mathcal{A}, \quad t \in [t_0, T], \tag{3}$$

where

$$\mathcal{A} = \{A(t) \in \mathbb{R}^{n \times n} : A(t) = \mathrm{diag}\, a, \ a = (a_1, \dots, a_n) \in \mathbf{A}_0\}, \tag{4}$$

$$\mathbf{A}_0 = \{a \in \mathbb{R}^n : \sum_{i=1}^{n} |a_i|^2 \leq 1\}.$$

It is assumed that control function $u(t) \in \mathbb{R}^n$ is Lebesgue measurable on $[t_0, T]$ and $u(t) \in \mathcal{U} = E(\hat{a}, \hat{Q})$ for all $t \in [t_0, T]$.

We suppose that the initial value $x_0 = x(-0)$ for the system (2) is unknown but belongs to a given set $\mathcal{X}_0 \in \mathrm{St}(p, \mathbb{R}^n)$, where $\mathcal{X}_0 \in \mathbb{R}^n$. The set $\mathcal{X}_0$ is a symmetric nondegenerate polytope $\mathcal{M}(p)$ with center $p \in \mathbb{R}^n$ and $2m$ faces ($m \geq n$). It is assumed that $\mathcal{M}(p)$ can be represented by the union of m parallelepipeds $\mathcal{P}(p, P_k)$

$$x_0 \in \mathcal{X}_0 = \mathcal{M}(p) = \bigcup_{k=1}^{m} \mathcal{P}(p, P_k), \tag{5}$$

where $P_k = \{p_k^1 \dots p_k^n\}$, p_k^i are the direction vectors for parallelepipeds $\mathcal{P}(p, P_k)$.

The control system (2)–(5) presents a model of an uncertain dynamic system with an unknown matrix and given inclusion descriptions $A(t) \in \mathcal{A}$, $x_0 \in \mathcal{X}_0$, and $u(t) \in \mathcal{U}$.

Let the function $x(\cdot) = x(\cdot; t_0, x_0, A(\cdot), u(\cdot))$ be a *solution* of the system (2) for initial state $x_0 \in \mathcal{X}_0$, a matrix $A(t) \in \mathcal{A}$ and admissible control $u(t) \in \mathcal{U}$. The *trajectory tube* $\mathcal{X}(\cdot)$ of the system (2) is defined as

$$\mathcal{X}(\cdot) = \mathcal{X}(\cdot\,;\, t_0, \mathcal{X}_0, \mathcal{A}, \mathcal{U}) = \bigcup \{x(\cdot) : \ x_0 \in \mathcal{X}_0, \ A(\cdot) \in \mathcal{A}, \ u(\cdot) \in \mathcal{U}\}$$

and the *reachable set* is the cross-section $\mathcal{X}(t)$ of this set for the time moment $t \in [t_0, T]$. It should be note that a reachable set has the following evolutionary property: $\mathcal{X}(t\,;\, t_0, \mathcal{X}_0, \mathcal{A}, \mathcal{U}) = \mathcal{X}(t\,;\, \tau, \mathcal{X}(\tau), \mathcal{A}, \mathcal{U})$ where $\tau \in [t_0, t]$.

The problem of the study is to find the internal and external ellipsoidal estimates (with respect to the inclusion of sets) of the reachable set $\mathcal{X}(t)$ $(t_0 < t \le T)$ for the bilinear system (2) by using the analysis of bilinear control systems with uncertainty on initial data.

3 Main Results

3.1 Ellipsoidal Estimates of the Initial Set

First we construct the ellipsoidal estimates of the initial set.

Lemma 1. [13] *For the polytope $\mathcal{M}(p)$ defined in (5) the following ellipsoidal estimates holds*

$$\bigcup_{k=1}^{m} E(p, D_k^-) \subseteq \mathcal{M}(p) \subseteq \bigcup_{k=1}^{m} E(p, D_k^+),$$
$$D_k^- = P_k P_k', \quad D_k^+ = nP_k P_k'. \tag{6}$$

Remark 1. If it is necessary to construct an estimate in the form of a single ellipsoid, then it is sufficient to find an external ellipsoidal estimate for the union of the ellipsoids in the inclusion (6). The algorithm of internal ellipsoidal estimation of the union of the ellipsoids is given in [20].

Example 1. Consider the symmetric polytope $\mathcal{M}(0)$ with vertices: $(4,0)$, $(2,3)$, $(-2,5)$, $(-3,1)$, $(-4,0)$, $(-2,-3)$, $(2,-5)$, $(3,-1)$. The octagon $\mathcal{M}(0)$ can be represented by the union of parallelograms $\mathcal{P}(0, P_k)$ $(k = 1, \dots, 4)$

$$\mathcal{M}(0) = \bigcup_{k=1}^{4} \mathcal{P}(0, P_k),$$
$$P_1 = \begin{pmatrix} 0 & 2 \\ 4 & -1 \end{pmatrix}, \quad P_2 = \begin{pmatrix} 3 & -1 \\ 1.5 & 1.5 \end{pmatrix},$$
$$P_3 = \begin{pmatrix} 3.5 & 1.5 \\ -0.5 & 0.5 \end{pmatrix}, \quad P_4 = \begin{pmatrix} 0.5 & -2.5 \\ 2 & 3 \end{pmatrix},$$

where P_k $(k = 1, \dots, 4)$ are the orientation matrices composed of direction vectors p_i. The set $\mathcal{M}(0)$ and its external and internal ellipsoidal estimates are shown in Fig. 1.

3.2 External Ellipsoidal Estimates

Note that the sets $\mathcal{X}(t)$ need not be convex for the bilinear system (2)–(5). However, these sets have other geometrical properties.

Assumption 1. (i) *For every $t \in [t_0, T]$ the inclusion $0 \in \mathcal{U}$ is true.* (ii) *The inclusion $0 \in \mathcal{X}_0$ is true.*

The following theorem is valid.

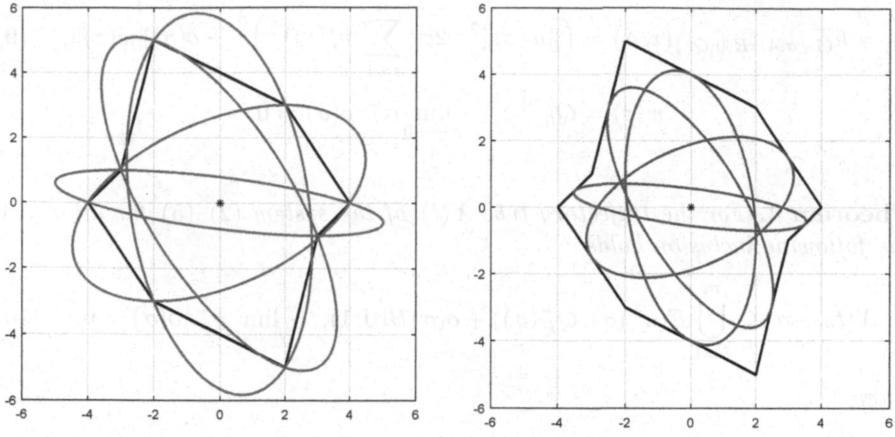

Fig. 1. Polytope $\mathcal{M}(0)$ and its external and internal ellipsoidal estimates.

Theorem 1. [10] *Under Assumption 1 the reachable sets* $\mathcal{X}(t)$ *are star-shaped and compact sets for all* $t \in [t_0, T]$ ($\mathcal{X}(t) \in \operatorname{St} \mathbb{R}^n$).

We need the following notation

$$\mathcal{M} * X = \{z \in \mathbb{R}^n : z = Mx,\ M \in \mathcal{M},\ x \in X\},$$

where $\mathcal{M} \in \operatorname{conv} \mathbb{R}^{n \times n}$, $X \in \operatorname{conv} \mathbb{R}^n$. Then the evolution equation that describes the dynamics of trajectory tubes has the following form.

Theorem 2. [5] *The trajectory tube* $\mathcal{X}(t)$ *of the bilinear differential system* (2)–(5) *is the unique solution to the evolution equation*

$$\lim_{\sigma \to +0} \sigma^{-1} h\big(\mathcal{X}(t+\sigma), (I+\sigma\mathcal{A})*\mathcal{X}(t)+\sigma\mathcal{U}\big) = 0, \quad \mathcal{X}(t_0) = \mathcal{X}_0, \quad t \in [t_0, T]. \quad (7)$$

Denote the *Minkowski function* of a set $M \in \operatorname{St} \mathbb{R}^n$ by

$$h_M(z) = \inf\{t > 0 : \ z \in tM,\ z \in \mathbb{R}^n\}.$$

Let $\rho(l|M)$ be the support function of a convex compact set $C \in \operatorname{conv} \mathbb{R}^n$, i.e.,

$$\rho(l|C) = \max\{(l, c) : \ c \in C,\ l \in \mathbb{R}^n\}.$$

Theorem 3. [5,12] For every $z \in \mathbb{R}^n$ such that $z_i \neq 0$ ($i = \overline{1, n}$) the following formula is true:

$$h_{(I+\sigma\mathcal{A}_0)*\mathcal{X}_0}(z, \sigma) = \min\Big\{ \max_{l \neq 0} \frac{1}{\rho(l|\mathcal{X}_0)} \sum_{i=1}^{n} \frac{l_i z_i}{1 + \sigma a_i} : \ a \in \mathbf{A}_0,\ i = \overline{1, n} \Big\}. \quad (8)$$

By using equality (8) for the given set $\mathcal{A}$ and if $\mathcal{X}_0 = E(0, Q_0)$ we get the Minkowski function of set $(I + \sigma\mathcal{A}) * E(0, Q_0)$ [12]

$$h_{(I+\sigma\mathcal{A})*E(0,Q_0)}(z,\sigma) = \Big(\|w(z)\|^2 - 2\sigma\big(\sum_{i=1}^{n} w_i^4(z)\big)^{\frac{1}{2}}\Big)^{\frac{1}{2}} + o(\sigma)\|w(z)\|, \qquad (9)$$

$$w(z) = Q_0^{\prime -\frac{1}{2}} z, \quad \lim_{\sigma\to+0} \sigma^{-1} o(\sigma) = 0.$$

Theorem 4. *For the trajectory tube $\mathcal{X}(t)$ of the system (2)–(5) for all $\sigma > 0$ the following inclusion holds*

$$\mathcal{X}(t_0+\sigma) \subseteq \bigcup_{k=1}^{m} E(a^+(\sigma), Q_k^+(\sigma)) + o(\sigma)B(0,1), \quad \lim_{\sigma\to+0} \sigma^{-1} o(\sigma) = 0, \qquad (10)$$

where

$$a^+(\sigma) = p + \sigma\hat{a},$$
$$Q_k^+(\sigma) = (q_1^{-1}+1)H_k(\sigma) + (q_1+1)\sigma^2\hat{Q},$$
$$H_k(\sigma) = (q_2^{-1}+1)\sigma^2 \operatorname{diag}\{(p_i)^2\} + (q_2+1)R_k(\sigma)D_k^+, \quad p = \{p_i\}$$
$$R_k(\sigma) = \max_z \frac{z'(D_k^+)^{-1}z}{\Big(h_{(I+\sigma\mathcal{A})*E(p,D_k^+)}(z,\sigma)\Big)}, \quad k = 1,\dots,m,$$

where $h_{(I+\sigma\mathcal{A})*E(p,D_k^+)}(z,\sigma)$ may be found by formula (8) and q_1 and q_2 are the unique positive root of the equations

$$\sum_{i=1}^{n} \frac{1}{q_1+\alpha_i} = \frac{n}{q_1(q_1+1)}, \quad \sum_{i=1}^{n} \frac{1}{q_2+\beta_i} = \frac{n}{q_2(q_2+1)},$$

with $\alpha_i \geq 0$ $(i=\overline{1,n})$ being the roots of the following equation $|H_k(\sigma) - \alpha\sigma^2\hat{Q}| = 0$ and $\beta_i \geq 0$ $(i=\overline{1,n})$ being the roots of the following equation

$$\prod_{i=1}^{n} \big(\sigma^2(p_i)^2 - \beta R_k^2(\sigma)\big) = 0.$$

Proof. From Theorem 2 we have the funnel equation for small σ $(t = t_0+\sigma)$

$$h\big(\mathcal{X}(t_0+\sigma), (I+\sigma\mathcal{A}) * \mathcal{X}_0\big) = o(\sigma), \quad \lim_{\sigma\to+0} \sigma^{-1} o(\sigma) = 0.$$

Note that

$$(I+\sigma\mathcal{A}) * E(p, D_k^+) = p + \sigma\mathcal{A} * p + (I+\sigma\mathcal{A}) * E(0, D_k^+).$$

The set $\mathcal{A} * p$ is convex, therefore

$$\rho(l|\mathcal{A}*p) = \max_{A\in\mathcal{A}} l'Ap = \big(\sum_{i=1}^{n} l_i^2(p_i)^2\big)^{\frac{1}{2}} = \rho\big(l|E(0, \operatorname{diag}\{(p_i)^2\})\big). \qquad (11)$$

Further we use the properties of the Minkowski function and the results of the Theorem 3. The inequality

$$h_{E(0,R_k(\sigma)D_k^+)}(z) \geq h_{(I+\sigma\mathcal{A})*E(0,D_k^+)}(z)$$

holds for

$$R_k(\sigma) = \max_z \frac{z'(D_k^+)^{-1}z}{\left(h_{(I+\sigma\mathcal{A})*E(p,D_k^+)}(z,\sigma)\right)}.$$

By direct calculation we obtain the following result

$$\begin{aligned}
&(I+\sigma\mathcal{A}) * E(p, D_k^+) + \sigma\mathcal{U} \\
&= E(p, \sigma^2 \operatorname{diag}\{(p_i)^2\}) + (I+\sigma\mathcal{A}) * E(0, D_k^+) + \sigma E(\hat{a}, \hat{Q}) \\
&\subseteq E(p, \sigma^2 \operatorname{diag}\{(p_i)^2\}) + E(0, R_k(\sigma)D_k^+) + \sigma E(\hat{a}, \hat{Q}), \\
&R_k(\sigma) = \max_z \frac{z'(D_k^+)^{-1}z}{\left(h_{(I+\sigma\mathcal{A})*E(p,D_k^+)}(z,\sigma)\right)}.
\end{aligned}$$

Based the procedure of external ellipsoidal estimate of the sum of the ellipsoids given in [3,9] and estimate for initial set (6)

$$\mathcal{X}_0 \subseteq \bigcup_{k=1}^{m} E(p, D_k^+)$$

we get the external estimate (10).

Algorithm 1. Introduce subsegments $[t_i, t_{i+1}]$ of the time segment $[t_0, T]$ where $t_i = t_0 + ih$ $(i = 1, \ldots, m)$, $t_m = T$, $h = (T - t_0)/m = \sigma$.

1. With applying of Lemma 1 find m ellipsoids $E(p, D_k^+)$, $k = \overline{1,m}$ for the given symmetric nondegenerate polytope $\mathcal{X}_0 = \mathcal{M}(p)$.
2. With applying Theorem 4 define the ellipsoids $E(a_k^1, Q_k^1) = E(a^+(\sigma), Q_k^+(\sigma))$ for each ellipsoid $E(p, D_k^+)$ $(k = \overline{1,m})$.
3. Consider the system on the next subsegment $[t_1, t_2]$ with $E(a_k^1, Q_k^1)$ as the initial ellipsoids at instant t_1.
4. If $t = T$ then end of the procedure otherwise the next step repeats the previous iterations.

The result of the process is the external estimate of the reachable set $\mathcal{X}(T)$ of the system (2).

3.3 Internal Ellipsoidal Estimates

The following theorem allows us to find an internal ellipsoidal estimate the reachable set of the bilinear control system (2)–(5).

Theorem 5. *For the trajectory tube $\mathcal{X}(t)$ of the system* (2)–(5) *for all $\sigma > 0$ the following inclusion holds*

$$\bigcup_{k=1}^{m} E(a^-(\sigma), Q_k^-(\sigma)) \subseteq \mathcal{X}(t_0 + \sigma) + o(\sigma)B(0,1), \quad \lim_{\sigma \to +0} \sigma^{-1} o(\sigma) = 0, \tag{12}$$

where

$$a^-(\sigma) = p + \sigma\hat{a},$$
$$Q_k^-(\sigma) = G_k(\sigma) + \sigma^2\hat{Q} + 2\sigma G_k(\sigma)^{\frac{1}{2}}(G_k(\sigma)^{-\frac{1}{2}}\hat{Q}G_k(\sigma)^{-\frac{1}{2}})^{\frac{1}{2}}G_k(\sigma)^{\frac{1}{2}},$$
$$G_k(\sigma) = \sigma^2 \operatorname{diag}\{(p_i)^2\} + r_k(\sigma)D_k^-$$
$$+2\sigma r_k(\sigma)\operatorname{diag}\{(p_i)\}(\operatorname{diag}\{p_i^{-1}\}D_k^- \operatorname{diag}\{p_i^{-1}\})^{1/2}\operatorname{diag}\{(p_i)\}, \quad p = \{p_i\},$$
$$r_k(\sigma) = \min_z \frac{z'(D_k^-)^{-1}z}{\left(h_{(I+\sigma\mathcal{A})*E(p,D_k^-)}(z,\sigma)\right)}, \quad k = 1,\ldots,m,$$

where $h_{(I+\sigma\mathcal{A})*E(p,D_k^-)}(z,\sigma)$ *may be found by formula* (8).

Proof. Consider the funnel Eq. (7). Note that

$$(I + \sigma\mathcal{A}) * E(p, D_k^-) = a_0 + \sigma\mathcal{A} * p + (I + \sigma\mathcal{A}) * E(0, D_k^-).$$

The following formulas may be derived by direct calculation by using the results of the Theorem 3, formulas (11) and Minkowski function for the ellipsoid

$$(I + \sigma\mathcal{A}) * E(p, D_k^-) + \sigma\mathcal{U}$$
$$= E(p, \sigma^2 \operatorname{diag}\{(p_i)^2\}) + (I + \sigma\mathcal{A}) * E(0, D_k^-) + \sigma E(\hat{a}, \hat{Q})$$
$$\supseteq E(p, \sigma^2 \operatorname{diag}\{(p_i)^2\}) + E(0, r_k(\sigma)D_k^-) + \sigma E(\hat{a}, \hat{Q}),$$
$$r_k(\sigma) = \min_z \frac{z'(D_k^-)^{-1}z}{\left(h_{(I+\sigma\mathcal{A})*E(p,D_k^-)}(z,\sigma)\right)}.$$

Based the procedure of internal ellipsoidal estimate of the sum of two ellipsoids given in [3,9] and estimate for initial set (6)

$$\bigcup_{k=1}^{m} E(p, D_k^-) \subseteq \mathcal{X}_0$$

we get the internal estimate (12).

Algorithm 2. Introduce subsegments $[t_i, t_{i+1}]$ of the time segment $[t_0, T]$ where $t_i = t_0 + ih$ $(i = 1, \ldots, m)$, $h = (T - t_0)/m = \sigma$, $t_m = T$.

1. Find m ellipsoids $E(p, D_k^-)$, $k = \overline{1,m}$ for the given symmetric nondegenerate polytope $\mathcal{X}_0 = \mathcal{M}(p)$ by Lemma 1.
2. Define the ellipsoids $E(a_k^1, Q_k^1){=}E(a^-(\sigma), Q_k^-(\sigma))$ for each ellipsoid $E(p, D_k^-)$ $(k = \overline{1,m})$ by Theorem 5.
3. Consider the system on the next subsegment $[t_1, t_2]$ with $E(a_k^1, Q_k^1)$ as the initial ellipsoids at instant t_1.
4. If $t = T$ then end of the procedure. Otherwise, the next step repeats the previous iterations.

At the end of the process, we will get the internal estimate of the reachable set $\mathcal{X}(T)$ of the system (2).

4 Numerical Simulation

The following example illustrates the main result of the study.

Example 2. Consider the following bilinear control system in $\mathbb{R}^2$

$$\begin{cases} \dot{x}_1 = & a_1 x_1 + u_1, \\ \dot{x}_2 = & a_2 x_2 + u_2, \quad 0 \le t \le 0.8. \end{cases}$$

Here the uncertain bounded matrix function $A \in \mathcal{A}$ where

$$\mathcal{A}=\{A : A = \operatorname{diag}\{a_1, a_2\},\, a_1^2 + a_2^2 \le 1\}.$$

The control function

$$u(t) \in \mathcal{U} = E(0, \hat{Q}), \quad \hat{Q} = \begin{pmatrix} 5 & 4 \\ 4 & 5 \end{pmatrix}.$$

The initial set $\mathcal{X}_0$ is the symmetric polytope $\mathcal{M}(0)$ with vertices: $(1.5, 0)$, $(0.5, 0.5)$, $(0, 1.5)$, $(-0.5, 0.5)$, $(-1.5, 0)$, $(-0.5, -0.5)$, $(0, -1.5)$, $(0.5, -0.5)$. For $\mathcal{M}(0)$ parallelepipeds $\mathcal{P}(0, P_i)$, $i = 1, 2, 3, 4$ were constructed,

$$\mathcal{X}_0 = \mathcal{M}(0) = \bigcup_{i=1}^{4} \mathcal{P}(0, P_i),$$

$$P_1 = \begin{pmatrix} -0.25 & 0.25 \\ 1 & 0.5 \end{pmatrix} \quad P_2 = \begin{pmatrix} 0.25 & 0.25 \\ 1 & -0.5 \end{pmatrix}$$

$$P_3 = \begin{pmatrix} 1 & -0.5 \\ 0.25 & 0.25 \end{pmatrix} \quad P_4 = \begin{pmatrix} 1 & 0.5 \\ 0.25 & 0.25 \end{pmatrix}.$$

The external and internal ellipsoidal estimates for $\mathcal{X}_0$ are given in Fig. 2. The trajectory tube $\mathcal{X}(t)$ is shown in the Fig. 3. This tube was constructed approximately with using results [15]. The external and internal ellipsoidal estimates of the reachable set $\mathcal{X}(t)$ under $t = 0.8$ are given in Fig. 4.

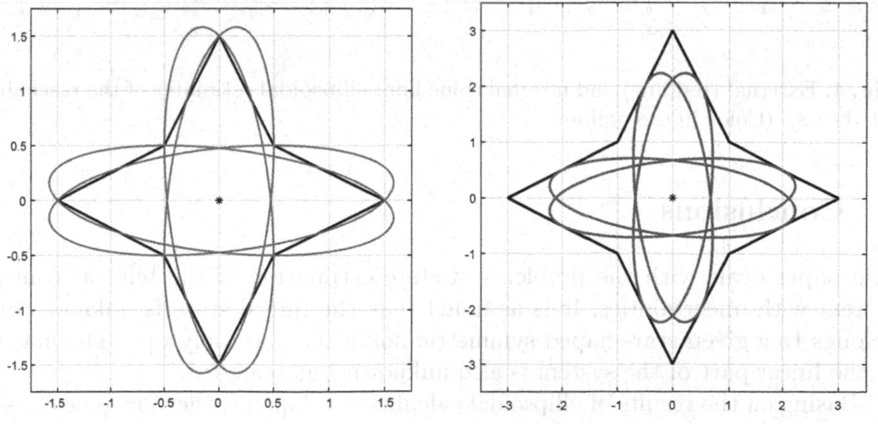

Fig. 2. Polytope $\mathcal{M}(0)$ and its external and internal ellipsoidal estimates.

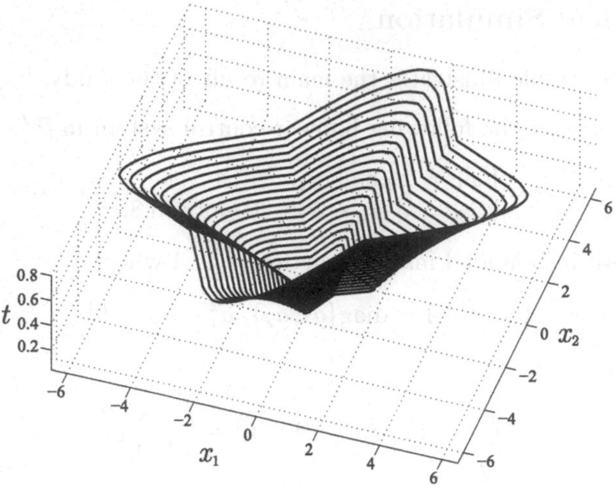

Fig. 3. The trajectory tube $\mathcal{X}(t)$, $0 < t \leq 0.8$.

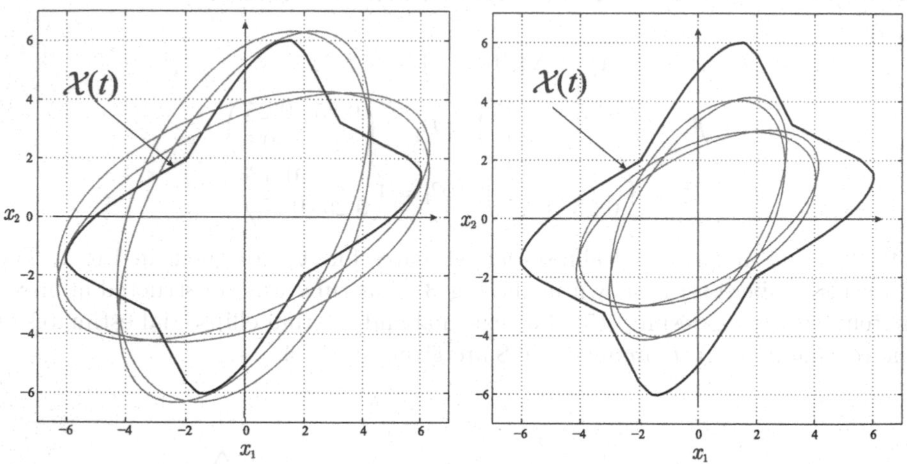

Fig. 4. External (red line) and internal (blue line) ellipsoidal estimates of the reachable set $\mathcal{X}(0.8)$. (Color figure online)

5 Conclusions

The paper deals with the problems of state estimation of the bilinear control system with uncertainties. It is assumed that the initial state is unknown but belongs to a given star-shaped symmetric nondegenerate polytope. The matrix in the linear part of the system is also unknown but bounded.

Basing on the results of ellipsoidal calculus developed earlier for some classes of uncertain systems the modified state estimation approach was presented. This approach uses the special constraints on the control and uncertainties and allows us to construct the external and internal ellipsoidal estimates of the reachable set.

References

1. Boscain, U., Chambrion, T., Sigalotti, M.: On some open questions in bilinear quantum control. In: European Control Conference (ECC), Zurich, Switzerland, 17–19 July 2013, pp. 2080–2085. IEEE Xplore Digital Library (2013). https://doi.org/10.23919/ECC.2013.6669238
2. Boussaïd, N., Caponigro, M., Chambrion, T.: Total variation of the control and energy of bilinear quantum systems. In: 52nd IEEE Conference on Decision and Control, Florence, Italy, 10–13 December 2013, pp. 3714–3719. IEEE Xplore Digital Library (2014). https://doi.org/10.1109/CDC.2013.6760455
3. Chernousko, F.L.: State Estimation for Dynamic Systems. CRC Press, Boca Raton (1994)
4. Filippova, T.F.: Set-valued dynamics in problems of mathematical theory of control processes. Int. J. Mod. Phys. Ser. B (IJMPB) **26**(25), 1–8 (2012)
5. Filippova, T.F., Lisin, D.V.: On the estimation of trajectory tubes of differential inclusions. Proc. Steklov Inst. Math. Probl. Control Dyn. Syst. (Suppl. 2), S28–S37 (2000)
6. Filippova, T.F., Matviychuk, O.G.: Estimates of reachable sets of control systems with bilinear-quadratic nonlinearities. Ural Math. J. **1**(1), 45–54 (2015). https://doi.org/10.15826/umj.2015.1.004
7. Kostousova, E.K.: On tight polyhedral estimates for reachable sets of linear differential systems. AIP Conf. Proc. **1493**, 579–586 (2012). https://doi.org/10.1063/1.4765545
8. Kurzhanski, A.B.: Control and Observation Under Conditions of Uncertainty. Nauka, Moscow (1977). [in Russian]
9. Kurzhanski, A.B., Valyi, I.: Ellipsoidal Calculus for Estimation and Control. Birkhäuser, Boston (1997)
10. Kurzhanski, A.B., Filippova, T.F.: On the theory of trajectory tubes – a mathematical formalism for uncertain dynamics, viability and control. In: Kurzhanski, A.B. (ed.) Advances in Nonlinear Dynamics and Control: a Report from Russia. Progress in Systems and Control Theory, vol. 17, pp. 22–188. Birkhäuser, Boston (1993)
11. Matviychuk, O.G.: Ellipsoidal estimates of reachable sets of impulsive control systems with bilinear uncertainty. Cybern. Phys. **5**(3), 96–104 (2016)
12. Matviychuk, O.G.: Internal ellipsoidal estimates for bilinear systems under uncertainty. AIP Conf. Proc. **1789**, 1–7 (2016). https://doi.org/10.1063/1.4968500. Article No. P. 060008
13. Matviychuk, O.G.: Estimates of the trajectory tubes of impulsive control systems. In: 14th International Conference "Stability and Oscillations of Nonlinear Control Systems" (Pyatnitskiy's Conference) (STAB), pp. 1–4. IEEE Xplore Digital Library (2018). https://doi.org/10.1109/STAB.2018.8408379
14. Matviychuk, O.G.: Estimation techniques for bilinear control systems. IFAC-PapersOnLine **51**(32), 877–882 (2018). https://doi.org/10.1016/j.ifacol.2018.11.434
15. Matviychuk, A.R., Matviychuk, O.G.: A method of approximate reachable set construction on the plane for a bilinear control system with uncertainty. AIP Conf. Proc. **2025**, 1–6 (2018). https://doi.org/10.1063/1.5064933. Article No. 100004
16. Mazurenko, S.: Partial differential equation for evolution of star-shaped reachability domains of differential inclusions. Set-Valued Var. Anal. **24**(2), 333–354 (2006). https://doi.org/10.1007/s11228-015-0345-4

17. Polyak, B.T., Nazin, S.A., Durieu, C., Walter, E.: Ellipsoidal parameter or state estimation under model uncertainty. Automatica **40**(7), 1171–1179 (2004). https://doi.org/10.1016/j.automatica.2004.02.014
18. Schweppe, F.: Uncertain Dynamic Systems. Prentice-Hall, Englewood Cliffs (1973)
19. Tidmore, F.E.: Extremal structure of star-shaped sets. Pac. J. Math. **29**(2), 461–465 (1969)
20. Vzdornova, O.G.: Ellipsoidal techniques in state estimation problem for linear impulsive control systems. In: Proceedings of the 3rd International Conference on Physics and Control (PhysCon 2007), Potsdam, Germany, 03–07 September 2007, pp. 1–6. Universitätsverlag Potsdam, Potsdam (2007). http://lib.physcon.ru/doc?id=5fc7ce762ef6
21. Walter, E., Pronzato, L.: Identification of Parametric Models from Experimental Data. Springer, Heidelberg (1997)

Problems of Hard Control for a Class of Degenerate Fractional Order Evolution Equations

Marina V. Plekhanova[1,2(✉)] and Guzel D. Baybulatova[2]

[1] South Ural State University, Lenin Av. 76, Chelyabinsk 454080, Russia
mariner79@mail.ru

[2] Chelyabinsk State University, Kashirin Brothers St. 129, Chelyabinsk 454001, Russia
baybulatova_g_d@mail.ru

Abstract. We find conditions of unique strong solution existence for the generalized Showalter—Sidorov problem to semilinear evolution equations with a degenerate operator at the highest fractional Gerasimov—Caputo derivative and with some constraint on the image of the nonlinear operator. Then we consider a class of optimal control problems for systems, whose dynamics is described by such equations endowed with the respective initial value conditions. Target functional is assumed not to take into account control costs. In such situation we used the additional condition of the admissible controls set boundedness. The obtained result of the initial problem unique solvability and properties of some functions spaces are applied to the proof of optimal control existence for such class of problems. Abstract results are applied to study of a control problem for a system, which is described by an initial-boundary value problem to a nonlinear partial differential equation, not solvable with respect to the highest time fractional derivative.

Keywords: Optimal control problem · Distributed control system · Semilinear degenerate evolution equation · Fractional derivative · Strong solution · Initial-boundary value problem

1 Introduction

Suppose that $\mathcal{X}, \mathcal{Y}, \mathcal{U}$ are Banach spaces, operators $L : \mathcal{X} \to \mathcal{Y}$, $\ker L \neq \{0\}$, $B : \mathcal{U} \to \mathcal{Y}$ are linear and continuous, $M : D_M \to \mathcal{Y}$ is linear and closed, D_M is dense in $\mathcal{X}$, $N : (t_0, T) \times \mathcal{X}^{m-1} \to \mathcal{Y}$. In this work an optimal control problem without taking into account control costs for a system, described by the equation

$$LD^\alpha x(t) = Mx(t) + N(t, x(t), x^{(1)}(t), \ldots, x^{(m-2)}(t)) + Bu(t), \ t \in (t_0, T), \quad (1)$$

with the fractional Gerasimov—Caputo [2,10] derivative D^α of the order $\alpha > 0$, is studied. Note that fractional differential equations and systems of such equations often arise in mathematical modeling of various real processes (see [11,13]

M. Khachay et al. (Eds.): MOTOR 2019, LNCS 11548, pp. 501–512, 2019.
https://doi.org/10.1007/978-3-030-22629-9_35

and the bibliographies there). This research continues the series of the first author's works on the solvability investigation of degenerate evolution equations and of optimal control problems for corresponding systems. Particularly, the existence of a unique strong solution for initial problems to degenerate fractional order equation in Banach spaces was researched in the linear case [20], in the semilinear case with the nonlinear operator N, not depending on the degeneration subspace elements [16,17,21], for the incomplete semilinear equation with restrictions on the image of the nonlinear operator $N(t, x(t))$ [17]. The existence of an optimal control in problems with compromise target functionals was investigated in [16–19] in the case of distributed control, in the work [19] for the start control problems. Obtained abstract results were applied to the problems for systems of equations of viscoelastic fluids dynamics [16,18,21], for pseudoparabolic equations [20] of time fractional order.

The presence of the degenerate ($\ker L \neq \{0\}$) linear operator L in (1) at the highest derivative does not allow to investigate the equation by classical methods [1,8,22]. We use the methods of the degenerate evolution equations from the works [5–7,14] of Fedorov and his co-authors. In this case, the original equation is represented as a system of a singular equation and an equation, resolved with respect to the fractional derivative. The paper considers the problem with the generalized Showalter—Sidorov initial conditions

$$(Px)^{(k)}(t_0) = x_k, \; k = 0, 1, \dots, m-1. \tag{2}$$

They mean, that the initial data are given only for the projection of the unknown function on the subspace $\mathcal{X}^1 = \mathrm{im} P$ without degeneration.

The results of this work differ from the other works results of the author in this direction firstly in the condition on a nonlinear operator. Here we use the constraints on the nonlinear operator N image, which are used before only for the incomplete nonlinear equation (1) [17]. The existence and the uniqueness of a strong solution for generalized Showalter—Sidorov problem (2) to nonlinear equation (1) is proved in the first part of this work.

Another feature of the present work is the form of the target functional

$$J_q(x, u) = \|x - x_d\|^q_{\mathcal{Q}_{\alpha,q}(t_0,T;\mathcal{X})} \to \inf, \tag{3}$$

which does not include control costs. Such control problems are often called problems of hard control [9]. When considering such problems, the properties of the coercivity of the target functional and its strict convexity (if any) are lost. The coercivity of the functional does not disappear, if we additionally require the boundedness of the set of admissible controls in the control space $\mathbf{U}$. Using these consideration, we proved the existence of an optimal control for the hard control problem to the system, which state is described by (1), (2). Obtained abstract results are illustrated on the example of the system, described by an initial-boundary value problem to a nonlinear partial differential equation, not solvable with respect to the time fractional derivative.

Other methods and approaches for optimal control problems study to fractional order systems, including systems, not solved with respect to the time derivative, can be found in [3,4,12,15,24] (see also the references lists there) and others.

2 Strong Solution of Semilinear Degenerate Equation

For $\beta > 0, t > 0$ we define functions $g_\beta(t) := t^{\beta-1}/\Gamma(\beta)$, where $\Gamma(\beta)$ is the Euler function at the point β. For convenience, we shall denote $\tilde{g}_\beta(t) := g_\beta(t - t_0)$.

The fractional Riemann—Liouville integral of order $\beta > 0$ is

$$J^\beta z(t) := \int_{t_0}^{t} \frac{(t-s)^{\beta-1}}{\Gamma(\beta)} z(s)ds, \quad t > t_0.$$

Let $m - 1 < \alpha \leq m \in \mathbf{N}$, $D^m := \frac{d^m}{dt^m}$ be the ordinary derivative of the integer order m. The fractional Gerasimov—Caputo derivative of order $\alpha > 0$ is defined as

$$D^\alpha z(t) := D^m J^{m-\alpha} \left(z(t) - \sum_{k=0}^{m-1} z^{(k)}(t_0) g_{k+1}(t - t_0) \right).$$

Let $\mathcal{Z}$ is a Banach space, operator $A \in \mathcal{L}(\mathcal{Z})$, i.e. it is a linear continuous operator from $\mathcal{Z}$ into $\mathcal{Z}$, $m - 1 < \alpha \leq m \in \mathbf{N}$. An operator $B : (t_0, T) \times \mathcal{Z}^m \to \mathcal{Z}$ is called Caratheodory mapping, if for any $z_0, z_1, \ldots, z_{m-1} \in \mathcal{Z}$ it is the measurable mapping on (t_0, T), and for almost all $t \in (t_0, T)$ it is continuous with respect to $z_0, z_1, \ldots, z_{m-1} \in \mathcal{Z}$.

For a constant $q > 1$ denote the space

$$\mathcal{Q}_{\alpha,q}(t_0, T; \mathcal{Z})$$
$$:= \left\{ z \in C^{m-1}([t_0, T]; \mathcal{Z}) : J^{m-\alpha} \left(z - \sum_{k=0}^{m-1} z^{(k)}(t_0) \tilde{g}_{k+1} \right) \in W_q^m(t_0, T; \mathcal{Z}) \right\}.$$

Consider the Cauchy problem

$$z^{(k)}(t_0) = z_k, \quad k = 0, 1, \ldots, m - 1, \tag{4}$$

for the semilinear equation

$$D^\alpha z(t) = Az(t) + B(t, z(t), z^{(1)}(t), \ldots, z^{(m-1)}(t)). \tag{5}$$

A strong solution of (4), (5) on (t_0, T) is a function $z \in \mathcal{Q}_{\alpha,q}(t_0, T; \mathcal{Z})$, for which conditions (4) and almost everywhere on (t_0, T) equality (5) hold.

The bar over a symbol will mean an ordered set of m elements with indexes from 0 to $m - 1$, for example, $\overline{z} = (z_0, z_1, \ldots, z_{m-1})$. A mapping $B : (t_0, T) \times \mathcal{Z}^m \to \mathcal{Z}$ is called uniformly Lipschitz continuous in $\overline{y}$, if there exists $l > 0$, such that the inequality $\|B(t, \overline{x}) - B(t, \overline{y})\|_\mathcal{Z} \leq l \sum_{k=0}^{m-1} \|x_k - y_k\|_\mathcal{Z}$ is true for almost all $t \in (t_0, T)$ and for all $\overline{x}, \overline{y} \in \mathcal{Z}^m$.

Theorem 1. [16]. *Let* $A \in \mathcal{L}(\mathcal{Z})$, $B : (t_0, T) \times \mathcal{Z}^m \to \mathcal{Z}$ *be Caratheodory mapping, uniformly Lipschitz continuous in* $\overline{y}$, $q \in (\max\{1, 1/\alpha\}, \infty)$, *for some* $\overline{v} \in \mathcal{Z}^m$ $B(\cdot, \overline{v}) \in L_q(t_0, T; \mathcal{Z})$. *Then for any* $z_0, z_1, \ldots, z_{m-1} \in \mathcal{Z}$ *the Cauchy problem* (4), (5) *has a unique strong solution on* (t_0, T).

We assume that $\mathcal{X}$, $\mathcal{Y}$ are Banach spaces, $L \in \mathcal{L}(\mathcal{X}; \mathcal{Y})$, i.e. it is a linear continuous operator from $\mathcal{X}$ into $\mathcal{Y}$, $\ker L \neq \{0\}$, $M \in \mathcal{C}l(\mathcal{X}; \mathcal{Y})$, i.e. it is a linear closed operator with a dense domain D_M in the space $\mathcal{X}$, which is acting into $\mathcal{Y}$. Define the L-resolvent set $\rho^L(M) = \{\mu \in \mathbf{C} : (\mu L - M)^{-1} \in \mathcal{L}(\mathcal{Y}; \mathcal{X})\}$ and the L-spectrum $\sigma^L(M) = \mathbf{C} \setminus \rho^L(M)$ of the operator M.

An operator M is said to be (L, σ)-bounded, if L-spectrum $\sigma^L(M)$ of operator M is bounded, i. e.

$$\exists a > 0 \quad \forall \mu \in \mathbf{C} \quad (|\mu| > a) \Rightarrow (\mu \in \rho^L(M)).$$

In the case of (L, σ)-boundedness of the operator M we can be define projectors P and Q on the spaces $\mathcal{X}$ and $\mathcal{Y}$ respectively. They have the forms

$$P := \frac{1}{2\pi i} \int_\gamma (\mu L - M)^{-1} L \, d\mu \in \mathcal{L}(\mathcal{X}), \; Q := \frac{1}{2\pi i} \int_\gamma L(\mu L - M)^{-1} \, d\mu \in \mathcal{L}(\mathcal{Y}),$$

where $\gamma := \{\mu \in \mathbf{C} : |\mu| = r > a\}$. Denote $\mathcal{X}^0 := \ker P$, $\mathcal{X}^1 := \mathrm{im} P$, $\mathcal{Y}^0 := \ker Q$, $\mathcal{Y}^1 := \mathrm{im} Q$. Then $\mathcal{X} = \mathcal{X}^0 \oplus \mathcal{X}^1$, $\mathcal{Y} = \mathcal{Y}^0 \oplus \mathcal{Y}^1$. By M_k (L_k) we denote the restriction of the operator M (L) onto $D_{M_k} := \mathcal{X}^k \cap D_M$ $(\mathcal{X}^k)$, $k = 0, 1$.

Theorem 2 [23, pp. 90, 91]. *Let an operator* M *be* (L, σ)*-bounded. Then*

(i) $M_1 \in \mathcal{L}(\mathcal{X}^1; \mathcal{Y}^1)$, $M_0 \in \mathcal{C}l(\mathcal{X}^0; \mathcal{Y}^0)$, $L_k \in \mathcal{L}(\mathcal{X}^k; \mathcal{Y}^k)$, $k = 0, 1$;
(ii) the operators $M_0^{-1} \in \mathcal{L}(\mathcal{Y}^0; \mathcal{X}^0)$, $L_1^{-1} \in \mathcal{L}(\mathcal{Y}^1; \mathcal{X}^1)$ *exist.*

Denote $G := M_0^{-1} L_0$. An operator M is called (L, p)-bounded at $p \in \mathbf{N}_0 := \mathbf{N} \cup \{0\}$, if G is a nilpotent operator of the power p.

Lemma 1. *Let* $H \in \mathcal{L}(\mathcal{X})$ *be a nilpotent operator of the power* $p \in \mathbf{N}_0$, *a function* $h : [t_0, T] \to \mathcal{X}$, $(HD^\alpha)^n h \in \mathcal{Q}_{\alpha,q}(t_0, T; \mathcal{X})$ *for* $n = 0, 1, \ldots, p$. *Then the equation* $HD^\alpha x(t) = x(t) + h(t)$ *has a unique strong solution. Moreover, it has the form* $x(t) = -\sum\limits_{n=0}^{p} (HD^\alpha)^n h(t)$.

Proof. Acting by the operator HD^α on the both sides of the equation, we get the equality $(HD^\alpha)^2 x(t) = x(t) + h(t) + HD^\alpha h(t)$. After p steps we have

$$(HD^\alpha)^{p+1} x = x + \sum_{n=0}^{p} (HD^\alpha)^n h.$$

Moreover, $(HD^\alpha)^{p+1} x = (D^\alpha)^{p+1} H^{p+1} x \equiv 0$ due to the nilpotency of the operator H. □

The semilinear equation

$$LD^{\alpha}x(t) = Mx(t) + N(t, x(t), x^{(1)}(t), \ldots, x^{(m-1)}(t)) + f(t). \tag{6}$$

will be called degenerate, since it is assumed before, that $\ker L \neq \{0\}$. A strong solution of Eq. (6) on (t_0, T) is $x \in C^{m-1}([t_0, T]; \mathcal{X}) \cap L_q(t_0, T; D_M)$, for which $J^{m-\alpha}\left(x - \sum\limits_{k=0}^{m-1} x^{(k)}(t_0)\tilde{g}_{k+1}\right) \in W_q^m(t_0, T; \mathcal{X})$, $q > 1$, and almost everywhere on (t_0, T) equality (6) is valid.

Consider the initial value problem

$$(Px)^{(k)}(t_0) = x_k, \; k = 0, 1, \ldots, m-1, \tag{7}$$

for Eq. (6). A strong solution of problem (6), (7) on the interval (t_0, T) is a solution of Eq. (6), such that conditions (7) are satisfied. Here we take into account that the smoothness of the function $Px(t) = L_1^{-1}QLx(t)$ is not less, than for $Lx(t)$.

Theorem 3. *Let $\alpha > 0$, $q > (\alpha - m + 1)^{-1}$, $p \in \mathbf{N}_0$, an operator M be (L, p)-bounded, an operator $N : (t_0, T) \times \mathcal{X}^m \to \mathcal{Y}$ be Caratheodory mapping, uniformly Lipschitz continuous in $\overline{v} \in \mathcal{X}^m$, for some $\overline{z} \in \mathcal{X}^m$ $N(\cdot, \overline{z}) \in L_q(t_0, T; \mathcal{Y})$, $\mathrm{im}N \subset \mathcal{Y}^1$, $Qf \in L_q(t_0, T; \mathcal{Y})$, $(GD^{\alpha})^n M_0^{-1}(I - Q)f \in \mathcal{Q}_{\alpha,q}(t_0, T; \mathcal{X})$ for $n = 0, 1, \ldots, p$; $x_0, x_1, \ldots, x_{m-1} \in \mathcal{X}^1$. Then problem (6), (7) has a unique strong solution on (t_0, T).*

Remark 1. If a function f is smooth enough, then all the conditions on f in Theorem 3 are satisfied.

Proof. If $\mathrm{im}N \subset \mathcal{Y}^1$, then $(I - Q)N \equiv 0$, $QN \equiv N$. In this case Eq. (6) after acting on its both sides by the operator $M_0^{-1}(I - Q)$ has the form

$$D^{\alpha}Gw(t) = w(t) + M_0^{-1}(I - Q)f(t),$$

where $w(t) = (I - P)x(t)$. By the nilpotency of the operator G and Lemma 1 this equation has a unique strong solution

$$w(t) = -\sum_{n=0}^{p}(D^{\alpha}G)^n M_0^{-1}(I - Q)f(t).$$

It remains to show the unique solvability of the problem

$$D^{\alpha}v(t) = S_1v(t) + L_1^{-1}Qf(t)$$
$$+ L_1^{-1}N(t, v(t) + w(t), v^{(1)}(t) + w^{(1)}(t), \ldots, v^{(m-1)}(t) + w^{(m-1)}(t)),$$
$$v^{(k)}(t_0) = Px_k, \; k = 0, 1, \ldots, m-1.$$

Here $S_1 = L_1^{-1}M_1$, $v(t) = Px(t)$. Since the operator

$$B(t, v_0, v_1, \ldots, v_{m-1})$$
$$= L_1^{-1}N(t, v_0 + w(t), v_1 + w^{(1)}(t), \ldots, v_{m-1} + w^{(m-1)}(t)) + L_1^{-1}Qf(t)$$

satisfies the conditions of Theorem 1, we get the required result from this theorem. □

3 Problems Without Control Costs

In this section we assume that $\mathcal{Y}$ is a Banach space, $\mathcal{X}, \mathcal{X}_1, \mathcal{U}$ are reflexive Banach spaces, $\mathcal{X}$ is compactly embedded in $\mathcal{X}_1$, $L \in \mathcal{L}(\mathcal{X};\mathcal{Y})$, $\ker L \neq \{0\}$, $M \in \mathcal{C}l(\mathcal{X};\mathcal{Y})$ is (L,p)-bounded operator, $N : (t_0,T) \times \mathcal{X}^{m-1} \to \mathcal{Y}$, $B \in \mathcal{L}(\mathcal{U};\mathcal{Y})$. Endow the domain D_M of the operator M with the graph norm, then D_M is the Banach space due to the closedness of M. Denote the space

$$\mathcal{Z}_{\alpha,q}(t_0,T;\mathcal{X}) = \{x \in L_q(t_0,T;D_M) \cap C^{m-1}([t_0,T];\mathcal{X}) :$$
$$J^{m-\alpha}\left(x - \sum_{k=0}^{m-1} x^{(k)}(t_0)\tilde{g}_{k+1}\right) \in W_q^m(t_0,T;\mathcal{X})\}, \quad q > 1.$$

Lemma 2 [16,19]. *$\mathcal{Q}_{\alpha,q}(t_0,T;\mathcal{X})$ and $\mathcal{Z}_{\alpha,q}(t_0,T;\mathcal{X})$ are Banach spaces with the norms $\|x\|_{\mathcal{Q}_{\alpha,q}(t_0,T;\mathcal{X})} = \|x\|_{C^{m-1}([t_0,T];\mathcal{X})} + \|D^\alpha x\|_{L_q(t_0,T;\mathcal{X})}$ and*

$$\|x\|_{\mathcal{Z}_{\alpha,q}(t_0,T;\mathcal{X})} = \|x\|_{L_q(t_0,T;D_M)} + \|x\|_{C^{m-1}([t_0,T];\mathcal{X})} + \|D^\alpha x\|_{L_q(t_0,T;\mathcal{X})}$$

respectively.

Consider the problem of hard control

$$LD^\alpha x(t) = Mx(t) + N(t, x(t), x^{(1)}(t), \ldots, x^{(m-2)}(t)) + Bu(t),\ t \in (t_0,T), \quad (8)$$

$$(Px)^{(k)}(t_0) = x_k,\ k = 0,1,\ldots,m-1, \quad (9)$$

$$u \in \mathcal{U}_\partial, \quad (10)$$

$$J_q(x,u) = \|x - x_d\|^q_{\mathcal{Q}_{\alpha,q}(t_0,T;\mathcal{X})} \to \inf, \quad (11)$$

where $m-1 < \alpha \leq m \in \mathbf{N}$, $x_d \in \mathcal{Q}_{\alpha,q}(t_0,T;\mathcal{X})$. Here $\mathcal{U}_\partial$ is the set of admissible controls.

Introduce the continuous operator $\gamma_0 : C([t_0,T];\mathcal{X}) \to \mathcal{X}$, $\gamma_0 x = x(t_0)$.

Set of pairs (x,u) will be called admissible pairs set $\mathcal{W}$ of problem (8)–(11), if $u \in \mathcal{U}_\partial$, $x \in \mathcal{Z}_{\alpha,q}(t_0,T;\mathcal{X})$ is a strong solution of (8), (9), $J(x,u) < \infty$. Problem (8)–(11) is a problem of finding of pairs $(\hat{x},\hat{u}) \in \mathcal{W}$, which minimize the cost functional, i. e. $J(\hat{x},\hat{u}) = \inf\limits_{(x,u)\in\mathcal{W}} J(x,u)$.

Lemma 3 [18]. *Let $\mathcal{X}_0$, $\mathcal{X}_1$ be reflexive Banach spaces, space $\mathcal{X}_0$ be compactly embedded in $\mathcal{X}_1$, $q \in (1,+\infty)$. Then for $m \in \mathbf{N}$ $W_q^m(t_0,T;\mathcal{X}_0)$ is compactly embedded in $W_q^{m-1}(t_0,T;\mathcal{X}_1)$.*

Theorem 4. *Let $\alpha > 1$, $q > (\alpha - m + 1)^{-1}$, operator M be (L,p)-bounded, $\mathcal{X}$ be compactly embedded in $\mathcal{X}_1$, the Banach space $\mathcal{Z}_{\alpha,q}(t_0,T;\mathcal{X})$ be reflexive, an operator $N : (t_0,T) \times \mathcal{X}_1^{m-1} \to \mathcal{Y}$ be Caratheodory mapping, uniformly Lipschitz continuous in $\overline{y} = (y_0, y_1, \ldots, y_{m-2}) \in \mathcal{X}_1^{m-1}$, for some $\overline{z} \in \mathcal{X}^{m-1}$ $N(\cdot,\overline{z}) \in L_q(t_0,T;\mathcal{Y})$, $N[(t_0,T)\times\mathcal{X}^{m-1}] \subset \mathcal{Y}^1$, $x_k \in \mathcal{X}^1$, $k = 0,1,\ldots,m-1$. Assume that $\mathcal{U}_\partial$ is a non-empty bounded closed convex subset in $L_q(t_0,T;\mathcal{U})$, for some $u_0 \in \mathcal{U}_\partial$ $(GD^\alpha)^l M_0^{-1}(I-Q)Bu_0 \in C^{m-1}([t_0,T];\mathcal{X})$, $D^\alpha(GD^\alpha)^l M_0^{-1}(I-Q)Bu_0 \in L_q(t_0,T;\mathcal{X})$ for $l = 0,1,\ldots,p$. Then problem (8)–(11) has a solution $(\hat{x},\hat{u}) \in \mathcal{Z}_{\alpha,q}(t_0,T;\mathcal{X}) \times \mathcal{U}_\partial$.*

Proof. We use Theorem 2.4 from the monograph [9] for the proof of an optimal control existence. Take spaces $\mathbf{Y} := \mathcal{Q}_{\alpha,q}(t_0,T;\mathcal{X})$, $\mathbf{Y}_1 := \mathcal{Z}_{\alpha,q}(t_0,T;\mathcal{X})$, $\mathbf{U} := L_q(t_0,T;\mathcal{U})$, $\mathbf{V} := L_q(t_0,T;\mathcal{Y}) \times \mathcal{X}^m$ and operators

$$\mathbf{L}(x,u) := (LD^\alpha x - Mx - Bu, \gamma_0 Px, \gamma_0(Px)^{(1)}, \dots, \gamma_0(Px)^{(m-1)}),$$

$$\mathbf{F}(z(\cdot)) := -(N(\cdot,x(\cdot),x^{(1)}(\cdot),\dots,x^{(m-2)}(\cdot)), x_0, x_1, \dots, x_{m-1}).$$

Since $\mathcal{X}$ is compactly embedded in $\mathcal{X}_1$, then the operator restriction $N|_{\mathcal{X}}$ satisfies all the conditions of Theorem 3. Hence by Theorem 3 the set $\mathcal{W}$ of admissible pairs is nonempty.

The continuity of the linear operator $\mathbf{L}$ from $\mathcal{Z}_{\alpha,q}(t_0,T;\mathcal{X}) \times L_q(t_0,T;\mathcal{U})$ to $L_q(t_0,T;\mathcal{Y}) \times \mathcal{X}^m$ follows from the inequalities

$$\begin{aligned}\|(LD^\alpha x - Mx - Bu, \gamma_0 Px, \gamma_0(Px)^{(1)}, \dots, \gamma_0(Px)^{(m-1)})\|_{L_q(t_0,T;\mathcal{Y})\times\mathcal{X}^m} \\ \le C_1\left(\|x\|_{\mathcal{Z}_{\alpha,q}(t_0,T;\mathcal{X})} + \|u\|_{L_q(t_0,T;\mathcal{U})} + \|x\|_{C^{m-1}([t_0,T];\mathcal{X})}\right) \\ \le C_2\|(x,u)\|_{\mathcal{Z}_{\alpha,q}(t_0,T;\mathcal{X})\times L_q(t_0,T;\mathcal{U})}.\end{aligned}$$

If $\|x_n - x\|_{\mathcal{Z}_{\alpha,q}(t_0,T;\mathcal{X})} \to 0$ as $n \to \infty$, due to the uniform Lipschitz continuity of the operator N we have

$$\begin{aligned}\|N(\cdot,x_n(\cdot),x_n^{(1)}(\cdot),\dots,x_n^{(m-2)}(\cdot)) - N(\cdot,x(\cdot),x^{(1)}(\cdot),\dots,x^{(m-2)}(\cdot))\|_{L_q(t_0,T;\mathcal{Y})} \\ \le C_1\sum_{k=0}^{m-2}\|x_n^{(k)} - x^{(k)}\|_{C([t_0,T];\mathcal{X}_1)} \le C_2\|x_n - x\|_{C^{m-2}([t_0,T];\mathcal{X})} \to 0\end{aligned}$$

as $n \to \infty$. This fact and the continuous embedding

$$\mathcal{Z}_{\alpha,q}(t_0,T;\mathcal{X}) \subset C^{m-2}([t_0,T];\mathcal{X})$$

imply the continuity of the operator $\mathbf{F} : \mathcal{Z}_{\alpha,q}(t_0,T;\mathcal{X}) \to \mathbf{V}$.

Choose $\mathbf{Y}_{-1} = W_q^{m-2}(t_0,T;\mathcal{X}_1)$ and check the remaining conditions of Theorem 2.4 [9]. By Lemma 3 the space $\mathcal{Z}_{\alpha,q}(t_0,T;\mathcal{X}) \subset W_q^{m-1}(t_0,T;\mathcal{X})$ is compactly embedded in $W_q^{m-2}(t_0,T;\mathcal{X}_1)$.

Due to the uniform Lipschitz continuity of the operator N for a linear continuous functional $v^* \in (L_q(t_0,T;\mathcal{X}_1))^*$ we have

$$\begin{aligned}|v^*(N(\cdot,x_n(\cdot),x_n^{(1)}(\cdot),\dots,x_n^{(m-2)}(\cdot)) - N(\cdot,x(\cdot),x^{(1)}(\cdot),\dots,x^{(m-2)}(\cdot)))| \\ \le C_1\|v^*\|_{(L_q(t_0,T;\mathcal{X}_1))^*}\|x_n - x\|_{W_q^{m-2}(t_0,T;\mathcal{X}_1)}.\end{aligned}$$

This reasoning allows us to conclude that the functional $w(\cdot) := v^*(\mathbf{F}(\cdot))$ can be continuously extended from $\mathcal{Z}_{\alpha,q}(t_0,T;\mathcal{X})$ onto $\mathbf{Y}_{-1}$.

For the pair $(x, u) \in \mathcal{W}$ we have

$$\|x\|_{\mathcal{Z}_{\alpha,q}(t_0,T;\mathcal{X})} + \|u\|_{L_q(t_0,T;\mathcal{U})} \le C_1\|x\|_{\mathcal{Q}_{\alpha,q}(t_0,T;\mathcal{X})}$$
$$+ \|Mx\|_{L_q(t_0,T;\mathcal{Y})} + \|u\|_{L_q(t_0,T;\mathcal{U})} = C_1\|x\|_{\mathcal{Q}_{\alpha,q}(t_0,T;\mathcal{X})} + \|u\|_{L_q(t_0,T;\mathcal{U})}$$
$$+ \|LD^\alpha x - N(\cdot, x(\cdot), x^{(1)}(\cdot), \dots, x^{(m-2)}(\cdot)) - Bu\|_{L_q(t_0,T;\mathcal{Y})}$$
$$\le C_2\|x\|_{\mathcal{Q}_{\alpha,q}(t_0,T;\mathcal{X})} + C_3\|u\|_{L_q(t_0,T;\mathcal{U})}$$
$$+ \|N(\cdot, x(\cdot), x^{(1)}(\cdot), \dots, x^{(m-2)}(\cdot))\|_{L_q(t_0,T;\mathcal{Y})}$$
$$\le C_3\|x\|_{\mathcal{Q}_{\alpha,q}(t_0,T;\mathcal{X})} + C_4 + \|N(\cdot, z_0, z_1, \dots, z_{m-2})\|_{L_q(t_0,T;\mathcal{Y})}$$
$$+ l\|x\|_{W_q^{m-2}(t_0,T;\mathcal{X})} + l(T-t_0)^{1/q}\sum_{k=0}^{m-2}\|z_k\|_{\mathcal{X}} \le C_3\|x\|_{\mathcal{Q}_{\alpha,q}(t_0,T;\mathcal{X})} + C_5.$$

Here we take into account that the set $\mathcal{U}_\partial$ is bounded in $L_q(t_0, T; \mathcal{U})$, the operator N is uniformly Lipschitz continuous, and $N(\cdot, z_0, z_1, \dots, z_{m-2}) \in L_q(t_0, T; \mathcal{Y})$. Thus, the functional J is coercive. □

4 Application

Consider the initial-boundary value problem

$$\left(\frac{\partial^2}{\partial s^2} + \beta\right)\left(\frac{\partial^k w}{\partial t^k}(s, t_0) - v_k(s)\right) = 0, \quad k = 0, 1, \dots, m-1, \quad s \in (0, \pi), \tag{12}$$

$$w(0, t) = w(\pi, t) = 0, \quad t \in (t_0, T), \tag{13}$$

for the model equation in $(0, \pi) \times (t_0, T)$

$$D_t^\alpha\left(\frac{\partial^2}{\partial s^2} + \beta\right)^2 w = \delta w + \sum_{n=0}^{m-2}\delta_n(t)\left(\frac{\partial^2}{\partial s^2} + \beta\right)\ln\left(1 + \left(\frac{\partial^n w}{\partial t^n}\right)^2\right) + u(s,t), \tag{14}$$

where D_t^α is the Gerasimov—Caputo derivative with respect to t, constants $\beta, \delta \in \mathbf{R}$, $m - 1 < \alpha \le m \in \mathbf{N}$, $\delta_n : (t_0, T) \to \mathbf{R}$, $n = 1, 2, \dots, m-2$.

Define Banach spaces

$$\mathcal{X} := \{v \in H^4(0, \pi) : v(0) = v(\pi) = v''(0) = v''(\pi) = 0\},$$
$$\mathcal{X}_1 := \{v \in H^2(0, \pi) : v(0) = v(\pi) = 0\}, \quad \mathcal{Y} = \mathcal{U} := L_2(0, \pi),$$

and operators

$$L := \left(\frac{\partial^2}{\partial s^2} + \beta\right)^2, \quad M := \delta I, \quad B = I,$$

$$N(t, x_0, x_1, \dots, x_{m-2}) := \sum_{n=0}^{m-2}\delta_n(t)\left(\frac{\partial^2}{\partial s^2} + \beta\right)\ln(1 + x_n^2),$$

$x_n \in \mathcal{X}_1$, $n = 0, \dots, m-2$.

We assume that $\beta = k_0^2$ for some $k_0 \in \mathbf{N}$, then $\ker L = \operatorname{span}\{\sin k_0 s\} \ne \{0\}$ and Eq. (14) is degenerate.

Theorem 5. *Let* $\alpha > 0$, $q > (\alpha - m + 1)^{-1}$, $\beta = k_0^2$ *for some* $k_0 \in \mathbf{N}$, *functions* $\delta_n : (t_0, T) \to \mathbf{R}$, $n = 0, 1, \dots, m-2$, *be measurable and essentially bounded,* $u \in L_q(t_0, T; L_2(0, \pi))$, $v_k \in \mathcal{X}$, $k = 0, 1, \dots, m-1$. *Then there exists a unique strong solution of problem* (12)–(14) *on* (t_0, T).

Proof. Here we have $(L, 0)$-bounded operator M due to [6, Theorem 8], hence $\ker L = \mathcal{X}^0$, $\mathrm{im} L = \mathcal{Y}^1$. We see that conditions (12) determine the initial data on the complement to $\ker L$, i. e. on $\mathcal{X}^1$, therefore, generalized Showalter—Sidorov conditions (9) are equivalent to conditions (12). Hence problem (12)–(14) is equivalent to problem (8), (9). We have also

$$\mathrm{im} N \subset \overline{\mathrm{span}}\{\sin ks : k \neq k_0\} = \mathrm{im} L = \mathcal{Y}^1$$

at almost all $t \in (t_0, T)$. Here the overline means the closure of the set in $L_2(0, \pi)$.

It is obvious, that $\mathcal{X}$ is compactly embedded in $\mathcal{X}_1$. Besides, $\mathcal{X}_1 \subset C^1[0, \pi]$, therefore, for any $x \in \mathcal{X}_1$ we have

$$(\ln(1 + x^2))'' = \frac{2xx''}{1 + x^2} + \frac{2x'^2(1 - x^2)}{(1 + x^2)^2} \in L_2(0, \pi),$$

$$\ln(1 + x^2(0)) = \ln(1 + x^2(\pi)) = 0,$$

hence $\ln(1 + x^2) \in \mathcal{X}_1$. Consequently, for every $\overline{x} = (x_0, x_1, \dots, x_{m-2}) \in \mathcal{X}_1^{m-1}$ and almost all $t \in (t_0, T)$ we have $N(t, \overline{x}) \in \mathcal{Y}$.

Denote for $z \in L_2(0, \pi)$ the Fourier coefficients

$$[z]_k = \sqrt{\frac{2}{\pi}} \int_0^{\pi} z(s) \sin ks ds, \quad k \in \mathbf{N},$$

with respect to the orthonormal basis $\left\{\sqrt{\frac{2}{\pi}} \sin ks : k \in \mathbf{N}\right\}$ in $L_2(0, \pi)$, then for any $z \in \mathcal{X}_1$ integration by parts implies the equality $[z'']_k = -k^2 [z]_k$. Therefore,

$$\|z\|_{H^2(0,\pi)}^2 = \|z\|_{L_2(0,\pi)}^2 + \|z''\|_{L_2(0,\pi)}^2 = \sum_{k=1}^{\infty} \left([z]_k^2 + [z'']_k^2\right) = \sum_{k=1}^{\infty} (1 + k^4)[z]_k^2.$$

Besides,

$$[\ln(1 + x^2) - \ln(1 + y^2)]_k = \sqrt{\frac{2}{\pi}} \int_0^{\pi} \sin ks \int_0^1 \frac{d}{d\theta} \ln(1 + (\theta x(s) + (1 - \theta) y(s))^2) d\theta ds$$

$$= \sqrt{\frac{2}{\pi}} \int_0^{\pi} \frac{2(\theta_0 x(s) + (1 - \theta_0) y(s))}{1 + (\theta_0 x(s) + (1 - \theta_0) y(s))^2} (x(s) - y(s)) \sin ks ds$$

$$= \frac{2(\theta_0 x(\xi) + (1 - \theta_0) y(\xi))}{1 + (\theta_0 x(\xi) + (1 - \theta_0) y(\xi))^2} [x - y]_k$$

at some $\theta_0 \in (0,1)$, $\xi \in (0,\pi)$ due to the mean value theorem, which is applied twice here. Since $\frac{2|x|}{1+x^2} \le 1$ for all $x \in \mathbf{R}$, we obtain the inequality

$$|[\ln(1+x^2) - \ln(1+y^2)]_k| \le |[x-y]_k|, \quad k \in \mathbf{N}.$$

Denote $b_n = \operatorname{ess\,sup}\{|\delta_n(t)| : t \in (t_0,T)\}$, $n = 0,1,\ldots,m-2$. For any $\overline{x},\overline{y} \in \mathcal{X}^{m-1}$ we have

$$\|N(t,\overline{x}) - N(t,\overline{y})\|_{\mathcal{Y}}^2 \le 2\sum_{n=0}^{m-2} b_n \left\| \left(\beta + \frac{d^2}{ds^2}\right)(\ln(1+x_n^2) - \ln(1+y_n^2)) \right\|_{L_2(0,\pi)}^2$$

$$\le C_1 \sum_{n=0}^{m-2} b_n \left\|\ln(1+x_n^2) - \ln(1+y_n^2)\right\|_{H^2(0,\pi)}^2$$

$$= C_1 \sum_{n=0}^{m-2} b_n \sum_{k=1}^{\infty} (1+k^4)[\ln(1+x_n^2) - \ln(1+y_n^2)]_k^2$$

$$\le C_1 \sum_{n=0}^{m-2} b_n \sum_{k=1}^{\infty} (1+k^4)[x_n - y_n]_k^2 = C_1 \sum_{n=0}^{m-2} b_n \|x-y\|_{H^2(0,\pi)}.$$

Thus, operator $N : (t_0,T) \times \mathcal{X}_1^{m-1} \to \mathcal{Y}$ is uniformly Lipschitz continuous in $\overline{x}$ and the Caratheodory mapping. Hence its restriction

$$N|_{(t_0,T)\times\mathcal{X}^{m-1}} : (t_0,T) \times \mathcal{X}^{m-1} \to \mathcal{Y}$$

has these properties, besides, $N(t,0,\ldots,0) \equiv 0 \in L_q(t_0,T;\mathcal{Y})$. By Theorem 3 we obtain the required statement. □

Consider the control problem

$$\|u\|_{L_q(t_0,T;L_2(0,\pi))} \le R, \tag{15}$$

$$\|w\|_{\mathcal{Q}_{\alpha,q}(t_0,T;H^2(0,\pi))}^q \to \inf \tag{16}$$

for system (12)–(14).

Theorem 6. *Let* $\alpha > 1$, $q > (\alpha - m + 1)^{-1}$, $\beta = k_0^2$ *for some* $k_0 \in \mathbf{N}$, *functions* $\delta_n : (t_0,T) \to \mathbf{R}$ *be measurable and essentially bounded,* $n = 0,1,\ldots,m-2$, $v_k \in \mathcal{X}$, $k = 0,1,\ldots,m-1$, *the Banach space* $\mathcal{Q}_{\alpha,q}(t_0,T;\mathcal{X})$ *be reflexive. Then there exists a solution of optimal control problem* (12)–(16).

Proof. Since the operator $M = \delta I$ is bounded, we have

$$\mathcal{Z}_{\alpha,q}(t_0,T;\mathcal{X}) = \mathcal{Q}_{\alpha,q}(t_0,T;\mathcal{X}).$$

Here the set of admissible controls

$$\mathcal{U}_\partial = \{u \in L_q((t_0,T);L_2(0,\pi)) : \|u\|_{L_q((t_0,T);L_2(0,\pi))} \le R\}$$

contains the function $u_0 \equiv 0$, which satisfies the conditions of Theorem 4. Besides, $\mathcal{U}_\partial$ is non-empty bounded closed convex subset in $L_q(t_0,T;\mathcal{U})$. Other conditions of that theorem are checked in the previous proof. By Theorem 4 we obtain the required result. □

5 Conclusion

The results can be used for the correct choice of formulation and parameters of the applied problems, for the development of numerical methods for solving problems, etc. In the future, the obtained results will allow us to investigate perturbed equations of the corresponding classes, as well as to proceed to the study of similar problems for more complex classes of fractional order evolution equations, in particular, of equations with the nonlinearity, which contains lower fractional derivatives.

Acknowledgements. The work is supported by Act 211 of Government of the Russian Federation, contract 02.A03.21.0011, and by Ministry of Education and Science of the Russian Federation, task No 1.6462.2017/BCh.

References

1. Bajlekova, E.G.: Fractional evolution equations in Banach spaces. Ph.D. thesis, University Press Facilities, Eindhoven University of Technology, Eindhoven (2001)
2. Caputo, M.: Linear model of dissipation whose Q is almost frequancy independent. Geophys. J. R. Astron. Soc. **13**, 529–539 (1967)
3. Debbouche, A., Nieto, J.J.: Sobolev type fractional abstract evolution equations with nonlocal conditions and optimal multi-controls. Appl. Math. Comput. **245**, 74–85 (2014)
4. Debbouche, A., Torres, D.F.M.: Sobolev type fractional dynamic equations and optimal multi-integral controls with fractional nonlocal conditions. Fract. Calc. Appl. Anal. **18**(1), 95–121 (2015)
5. Fedorov, V.E., Davydov, P.N.: On nonlocal solutions of semilinear equations of the Sobolev type. Differ. Equ. **49**(3), 338–347 (2013)
6. Fedorov, V.E., Gordievskikh, D.M.: Resolving operators of degenerate evolution equations with fractional derivative with respect to time. Russ. Math. **59**(1), 60–70 (2015)
7. Fedorov, V.E., Gordievskikh, D.M., Plekhanova, M.V.: Equations in Banach spaces with a degenerate operator under a fractional derivative. Differ. Equ. **51**(10), 1360–1368 (2015)
8. Fedorov, V.E., Kostić, M.: On a class of abstract degenerate multi-term fractional differential equations in locally convex spaces. Eurasian Math. J. **9**(3), 33–57 (2018)
9. Fursikov, A.V.: Optimal Control of Distributed Systems: Theory and Applications. Translations of Mathematical Monographs, vol. 187. AMS, Providence (1999)
10. Gerasimov, A.N.: Generalization of the linear laws of deformation and their application to the problems of internal friction. Appl. Math. Mech. **12**, 529–539 (1948). (In Russian)
11. Hilfer, R.: Applications of Fractional Calculus in Physics. WSPC, Singapore (2000)
12. Kamocki, R.: On the existence of optimal solutions to fractional optimal control problems. Appl. Math. Comput. **235**, 94–104 (2014)
13. Kilbas, A.A., Srivastava, H.M., Trujillo, J.J.: Theory and Applications of Fractional Differential Equations. Elsevier Science Publishing, Amsterdam, Boston, Heidelberg (2006)
14. Kostić, M., Fedorov, V.E.: Degenerate fractional differential equations in locally convex spaces with a σ-regular pair of operators. Ufa Math. J. **8**(4), 98–110 (2016)

15. Mophoua, G.M., Guérékata, G.M.: Optimal control of a fractional diffusion equation with state constraints. Comput. Math. Appl. **62**(3), 1413–1426 (2011)
16. Plekhanova, M.V.: Degenerate distributed control systems with fractional time derivative. Ural. Math. J. **2**(2), 58–71 (2016)
17. Plekhanova, M.V.: Distributed control problems for a class of degenerate semilinear evolution equations. J. Comput. Appl. Math. **312**, 39–46 (2017)
18. Plekhanova, M.V.: Optimal control existence for degenerate infinite dimensional systems of fractional order. IFAC-PapersOnLine **51**(32), 669–674 (2018). 17th IFAC Workshop on Control Applications of Optimization CAO 2018, Yekaterinburg, Russia, 15–19 October 2018
19. Plekhanova, M.V.: Solvability of control problems for degenerate evolution equations of fractional order. J. Comput. Appl. Math. **2**(1), 53–65 (2017)
20. Plekhanova, M.V.: Start control problems for fractional order evolution equations. Chelyabinsk Phys. Math. J. **1**(3), 15–36 (2016)
21. Plekhanova, M.V.: Strong solutions to nonlinear degenerate fractional order evolution equations. J. Math. Sci. **230**(1), 146–158 (2018)
22. Prüss, J.: Evolutionary Integral Equations and Applications. Springer, Basel (1993). https://doi.org/10.1007/978-3-0348-8570-6
23. Sviridyuk, G.A., Fedorov, V.E.: Linear Sobolev Type Equations and Degenerate Semigroups of Operators. VSP, Utrecht, Boston (2003)
24. Zhou, Y.: Fractional Evolution Equations and Inclusions: Analysis and Control. Elseiver, Amsterdam (2016)

Feedback Optimality Conditions with Weakly Invariant Functions for Nonlinear Problems of Impulsive Control

Olga Samsonyuk, Stepan Sorokin(✉), and Maxim Staritsyn

Matrosov Institute for System Dynamics and Control Theory, Irkutsk, Russia
olga.samsonyuk@gmail.com, sorsp@mail.ru, starmax@icc.ru

Abstract. We consider a broad class of optimal control problems for nonlinear measure-driven equations. For such problems, we propose necessary optimality conditions, which are based on a specific procedure of "feedback variation" of a given, reference impulsive control. The approach is based on using impulsive feedback controls designed by means of "weakly invariant functions". The concept of weakly invariant function generalizes the notion of weakly monotone function. In the paper, we discuss the advantages of this approach and some perspectives of designing, on its base, nonlocal numeric algorithms for optimal impulsive control.

Keywords: Measure differential equations · Impulsive control · Feedback control · Optimality condition · Functions of Lyapunov type

1 Introduction

The work lays in the vein of optimal impulsive control theory—the area of dynamic optimization, where the state trajectories can be discontinuous due to "shock impacts" produced by Dirac-type distributions or signed measures, playing the part of control inputs [1,3,4,9,13–15,17,19,21].

In this paper, we extend the ideas [7,8,10,11,22,23] towards deriving new versions of "feedback" necessary optimality conditions for impulsive control problems with states of bounded variation. The approach is, to some extent, in tune with the dynamic programming [12]. At the same time, it does not require any global information (exact or viscosity solution to the Hamilton-Jacobi equation). Instead, it only operates with the known reference impulsive process, and, being applied iteratively, generates a nonlocal algorithm for optimal impulsive control. In order to discard ("improve") a reference process, we shall apply feedback controls of a specific "extremal" structure, whose construction appeals to the concept of *weakly invariant* (with respect to the impulsive system) *function*—a generalization of the more familiar notion of weakly monotone function [5,6,24].

Partially supported by the Russian Foundation for Basic Research, projects nos 18-31-20030, 18-31-00425, 17-01-00733.

M. Khachay et al. (Eds.): MOTOR 2019, LNCS 11548, pp. 513–526, 2019.
https://doi.org/10.1007/978-3-030-22629-9_36

2 Problem Statement and Preliminaries

Assume that we are given the following data: a finite time interval $T = [a, b] \subset \mathbb{R}$, a compact set $U \subset \mathbb{R}^m$, a closed convex cone $K \subseteq \mathbb{R}^r$, and a vector $x_0 \in \mathbb{R}^n$. Consider the following optimal impulsive control problem (P):

$$\text{Minimize } J = l(x(b)) \text{ subject to} \tag{1}$$

$$dx(t) = f(t, x(t), u(t))\, dt + G(t, x(t))\, \pi(dt), \quad x(a^-) = x_0; \tag{2}$$

$$u \in L^\infty(T, U), \quad \pi \in \mathcal{W}(T, K). \tag{3}$$

Here, by $x(a^-)$ we denote the left one-sided limit of a function x at a point $t = a$.

In (P), we operate with two principally different types of input signals: u is a "usual" control, involved in the drift of our dynamical system, while the term π presents an *impulsive control* produced by a vector-valued Borel measure, whose action is responsible for the jumps of the state x. An impulsive control is a collection

$$\pi = \big(\mu, S, \{d_s, \omega_s(\cdot)\}_{s \in S}\big),$$

containing the following objects: (i) a K-valued Borel measure μ on T; (ii) a finite or countable subset $S \supseteq S_{\mathrm{d}}(\mu) \doteq \big\{s \in T \,\big|\, \mu(\{s\}) \neq 0\big\}$ of the interval T; (iii) a collection of real numbers d_s such that $d_s \geq \big\|\mu(\{s\})\big\|_1$ for all $s \in S$, and $\sum_{s \in S} d_s < \infty$ ($\|\cdot\|_1$ denotes the Manhattan norm in $\mathbb{R}^r$), and (iv) Borel measurable functions $\omega_s(\cdot) : [0, d_s] \to co\, K_1$, $s \in S$, with the property

$$\int_0^{d_s} \omega_s(\tau)\, d\tau = \mu(\{s\}).$$

Here, $K_1 \doteq \{v \in K \mid \|v\|_1 = 1\}$ and $co\, A$ denotes the convex hull of a set A. The set of impulsive controls is denoted by $\mathcal{W}(T, K)$. Note that the cone K establishes the constraint on admissible "directions" of impulsive actions.

Equations of the form (2), called *measure differential equations*, give a conventional but rather symbolic representation of impulsive dynamic processes. In fact, (2) performs certain limit version (impulsive relaxation) of an ordinary control system, which is affine in an "unbounded" (non-compact-valued) control input. To become precise, the measure differential equation should be interpreted as the following relations:

$$\begin{aligned} x(t) = {} & x_0 + \int_a^t f(\tau, x(\tau), u(\tau))\, d\tau + \int_a^t G(\tau, x(\tau))\, \mu_{\mathrm{c}}(d\tau) \\ & + \sum_{s \in S,\ s \leq t} \big(z^s(d_s) - x(s^-)\big), \ t \in T, \end{aligned} \tag{4}$$

$$\frac{dz^s(\tau)}{d\tau} = G(s, z^s(\tau))\, \omega_s(\tau), \quad z^s(0) = x(s^-), \quad \text{for } \mathcal{L}\text{-a.e. } \tau \in [0, d_s] \text{ and all } s \in S. \tag{5}$$

Here, $\mathcal{L}$ stands for the Lebesgue measure on the real line, and μ_c is the continuous part of measure μ.

Under the following (rather standard) assumptions (H_1)–(H_3), a solution $x : T \mapsto \mathbb{R}^n$ to (4), (5) is correctly defined for any $(u, \pi) \in L_\infty(T, U) \times \mathcal{W}(T, K)$ as a (unique) right continuous on $[a, b)$ function of bounded variation ($x \in BV_r(T, \mathbb{R}^n)$) [18].

(H_1) The function $l : \mathbb{R}^n \mapsto \mathbb{R}$ is continuous.
(H_2) The functions $f : T \times \mathbb{R}^n \times U \mapsto \mathbb{R}^n$, $G : T \times \mathbb{R}^n \mapsto \mathbb{R}^{n\times r}$ are continuous and locally Lipschitz continuous in x.
(H_3) The set $f(t, x, U) \doteq \{f(t, x, u) \mid u \in U\}$ is convex for all $(t, x) \in T \times \mathbb{R}^n$.

Triples $\sigma = (x, u, \pi)$ satisfying conditions (3)–(5), are called *feasible processes* of problem (P), and Σ denotes the set of all feasible processes.

Given $\pi \in \mathcal{W}(T, K)$, we introduce the function $V = V[\pi] : T \to \mathbb{R}_+$ as:

$$V(t) = |\mu_c|([t, b]) + \sum_{s \geq t,\, s \in S} d_s, \quad t \in [a, b), \quad V(b) = 0, \tag{6}$$

where $|\mu|$ denotes the total variation of μ, $\mathbb{R}_+ \doteq \{v \in \mathbb{R} \mid v \geq 0\}$. The value $V(t)$ characterizes the "resource" of impulsive control on the time interval $[t, b]$ (or rather, the remaining energy of the guide). In what follows, it will be convenient to consider our impulsive system (3)–(5) together with (6) (i.e. to treat V as an extra state variable). Furthermore, in some cases, problem (P) is naturally weighted by the constraint

$$V(a) \leq M \tag{7}$$

with given $M \geq 0$. Note that, in this case, (P) does have a solution [18].

In the next section, we propose a notion of weak invariance of a function with respect to (w.r.t.) the impulsive system. This property will be formulated in terms of *supplemented trajectories* to be introduced below: Taken $u \in L^\infty(T, U)$ and $\pi \in \mathcal{W}(T, K)$, let x, $\{z^s\}_{s\in S}$, and V be the associated solutions to (4)–(6). We define the set-valued function X_V, called a supplemented trajectory, as follows:

(i) $X_V(t) = \big(x(t), V(t)\big)$, if $t \in T \setminus S$, and
(ii) $X_V(s) = \big\{\big(z^s(\tau), V(s^-) - \tau\big) \mid \tau \in [0, d_s]\big\}$, if $t = s \in S$.

The set of supplemented trajectories in problem (P) is denoted by $\mathcal{X}$. The graph of X_V on T is defined as $\operatorname{graph}_T X_V \doteq \big\{(t, x, V) \mid t \in T,\ (x, V) \in X_V(t)\big\}$.

2.1 Weakly Invariant Functions of the Lyapunov Type

Given a continuous function $\varphi : T \times \mathbb{R}^n \times \mathbb{R}_+ \mapsto \mathbb{R}$, denote

$$Q_\varphi \doteq \big\{(t, x, V) \in T \times \mathbb{R}^n \times \mathbb{R}_+ \,\big|\, \varphi(t, x, V) \leq 0\big\}. \tag{8}$$

Below, we introduce a specific class of the Lyapunov-type functions associated to the impulsive dynamics (3)–(6). These functions are called by us "weakly invariant", which reflects the following characteristic property: the 0-sublevel set Q_φ of such a function φ should be weakly invariant w.r.t. our control system.

Definition 1. *We say that φ is* weakly invariant *iff the set Q_φ is weakly invariant, i.e., for any $(t_\alpha, x_\alpha, V_\alpha) \in Q_\varphi$, there is $X_V \in \mathcal{X}$ with $X_V(t_\alpha-) = (x_\alpha, V_\alpha)$ such that* $\operatorname{graph}_{[t_\alpha,b]} X_V \subset Q_\varphi$.

We note that any function, which is weakly monotone w.r.t. the control system (3)–(6), is also weakly invariant.

Now we shall present a constructive criteria for the weak invariance of a closed set $Q \subset \mathbb{R}^{n+2}$ w.r.t. the control system (3)–(6).

First, recall the notion [6] of proximal normal cone: Given a closed set $A \subset \mathbb{R}^k$ and $x \in A$, a vector $\zeta \in \mathbb{R}^k$ is said to be proximal normal to A at a point x iff there exists $\alpha > 0$ such that $d_A(x + \alpha\zeta) = \alpha\|\zeta\|$, where $d_A(y) \doteq \inf_{x\in A} \|y - x\|$. The set $N_A^P(x)$ of all proximal normals ζ is called the proximal normal cone to A at x.

Introduce the functions

$$\overline{h}_0(t, x, \psi_1, \psi_2) = \psi_1 + \min_{u\in U} \langle\psi, f(t, x, u)\rangle,$$

$$\overline{h}_1(t, x, \psi_1, \psi_2) = \psi_1 + \min_{v\in K_1} \langle\psi, G(t, x)\, v\rangle,$$

and denote

$$\begin{aligned} Q_{[a,b)} &= Q \cap \big([a, b) \times \mathbb{R}^n \times (0, +\infty)\big),\\ Q_{[a,b]V_0} &= \big\{(t, x) \in [a, b] \times \mathbb{R}^n \mid (t, x, 0) \in Q\big\},\\ Q_{V_0} &= \big\{(t, x) \in (a, b) \times \mathbb{R}^n \mid (t, x, 0) \in Q\big\},\\ Q_t &= \big\{(x, V) \in \mathbb{R}^n \times (0, +\infty) \mid (t, x, V) \in Q\big\}. \end{aligned}$$

Assumed that $Q_{[a,b]V_0} = \limsup_{V\downarrow 0} Q_{[a,b]V}$ and $\{x \mid (b, x, 0) \in Q\} \neq \varnothing$, consider the following condition, which characterizes the set Q near a point (t, x, V) w.r.t the impulsive dynamics:

Condition (A): for all $\zeta = (\zeta_t, \zeta_x, \zeta_V) \in N^P_{\overline{Q}_{[a,b)}}(t, x, V)$ and $(t, x, y) \in Q_{[a,b)}$ it holds:

$$\min_{\substack{\omega_0,\,\omega_1\geq 0\\ \omega_0+\omega_1=1}} \big\{\overline{h}_0(t, x, \zeta_t, \zeta_x)\, \omega_0 + \overline{h}_1(t, x, -\zeta_V, \zeta_x)\, \omega_1\big\} \leq 0;$$

for all $(\zeta_t, \zeta_x) \in N^P_{\overline{Q}_{V_0}}(t, x)$ and $(t, x) \in Q_{V_0}$,

$$\overline{h}_0(t, x, \zeta_t, \zeta_x) \leq 0;$$

and, for all $(\zeta_x, \zeta_V) \in N^P_{\overline{Q}_b}(x, V)$ and $(x, V) \in Q_b$,

$$\overline{h}_1(b, x, -\zeta_V, \zeta_x) \leq 0.$$

Here, $\overline{Q}_{[a,b)}$, $\overline{Q}_t$, and $\overline{Q}_{V_0}$ are the closures of the sets $Q_{[a,b)}$, Q_t, and Q_{V_0}, respectively.

In [20], it is proved that Condition (A) is equivalent to the weak invariance of a closed set $Q \subset \mathbb{R}^{n+2}$ w.r.t. the control system (3)–(6).

2.2 Time Reparameterization and Impulsive Feedback Controls

Let us introduce the following variational problem $(P_{\rm a})$, which is an ordinary counterpart of the impulsive control problem (P). This problem is stated on processes of the so-called *space-time* system $(S_{\rm a})$, obtained from the measure differential Eqs. (3)–(6) through the standard discontinuous time reparameterization technique [16–19,21]. Problem $(P_{\rm a})$ takes the form:

$$\text{Minimize } \hat{J} = l(y(\tau_1)) \text{ subject to}$$

$$\frac{d}{d\tau}\eta \doteq \eta'(\tau) = \omega_0(\tau), \quad \eta(0) = a, \quad \eta(\tau_1) = b,$$

$$y'(\tau) = f\big(\eta(\tau), y(\tau), \nu(\tau)\big)\,\omega_0(\tau) + G\big(\eta(\tau), y(\tau)\big)\,\omega(\tau), \quad y(0) = x_0,$$

$$m'(\tau) = \omega_0(\tau) - 1, \quad m(\tau_1) = 0,$$

$$\nu(\tau) \in U, \quad (\omega_0(\tau), \omega(\tau)) \in co\,\tilde{K}_1 \quad \text{for } \mathcal{L}\text{-a.e. } \tau \in [0, \tau_1].$$

Here, $(\eta, y, m) \in W^{1,1}([0,\tau_1], \mathbb{R}^{n+2})$ are new states, and $\nu \in L^\infty([0,\tau_1], \mathbb{R}^m)$, $(\omega_0, \omega) \in L^\infty([0,\tau_1], \mathbb{R}^{r+1})$ are controls;

$$\tilde{K}_1 \doteq \{(\omega_0, \omega) \in [0,1] \times K \mid \omega_0 + \|\omega\|_1 = 1\}.$$

Furthermore, if constraint (7) on the resource of impulsive control is imposed in the original system, than $\tau_1 \leq b - a + M$.

Note that the set $\Sigma_{\rm a}$ of processes $\rho = (\tau_1, \eta, y, m, \nu, \omega_0, \omega)$ that are feasible for $(P_{\rm a})$ is in one-to-one correspondence with Σ [17], and $\min\limits_{\sigma\in\Sigma} J(\sigma) = \min\limits_{\rho\in\Sigma_{\rm a}} J(\rho)$ (in other words, problems (P) and $(P_{\rm a})$ are equivalent one to another).

Let $\tau_1 \geq 0$. Now we shall fix an appropriate class of *time reparameterizations*, which are involved in the notion of feedback control. By the time reparametrization we mean a nondecreasing Lipschitz continuous function $\eta : [0, \tau_1] \mapsto [a, b]$ such that $\eta(0) = a$ and $\eta(\tau_1) = b$, and denote by $\mathcal{T}_{\tau_1}$ the set of time reparametrizations enjoying the property: $\eta'(\theta) \in [0,1]$ $\mathcal{L}$-a.e. on $[0, \tau_1]$. The feedback control is a collection $(\tau_1, \eta(\cdot), \nu(\tau, y), \omega(\tau, y))$, where

$$\eta(\cdot) \in \mathcal{T}_{\tau_1}, \tag{9}$$

$$\nu(\tau, y) \in U, \tag{10}$$

$$\omega(\tau, y) \in \mathcal{V}(\tau; \eta) \doteq \{v \in K \mid (\eta'(\tau), v) \in co\,\tilde{K}_1\} \tag{11}$$

(cf. [2]). Note that any feedback control produces at least one sampling (Krasovskii-Subbotin/Euler) solution of the closed-looped system $(S_{\rm a})$ [6], and this sampling solution satisfies terminal constraint $\eta(\tau_1) = b$.

3 Feedback Necessary Optimality Conditions

In this section, we formulate the main result—the feedback necessary conditions for optimality,—and present some illustrative cases demonstrating the machinery of our approach along with its key features.

Let $\bar{\sigma} = (\bar{x}, \bar{u}, \bar{\pi}) \in \Sigma$ be a reference control process in problem (P), and $\bar{g} = (\bar{\tau}_1, \bar{\eta}, \bar{y}, \bar{m}, \bar{\nu}, \bar{\omega}_0, \bar{\omega})$ the associated process of $(P_{\rm a})$. We abbreviate $\bar{x}_1 \doteq \bar{x}(b)$. Consider a function $\varphi(t, x, V)$, which is weakly invariant w.r.t. system (3)–(5), and assume that i) $\varphi(a, x_0, V) \leq 0$ for all $V \in \mathbb{R}_+$ and ii) $\varphi(b, x, 0) \geq l(x) - l(\bar{x}_1)$. Then, $\varphi(\eta, y, m)$ is weakly invariant w.r.t. system $(S_{\rm a})$. Moreover, there exists $g = (\tau_1, \eta, y, m, \nu, \omega_0, \omega) \in \Sigma_{\rm a}$ such that the map $\tau \mapsto \varphi(\eta(\tau), y(\tau), m(\tau))$ is non-positive on $[0, \tau_1]$. One concludes that

$$\varphi(b, y(\tau_1), 0) \leq 0, \text{ and}$$

$$l(y(\tau_1)) - l(\bar{x}_1) \leq 0. \tag{12}$$

In view of this, one easily derives the following (variational) necessary optimality condition for problem (P):

Theorem 1. *Assume that $\bar{\sigma}$ is globally optimal for (P). Then the associated process $\bar{g}$ is a minimizer for $(P_{\rm a})$, and the inequality (12) does not hold strictly.*

This theorem, looking rather straightforward, shall be handled in its counterpositive version, i.e., as a *sufficient condition for non-optimality.* Then, applied iteratively, it is trivially turned into the following conceptual algorithm: Let an appropriate φ be chosen. Fix τ_1 and $\eta(\cdot)$. Design a feedback control $(\tau_1, \eta(\cdot), \nu(\tau, y), \omega(\tau, y))$, satisfying conditions (9)–(11) in the way [6] (using the set $\Omega_\varphi(t, x, V)$ to be defined below, or the respective criterion for the invariance w.r.t $(S_{\rm a})$), such that φ, contracted on an associated sampling solution, stays within the set Q_φ (this is always possible due to the weak invariance of φ).

Here, given a weakly invariant function φ and $(t, x, V) \in T \times \mathbb{R}^n \times \mathbb{R}_+$, $\Omega_\varphi(t, x, V)$ denotes the set of triples (u, ω_0, ω) such that vector $(u, \omega_0, \omega_1, v)$ with $\omega_1 = \|\omega\|_1$ and $v = \begin{cases} \omega/\|\omega\|_1, & \|\omega\|_1 \neq 0, \\ 0, & \|\omega\|_1 = 0 \end{cases}$ brings the minima in the left-hand sides of the respective inequalities in Condition (A), applied to the set $Q = Q_\varphi$ at the point (t, x, V).

To illustrate, how the formulated scheme works in practice, we propose three simple but eloquent cases, where an appropriate weakly invariant (for $(P_{\rm a})$) function φ can be chosen in the simplest state-linear form

$$\varphi(\tau, y) = \big(\psi(\tau) + l_y(\bar{y}_1)\big)\big(\bar{y}(\tau) - y\big) + l(y) - l\big(\bar{y}(\tau)\big). \tag{13}$$

Here, $\bar{y}_1 \doteq \bar{y}(\bar{\tau}_1)$, and ψ is the adjoint of $\bar{g}$ in the sense of the Maximum Principle. Note that this situation brings us to the paradigm of the so-called *feedback minimum principle* [7,8].

Example 1. Minimize $J[\sigma] = \langle x(t_1), Ax(t_1)\rangle$, $A = \begin{pmatrix} -5 & 3 \\ 3 & -5 \end{pmatrix}$, subject to

$$x_1(t) = \int_0^t \mu_{c1}(d\tau) + \sum_{s \in S,\, s \leq t} \big(z_1^s(d_s) - x_1(s^-)\big), \quad x_1(0) = 0, \tag{14}$$

$$x_2(t) = \int_0^t x_1(\tau)\, \mu_{c2}(d\tau) + \sum_{s \in S,\, s \leq t} \big(z_2^s(d_s) - x_2(s^-)\big), \quad x_2(0) = 0, \tag{15}$$

$$\frac{z_1^s(\tau)}{d\tau} = \omega_1^s(\tau), \quad z_1^s(0) = x_1(s^-), \tag{16}$$

$$\frac{z_2^s(\tau)}{d\tau} = z_1^s(\tau)\,\omega_2^s(\tau), \quad z_2^s(0) = x_2(s^-), \tag{17}$$

$$\left|\omega_1^s(\tau)\right| + \left|\omega_2^s(\tau)\right| \le 1 \quad \text{for } s \in S,\ \mathcal{L}\text{- a.e. } \tau \in [0, d_s], \tag{18}$$

$$|\mu_c|\big([0,t_1]\big) + \sum_{s\in S} d_s \le M. \tag{19}$$

This problem is characterized by infinitely many extrema of the impulsive Maximum Principle [18]; the terminal states of any extremal trajectory belong to the four-point set presented on Fig. 1 (the two green points correspond to all local extrema, and the two red ones—to global solutions). Notice that the reachable set $\mathcal{R}(t_1)$ of system (14)–(19) at any time moment $t_1 \ge 0$ can be represented as follows:

$$\mathcal{R}(t_1) = \big\{(x_1, x_2) \mid \varphi_{1,2}(x_1, x_2, M) \le 0\big\},$$

where $\varphi_{1,2}$ are strongly decreasing functions of the form

$$\varphi_1(x_1, x_2, V) = \begin{cases} x_2 + (V + x_1)x_1, & x_1 \le -V/3, \\ x_2 - \dfrac{(V - x_1)^2}{8}, & x_1 \in \big[-V/3, 0\big], \\ x_2 - \dfrac{(V + x_1)^2}{8}, & x_1 \in \big[0, V/3\big], \\ x_2 - (V - x_1)x_1, & x_1 \ge V/3, \text{ and} \end{cases}$$

$$\varphi_2(x_1, x_2, V) = \varphi_1(x_1, -x_2, V).$$

Let $M = 3$. To illustrate the point, among all extremal controls, we choose the four ones with the property: $|\mu_c| = 0$ and $S = \{0\}$. For the respective control processes, we apply Theorem 1. We are aimed at discarding non-optimal extrema $\widetilde{\sigma}^{1,2}$ produced by controls

$$\widetilde{\omega}^1(\tau) = \begin{cases} (1,0), \tau \in [0,\tau^*), \\ (0,1), \tau \in [\tau^*, 3] \end{cases} \text{ and } \widetilde{\omega}^2(\tau) = \begin{cases} (-1,0), \tau \in [0,\tau^*), \\ (0,1), \quad \tau \in [\tau^*, 3], \end{cases} \quad \tau^* = 1.6,$$

with $J[\widetilde{\sigma}^{1,2}] = -16,384$. Note that global solutions $\overline{\sigma}^{1,2}$ correspond to the inputs

$$\overline{\omega}^1(\tau) = \begin{cases} (1,0), \quad \tau \in [0,2), \\ (0,-1), \tau \in [2,3], \end{cases} \quad \overline{\omega}^2(\tau) = \begin{cases} (-1,0), \tau \in [0,2), \\ (0,-1), \tau \in [2,3], \end{cases}$$

and give $J[\overline{\sigma}^{1,2}] = -64$.

The Pontryagin function takes the form $H = \psi_1\,\omega_1 + \psi_2\,y_1\,\omega_2$, and the adjoint system looks as follows:

$$\dot{\psi}_1 = -\psi_2\,\omega_2, \quad \psi_1(3) = 10\,y_1(3) - 6\,y_2(3), \quad \psi_2 \equiv -6\,y_1(3) + 10\,y_2(3).$$

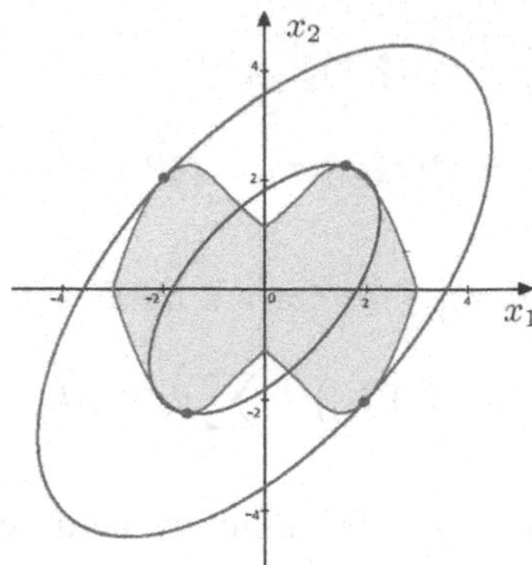

Fig. 1. Example 1: The level lines of the cost function and the reachable set $\mathcal{R}(t_1)$ for $M = 3$ and any $t_1 \geq 0$.

Consider a non-optimal (local) extrema $\widetilde{\sigma}^1$, and introduce the functions g_i generated by $\varphi \doteq \varphi(\widetilde{\sigma}^1)$ from (13):

$$g_1 = g_1(\tau, y; \widetilde{\psi}_1^1) \doteq \varphi_{y_1} = \widetilde{\psi}_1^1(\tau) + l_{y_1}(y) - l_{y_1}(\widetilde{y}^1),$$

$$g_2 = g_2(\tau, y; \widetilde{\psi}_2^1) \doteq \varphi_{y_2} y_1 = \left(\widetilde{\psi}_2^1(\tau) + l_{y_2}(y) - l_{y_2}(\widetilde{y}^1)\right) y_1.$$

Then, the respective H-extremal set-valued function (which corresponds to the set $\Omega_\varphi(t, x, V)$) has the form

$$\Omega_\varphi[g_1, g_2] = \begin{cases} (0, -1), & g_2 < -|g_1|, \\ (0, 1), & g_2 > |g_1|, \\ (1, 0), & g_1 > |g_2|, \\ (-1, 0), & g_1 < -|g_2|, \\ (-\lambda, \lambda - 1),\ \lambda \in [0, 1], & g_2 = g_1 < 0, \\ (\lambda, 1 - \lambda),\ \lambda \in [0, 1], & g_2 = g_1 > 0, \\ (\lambda, \lambda - 1),\ \lambda \in [0, 1], & g_2 = -g_1 < 0, \\ (-\lambda, 1 - \lambda),\ \lambda \in [0, 1], & g_2 = -g_1 > 0, \\ g_1 = g_2 = 0 & \omega^* \text{ is any admissible.} \end{cases}$$

Taken any selection of this multifunction such that $\omega^* = (1, 0)$ for $g_2 = g_1 > 0$, one generates a new process with

$$\hat{\omega}(\tau) = \begin{cases} (1, 0), & \tau \in [0, \hat{\tau}), \\ (0, -1), & \tau \in [\hat{\tau}, 3], \end{cases} \quad \hat{\tau} = \frac{274 + \sqrt{96196}}{220}, \text{ and } J[\hat{\sigma}] \approx -53,835.$$

Few consecutive iterations lead to a global solution, see Table 1.

Example 2. Consider a version of Example 1, where the cost functional is linear in $x(t_1)$, while the component μ_2 of the control measure is non-negative.

Table 1. Example 1: Iterations of the feedback optimality condition (values of the functional, and improvement w.r.t. the previous iteration)

Iteration	Problem value	% of improvement
0	−16,384	–
1	−53,835	228,583
2	−60,058	11,558
3	−63,464	5,671
4	−63,716	0,397
5	−63,930	0,335
6	−63,974	0,069
7	−63,988	0,022
8	−63,997	0,013
9	−63,997	1,25005E–05
10	−64	0,004

For simplicity, we describe the model through its conventional prototype:

$$J = c_1 x_1(t_1) + c_2 x_2(t_1) \to \inf; \tag{20}$$

$$\dot{x}_1 = v_1, \qquad x_1(0) = 0, \tag{21}$$

$$\dot{x}_2 = x_1 v_2, \qquad x_2(0) = 0, \tag{22}$$

$$v_1(t) \in \mathbb{R}, \quad v_2(t) \geq 0, \quad \int_0^{t_1} \big(|v_1(t)| + v_2(t)\big)\, dt \leq M. \tag{23}$$

Let $M = 5$. Reachable sets of the respective impulsive system are depicted on Fig. 2. Again, we study the optimal impulsive control problem by using its space-time representation. We have:

$$H = \psi_1\, \omega_1 - c_2\, y_1\, \omega_2,$$

$$\dot{\psi}_1 = c_2\, \omega_2, \quad \psi_1(M) = -c_1, \quad \psi_2 \equiv -c_2;$$

$$g_1 = g_1(\psi_1) \doteq \psi_1, \quad g_2 = g_2(y_1) \doteq -c_2\, y_1,$$

and the H-extremal controls can be described as follows:

$$\begin{aligned}
&\Omega_1 : g_2 > |g_1| && \Rightarrow \omega^* = (0, 1),\\
&\Omega_2 : g_2 < -g_1,\ g_1 < 0 && \Rightarrow \omega^* = (-1, 0),\\
&\Omega_3 : g_2 < g_1,\ g_1 > 0 && \Rightarrow \omega^* = (1, 0),\\
&\Omega_4 : g_2 = g_1,\ g_1 > 0 && \Rightarrow \omega^* = (\lambda, 1-\lambda),\ \lambda \in [0,1],\\
&\Omega_5 : g_2 = -g_1,\ g_1 < 0 && \Rightarrow \omega^* = (-\lambda, 1-\lambda),\ \lambda \in [0,1],\\
&\Omega_6 : g_1 = 0,\ g_2 < 0 && \Rightarrow \omega^* = (2\lambda - 1, 0),\ \lambda \in [0,1],\\
&\Omega_7 : g_1 = g_2 = 0 && \Rightarrow \omega^* \text{ is any admissible.}
\end{aligned}$$

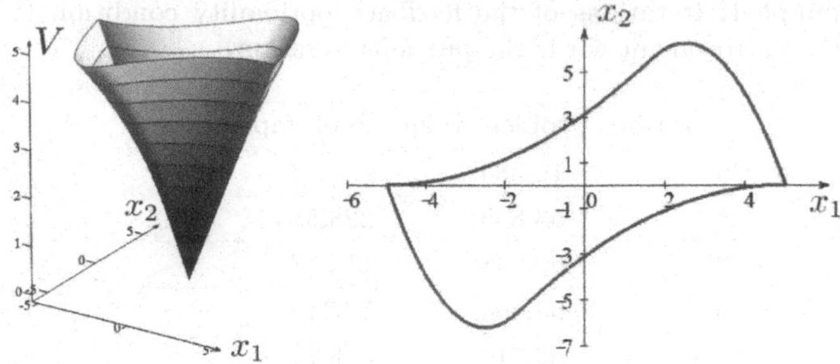

Fig. 2. Example 2: The reachable sets of the impulsive system for $M = 5$, $t_1 \geq 0$.

(1) Let $c_1, c_2 > 0$. Then, simple calculations give the following characterization of the optimal solution:

(a) if $M < \dfrac{c_1}{c_2}$, then $\bar{\omega} \equiv (-1, 0)$, $\begin{array}{l} \bar{y}_1(\tau) = -\tau, \\ \bar{y}_2(\tau) \equiv 0, \\ \bar{\psi}_1(\tau) \equiv -c_1, \end{array}$ $\tilde{J}(\bar{\sigma}) = -M$;

(b) if $M > \dfrac{c_1}{c_2}$, then $\bar{\omega}(\tau) = \begin{cases} (-1, 0), \ \tau \in \Delta_1 \doteq [0, \tau^*], \\ (0, 1), \quad \tau \in \Delta_2 \doteq (\tau^*, M], \end{cases}$ $\tau^* = \dfrac{c_1 + c_2 M}{2c_2}$,

$$\bar{y}_1(\tau) = \begin{cases} -\tau, & \tau \in \Delta_1, \\ -\dfrac{c_1 + c_2 M}{2c_2}, & \tau \in \Delta_2, \end{cases}$$

$$\bar{y}_2(\tau) = \begin{cases} 0, & \tau \in \Delta_1, \\ -\dfrac{c_1 + c_2 M}{2c_2}\tau + \dfrac{(c_1 + c_2 M)^2}{4c_2^2}, & \tau \in \Delta_2, \end{cases}$$

$$\bar{\psi}_1(\tau) = \begin{cases} -\dfrac{c_1 + c_2 M}{2}, & \tau \in \Delta_1 \doteq [0, \tau^*], \\ c_2(\tau - M) - c_1, & \tau \in \Delta_2 \doteq (\tau^*, M], \end{cases}$$

$$\tilde{J}(\bar{\sigma}) = -\frac{c_1(c_1 + c_2 M)}{2c_2} + \frac{(c_1 + c_2 M)^2}{4c_2} - \frac{(c_1 + c_2 M)M}{2}.$$

For example, put $c_1 = c_2 = 1$, $M = 2$ (case b), and let $\omega^0 \equiv (0, 0)$, $y_1^0 = x_2^0 \equiv 0$, $\psi_1^0 \equiv -1$, $\tilde{J}^0 = 0$. Then $g_1 \equiv -1$, $g_2 = -y_1$. Consider the following feedback control

$$\omega(\tau, y) = \begin{cases} (-1, 0), & \text{in } \Omega_2, \\ (0, 1), & \text{in } \Omega_1 \cup \Omega_5, \\ \text{any admissible, otherwise.} \end{cases}$$

Theorem 1 gives:

$$\tilde{\omega}(\tau) = \begin{cases} (-1, 0), & \tau \in [0, 1], \\ (0, 1), & \tau \in (1, 2], \end{cases} \quad \text{with } \tilde{\psi}(\tau) = \begin{cases} -2, & \tau \in [0, 1], \\ t - 3, & \tau \in (1, 2], \end{cases} \quad \tilde{J}[\tilde{\omega}] = -2,$$

and the second iteration $\tilde{\omega}$ leads to an optimal control

$$\bar{\omega}(\tau) = \begin{cases} (-1,0), \, \tau \in [0,3/2], \\ (0,1), \quad \tau \in (3/2,2] \end{cases} \quad \text{with } \tilde{J}[\bar{\sigma}] = -\frac{9}{4}.$$

(2) A more interesting situation appears when $c_1 > 0$, $c_2 < 0$. Now, our problem admits up to two extrema. Indeed, taken $c_1 = 1$, $c_2 = -1$, $M = 5$, we obtain two extremal processes:

– a (strong) local minimum: $\tilde{\omega} = \begin{cases} (1,0), \tau \in \Delta_1 \doteq [0,2], \\ (0,1), \tau \in \Delta_2 \doteq (2,5], \end{cases}$

$$\tilde{y}_1(\tau) = \begin{cases} \tau, \tau \in \Delta_1, \\ 2, \tau \in \Delta_2, \end{cases} \quad \tilde{y}_2(\tau) = \begin{cases} 0, & \tau \in \Delta_1, \\ 2t-4, \tau \in \Delta_2, \end{cases}$$

$$\tilde{\psi}_1(\tau) = \begin{cases} 2, & \tau \in \Delta_1, \\ 4-\tau, \tau \in \Delta_2, \end{cases} \quad \tilde{J}[\tilde{\omega}] = -4,$$

and

– a global minimum: $\bar{\omega} \equiv (-1,0)$, $\bar{y}_1(\tau) = -\tau$, $\bar{x}_2(\tau) \equiv 0$, $\bar{\psi}_1(\tau) \equiv -1$ with $\tilde{J}[\bar{\sigma}] = -5$.

Thus, any initial process satisfying $\int_0^M \omega_2(\tau)\,d\tau < 1$ could be discarded (improved) in just one iteration.

Let $\omega^0 \equiv (0,1)$, $y_1^0 = y_2^0 \equiv 0$, $\psi_1(\tau) = 4-\tau$. All extremal feedbacks generate the following parametrized family of open-loop controls:

$$\tilde{\omega} = \tilde{\omega}(\tau,\tau^*) = \begin{cases} (1,0), & \tau \in [0,2), \\ (-1,0), \tau \in [2,\tau^*), \\ (0,1), & \tau \in [\tau^*,5], \end{cases} \quad \tau^* \in [2,5].$$

Taken $\tau^* = 2$, one comes to a point of local minimum, while, for $\tau^* = 5$, one obtains a control with $\tilde{\omega}_2 \equiv 0$, and the second iteration leads to the optimal solution.

Example 3. Consider an impulsive relaxation of the following ordinary variational problem:

$$J = x_1(1) + 2x_2(1) - 3x_3(1) \to \inf;$$
$$\dot{x}_1 = f_1(t) + (ax_2 + b)v_1, \quad x_1(0) = x_{10},$$
$$\dot{x}_2 = f_2(t) + (cx_1 + d)v_2, \quad x_2(0) = x_{20},$$
$$\dot{x}_3 = v_1 + v_2, \quad x_3(0) = 0, \; x_3(1) \le M,$$
$$v_1(t) \ge 0, \quad v_2(t) \ge 0, \quad t \in [0,t_1].$$

Here, $a = 2$, $b = 1$, $c = 1$, $d = 3$, $M = 2$, $x_0 = (1,1)$, $f(t) = (1,1)$. Figure 3 demonstrates a cross-section of the reachable set of the associated measure differential equation, for $x_3 = 2$, $t_1 = 2$.

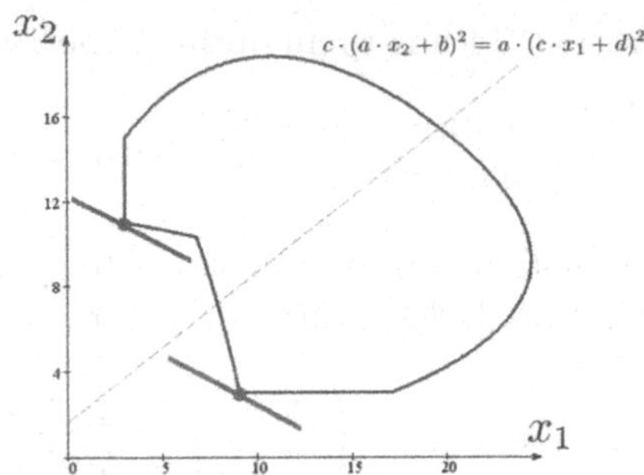

Fig. 3. Example 3: The cross-section of the reachable set for $x_3 = 2$, $t_1 = 2$.

In this example, the impulsive Maximum Principle [18] extracts two extremal process; the terminal values of the extremal trajectories are located at the points, depicted on Fig. 3: the green point (σ_1) corresponds to a non-optimal extremum, and the red one (σ_2) presents the global solution.

By applying Theorem 1 with $\varphi(\tau, y) = \psi(\tau)\big(\bar{y}(\tau) - y\big)$ ($\bar{y}$ and ψ are the phase and adjoint states corresponding to σ_1, respectively), we discard the non-optimal process σ_1. Furthermore, extremal feedback controls generated by φ provide the property of descent (control improvement) w.r.t. the cost functional. By the arguments similar to Examples 1, 2, one finally reaches the globally optimal process σ_2, as above.

4 Conclusion

As we have shown, the proposed feedback optimality conditions enjoy a constructive feature: for impulsive processes, there are known criteria for weak invariance (weak monotonicity) [20], which let one to design the feedback controls generating "viable" trajectories, together with respective open-loop controls. By an appropriate choice of function φ, one can construct an admissible control process, whose cost is less or equal to the problem value on the reference process. Furthermore, as multiple (even rather pathological) cases show, the effect of such feedback variations can be quite essential; sometimes it serves to obtain a global solution in few iterations, starting from a local extremum.

References

1. Bressan, A., Rampazzo, F.: Impulsive control systems without commutativity assumptions. J. Optim. Theory Appl. **81**(3), 435–457 (1994). https://doi.org/10.1007/BF02193094
2. Bressan, A., Mazzola, M.: Graph completions for impulsive feedback controls. J. Math. Anal. Appl. **412**(2), 976–988 (2014). https://doi.org/10.1016/j.jmaa.2013.11.015

3. Bressan, A., Piccoli, B.: Introduction to the mathematical theory of control, AIMS Series on Applied Mathematics, vol. 2. American Institute of Mathematical Sciences (AIMS), Springfield (2007)
4. Bressan, A., Rampazzo, F.: On differential systems with quadratic impulses and their applications to Lagrangian mechanics. SIAM J. Control Optim. **31**(5), 1205–1220 (1993). https://doi.org/10.1137/0331057
5. Clarke, F.H., Ledyaev, Yu. S., Stern, R.J., Wolenski, R.R.: Nonsmooth Analysis and Control Theory. Graduate Texts in Mathematics, vol. 178. Springer, New York (1998). https://doi.org/10.1007/b97650
6. Clarke, F.: Functional Analysis, Calculus of Variations and Optimal Control. Graduate Texts in Mathematics, vol. 264. Springer, London (2013). https://doi.org/10.1007/978-1-4471-4820-3
7. Dykhta, V.: Variational necessary optimality conditions with feedback descent controls for optimal control problems. Dokl. Math. **91**(3), 394–396 (2015)
8. Dykhta, V.: Positional strengthenings of the maximum principle and sufficient optimality conditions. Proc. Steklov Inst. Math. **293**(1), S43–S57 (2016)
9. Dykhta, V., Samsonyuk, O.: Optimal Impulsive Control with Applications, 2nd edn. Fizmatlit, Moscow (2003)
10. Dykhta, V., Samsonyuk, O.: Hamilton-Jacobi Inequalities and Variational Optimality Conditions. Irkutsk state university, Irkutsk (2015)
11. Dykhta, V., Samsonyuk, O.: Optimality conditions with feedback controls for optimal impulsive control problems. IFAC-PapersOnLine **51**(32), 509–514 (2018)
12. Fraga, S.L., Pereira, F.L.: Hamilton-Jacobi-Bellman equation and feedback synthesis for impulsive control. IEEE Trans. Autom. Control **57**(1), 244–249 (2012). https://doi.org/10.1109/TAC.2011.2167822
13. Gurman, V.: The Extension Principle in Optimal Control Problems, 2nd edn. Fizmatlit, Moscow (1997)
14. Karamzin, D., Oliveira, V., Pereira, F., Silva, G.: On some extension of optimal control theory. Eur. J. Control **20**(6), 284–291 (2014)
15. Karamzin, D., Oliveira, V., Pereira, F., Silva, G.: On the properness of the extension of dynamic optimization problems to allow impulsive controls. ESAIM Control Optim. Calc. Var. **21**(3), 857–875 (2015)
16. Miller, B.: The generalized solutions of nonlinear optimization problems with impulse control. SIAM J. Control Optim. **34**, 1420–1440 (1996)
17. Miller, B., Rubinovich, E.: Discontinuous solutions in the optimal control problems and their representation by singular space-time transformations. Autom. Remote Control **74**, 1969–2006 (2013)
18. Miller, B.M., Rubinovich, E.Y.: Impulsive Control in Continuous and Discrete-Continuous Systems. Kluwer Academic/Plenum Publishers, New York (2003). http://dx.doi.org/10.1007/978-1-4615-0095-7
19. Motta, M., Rampazzo, F.: Space-time trajectories of nonlinear systems driven by ordinary and impulsive controls. Differ. Integr. Equ. **8**, 269–288 (1995)
20. Samsonyuk, O.: Invariant sets for nonlinear impulsive control systems. Autom. Remote Control **76**(3), 405–418 (2015)
21. Sesekin, A., Zavalishchin, S.: Dynamic Impulse Systems: Theory and Applications. Kluwer Academic Publishers, Dordrecht (1997)

22. Sorokin, S., Staritsyn, M.: Feedback necessary optimality conditions for a class of terminally constrained state-linear variational problems inspired by impulsive control. Numer. Algebra Control Optim. **7**(2), 201–210 (2017)
23. Staritsyn, M., Sorokin, S.: On feedback strengthening of the maximum principle for measure differential equations. J. Glob. Optim. (2019). https://doi.org/10.1007/s10898-018-00732-3
24. Vinter, R.: Optimal Control. Birkhauser, Berlin (2000)

Data Mining and Computational Geometry

Semi-supervised Classification Using Multiple Clustering and Low-Rank Matrix Operations

Vladimir Berikov[1,2](✉)

[1] Sobolev Institute of Mathematics, Koptyug pr. 4, 630090 Novosibirsk, Russia
berikov@math.nsc.ru

[2] Novosibirsk State University, Universitetsky pr. 1, 630090 Novosibirsk, Russia

Abstract. This paper proposes a semi-supervised classification method which combines machine learning regularization framework and cluster ensemble approach. We use the low-rank decomposition of the co-association matrix of the ensemble to significantly speed up calculations and save memory. Numerical experiments using Monte Carlo approach demonstrate the efficiency of the proposed method.

Keywords: Semi-supervised classification · Cluster ensemble · Co-association matrix · Regularization · Low-rank matrix decomposition

1 Introduction

In pattern recognition problems, it is required to classify objects characterized by a set of features into several classes (patterns) obtaining an optimal value of a certain quality functional (e.g., minimize an estimate of misclassification probability). The creation of the classifier is based on the analysis of the learning sample consisting of precedents, i.e., objects for which their class labels are identified. In the basic formulation of the problem, the labels are known for all objects in the sample (*fully supervised classification*).

This paper considers another variant of the problem: so-called *semi-supervised classification*. In this task, class labels are defined only for a small part of the sample. It is required to classify the existed unlabeled objects (*transductive learning*) or find a decision rule for classifying any new objects from the statistical population (*inductive learning*). In this paper, we consider the transductive learning problem.

Semi-supervised classification is important because the procedure of class registration can be rather costly. For example, measurement of vegetation type in remote sensing [1] requires expensive field research, therefore labels can be attributed to only a small part of pixels. To improve the accuracy of prediction, it is necessary to utilize information from both labeled and unlabeled data.

M. Khachay et al. (Eds.): MOTOR 2019, LNCS 11548, pp. 529–540, 2019.
https://doi.org/10.1007/978-3-030-22629-9_37

A number of approaches and methods in semi-supervised classification exist (see, e.g., [2]). At the present time, such methods as heuristic self-training, probabilistic mixture decomposition, transductive support vector machine, and graph-based Laplacian Regularization are widely used.

Unsupervised learning (clustering) is applied in the semi-supervised classification as an instrument of knowledge extraction from unlabeled data. Ensemble clustering is a way of obtaining robust clustering decisions, especially in the case of uncertainty in the data structure. This methodology aims at finding consensus decision from multiple partition variants [5]. Properly organized ensemble (even composed of "weak" classifiers), as a rule, significantly improves the overall clustering quality.

Different schemes of application of ensemble clustering in semi-supervised classification have been proposed in [6,7]. The suggested methods are based on evidence showing that usage of averaged co-association matrix in the role of similarity matrix often improves the accuracy of decisions. This observation was supported with theoretical analysis: it was proved in [7] that the probability of classification error decreases with increase in ensemble size under some regularity conditions.

In this paper, we propose a novel semi-supervised classification method using graph Laplacian regularization and cluster ensemble methodology. Graph regularization (also known as manifold regularization) postulates that if two data points belong to the same manifold, it is likely they have the same labels. A graph Laplacian is used to measure the smoothness of the classifications in the data manifold comprising both labeled and unlabeled data [8–10].

We suggest a low-rank decomposition of averaged co-association matrix to reduce numerical cost and storage of the method.

In the rest of the paper, we give a short overview of related methods and describe the details of the suggested method. Numerical experiments with artificial and real datasets are presented. Finally, concluding remarks are given.

2 Basic Preliminaries

Let a dataset $\mathbf{X} = \{x_1, \ldots, x_n\}$ be given, where $x_i \in \mathbf{R}^d$ is feature vector, d is dimensionality of feature space $X = (X_1, \ldots, X_d)$. In a fully supervised classification, we are given an additional set $Y = \{y_1, \ldots, y_n\}$ of class labels, $y_i \in D_Y$, where $D_Y = \{c_1, \ldots, c_K\}$ is target feature domain, i.e., unordered set of categorical values (classes). Using this information, it is necessary to find a classifier $y = f(x)$ for predicting target feature labels for any new data point $x \in \mathbf{R}^d$ from the same statistical population. The function should be optimal in some sense, e.g., give minimal value to the expected losses.

In an unsupervised learning setting, the target feature values are not provided. The problem of cluster analysis, which is an important direction in unsupervised classification, consists in finding a partition $P = \{C_1, \ldots, C_K\}$ of $\mathbf{X}$ on a relatively small number of homogeneous clusters describing the structure of data. As a criterion of homogeneity, it is possible to use a function dependent on the scatter of observations within groups and the distances between clusters. The desired number of clusters is either a predefined parameter or should be found in the best way.

In a semi-supervised transductive classification problem, the target feature labels are known only for a part of the data set $\mathbf{X}_1 \subset \mathbf{X}$. It is possible to assume that $\mathbf{X}_1 = \{x_1, \ldots, x_{n_1}\}$, and the unlabeled part is $\mathbf{X}_0 = \{x_{n_1+1}, \ldots, x_n\}$. The set of labels for points from $\mathbf{X}_1$ is denoted by $\mathbf{Y}_1 = \{y_1, \ldots, y_{n_1}\}$. It is required to predict labels $\mathbf{Y}_0 = (y_{n_1+1}, \ldots, y_n)$ in some best way for a given unlabeled sample $\mathbf{X}_0$.

3 Overview of Methods

Consider some of the commonly used approaches in semi-supervised classification.

3.1 Heuristic Self-training

In this approach, some basic supervised classification algorithm is used. At the first step, the algorithm is trained on the labeled sample and then classifies the unlabeled part. For each classified object, a recognition quality score is calculated (for example, the distance to the separating hyperplane). In the next step, those observations for which the quality estimates are above a certain predetermined threshold are excluded from the set $\mathbf{X}_0$ and added to $\mathbf{X}_1$, and their labels are included to $\mathbf{Y}_1$. Then the basic algorithm is again used for training on the updated labeled sample and classification the remaining unlabeled part. The process is repeated until no unlabeled points remain.

Methods based on the described heuristic procedure, as a rule, are quite effective, but theoretical analysis of their properties is a difficult task.

3.2 Probabilistic Approach

When using this approach, it is assumed that for each class c_k some distribution $p_k(x|\theta_k)$ is specified on feature space, where θ_k is a vector of parameters, $k = 1, \ldots, K$. It is assumed that the general form of the distribution is known (for example, multidimensional normal), and its parameters need to be estimated with the sample. Denote $\theta = (\theta_1, \ldots, \theta_K)$, and let $q = (q_1, \ldots, q_K)$ be a set of a priori probabilities of classes, where $q_1 + \cdots + q_K = 1$. Then for a labeled point $x_i \in \mathbf{X}_1$, for which $y_i = c_k$, by the Chain rule we obtain:

$$p(x_i, y_i|\theta) = q_k p_k(x_i|\theta_k).$$

According to the Law of Total Probability, for unlabeled point $x_i \in \mathbf{X}_0$ the following is true:

$$p(x_i|\theta) = \sum_{k=1}^{K} q_k p_k(x_i|\theta_k).$$

One may consider the problem of maximizing the likelihood function:

$$(q^*, \theta^*) = \arg\max_{q,\theta} \left\{ \sum_{x_i \in \mathbf{X}_1} \log p(x_i, y_i|\theta) + \sum_{x_i \in \mathbf{X}_0} \log p(x_i|\theta) \right\},$$

For the solution of the problem, iterative algorithms have been developed (for example, EM algorithm for analysis of distribution mixture [2]). After finding optimal $q^* = (q_1^*, \dots q_K^*)$, $\theta^* = (\theta_1^*, \dots, \theta_K^*)$, the unlabeled objects are classified according to the Bayes formula:

$$f(x_i) = c_{k*}, \text{ where } k^* = \arg\max_k \{q_k^* p_k(x_i|\theta_k^*)\}, \ i = n_1 + 1, \dots, n.$$

The disadvantage of this approach is that under significant violation of the assumptions on the distribution model, the found solutions will have large error.

3.3 Transductive Support Vector Machine

The methods of this type are based on support vector machine (SVM) methodology for the fully supervised two-class recognition problem (which can be extended to the multi-class case). In the basic formulation (binary classification), it is required to find the direction of the separating hyperplane, for which the width of the margin separating the classes is maximum. The input of the algorithm is a training sample $\mathbf{X}$ with class labels $\mathbf{Y} = \{y_1, \dots, y_n\}$, $y_i \in \{-1, +1\}$, $i = 1, \dots, n$. In the case of linear separability of classes, there are an infinite number of separating hyperplanes. It is reasonable to choose a hyperplane, the distance from which to both classes is maximal. The points lying on the border of the separating margin are called support vectors.

The hyperplane equation can be represented as $\langle w, x \rangle + b = 0$, where $\langle \cdot, \cdot \rangle$ is a scalar product, w is a normal vector, and b is an auxiliary parameter. The support vector method builds a decision function in the form

$$f(x) = sign(\sum_{i=1}^{n} \alpha_i y_i \langle x_i, x \rangle + b)$$

where $\alpha_1, \dots, \alpha_n \geq 0$ are some parameters; at the same time, the hyperplane coefficients are normalized so that $\langle w, x_i \rangle + b = 1$ for the support vectors. It is important to note that the summation goes only over the support vectors for which $\alpha_i \neq 0$.

For transductive support vector machine, it is required to find a hyperplane which separates with the maximum margin not only labeled points from $\mathbf{X}_1$, but also the unlabeled points from $\mathbf{X}_0$. Thus, the hyperplane must be located in the region with the lowest density. The problem of finding the optimal hyperplane can be formulated as follows:

$$\begin{gathered}\text{find } \mathbf{Y}_0, w, b, \xi : \tfrac{1}{2}||w||^2 + C\sum_i \xi_i \to \min_{\mathbf{Y}_0, w, b, \xi} \\ \text{s.t. } y_i(\langle w, x_i \rangle + b) \geq 1 - \xi_i, \ i = 1, \dots, n, \\ \xi_i \geq 0, \ i = 1, \dots, n.\end{gathered}$$

Here $\xi_1, \ldots, \xi_n$ are variables denoting penalty for violation margin boundaries, $C \geq 0$ is a given parameter. Thus, we need to maximize the width of separating margin (it can be shown that this requirement is equivalent to minimizing $||w||^2$), as well as minimize the total penalty for violating its boundaries.

Algorithms for the approximate solution of this problem exist (see [2]). For the solution, the corresponding dual problem is solved with respect to parameters $\alpha_1, \ldots, \alpha_n$.

With linearly non-separable classes, it is possible to find a transform $\varphi : X \to X'$ of the original feature space X to a new space X' of higher dimensionality by use of some kernel function. In the new space, the objects can already be linearly separable.

It is known that the optimization problem for transductive support vector machine is not convex, and existing algorithms for its approximate solution have a polynomial complexity depending on the number of observations. Therefore, this approach is applicable to samples of relatively small size (about a thousand observations).

3.4 Graph Laplacian Regularization

Consider weighted non-oriented complete graph $G = (V, E)$, in which the set of vertices V corresponds to points from $\mathbf{X}$, and the set of edges E corresponds to pairs (x_i, x_j), $i, j = 1, \ldots, n$, $i \neq j$. Each edge (x_i, x_j) is associated with a non-negative weight W_{ij} (the degree of similarity between the points). For example, the weight can be determined using the RBF kernel:

$$W_{ij} = \exp\left(-\frac{||x_i - x_j||}{2\sigma^2}\right)$$

where σ is a given parameter.

Let us introduce a diagonal matrix D with elements $D_{ii} = \sum_j W_{ij}$. Matrix

$$L = D^{-1/2} W D^{-1/2}$$

is called normalized graph Laplacian. It has a dimension $n \times n$; elements of the matrix are: $L_{ij} = \frac{W_{ij}}{\sqrt{D_{ii}}\sqrt{D_{jj}}}$. Other types of graph Laplacian are also used: standard Laplacian [3] $L_s = D - W$, and graph Laplacian associated with random walk [4] $L_{rw} = D^{-1}W$.

Let $Y_i = (Y_{i1}, \ldots, Y_{iK})$ denote a Boolean vector of observable class labels: $Y_{ik} = \mathbb{I}[y_i = c_k]$, where $\mathbb{I}[\cdot]$ is an indicator function: $\mathbb{I}[true] = 1$, $\mathbb{I}[false] = 0$, $i = 1, \ldots, n_1$, $k = 1, \ldots, K$. Denote by $F_i = (F_{i1}, \ldots, F_{iK})$ classification vector with elements equal to the estimated degrees of belonging of point x_i to classes $c_1, \ldots, c_K$, and let $F = (F_1, \ldots, F_n)^T$ denote classification matrix.

Consider the following optimization problem:

$$\begin{aligned} &\text{find } F^* = \arg \min_{F \in R^{n \times K}} Q(F) \\ &= \frac{1}{2}\left(\sum_{x_i \in X_1} ||F_i - Y_i||^2 + \beta \sum_{x_i, x_j \in \mathbf{X}} W_{ij} \left\| \frac{F_i}{\sqrt{D_{ii}}} - \frac{F_j}{\sqrt{D_{jj}}} \right\|^2 \right), \; s.t. \; F \geq 0 \end{aligned} \quad (1)$$

where $\beta > 0$ is a regularization parameter. The first sum in the right side of (1) is aimed to reduce the fitting error on labeled data; the second component plays the role of a smoothing function: its minimization means that if two points x_i, x_j (either labeled or unlabeled) are similar, their classification vectors should not be very different.

It is known that the optimized function is convex. To find the optimal solution, we differentiate (1), and after simple transformations with normalized graph Laplacian get:

$$\frac{\partial Q}{\partial F_{ik}}|_{F_{ik}=F^*_{ik}} = F^*_{ik} - Y_{ik} + \beta F^*_{ik} - \beta L_{i\cdot} F^*_{\cdot k} = 0, \quad i = 1, \ldots, n_1, \tag{2}$$

$$\frac{\partial Q}{\partial F_{ik}}|_{F_{ik}=F^*_{ik}} = \beta F^*_{ik} - \beta L_{i\cdot} F^*_{\cdot k} = 0, \quad i = n_1 + 1, \ldots, n \tag{3}$$

where $L_{i\cdot}$, $F^*_{\cdot k}$ are ith row of matrix L and kth column of the matrix F^*, respectively, $i = 1, \ldots, n_1,\ k = 1, \ldots, K$.

Denote by $Y_{1,0}$ the following matrix:

$$Y_{1,0} = (Y_1, \ldots, Y_{n_1}, \underbrace{0, \ldots, 0}_{n-n_1})^T$$

of dimensionality $n \times K$, and by G diagonal matrix:

$$G = diag(G_{11} \ldots, G_{nn}),\ G_{ii} = \begin{cases} \beta+1, & i=1,\ldots,n_1 \\ \beta, & i=n_1+1,\ldots,n \end{cases} \tag{4}$$

of dimensionality $n \times n$. Then one may rewrite (2), (3) in matrix form:

$$(G - \beta L)F^* = Y_{1,0}, \tag{5}$$

hence

$$F^* = (G - \beta L)^{-1} Y_{1,0} \tag{6}$$

if the inverse matrix exists (note that it is always possible to choose regularization parameter β to ensure the well-posedness of the problem).

To find F^*, one may use iterative matrix inversion methods [11]. The authors of [9] describe the iterative algorithm (*Label Spreading*) used to solve a problem similar to (6). One may also apply the existing methods of solving systems of linear algebraic equations for (5), where each system is determined for corresponding columns of matrices F^* and $Y_{1,0}$. After calculating F^*, the final classification is determined as

$$y_i = c_{k*}, \text{ where } k^* = \arg\max_{k=1,\ldots,K} F^*_{ik},\ i = n_1 + 1, \ldots, n. \tag{7}$$

A limitation of this approach is that one needs to keep in memory non-sparse graph Laplacian matrix of dimensionality $n \times n$, and also a large cost of matrix operations.

4 Proposed Method

The method proposed in this paper is based on a combination of ensemble clustering and graph Laplacian regularization. We use averaged co-association matrix of clustering ensemble [12] as similarity matrix in (1).

This replacement has a number of reasons. First of all, the co-association matrix defines semi-metric on observations space [13]. Thus the frequencies of assigning pairs of objects to the same clusters can be viewed as measures of similarity between data points. Second, one may believe that objects from a dense region in feature space share their class labels with larger probability, even though the region has complicated form (elongated or strip-like cluster) and the Euclidian distance between points is large. Viewed from this angle, such points are similar to each other.

Usually averaged co-association matrix is calculated in the process of cluster ensemble formation and requires a memory of quadratic size. However, it is possible to reduce the requirement by low-rank matrix decomposition.

4.1 Cluster Ensemble and Low-Rank Decomposition of the Co-association Matrix

Let us consider a set of partition variants $\{P_l\}_{l=1}^r$, where $P_l = \{C_{l,1}, \ldots, C_{l,K_l}\}$, $C_{l,k} \subset \mathbf{X}$, $C_{l,k} \bigcap C_{l,k'} = \varnothing$, K_l is the number of clusters in lth partition. For each P_l we determine matrix $H_l = (h_l(i,j))_{i,j=1}^n$ with elements indicating whether a pair x_i, x_j belong to the same cluster in lth variant or not: $h_l(i,j) = \mathbb{I}[c_l(x_i) = c_l(x_j)]$, $c_l(x)$ is the cluster label assigned to x. The weighted averaged co-association matrix (WACM) is defined as follows:

$$H = (H(i,j))_{i,j=1}^n,\ H(i,j) = \sum_{l=1}^r w_l H_l(i,j)$$

where $w_1, \ldots, w_r$ are weights of ensemble elements, $w_l \geq 0, \sum w_l = 1$. The weights should reflect the "importance" of base clustering variants in the ensemble and be dependent on some evaluation function Γ (cluster validity index, diversity measure) [13]: $w_l = \gamma_l / \sum\limits_{l'} \gamma_{l'}$, where $\gamma_l = \Gamma(l)$ is an estimate of clustering quality for the lth partition (we assume that a larger value of Γ indicates better quality).

The following obvious property of WACM allows increasing the processing speed.

Proposition 1. Weighted averaged co-association matrix admits low-rank decomposition in the form:

$$H = BB^T,\ B = [B_1 B_2 \ldots B_r] \tag{8}$$

where B is a block matrix, $B_l = \sqrt{w_l}\, A_l$, A_l is $n \times K_l$ cluster assignment matrix for lth partition: $A_l(i,k) = \mathbb{I}[c(x_i) = k]$, $i = 1, \ldots, n$, $k = 1, \ldots, K_l$.

As a rule, $m = \sum_l K_l \ll n$, thus (8) gives us an opportunity of saving memory by storing $n \times m$ sparse matrix instead of full $n \times n$ co-association matrix. The complexity of matrix-vector multiplication $H \cdot x$ is decreased from $O(n^2)$ to $O(nm)$.

4.2 Cluster Ensemble and Graph Laplacian Regularization

Let us consider normalized graph Laplacian in the form: $\tilde{L} = \tilde{D}^{-1/2} H \tilde{D}^{-1/2}$ where $\tilde{D} = \mathrm{diag}(\tilde{D}'_{11}, \ldots, \tilde{D}_{nn})$, $\tilde{D}_{ii} = \sum\limits_j H(i,j)$. We have:

$$\tilde{D}_{ii} = \sum_{j=1}^{n} \sum_{l=1}^{r} w_l \sum_{k=1}^{K_l} A_l(i,k) A_l(j,k)$$
$$= \sum_{l=1}^{r} w_l \sum_{k=1}^{K_l} A_l(i,k) \sum_{j=1}^{n} A_l(j,k) = \sum_{l=1}^{r} w_l n_l(i) \tag{9}$$

where $n_l(i)$ is the size of the cluster which includes the point x_i in lth partition variant.

Substituting $\tilde{L}$ in (6), we obtain a cluster ensemble-based classification matrix:

$$F^{**} = (G - \beta \tilde{L})^{-1} \, Y_{1,0}. \tag{10}$$

Using low-rank decomposition, this expression can be transformed into the form which involves more efficient matrix operations.

Denote $U = \tilde{D}^{-1/2} B$. From (8) and (10) we get:

$$F^{**} = (G - \beta U U^T)^{-1} \, Y_{1,0}.$$

In linear algebra, the following Woodbury matrix identity is known:

$$(G + UV)^{-1} = G^{-1} - G^{-1} U (I + V G^{-1} U)^{-1} V G^{-1}$$

where $G \in \mathbf{R}^{n \times n}$ is invertible matrix, $U \in \mathbf{R}^{n \times m}$ and $V \in \mathbf{R}^{m \times n}$. From (4) it follows that

$$G^{-1} = \mathrm{diag}(1/G_{11}, \ldots, 1/G_{nn}).$$

Now it is possible to formulate the following statement:

Proposition 2. Cluster ensemble-based classification matrix (10) *can be calculated using low-rank decomposition as follows:*

$$F^{**} = (G^{-1} + \beta G^{-1} U (I - \beta U^T G^{-1} U)^{-1} U^T G^{-1}) \, Y_{1,0}. \tag{11}$$

A remarkable fact is that in (11) one has to invert significantly smaller $m \times m$ dimensional matrix instead of $n \times n$ dimensional in (10). The overall computational complexity of (11) is $O(nm + m^3)$.

The scheme of the suggested semi-supervised classification algorithm based on graph Laplacian and low-rank decomposition of a co-association matrix (SSC-LR) is as follows.

Algorithm SSC-LR
Input:
$\mathbf{X}$: dataset including both labeled $\mathbf{X}_1$ and unlabeled samples $\mathbf{X}_0$;
$\mathbf{Y}_1$: set of known labels;
r: number of runs for base clustering algorithm;
$\boldsymbol{\Omega}$: set of parameters (working conditions) of the clustering algorithm.
Output:
$\mathbf{Y}_0$: predicted class labels for objects from $\mathbf{X}_0$.
Steps:
1. Generate r variants of clustering partition for working parameters randomly chosen from $\boldsymbol{\Omega}$; calculate weights $w_1, \ldots, w_r$ of variants.
2. Find normalized graph Laplacian in low-rank representation using matrices B in (8) and $\tilde{D}$ in (9);
3. Determine predicted classification matrix according to (11);
4. Calculate label assignments $\mathbf{Y}_0$ using (7).
end.

In the computer implementation of SSC-LR, we use K-means as base clustering algorithm which has linear complexity with respect to data dimensions.

5 Numerical Experiments

In this section, we describe numerical experiments with the proposed SSC-LR algorithm. The aim of experiments is to confirm the usefulness of involving cluster ensemble for similarity matrix estimation in semi-supervised classification. We experimentally evaluate the classification quality on one synthetic and one real-life example.

In the first example, we consider data sets generated from a mixture of three multidimensional normal distributions $\mathcal{N}(a_i, \sigma_X I)$ under equal weights; $a_i \in \mathbf{R}^d$, $i = 1, 2, 3$, $d = 8$, σ_X is a parameter, $a_1 = (0, 0, \ldots, 0)^T$, $a_2 = (5, 5, \ldots, 5)^T$, $a_3 = (-5, 5, \ldots, 5)^T$. To study the robustness of the algorithm, we also generate two independent random variables following uniform distribution $\mathcal{U}(0, 5)$ and use them as additional "noisy" features.

In Monte Carlo modeling, we repeatedly generate samples of size n according to the given distribution mixture. In the experiment, 10% of the points selected at random from each component compose the labeled sample; the remaining ones are included in the unlabeled part.

The ensemble variants are designed by random initialization of centroids (selected data points) in K-means; the number of clusters equals three. The ensemble size is $r = 10$. The wights of ensemble elements are the same: $w_l \equiv 1/r$. The regularization parameter $\beta = 0.1$.

For the comparison purposes, we consider the method (denoted as SSC-RBF) which uses standard similarity matrix evaluated with RBF kernel (best results were obtained with RBF parameter $\sigma = 4$), where the output classifications are calculated according to (6) and (7). The accuracy of classification is evaluated by comparison of the predictions with true class labels for $\mathbf{X}_0$ (unknown in the stage of classifier design).

To make the comparison results more statistically sound, we have averaged accuracy estimations over 40 Monte Carlo repetitions and compare the results by paired two-sample Student's t-test.

Table 1 presents the results of experiments. In addition to the averaged classification accuracy, the table shows the averaged execution times for the algorithms (working on dual-core Intel Core i5 processor with a clock frequency of 2.8 GHz and 4 GB RAM). For SSC-LR, we separately indicate ensemble generation time t_{ens} and low-rank matrix operation time t_{matr} (in seconds). The obtained p-values for Student's t-test are also taken into account. A p-value less than the given significance level indicates a statistically significant difference between the performance estimates.

Table 1. Results of experiments with a mixture of three distributions. Significantly larger accuracy estimates (p-value $< 10^{-5}$) are in bold. For $n = 10^5$ and $n = 10^6$, SSC-RBF failed due to unacceptable memory demands.

n	σ_X	SSC-LR			SSC-RBF	
		Accuracy	t_{ens} (sec)	t_{matr} (sec)	Accuracy	Time (sec)
1000	2	0.999	0.07	0.006	0.999	0.26
	3	0.988	0.08	0.01	0.987	0.26
	4	**0.953**	0.06	0.05	0.942	0.26
3000	2	0.999	0.08	0.02	0.999	4.02
	3	0.988	0.08	0.02	0.988	4.09
	4	**0.952**	0.09	0.02	0.949	4.08
7000	2	0.999	0.75	0.12	0.999	44.7
	3	0.989	0.78	0.13	0.989	47.1
	4	**0.953**	0.28	0.08	0.951	44.1
10^5	2	0.999	1.87	0.69	-	-
10^6	2	0.999	27.2	8.28	-	-

The results show that the proposed algorithm has comparable or even better classification accuracy than SSC-RBF. At the same time, it works much faster. For a large volume of data ($n = 10^5$, $n = 10^6$) only SSC-LR is able to find a solution because SSC-RBF has refused to work due to unacceptable memory demands (74.5 GB and 7450.6 GB correspondingly).

In the second example, we consider Cardiotocography Data Set [14]. Cardiotocography is a simultaneous recording of Fetal Heart Rate (FHR) and

Uterine Contractions (UC) and it is one of the most common diagnostic techniques to evaluate maternal and fetal well-being during pregnancy and before delivery. Fetal cardiotocograms were automatically processed and the respective diagnostic features were measured. The dataset includes a total of 2126 observations of which is 1655 normal, 295 suspicious and 176 pathologic samples which indicate the existing of fetal distress. 22 numerical features are FHR baseline (beats per minute), number of accelerations per second, number of fetal movements per second, number of UC per second, etc.

The following experiment's settings are used. The volume of the labeled sample is 10% of overall data; the cluster ensemble architecture is the same as in the previous example: K-means base algorithm with 10 clusters; ensemble size $r = 10$. Parameter $\beta = 0.1$; SSC-RBF parameter $\sigma = 4$. The number of generations of labeled samples is 40.

As a result of modeling, the averaged accuracy rate for SSC-LR equals 0.89. For SSC-RBF, the averaged accuracy is 0.80. The p-value less than 0.0001 indicates a statistically significant difference between the quality estimates.

6 Conclusion

This work has introduced a semi-supervised classification method which combines graph Laplacian regularization and multiple clustering methodologies. Low-rank decomposition of co-association matrix gave us a possibility to speed up calculations and save memory from cubic to linear.

There are a number of arguments for the usefulness of ensemble clustering in semi-supervised classification. Ensemble decisions allow us to restore more accurately metric relations between objects under the existence of complex data structures. The obtained co-association matrix depends on the outputs of clustering algorithms and is less noise-addicted than conventional similarity matrices. Clustering with a sufficiently large number of clusters can be viewed as Learning Vector Quantization known for lowering the average distortion in data.

The efficiency of the suggested SSC-LR algorithm was confirmed experimentally. Monte Carlo experiments have demonstrated comparable or even statistically significantly better accuracy estimates and a considerable decrease in running time for SSC-LR in comparison with analogous SSC-RBF algorithm based on standard similarity matrix with RBF kernel.

We plan to continue studying the theoretical properties of the proposed method, as well as to perform a detailed comparison with other state-of-the-art methods. Applications of the method in various fields are also planned, especially for hyperspectral imagery classification and analysis of genetic sequences.

Acknowledgments. The work was supported by the program of Fundamental Scientific Researches of the RAS, project 0314-2019-0015 of the Sobolev Institute of mathematics. The research was partly supported by RFBR grants 18-07-00600, 18-29-09041mk and partly by the Russian Ministry of Science and Higher Education under Project 5-100. The author thanks anonymous reviewers for helpful comments.

References

1. Camps-Valls, G., Marsheva, T., Zhou, D.: Semi-supervised graph-based hyperspectral image classification. IEEE Trans. Geosci. Remote Sens. **45**(10), 3044–3054 (2007)
2. Zhu, X.: Semi-supervised learning literature survey. Technical report. Department of Computer Science, University of Wisconsin, Madison. N. 1530 (2008)
3. Zhu, X., Ghahramani, Z., Lafferty, J.D.: Semi-supervised learning using Gaussian fields and harmonic functions. In: Proceedings of the 20th International Conference on Machine Learning (ICML 2003), pp. 912–919 (2003)
4. Avrachenkov, K., Mishenin, A., Goncalves, P., Sokol, M.: Generalized optimization framework for graph-based semi-supervised learning. In: Proceedings of the 2012 SIAM International Conference on Data Mining, pp. 966–974 (2012)
5. Boongoen, T., Iam-On, N.: Cluster ensembles: a survey of approaches with recent extensions and applications. Comput. Sci. Rev. **28**, 1–25 (2018)
6. Yu, G.X., Feng, L., Yao, G.J., Wang, J.: Semi-supervised classification using multiple clusterings. Pattern Recognit. Image Anal. **26**(4), 681–687 (2016)
7. Berikov, V., Karaev, N., Tewari, A.: Semi-supervised classification with cluster ensemble. In: 2017 International Multi-Conference Engineering, Computer and Information Sciences (SIBIRCON), pp. 245–250. IEEE (2017)
8. Wu, M., Scholkopf, B.: Transductive classification via local learning regularization. In: Artificial Intelligence and Statistics, pp. 628–635 (2007)
9. Zhou, D., Bousquet, O., Lal, T., Weston, J., Scholkopf, B.: Learning with local and global consistency. In: Advances in Neural Information Processing Systems, pp. 321–328 (2003)
10. Belkin, M., Niyogi, P., Sindhwani, V.: Manifold regularization: a geometric framework for learning from labeled and unlabeled examples. J. Mach. Learn. Res. **7**, 2399–2434 (2006)
11. Tikhonov, A.N., Goncharsky, A., Stepanov, V.V., Yagola, A.G.: Numerical Methods for the Solution of Ill-posed Problems. Springer, Dordrecht (2013). https://doi.org/10.1007/978-94-015-8480-7
12. Fred, A., Jain, A.: Combining multiple clusterings using evidence accumulation. IEEE Trans. Pattern Anal. Mach. Intell. **27**(6), 835–850 (2005)
13. Berikov, V.B.: Construction of an optimal collective decision in cluster analysis on the basis of an averaged co-association matrix and cluster validity indices. Pattern Recognit. Image Anal. **27**(2), 153–165 (2017)
14. http://archive.ics.uci.edu/ml/datasets/Cardiotocography. Accessed 22 Feb 2019

Maximum Diversity Problem with Squared Euclidean Distance

Anton V. Eremeev[1,2(✉)], Alexander V. Kel'manov[1,3], Mikhail Y. Kovalyov[4], and Artem V. Pyatkin[1,3]

[1] Sobolev Institute of Mathematics, Novosibirsk, Russia
[2] Dostoevsky Omsk State University, Omsk, Russia
eremeev@ofim.oscsbras.ru
[3] Novosibirsk State University, Novosibirsk, Russia
[4] United Institute of Informatics Problems, Minsk, Belarus

Abstract. In this paper we consider the following Maximum Diversity Subset problem. Given a set of points in Euclidean space, find a subset of size M maximizing the squared Euclidean distances between the chosen points. We propose an exact dynamic programming algorithm for the case of integer input data. If the dimension of the Euclidean space is bounded by a constant, the algorithm has a pseudo-polynomial time complexity. Using this algorithm, we develop an FPTAS for the special case where the dimension of the Euclidean space is bounded by a constant. We also propose a new proof of strong NP-hardness of the problem in the general case.

Keywords: Euclidean space · Subset of points · Given size · Maximum variance · Strong NP-hardness · Integer instance · Exact algorithm · Fixed space dimension · Pseudo-polynomial time

1 Introduction

The subject of this research is the well-known Maximum Diversity Problem, assuming the squared Euclidean distance as a distance measure: given N points in Euclidean space, choose a subset of size M to maximize the sum of squared Euclidean distances between all points of the subset. It is also known as *max-sum* Dispersion Problem. The problem is strongly NP-hard [4].

Our goals are to study the possibility of finding approximate solutions to the problem, to develop an exact algorithm for the special case of the Maximum Diversity Problem where all points have integer coordinates, and to analyse the complexity of this algorithm.

Suppose an objective function $f(x)$ of a maximization problem is non-negative (as in the case of the Maximum Diversity Problem). A feasible solution x is called a $(1-\varepsilon)$-*approximate* solution if it satisfies the inequality $f(x) \geq (1-\varepsilon)f^*$, where $0 < \varepsilon < 1$ and f^* is the optimal objective function value.

M. Khachay et al. (Eds.): MOTOR 2019, LNCS 11548, pp. 541–551, 2019.
https://doi.org/10.1007/978-3-030-22629-9_38

An algorithm is called a $(1-\varepsilon)$-*approximation algorithm* if in a polynomial time it outputs a $(1-\varepsilon)$-approximate solution for every solvable problem instance. A family of $(1-\varepsilon)$-approximation algorithms parameterized by $\varepsilon > 0$, such that the time complexity of these algorithms is polynomially bounded by $1/\varepsilon$ and by the problem instance length is called *a fully polynomial-time approximation scheme (FPTAS)*.

The paper is organized as follows. In Sect. 2 we introduce the Maximum Diversity problem and survey known algorithmic results on this and related problems. In Sect. 3 we suggest a new proof of strong NP-hardness of the problem. Section 4 contains some supplementary results, related to the properties of the optimization criterion. In Sect. 5 we suggest an exact algorithm for the case of integer input data and study its complexity. Section 6 presents an FPTAS for the case when the dimension of the Euclidean space is bounded by a constant. Concluding remarks are given in Sect. 7.

2 Problem Formulation, Related Problems and Known Results

The Maximum Diversity Problem (later on referred to as Problem 1) considered in this paper has the following formulation.

Problem 1.

Input: an N-element set Y of points in k-dimensional Euclidean space $\mathbb{R}^k$ and a positive integer $M \leq N$.

Problem: find a subset $C \subseteq Y$ of points such that the objective function

$$h(C) := \sum_{y\in C}\sum_{z\in C} ||z-y||^2 \tag{1}$$

is maximal under the constraint $|C| = M$ on the size of the subset C.

Without loss of generality we can assume that Y contains at least one non-zero vector (otherwise the problem is trivial).

Note that for any nonempty set C of points in Euclidean space the following equations are true (see e.g. [5]):

$$f(C) := \sum_{z\in C} ||z-\bar{z}(C)||^2 = \frac{1}{2|C|}\sum_{z\in C}\sum_{y\in C} ||z-y||^2 = \frac{1}{2|C|}h(C). \tag{2}$$

Here and below, $\bar{z}(C) := \frac{1}{|C|}\sum_{z\in C} z$ denotes the centroid of a set C.

Therefore the optimum of the Problem 1 is attained on the same subsets as the optimum of $\max\{f(C) \mid |C| = M, C \subseteq Y\}$ and these optimization problems are equivalent.

Problem 1 has numerous applications in ecological, medical or social sciences, in animal and plant genetics [17]. In particular, if the given points correspond to people so that the coordinates of points are equal to some characteristics of these people, then the Maximum Diversity Problem may be treated as a problem of finding a maximally diverse group of people of a given size (e.g. for composition of medical crews [2] or in immigration control [15]). In evolutionary algorithms for single-objective or multi-objective optimization in $\mathbb{R}^k$, maximally diverse subset of tentative solutions may be beneficial in selection for the next population.

The complexity and algorithmic issues of Problem 1 are addressed in a number of publications. Kuo, Glover and Dhir [14] have proposed several integer programming formulations for this problem and showed that a more general problem, where the distances between the points are arbitrary positive numbers, is strongly NP-hard. A generalization of Problem 1 to the distance spaces of negative type is considered by Cevallos, Eisenbrand, and Zenklusen [4], where a polynomial-time approximation scheme (PTAS) has been developed and the strong NP-hardness of the problem is established. Cevallos, Eisenbrand and Morell in [3] (along with other results) propose a proof of NP-hardness of Problem 1 in the case of 3-dimensional Euclidean space and give a survey of the current state of the art in theoretical analysis of Problem 1 and other diversity maximization problems.

Note that changing the optimization direction from maximization to minimization converts Problem 1 into the M-Variance problem [1]. Strong NP-hardness of the M-Variance problem is established in [9]. In the same paper, it is shown that there does not exist a fully polynomial time approximation scheme (FPTAS) for this problem unless P = NP. The exact algorithms with time complexity $\mathcal{O}(kN^{k+1})$ were proposed in [1,19]. If the space dimension k is fixed, these algorithms are polynomial and their time complexity is $\mathcal{O}(N^{k+1})$. An exact algorithm for the case of integer inputs was presented in [10]. The time complexity of the algorithm is $\mathcal{O}(kN(2MB+1)^k)$, where B is the maximum absolute coordinate value in the input set. If the space dimension is fixed, the algorithm is pseudo-polynomial and its time complexity is $\mathcal{O}(N(MB)^k)$. In [11], a 2-approximation polynomial algorithm with time complexity $\mathcal{O}(kN^2)$ was presented for the general case of the problem. A polynomial time approximation scheme (PTAS) was proposed in [18]. The time complexity of the scheme is $\mathcal{O}(kN^{2/\varepsilon+1}(9/\varepsilon)^{3/\varepsilon})$, where $\varepsilon > 0$ is a relative error.

In [12], the algorithm was proposed which allows finding a $(1+\varepsilon)$-approximate solution in $\mathcal{O}(kN^2(2\sqrt{k}M/\varepsilon+2)^k)$ time for given $\varepsilon \in (0,1)$. For fixed space dimension k, the algorithm runs in $\mathcal{O}(N^2(M/\varepsilon)^k)$ time and implements an FPTAS.

An improved approximation scheme that allows finding a $(1+\varepsilon)$-approximate solution in $\mathcal{O}\left(kN^2(\sqrt{\frac{2k}{\varepsilon}}+2)^k\right)$ time was proposed in [13]. If the space dimension is fixed, the algorithm implements an FPTAS, since its time complexity in this case is $\mathcal{O}(N^2(1/\varepsilon)^{k/2})$. In the same work, an improved approximation scheme were proposed. The time complexity of this scheme is $\mathcal{O}\left(\sqrt{k}N^2\left(\frac{\pi e}{2}\right)^{k/2}\left(\sqrt{\frac{2}{\varepsilon}}+2\right)^k\right)$. In the case of dimension $k = \mathcal{O}(\log N)$, the improved scheme remains polynomial. In this case it implements a PTAS with $\mathcal{O}\left(N^{c\,(1.05+\log(2+\sqrt{\frac{2}{\varepsilon}}))}\right)$ time, where c is a positive constant.

In [8], a parameterized randomized algorithm for the M-Variance problem was proposed. For given upper bounds on the relative error and failure probability, the parameter value is defined for which the algorithm finds approximate solutions in a polynomial time. The conditions are found under which these algorithms are asymptotically exact and have the time complexity that is linear in the space dimension and quadratic in the size of the input set.

3 New Proof of Strong NP-hardness

Theorem 9 in [4] implies that Problem 1 is NP-hard in the strong sense because the square of Euclidean norm induces a distance space of negative type. Nevertheless, for the sake of completeness, we provide a direct proof of this fact since it is very short and simple.

It is known [16] that the classic NP-complete Independent Set problem [6] remains NP-complete for regular graphs:

Independent Set in a Regular Graph. Given a regular graph G of degree d and a positive integer M, find whether this graph contains a vertex subset of cardinality M such that every two vertices of this subset are not connected by an edge.

Theorem 1. *Problem 1 is NP-hard in the strong sense.*

Let G be a regular graph of degree d with N vertices and $q = dN/2$ edges. Construct the following instance of Problem 1. Put $k = q$ and assign to every vertex of the graph G a k-dimensional vector y whose i-th coordinate is 1 if the edge i is incident with this vertex, and is 0 otherwise. Then, for a pair of vectors y and z from $Y = \{y_1, \dots, y_N\}$, clearly, $\|y - z\|^2 = 2d - 2$, if the vertices of G corresponding to y and z are adjacent, and $\|y - z\|^2 = 2d$, otherwise. So, the optimum of Problem 1 is at least $2dM(M-1)$ iff G contains an independent set of size M. □

Note that the instance of Problem 1 constructed in Theorem 1 has integer (indeed, Boolean) input vectors, and therefore, the objective function takes only integer values. Besides, the value of objective function is bounded by a polynomial in the input size. Therefore (see e.g. [6]), unless P = NP, the general case of Problem 1 does not admit a fully polynomial time approximation scheme (FPTAS).

4 Properties of the Objective Function

Denote by C^* the optimal solution of the Problem 1. The following "folklore" result is well-known. It gives an expression of the total quadratic cluster spread with respect to any given point in terms of the cluster centroid.

Lemma 1. *For an arbitrary point $x \in \mathbb{R}^k$ and a finite set $C \subset \mathbb{R}^k$, we have the equality*

$$\sum_{z\in C} ||z-x||^2 = \sum_{z\in C} ||z-\overline{z}(C)||^2 + |C| \cdot ||x-\overline{z}(C)||^2, \tag{3}$$

where $\overline{z}$ is the centroid of C.

The above mentioned equality (2) relating the objective functions $f(C)$ and $h(C)$ follows from Lemma 1 by summation over $x \in C$. A number of results in [10,12,13] for the minimization version of the problem exploit Lemma 1 as well.

The next lemma provides another expression for the objective function $f(C)$, which will be useful for development of an exact algorithm based on dynamic programming in Sect. 5. Here we use the Boolean programming formulation, where binary variables $x_1, \ldots, x_N$ give a natural representation of a subset $C \subseteq Y$, assuming $x_j = 1$ if point $y_j \in C$ and $x_j = 0$ otherwise, $j = 1, \ldots, N$. Let $x = (x_1, \ldots, x_N)$ and $f_N(x) = f(C)$, where C is represented by x.

Let $y_j = (a_{1j}, a_{2j} \ldots, a_{kj})$, $j = 1, \ldots, N$. Then Problem 1 is equivalent to the following Boolean linear programming problem.

Problem 2.

$$\max f_N(x) = \max \sum_{j=1}^{N} \sum_{r=1}^{k} x_j \left(a_{rj} - \frac{\sum_{i=1}^{N} a_{ri} x_i}{M} \right)^2, \tag{4}$$

subject to

$$\sum_{j=1}^{N} x_j = M, \tag{5}$$

$$x_j \in \{0,1\}, j = 1, \ldots, N. \tag{6}$$

Lemma 2. *If $\sum_{j=1}^{N} x_j = M$ then*

$$f_N(x) = \sum_{j=1}^{N} \frac{x_j}{M} \sum_{r=1}^{k} \Big((M-1)a_{rj}^2 - 2a_{rj} \sum_{i=1}^{j-1} a_{ri} x_i \Big). \tag{7}$$

Proof. We can rewrite the objective function (4) as

$$\begin{aligned}
f_N(x) &= \sum_{r=1}^{k}\sum_{j=1}^{N} x_j\Big(a_{rj} - \frac{\sum_{i=1}^{N} a_{ri}x_i}{M}\Big)^2 \\
&= \sum_{r=1}^{k}\sum_{j=1}^{N}\Big(a_{rj}^2 x_j - \tfrac{2}{M}a_{rj}x_j\sum_{i=1}^{N} a_{ri}x_i + \tfrac{1}{M^2}x_j\sum_{i=1}^{N} a_{ri}^2 x_i \\
&\quad + \tfrac{2}{M^2}x_j\sum_{i=1}^{N} a_{ri}x_i\sum_{h=1}^{i-1} a_{rh}x_h\Big) \\
&= \sum_{r=1}^{k}\Big(\sum_{j=1}^{N} a_{rj}^2 x_j - \tfrac{2}{M}\sum_{j=1}^{N} a_{rj}x_j\sum_{i=1}^{N} a_{ri}x_i + \tfrac{1}{M^2}\sum_{j=1}^{N} x_j\sum_{i=1}^{N} a_{ri}^2 x_i \\
&\quad + \tfrac{2}{M^2}\sum_{j=1}^{N} x_j\sum_{i=1}^{N} a_{ri}x_i\sum_{h=1}^{i-1} a_{rh}x_h\Big) \\
&= \sum_{r=1}^{k}\Big(\sum_{j=1}^{N} a_{rj}^2 x_j - \tfrac{2}{M}\sum_{j=1}^{N} a_{rj}^2 x_j - \tfrac{4}{M}\sum_{j=1}^{N} a_{rj}x_j\sum_{i=1}^{j-1} a_{ri}x_i + \tfrac{1}{M}\sum_{i=1}^{N} a_{ri}^2 x_i \\
&\quad + \tfrac{2}{M}\sum_{i=1}^{N} a_{ri}x_i\sum_{h=1}^{i-1} a_{rh}x_h\Big) \\
&= \sum_{r=1}^{k}\Big(\tfrac{M-1}{M}\sum_{j=1}^{N} a_{rj}^2 x_j - \tfrac{2}{M}\sum_{j=1}^{N} a_{rj}x_j\sum_{i=1}^{j-1} a_{ri}x_i\Big) \\
&= \sum_{j=1}^{N}\tfrac{x_j}{M}\sum_{r=1}^{k}\Big((M-1)a_{rj}^2 - 2a_{rj}\sum_{i=1}^{j-1} a_{ri}x_i\Big).
\end{aligned}$$

□

5 Exact Algorithm for Integer Instances

Our exact algorithm for instances with integer coordinates is based on dynamic programming. Consider the Boolean programming formulation (4)–(6) and introduce functions $s_j(\cdot)$ for partial sums in (7):

$$s_j(x_1,\ldots,x_j) = \frac{x_j}{M}\sum_{r=1}^{k}\Big((M-1)a_{rj}^2 - 2a_{rj}\sum_{i=1}^{j-1} a_{ri}x_i\Big),\ j=1,\ldots,N.$$

Then we have $f_N(x) = \sum_{j=1}^{N} s_j(x_1,\ldots,x_j)$.

Let $F_j(m, A_1,\ldots,A_k)$ be the maximum diversity for partial solutions $(x_1,\ldots,x_j)$ such that exactly $m \le M$ components among $x_1,\ldots,x_j$ are equal to 1 and $\sum_{i=1}^{j-1} a_{ri}x_i = A_r$ for every $r = 1,\ldots,k$. Formally,

$$F_j(m, A_1,\ldots,A_k) = \max\sum_{i=1}^{j} s_i(x_1,\ldots,x_i), \tag{8}$$

subject to

$$\sum_{i=1}^{j} x_j = m, \tag{9}$$

$$\sum_{i=1}^{j-1} a_{ri} x_i = A_r, \ \ r = 1, \dots, k, \tag{10}$$

$$x_i \in \{0,1\}, i = 1, \dots, j. \tag{11}$$

In the case of integer inputs $y_j \in \mathbb{Z}^k$, $j = 1, \dots, N$, the partial sums $\sum_{i=1}^{j-1} a_{ri} x_i$ take only integer values from $[-B, B]$ (recall that B denotes the maximum absolute coordinate value in the input set) and we have the Bellman Equation:

$$F_j(m, A_1, \dots, A_k) = \max_{x_j \in \{0,1\}} \begin{cases} F_{j-1}(m, A_1, \dots, A_k), & \text{if } x_j = 0, \\ F_{j-1}(m-1, A_1 - a_{1j}, \dots, A_k - a_{kj}) + \\ \quad + \frac{1}{M} \sum_{r=1}^{k} \left((M-1) a_{rj}^2 - 2 a_{rj} A_r \right), & \text{otherwise.} \end{cases}$$

Our exact algorithm for instances with integer coordinates of the input points works as follows. Put $A := \max_{1 \le r \le k} \sum_{j=1}^{N} a_{rj} \le BN$. First, compute recursively the set of values $F_j(m, A_1, \dots, A_k)$ for all $j = 1, \dots, N$, $m = 1, \dots, M$, and $A_1 = -A, \dots, A$, where they are defined (otherwise assume $F_j(m, A_1, \dots, A_k) = 0$). The set of binary vectors $x_1, \dots, x_j$ corresponding to each $F_j(m, A_1, \dots, A_k)$ can be easily back-tracked. Then, compute

$$f_N(x^*) = \max_{x \in \{0,1\}^N} \sum_{j=1}^{N} s_j(x_1, \dots, x_j) = \max_{(A_1, \dots, A_k) \in [-A, A]^k} F_N(M, A_1, \dots, A_k).$$

Output a subset $C = \cup_{j : x_j^* = 1} \{y_j\}$ as a solution to the problem. This algorithm is called Algorithm DP in what follows.

Clearly, Lemma 3 below establishes a relation between optimal and algorithmic solutions.

Lemma 3. *Suppose that components of all points in Y are integers from the interval $[-B, B]$. Then the solution found by the Algorithm DP has the objective value $f_{DP} = f(C^*) = \frac{1}{2M} h(C^*)$.*

The following theorem establishes the time complexity and correctness of the Algorithm DP.

Theorem 2. *If components of all points in Y have integer values in the interval $[-B, B]$, then Algorithm DP finds an optimal solution to Problem 2 in time $\mathcal{O}(MN(2A+1)^k)$.*

Algorithm DP is pseudo-polynomial for a fixed k of the space dimension since the time complexity of this algorithm is $\mathcal{O}(MN(BN)^k)$ that is polynomially bounded in terms of problem dimension N and the value of B. This is expressed in the following

Corollary 1. *If components of all points in Y are integers from the interval $[-B, B]$ and k is bounded above by a constant, then Algorithm DP finds an optimal solution to Problem 1 in pseudo-polynomial time $\mathcal{O}(MN(BN)^k)$.*

6 Approximation Scheme

In order to obtain an FPTAS in the case of fixed space dimension k, we can use the well-known *rounding the input* technique (see e.g. [7] or [6], Chapter 6).

For each subset $C \subset Y$ denote by $J(C) \subset \{1, \ldots, N\}$ the set of indices j of all elements y_j from C.

First of all, preprocess the input data in such a way that all coordinates become non-negative and, moreover, $\min_j a_{ij} = 0$ for each $i = 1, \ldots, k$. Recall that a_{ij} is the i-th coordinate of the vector y_j. This can be done by an affine transformation that, clearly, changes neither the value of the objective function nor the set of indices $J^* := J(C^*)$ of the optimal solution.

Given an instance of Problem 1, i.e. an N-element set $Y \subset \mathbb{R}^k$ and an integer $M \leq N$, modify the instance, replacing each coordinate $a_{ij}, i = 1, \ldots, k$ of every vector $y_j, j = 1 \ldots, N$ by the new value $\hat{a}_{ij} := \lfloor a_{ij}/K \rfloor$ for some appropriate $K > 0$ which will be chosen later. Let $\hat{y}_j = (\hat{a}_{1j}, \ldots, \hat{a}_{kj})$, $j = 1, \ldots, N$. Suppose that B is the maximum absolute coordinate value in the input set. The time complexity of Algorithm DP on the resulting instance with integer input data is $\mathcal{O}(MN(NB/K)^k)$ by Corollary 1. Let the Algorithm DP output a set $\hat{C} \subseteq \{\hat{y}_1, \ldots, \hat{y}_N\} \subset \mathbb{Z}^k$, where $|\hat{C}| = M$ and $h(\hat{C})$ is the optimal value of the objective function of the rounded instance.

Note that for every $j \neq j'$ the squared distance $||\hat{y}_j - \hat{y}_{j'}||^2$ is at least

$$\sum_{i=1}^{k} \left(\frac{|a_{ij} - a_{ij'}|}{K} - 1 \right)^2 = \sum_{i=1}^{k} \left(\frac{a_{ij}}{K} - \frac{a_{ij'}}{K} \right)^2 - 2 \sum_{i=1}^{k} \frac{|a_{ij} - a_{ij'}|}{K} + k.$$

So,

$$\begin{aligned} ||y_j - y_{j'}||^2 &\leq K^2 ||\hat{y}_j - \hat{y}_{j'}||^2 + 2kK \max_i |a_{ij} - a_{ij'}| - kK^2 \\ &\leq K^2 ||\hat{y}_j - \hat{y}_{j'}||^2 + 2kKB. \qquad (12) \end{aligned}$$

Denote by $Z(J)$ the subset of vectors $\{\hat{y}_j,\ j \in J\}$, of the rounded instance defined by the set of indices $J \subseteq \{1, ..., N\}$. Since $|C^*| = M$, we have the following inequality

$$h(C^*) \leq K^2 h(Z(J^*)) + 2kKBM^2, \qquad (13)$$

giving us the following relation between the optimum value $h(C^*)$ and the optimum value $h(\hat{C})$ for the modified instance:

$$h(C^*) \leq K^2 h(\hat{C}) + 2kKBM^2. \qquad (14)$$

Similarly to (12), for every $j \neq j'$ the squared distance $||\hat{y}_j - \hat{y}_{j'}||^2$ is at most

$$\sum_{i=1}^{k} \left(\frac{|a_{ij} - a_{ij'}|}{K} + 1 \right)^2 = \sum_{i=1}^{k} \left(\frac{a_{ij}}{K} - \frac{a_{ij'}}{K} \right)^2 + 2 \sum_{i=1}^{k} \frac{|a_{ij} - a_{ij'}|}{K} + k,$$

i.e.

$$||y_j - y_{j'}||^2 \geq K^2||\hat{y}_j - \hat{y}_{j'}||^2 - 2kKB - kK^2, \tag{15}$$

which means that

$$K^2 h(\hat{C}) \leq \sum_{j \in J'} \sum_{j' \in J'} ||y_j - y_{j'}||^2 + (2KB + K^2)kM^2, \tag{16}$$

where $J' = J(\hat{C})$. So, for an approximate solution $C := \cup_{j \in J'} \{y_j\}$ of the original instance the value $h(C) = \sum_{j \in J'} \sum_{j' \in J'} ||y_j - y_{j'}||^2$ differs from $K^2 h(\hat{C})$ by at most $(2KB + K^2)kM^2$ due to (16). Combining this with (14), we conclude that

$$h(C^*) - h(C) \leq (4KB + K^2)kM^2. \tag{17}$$

Given a precision parameter $\varepsilon \in (0, 1)$, choose

$$K := \frac{3B}{2^4 k^2 M^4 (\varepsilon^{-1} + 1)^2}.$$

In what follows, Algorithm DP applied to the rounded input data with this K outputting the feasible solution C defined above is referred to as Algorithm A_ε.

The approximation algorithm A_ε has the time complexity $\mathcal{O}\Big(MN(NM^4 (\varepsilon^{-1} + 1)^2)^k\Big)$ which is polynomial in N and ε^{-1}, if $k = O(1)$. Moreover, in view of (17) it holds that

$$h(C^*) - h(C) \leq (4KB + K^2)kM^2 < \frac{3B^2}{4kM^2(\varepsilon^{-1}+1)^2} + \frac{9B^2}{k^3 2^8 M^6 (\varepsilon^{-1}+1)^4}$$
$$= \frac{B^2}{\varepsilon^{-1}+1}\left(\tfrac{3}{4}A + \tfrac{9}{256}A^3\right),$$

where $A := \frac{1}{kM^2(\varepsilon^{-1}+1)} < 1/2$. So, $\frac{3}{4}A + \frac{9}{256}A^3 < 1$ and

$$h(C) > h(C^*) - \frac{B^2}{\varepsilon^{-1} + 1}. \tag{18}$$

Now note that $h(C^*) \geq D^2$, where $D := \max_{j \neq j'} ||y_j - y_{j'}||$ is the distance between the most remote points of the input set. Due to the preprocessing, we have $D \geq B$ and $h(C^*) \geq D^2 \geq B^2$.

Therefore, by (18) it holds that

$$h(C) > B^2 - \frac{B^2}{\varepsilon^{-1} + 1}. \tag{19}$$

Using (18) and (19) we get the approximation guarantee

$$\frac{h(C^*)}{h(C)} \leq \frac{h(C) + \frac{B^2}{\varepsilon^{-1}+1}}{h(C)} < 1 + \frac{\frac{B^2}{\varepsilon^{-1}+1}}{B^2 - \frac{B^2}{\varepsilon^{-1}+1}} = 1 + \varepsilon. \tag{20}$$

Therefore the following theorem holds.

Theorem 3. *In the case of $k = \mathcal{O}(1)$, Algorithm A_ε constitutes an FPTAS for Problem 1 with the time complexity $\mathcal{O}\Big(N^{k+1} M^{4k+1} \varepsilon^{-2k}\Big)$.*

7 Conclusions

The Maximum Diversity Subset problem is considered, where the optimization criterion is to maximize the sum of squared Euclidean distances between all points of the chosen subset. The main results of this paper are obtained for the case when the dimension of the space is not a part of the input (i.e. the dimension is fixed or bounded by a constant). In particular we show that in this special case, the Maximum Diversity Subset problem is solvable in a pseudo-polynomial time, if the coordinates of the input points are all integer. Using this fact, a fully polynomial time approximation scheme is proposed for the case when the dimension of the space is not a part of the input.

Further "positive" results for the Maximum Diversity Subset problem with squared Euclidean distances may be expected for the special cases of Euclidean space dimensions 1 and 2.

Acknowledgements. The study presented in Sect. 4 was supported by the Russian Foundation for Basic Research project 19-01-00308. The study presented in Sect. 6 was supported by the Russian Academy of Science (the Program of basic research), projects 0314-2019-0015, 0314-2019-0019, and in Sect. 3 by the Russian Ministry of Science and Education under the 5-100 Excellence Programme.

References

1. Aggarwal, H., Imai, N., Katoh, N., Suri, S.: Finding k points with minimum diameter and related problems. J. Algorithms **12**(1), 38–56 (1991)
2. Aringhieri, R.: Composing medical crews with equity and efficiency. Cent. Eur. J. Oper. Res. **17**(3), 343–357 (2009). https://doi.org/10.1007/s10100-009-0093-3
3. Cevallos, A., Eisenbrand, F., Morell, S.: Diversity maximization in doubling metrics. In: Proceedings of 29th International Symposium on Algorithms and Computation (ISAAC 2018). LIPIcs, vol. 123, pp. 33:1–33:12. Schloss Dagstuhl-Leibniz-Zentrum fuer Informatik, Dagstuhl (2016). https://doi.org/10.4230/LIPIcs.ISAAC.2018.33
4. Cevallos, A., Eisenbrand, F., Zenklusen, R.: Max-sum diversity via convex programming. In: 32nd Annual Symposium on Computational Geometry (SoCG). LIPIcs, vol. 51, pp. 26:1–26:14. Schloss Dagstuhl-Leibniz-Zentrum fuer Informatik, Dagstuhl (2016). https://doi.org/10.4230/LIPIcs.SoCG.2016.26
5. Edwards, A.W.F., Cavalli-Sforza, L.L.: A method for cluster analysis. Biometrics **21**, 362–375 (1965)
6. Garey, M.R., Johnson, D.S.: Computers and intractability. A guide to the Theory of NP-Completeness. W.H. Freeman and Company, San Francisco (1979)
7. Ibarra, O., Kim, C.E.: Fast approximation algorithms for the knapsack and sum of subset problems. J. ACM **22**, 463–468 (1975)
8. Kel'manov, A., Khandeev, V., Panasenko, A.: Randomized algorithms for some clustering problems. In: Eremeev, A., Khachay, M., Kochetov, Y., Pardalos, P. (eds.) OPTA 2018. CCIS, vol. 871, pp. 109–119. Springer, Cham (2018). https://doi.org/10.1007/978-3-319-93800-4_9
9. Kel'manov, A.V., Pyatkin, A.V.: NP-completeness of some problems of choosing a vector subset. J. Appl. Ind. Math. **5**(3), 352–357 (2011)

10. Kel'manov, A.V., Romanchenko, S.M.: Pseudopolynomial algorithms for certain computationally hard vector subset and cluster analysis problems. Autom. Remote Control **73**(2), 349–354 (2012)
11. Kel'manov, A.V., Romanchenko, S.M.: An approximation algorithm for solving a problem of search for a vector subset. J. Appl. Ind. Math. **6**(1), 90–96 (2012)
12. Kel'manov, A.V., Romanchenko, S.M.: An FPTAS for a vector subset search problem. J. Appl. Ind. Math. **8**(3), 329–336 (2014)
13. Kel'manov, A., Motkova, A., Shenmaier, V.: An approximation scheme for a weighted two-cluster partition problem. In: van der Aalst, W.M.P., et al. (eds.) AIST 2017. LNCS, vol. 10716, pp. 323–333. Springer, Cham (2018). https://doi.org/10.1007/978-3-319-73013-4_30
14. Kuo, C.C., Glover, F., Dhir, K.S.: Analyzing and modeling the maximum diversity problem by zero-one programming. Decis. Sci. **24**(6), 1171–1185 (1993)
15. McConnell, S.: The new battle over immigration. Fortune **117**(10), 89–102 (1988)
16. Papadimitriou, C.H.: Computational Complexity. Addison-Wesley, New York (1994)
17. Porter, W.M., Eawal, K.M., Rachie, K.O., Wien, H.C., Willians, R.C.: Cowpea germplasm catalog No. 1. International Institute of Tropical Agriculture, Ibadan, Nigeria (1975)
18. Shenmaier, V.V.: An approximation scheme for a problem of search for a vector subset. J. Appl. Ind. Math. **6**(3), 381–386 (2012)
19. Shenmaier, V.V.: Solving some vector subset problems by Voronoi diagrams. J. Appl. Ind. Math. **10**(4), 560–566 (2016)

Estimation of the Necessary Sample Size for Approximation of Stochastic Optimization Problems with Probabilistic Criteria

Sergey V. Ivanov(✉) and Irina D. Zhenevskaya

Moscow Aviation Institute (National Research University), Volokolamskoe Shosse, 4, A-80, GSP-3, 125993 Moscow, Russia
sergeyivanov89@mail.ru, genevskaia@gmail.com

Abstract. Stochastic optimization problems with probabilistic and quantile objective functions are considered. The probability objective function is defined as the probability that the value of losses does not exceed a fixed level. The quantile function is defined as the minimal value of losses that cannot be exceeded with a fixed probability. Sample approximations of the considered problems are formulated. A method to estimate the accuracy of the approximation of the probability maximization and quantile minimization is described for the case of a finite set of feasible strategies. Based on this method, we estimate the necessary sample size to obtain (with a given probability) an epsilon-optimal strategy to the original problems by solving their approximations in the cases of finite set of feasible strategies. Also, the necessary sample size is obtained for the probability maximization in the case of a bounded infinite set of feasible strategies and a Lipschitz continuous probability function.

Keywords: Stochastic optimization · Sample approximation · Probability function · Quantile function

1 Introduction

Problems of stochastic programming with probabilistic and quantile objective functions are encountered in many applied problems, where special attention is paid to the reliability of the system. These problems and methods for solving them are well covered in [1].

In this paper, we research the sample average approximation (SAA) method for solving stochastic programming problems with probabilistic and quantile objective functions. This method is based on statistical estimation of the objective function. For the expectation objective function, the convergence of the SAA method is proved in [2]. The SAA method was applied to stochastic programming problems with probabilistic constraints in [3], where the convergence of the method was proved for a special case of the problem. In [4], the possibility of approximating a stochastic programming problem with probabilistic and

M. Khachay et al. (Eds.): MOTOR 2019, LNCS 11548, pp. 552–564, 2019.
https://doi.org/10.1007/978-3-030-22629-9_39

quantile objective function was investigated. The hypo-convergence of sampling probability functions is proved, which in turn guarantees the convergence of the approximation of the probability maximization and quantile minimization problems with respect to the value of the objective function and with respect to the optimization strategy.

From recent works on the SAA, [5] can be noted, where general approach to study approximations of stochastic programming problems is suggested. In [6], confidence bounds on the optimal objective value are constructed.

When the SAA method is applied, the reduced problem can be considered as a stochastic optimization problem with discrete distribution of the vector of random parameters. These problems can be reduced to mixed integer programming problems [7], which can be solved by using available software.

To apply the SAA method, it is useful to know the quality of the obtained approximate solution. In [8–10], in the case of a finite set of feasible strategies, an estimate of the required sample size was obtained to approximate an expectation minimization problem. This result was extended for the case of a Lipschitz continuous expectation function in [8]. To estimate the required number of realizations of the random vector, the exponential estimation of large deviations was used. In [11], the rate of convergence is studied for stochastic programming problems with probabilistic constraints.

This paper presents sample size estimates for problems of stochastic programming with probabilistic and quantile objective function. For the probability maximization, we consider cases of finite and bounded set of feasible strategies. For the quantile minimization, the case of finite set of feasible strategies is considered.

2 Statement

Let $(\mathcal{X}, \mathcal{F}, \mathbf{P})$ be a complete probability space. Let X be a random vector defined on this probability space. For simplicity, we assume that $\mathcal{X} \subset \mathbb{R}^m$ is a closed set.

Let us denote by $\Phi(\cdot)\colon U \times \mathcal{X} \to (-\infty, +\infty)$ a loss function, where $U \subset \mathbb{R}^n$ is a nonempty compact set of strategies u. We assume that the function $(u, x) \mapsto \Phi(u, x)$ is lower semi-continuous in $u \in U$ and $\mathcal{B}(U) \times \mathcal{F}$-measurable in $x \in \mathcal{X}$, where $\mathcal{B}(U)$ is the Borel σ-algebra of subsets U. These conditions guarantees that the function $(u, x) \mapsto \Phi(u, x)$ is a normal integrand [12].

Let us introduce the probability function

$$P_\varphi(u) \triangleq \mathbf{P}\{\Phi(u, X) \leqslant \varphi\},$$

and quantile function

$$\varphi_\alpha(u) \triangleq \min\{\varphi : P_\varphi(u) \geqslant \alpha\},$$

where $\alpha \in (0, P^*)$ is a given reliability level,

$$P^* = \sup_{u \in U} \mathbf{P}\{\Phi(u, X) < +\infty\}.$$

We consider the probability maximization problem

$$\alpha^* \triangleq \sup_{u \in U} P_\varphi(u) \tag{1}$$

and the quantile minimization problem

$$\varphi^* \triangleq \inf_{u \in U} \varphi_\alpha(u). \tag{2}$$

The sets of ε-optimal solutions to problems (1) and (2) are denoted by

$$U_\varphi^\varepsilon \triangleq \{u \in U \colon P_\varphi(u) \geq \alpha^* - \varepsilon\},$$
$$V_\alpha^\varepsilon \triangleq \{u \in U \colon \varphi_\alpha(u) \leq \varphi^* + \varepsilon\}$$

respectively.

3 Sample Approximation

Let $X_1, \ldots, X_N$ be a sample generated by random vector X, i.e., random vectors X_k, $k = \overline{1, N}$, are independent identically distributed with distribution function $F(x) = \mathbf{P}\{X \leqslant x\}$. We assume that the sample is defined on a complete probability space $(\Omega, \mathcal{F}', \mathbf{P}')$. This probability space may differ from the space $(\mathcal{X}, \mathcal{F}, \mathbf{P})$. However, below we will use the same letter $\mathbf{P}$ for the probability $\mathbf{P}'$, because it is clear which probability space is considered. Then we can write the frequency of the event $\{\Phi(u, X) \leqslant \varphi\}$ as

$$P_\varphi^{(N)} \triangleq \frac{1}{N} \sum_{k=1}^{N} \chi_{(-\infty, \varphi]}(\Phi(u, X_k)). \tag{3}$$

where

$$\chi_A(x) \triangleq \begin{cases} 0, & x \in A; \\ 1, & x \notin A. \end{cases}$$

By using (3), the quantile function can be estimated by

$$\varphi_\alpha^{(N)}(u) \triangleq \min\{\varphi \colon P_\varphi^{(N)}(u) \geq \alpha\}. \tag{4}$$

The sample approximation of the probability maximization problem is formulated as

$$\hat{U}_\varphi^{(N)} \triangleq \operatorname{Arg}\max_{u \in U} P_\varphi^{(N)}(u), \ N \in \mathbb{N}; \tag{5}$$

and the sample approximation of the quantile minimization problem is formulated as

$$\hat{V}_\alpha^{(N)} \triangleq \operatorname{Arg}\min_{u \in U} \varphi_\alpha^{(N)}(u), \ N \in \mathbb{N}. \tag{6}$$

From [4] it follows that quantile function (4) coincides with the order statistics of sample values $\{\Phi_k\}_{k=1}^{N}$ for the random variable $\Phi \triangleq \Phi(u, X)$ that has index $\lceil \alpha N \rceil$ and is called the sample quantile, where $\lceil x \rceil \triangleq \min\{k \in \mathbb{N} \colon x \leqslant k\}$.

4 Estimation of the Necessary Sample Size for Probability Maximization

In [4] the convergence of the approximating stochastic programming problems (5) and (6). These results, however, do not show the quality of solutions for a given sample of size N. In this section, we find upper bounds on the necessary sample size to consider a solution to (5) as an approximate solution to problem (1).

4.1 Case of Finite Set of Feasible Strategies

Let us begin with the case when the set U is finite. Its cardinality is denoted by $|U|$.

We consider the event

$$\left\{\hat{U}_\varphi^{(N)} \not\subset U_\varphi^\varepsilon\right\} = \bigcup_{u\in U\setminus U_\varphi^\varepsilon} \bigcap_{y\in U} \left\{\hat{P}_\varphi^{(N)}(u) \geqslant \hat{P}_\varphi^{(N)}(y)\right\}.$$

The event $\left\{\hat{U}_\varphi^{(N)} \not\subset U_\varphi^\varepsilon\right\}$ means that there exists an optimal solution $u_\varphi^{(N)}$ to the approximation problem (5) such that $u_\varphi^{(N)}$ is not ε-optimal solution to the true problem (1).

Then, given that the set U is finite, we can find an upper bound for the probability

$$\mathbf{P}\left\{\hat{U}_\varphi^{(N)} \not\subset U_\varphi^\varepsilon\right\} \le \sum_{u\in U\setminus U_\varphi^\varepsilon} \mathbf{P}\left(\bigcap_{y\in U}\left\{\hat{P}_\varphi^{(N)}(u) \ge \hat{P}_\varphi^{(N)}(y)\right\}\right). \tag{7}$$

Let u^* be an optimal solution to the true problem (1). If there several optimal solution to problem (1), then u^* can be taken arbitrarily. It follows from (7) that

$$\begin{aligned}\mathbf{P}\left\{\hat{U}_\varphi^{(N)} \not\subset U_\varphi^\varepsilon\right\} &\le \sum_{u\in U\setminus U_\varphi^\varepsilon} \mathbf{P}\left\{P_\varphi^{(N)}(u) \ge P_\varphi^{(N)}(u^*)\right\}\\ &\le |U| \max_{u\in U\setminus U_\varphi^\varepsilon} \mathbf{P}\left\{\hat{P}_\varphi^{(N)}(u) \ge \hat{P}_\varphi^{(N)}(u^*)\right\}.\end{aligned} \tag{8}$$

Let us introduce the random variables

$$\xi_k = \chi_{(-\infty,\varphi]}(\Phi(u, X_k)) - \chi_{(-\infty,\varphi]}(\Phi(u^*, X_k)).$$

Notice that ξ_k are independent. Then we can write

$$\mathbf{P}\left\{\hat{P}_\varphi^{(N)}(u) \ge \hat{P}_\varphi^{(N)}(u^*)\right\} = \mathbf{P}\left\{\sum_{k=1}^N \xi_k \ge 0\right\} = \mathbf{P}\left\{\exp\left(t\sum_{k=1}^N \xi_k\right) \ge 1\right\},$$

where $t > 0$. By Chebyshev's inequality,

$$\mathbf{P}\left\{\exp\left(t\sum_{k=1}^{N}\xi_k\right) \geq 1\right\} \leq \mathbf{E}\left[\exp\left(t\sum_{k=1}^{N}\xi_k\right)\right] = (M(t))^N, \tag{9}$$

where

$$M(t) \triangleq \mathbf{E}\left[\exp\left(t\xi_1\right)\right] \tag{10}$$

is the moment-generating function of the random variable ξ_1.

Lemma 1. *Let $M(t)$ be the function defined by (10). Let $0 < \varepsilon < \alpha^*$. Then*

$$\inf_{t>0} M(t) \leq \sqrt{1-\varepsilon^2}.$$

If $\alpha^* \leq \frac{1+\varepsilon}{2}$, then

$$\inf_{t>0} M(t) \leq 2\sqrt{(\alpha^*-\varepsilon)\alpha^*} + 1 + \varepsilon - 2\alpha^* \leq \sqrt{1-\varepsilon^2}.$$

Proof. Let us introduce the events

$$A \triangleq \{\Phi(u, X) \leq \varphi\},$$
$$B \triangleq \{\Phi(u^*, X) \leq \varphi\}.$$

Then

$$M(t) = p_+e^t + p_-e^{-t} + 1 - p_+ - p_-,$$

where

$$p_+ = \mathbf{P}(A \cap \overline{B}), \quad p_- = \mathbf{P}(\overline{A} \cap B). \tag{11}$$

Since u^* is an optimal solution to the true problem (1) and $u \in U \setminus U^{\varepsilon}_{\varphi}$, it holds that

$$\mathbf{P}(A) \leq \alpha^* - \varepsilon, \quad \mathbf{P}(B) = \alpha^*. \tag{12}$$

From (11) and (12), it follows that

$$p_+ \in [0, \alpha^* - \varepsilon], \quad p_- \in [\varepsilon, \alpha^*]. \tag{13}$$

Therefore,

$$\varepsilon \leq \mathbf{P}(B) - \mathbf{P}(A) = \mathbf{P}(\overline{A} \cap B) + \mathbf{P}(A \cap B) - (\mathbf{P}(A \cap \overline{B}) + \mathbf{P}(A \cap B)) = p_- - p_+. \tag{14}$$

The function $t \mapsto M(t)$ is convex. From the optimality conditions, we obtain

$$\operatorname{Arg}\min_{t>0} M(t) = \left\{\frac{1}{2}\ln\frac{p_-}{p_+}\right\}$$

if $p_+ > 0$. From (14), it follows that

$$\frac{1}{2}\ln\frac{p_-}{p_+} > 0.$$

Thus,

$$Q(p_+, p_-) = \inf_{t>0} M(t) = 2\sqrt{p_- p_+} + 1 - p_+ - p_-. \tag{15}$$

If $p_+ = 0$, then equality (15) holds too.

The function $(p_+, p_-) \mapsto Q(p_+, p_-)$ is concave (see, for example, [13, P. 74]). Let us find

$$\max_{p_+, p_-} Q(p_+, p_-)$$

subject to (13), (14), and (15). Since the unconditional maximum of $Q(p_+, p_-)$ is attained when $p_+ = p_-$, the conditional maximum is attained when the constraint (14) is active. Taking into account the constraint $p_+ + p_- \leq 1$, it is easy to see that

$$\inf_{t>0} M(t) \leq$$

$$\max_{p_+, p_-} \{Q(p_+, p_-) \mid p_- - p_+ = \varepsilon,\ p_+ + p_- \leq 1,\ p_+ \in [0, \alpha^* - \varepsilon],\ p_- \in [\varepsilon, \alpha^*]\}$$

$$= \max_{p_+} \left\{2\sqrt{p_+(p_+ + \varepsilon)} + 1 - 2p_+ - \varepsilon \mid p_+ + p_+ + \varepsilon \leq 1,\ p_+ \in [0, \alpha^* - \varepsilon]\right\}$$

$$= \begin{cases} 2\sqrt{\frac{1-\varepsilon}{2} \cdot \frac{1+\varepsilon}{2}} + 1 - (1-\varepsilon) - \varepsilon = \sqrt{1-\varepsilon^2} & \text{if } \alpha^* - \varepsilon > \frac{1-\varepsilon}{2}, \\ 2\sqrt{(\alpha^* - \varepsilon)\alpha^*} + 1 + \varepsilon - 2\alpha^* \leq \sqrt{1-\varepsilon^2} & \text{if } \alpha^* - \varepsilon \leq \frac{1-\varepsilon}{2}. \end{cases}$$

Thus, Lemma 1 is proved.

Let us prove a theorem on the necessary sample size to approximate the true problem (1).

Theorem 1. *Let $\beta \in (0,1)$. If the set U is finite and*

$$N \geq 2\frac{\ln|U| - \ln(1-\beta)}{|\ln(1-\varepsilon^2)|}, \tag{16}$$

then

$$\mathbf{P}\left\{\hat{U}_\varphi^{(N)} \subset U_\varphi^\varepsilon\right\} \geq \beta. \tag{17}$$

Moreover, if it is known that $\alpha^* \leq \frac{1+\varepsilon}{2}$, then inequality (17) holds if

$$N \geq \frac{\ln|U| - \ln(1-\beta)}{\left|\ln\left(2\sqrt{(\alpha^* - \varepsilon)\alpha^*} + 1 + \varepsilon - 2\alpha^*\right)\right|}. \tag{18}$$

Proof. First, let us consider the case $\alpha^* \leq \varepsilon$. Then, it is obvious that

$$\mathbf{P}\left\{\hat{U}_\varphi^{(N)} \subset U_\varphi^\varepsilon\right\} = 1,$$

hence, the assertion of the theorem is true.

If $\alpha^* > \varepsilon$ and $\alpha^* \leq \frac{1+\varepsilon}{2}$, then, from (8), (9), and Lemma 1, it follows that

$$\mathbf{P}\left\{\hat{U}_\varphi^{(N)} \not\subset U_\varphi^\varepsilon\right\} \leq \inf_{t>0} |U|(M(t))^N \leq \left(2\sqrt{(\alpha^* - \varepsilon)\alpha^*} + 1 + \varepsilon - 2\alpha^*\right)^N.$$

Thus, inequality (17) holds if

$$|U|\left(2\sqrt{(\alpha^*-\varepsilon)\alpha^*}+1+\varepsilon-2\alpha^*\right)^N \leq 1-\beta \Leftrightarrow$$
$$N \geq \frac{\ln(1-\beta)-\ln|U|}{\ln\left(2\sqrt{(\alpha^*-\varepsilon)\alpha^*}+1+\varepsilon-2\alpha^*\right)} = \frac{\ln|U|-\ln(1-\beta)}{\left|\ln\left(2\sqrt{(\alpha^*-\varepsilon)\alpha^*}+1+\varepsilon-2\alpha^*\right)\right|}.$$

Since

$$2\sqrt{(\alpha^*-\varepsilon)\alpha^*}+1+\varepsilon-2\alpha^* \leq \sqrt{1-\varepsilon^2},$$

we obtain that inequality (17) holds for

$$N \geq \frac{\ln|U|-\ln(1-\beta)}{\left|\ln\sqrt{1-\varepsilon^2}\right|} = 2\frac{\ln|U|-\ln(1-\beta)}{|\ln(1-\varepsilon^2)|}.$$

In the case when $\alpha^* > \varepsilon$ and $\alpha^* > \frac{1+\varepsilon}{2}$, the theorem is proved in the same manner. Theorem 1 is proved.

Remark 1. In [9, Theorem 5.17], a result similar to Theorem 1 is proved for the maximization of the expectation function. By applying this result to problem (1), it can be obtained that inequality (17) holds for

$$N \geq 2\frac{\ln|U|-\ln(1-\beta)}{\varepsilon^2}.$$

It is easy to check that

$$\varepsilon^2 < \left|\ln\left(1-\varepsilon^2\right)\right|$$

for $\varepsilon \in (0,1)$. Thus, the sample estimate (16) improves the result in [8] for maximization of the probability function.

Remark 2. To apply the sample estimate (18), we need to know exact solution to problem (1). However, the sample approximation is construct to estimate α^*. Sometimes, it possible to find an upper bound $\bar{\alpha} \geq \alpha^*$. If $\bar{\alpha} \leq \frac{1+\varepsilon}{2}$, then we can improve the sample estimate (16). It is guaranteed that inequality (17) holds if

$$N \geq \frac{\ln|U|-\ln(1-\beta)}{\left|\ln\left(2\sqrt{(\bar{\alpha}-\varepsilon)\bar{\alpha}}+1+\varepsilon-2\bar{\alpha}\right)\right|}.$$

4.2 Case of Bounded Set of Feasible Strategies

Let us consider the case when U is a bounded, not necessarily finite, subset of $\mathbb{R}^n$. The diameter of U is denoted by

$$D \triangleq \sup_{u,v\in U} \|u-v\|,$$

where the norm $\|u\| = \max\{|u_1|,\ldots,|u_n|\}$ is used.

We suppose that the probability function $u \mapsto P_\varphi(u)$ is Lipschitz continuous on U with a Lipschitz constant L, i.e.

$$|P_\varphi(u) - P_\varphi(v)| \leq L\|u - v\| \tag{19}$$

for all $u, v \in U$.

Theorem 2. *Let $\beta \in (0,1)$. If the assumption (19) is satisfied and*

$$N \geq 2 \inf_{\gamma \in (0,1)} \frac{n \ln \left\lceil \frac{DL}{(1-\gamma)\varepsilon} \right\rceil - \ln(1-\beta)}{|\ln(1-\gamma^2\varepsilon^2)|}, \tag{20}$$

then

$$\mathbf{P}\left\{\hat{U}_\varphi^{(N)} \subset U_\varphi^\varepsilon\right\} \geq \beta. \tag{21}$$

Proof. First, let us check that the event $\left\{\hat{U}_\varphi^{(N)} \subset U_\varphi^\varepsilon\right\} \in \mathcal{F}'$. Since the function $(u,x) \mapsto \Phi(u,x)$ is a normal integrand, the function $u \mapsto P_\varphi(u)$ is upper semi-continuous and the set U_φ^ε is compact [4]. Also, the $u \mapsto \hat{P}_\varphi^{(N)}(u)$ is upper semi-continuous for all fixed realizations of the sample. The compactness of U and semi-continuity of these function imply that the supremum

$$\sup_{u \in U} \hat{P}_\varphi^{(N)}(u)$$

is attained and is a measurable function of the sample. Thus, the considered event can be represented as

$$\begin{aligned}\left\{\hat{U}_\varphi^{(N)} \subset U_\varphi^\varepsilon\right\} &= \bigcap_{u \in U \setminus U_\varphi^\varepsilon} \left\{\hat{P}_\varphi^{(N)}(u) < \sup_{v \in U} \hat{P}_\varphi^{(N)}(v)\right\} \\ &= \bigcap_{k \in \mathbb{N}} \left\{\sup_{u \in U_k} \hat{P}_\varphi^{(N)}(u) < \sup_{u \in U} \hat{P}_\varphi^{(N)}(u)\right\},\end{aligned}$$

where

$$U_k = \left\{u \in U\colon \inf_{v \in U_\varphi^\varepsilon} \|u - v\| \geq \frac{1}{k}\right\}.$$

The set U_k is compact, so the function

$$(x_1, \dots, x_N) \mapsto \sup_{u \in U_k} \hat{P}_\varphi^{(N)}(u)$$

is measurable and, hence,

$$\left\{\hat{U}_\varphi^{(N)} \subset U_\varphi^\varepsilon\right\} \in \mathcal{F}'.$$

Let $\tilde{U}$ be a finite subset of U. Let

$$\nu \triangleq \sup_{u \in U} \inf_{v \in \tilde{U}} \|u - v\|.$$

The value of ν shows the maximal distance between an arbitrary point of U and the nearest point of $\tilde{U}$. The set $\tilde{U}$ can be selected in such a way that

$$|\tilde{U}| \leq \left\lceil \frac{D}{\nu} \right\rceil^n . \tag{22}$$

It will be assumed that condition (22) is satisfied.

Let

$$\tilde{U}_\varphi^\varepsilon \triangleq \{u \in \tilde{U} \colon P_\varphi(u) \geq \alpha_{\tilde{U}}^* - \varepsilon\},$$

where

$$\alpha_{\tilde{U}}^* \triangleq \sup_{u \in \tilde{U}} P_\varphi(u).$$

Since the function $u \mapsto P_\varphi(u)$ is Lipschitz continuous, the condition $u \in \tilde{U}_\varphi^\varepsilon$ implies $u \in U_\varphi^{\varepsilon + L\nu}$. If $\gamma \in (0,1)$ is a fixed number and $L\nu = (1-\gamma)\varepsilon$, then

$$\left\{\hat{U}_\varphi^{(N)} \subset \tilde{U}_\varphi^{\gamma\varepsilon}\right\} \subset \left\{\hat{U}_\varphi^{(N)} \subset U_\varphi^\varepsilon\right\}. \tag{23}$$

From (23) and Theorem 1, it follows that

$$\mathbf{P}\left\{\hat{U}_\varphi^{(N)} \subset U_\varphi^\varepsilon\right\} \geq \mathbf{P}\left\{\hat{U}_\varphi^{(N)} \subset \tilde{U}_\varphi^{\gamma\varepsilon}\right\} \geq \beta$$

if

$$N \geq 2\frac{\ln|\tilde{U}| - \ln(1-\beta)}{|\ln(1-\gamma^2\varepsilon^2)|} \leq 2\frac{\ln\left\lceil \frac{D}{\nu}\right\rceil^n - \ln(1-\beta)}{|\ln(1-\gamma^2\varepsilon^2)|} = 2\frac{n\ln\left\lceil \frac{DL}{(1-\gamma)\varepsilon}\right\rceil - \ln(1-\beta)}{|\ln(1-\gamma^2\varepsilon^2)|}.$$

Since γ is selected arbitrarily, the theorem is proved.

Remark 3. A similar result for minimization of the expectation function is proved in [8]. This result can be obtained from Theorem 2 if $t = 1/2$. Additional optimization in $t \in (0,1)$ can improve the result [8] for the special case of probability maximization. If the exact value of the infimum is difficult to find, then the sample estimate N can be found by substituting several values of $t \in (0,1)$ into (20). The minimal value of the obtained numbers can be set as the sample estimate N.

Remark 4. To apply the sample estimate (20), the Lipschitz constant is need to know. If the function $u \mapsto P_\varphi(u)$ is continuously differentiable on the set U, then, from the mean value theorem, it follows that

$$|P_\varphi(u) - P_\varphi(v)| \leqslant \sup_{w \in U} \|\nabla P_\varphi(w)\| \|u - v\|.$$

Therefore, we can take $L = \sup_{w \in U} \|\nabla P_\varphi(w)\|$. Methods to find the gradient of the probability function are described in [14,15].

5 Estimation of the Necessary Sample Size for Quantile Minimization

In this section, we find upper bounds on the necessary sample size to consider a solution to (6) as an approximate solution to problem (2). We assume that the set U is finite.

We suppose that the following assumption is satisfied.

Assumption 1. *The random variable $\Phi(u, X)$ is absolutely continuous for all $u \in U$ with probability density function $p_u(\cdot)$ continuous at the point $\varphi_\alpha(u)$ with its neighborhood. Also, there exists a number $C > 0$ such that*

$$\min_{u \in U} p_u(\varphi_\alpha(u)) > C.$$

As for the probability function, consider the event

$$\left\{\hat{V}_\alpha^{(N)} \not\subset V_\alpha^\varepsilon\right\} = \bigcup_{u \in U \setminus V_\alpha^\varepsilon} \bigcap_{y \in U} \left\{\hat{\varphi}_\alpha^{(N)}(u) \leqslant \hat{\varphi}_\alpha^{(N)}(y)\right\}.$$

Then we can find an upper bound for the probability

$$\mathbf{P}\left\{\hat{V}_\alpha^{(N)} \not\subset V_\alpha^\varepsilon\right\} \leq \sum_{u \in U \setminus V_\alpha^\varepsilon} \mathbf{P}\left(\bigcap_{y \in U} \left\{\hat{\varphi}_\alpha^{(N)}(u) \leqslant \hat{\varphi}_\alpha^{(N)}(y)\right\}\right). \tag{24}$$

Let us fix an optimal solution to the true problem (2) u^*. From (24), we obtain

$$\begin{aligned}\mathbf{P}\left\{\hat{V}_\alpha^{(N)} \not\subset V_\alpha^\varepsilon\right\} &\leq \sum_{u \in U \setminus V_\alpha^\varepsilon} \mathbf{P}\left\{\hat{\varphi}_\alpha^{(N)}(u) \leqslant \hat{\varphi}_\alpha^{(N)}(u^*)\right\} \\ &\leq |U| \max_{u \in U \setminus V_\alpha^\varepsilon} \mathbf{P}\left\{\hat{\varphi}_\alpha^{(N)}(u) \leqslant \hat{\varphi}_\alpha^{(N)}(u^*)\right\}. \end{aligned} \tag{25}$$

Let us define the random variables

$$\eta_N = \hat{\varphi}_\alpha^{(N)}(u^*) - \hat{\varphi}_\alpha^{(N)}(u), \quad N \in \mathbb{N}.$$

Notice that the random variables $\varphi_\alpha^{(N)}(u^*)$ and $\varphi_\alpha^{(N)}(u)$ can be dependent. We need to find an upper bound on the probability

$$\mathbf{P}\{\eta_N \geq 0\}.$$

By the Mosteller theorem [16], the distribution of the order statistics $\varphi_\alpha^{(N)}(u^*)$ and $\varphi_\alpha^{(N)}(u)$ converges to a normal distribution:

$$\sqrt{N}\left(\varphi_\alpha^{(N)}(u^*) - \varphi^*\right) \xrightarrow{d} Z_{u^*} \sim \mathcal{N}\left(0, \frac{\alpha(1-\alpha)}{p_{u^*}^2(\varphi^*)}\right),$$

$$\sqrt{N}\left(\varphi_\alpha^{(N)}(u) - \varphi_\alpha(u)\right) \xrightarrow{d} Z_u \sim \mathcal{N}\left(0, \frac{\alpha(1-\alpha)}{p_u^2(\varphi_\alpha(u))}\right).$$

Therefore, we can write

$$\lim_{N\to\infty} \mathbf{E}\eta_N \leq -\varepsilon,$$

$$\limsup_{N\to\infty} \mathbf{var}\left[\eta_N\sqrt{N}\right] < \frac{4\alpha(1-\alpha)}{C}.$$

Thus, for any $0 < \varepsilon' < \varepsilon$, there exists $\tilde{N}(\varepsilon') \in \mathbb{N}$ such that

$$\mathbf{E}\eta_N \leq -\varepsilon + \varepsilon', \tag{26}$$

$$\mathbf{var}\left[\eta_N\right] < \frac{4\alpha(1-\alpha)}{CN} \tag{27}$$

for all $N > \tilde{N}(\varepsilon')$. We would like to notice that $\tilde{N}(\varepsilon')$ can depend on u. So, the maximal value of $\tilde{N}(\varepsilon')$ should be taken.

By Cantelli's inequality, for $N > \tilde{N}(\varepsilon')$,

$$\mathbf{P}\{\eta_N \geq 0\} \leq \mathbf{P}\{\eta_N - \mathbf{E}\eta_N \geq \varepsilon - \varepsilon'\} \leq \frac{\mathbf{var}\left[\eta_N\right]}{\mathbf{var}\left[\eta_N\right] + (\varepsilon-\varepsilon')^2}$$
$$< \frac{4\alpha(1-\alpha)}{4\alpha(1-\alpha) + (\varepsilon-\varepsilon')^2 CN}.$$

Thus, from (25), the theorem follows.

Theorem 3. *Let U be a finite set, $\beta \in (0,1)$. Assumption 1 is supposed to be satisfied. Then*

$$\mathbf{P}\left\{\hat{V}_\alpha^{(N)} \not\subset V_\alpha^\varepsilon\right\} \leq |U|\frac{4\alpha(1-\alpha)}{4\alpha(1-\alpha) + (\varepsilon-\varepsilon')^2 CN}. \tag{28}$$

for sufficiently large N.

Now, we can obtain the corollary from Theorem 3.

Corollary 1. *Let assumptions of Theorem 3 be satisfied. Then*

$$\mathbf{P}\left\{\hat{V}_\alpha^{(N)} \subset V_\alpha^\varepsilon\right\} \geq \beta$$

if

$$N \geq \frac{4\alpha(1-\alpha)(|U| + \beta - 1)}{(1-\beta)(\varepsilon-\varepsilon')^2 C} \tag{29}$$

and $N > \tilde{N}(\varepsilon')$.

Proof. From (28), it follows that the assertion of the theorem is true if

$$|U|\frac{4\alpha(1-\alpha)}{4\alpha(1-\alpha) + (\varepsilon-\varepsilon')^2 CN} \leq 1 - \beta.$$

By solving this inequality, we obtain (29). The corollary is proved.

Remark 5. Unfortunately, it is difficult find bounds on the value $\tilde{N}(\varepsilon')$. To use the estimate (29), inequalities (26) and (27) should be checked. It can be made by statistical methods.

6 Conclusion

In the paper, sample size estimates for approximation of stochastic optimization problems with probabilistic and quantile objective functions are obtained. These estimates are quite rough for practical use, but they allow us to judge the complexity of the solution to the original problem. In future research, it is planned to improve this result for special cases of stochastic optimization problems. We hope that it is possible to describe a class of problems for which exponential bounds can be obtained instead of (28). Also, a more general case of quantile minimization problem on a bounded set should be studied.

Acknowledgements. The work is supported by the Russian Foundation for Basic Research (project 19-07-00436).

References

1. Kibzun, A.I., Kan, Y.S.: Stochastic Programming Problems with Probability and Quantile Functions. Wiley, Chichester, New York, Brisbane, Toronto, Singapore (1996)
2. Artstein, Z., Wets, R.J.-B.: Consistency of minimizers and the SLLN for stochastic programs. J. Convex Anal. **2**, 1–17 (1996)
3. Pagnoncelli, B.K., Ahmed, S., Shapiro, A.: Sample average approximation method for chance constrained programming: theory and applications. J. Optim. Theory Appl. **142**, 399–416 (2009). https://doi.org/10.1007/s10957-009-9523-6
4. Kibzun, A.I., Ivanov, S.V.: On the convergence of sample approximations for stochastic programming problems with probabilistic criteria. Automat. Remote Control. **79**(2), 216–228 (2018). https://doi.org/10.1134/S0005117918020029
5. Hess, C., Seri, R.: Generic consistency for approximate stochastic programming and statistical problems. SIAM J. Optim. **29**(1), 290–317 (2019). https://doi.org/10.1137/17M1156769
6. Guigues, V., Juditsky, A., Nemirovski, A.: Non-asymptotic confidence bounds for the optimal value of a stochastic program. Optim. Method. Softw. **32**(5), 1033–1058 (2017). https://doi.org/10.1080/10556788.2017.1350177
7. Kibzun, A.I., Naumov, A.V., Norkin, V.I.: On reducing a quantile optimization problem with discrete distribution to a mixed integer programming problem. Automat. Remote Control. **74**(6), 951–967 (2013). https://doi.org/10.1134/S0005117913060064
8. Shapiro, A.: Monte Carlo sampling methods. In: Ruszczyński, A., Shapiro, A. (eds.) Handbooks in OR Handbooks in Operations Research and Management Science & MS, North-Holland, Dordrecht, The Netherlands, vol. 10, pp. 353–425 (2003). https://doi.org/10.1016/S0927-0507(03)10006-0
9. Shapiro, A., Dentcheva, D., Ruszczyński, A.: Lectures on Stochastic Programming. Society for Industrial and Applied Mathematics, Philadelphia (2009)
10. Kleywegt, A.J., Shapiro, A., Homem-De-Mello, T.: The sample average approximation method for stochastic discrete optimization. SIAM J. Optim. **12**(2), 479–502 (2001). https://doi.org/10.1137/S1052623499363220
11. Luedtke, J., Ahmed, S.: A sample approximation approach for optimization with probabilistic constraints. SIAM J. Optim. **19**(2), 674–699 (2008). https://doi.org/10.1137/070702928

12. Rockafellar, R.T., Wets, R.J.-B.: Variational Analysis. Springer, Heidelberg (2009). https://doi.org/10.1007/978-3-642-02431-3
13. Boyd, S., Vandenberghe, L.: Convex Optimization. Cambridge University Press, New York (2004)
14. Kibzun, A., Uryasev, S.: Differentiability of probability function. Stochast. Anal. Appl. **16**(6), 1101–1128 (1998). https://doi.org/10.1080/07362999808809581
15. Van Ackooij, W., Henrion, R.: Gradient formulae for nonlinear probabilistic constraints with Gaussian and Gaussian-like distributions. SIAM J. Optim. **24**(4), 1864–1889 (2014). https://doi.org/10.1137/130922689
16. Mosteller, F.: One some useful inefficient statistics. Ann. Math. Stat. **17**, 317–408 (1946)

Approximation Algorithms for Piercing Special Families of Hippodromes: An Extended Abstract

Konstantin Kobylkin[1,2](✉) and Irina Dryakhlova[2]

[1] Krasovsky Institute of Mathematics and Mechanics, Ural Branch of RAS, Sophya Kovalevskaya str. 16, 620108 Ekaterinburg, Russia
kobylkinks@gmail.com

[2] Ural Federal University, Mira str. 19, 620002 Ekaterinburg, Russia
iadryahlova@gmail.com

Abstract. Polynomial time approximation algorithms are proposed with constant approximation factors for a problem of computing the smallest cardinality set of identical disks whose union intersects each segment from a given set E of n straight line segments on the plane. This problem has important applications in operations research, namely in wireless and road network analysis. It is equivalent to finding the least cardinality piercing (or hitting) set for the corresponding family of n Euclidean r-neighborhoods of straight line segments of E on the plane, which are called r-hippodromes in the literature. When the number of distinct orientations is upper bounded by k of segments from E, a simple $O(n \log n)$-time $4k$-approximate algorithm is known for this problem. Besides, when E contains arbitrary straight line segments, overlapping at most at their endpoints, $O(n^4 \log n)$-time 100-approximate algorithm is given recently. In the present paper, simple approximation algorithms are proposed with small approximation factors for E, being edge set of some special plane graphs of interest in road network applications; here the number of distinct orientations of straight line segments from E can be arbitrarily large. More precisely, $O(n^2)$-time approximation algorithms are constructed for edge sets of either Gabriel or relative neighborhood graphs or of Euclidean minimum spanning trees with factors of 14, 12 and 10 respectively. These algorithms are much faster, more accurate and conceptually much simpler than the aforementioned 100-approximate algorithm for the general case of the problem on edge sets of arbitrary plane graphs.

Keywords: Operations research · Computational geometry · Approximation algorithms · Geometric piercing problem · Straight line segment · Hippodrome · Proximity graph

1 Introduction

Placement of geometric objects (or facilities) on the plane is a widely studied problem at the intersection of computational geometry and operations research,

M. Khachay et al. (Eds.): MOTOR 2019, LNCS 11548, pp. 565–580, 2019.
https://doi.org/10.1007/978-3-030-22629-9_40

known as the facility location problem in the literature. In its classical setting, an arbitrary fixed number of objects is to be placed, say, production plants, inventories or markets such that the average (or the maximum) distance from clients (e.g. from customers or from roads) to their nearest object is the minimum possible. Here feasible locations of objects to place are given by a set F of simply shaped geometric objects like disks or rectangles on the plane whereas clients are given by a set K, which contains points or straight line segments. In the more involved problem setting the sum is minimized of transportation costs and placement fees. There is also a subclass of facility location problems called set covering problems in which one needs to either cover or intersect segments from K with the least cardinality subset of objects from F. Such settings might model placement of markets at the vicinity of potential customers and road monitoring with sensors, having bounded sensing area. In this paper, small guaranteed constant factor approximation algorithms are designed for the following facility location problem from the class of set covering problems:

Intersecting Plane Graph with Disks (IPGD): given a fixed constant $r > 0$ and a straight line embedding $G = (V, E)$ of a simple planar graph with n straight line edges, which are allowed to intersect at most at their endpoints, find the smallest cardinality subset $H \subset \mathbb{R}^2$ of points (or disk centers) such that each $e \in E$ is within Euclidean distance r from some point $x = x(e) \in H$; equivalently, a disk of radius r centered at x intersects e. Each isolated vertex $v \in V$ is considered a zero-length segment $e_v \in E$. Moreover, the vertex set V is assumed to be in general position.

Below the term "plane graph" is used to denote any straight line embedding of a planar graph whose (straight line) edges intersect at most at their endpoints.

The IPGD problem is a special case of the well-known geometric piercing (or, more generally, hitting set) problem on the plane. In its general setting the latter problem is formulated as follows: one is to find the least cardinality subset $H \subset \mathbb{R}^2$ such that $H \cap R \neq \varnothing$ for every $R \in \mathcal{R}$, where $\mathcal{R}$ is a given family of subsets of $\mathbb{R}^2$ also called *objects*. Below a set $H \subset \mathbb{R}^2$ is called a *piercing* set for $\mathcal{R}$ when $H \cap R \neq \varnothing$ for every $R \in \mathcal{R}$.

Obviously, the IPGD problem is equivalent to the piercing problem for the family $\mathcal{N}_r(E) = \{N_r(e) : e \in E\}$, where $N_r(e) = \{x \in \mathbb{R}^2 : d(x, e) \leq r\}$ and $d(x, e)$ is Euclidean distance between a point $x \in \mathbb{R}^2$ and a segment $e \in E$. Objects from $\mathcal{N}_r(E)$ are called hippodromes or, more precisely, r-hippodromes in the literature.

Design of approximation algorithms for the IPGD problem above finds its applications in operations research, more specifically, in optimal sensor placement e.g. for road monitoring. Suppose a road network is to be monitored using identical sensors with a circular sensing area. Geometrically, its roads can be modeled by piecewise linear arcs on the plane. One can split these arcs into chains of elementary straight line segments such that any two of the resulting elementary segments intersect at most at their endpoints. When full road network surveillance is costly, it may be a good approach to place the minimum number of sensors such that each piece of every road (represented by an elementary segment) is partially covered by sensing area of some of the placed sensors.

This approach leads to a geometric combinatorial optimization model, which coincides with the IPGD problem.

1.1 Basic Definitions and Notations

In this paper a polynomial time and space algorithm (denote it by $\mathcal{A}$) for the IPGD problem is called *f-approximate* (or *having an approximation factor of f*), if for any constant $r > 0$ and any plane graph $G = (V, E)$ from a given graph class the following inequality holds true uniformly within that class:

$$\frac{|H_{\mathcal{A}}(G, r)|}{\mathrm{OPT}(G, r)} \leq f,$$

where $H_{\mathcal{A}}(G, r)$ is a feasible solution to the IPGD problem for the graph G and the radius r, output by $\mathcal{A}$, and $\mathrm{OPT} = \mathrm{OPT}(G, r)$ is the optimum of the IPGD problem for G and r. Below f is assumed to be some absolute constant, thus, being r-independent.

Let α and β be computable functions with positive integer argument m. Below the standard notation $\alpha(m) = O(\beta(m))$ is used, reporting that there is a constant $D > 0$ such that $|\alpha(m)| \leq D|\beta(m)|$ for any sufficiently large m.

1.2 Our Results and Related Work

The IPGD problem generalizes a classical problem of covering a given n-point set E on the plane with the minimum number of identical disks on the case of sets of non-zero length segments. Therefore the IPGD problem is NP-hard [5,8]. Apparently, [5] is the first work to tackle a close problem in which segments of E generally overlap by their relative interiors and are restricted to have their orientations parallel to any of two coordinate axes. A simple 8-approximate algorithm is built for this latter problem, working in $O(n \log n)$ time and $O(n \log n)$ space. This algorithm can be easily extended to $4k$-approximate algorithm for the case of at most k distinct orientations of segments from E.

In the most general case, allowing segments to overlap by their relative interiors and admitting segment sets with arbitrarily large number of distinct orientations, it is unlikely [2] that an $O(1)$-approximate algorithm exists, at least, based on the known algorithmic paradigms. However, when segments from E are allowed to intersect at most at their endpoints (which coincides with the general IPGD problem setting), 100-approximate algorithm [9] is known which takes $O(n^4 \log n)$ time and $O(n^2 \log n)$ space. It follows from results of the earlier paper [14] that an $O(1)$-approximate algorithm exists for the IPGD problem, though with huge constant upper bound on its approximation factor.

Besides, one can assume (without loss of generality) that $\mathcal{N}_r(E)$ is a set of closed convex pseudo-disks, i.e. $|\mathrm{bd}\, N_1 \cap \mathrm{bd}\, N_2| \leq 2$ for any distinct $N_1, N_2 \in \mathcal{N}_r(E)$ and both $N_1 \backslash N_2$ and $N_2 \backslash N_1$ are connected, where $\mathrm{bd}\, N$ denotes boundary of a set $N \subset \mathbb{R}^2$. Indeed, as straight line segments from E intersect at most at their endpoints, segments of E can be slightly shifted to become pairwise disjoint and non-parallel while keeping all nonempty intersections of subsets of objects from $\mathcal{N}_r(E)$ with some slightly larger r. For two non-overlapping segments e and e' it can be understood that $|\mathrm{bd}\, N_r(e) \cap \mathrm{bd}\, N_r(e')| \leq 2$ because Euclidean distance grows strictly monotonically from e (or from e') to a point of the curve $\chi(e,e') = \{x \in \mathbb{R}^2 : d(x,e) = d(x,e')\}$[1] as that point moves along $\chi(e,e')$ in any of two opposite directions starting from midpoint of the segment which joins closest points of e and e'; here also $\mathrm{bd}\, N_r(e) \cap \mathrm{bd}\, N_r(e') \subset \chi(e,e')$. This implies that $N_r(e) \backslash N_r(e')$ is connected.

It follows from the aforementioned assumption that the IPGD problem admits PTAS [13] (see also [12]) of large time complexity. Moreover, this assumption gives that a 4-approximate local search based algorithm can be constructed [3] for the IPGD problem taking $O(n^{18})$ time.

Thus, for the general setting of the IPGD problem with arbitrary straight line segments, overlapping at most at their endpoints, there is an ugly tradeoff between accuracy estimates and time complexity of the known approximation algorithms. Namely, achieving close to 1 approximation factor (which results in high accuracy of produced feasible solutions) requires high computational cost whereas only loose approximation factor can be guaranteed by known algorithms with modest time complexity. This situation is very typical in geometric piercing problems for sets of objects of more or less sophisticated shape.

Under those circumstances, apparently the best one can hope is to provably approximate the IPGD problem well in some special cases, arising in various applications. In the present paper approximation algorithms are proposed combining both high accuracy and low complexity for the IPGD problem considered on special configurations of straight line segments forming edge sets of special graphs, called proximity graphs in the literature. These algorithms are conceptually much simpler than the algorithm from [9], designed for the general setting of the IPGD problem.

Let us give some definitions. Let V be a finite point set in general position on the plane. Assuming that no 4 points of V are cocircular, a plane graph $G = (V, E)$ is called a *Gabriel* graph [11] when $[u, v] \in E$ iff intersection of V is empty with interior of the disk with diameter $[u, v]$. Under the same assumption a plane graph $G = (V, E)$ is called a *relative neighborhood* graph [7] when $[u, v] \in E$ iff $\max\{d(u, w), d(v, w)\} \geq d(u, v)$ for any $w \in V \backslash \{u, v\}$.

A plane graph $G = (V, E)$ is called *Euclidean minimum spanning tree* if G coincides with the smallest weight spanning tree for the complete weighted graph whose vertices are at points of V whereas edge weights are given by Euclidean distances between edge endpoints. It can be shown that each Gabriel graph on V contains some relative neighborhood graph on V, which itself contains some Euclidean minimum spanning tree on the set V.

All types of proximity graphs defined above appear in a variety of network applications. For example, they represent convenient network topologies, simplifying routing and control in geographical (e.g. wireless) networks.

---

[1] The curve $\chi(e, e')$ is composed of pieces of straight lines and parabolas.

In [8] for $r > 0$ NP-hardness is proved of the IPGD problem for any of these three classes of proximity graphs. In the present paper, extending work [10], a 14-approximate algorithm is proposed for the IPGD problem considered on the class of Gabriel graphs, 12-approximate algorithm—for relative neighborhood graphs and 10- approximate algorithm—for Euclidean minimum spanning trees. These algorithms have identical order $O(n\text{OPT})$ of time complexity and $O(n)$ space cost. Furthermore, our algorithm for Gabriel graphs outperforms the algorithm from [9] designed for this type of graphs both with respect to its time and space cost.

2 Main Ideas of the Proposed Algorithms

Ideas beneath our $O(1)$-approximate algorithms can be summarized in the following algorithm and two underlying definitions.

Definition 1. *A subset $\mathcal{I} \subseteq \mathcal{N}_r(E)$ is called a maximal (with respect to inclusion) independent set in $\mathcal{N}_r(E)$, if $I \cap I' = \varnothing$ for any $I, I' \in \mathcal{I}$, and for any $N \in \mathcal{N}_r(E)$ there is some $I \in \mathcal{I}$ with $N \cap I \neq \varnothing$.*

Definition 2. *Let $G = (V, E)$ be a plane graph and $C > 0$ be some absolute constant. An edge $e \in E$ is called C-coverable with respect to E, if for any constant $\rho > 0$ one can construct at most C-point piercing set $U(e, E) \subset \mathbb{R}^2$ for $\mathcal{N}_e(E) = \{N \in \mathcal{N}_\rho(E) : N \cap N_\rho(e) \neq \varnothing\}$ in polynomial time with respect to $|\mathcal{N}_e(E)|$.*

Consider the following basic algorithm:

COVERING EDGES OF SPECIAL PLANE GRAPHS WITH EQUAL DISKS.

Input: a constant $r > 0$ and a plane graph $G = (V, E)$;
Output: an approximate solution H of the IPGD problem for G and r.

---

1. $E' := \varnothing, E_0 := E$ and $H := \varnothing$;
2. while $E_0 \neq \varnothing$, repeat steps 3-4:
3. choose an arbitrary C-coverable edge $e^* \in E_0$ (with respect to E_0) and construct a piercing set $U(e^*, E_0)$ of at most C points for $\mathcal{N}_{e^*}(E_0)$, applying some auxiliary procedure;
4. set $E_0 := E_0 \backslash \{e \in E_0 : N_r(e) \cap U(e^*, E_0) \neq \varnothing\}$, $E' := E' \cup \{e^*\}$ and $H := H \cup U(e^*, E_0)$;
5. return H.

To give a comprehensive description of work of the algorithm above for some class $\mathcal{G}$ of plane graphs, one should specify an algorithmic way to choose C-coverable edge $e^* \in E_0$ for any subset $E_0 \subseteq E$ of edge set E of any graph from $\mathcal{G}$. Besides, one should implement an auxiliary procedure, which, given an arbitrary $E_0 \subseteq E$ and a C-coverable edge $e \in E_0$, seeks a piercing set $U(e, E_0)$ for $\mathcal{N}_e(E_0)$ of size at most C. Here it is assumed that such a procedure accepts an edge $e \in E_0$, a constant $r > 0$ and a family $\mathcal{N}_e(E_0)$ as its input.

In the sequel notation $\mathcal{N}_e(E_0)$ is only used when one needs to emphasize that $\mathcal{N}_e(E_0)$ contains those ρ-hippodromes, $N_\rho(e)$, whose underlying segments are from some subset $E_0 \subseteq E$, which is generally not equal to E. If $E_0 = E$, a simpler notation $\mathcal{N}_e$ is applied.

In the statement below a sufficient condition is given under which the Covering edges of special plane graphs with equal disks algorithm is C-approximate.

Statement 1. *Let $C > 0$ be some constant and $\mathcal{G}$ be a class of plane graphs such that for any $G = (V, E) \in \mathcal{G}$ and any $E_0 \subseteq E$ an edge $e^* \in E_0$ can be found in (polynomial) $O(\varphi(|E_0|))$ time and linear space cost, which is C-coverable with respect to E_0. Let $O(\xi(|\mathcal{N}_{e^*}(E_0)|))$ also be time complexity of an auxiliary procedure with linear space cost, which seeks a piercing set for $\mathcal{N}_{e^*}(E_0)$ of size at most C for any $\rho > 0$. Then the* Covering edges of special plane graphs with equal disks *algorithm is C-approximate for the* IPGD *problem within the class $\mathcal{G}$. Its time complexity is of the order $O((n + \varphi(n) + \xi(n))\mathrm{OPT}(G, r))$ whereas its space cost is $O(n)$ for any $G \in \mathcal{G}$ and $r > 0$, where $n = |E|$.*

Proof. As the set $\mathcal{N}_r(E')$ becomes maximal independent in $\mathcal{N}_r(E)$ to the step 5 of the Covering edges of special plane graphs with equal disks algorithm, $|E'| \leq \mathrm{OPT}$ and $|H| \leq C|E'| \leq C\mathrm{OPT}$. Trivial arguments lead to claimed algorithm complexity bounds. □

To build $O(1)$-approximate algorithms for the IPGD problem, relying on the Covering edges of special plane graphs with equal disks algorithm, one should implement efficient search of C-coverable edges. Below a sufficient condition is established for edge C-coverability. Given any $\mathcal{N} \subseteq \mathcal{N}_r(E)$ let $\mathcal{E}(\mathcal{N}) \subseteq E$ be such that $\mathcal{N}_r(\mathcal{E}(\mathcal{N})) = \mathcal{N}$.

Statement 2 *Let $G = (V, E)$ be a plane graph and $C > 0$ be some absolute constant. Then an edge $e^* \in E$ is C-coverable (with respect to E), if there is an auxiliary procedure, which for any $\rho > 0$ efficiently seeks a point set $U(e^*, E) \subset \mathbb{R}^2$ of size at most C such that*

$$e \cap \left(N_{2\rho}(e^*) \cap \bigcup_{u \in U(e^*, E)} N_\rho(u) \right) \neq \varnothing$$

for any $e \in \mathcal{E}(\mathcal{N}_{e^*}(E))$.

The statement proof relies on the fact that $e \cap N_{2\rho}(e^*) \neq \varnothing$ for any $e \in \mathcal{E}(\mathcal{N}_{e^*}(E))$.

Without additional restrictions either on mutual location of edges and vertices of plane graphs in some more or less non-trivial graph class $\mathcal{G}$, or, perhaps, on either orientation or length of these edges, it is quite involved task to prove that the sufficient condition above holds true within $\mathcal{G}$ for some constant C.

An Idea to Implement Step 3 of the Basic Algorithm. For Euclidean minimum spanning trees, Gabriel and relative neighborhood graphs, it follows from their definitions that vertices and edges are located on the plane in some very specific way of graphs of either type. More precisely, for each of the aforementioned three graph types the following property holds true: for any subset $E_0 \subseteq E$ of edge set E of any graph of an arbitrary type one can find such an edge $e^* \in E_0$ that $M(e^*) \cap V_0 = \varnothing$, where V_0 is the endpoint set for segments from E_0, orientation and location of object $M(e^*)$ is defined by e^* whereas shape of $M(e^*)$ is defined by the graph type.

For any of the aforementioned three graph types, due to specifics of shape and mutual location of sets $M(e^*)$ and $N_{2r}(e^*)$, it can be guaranteed that (see subsequent sections) each segment from $\mathcal{E}(\mathcal{N}_{e^*}(E_0))$ must intersect $N_{2r}(e^*)\backslash M(e^*)$. Due to the Statement 2, applied for $\rho := r$, in order to get a piercing set for $\mathcal{N}_{e^*}(E_0)$ it is sufficient to compute a point set $U(e^*, E_0)$ of small constant size such that

$$N_{2r}(e^*)\backslash M(e^*) \subset \bigcup_{u \in U(e^*, E_0)} N_r(u).$$

Thus, to build a piercing set for $\mathcal{N}_{e^*}(E_0)$ one can compute a "good" partial cover of $N_{2r}(e^*)$ with small number of radius r disks, which does not depend on the location of segments from $\mathcal{E}(\mathcal{N}_{e^*}(E_0))$ with respect to e^*. In the next three sections, procedures are described for each of the aforementioned three graph types to construct such partial covers.

Finally, one can design an approximation algorithm for those graphs by applying the COVERING EDGES OF SPECIAL PLANE GRAPHS WITH EQUAL DISKS algorithm, evoking search at its step 3 of a cover of complement of $2r$-hippodrome of some special edge $e^* \in E_0$ to the set $M(e^*)$ with at most C radius r disks for a graph type specific constant C. Moreover, one applies a graph type specific heuristic at this step to choose C-coverable edge e^*, trying to achieve the smallest possible constant C.

This straightforward approach is conceptually simpler than the approach of work [9] tackling the general setting of the IPGD problem, which relies on tangled machinery of epsilon nets [1]. From one hand, it gives a significant cut in upper bounds for guaranteed approximation factor, being compared with direct application of the epsilon net based approach. From the other hand, our approach has some drawbacks as size of the built cover of $N_{2r}(e^*)\backslash M(e^*)$ does not depend on $|\mathcal{E}(\mathcal{N}_{e^*}(E_0))|$ and mutual location of segments from $\mathcal{E}(\mathcal{N}_{e^*}(E_0))$ with respect to e^*. This may result in larger piercing set size than that obtained e.g. after taking a single point from each object of $N_{e^*}(E_0)$.

Review of Related Approaches. Several examples are known in the literature of designing $O(1)$-approximate algorithms for close problem settings, which rely on similar approaches. In distinction to the IPGD problem in those settings segments are allowed to intersect by their relative interiors. Moreover, some restrictions are also imposed therein on either orientations or lengths of segments from E. For example, for identically oriented segments, say, parallel to x-axis, one can design a 4-approximate algorithm [5], which is similar to the COVERING

EDGES OF SPECIAL PLANE GRAPHS WITH EQUAL DISKS algorithm: segments from E are sorted with respect to x-coordinates of their right endpoints, the segment $e^* \in E_0$ is chosen with the least x-coordinate of its right endpoint; segments from $\mathcal{E}(\mathcal{N}_{e^*}(E_0))$ must intersect the half-disk of radius $2r$ centered at that endpoint; this half-disk can be covered by at most 4 radius r disks. In [4] the same idea is used in the case, where E consists of zero-length segments.

As follows from [6], the COVERING EDGES OF SPECIAL PLANE GRAPHS WITH EQUAL DISKS algorithm is C-approximate in the class $\mathcal{G}_\lambda$, which consists of those plane graphs $G = (V, E)$ such that each edge $e \in E$ is of Euclidean length of at most λr, where $\lambda > 0$ depends only on the class $\mathcal{G}_\lambda$ and not on G. Here one has $C = O(\lambda)$.

3 12-Approximate Algorithm for Relative Neighborhood Graphs

In this section, an $O(1)$-approximate algorithm is designed for the IPGD problem in the class of relative neighborhood graphs, which is based on applying the COVERING EDGES OF SPECIAL PLANE GRAPHS WITH EQUAL DISKS algorithm. The idea from the previous section is used to implement its step 3, relying on computing a cover of $N_{2r}(e)\backslash M(e)$ by a few radius r disks for an arbitrarily chosen edge $e \in E_0$ and some special set $M(e)$. The following characteristic property of relative neighborhood graphs is formulated below, which defines a shape of $M(e)$ specific to graphs of this type.

Observation 1. Let $G = (V, E)$ be a relative neighborhood graph, $e = [u_1, u_2] \in E$ and $M_{\text{RNG}}(e) = \text{int}\,(N_{2\Delta}(u_1) \cap N_{2\Delta}(u_2))$, where $\Delta = d(u_1, u_2)/2 > 0$ and $\text{int}\, N$ denotes interior of set $N \subseteq \mathbb{R}^2$. Then $M_{\text{RNG}}(e) \cap V = \varnothing$.

The lemma below provides a sufficient condition for edge C-coverability, which is specific to relative neighborhood graphs. It is weaker than that from the Statement 2, but follows the aforementioned idea by describing "good" partial covers with radius r disks of $2r$-hippodromes, induced by edges of relative neighborhood graphs.

Lemma 1. *Let $G = (V, E)$ be a relative neighborhood graph and $e \in E$. Then the edge e is C-coverable (with respect to E), if for any $\rho > 0$ a point set $U(e) \subset \mathbb{R}^2$ can be found in polynomial time of size at most C such that*

$$N_{2\rho}(e)\backslash M_{\text{RNG}}(e) \subset \bigcup_{u \in U(e)} N_\rho(u). \quad (1)$$

The lemma proof is based on the fact that $g \cap N_{2\rho}(e) \neq \varnothing$ for any $g \in \mathcal{E}(\mathcal{N}_e)$. Moreover, it strongly relies on the assumption that relative neighborhood graphs are plane.

Thus, to describe work of the COVERING EDGES OF SPECIAL PLANE GRAPHS WITH EQUAL DISKS algorithm in the class of relative neighborhood graphs, it is sufficient to formulate a procedure, which, given an arbitrary edge $e \in E$ of any graph $G = (V, E)$ of this type, seeks a point set $U(e)$ of size at most C such that (1) is hold for e and $U(e)$ with $\rho = r$. Below a procedure is given, seeking such a point set $U(e)$ with $C = 12$.

PARTIAL r-DISK COVER SEARCH FOR $2r$-HIPPODROMES ON RNG EDGES

Input: a constant $r > 0$ and an edge $e = [u_1, u_2]$ of an arbitrary relative neighborhood graph;
Output: $U_{\text{RNG}}(e) \subset \mathbb{R}^2$ such that (1) is hold for e and $U(e) = U_{\text{RNG}}(e)$ with $\rho = r$;

---

1. for each $s \in \{1, 2\}$ construct a regular hexagon inscribed in $N_{2r}(u_s)$, whose orientation is such that the straight line through e contains a pair of vertices of that hexagon; form a 7-point set V_s, which contains u_s and midpoints of the hexagon sides (of length $2r$);
2. for each $s \in \{1, 2\}$ choose a subset $U_s \subset V_s$ with $|U_s| = 5$ and $T_s \subset \bigcup_{u \in U_s} N_r(u)$, where $N_{2r}(e) = T_1 \cup T_2 \cup R$ for some rectangle R and two closed half-disks T_1 and T_2 of radii $2r$ centered at u_1 and u_2 respectively;
3. if either $\Delta \geq \frac{(2\sqrt{3}-1)r}{2\sqrt{4\sqrt{3}-6}}$ or $\Delta \in (0, r/2]$, return $U_{\text{RNG}}(e) := U_1 \cup U_2$;
4. for $\Delta \in \left(r/2, \frac{(2\sqrt{3}-1)r}{2\sqrt{4\sqrt{3}-6}}\right)$ choose those points $v_{s1}, v_{s2} \in U_s$, which are symmetric with respect to u_s, $s = 1, 2$; set $a_i = \frac{v_{1i}+v_{2i}}{2}$, where v_{1i} and v_{2i} lie on the same side with respect to the straight line through e, $i = 1, 2$;
5. return $U_{\text{RNG}}(e) := U_1 \cup U_2 \cup \{a_1, a_2\}$.

The following lemma reports on the efficiency of the procedure above.

Lemma 2. *Let e be an edge in the input of the* PARTIAL r-DISK COVER SEARCH FOR $2r$-HIPPODROMES ON RNG EDGES *procedure, whereas $U_{\text{RNG}}(e)$ is its output. Then the inclusion (1) holds for e and $U_{\text{RNG}}(e)$, where $|U_{\text{RNG}}(e)| \leq 12$.*

Proof. Below a proof sketch is provided.

To prove (1) for e and $U_{\text{RNG}}(e)$ consider segments $f_1(e)$ and $f_2(e)$, which are parallel and equal to e, such that

$$\text{bd}\, N_{2r}(e) = S_1 \cup S_2 \cup f_1(e) \cup f_2(e),$$

where S_s is the radius $2r$ half-circle on the boundary of T_s and $f_i(e)$ lies at the same side with respect to the straight line through e as the segment $[v_{1i}, v_{2i}]$ does. Obviously, (1) holds if $f_i(e) \cap M_{\text{RNG}}(e) \cap N_r(v_{si}) \neq \varnothing$ for every $s = 1, 2$. The latter situation occurs when $\Delta \geq \frac{(2\sqrt{3}-1)r}{2\sqrt{4\sqrt{3}-6}}$.

In the case $\Delta \in (0, r/2]$ one gets the inclusion

$$N_{2r}(e) \subset \bigcup_{u \in U_{\text{RNG}}(e)} N_r(u). \tag{2}$$

For the case $\Delta \in \left(\frac{r}{2}, \frac{(2\sqrt{3}-1)r}{2\sqrt{4\sqrt{3}-6}}\right)$ let $u_{si} = f_i(e) \cap \operatorname{bd} N_r(v_{si})$. In the subcase $\Delta \in \left[r, \frac{(2\sqrt{3}-1)r}{2\sqrt{4\sqrt{3}-6}}\right)$ (1) holds if $u_{si}, c_{si} \in N_r(a_i)$, where $c_{si} = R \cap \operatorname{bd} N_r(v_{si}) \cap \operatorname{bd} N_{2r}(u_s)$, $i, s = 1, 2$. The latter can be proved with a few lines of algebra.

For the subcase $\Delta \in (r/2, r)$ let $c'_{si} = R \cap \operatorname{bd} N_r(v_{si}) \cap \operatorname{bd} N_r(u_s)$. Here it is enough to prove the inclusion $u_{si}, c'_{si} \in N_r(a_i)$ for $i, s = 1, 2$, to establish (1). □

Taking Lemmas 1, 2 and the Statement 1 into account, one can give the following approximation algorithm for the IPGD problem in the class of relative neighborhood graphs.

Theorem 3. *If one chooses an arbitrary edge $e \in E_0$ at the step 3 of the* Covering edges of special plane graphs with equal disks *algorithm and performs the* Partial r-disk cover search for $2r$-hippodromes on RNG edges *procedure for that edge, this algorithm becomes 12-approximate for the* IPGD *problem in the class of relative neighborhood graphs with $O(n\,\mathrm{OPT})$ time complexity and $O(n)$ space cost, where $n = |E|$.*

4 10-Approximate Algorithm for Euclidean Minimum Spanning Trees

As every Euclidean minimum spanning tree on any point set V in general position is a subgraph of some relative neighborhood graph on V, the 12-approximate algorithm is applicable from the previous section for the IPGD problem in the class of such trees. In this section, it is shown that there is a 10-approximate algorithm for the IPGD problem in the class of Euclidean minimum spanning trees for any $r > 0$. It is curious that for $r = 0$ the IPGD problem coincides with the classical Vertex Cover problem, which is polynomially solvable in the class of arbitrary trees.

Our 10-approximate algorithm uses the Covering edges of special plane graphs with equal disks algorithm. It also employs the aforementioned idea of building a cover of size at most 10 at its step 3 of the complement of $2r$-hippodrome of some edge e to a special set $M_{\mathrm{EMST}}(e)$, whose shape is defined in the Lemma 3 below. Here a smaller value of the coverability parameter C is gained due to the fact that $M_{\mathrm{RNG}}(e) \subset M_{\mathrm{EMST}}(e)$.

The observation below follows from definition of Euclidean minimum spanning trees.

Observation 2. Let $G = (V, E)$ be Euclidean minimum spanning tree with a root $v_0 \in V$, $\mathrm{depth}(u) = \mathrm{depth}(u|v_0, G)$ be the (graph-theoretic) distance in G from v_0 to an arbitrary $u \in V$ and $V(u|v_0)$ be the subset of those vertices $w \in V$ such that the shortest path in G from w to v_0 (with respect to the number of its edges) passes through u. If an edge $e = [u_1, u_2] \in E$ is such that $\mathrm{depth}(u_1) = \mathrm{depth}(u_2) - 1$, then $\operatorname{int} N_{2\Delta}(u_2) \cap (V \backslash V(u_2|v_0)) = \varnothing$, where $\Delta = d(u_1, u_2)/2$.

One can formulate a sufficient condition for edges of Euclidean minimum spanning trees to be 10-coverable, being an analog of the Lemma 1, specific to such trees. Notation is kept of the Observation 2.

Lemma 3. *Let* $G_0 = (V_0, E_0)$ *be a subgraph without isolated vertices of Euclidean minimum spanning tree* $G = (V, E)$*. Let* $\mathrm{depth}(\cdot|v_0)$ *be a distance function on* V *with respect to a chosen* $v_0 \in V$ *as defined in the Observation 2. Then an edge* $e^* = [u_1, u_2] \in E_0$ *is 10-coverable with respect to* E_0*, if* $u_2 \in \operatorname{Arg}\max\limits_{u \in V_0} \mathrm{depth}(u)$ *and for any constant* $\rho > 0$ *a point set* $U(e^*) \subset \mathbb{R}^2$ *can be found in polynomial time of size at most* 10 *such that*

$$N_{2\rho}(e^*) \backslash M_{\mathrm{EMST}}(e^*) \subset \bigcup_{u \in U(e^*)} N_\rho(u), \qquad (3)$$

where $M_{\mathrm{EMST}}(e^*) = \mathrm{int}\, N_{2\Delta}(u_2) \backslash S_2$ *and* S_2 *is the closed half-circle of radius* 2ρ *centered at* u_2 *on the boundary of* $N_{2\rho}(e^*)$*.*

Proof. Make use of the Observation 2. For $\Delta \leq \rho$ one can prove (3) analogously to proof of the Lemma 1, applied for the set E_0 instead of E.

For $\Delta > \rho$, from the inclusion $u_2 \in \operatorname{Arg}\max\limits_{u \in V_0} \mathrm{depth}(u)$ it follows that none of edges from E_0 is incident to u_2 except for e^*. As segments from E_0 intersect at most at their endpoints, one gets $e \cap (N_{2\rho}(e^*) \backslash \mathrm{int} N_{2\Delta}(u_2)) \neq \varnothing$ for any $e \in \mathcal{E}(\mathcal{N}_{e^*}(E_0))$ with $e \cap S_2 = \varnothing$. $\square$

To describe the Covering edges of special plane graphs with equal disks algorithm work for an arbitrary Euclidean minimum spanning tree $G = (V, E)$ based on the Lemma 3, one should first implement a heuristic of choice of 10-coverable edge at its step 3, implied by this lemma. To simplify its implementation, a breadth-first search is preliminarily performed on G in $O(|E|)$ time at step 1 of the algorithm to compute a distance function $\mathrm{depth}(\cdot) = \mathrm{depth}(\cdot|v_0, G)$ for an arbitrarily chosen $v_0 \in V$ as defined in the Observation 2. To its step 3 a subgraph $G_0 = (V_0, E_0)$ is constructed of G without isolated vertices. One is ready to give the following heuristic to choose a 10-coverable edge in G_0.

Choice of 10-Coverable Edge. At the algorithm step 3 in $O(|E_0|)$ time one chooses an edge $e^* = [u_1, u_2] \in E_0$ such that $u_2 \in \operatorname{Arg}\max\limits_{u \in V_0} \mathrm{depth}(u)$.

The following auxiliary procedure can be formulated to construct a point set $U(e^*)$ of size at most 10, for which the inclusion (3) holds true for $\rho = r$ with respect to the segment e^* chosen according to the Choice of 10-coverable edge heuristic. In the procedure pseudocode some notations are kept of the Partial r-disk cover search for $2r$-hippodromes on RNG edges procedure and of proof of the Lemma 2.

PARTIAL r-DISK COVER SEARCH FOR $2r$-HIPPODROMES ON EMST EDGES.

Input: a constant $r > 0$ and an edge $e^* = [u_1, u_2]$ of an arbitrary subgraph $G_0 = (V_0, E_0)$ without isolated vertices of Euclidean minimum spanning tree such that $u_2 \in \operatorname{Arg}\max_{u \in V_0} \operatorname{depth}(u)$;

Output: $U_{\mathrm{EMST}}(e^*) \subset \mathbb{R}^2$ such that (3) is hold for e^* and $U(e^*) = U_{\mathrm{EMST}}(e^*)$ with $\rho = r$;

---

1. compute a set U_1 in the same manner as in the PARTIAL r-DISK COVER SEARCH FOR $2r$-HIPPODROMES ON RNG EDGES procedure;
2. for $\Delta \in (0, r/2]$ return $U_{\mathrm{EMST}}(e^*) := U_1 \cup U_2$, where U_2 is built as in the PARTIAL r-DISK COVER SEARCH FOR $2r$-HIPPODROMES ON RNG EDGES procedure;
3. consider a hexagon D inscribed in $N_{2r}(u_2)$, whose orientation is such that exactly three sides of D are contained in T_2; compute a set U_2' of midpoints of those three sides of D, which lie completely in T_2;
4. if $\Delta \geq \frac{(2\sqrt{3}-1)r}{2\sqrt{4\sqrt{3}-6}}$, return $U_{\mathrm{EMST}}(e^*) := U_1 \cup U_2'$;
5. for $\Delta \in \left[r, \frac{(2\sqrt{3}-1)r}{2\sqrt{4\sqrt{3}-6}}\right)$ set $u_{1i} = f_i(e^*) \cap \operatorname{bd} N_r(v_{1i})$, $z_i := f_i(e^*) \cap \operatorname{bd} N_{2\Delta}(u_2)$ and $a_i' := \frac{u_{1i}+z_i}{2}$, where $f_1(e^*)$ and $f_2(e^*)$ are segments, which are parallel and identical to e^*, such that $\operatorname{bd} N_{2r}(e^*) = S_1 \cup S_2 \cup f_1(e^*) \cup f_2(e^*)$ and S_s is the radius $2r$ half-circle on the boundary of T_s, $i, s = 1, 2$;
6. for $\Delta \in (r/2, r)$ introduce a (rectangular) coordinate system with its origin at u_2, whose x-axis is along e^* and y-axis is towards $f_i(e^*)$; set $a_i' = \frac{u_{1i}+b_i}{2}$, where $b_i = (0, 2\Delta)$;
7. return $U_{\mathrm{EMST}}(e^*) := U_1 \cup U_2' \cup \{a_1', a_2'\}$.

The following lemma reports on the PARTIAL r-DISK COVER SEARCH FOR $2r$-HIPPODROMES ON EMST EDGES procedure efficiency.

Lemma 4. *Let $e^* = [u_1, u_2]$ be an input of the* PARTIAL r-DISK COVER SEARCH FOR $2r$-HIPPODROMES ON EMST EDGES *procedure. Then it returns a point set $U_{\mathrm{EMST}}(e^*)$ of size at most 10 such that (3) holds true for e^* and $U_{\mathrm{EMST}}(e^*)$.*

Proof. Below a proof sketch is provided.

Cases $\Delta \geq \frac{(2\sqrt{3}-1)r}{2\sqrt{4\sqrt{3}-6}}$ and $\Delta \in (0, r/2]$ are proved using the same arguments as in the Lemma 2 proof.

For $\Delta \in \left[r, \frac{(2\sqrt{3}-1)r}{2\sqrt{4\sqrt{3}-6}}\right)$, it is sufficient to establish that points u_{1i}, z_i and $c_{1i} = R \cap \operatorname{bd} N_r(v_{1i}) \cap \operatorname{bd} N_{2r}(u_1)$ lie in $N_r(a_i')$ to prove (3). This can be done using a few lines of algebra.

For $\Delta \in (r/2, r)$ let $z_i' = (0, 2r)$ and c_i' be the point at the intersection $\operatorname{bd} N_{2\Delta}(u_2) \cap \operatorname{bd} N_r(v_{1i})$ with the largest y-coordinate. To prove (3) it is enough to establish the inclusion $\{u_{1i}, z_i', b_i, c_i'\} \subset N_r(a_i')$. $\square$

The following 10-approximate algorithm results for the class of Euclidean minimum spanning trees from Lemmas 3 and 4.

Theorem 4. *The* COVERING EDGES OF SPECIAL PLANE GRAPHS WITH EQUAL DISKS *algorithm becomes 10-approximate for the* IPGD *problem in the class of Euclidean minimum spanning trees when at its step 3 the* CHOICE OF 10-COVERABLE EDGE *heuristic is used to choose $e^* \in E_0$ and the* PARTIAL r-DISK COVER SEARCH FOR $2r$-HIPPODROMES ON EMST EDGES *procedure is applied to compute a point set, defining a cover of $N_{2r}(e^*)\backslash M_{\mathrm{EMST}}(e^*)$. This algorithm takes $O(n\,\mathrm{OPT})$ time and $O(n)$ space, where $n = |E|$.*

5 14-Approximate Algorithm for Gabriel Graphs

In this section, a 14-approximate algorithm is built for the IPGD problem in the class of Gabriel graphs based on the COVERING EDGES OF SPECIAL PLANE GRAPHS WITH EQUAL DISKS algorithm, evoking a call of an auxiliary procedure to get a cover of $N_{2r}(e)\backslash M_G(e)$ for an arbitrary edge $e \in E_0$ chosen at the step 3 of the latter algorithm, where $M_G(e)$ is defined in the following sufficient condition of 14-coverability of edges of Gabriel graphs. Its proof is analogous to the Lemma 1 proof.

Lemma 5. *Let $G = (V, E)$ be a Gabriel graph, $e = [u_1, u_2] \in E$, $M_G(e) = \mathrm{int}\, N_\Delta\left(\frac{u_1+u_2}{2}\right)$ and $\Delta = \frac{d(u_1,u_2)}{2} > 0$. Then the edge e is 14-coverable, if for any $\rho > 0$ a point set $U(e) \subset \mathbb{R}^2$ can be found of size at most 14 in polynomial time such that*

$$N_{2\rho}(e)\backslash M_G(e) \subset \bigcup_{u\in U(e)} N_\rho(u). \tag{4}$$

Keeping notation from both PARTIAL r-DISK COVER SEARCH FOR $2r$-HIPPODROMES ON RNG EDGES and PARTIAL r-DISK COVER SEARCH FOR $2r$-HIPPODROMES ON EMST EDGES procedures, one can formulate an analogous procedure for Gabriel graphs.

PARTIAL r-DISK COVER SEARCH FOR $2r$-HIPPODROMES ON GG EDGES.

Input: a constant $r > 0$ and an edge $e = [u_1, u_2]$ of an arbitrary Gabriel graph;
Output: $U_{\mathrm{G}}(e) \subset \mathbb{R}^2$ such that (4) is hold for e and $U(e) = U_{\mathrm{G}}(e)$ with $\rho = r$;

---

1. compute sets U_s, $s = 1, 2$, as in the PARTIAL r-DISK COVER SEARCH FOR $2r$-HIPPODROMES ON RNG EDGES procedure;
2. if either $\Delta \geq \frac{(2\sqrt{3}-1)r}{\sqrt{4\sqrt{3}-6}}$ or $\Delta \in (0, r/2]$, return $U_{\mathrm{G}}(e) := U_1 \cup U_2$;
3. for $\Delta \in \left[2r, \frac{(2\sqrt{3}-1)r}{\sqrt{4\sqrt{3}-6}}\right)$ set $u_0 := \frac{u_1+u_2}{2}$ and construct two points z_{1i} and z_{2i} at the intersection $f_i(e) \cap \mathrm{bd}\, M_{\mathrm{G}}(e)$, where z_{si} is closer to u_{si} than the point $z_{(3-s)i}$ is; set $a_{si} = \frac{u_{si}+z_{si}}{2}$, $i, s = 1, 2$;
4. if $\Delta \in (r, 2r)$, consider a (rectangular) coordinate system with the origin at u_s whose x-axis is along e and y-axis is perpendicular to x-axis, being directed towards $f_i(e)$; set $b_i = (\Delta, \Delta)$ and $a_{si} = \frac{u_{si}+b_i}{2}$, $i, s = 1, 2$;

5. for $\Delta \in (r/2, r]$ set $a_{1i} := (\Delta, \sqrt{3}r)$ and $a_{2i} := \left(\Delta, \frac{\sqrt{3}r}{2}\right)$, $i = 1, 2$;
6. return $U_G(e) := U_1 \cup U_2 \cup \{a_{si}\}_{s,i=1,2}$;

The lemma below guarantees the efficiency of the PARTIAL r-DISK COVER SEARCH FOR $2r$-HIPPODROMES ON GG EDGES procedure.

Lemma 6. *Let e and $U_G(e)$ be an input and an output of the* PARTIAL r-DISK COVER SEARCH FOR $2r$-HIPPODROMES ON GG EDGES *procedure respectively. Then $|U_G(e)| \leq 14$ and the inclusion (4) holds true for e and $U_G(e)$.*

Proof. Below a proof sketch is provided.

For the case $\Delta \geq \frac{(2\sqrt{3}-1)r}{\sqrt{4\sqrt{3}-6}}$ the proof is analogous to that from the Lemma 2, replacing 2Δ with Δ. The case $\Delta \leq r/2$ is considered in the same way as in the proof of that lemma.

In cases $\Delta \in \left[2r, \frac{(2\sqrt{3}-1)r}{\sqrt{4\sqrt{3}-6}}\right)$ and $\Delta \in (r, 2r)$ the proof is analogous to the Lemma 4 proof, replacing 2Δ with Δ.

In the case $\Delta \in (r/2, r]$ to establish (4) it is enough to prove inclusions $\text{conv}\,\{u_{1i}, u_{2i}, c'_{1i}, c'_{2i}\} \subset N_r(a_{1i})$ and $\text{conv}\,\{c'_{1i}, c'_{2i}, u_0\} \subset N_r(a_{2i})$. □

One is ready to give the following approximation algorithm for the IPGD problem in the class of Gabriel graphs.

Theorem 5. *The* COVERING EDGES OF SPECIAL PLANE GRAPHS WITH EQUAL DISKS *algorithm becomes 14-approximate for the class of Gabriel graphs, when it chooses an arbitrary edge $e \in E_0$ at its step 3 and performs the* PARTIAL r-DISK COVER SEARCH FOR $2r$-HIPPODROMES ON GG EDGES *procedure for e. It has $O(n\,\text{OPT})$ time complexity and $O(n)$ space cost, where $n = |E|$.*

6 Tightness Analysis for Upper Bounds on Approximation Factors of Our Algorithms

In this section a series of the IPGD problem instances is given for which constants in upper bounds on approximation factors of our algorithms do not strongly deviate from the ratio $\frac{|H(G,r)|}{\text{OPT}(G,r)}$, where $H(G,r)$ is a feasible solution to the IPGD problem for a graph G and a radius $r > 0$, output by our algorithms, and $\text{OPT} = \text{OPT}(G,r)$ is its optimum.

For an arbitrary plane graph G, choose $r > 0$ such a small that the IPGD problem for G and r can be considered equivalent to the VERTEX COVER problem on G. In this case the segment set E', which the COVERING EDGES OF SPECIAL PLANE GRAPHS WITH EQUAL DISKS algorithm constructs to its step 5, is, in fact, a maximal (with respect to inclusion) matching in G and $|E'| \leq \text{OPT} \leq 2|E'|$. As a consequence, the ratio $\frac{|H(G,r)|}{\text{OPT}(G,r)}$ is at least 5, 5 and 4 for our algorithms on Gabriel, relative neighborhood graphs and Euclidean minimum spanning trees respectively.

7 Conclusion

In the present paper, approximation algorithms are designed with constant approximation factors for the problem of intersecting sets of edges of special plane graphs with the least number of identical disks, which generalizes on the case of sets of non-zero length segments the classical NP-hard problem of covering a given finite point set on the plane with the fewest number of identical disks. More precisely, based on specifics of mutual location of edges and vertices of Gabriel, relative neighborhood graphs and Euclidean minimum spanning trees approximation algorithms are built with factors of 14, 12 and 10 respectively, combining modest complexity with good accuracy. Namely, their approximation factors and time complexity bounds are smaller than those for known algorithms designed for close problems (see e.g. [9] for algorithms on Gabriel graphs and on arbitrary plane graphs). Our algorithms can be used in various network applications.

References

1. Agarwal, P., Pan, J.: Near-linear algorithms for geometric hitting sets and set covers. In: Proceedings of the 30th Annual Symposium on Computational Geometry, pp. 271–279. ACM, New York (2014)
2. Alon, N.: A non-linear lower bound for planar epsilon-nets. Discrete Comput. Geom. **47**(2), 235–244 (2011)
3. Antunes, D., Mathieu, C. and Mustafa, N.: Combinatorics of local search: an optimal 4-local Hall's theorem for planar graphs. In: Proceedings of the 25th Annual European Symposium on Algorithms, vol. 87, pp. 8:1–8:13. Schloss Dagstuhl-Leibniz-Zentrum fuer Informatik, Dagstuhl (2017)
4. Biniaz, A., Liu, P., Maheshwari, A., Smid, M.: Approximation algorithms for the unit disk cover problem in 2D and 3D. Comput. Geom. **60**, 8–18 (2017)
5. Dash, D., Bishnu, A., Gupta, A., Nandy, S.: Approximation algorithms for deployment of sensors for line segment coverage in wireless sensor networks. Wirel. Netw. **19**(5), 857–870 (2012)
6. Efrat, A., Katz, M., Nielsen, F., Sharir, M.: Dynamic data structures for fat objects and their applications. Comput. Geom. **15**(4), 215–227 (2000)
7. Jaromczyk, J., Toussaint, G.: Relative neighborhood graphs and their relatives. Proc. IEEE **80**(9), 1502–1517 (1992)
8. Kobylkin, K.: Stabbing line segments with disks: complexity and approximation algorithms. In: van der Aalst, W.M.P., Ignatov, D.I., Khachay, M., Kuznetsov, S.O., Lempitsky, V., Lomazova, I.A., Loukachevitch, N., Napoli, A., Panchenko, A., Pardalos, P.M., Savchenko, A.V., Wasserman, S. (eds.) AIST 2017. LNCS, vol. 10716, pp. 356–367. Springer, Cham (2018). https://doi.org/10.1007/978-3-319-73013-4_33
9. Kobylkin, K.: Constant factor approximation for intersecting line segments with disks. In: Battiti, R., Brunato, M., Kotsireas, I., Pardalos, P.M. (eds.) LION 12 2018. LNCS, vol. 11353, pp. 447–454. Springer, Cham (2019). https://doi.org/10.1007/978-3-030-05348-2_39
10. Kobylkin, K.: Approximation algorithms with guaranteed performance for the intersection of edge sets of some metric graphs with equal disks. Trudy Instituta Matematiki i Mekhaniki UrO RAN **25**(1), 62–77 (2019)

11. Matula, D., Sokal, R.: Properties of Gabriel graphs relevant to geographic variation research and the clustering of points in the plane. Geogr. Anal. **12**(3), 205–222 (1980)
12. Madireddy, R. and Mudgal, A.: Stabbing line segments with disks and related problems. In: Proceedings of the 28th Canadian Conference on Computational Geometry, pp. 201–207. Simon Fraser University, Vancouver, Canada (2016)
13. Mustafa, N., Ray, S.: Improved results on geometric hitting set problems. Discrete Comput. Geom. **44**(4), 883–895 (2010)
14. Pyrga, E., Ray, S.: New existence proofs for ε-nets. In: Proceedings of the 24th Annual Symposium on Computational Geometry, pp. 199–207. ACM, New York (2008)

A PTAS for One Cardinality-Weighted 2-Clustering Problem

Anna Panasenko[1,2](✉)

[1] Sobolev Institute of Mathematics, 4 Koptyug Avenue, 630090 Novosibirsk, Russia
a.v.panasenko@math.nsc.ru

[2] Novosibirsk State University, 2 Pirogova Street, 630090 Novosibirsk, Russia

Abstract. We consider one strongly NP-hard problem of clustering a finite set of points in Euclidean space. In this problem, we need to partition a finite set of points into two clusters minimizing the sum over both clusters of the weighted intracluster sums. Each of these sums is the sum of squared distances between the elements of the cluster and their center. The center of the one cluster is unknown and determined as the centroid, while the center of the other one is fixed at the origin. The weight factors for both intracluster sums are the given sizes of the clusters. In this paper, we present an approximation algorithm for the problem and prove that it is a polynomial-time approximation scheme (PTAS).

Keywords: Euclidean space · Weighted 2-clustering · Quadratic variation · NP-hardness · Approximation algorithm · PTAS

1 Introduction

The subject of this study is one strongly NP-hard cardinality-weighted 2-clustering problem of a finite set of points in Euclidean space. Our goal is to substantiate an approximation algorithm for this problem and show that it implements a PTAS.

Our research is motivated by the fact that the problem under consideration has been poorly studied in the algorithmic direction and by the problem importance in some applications, for example, in Data Analysis, and in Data mining.

The paper has the following structure. Section 2 contains the problem formulation, related problems, and known results. In the same section, we announce our result. We formulate and prove the basics of the algorithm in Sect. 3. The approximation algorithm is presented in Sect. 4. Also in Sect. 4, we show that our algorithm implements a PTAS.

The study presented in Sects. 2 and 3 was supported by the Russian Foundation for Basic Research, project 18-31-00398. The study presented in the other sections was supported by the Russian Academy of Science (the Program of basic research), project 0314-2019-0015, and by the Russian Ministry of Science and Education under the 5-100 Excellence Programme.

M. Khachay et al. (Eds.): MOTOR 2019, LNCS 11548, pp. 581–592, 2019.
https://doi.org/10.1007/978-3-030-22629-9_41

2 Problem Formulation and Related Problems, Known and Obtained Results

Everywhere below $\mathbb{R}$ denotes the set of real numbers, $\|\cdot\|$ denotes the Euclidean norm, and $\langle\cdot,\cdot\rangle$ denotes the scalar product.

We consider the following

Problem 1 (*Cardinality-weighted variance-based* 2*-clustering with given center*). *Given* an N-element set $\mathcal{Y}$ of points in $\mathbb{R}^d$, and a positive integer number M. *Find* a partition of $\mathcal{Y}$ into two non-empty clusters $\mathcal{C}$ and $\mathcal{Y}\setminus\mathcal{C}$ such that

$$f(\mathcal{C}) = |\mathcal{C}|\sum_{y\in\mathcal{C}}\|y-\overline{y}(\mathcal{C})\|^2 + |\mathcal{Y}\setminus\mathcal{C}|\sum_{y\in\mathcal{Y}\setminus\mathcal{C}}\|y\|^2 \to \min, \tag{1}$$

where $\overline{y}(\mathcal{C}) = \frac{1}{|\mathcal{C}|}\sum\limits_{y\in\mathcal{C}} y$ is the centroid of $\mathcal{C}$, subject to constraint $|\mathcal{C}| = M$.

This optimization problem of geometric data (i.e., $\mathcal{Y}$) approximation by the clusters and the algorithms for its solution are important for Data mining [1,2]. It is known that, in this applied field of testing hypotheses about the data structure, efficient cluster approximation algorithms are the main mathematical tools.

Besides, this problem simulates the following applied problem. We have a set $\mathcal{Y}$ of N measurement results for d characteristics of some object in two different states (active and passive, for example). Each measurement has an error and nobody knows the correspondence between the elements of the input set and the states. In addition, it is known that exactly M times the object was in the active state (or the probability of the active state is $\frac{M}{N}$). It requires to find 2-partition of the input set and evaluate the object characteristics (i.e., $\overline{y}(\mathcal{C})$ in accordance with (1)).

The applied problem is typical, in particular, for medical and technical applications. In these applications, the objects might be presented by patients and technical devices, and the states are healthy or sick, serviceable or malfunctioning.

One can easily check that only in the particular case of Problem 1, when $2M = N$, the optimal clusters are separated by a hyperplane. In other cases, the separating surface is non-linear. It is known that the construction of optimal separating surfaces (i.e. optimal classifiers) is important for Pattern recognition and Machine learning [3,4].

Problem 1 is closely related to the well-known *Min-sum 2-clustering* problem. In this problem, it is required to find a partition of $\mathcal{Y}$ into two clusters so as to minimize the sum

$$|\mathcal{C}|\sum_{y\in\mathcal{C}}\|y-\overline{y}(\mathcal{C})\|^2 + |\mathcal{Y}\setminus\mathcal{C}|\sum_{y\in\mathcal{Y}\setminus\mathcal{C}}\|y-\overline{y}(\mathcal{Y}\setminus\mathcal{C})\|^2.$$

Both centroids of the clusters are unknown in this problem. *Min-sum 2-clustering* problem is equivalent to another well-known *Min-sum all-pairs 2-clustering* problem minimizing the sum

$$\sum_{x\in\mathcal{C}}\sum_{z\in\mathcal{C}}\|x-z\|^2 + \sum_{x\in\mathcal{Y}\setminus\mathcal{C}}\sum_{z\in\mathcal{Y}\setminus\mathcal{C}}\|x-z\|^2.$$

The NP-hardness of the general metric case of these problems was shown earlier in [5,6]. The strong NP-hardness of the Euclidean case was proved in [7,8].

There are some approximation results for these problems [6,9–12], but they can not be applied to Problem 1 because these problems are not equivalent.

It was proved in [7,8] that Problem 1 is the strongly NP-hard one. The following algorithmic results were obtained for this problem.

First of all, recall that a number of algorithmic results [13–19] were obtained for the particular case of Problem 1 when $2M = N$.

Further, in [20], an exact pseudo-polynomial algorithm was constructed for the case of integer components of the input points and fixed dimension d of the space. The running time of this algorithm is $\mathcal{O}(N(MD)^d)$, where D is the maximum absolute value of coordinates of the input points.

An approximation scheme that allows one to find $(1+\varepsilon)$-approximate solution in $\mathcal{O}\left(dN^2(\sqrt{\frac{2d}{\varepsilon}}+2)^d\right)$ time was proposed in [21]. It implements an FPTAS in the case of the fixed space dimension.

Moreover, the modification of this algorithm [22] with improved time complexity: $\mathcal{O}\left(\sqrt{d}N^2(\frac{\pi e}{2})^{d/2}(\sqrt{\frac{2}{\varepsilon}}+2)^d\right)$, was proposed. The algorithm implements an FPTAS in the case of fixed space dimension and remains polynomial for instances of dimension $\mathcal{O}(\log n)$. In this case, it implements a PTAS with $\mathcal{O}\left(N^{C\,(1.05+\log(2+\sqrt{\frac{2}{\varepsilon}}))}\right)$ time, where C is a positive constant.

An approximation algorithm that allows one to find a 2-approximate solution to the problem in $\mathcal{O}\left(dN^2\right)$ time was constructed in [23].

In [24], a randomized algorithm was constructed. It allows one to find $(1+\varepsilon)$-approximate solution with probability not less than $1-\gamma$ in $\mathcal{O}(dN)$ time for an established parameter value, a given relative error ε and fixed γ. The conditions are found under which the algorithm is asymptotically exact and runs in $\mathcal{O}(dN^2)$ time.

In this paper, we present an approximation algorithm with parameters s and t that allows one to find a $(1/t+8\zeta(t,s))$-approximate solution, where $\zeta(t,s)=\sqrt{t-1}/s+(t-1)/s^2$. It is proved that the algorithm is a PTAS with a relative error of $\varepsilon>0$ when $t=2/\varepsilon$ and $s=9t^{3/2}$. The time complexity in this case is $O(dN^{2/\varepsilon+1}((9/\varepsilon)^{3/\varepsilon}+2/\varepsilon))$.

Our algorithm is based on the approach presented in [14,25] and develops it.

3 Basics of the Algorithm

In this section, we formulate and prove some statements which are necessary for substantiation of our algorithms. The proof of the following lemma can be found in [20].

Lemma 1. *Let*

$$S(\mathcal{C},x)=|\mathcal{C}|\sum_{y\in\mathcal{C}}\|y-x\|^2+|\mathcal{Y}\setminus\mathcal{C}|\sum_{y\in\mathcal{Y}\setminus\mathcal{C}}\|y\|^2,\ \mathcal{C}\subseteq\mathcal{Y},\ x\in\mathbb{R}^d. \qquad (2)$$

Then the next statements are true:

(1) for any nonempty fixed set $\mathcal{C} \subseteq \mathcal{Y}$, the minimum of the function $S(\mathcal{C}, x)$ over $x \in \mathbb{R}^d$ is reached at the point $\overline{y}(\mathcal{C}) = \frac{1}{|\mathcal{C}|} \sum_{y \in \mathcal{C}} y$;

(2) if $|\mathcal{C}| = M = const$, then for any fixed point $x \in \mathbb{R}^d$ the minimum of function $S(\mathcal{C}, x)$ over $\mathcal{C} \subseteq \mathcal{Y}$ is reached at the subset $\mathcal{B}^x$ that consists of M points of the set $\mathcal{Y}$, at which the function

$$h^x(y) = (2M - N) \|y\|^2 - 2M \langle y, x \rangle, \ \ y \in \mathcal{Y}, \tag{3}$$

has the smallest values.

Lemma 1 allows us to find the optimal solution of some auxiliary problem when the point $x \in \mathbb{R}^d$ is fixed. This auxiliary problem is to minimize the objective function (2) instead of (1), and that is the only difference from Problem 1.

Let us find out the necessary conditions for choosing x so as to construct the solution of the auxiliary problem that is good enough in some sense for Problem 1.

Let $\varepsilon(\mathcal{C}, y)$ denotes the relative error of the feasible solution y and $\mathcal{C}$ of the auxiliary problem:

$$\varepsilon(\mathcal{C}, y) = \frac{S(\mathcal{C}, y) - f(\mathcal{C})}{f(\mathcal{C})}.$$

Everywhere below $\mathcal{C}^*$ denotes the optimal solution of Problem 1. Note that if $\mathcal{B}^y$ is the set from Lemma 1 for the fixed point y, then the value $\varepsilon(\mathcal{C}^*, y)$ is an upper estimate of the relative error of the solution $\mathcal{B}^y$ of Problem 1. Indeed, by Lemma 1,

$$f(\mathcal{B}^y) = S(\mathcal{B}^y, \overline{y}(\mathcal{B}^y)) \leq^{L1.1} S(\mathcal{B}^y, y) \leq^{L1.2} S(\mathcal{C}^*, y),$$

and, therefore,

$$\frac{f(\mathcal{B}^y) - f(\mathcal{C}^*)}{f(\mathcal{C}^*)} \leq \frac{S(\mathcal{C}^*, y) - f(\mathcal{C}^*)}{f(\mathcal{C}^*)} = \varepsilon(\mathcal{C}^*, y).$$

The proof of the following well-known lemma is presented in many publications (see, for example, [26]).

Lemma 2. *For an arbitrary point $x \in \mathbb{R}^d$, a finite set $\mathcal{Z} \subset \mathbb{R}^d$ and $\overline{z} = \frac{1}{|\mathcal{Z}|} \sum_{z \in \mathcal{Z}} z$ ($\overline{z}$ is the centroid of $\mathcal{Z}$), it is true that*

$$\sum_{z \in \mathcal{Z}} \|z - x\|^2 = \sum_{z \in \mathcal{Z}} \|z - \overline{z}\|^2 + |\mathcal{Z}| \cdot \|x - \overline{z}\|^2.$$

Applying Lemma 2 for objective function (2) and $|\mathcal{C}| = M$, we get the next equality:

$$S(\mathcal{C}, x) = f(\mathcal{C}) + M^2 \|x - \overline{y}(\mathcal{C})\|^2, \tag{4}$$

and, therefore,

$$\varepsilon(\mathcal{C}, y) = \frac{S(\mathcal{C}, y) - f(\mathcal{C})}{f(\mathcal{C})} = \frac{M^2\|y - \overline{y}(\mathcal{C})\|^2}{f(\mathcal{C})}.$$

The statements of the following lemma and the theorem are completely identical with those in [25]. The proofs are similar but have some important differences. We omit the proof of the following lemma and present the proof of the theorem. It is enough in order to demonstrate the differences.

Lemma 3. *Let $\mathcal{C}$ be an arbitrary subset of $\mathcal{Y}$ of cardinality M. Let x be an arbitrary point in Euclidean space and let $y = y(x, \mathcal{C})$ be the point closest to $\overline{y}(\mathcal{C})$ among those lying on the beams from x to all points of the cluster $\mathcal{C}$. Then*

$$\varepsilon(\mathcal{C}, y) \leq \frac{\varepsilon(\mathcal{C}, x)}{1 + \varepsilon(\mathcal{C}, x)}.$$

Theorem 1. *Let $\mathcal{C}$ be an arbitrary subset of $\mathcal{Y}$ of cardinality M, $1 \leq t \leq M$. Then a linear span of some subset of $\mathcal{C}$ of cardinality t contains a point y_t such that $\varepsilon(\mathcal{C}, y_t) \leq 1/t$.*

Proof. Prove the theorem by induction on t.

Base case: $t = 1$. Let $y_1 = \arg\min_{y\in\mathcal{C}} \|y - \overline{y}(\mathcal{C})\|^2$ be the point from the subset $\mathcal{C}$ closest to its centroid. Then

$$\begin{aligned} S(\mathcal{C}, \overline{y}(\mathcal{C})) &= M\sum_{y\in\mathcal{C}} \|y - \overline{y}(\mathcal{C})\|^2 + (N - M)\sum_{y\in\mathcal{Y}\setminus\mathcal{C}} \|y\|^2 \\ &\geq M\sum_{y\in\mathcal{C}} \|y - \overline{y}(\mathcal{C})\|^2 \geq M^2\|y_1 - \overline{y}(\mathcal{C})\|^2. \end{aligned}$$

Hence,

$$S(\mathcal{C}, y_1) = S(\mathcal{C}, \overline{y}(\mathcal{C})) + M^2\|y_1 - \overline{y}(\mathcal{C})\|^2 \leq 2S(\mathcal{C}, \overline{y}(\mathcal{C})) = 2f(\mathcal{C})$$

and

$$\varepsilon(\mathcal{C}, y_1) = \frac{S(\mathcal{C}, y_1) - f(\mathcal{C})}{f(\mathcal{C})} \leq 1.$$

Induction step: Consider as y_{t+1} the point $y(y_t, \mathcal{C})$ of Lemma 3. It is contained in the linear span of some subset of $\mathcal{C}$ of cardinality $t+1$. In this case, by Lemma 3, we have

$$\varepsilon(\mathcal{C}, y_{t+1}) \leq \frac{\varepsilon(\mathcal{C}, y_t)}{1 + \varepsilon(\mathcal{C}, y_t)}.$$

By induction, $\varepsilon(\mathcal{C}, y_t) \leq 1/t$, hence

$$\varepsilon(\mathcal{C}, y_{t+1}) \leq \frac{1/t}{1 + 1/t} = \frac{1}{t+1}.$$

□

Remark 1. For $t \geq M$ Theorem 1 also holds since $\overline{y}(\mathcal{C})$ is contained in the linear span of $\mathcal{C}$ of the size M and, in this case, $\varepsilon(\mathcal{C}, \overline{y}(\mathcal{C})) = 0 \leq 1/t$.

Theorem 1 and Remark 1 guarantee that if we consider the points of the linear spans of all t-tuples of $\mathcal{Y}$ as a local center then one of them leads to the solution with a relative error of $1/t$.

Lemma 4. *Let $1 \leq t \leq M$, and let points $y_1, \dots, y_t$ be constructed in series as in Theorem 1 applied to $\mathcal{C}^*$. Let $f_1 = \min_{y \in \mathcal{Y}} f(\mathcal{B}^y)$, where $\mathcal{B}^y$ is constructed as in Lemma 1. Then*

$$\|y_t - \overline{y}(\mathcal{C}^*)\| \leq \sqrt{\frac{f_1}{M^2}}.$$

Proof. The geometrical considerations imply

$$\|y_t - \overline{y}(\mathcal{C}^*)\| \leq \|y_1 - \overline{y}(\mathcal{C}^*)\|. \tag{5}$$

Since $y_1 = \arg\min_{y \in \mathcal{C}^*} \|y - \overline{y}(\mathcal{C}^*)\|^2$:

$$f(\mathcal{C}^*) = S(\mathcal{C}^*, \overline{y}(\mathcal{C}^*)) \geq M \sum_{y \in \mathcal{C}^*} \|y - \overline{y}(\mathcal{C}^*)\|^2 \geq M^2 \|y_1 - \overline{y}(\mathcal{C}^*)\|^2,$$

and, therefore,

$$\|y_1 - \overline{y}(\mathcal{C}^*)\| \leq \sqrt{\frac{f(\mathcal{C}^*)}{M^2}},$$

hence, by inequality (5),

$$\|y_t - \overline{y}(\mathcal{C}^*)\| \leq^{(5)} \|y_1 - \overline{y}(\mathcal{C}^*)\| \leq \sqrt{\frac{f(\mathcal{C}^*)}{M^2}} \leq \sqrt{\frac{f_1}{M^2}},$$

since $f(\mathcal{C}^*) \leq f_1$. We get the desired inequality. □

Remark 2. For $t \geq M$ Lemma 4 also holds since in this case $y_t = \overline{y}(\mathcal{C}^*)$ by Remark 1.

Remark 3. Note that the distance from y_1 to y_t in Lemma 4 is not greater than $2\sqrt{\frac{f_1}{M^2}}$ by triangle inequality. Hence, y_t is somewhere in the ball centered at y_1 with the radius $2\sqrt{\frac{f_1}{M^2}}$. Since $y_1 = \arg\min_{y \in \mathcal{C}^*} \|y - \overline{y}(\mathcal{C}^*)\|^2$, it is enough to go through all the points from $\mathcal{Y}$ to meet y_1.

Lemma 5. *Let the conditions of Lemma 4 hold and $s > 0$ be some integer number. Let y'_t be the point of $\mathbb{R}^d$ such that $\|y'_t - y_t\| \leq h\sqrt{t-1}/2$, where $h = 4\sqrt{\frac{f_1}{M^2}}/s$. Then,*

$$\varepsilon(\mathcal{C}^*, y'_t) \leq \varepsilon(\mathcal{C}^*, y_t) + 8\zeta(t, s),\ \zeta(t, s) = \sqrt{t-1}/s + (t-1)/s^2. \tag{6}$$

Proof. Let v denote $h\sqrt{t-1}/2$. From (4), conditions of the lemma and the triangle inequality we obtain

$$S(\mathcal{C}^*, y_t') - S(\mathcal{C}^*, y_t) = M^2(\|y_t' - y_t + y_t - \overline{y}(\mathcal{C}^*)\|^2 - \|y_t - \overline{y}(\mathcal{C}^*)\|^2) \\ \leq M^2((a+v)^2 - a^2),$$

where $a = \|y_t - \overline{y}(\mathcal{C}^*)\|$. By Lemma 4 we have $a \leq \sqrt{\frac{f_1}{M^2}}$ (let A denote $\sqrt{\frac{f_1}{M^2}}$) and, therefore,

$$(a+v)^2 - a^2 \leq 2Av + v^2 = A\sqrt{t-1}h + (t-1)h^2/4 \\ = 4A^2\sqrt{t-1}/s + 4A^2(t-1)/s^2 = 4A^2\zeta(t,s).$$

Thus,

$$S(\mathcal{C}^*, y_t') - S(\mathcal{C}^*, y_t) \leq 4M^2A^2\zeta(t,s) = 4f_1\zeta(t,s).$$

In [23] was shown that $f_1 \leq 2f(\mathcal{C}^*)$. Hence,

$$S(\mathcal{C}^*, y_t') - S(\mathcal{C}^*, y_t) \leq 8f(\mathcal{C}^*)\zeta(t,s).$$

Dividing the expression by $f(\mathcal{C}^*)$, we obtain the required inequality:

$$\frac{S(\mathcal{C}^*, y_t') - f(\mathcal{C}^*) + f(\mathcal{C}^*) - S(\mathcal{C}^*, y_t)}{f(\mathcal{C}^*)} = \varepsilon(\mathcal{C}^*, y_t') - \varepsilon(\mathcal{C}^*, y_t) \leq 8\zeta(t,s).$$

The proof of Lemma 5 is complete. □

Remark 4. For $t \geq M$ Lemma 5 also holds by Remark 2.

4 Approximation Algorithm

We present the approximation algorithm for Problem 1 in this section. The main idea of this algorithm can be described as follows. For each point u of the input set $\mathcal{Y}$, we form the subsets $\{x_1, \ldots, x_q\} \subseteq (\mathcal{Y} \setminus \{u\})$ of the other points with the size q fixed. For each subset, we construct a domain (spherical with the radius $H + h\sqrt{q}/2$ centered at u, where H and h will be defined below) so that the center of the desired subset necessarily belongs to one of these domains. We generate, using given (as input) parameters, a lattice (a grid) that discretizes the domain with a uniform step h in all directions. These directions are defined with the help of the Gram-Schmidt process applied to the vectors $x_1 - u, \ldots, x_q - u$ (note that the dimension can be less than q). This kind of constructing allows us to have the nodes which belong to the linear span of the current subset. For each lattice node, a subset of M points from the input set that have the smallest values of the function (3) is formed. The resulting set is declared as a solution candidate. The candidate that minimizes the objective function is chosen to be the final solution.

Let $s, t > 0$ be the integer parameters of the algorithm. Let $f_1 = \min_{y \in \mathcal{Y}} f(\mathcal{B}^y)$, where $\mathcal{B}^y$ is constructed as in Lemma 1. Let

$$A = \sqrt{\frac{f_1}{M^2}}, \quad H = 2A, \quad h = 4A/s, \quad q = \min\{M, d+1, t\} - 1. \tag{7}$$

For an arbitrary point $u \in \mathcal{Y}$, positive numbers h and H, and an arbitrary subset $\{x_1, \ldots, x_q\} \subseteq (\mathcal{Y} \setminus \{u\})$, we define a grid $\mathcal{D}(u, x_1, \ldots, x_q, h, H)$ by the following way. We apply the Gram-Shmidt process to the vectors $\{x_1 - u, \ldots, x_q - u\}$ and use the result as the directions for the grid. Then we construct the desired grid of size $2H$ centered at the point u with node spacing h.

For each $y \in \mathcal{Y}$ let $R = H + \frac{h\sqrt{q}}{2}$. Let us construct the lattice

$$\mathcal{D}_R(y, x_1, \ldots, x_q, h, H + h/2) = \mathcal{D}(y, x_1, \ldots, x_q, h, H + h/2) \cap B(y, R), \tag{8}$$

where $B(y, R) = \{x \in \mathbb{R}^d |\ \|x - y\| \leq R\}$ is the ball of radius R and center y.

The step-by-step description looks like as follows.

Algorithm $\mathcal{A}$.

Input: a set $\mathcal{Y}$, positive integers M, s, t.

Step 0. $f_1 := \infty$.

For each point $y \in \mathcal{Y}$ Steps 1–3 are executed:

Step 1. Compute the values $h^y(z), z \in \mathcal{Y}$, using formula (3). Find an M-element subset $\mathcal{B}^y \subseteq \mathcal{Y}$ with the smallest values $h^y(z)$. Compute $f(\mathcal{B}^y)$ using formula (1).

Step 2. If $f(\mathcal{B}^y) = 0$ then put $\mathcal{C}_\mathcal{A} = \mathcal{B}^y$; exit.

Step 3. If $f(\mathcal{B}^y) < f_1$ then put $f_1 = f(\mathcal{B}^y)$.

Step 4. Compute the values A, H, h, q using formulae (7). Compute $R = H + \frac{h\sqrt{q}}{2}$.

For each point $u \in \mathcal{Y}$ and for each subset $\{x_1, \ldots, x_q\} \subseteq (\mathcal{Y} \setminus \{u\})$, Steps 5–7 are executed.

Step 5. Apply the Gram-Shmidt process to the vectors $x_1 - u, \ldots, x_q - u$.

Step 6. Construct the grip $\mathcal{D}_R(u, x_1, \ldots, x_q, h, H + h/2)$ using formula (8).

Step 7. For each node y of the grid $\mathcal{D}_R(u, x_1, \ldots, x_q, h, H + h/2)$, compute the values $h^y(z), z \in \mathcal{Y}$, using formula (3). Find and remember an M-element subset $\mathcal{B}^y \subseteq \mathcal{Y}$ with the smallest values $h^y(z)$. Compute and remember $f(\mathcal{B}^y)$ using formula (1).

Step 8. In the family W of candidate sets constructed above choose as a solution $\mathcal{C}_\mathcal{A}$ the set $\mathcal{B}^x$ with minimal value $f(\mathcal{B}^x)$, where

$$W = \{\mathcal{B}^y : y \in \mathcal{D}_R(u, x_1, \ldots, x_q, h, H + h/2), u \in \mathcal{Y}, \{x_1, \ldots, x_q\} \subseteq (\mathcal{Y} \setminus \{u\})\}.$$

Output: The set $\mathcal{C}_\mathcal{A}$.

The following two lemmas were proved in [22]. We can apply these lemmas to our case with $d = q$ and grid $\mathcal{D}_R(u, x_1, \ldots, x_q, h, H + h/2)$.

Lemma 6. *For an arbitrary point x of $B(y, H)$, $y \in \mathcal{Y}$, the distance from x to the closest node of the grid $\mathcal{D}_R(y, h, H + h/2)$ does not exceed the value $\frac{h\sqrt{d}}{2}$.*

Lemma 7. *For an arbitrary point $y \in \mathcal{Y}$ the cardinality of the lattice $\mathcal{D}_R(y, h, H + h/2)$ does not exceed the value*

$$\frac{1}{\sqrt{\pi d}}\Big(\frac{2\pi e}{d}\Big)^{d/2}\Big(\frac{H}{h} + \sqrt{d}\Big)^d.$$

Remark 5. The cardinality of the lattice $\mathcal{D}_R(u, x_1, \ldots, x_q, h, H + h/2)$ does not exceed the value

$$L_R = \frac{1}{\sqrt{\pi q}}\Big(\frac{2\pi e}{q}\Big)^{q/2}\Big(\frac{s}{2A} + \sqrt{q}\Big)^q.$$

We know that $q \leq d$, so if the space dimension d is fixed, then we get $\mathcal{O}(L_R) \sim \mathcal{O}(s^d)$.

Theorem 2. *For any fixed $s, t > 0$ Algorithm $\mathcal{A}$ finds $(1/t + 8\zeta(t, s))$-approximate solution of Problem 1, where $\zeta(t, s) = \sqrt{t-1}/s + (t-1)/s^2$, in*

$$\mathcal{O}\left(dN^2\binom{N-1}{q}(L_R + q)\right) \tag{9}$$

time, where $q = \min\{M, d + 1, t\} - 1$.

Proof. Let us bound the approximation factor of the algorithm. It is obvious by Remark 3, that the algorithm once uses the point $y_1 \in \mathcal{Y}$ (the $2A$-neighbourhood of which contains y_t from Lemmas 4 and 5) as the center of the constructed grid. Point y_t from Lemmas 4 and 5 is inside the ball $B(y_1, H)$. Hence, by Lemma 6, the distance from y_t to the closest node (let us denote this node by y_t') of the considered grid does not exceed the value $h\sqrt{q}/2$ and so does not exceed the value $h\sqrt{t-1}/2$, since $q \leq t - 1$. So, we can apply Lemma 5 to the point y_t' and get inequality (6). And then we can evaluate the right-hand side of the inequality with the help of Theorem 1 because:

1. If $q = t - 1$ and $t \leq M$, $t \leq d + 1$, then the algorithm meets the linear span of each possible subset with cardinality t. And so it meets the desired one from Theorem 1.
2. If $q = d$ and $d + 1 \leq t$, $d + 1 \leq M$, then for arbitrary $x_i, u \in \mathcal{Y}$, the linear span of vectors $\{x_1 - u, \ldots, x_t - u\}$ coincides with the linear span of vectors $\{y_1 - u, \ldots, y_d - u\}$, where $\{y_1 - u, \ldots, y_d - u\}$ is some subset of $\{x_1 - u, \ldots, x_t - u\}$. So, Steps 5–7 would have the same results if we use t instead of $q = d$, and so the algorithm meets the desired one linear span from Theorem 1.
3. If $q = M - 1$ and $M \leq d + 1$, $M \leq t$, then Remark 1 holds.

Let $\mathcal{C}_{\mathcal{A}}$ be the set produced by algorithm $\mathcal{A}$. Applying Lemma 5 and Theorem 1, we get the following chain of inequalities:

$$\frac{f(\mathcal{C}_{\mathcal{A}}) - f(\mathcal{C}^*)}{f(\mathcal{C}^*)} \leq \frac{f(\mathcal{B}^{y_t'}) - f(\mathcal{C}^*)}{f(\mathcal{C}^*)} \leq \varepsilon(\mathcal{C}^*, y_t') \leq^{L5}$$
$$\leq^{L5} \varepsilon(\mathcal{C}^*, y_t) + 8\zeta(t, s) \leq^{Th1} 1/t + 8\zeta(t, s).$$

Let us evaluate the time complexity of the algorithm.

Step 0 takes $\mathcal{O}(1)$ time.

Then Steps 1–3 are executed for N times. Step 1 requires $\mathcal{O}(dN)$ operations: calculation of $h^y(z)$ requires at most $\mathcal{O}(dN)$-time; finding the M smallest elements in the set of N elements requires $\mathcal{O}(N)$ operations (for example, using the algorithm of finding the n-th smallest value in an unordered array [27]); computation of the value $f(\mathcal{B}^y)$ takes $\mathcal{O}(dN)$ time. Steps 2 and 3 are executed in $\mathcal{O}(d)$ operations.

Step 4 requires $\mathcal{O}(d)$ operations.

Steps 5–7 are executed for $N\binom{N-1}{q}$ times. Step 5 takes $\mathcal{O}(dq^2)$ time [28]. Step 6 requires at most $\mathcal{O}(dL_R)$-time. Step 7 takes $\mathcal{O}(dNL_R)$ time. The total time complexity of these steps is $dN\binom{N-1}{q}(q^2 + NL_R)$. Since $q \leq N$, we can evaluate the time complexity by $dN^2\binom{N-1}{q}(q + L_R)$.

Step 8 requires $N\binom{N-1}{q}L_R$ operations and the total time complexity of all Steps is

$$\mathcal{O}\left(dN^2\binom{N-1}{q}(L_R + q)\right).$$

□

Remark 6. Note some rough estimates:

$$N\binom{N-1}{q} \leq N(N-1)^q \leq N^{q+1} \leq N^t; \tag{10}$$

$$L_R \leq (2H/h+2)^q = (s/A+2)^q \leq const_1 s^q \leq const_1 s^{t-1}.$$

It implies that we can evaluate the time complexity of the algorithm by $\mathcal{O}(dN^{t+1}(s^{t-1} + q))$. This value coincides with the one from [25] when $q \leq s^{t-1}$ (that is correct if $s \geq 2$).

Property 1. In the case of $t = 2/\varepsilon$, where $\varepsilon > 0$, and $s = 9t^{3/2}$, the algorithm can solve the problem in time $O(dN^{2/\varepsilon+1}((9/\varepsilon)^{3/\varepsilon} + 2/\varepsilon - 1))$ with the relative error of ε.

Indeed, chosen s, we have

$$\zeta(t,s) \leq \frac{1}{9t} + \frac{1}{81t^2} \leq \frac{1}{8t}.$$

Hence, the relative error of the algorithm does not exceed $2/t = \varepsilon$. The estimate of the running time follows from

$$s^{t-1} = (9t^{3/2})^{t-1} \leq (9(2/\varepsilon)^{3/2})^{2/\varepsilon} \leq (9/\varepsilon)^{3/\varepsilon};$$

$$q \leq t-1 = 2/\varepsilon - 1.$$

Thus, the PTAS is obtained.

Moreover, it is true that:

Property 2. In the case of $t = 2/\varepsilon$, where $\varepsilon > 0$, $s = 9t^{3/2}$, and the fixed space dimension d (or bounded by a constant), the algorithm can solve the problem in time $\mathcal{O}(N^{d+2}\varepsilon^{-3d/2})$ with the relative error of ε.

Indeed, the value ε of the relative error is obvious by Property 1. The time complexity can be evaluated by $\mathcal{O}(N^{q+2}s^d)$ by Remark 5, (9) and (10). Since $q \leq d$, $s = 9t^{3/2}$ and $t = 2/\varepsilon$, the algorithm can solve the problem in $\mathcal{O}(N^{d+2}\varepsilon^{-3d/2})$-time.

5 Conclusion

In this paper, we presented an approximation algorithm for one Euclidean cardinality-weighted 2-clustering problem for a finite set of points. Our algorithm on the one hand based on an adaptive-grid-approach and on the other hand based on the geometry of the linear spans and the Gram-Schmidt process. It was proved that the algorithm is a PTAS for some parameters values. This algorithm is the first PTAS for the considered clustering problem.

References

1. Aggarwal, C.C.: Data Mining: The Textbook. Springer, Cham (2015). https://doi.org/10.1007/978-3-319-14142-8
2. Hastie, T., Tibshirani, R., Friedman, J.: The Elements of Statistical Learning: Data Mining, Inference, and Prediction. Springer, New York (2009). https://doi.org/10.1007/978-0-387-84858-7
3. Bishop, C.M.: Pattern Recognition and Machine Learning. Springer, New York (2006)
4. James, G., Witten, D., Hastie, T., Tibshirani, R.: An Introduction to Statistical Learning. Springer, New York (2013). https://doi.org/10.1007/978-1-4614-7138-7
5. Brucker, P.: On the complexity of clustering problems. In: Henn, R., et al. (eds.) Optimization and Operations Research. LNE, vol. 157, pp. 45–54. Springer, Heidelberg (1978). https://doi.org/10.1007/978-3-642-95322-4_5
6. Sahni, S., Gonzalez, T.: P-Complete Approximation Problems. J. ACM **23**, 555–566 (1976)
7. Kel'manov, A.V., Pyatkin, A.V.: NP-hardness of some quadratic Euclidean 2-clustering problems. Doklady Math. **92**(2), 634–637 (2015)
8. Kel'manov, A.V., Pyatkin, A.V.: On the complexity of some quadratic Euclidean 2-clustering problems. Comput. Math. Math. Phys. **56**(3), 491–497 (2016)
9. de la Vega F., Karpinski M., Kenyon C., Rabani Y.: Polynomial Time Approximation Schemes for Metric Min-Sum Clustering. Electronic Colloquium on Computational Complexity (ECCC), Report 25 (2002)
10. de la Vega, F., Kenyon, C.: A randomized approximation scheme for metric max-cut. J. Comput. Syst. Sci. **63**, 531–541 (2001)
11. Hasegawa, S., Imai, H., Inaba, M., Katoh, N., Nakano, J.: Efficient algorithms for variance-based k-clustering. In: Proceedings of the 1st Pacific Conference on Computer Graphics and Applications, Pacific Graphics 1993, Seoul, Korea, vol. 1, pp. 75–89. World Scientific, River Edge (1993)

12. Inaba, M., Katoh, N., Imai, H.: Applications of weighted Voronoi diagrams and randomization to variance-based k-clustering: (extended abstract). In: SCG 1994 Proceedings of the Tenth Annual Symposium on Computational Geometry, Stony Brook, NY, USA, 6–8 June 1994, pp. 332–339. ACM, New York (1994)
13. Dolgushev, A.V., Kel'manov, A.V.: An approximation algorithm for solving a problem of cluster analysis. J. Appl. Indust. Math. **5**(4), 551–558 (2011)
14. Dolgushev, A.V., Kel'manov, A.V., Shenmaier, V.V.: Polynomial-time approximation scheme for a problem of partitioning a finite set into two clusters. Proc. Steklov Inst. Math. **295**(1), 47–56 (2016)
15. Gimadi, E.K., Pyatkin, A.V., Rykov, I.A.: On polynomial solvability of some problems of a vector subset choice in a Euclidean space of fixed dimension. J. Appl. Ind. Math. **4**(1), 48–53 (2010)
16. Shenmaier, V.V.: Solving some vector subset problems by Voronoi diagrams. J. Appl. Ind. Math. **10**(4), 560–566 (2016)
17. Kel'manov, A.V., Khandeev, V.I.: A randomized algorithm for two-cluster partition of a set of vectors. Comput. Math. Math. Phys. **55**(2), 330–339 (2015)
18. Kel'manov, A.V., Khandeev, V.I.: An exact pseudopolynomial algorithm for a problem of the two-cluster partitioning of a set of vectors. J. Appl. Indust. Math. **9**(4), 497–502 (2015)
19. Kel'manov, A.V., Khandeev, V.I.: Fully polynomial-time approximation scheme for a special case of a quadratic Euclidean 2-clustering problem. Comput. Math. Math. Phys. **56**(2), 334–341 (2016)
20. Kel'manov, A.V., Motkova, A.V.: Exact pseudopolynomial algorithms for a balanced 2-clustering problem. J. Appl. Ind. Math. **10**(3), 349–355 (2016)
21. Kel'manov, A., Motkova, A.: A fully polynomial-time approximation scheme for a special case of a balanced 2-clustering problem. In: Kochetov, Y., Khachay, M., Beresnev, V., Nurminski, E., Pardalos, P. (eds.) DOOR 2016. LNCS, vol. 9869, pp. 182–192. Springer, Cham (2016). https://doi.org/10.1007/978-3-319-44914-2_15
22. Kel'manov, A., Motkova, A., Shenmaier, V.: An approximation scheme for a weighted two-cluster partition problem. In: van der Aalst, W.M.P., Ignatov, D.I., Khachay, M., Kuznetsov, S.O., Lempitsky, V., Lomazova, I.A., Loukachevitch, N., Napoli, A., Panchenko, A., Pardalos, P.M., Savchenko, A.V., Wasserman, S. (eds.) AIST 2017. LNCS, vol. 10716, pp. 323–333. Springer, Cham (2018). https://doi.org/10.1007/978-3-319-73013-4_30
23. Kel'manov, A.V., Motkova, A.V.: Polynomial-time approximation algorithm for the problem of cardinality-weighted variance-based 2-clustering with a given center. Comp. Math. Math. Phys. **58**(1), 130–136 (2018)
24. Kel'manov, A., Khandeev, V., Panasenko, A.: Randomized algorithms for some clustering problems. In: Eremeev, A., Khachay, M., Kochetov, Y., Pardalos, P. (eds.) OPTA 2018. CCIS, vol. 871, pp. 109–119. Springer, Cham (2018). https://doi.org/10.1007/978-3-319-93800-4_9
25. Shenmaier, V.V.: An approximation scheme for a problem of search for a vector subset. J. Appl. Ind. Math. **6**(3), 381–386 (2012)
26. Kel'manov, A.V., Romanchenko, S.M.: An approximation algorithm for solving a problem of search for a vector subset. J. Appl. Ind. Math. **6**(1), 90–96 (2012)
27. Wirth, N.: Algorithms + Data Structures = Programs. Prentice Hall, New Jersey (1976)
28. Golub, G.H., Van Loan, C.F.: Matrix Computations. Johns Hopkins Studies in the Mathematical Sciences. Johns Hopkins University Press, Baltimore (1996)

Games and Mathematical Economics

On a Single-Type Differential Game with a Non-convex Terminal Set

Igor' V. Izmest'ev([✉]) and Viktor I. Ukhobotov

Chelyabinsk State University, Br. Kashirinykh str., 129, 454001 Chelyabinsk, Russia
j748e8@gmail.com, ukh@csu.ru

Abstract. We consider the problem of controlling a rod attached to a rotor. A rotating flywheel is attached to one end of the rod. The rotor is controlled by the first player. The flywheel is controlled by the second player. The goal of the first player is to bring the rotor to a vertical position at a given time. The goal of the second player is the opposite. This problem is an example of a more general linear differential game with a one-dimensional aim. Using a linear change of variables, this problem is reduced to a single-type one-dimensional differential game with a non-convex terminal set, for which we have found the necessary and sufficient conditions of termination and constructed the corresponding controls of the players.

Keywords: Control · Differential game · Terminal set · Flywheel

1 Introduction

The linear differential game with fixed terminal time, using a linear change of variables [5], can be reduced to the form when on the right-hand side of the new equations there is only the sum of player controls whose values belong to given sets depending on the time. In the case that in a linear differential game the payoff is the value of the modulus of a linear function of the phase vector at a given time moment, a linear change of variables leads to a single-type differential game when the sets of player controls values are time-dependent segments. In a more general case, such problems are characterized by the fact that player controls vectograms are balls whose radii depend on time. Such dynamics after a change of variable arises in well-known differential games "isotropic missiles" [3], control example Pontryagin [9]. For such differential games, in the case when the terminal set is a ball of a given radius, the form of an alternating integral is found in [9]. In [10], optimal positional strategies for players were constructed. In [12] the form of alternating integral for single-type games with an arbitrary convex closed terminal set is found and optimal positional controls of players are constructed.

The problems of controlling oscillatory mechanical systems are actual. When conducting research and educational process in the field of mechatronics and

This work was funded by the Russian Science Foundation (project no. 19-11-00105).

M. Khachay et al. (Eds.): MOTOR 2019, LNCS 11548, pp. 595–606, 2019.
https://doi.org/10.1007/978-3-030-22629-9_42

automatic control systems, laboratory pendulum training and laboratory installations are widely used. In [1,2], problems of controlling the system, which consists of a physical pendulum with a flywheel at the end, were investigated. In the works [7,14], a numerical method developed by the authors was applied to solve this problem and computer simulation was carried out.

In this paper, we consider a game problem, in which the first player controls a rod attached to a rotor. A rotating flywheel is attached to one end of the rod. The flywheel is controlled by the second player. The goal of the first player is to bring the rotor to a vertical position at a given time. The goal of the second player is the opposite. This problem is an example of a more general linear differential game with a one-dimensional aim, which is determined using the modulus of a linear function of the phase vector. Using a linear change of variables, this linear differential game reduced to a single-type one-dimensional differential game. The terminal set in this game is the union of an infinite number of disjoint segments.

2 Introductory Example

The rotor axis of the first electric motor passes through the point O perpendicular to the plane of the figure (see Fig. 1). The rod AB is rigidly attached to the axis of the rotor so that it can rotate together with the rotor about its axis in the plane of the figure.

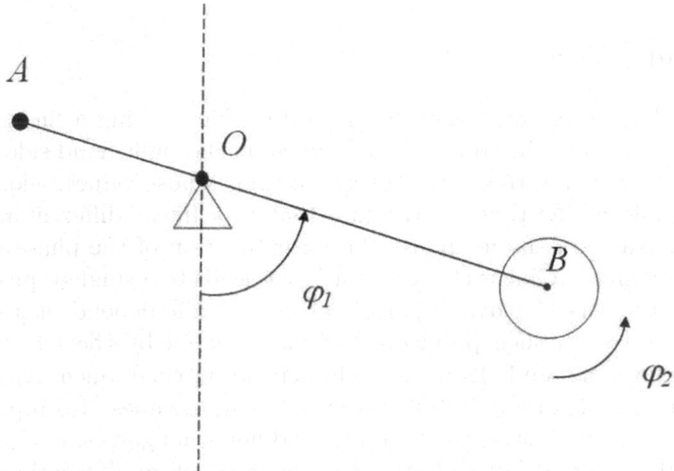

Fig. 1. The game problem of controlling rod AB with a rotating flywheel, which is attached to it at point B.

Rotary flywheel symmetrical about its axis is mounted so that its center is located at the point B. The flywheel can rotate about an axis passing through B perpendicular to the plane of the figure. The axis of rotation of the flywheel is the axis of the rotor of the second electric motor. The total mass of the flywheel and the second electric motor is M, and the total moment of inertia of the flywheel and the second electric motor relative to their axis of rotation is J_*.

The center of mass of the system consisting of the rod AB, the rotor of the first electric motor, the flywheel, and the second electric motor is at the point O, and the moment of inertia of this mechanical system relative to the axis of rotation passing through the point O is equal to J. Distance $OB = L$.

The kinetic energy of the system is equal to

$$T = \frac{1}{2}\left(J + ML^2\right)\dot{\phi}_1^2 + \frac{J_*}{2}(\dot{\phi}_1 + \dot{\phi}_2)^2. \tag{1}$$

Calculate the virtual work $\delta A = N_1\delta\phi_1 + N_2\delta\phi_2$ of external forces acting on the system. Here, N_i is the moment of electromagnetic forces applied to the rotor of i–th motor on the side of its stator, $i = 1, 2$. Neglecting the inductance in the rotor circuit, we assume [2,7,14]

$$N_i = c_i w_i - b_i\dot{\phi}_i, \quad c_i > 0, \quad b_i > 0, \quad i = 1, 2. \tag{2}$$

Here, w_i is the voltage supplied to i–th motor, and it is bounded $|w_i| \leq w_i^0$, $i = 1, 2$. The product $b_i\dot{\phi}_i$ describes moment of forces, which arise because of counter-emf.

Write down Lagrange equations, which describe the motion of the system

$$\frac{d}{dt}\frac{\partial T}{\partial \dot{\phi}_i} - \frac{\partial T}{\partial \phi_i} = N_i, \quad i = 1, 2.$$

Formulas (1) and (2) imply that

$$(J + ML^2)\ddot{\phi}_1 + J_*(\ddot{\phi}_1 + \ddot{\phi}_2) = c_1 w_1 - b_1\dot{\phi}_1,$$

$$J_*(\ddot{\phi}_1 + \ddot{\phi}_2) = c_2 w_2 - b_2\dot{\phi}_2.$$

From here we obtain that

$$\ddot{\phi}_1 = -\frac{b_1}{J + ML^2}\dot{\phi}_1 + \frac{b_2}{J + ML^2}\dot{\phi}_2 + \frac{c_1}{J + ML^2}w_1 - \frac{c_2}{J + ML^2}w_2,$$

$$\ddot{\phi}_2 = \frac{b_1}{J + ML^2}\dot{\phi}_1 - \left(\frac{b_2}{J + ML^2} + \frac{b_2}{J_*}\right)\dot{\phi}_2 - \frac{c_1}{J + ML^2}w_1 + \left(\frac{c_2}{J + ML^2} + \frac{c_2}{J_*}\right)w_2. \tag{3}$$

The goal of the first player is the fulfilment of the inequality

$$\min_{i \in I} |\phi_1(p) - 2\pi i| \leq \varepsilon, \tag{4}$$

where $0 \leq \varepsilon < \pi$, $I = 0, \pm 1, \pm 2, \pm 3, \ldots$. Note that (4) is the condition of the rod deviation from the vertical position is not more than on ε, taking into account the periodicity. The second player has the opposite goal.

Making the change of variables

$$x_1 = \phi_1, \quad x_2 = \dot{\phi}_1, \quad x_3 = \phi_2, \quad x_4 = \dot{\phi}_2$$

and denoting

$$a_1 = \frac{b_1}{J + ML^2}, \quad a_2 = \frac{b_2}{J + ML^2}, \quad a_3 = \frac{b_2}{J + ML^2} + \frac{b_2}{J_*},$$

$$\gamma = \frac{c_1}{J + ML^2}, \quad \delta_1 = \frac{c_2}{J + ML^2}, \quad \delta_2 = \frac{c_2}{J + ML^2} + \frac{c_2}{J_*},$$

we write the system (3) in the following form:

$$\dot{x} = A(t)x - \xi + \eta,$$

where

$$x = \begin{pmatrix} x_1 \\ x_2 \\ x_3 \\ x_4 \end{pmatrix}, \quad A(t) = \begin{pmatrix} 0 & 1 & 0 & 0 \\ 0 & -a_1 & 0 & a_2 \\ 0 & 0 & 0 & 1 \\ 0 & a_1 & 0 & -a_3 \end{pmatrix}, \tag{5}$$

$$\xi = \begin{pmatrix} 0 \\ -\gamma w_1 \\ 0 \\ \gamma w_1 \end{pmatrix}, \quad \eta = \begin{pmatrix} 0 \\ -\delta_1 w_2 \\ 0 \\ \delta_2 w_2 \end{pmatrix}. \tag{6}$$

3 Reduction to a Single-Type Problem

Consider the differential game

$$\dot{x} = A(t)x - \xi + \eta, \quad x(t_0) = x_0; \quad x \in \mathbb{R}^n, \quad t \leq p. \tag{7}$$

Here, control of the first player is $\xi \in W \subset \mathbb{R}^n$, control of the second player is $\eta \in F \subset \mathbb{R}^n$, where W and F are connected compacts; $A(t)$ is a matrix with corresponding dimension whose elements are continuous for $t_0 \leq t \leq p$ functions.

Vector $\psi_0 \in \mathbb{R}^n$ and numbers $\alpha, \varepsilon \in \mathbb{R}$ such that $0 \leq 2\varepsilon < \alpha$ are given. The goal of the first player is the fulfilment of the inequality

$$\min_{i \in I} |\langle \psi_0, x(p) \rangle - i\alpha| \leq \varepsilon. \tag{8}$$

Here, $\langle \cdot, \cdot \rangle$ denotes the scalar product in $\mathbb{R}^n$. The second player has the opposite goal. Note that condition (4) is an example of condition (8) with

$$\psi_0 = \begin{pmatrix} 1 \\ 0 \\ 0 \\ 0 \end{pmatrix}, \quad \alpha = 2\pi.$$

Denote by $\psi(t)$ the solution of the Cauchy problem

$$\dot{\psi}(t) = -A^*(t)\psi(t), \quad \psi(p) = \psi_0; \quad t \leq p. \tag{9}$$

Here, $A^*(t)$ denotes the transposed of matrix $A(t)$.

Denote

$$a_-(t) = \min_\xi \langle \psi(t), \xi \rangle, \quad a_+(t) = \max_\xi \langle \psi(t), \xi \rangle, \quad \xi \in W;$$

$$b_-(t) = \min_\eta \langle \psi(t), \eta \rangle, \quad b_+(t) = \max_\eta \langle \psi(t), \eta \rangle, \quad \eta \in F.$$

Note that these functions are continuous [8, p. 84, Lemma II.3.5].

Then connectivity of compacts W and F imply [6, pp. 333–334, Theorem 4] that

$$\langle \psi(t), \xi \rangle = \frac{a_+(t) + a_-(t)}{2} + a(t)u, \quad |u| \le 1, \quad a(t) = \frac{a_+(t) - a_-(t)}{2} \ge 0;$$

$$\langle \psi(t), \eta \rangle = \frac{b_+(t) + b_-(t)}{2} + b(t)v, \quad |v| \le 1, \quad b(t) = \frac{b_+(t) - b_-(t)}{2} \ge 0.$$

Introduce a new one-dimensional variable

$$z = \langle \psi(t), x \rangle + \frac{1}{2} \int_t^p (b_+(r) + b_-(r) - a_+(r) - a_-(r)) dr.$$

Differentiate z:

$$\dot{z} = \langle -A^*(t)\psi(t), x \rangle + \langle \psi(t), A(t)x - \xi + \eta \rangle + \frac{1}{2}(a_+(t) + a_-(t) - b_+(t) - b_-(t)).$$

Given equality $\langle \psi(t), A(t)x \rangle = \langle A^*(t)\psi(t), x \rangle$, the problem (7), (8) can be written as follows

$$\dot{z} = -a(t)u + b(t)v, \quad |u| \le 1, \quad |v| \le 1, \quad \min_{i \in I} |z(p) - i\alpha| \le \varepsilon.$$

Here, the terminal condition can also be written as

$$z(p) \in \bigcup_{i \in I} [i\alpha - \varepsilon, i\alpha + \varepsilon],$$

where set from the right side of inclusion is the union of an infinite number of disjoint segments.

4 Problem Statement

The motion of $z \in \mathbb{R}$ occurs according to the rule

$$\dot{z} = -a(t)u + b(t)v, \quad |u| \le 1, \quad |v| \le 1, \quad t \le p. \tag{10}$$

For the completeness of the exposition we assume that the functions $a(t) \ge 0$ and $b(t) \ge 0$ are summable on each segment of the semiaxis $(-\infty, p]$.

The numbers $\alpha, \varepsilon \in \mathbb{R}$ are given such that $0 \le 2\varepsilon < \alpha$. The goal of the first player, which chooses the control u, is to lead point z at time moment p to terminal set Z:

$$z(p) \in Z = \bigcup_{i \in I} [i\alpha - \varepsilon, i\alpha + \varepsilon]. \tag{11}$$

The goal of the second player, which chooses the control v, is the opposite.

Admissible controls of players are arbitrary function, which satisfy inequalities

$$|u(t,z)| \leq 1, \quad |v(t,z)| \leq 1, \quad t \leq p, \quad z \in \mathbb{R}. \tag{12}$$

Fix the initial state $t_0 < p$, $z(t_0) \in \mathbb{R}$ and the time moment $t_0 < t_* \leq p$. Take partition

$$\omega : t_0 < t_1 < \ldots < t_i < t_{i+1} < \ldots < t_k < t_{k+1} = t_*$$

with diameter $d(\omega) = \max(t_{i+1} - t_i)$, $i = \overline{0,k}$. Construct polygonal line for Eq. (10)

$$z_\omega(t) = z_\omega(t_i) - \left(\int_{t_i}^{t} a(r)dr\right) u(t_i, z_\omega(t_i)) + \left(\int_{t_i}^{t} b(r)dr\right) v(t_i, z_\omega(t_i)) \tag{13}$$

for $t_i < t \leq t_{i+1}$. Here, $z_\omega(t_0) = z(t_0)$.

It can be shown that

$$|z_\omega(\tau) - z_\omega(t)| \leq \int_t^\tau (a(r) + b(r))dr \text{ for } t_0 \leq t < \tau \leq t_*.$$

This equality and absolute continuity theorem on the Lebesgue integral [4, p. 282] imply that the family of these polygonal line (13), which are determined on the segment $[t_0, t_*]$, is uniformly bounded and equicontinuous [11, p. 56]. By Arzela's theorem [4, p. 104], from any sequence of the polygonal lines (13) we can select a subsequence $z_{\omega_m}(t)$ with diameter $d(\omega_m) \to 0$ that converges uniformly on the segment $[t_0, t_*]$ to some function $z(t)$.

The motion of the system $z(t)$ realized with admissible controls (12) from the initial state $z(t_0)$ is defined as any uniform limit of the sequence of the polygonal lines (13), for which diameters of partition tend to zero.

5 Necessary and Sufficient Conditions of Termination in a Single-Type Problem

Define function

$$f(t) = \int_t^p (a(r) - b(r))dr$$

for $t \leq p$ and denote

$$q_1 = \inf\{t < p : \varepsilon + f(\tau) < \alpha - \varepsilon - f(\tau) \text{ for all } t < \tau \leq p\},$$

$$q_2 = \inf\{t < p : 0 \leq \varepsilon + f(\tau) \text{ for all } t < \tau \leq p\}.$$

Define set $W(t)$ for $t \leq p$ as follows:

$$W(t) = \begin{cases} \bigcup_{i \in I}[i\alpha - \varepsilon - f(t), i\alpha + \varepsilon + f(t)] & \text{for } \max(q_1, q_2) \leq t \leq p, \\ \mathbb{R} & \text{for } t < q_1, \quad q_2 < q_1 \\ \emptyset & \text{for } t < q_2, \quad q_1 < q_2. \end{cases}$$

Here, $\emptyset$ denotes empty set.

5.1 Solution of Pursuit Problem

We show that if $z(t_0) \in W(t_0)$, then there is a control of the first player, which, with any admissible control of the second player, guarantees inclusion $z(p) \in Z$ for any realized motion $z(t)$.

Case 1. Let $\max(q_1, q_2) \le t_0 \le p$, and the inclusion

$$z(t_0) \in [i\alpha - \varepsilon - f(t_0), i\alpha + \varepsilon + f(t_0)]$$

holds for some $i \in I$.

Consider the differential game (10) with the terminal condition

$$|z(p)| \le \varepsilon, \quad \varepsilon \ge 0. \tag{14}$$

For this problem, the following lemma is performed.

Lemma 1 ([10]). *Let the initial state t_0, $z(t_0)$ be such that*

$$q_2 \le t_0 \le p \quad \text{and} \quad |z(t_0)| \le \varepsilon + f(t_0).$$

Then the control of the first player $u(t, z) = \text{sign } z$ with any admissible control of the second player guarantees the fulfilment of inequality

$$|z(\tau)| \le \varepsilon + f(\tau), \quad t < \tau \le p$$

for any realized motion $z(t)$.

Now we consider the case when the terminal condition in the differential game (10) has the form

$$i\alpha - \varepsilon \le z(p) \le i\alpha + \varepsilon \tag{15}$$

for some $i \in I$.

Let's change the variable

$$z_* = z - i\alpha.$$

Then (15) takes the form

$$|z_*(p)| \le \varepsilon.$$

Thus, in the new coordinates we obtain the differential game (10), (14).

Making the inverse change of variables, from Lemma 1 we obtain the following necessary and sufficient conditions of termination in the differential game (10), (15)

$$q_2 \le t_0 \le p \quad \text{and} \quad i\alpha - \varepsilon - f(t_0) \le z(t_0) \le i\alpha + \varepsilon + f(t_0),$$

and optimal control of the first player takes the form

$$u(t, z) = \text{sign}\,(z - i\alpha)\,. \tag{16}$$

The number $i \in I$ in formula (16) can be calculated as a solution of the minimization problem $\min\limits_{i\in I}|z - i\alpha|$.

Case 2. Let $t_0 < q_1$ and $q_2 < q_1$. According to the definition of q_1, equality

$$\varepsilon + f(q_1) = \alpha - \varepsilon - f(q_1),$$

holds, and, therefore,

$$i\alpha + \varepsilon + f(q_1) = (i+1)\alpha - \varepsilon - f(q_1)$$

holds for any $i \in I$. Thus, for any control of players on segment $[t_0, q_1]$, inclusion

$$z(q_1) \in [i\alpha - \varepsilon - f(q_1), i\alpha + \varepsilon + f(q_1)]$$

holds for some $i \in I$. From here we fall into the condition of case 1.

5.2 Solution of Evasion Problem

We show that if $z(t_0) \notin W(t_0)$, then there is a second player control, which, with any admissible control of the first player, guarantees inclusion $z(p) \notin Z$ for any realized motion $z(t)$.

Case 1. Let $\max(q_1, q_2) \le t_0 \le p$, and inequalities

$$i\alpha + \varepsilon + f(t_0) < z(t_0) < (i+1)\alpha - \varepsilon - f(t_0) \tag{17}$$

are fulfilled for some $i \in I$.

For $\varepsilon_1 > 0$ we denote

$$t(\varepsilon_1) = \inf\{t < p : \varepsilon_1 > f(\tau) \text{ for all } t < \tau \le p\}.$$

Lemma 2 ([13]). *Let the second player chooses control $v(t, z) = -\text{sign } z$. Then for any motion $z(t)$ and for any numbers $t(\varepsilon_1) \le t_* < t^* \le p$, the following condition is satisfied:*

$$\textit{if } |z(t_*)| < \varepsilon_1 - f(t_*), \textit{ then } |z(t^*)| < \varepsilon_1 - f(t^*).$$

Further, by making the change of variable

$$z_* = z - (i + 0.5)\alpha,$$

rewrite (17) as follows

$$|z_*(t_0)| < 0.5\alpha - \varepsilon - f(t_0).$$

Note that for $\varepsilon_1 = 0.5\alpha - \varepsilon > 0$ the definitions of q_1 and $t(\varepsilon_1)$ are the same. Using Lemma 2 and the inverse change of variables, we obtain that the control of the second player

$$v(t, z) = -\text{sign}\,(z - (i + 0.5)\alpha) \tag{18}$$

guarantees fulfilment of inequalities $i\alpha+\varepsilon < z(p) < (i+1)\alpha-\varepsilon$. Thus, inclusion (11) is not fulfilled.

Case 2. Let $t_0 < q_2$ and $q_1 < q_2$. According to the definition of q_2, there is time moment $q_1 < t_* < q_2$ such that

$$\varepsilon + f(\tau) < 0$$

for all $t_* \le \tau < q_2$. Thus, for any player controls on segment $[t_0, t_*]$, inequalities

$$i\alpha + \varepsilon + f(t_*) < z(t_*) < (i+1)\alpha - \varepsilon - f(t_*)$$

are satisfied for some $i \in I$. Further proof is similar to case 1.

6 Example

Let's return to the introductory example. Given the form of the matrix (5), write down the system of equations (9)

$$\dot{\psi}_1 = 0, \quad \dot{\psi}_2 = -\psi_1 + a_1\psi_2 - a_1\psi_4, \quad \dot{\psi}_3 = 0, \quad \dot{\psi}_4 = -a_2\psi_2 - \psi_3 + a_3\psi_4$$

with initial conditions

$$\psi_1(p) = 1, \quad \psi_2(p) = \psi_3(p) = \psi_4(p) = 0.$$

Hence, $\psi_1(t) = 1$, $\psi_3(t) = 0$. Therefore,

$$\dot{\psi}_2 = a_1\psi_2 - a_1\psi_4 - 1, \quad \dot{\psi}_4 = -a_2\psi_2 + a_3\psi_4. \tag{19}$$

Since the numbers $a_i > 0$, $i = 1, 2, 3$, then the characteristic equation

$$\begin{vmatrix} a_1 - \lambda & -a_1 \\ -a_2 & a_3 - \lambda \end{vmatrix} = (\lambda - a_1)(\lambda - a_3) - a_1a_2 = 0$$

has two real roots

$$\lambda_{1,2} = \frac{(a_1 + a_3) \pm \sqrt{(a_1 - a_3)^2 + 4a_1a_2}}{2}.$$

Then the general solution of system (19) takes the following form:

$$\psi_2(t) = c_1a_1e^{\lambda_1 t} + c_2a_1e^{\lambda_2 t} - \frac{a_3}{a_1(a_3 - a_2)},$$

$$\psi_4(t) = c_1(a_1 - \lambda_1)e^{\lambda_1 t} + c_2(a_1 - \lambda_2)e^{\lambda_2 t} - \frac{a_2}{a_1(a_3 - a_2)}.$$

Note that $a_3 > a_2$. Assume $c_i = \beta_i e^{-\lambda_i p}$, $i = 1, 2$. From condition $\psi_2(p) = \psi_4(p) = 0$ we obtain the equations for calculating the numbers β_i:

$$\beta_1 + \beta_2 = \frac{a_3}{a_1^2(a_3 - a_2)}, \quad \lambda_1\beta_1 + \lambda_2\beta_2 = \frac{1}{a_1}.$$

Thus,

$$\psi_1(t) = 1, \quad \psi_2(t) = \beta_1 a_1 e^{-(p-t)\lambda_1} + \beta_2 a_1 e^{-(p-t)\lambda_2} - \frac{a_3}{a_1(a_3 - a_2)}, \quad \psi_3(t) = 0,$$

$$\psi_4(t) = \beta_1(a_1 - \lambda_1)e^{-(p-t)\lambda_1} + \beta_2(a_1 - \lambda_2)e^{-(p-t)\lambda_2} - \frac{a_2}{a_1(a_3 - a_2)}. \tag{20}$$

Formulas (6) imply that the sets W and F are symmetric in the example. Using this, it can be shown that

$$a_+(t) = -a_-(t), \quad b_+(t) = -b_-(t) \quad \text{for all} \quad t \leq p. \tag{21}$$

Therefore,

$$a(t) = \max_{\xi} \langle \psi(t), \xi \rangle, \quad \xi \in W; \quad b(t) = \max_{\eta} \langle \psi(t), \eta \rangle, \quad \eta \in F.$$

These formulas and (6) imply that

$$\langle \psi(t), \xi \rangle = \gamma(-\psi_2(t) + \psi_4(t))w_1, \quad a(t) = \gamma w_1^0 \left| \psi_4(t) - \psi_2(t) \right|;$$

$$\langle \psi(t), \eta \rangle = (-\delta_1 \psi_2(t) + \delta_2 \psi_4(t))w_2, \quad b(t) = w_2^0 \left| \delta_2 \psi_4(t) - \delta_1 \psi_2(t) \right|.$$

Using (21), we obtain that the variable z takes the form

$$z = \langle \psi(t), x \rangle.$$

This equality and (20) imply that

$$z = x_1 + (\beta_1 a_1 x_2 + \beta_1(a_1 - \lambda_1)x_4)e^{-(p-t)\lambda_1}$$

$$+(\beta_2 a_1 x_2 + \beta_2(a_1 - \lambda_2)x_4)e^{-(p-t)\lambda_2} - \frac{a_3 x_2 + a_2 x_4}{a_1(a_3 - a_2)}.$$

Note that z and, consequently, the optimal controls of the players do not depend on $x_3 = \phi_2$, i.e. on the angle of rotation of the flywheel with the center at point B.

The optimal player controls (16) and (18) can be written as follows:

$$u(t, z) = \text{sign}(z - i\pi),$$

where $i \in I$ gives a minimum to the expression $|z - i\pi|$;

$$v(t, z) = -\text{sign}(z - (i + 0.5)\pi),$$

where $i \in I$ gives a minimum to the expression $|z - (i + 0.5)\pi|$.

7 Conclusion

In this paper, we consider the game problem, in which the first player controls the rod attached to the rotor of the electric motor. A rotating flywheel controlled by the second player is attached to one end of the rod. The goal of the first player is to bring the rotor to a vertical position at a given time. The goal of the second player is the opposite. We consider this problem as an example of a more general linear differential game with a one-dimensional aim, which is determined using the modulus of a linear function of the phase vector. Using a linear change of variables, this problem is reduced to a single-type one-dimensional differential game. The terminal set in this game is the union of an infinite number of disjoint segments. This terminal set has the meaning of ε-neighbourhood of the target position of the system, taking into account the periodicity.

Based on our previous results, we find the necessary and sufficient conditions of termination and constructed the corresponding controls of the players in the single-type differential game. Then, using the obtained results, we write down the optimal controls of the players in the original game problem of controlling the rod.

The obtained results can find applications in the design of control algorithms for complex oscillatory mechanical systems. For example, we can consider the behaviour of some elements of such systems as realizations of uncontrolled disturbances, the values of which belong given sets. Mathematical modelling of control in such problems based on an approach that prescribes disturbance behaviour that degrades the quality indicator, in accordance with which control is modelled. This approach leads to the consideration of the problem of constructing a control in the framework of the theory of differential games.

References

1. Andrievsky, B.R.: Global stabilization of the unstable reaction-wheel pendulum. Control Big Syst. **24**, 258–280 (2009). (in Russian)
2. Beznos, A.V., Grishin, A.A., Lenskiy, A.V., Okhozimskiy, D.E., Formalskiy, A.M.: The control of pendulum using flywheel. In: Workshop on Theoretical and Applied Mechanics, pp. 170–195. Publishing of Moscow State University, Moscow (2009). (in Russian)
3. Isaacs, R.: Differential Games: A Mathematical Theory with Applications to Warfare and Pursuit, Control and Optimization. Wiley, New York (1965)
4. Kolmogorov, A.N., Fomin, S.V.: Elements of the Theory of Functions and Functional Analysis. Nauka Publ., Moscow (1972). (in Russian)
5. Krasovskii, N.N., Subbotin, A.I.: Positional Differential Games. Nauka Publ., Moscow (1974). (in Russian)
6. Kudryavtsev, L.D.: A Course of Mathematical Analysis, vol. 1. Vysshaya shkola Publ., Moscow (1981). (in Russian)
7. Matviychuk, A.R., Ukhobotov, V.I., Ushakov, A.V., Ushakov, V.N.: The approach problem of a nonlinear controlled system in a finite time interval. J. Appl. Math. Mech. **81**(2), 114–128 (2017). https://doi.org/10.1016/j.jappmathmech.2017.08.005

8. Pshenichnyi, B.N.: Convex Analysis and Extremal Problems. Nauka Publ., Moscow (1980). (in Russian)
9. Pontryagin, L.S.: Linear differential games of pursuit. Math. USSR-Sbornik **40**(3), 285–303 (1981). https://doi.org/10.1070/SM1981v040n03ABEH001815
10. Ukhobotov, V.I.: Synthesis of control in single-type differential games with fixed time. Bull. Chelyabinsk Univ. **1**, 178–184 (1996). (in Russian)
11. Ukhobotov, V.I.: Method of One-Dimensional Design in Linear Differential Games with Integral Constraints: Study Guide. Publishing of Chelyabinsk State University, Chelyabinsk (2005). (in Russian)
12. Ukhobotov, V.I.: The same type of differential games with convex purpose. Proc. Inst. Math. Mech. Ural. Branch Russ. Acad. Sci. **16**(5), 196–204 (2010). (in Russian)
13. Ukhobotov, V.I., Izmest'ev I.V.: Single-type differential games with a terminal set in the form of a ring. In: Systems Dynamics and Control Process, Proceedings of the International Conference, Dedicated to the 90th Anniversary of Acad. N.N. Krasovskiy, Ekaterinburg, 15–20 September 2014, pp. 325–332. Publishing House of the UMC UPI, Ekaterinburg, Russia (2015). (in Russian)
14. Ushakov, V.N., Ukhobotov, V.I., Ushakov, A.V., Parshikov, G.V.: On solution of control problems for nonlinear systems on finite time interval. IFAC-PapersOnLine **49**(18), 380–385 (2016). https://doi.org/10.1070/10.1016/j.ifacol.2016.10.195

General Limit Value for Stationary Nash Equilibrium

Dmitry Khlopin(✉)

Krasovskii Institute of Mathematics and Mechanics, Yekaterinburg 620990, Russia
khlopin@imm.uran.ru

Abstract. We analyze the uniform asymptotics of the equilibrium value (as a function of initial state) in the case when its payoffs are averaged with respect to a density that depends on some scale parameter and this parameter tends to zero; for example, the Cesàro and Abel averages as payoffs for the uniform and the exponential densities, respectively. We also investigate the robustness of this asymptotics of the equilibrium value with respect to the choice of distribution when its scale parameter is small enough. We establish the class of densities such that the existence of the asymptotics of the equilibrium value for some density guarantees the same asymptotics for a piecewise-continuous density; in particular, this class includes the uniform, exponential, and rational densities. By reducing the general n-person dynamic games to mappings that assigns to each payoff its corresponding equilibrium value, we gain an ability to consider dynamic games in continuous and discrete time, both in deterministic and stochastic settings.

Keywords: n-person game · Dynamic game · Uniform value · Stationary Nash equilibrium · Tauberian theorem · Weighted payoffs

Introduction

In this paper, we study the asymptotic properties of dynamic games' equilibrium value with the payoff taken in the form of an average with respect to the given discount function (density) as the time scale decreases (the horizon gets larger and larger).

In dynamic optimization, ever so often, the payoff can be averaged over time with respect to a certain probability distribution. In this case, to a realization of the process—some function $t \mapsto z(t)$—in addition to the running cost—a function $t \mapsto g(z(t))$—one takes a payoff in the form of a certain average of the running cost,

$$\int_0^\infty \rho(t) g(z(t))\, dt,$$

with respect to a certain discount function, a density ρ. Most often, when the problem is considered on infinite horizon, the potential infinity of the interval is

Supported by the Russian Science Foundation (project no. 17-11-01093).

M. Khachay et al. (Eds.): MOTOR 2019, LNCS 11548, pp. 607–619, 2019.
https://doi.org/10.1007/978-3-030-22629-9_43

emulated by considering the problems where the payoff is taken over increasingly large intervals $[0,T]$ or in view of increasingly small discount λ; then, the limits of these problems are studied if such exist. Thus, effectively, one considers the asymptotic behavior of the equilibrium value for the payoffs

$$\int_0^\infty \lambda\rho(\lambda t)g(z(t))\,dt \tag{1}$$

as the scale parameter λ tends to zero, for instance, for the densities of the uniform $\rho(t) = 1_{[0,1]}(t)$ (Cesàro mean) and exponential $\rho(t) = e^{-t}$ (Abel mean) distributions, respectively.

Existence of a limit of equilibrium value with respect to density means that the equilibrium value is robust with respect to the scale parameter λ if this parameter is sufficiently small. In particular, in the stochastic statement, this limit (the asymptotic value) is customarily considered the equilibrium value when the planning horizon is infinite [3]. In these statements, one could often obtain, in addition, an asymptotically optimal strategy guaranteeing a payoff that is close to the optimal one (uniform value)—for sufficiently small scale parameter [18]. In this paper, we will have to assume the existence of the equilibrium value asymptotics in the sense of the uniform approach. Unfortunately, the existence of the uniform equilibrium in nonzero-sum stochastic games was proven only in special cases, see [7,22]; a lonely example in differential games was obtained in [12]. In the case of zero-sum games, the conditions of existence of uniform strategy are known well enough, see, for example [2,4–6,15,16].

Assuming the existence of these limits of equilibrium values for a certain density family (for example, the Abel mean), we can consider the same game with another density family (for example, the Cesàro mean). The theorems relating the asymptotics of different densities with each other are called Tauberian theorems. For a variety of two-person zero-sum dynamic game and optimal control statements, it is possible to obtain the following Tauberian theorem: the uniform convergence of the equilibrium value for the running costs averaged with respect to the uniform and/or exponential distributions guarantees the uniform convergence to the same limit for the other distribution [9,14,18,24]. Later, a Tauberian theorem for all two-person zero-sum games satisfying the Dynamic Programming Principle was proved [11]; the case of the stationary-like uniform Nash equilibrium was considered in [12].

Then, robustness can be stated in a more broad form: find a sufficient condition on the distribution family ϱ_λ under which the uniform convergence of equilibrium values with the payoffs averaged in view of Cesàro/Abel means (or in view of some other density family) implies the uniform convergence to the same limit of the equilibrium values for all distributions. For example, in paper [19], for discrete-time control processes, it was proved that the uniform convergence of the equilibrium values for Cesàro/Abel means automatically implies the uniform convergence (to the same limit) of the equilibrium values in view of payoffs (1) for arbitrary nonincreasing density ρ (since density could be rendered as a convex combination of Cesàro means). This result may be significantly improved upon, at the very least, for the control systems on compact invariant sets under nonexpansive dynamics assumption; however, without additional assumptions, this may already fail on Markov discrete-time control processes (see [25]). In the general case of zero-sum two-person game, the corresponding general Tauberian theorem was proved as a consequence of the Dynamic Programming Principle in [13]. In this article, we extend this result to the stationary-like uniform Nash equilibrium.

We consider the uniform equilibrium values' robustness under the choice of the scale parameter and the choice of density. Previously, this robustness could be verified with only the exponential or uniform densities as initial; the same result was obtained in [8,19,21,25] for different zero-sum game conceptions. A part of the reason for considering the uniform or exponential families first is the fact that they are invariant under scaling; however, other densities were also considered (see, for instance, [17]). In [13], it is stated that the polynomial densities of the kind $(A+Bt)^\gamma$ can be considered as initial density. Here, we improve this result, allowing one to consider a rational density. From the existence of a uniform equilibrium value with respect to a rational, uniform, or exponential densities, it follows that this value is the common uniform equilibrium value with respect to a continuous density.

The structure of the paper is as follows. In Sect. 1, we give the general statement and definitions, in particular, the definitions of uniform equilibrium value and a stationary-like strategy profile; further, we describe and discuss the needed assumptions. The main result, Theorem 1, the Tauberian theorem for the uniform equilibria with arbitrary densities, is formulated in Sect. 2, and an example is presented. The next section is devoted to the proof of Theorem 1; this proof is by reduction to [13, Theorem 6] in the statement from [12].

1 The Game Model

1.1 The Dynamics

Dynamic System. Assume the following items are given:

- a nonempty set Ω of states;
- the set of players $I \triangleq \{1, 2, \ldots, n\}$;
- a nonempty set $\mathbb{K}$ of processes, which are maps from $\mathbb{R}_+$ to Ω;
- a running cost $g^i : \Omega \mapsto [0, 1]$ for every player $i \in I$;

Assume that the map $t \mapsto g^i(z(t))$ is Borel measurable for every process $z \in \mathbb{K}$ and every player $i \in I$.

For every player $i \in I$, fix a nonempty strategy set $\mathcal{S}_i$. Let $\mathcal{S} \triangleq \prod_{i\in I} \mathcal{S}_i$ be the set of all strategy profiles. For each $\omega \in \Omega$, each strategy profile $s \in \mathcal{S}$ generates a unique process $z[\omega, s] \in \mathbb{K}$.

Let $s^{-i} \triangleq (s^1, \ldots, s^{i-1}, s^{i+1}, \ldots, s^n)$ for all $i \in I$ and let $s = (s^1, \ldots, s^n) \in \mathcal{S}$; furthermore, for every $s, \hat{s} \in \mathcal{S}$, set $[\hat{s}^i, s^{-i}] \equiv (s^1, \ldots, s^{i-1}, \hat{s}^i, s^{i+1}, \ldots, s^n)$, $[s^i, s^{-i}] \equiv s$ for all $i \in I$.

1.2 Densities and Weighted Payoffs

Let $\mathfrak{D}$ be the set of all Borel measurable maps $\varrho\colon \mathbb{R}_+ \to \mathbb{R}_+$ such that

$$\int_0^\infty \varrho(t)\,dt = \lim_{T\uparrow\infty} \int_0^T \varrho(t)\,dt = 1.$$

The elements of $\mathfrak{D}$ will be called densities.

For each density $\varrho \in \mathfrak{D}$ and each number $r \in (0,1)$, the quantile $q[\varrho](r)$ is uniquely defined as the minimum root of

$$\int_0^{q[\varrho](r)} \varrho(t)\,dt = r.$$

For a density $\varrho \in \mathfrak{D}$ and a positive T with the property $\int_T^\infty \varrho(t)\,dt > 0$ and them only, it is possible to introduce the density ϱ^T_{shift} by the following rule:

$$\varrho^T_{\text{shift}}(t) = \frac{\varrho(t+T)}{\int_T^\infty \varrho(t)\,dt} \qquad \forall t \geq 0.$$

For each density $\varrho \in \mathfrak{D}$ and arbitrary $\lambda > 0$, define the density $\varrho^\lambda_{\text{scale}}$ by the rule

$$\varrho^\lambda_{\text{scale}}(t) = \lambda\varrho(\lambda t) \qquad \forall t \geq 0.$$

To a density ϱ, assign a payoff profile $\mathtt{c}[\varrho] : \mathbb{K} \to \mathbb{R}^n$ by the following rule:

$$\mathtt{c}^i[\varrho](z) \triangleq \int_0^\infty \varrho(t) g^i(z(t))\,dt, \quad \forall i \in I, z \in \mathbb{K}.$$

1.3 On Accepted Value and Best-Reply Value

Let there be given a payoff profile c. Let all players pick a strategy profile $s_* \in \mathcal{S}$. Then, each player must take the accepted value

$$\mathbb{V}^{i,\flat}[c^i](\omega) \triangleq c^i\big(z[\omega, s_*]\big) \qquad \forall \omega \in \Omega.$$

However, each player could hope to come up with his own best-reply value

$$\mathbb{V}^{i,\sharp}[c^i](\omega) \triangleq \sup_{s^i \in \mathcal{S}_i} c^i\big(z[\omega, s^i, s_*^{-i}]\big) \qquad \forall \omega \in \Omega$$

assuming other players keep s_*.

Recall that a strategy profile $s_* \in \mathcal{S}^n$ is a Nash equilibrium for a payoff c with an initial state $\omega \in \Omega$ if $\mathbb{V}^{i,\sharp}[c^i](\omega) = c^i(z[\omega, s_*])$ for all $i \in I$; moreover, a strategy profile $s_* : \mathbb{R} \to \mathcal{S}$ is a Nash ε-equilibrium for c with an initial state $\omega \in \Omega$ if $\mathbb{V}^{i,\sharp}[c^i](\omega) \leq c^i(z[\omega, s_*]) + \varepsilon$ for all $i \in I$. Thus, the best-reply value and the accepted value coincide for each player in the case of Nash equilibrium, and are ε-closed in the case of ε-Nash equilibrium.

Usually, Nash equilibrium is defined only for the fixed initial state and fixed payoff profile c. Nonetheless, in this article we deal with the asymptotics of the equilibrium value (as a map from state space) for weighted payoff families $(c_\lambda)_{\lambda>0}$ as $\lambda \downarrow 0$ for all initial states. These asymptotics are linked with the corresponding Tauberian theorem, and these theorems have to consider the equilibrium value for all initial states; in particular, for the simplest case—a control problem—we have to require the uniform convergence, uniform on a given invariant set; otherwise, see the corresponding counterexample in [20]. By this reason, we also have to consider the uniform case, i.e., we assume that some strategy profile $s_* \in \mathbb{S}$ is a Nash ε-equilibrium for $(c_\lambda)_{\lambda>0}$ and all initial states $\omega \in \Omega$ if λ is sufficiently small. Here, we assume that s_* is independent of the initial state ω and the choice of λ. The first requirement is not essential; if we wanted it, we could consider the sets of all functions $s : \Omega \to \mathbb{S}_i$ as new sets $\mathbb{S}_i$. The second requirement is important and essential, the uniform approach (see Introduction) is based on the existence of a Nash equilibrium robust under the choice of sufficiently large interval $[0, T]$ and/or of sufficiently small discount rate λ.

Definition 1. *Let us say that a strategy profile $s_* \in \mathbb{S}$ is a uniform Nash equilibrium for the payoff profile family $(c_\lambda)_{\lambda>0}$ if the limit*

$$\lim_{\lambda\downarrow 0}\left|\mathbb{V}^{i,\sharp}[c^i_\lambda](\omega) - c^i_\lambda\big(z[\omega, s_*]\big)\right| = 0 \qquad \forall \omega \in \Omega, i \in I \tag{2}$$

exists and is uniform in $\omega \in \Omega$.

Definition 2. *Let us say that a function $U_* = (U^1_*, \ldots, U^n_*) : \Omega \to \mathbb{R}^n$ is the uniform equilibrium value of $s_* \in \mathbb{S}$ for the payoff profile family $(c_\lambda)_{\lambda>0}$ as $\lambda \downarrow 0$ if $s_* \in \mathbb{S}$ is a uniform Nash equilibrium for this payoff family and the limit*

$$\lim_{\lambda\downarrow 0}\mathbb{V}^{i,\sharp}[c^i_\lambda](\omega) = U^i_*(\omega) \qquad \forall \omega \in \Omega$$

exists and is uniform in $\omega \in \Omega$ for all $i \in I$.

It would be instructive to assume just (2), however, the author is unaware of such a Tauberian theorem in game statements. Hereinafter we can assume the existence of the uniform equilibrium value for a given payoff family. In Sect. 2, we apply the Tauberian theorem proved in this paper to the uniform equilibrium from [12].

1.4 Concatenations and Stationary-Like Strategies

We must also impose two conditions on strategies and processes. Similarly to [12], for Nash equilibrium, consider merely stationary-like strategy profiles.

Definition 3. *We will say that a strategy profile $s_* \in \mathbb{S}$ is stationary-like if*

$$z\big[z[\omega, s_*](\Delta), s_*\big](t) = z[\omega, s_*](t + \Delta) \quad \forall \omega \in \Omega, \Delta, t \geq 0. \tag{3}$$

Note that this condition on s_* is very strong; on the other side, we will hope that it could be weakened by the method of [10, Sect. 2], by applying multi-valued strategies. For differential games, similar constructions were applied in [1,9].

For the second condition, we need the concatenation of processes. For all $\tau \in \mathbb{R}_+, z', z'' \in \mathbb{K}$ with the property $z'(\tau) = z''(0)$ and them only, we define their concatenation, a map from $\mathbb{R}_+$ to Ω, as follows:

$$(z' \diamond_\tau z'')(t) \triangleq \begin{cases} z'(t), & 0 \leq t \leq \tau; \\ z''(t-\tau), & t > \tau. \end{cases}$$

Recall the conditions of the Tauberian theorem for a most simple one-person game, in particular, for a control problem. In [20], it was considered that the set of processes was closed with respect to concatenation. Later (see [10, Sect. 7.2]), it was shown that the restriction of an admissible process to a right-infinite interval must also be an admissible process. Similarly to [12], we require these conditions for the set of all processes generated by s^* and for the set of all processes generated by $(s_*^{-i}, \hat{s}^i)$. We assume the following condition on a stationary-like strategy profile $s_* \in \mathcal{S}$: for all $i \in I, \tau \in \mathbb{R}$,

$$\{z[\omega, s^i, s_*^{-i}] \mid s^i \in \mathcal{S}_i\} = \{\check{z} \diamond_\tau z[\check{z}(\tau), \hat{s}^i, s_*^{-i}] \mid \hat{s}^i, s^i \in \mathcal{S}_i, \check{z} = z[\omega, s^i, s_*^{-i}]\}. \quad (4)$$

This condition guarantees that every player $i \in I$ could declare his strategy stepwise for a time interval, not necessarily for the whole positive semiaxis, and this stepwise declaration of Nash equilibrium would generate this Nash equilibrium just as well. Condition (4) is sufficiently weak and it is similar to the conditions of the Tauberian theorem [20], [10, (7.1)] for one-person game, and condition [12, (5)]. This condition could also be relaxed in two ways. First, similarly to [12, (4)], we could assume only the inclusion "$\subset$" instead of the equality "$=$" in (4). Second, we could consider only natural τ in (4) for discrete-time statements, i.e., each process z would satisfy $z(n+r) = z(n)$ for natural n and $r \in (0,1)$. These modifications will have required nothing more than additional cumbersome references to proofs from [13]. The proof of the result below will be much more simple.

2 Tauberian Theorem

Set the exponential and uniform density families as follows:

$$\varpi_\lambda(t) = \lambda \cdot 1_{[0,1/\lambda]}(t), \quad \pi_\lambda(t) = \lambda \cdot e^{-\lambda t}, \qquad \forall \lambda > 0, t \geq 0.$$

A density $\varrho \in \mathfrak{D}$ is called rational if ϱ is a nonnegative rational function.

Theorem 1. *Let a stationary-like strategy profile $s_* \in \mathcal{S}$ satisfy (4).*

Then, for a given arbitrary function $U_* : \Omega \to \mathbb{R}^n$, the following conditions are equivalent:

(c) *for all piecewise continuous on $(0, \infty)$ densities $\mu \in \mathfrak{D}$, U_* is the uniform equilibrium value of s^* for the payoff profile family $(\mathtt{C}[\mu^\lambda_{scale}])_{\lambda>0}$;*

(u) U_* *is the uniform equilibrium value of* s^* *for the family* $\left(\mathtt{C}[\varpi_\lambda]\right)_{\lambda>0}$;
(e) U_* *is the uniform equilibrium value of* s^* *for the family* $\left(\mathtt{C}[\pi_\lambda]\right)_{\lambda>0}$;
(p) for a certain rational density $\rho\in\mathfrak{D}$, U_* *is the uniform equilibrium value of* s^* *for the payoff profile family* $\left(\mathtt{C}[\rho^{1/\lambda}_{shift}]\right)_{\lambda>0}$;
(q) for all rational densities $\rho\in\mathfrak{D}$, U_* *is the uniform equilibrium value of* s^* *for the payoff profile family* $\left(\mathtt{C}[\rho^{1/\lambda}_{shift}]\right)_{\lambda>0}$.

In [12] for uniform and exponential densities there was considered a uniform stationary Nash equilibrium in a nonlinear modification of the Lanchester model of competition of two firms ($i=1,2$) proposed in [23]. Let us extend this result over all densities.

Example 1. For each density μ, consider the game

$$\dot x_1=u_1\sqrt{1-x_1}-u_2\sqrt{x_1},\quad \dot x_2=u_2\sqrt{1-x_2}-u_1\sqrt{x_2}, \tag{5}$$

$$x_1(0)=y_{*1}>0,\ x_2(0)=y_{*2}>0,\ y_{*1}+y_{*2}=1, u_1,u_2\geq 0, \tag{6}$$

$$\mathtt{C}^i[\mu](x_1,x_2,u_1,u_2)\triangleq\int_0^\infty\mu(t)\big[q_ix_i(t)-\frac{1}{4}(u_i(t))^2\big]\,dt\quad\forall i\in\{1,2\}. \tag{7}$$

Here x_i is the market share of firm i, u_i is advertising rate of firm i, q_i is the unit margin of firm i, and the payoff (7) is typical for Lanchester-type model with constant potential market. It is easy to see that (6) guarantees $x_1(t)+x_2(t)\equiv 1$.

Fix a number $Q\geq 2\sqrt{q_1+q_2}$. Let $\mathbb{S}_1=\mathbb{S}_2$ be the set of all bounded maps $\mathbb{R}_+\times(0,1)\ni(t,x)\mapsto s(t,x)\in[0,Q]$ such that (1) $s(t,\cdot)$ are Lipschitz continuous on any compact subset of $(0,1)$ for all $t\geq 0$; (2) $s(\cdot,x)$ are Borel measurable for all $x\in(0,1)$; (3) there exists a positive $\varkappa$ such that $s(t,x)>\varkappa$ if $x<\varkappa$.

Paper [23] thoroughly investigated Nash equilibria of the game (5)–(7) with exponential densities π_{r_1},π_{r_2}; thanks to [23, Theorem 5], the unique feedback Nash equilibrium for this game is the following pair of stationary strategies:

$$\hat{s}^i_{r_1,r_2}(x_i)=2a_i\sqrt{1-x_i},\qquad\{i,j\}=\{1,2\}, \tag{8}$$

where (a_1,a_2) is the unique positive solution of the system

$$a_ir_i+a_i(a_i+a_j)=q_i,\qquad\{i,j\}=\{1,2\}.$$

Similarly [12], put $Y_0\triangleq\{(y_1,y_2)\,|\,y_1+y_2=1,y_1,y_2>0\}$. $\Omega\triangleq Y_0\times[0,Q]^2$, $\mathbb{K}\triangleq C(\mathbb{R}_+,Y_0)\times B(\mathbb{R}_+,[0,Q]^2)$. Define the running cost $g^i(\omega)=\frac{1}{2}+\frac{1}{2Q^2}\big[q_iy_i-\frac{1}{4}(u_i)^2\big]\in(0,1)$ for all initial data $\omega=(y_1,y_2,u_1,u_2)$. For each $\omega=(y_{1*},y_{2*},u_1,u_2)\in\Omega$ and strategy profile $s=(s^1,s^2)\in\mathbb{S}_1\times\mathbb{S}_2$ we set

$$z[\omega,u](t)=(x_1(t),x_2(t),u_1(t),u_2(t)),$$

where (x_1,x_2) is the solution of (5) with the initial conditions $(x_1,x_2)(0)=(y_{1*},y_{2*})$ and controls $u_1(\cdot)=s^1(\cdot,x_1(\cdot)),u_2(\cdot)=s^2(\cdot,x_2(\cdot))$.

Then we construct the basic statement of Sect. 1.

Let (b_1, b_2) be the unique positive solution of the system

$$b_i(b_i + b_j) = q_i, \ \{i, j\} = \{1, 2\}.$$

Define the strategy profile $\hat{s}_{*o} = (\hat{s}^1_{*o}, \hat{s}^2_{*o})$ by (8) with $(a_1, a_2) = (b_1, b_2)$.

In [12] it was shown that the strategy profile $\hat{s}_{*o} = (\hat{s}^1_{*o}, \hat{s}^2_{*o})$, defined by (8) with $(a_1, a_2) = (b_1, b_2)$, is the uniform feedback Nash equilibrium for game (5)–(7) with exponential density family $(\pi_\lambda)_{\lambda>0}$ (and, as consequence, uniform density family $(\varpi_\lambda)_{\lambda>0}$). Then, Theorem 1 guarantees that for all piecewise continuous on $(0, \infty)$ densities $\mu \in \mathfrak{D}$ the strategy profile $\hat{s}_{*o} = (\hat{s}^1_{*o}, \hat{s}^2_{*o})$ is the common uniform feedback Nash equilibrium for game (5)–(7) with the density family $\mu^\lambda_{\text{scale}}$.

3 The Proof of Theorem 1

The proof of this theorem will be reduced to [13, Theorem 6]. At the beginning, we must verify the main conditions of [13, Theorem 6] for the accepted and best reply values. In [12], it was proved that the maps $c \mapsto \mathbb{V}^{i,\sharp}[c]$ and $c \mapsto \mathbb{V}^{i,\flat}[c]$ are game value maps. Further, in the proof of [12, Theorem 1], the Dynamic Programming Principle was established for these values with respect to exponential and uniform densities. To apply [13, Theorem 6], we will only need to verify this principle with respect to an arbitrary density.

Fix a density $\rho \in \mathfrak{D}$ and some positive number T such that $\int_0^T \rho(t)\,dt < 1$. Consider two payoff profiles: for all $i \in I, z \in \mathbb{K}$,

$$c^{i,\flat}[\rho](z) \triangleq \int_0^T g^i(z(t))\,dt + \int_T^\infty \rho(t)\,dt \mathbb{V}^{i,\flat}\big[\mathtt{c}^i[\rho^T_{\text{shift}}]\big](z(T)),$$

$$c^{i,\sharp}[\rho](z) \triangleq \int_0^T g^i(z(t))\,dt + \int_T^\infty \rho(t)\,dt \mathbb{V}^{i,\sharp}\big[\mathtt{c}^i[\rho^T_{\text{shift}}]\big](z(T)).$$

We must prove that $\mathbb{V}^{i,\flat}\big[\mathtt{c}^i[\rho]\big] \equiv \mathbb{V}^{i,\flat}\big[c^{i,\flat}\big]$ and $\mathbb{V}^{i,\sharp}\big[\mathtt{c}^i[\rho]\big] \equiv \mathbb{V}^{i,\sharp}\big[c^{i,\sharp}\big]$.

Fix some initial data $\omega \in \Omega$; define $\bar{z} \triangleq z[\omega, s_*]$. Then,

$$\begin{aligned}
\mathbb{V}^{i,\flat}\big[\mathtt{c}^i[\rho]\big](\omega) &= \mathtt{c}^i[\rho]\big[z[\omega, s_*]\big] \\
&= \int_0^T \rho(t) g^i\big(\bar{z}(t)\big)\,dt + \int_0^\infty \rho(t+T) g^i\big(\bar{z}(t+T)\big)\,dt \\
&\overset{(3)}{=} \int_0^T g^i\big(\bar{z}(t)\big)\,dt + \int_0^\infty \rho(t+T) g^i\big(z[\bar{z}(T), s_*](t)\big)\,dt \\
&= \int_0^T g^i\big(\bar{z}(t)\big)\,dt + \int_T^\infty \rho(t)\,dt \int_0^T \rho^T_{\text{shift}}(t) g^i\big(z[\bar{z}(T), s_*](t)\big)\,dt \\
&= \int_0^T g^i\big(z[\omega, s_*](t)\big)\,dt + \int_T^\infty \rho(t)\,dt \mathbb{V}^{i,\flat}\big[\mathtt{c}^i[\rho^T_{\text{shift}}]\big]\big(z[\omega, s_*](T)\big) \\
&= c^{i,\flat}\big(z[\omega, s_*]\big) = \mathbb{V}^{i,\flat}\big[c^{i,\flat}\big](\omega).
\end{aligned}$$

For the best reply value, we also have

$$\begin{aligned}
\mathbb{V}^{i,\sharp}\big[\mathtt{c}^i[\rho]\big](\omega) &= \sup_{s^i\in\mathcal{S}_i}\int_0^\infty \rho(t)g^i(z[\omega,s^i,s_*^{-i}](t))\,dt\\
&\overset{(4)}{=} \sup_{s^i,\hat{s}^i\in\mathcal{S}_i,\check{z}\triangleq z[\omega,s^i,s_*^{-i}]}\int_0^\infty \rho(t)g^i\big((\check{z}\diamond_T z[\check{z}(T),\hat{s}^i,s_*^{-i}])(t)\big)\,dt\\
&= \sup_{s^i,\hat{s}^i\in\mathcal{S}_i,\check{z}\triangleq z[\omega,s^i,s_*^{-i}]}\Bigg[\int_0^T \rho(t)g^i(\check{z}(t))\,dt\\
&\qquad+\int_0^\infty \rho(t+T)g^i\Big(z\big[\check{z}(T),\hat{s}^i,s_*^{-i}\big](t)\Big)\,dt\Bigg]\\
&= \sup_{s^i\in\mathcal{S}_i,\check{z}\triangleq z[\omega,s^i,s_*^{-i}]}\Bigg[\int_0^T \rho(t)g^i(\check{z}(t))\,dt\\
&\qquad+\int_T^\infty \rho(t)\,dt\sup_{\hat{s}^i\in\mathcal{S}_i}\int_0^T \rho^T_{\mathrm{shift}}(t)g^i\big(z\big[\check{z}(T),\hat{s}^i,s_*^{-i}\big](t)\big)\,dt\Bigg]\\
&= \sup_{s^i\in\mathcal{S}_i}\Bigg[\int_0^T \rho(t)g^i(z[\omega,s^i,s_*^{-i}](t))\,dt\\
&\qquad+\int_T^\infty \rho(t)\,dt\;\mathbb{V}^{i,\sharp}\big[\mathtt{c}^i[\rho^T_{\mathrm{shift}}]\big]\big(z[\omega,s^i,s_*^{-i}](T)\big)\Bigg]\\
&= \sup_{s^i\in\mathcal{S}_i} c^{i,\sharp}\big(z[\omega,s^i,s_*^{-i}]\big) = \mathbb{V}^{i,\sharp}[c^{i,\sharp}](\omega).
\end{aligned}$$

So, the accepted value and the best reply value satisfy the Dynamic Programming Principle. Thus, we now have all the main conditions of [13, Theorem 6].

We prove Theorem 1 by the following scheme: $(u)\Leftrightarrow(e)$, $(c)\Rightarrow(e)$, $(e)\Rightarrow(c)$, $(e)\Rightarrow(q)$, $(q)\Rightarrow(p)$, $(p)\Rightarrow(e)$.

$(c)\Leftrightarrow(e)$. This implication is evident because $(\pi_1)^\lambda_{\mathrm{scale}}=\pi^\lambda_{\mathrm{scale}}$ and $(\varpi_1)^\lambda_{\mathrm{scale}}=\varpi^\lambda_{\mathrm{scale}}$ for all $\lambda>0$.

$(u)\Leftrightarrow(e)$. This implication was proved as [12, Theorem 1, $(u)\Leftrightarrow(e)$].

$(u)\Rightarrow(c)$. Assume (e). Then, $\mathbb{V}^{i,\sharp}\big[\mathtt{c}^i[\pi_\lambda]\big]$ and $\mathbb{V}^{i,\flat}\big[\mathtt{c}^i[\pi_\lambda]\big]$ uniformly on Ω converge to U^i_*. Applying [13, Theorem 1, $(e)\Rightarrow(c)$] to $\mathbb{V}^{i,\sharp}$, we see that $\mathbb{V}^{i,\sharp}\big[\mathtt{c}[\mu^\lambda_{\mathrm{scale}}]\big]$ uniformly on Ω converges to U^i_* for all piecewise continuous on $(0,\infty)$ densities $\mu\in\mathfrak{D}$. Applying this implication to $\mathbb{V}^{i,\flat}$, we obtain the same for $\mathbb{V}^{i,\flat}\big[\mathtt{c}[\mu^\lambda_{\mathrm{scale}}]\big]$. Then, $\mathbb{V}^{i,\sharp}\big[\mathtt{c}^i[\mu^\lambda_{\mathrm{scale}}]\big]-\mathbb{V}^{i,\flat}\big[\mathtt{c}^i[\mu^\lambda_{\mathrm{scale}}]\big]$ uniformly on Ω converges to zero, s^* is uniform Nash equilibrium, and U_* is the uniform equilibrium value of $s_*\in\mathcal{S}$ for all families $(\mu^\lambda_{\mathrm{scale}})_{\lambda>0}$. So, (c) is proved.

$(e)\Rightarrow(q)$. Assume (e). Fix a rational density $\rho(t)=P(t)/Q(t)$. Clearly,

$$\lim_{t\to\infty}\rho(t)=\lim_{t\to\infty}P(t)/Q(t)=0+.$$

Further, $\rho(T)$ and $\int_T^\infty \rho(t)dt$ tend to 0 as $T \to \infty$. Hence, $\frac{d}{dT}\Big(\int_T^\infty \rho(t)dt\Big) = -\rho(T)$ and, by the L'Hôpital rule, we have

$$\lim_{T\uparrow\infty} \frac{\rho(T)}{\int_0^T \rho(t)\,dt} = \lim_{T\uparrow\infty} \frac{d\rho(T)}{dT\,\rho(T)} = \lim_{T\uparrow\infty} \Big[\frac{dP(T)}{dT\,P(T)} - \frac{dQ(T)}{dT\,Q(T)}\Big] = 0. \quad (9)$$

Moreover, we can choose some natural k and positive A and B such that, for all positive λ and T,

$$\begin{aligned}\int_T^\infty \rho(t)\,dt &= AT^{-k}(1+O(1/T)),\\ \rho(T) &= BT^{-k-1}(1+O(1/T)),\\ q[\rho](1-\lambda) &= (\lambda/A)^{-\frac{1}{k}}(1+O(\lambda^{\frac{1}{k}})),\\ \rho(q[\rho](1-\lambda)) &= B\lambda^{\frac{k+1}{k}}(1+O(\lambda^{\frac{k+1}{k}})).\end{aligned} \quad (10)$$

Since ρ has a finite number of local maximums, we can assume that ρ is decreasing on the positive semiaxis; otherwise, we can consider $\rho^T_{\rm shift}$ for a sufficiently large T.

Also, note that existence of a uniform equilibrium value for the density family $(\rho^{1/\lambda}_{\rm shift})_{\lambda>0}$ implies the same for the density family $(\rho^{q[\rho](1-\lambda)}_{\rm shift})_{\lambda>0}$.

We again reduce the proof to a certain implication from [13]: to [13, Theorem 6, $(e) \Rightarrow (v)$]. Condition (v) means the uniform on Ω convergence of $\mathbb{V}^{i,\flat}\big[\mathtt{c}^i[\nu_\lambda]\big]$ and $\mathbb{V}^{i,\sharp}\big[\mathtt{c}^i[\nu_\lambda]$ to U^i_* with respect to an arbitrary density family $(\nu_\lambda)_{\lambda>0}$ satisfying

$$\lim_{\lambda\downarrow 0}\sup_{t>0} \nu_\lambda(t) = 0$$

$$\sup_{\lambda\in(0,1)} V_0^{q[\nu_\lambda](1-\varepsilon)}[\nu_\lambda] q[\nu_\lambda](1-\varepsilon) < +\infty, \qquad \forall\varepsilon\in(0,1),$$

here $V_a^b[y]$ is the total variation of a function y in an interval $[a,b)$.

So, we must require this condition with $\nu_\lambda = \rho^{q[\rho](1-\lambda)}_{\rm shift}$ $(\lambda>0)$;, now, since ρ is decreasing, $\nu_\lambda(0)$ becomes the total variation of ν_λ in $[0,\infty)$, and it is sufficient to require

$$\lim_{\lambda\downarrow 0} \rho^{q[\rho](1-\lambda)}_{\rm shift}(0) = 0$$

$$\sup_{\lambda\in(0,1)} \rho^{q[\rho](1-\lambda)}_{\rm shift}(0) q[\rho^{1/\lambda}_{\rm shift}](1-\varepsilon) < +\infty, \qquad \forall\varepsilon\in(0,1),$$

i.e.,

$$\lim_{T\uparrow\infty} \frac{\rho(T)}{\int_T^\infty \rho(t)\,dt} = 0,$$

$$\sup_{\lambda\in(0,1)} \frac{\rho(q[\rho](1-\lambda))}{\lambda}\Big[q[\rho](1-\lambda\varepsilon) - q[\rho](1-\lambda)\Big] < +\infty, \qquad \forall\varepsilon\in(0,1).$$

Note that the first part of this condition was proved in (9). With respect to the second part of condition, we note that

$$\lim_{\lambda\uparrow 1}\frac{\rho(q[\rho](1-\lambda))}{\lambda}\Big[q[\rho](1-\lambda\varepsilon)-q[\rho](1-\lambda)\Big]=\rho(0)q[\rho](1-\varepsilon);$$

$$\lim_{\lambda\downarrow 0}\frac{\rho(q[\rho](1-\lambda))}{\lambda}\Big[q[\rho](1-\lambda\varepsilon)-q[\rho](1-\lambda)\Big]=B\lambda^{1/k}\Big[\Big(\frac{\lambda\varepsilon}{A}\Big)^{-1/k}-\Big(\frac{\lambda}{A}\Big)^{-1/k}\Big]$$
$$=BA^{1/k}(\varepsilon^{-1/k}-1).$$

By applying the continuity of $\frac{\rho(q[\rho](1-\lambda))}{\lambda}\Big[q[\rho](1-\lambda\varepsilon)-q[\rho](1-\lambda)\Big]$ in λ, we have verified the second part of the condition.

So, the condition [13, Theorem 6, (v)] is verified, and condition (q) is proved.

$(q) \Rightarrow (p)$. This implication is clear.

$(p) \Rightarrow (e)$. Assume (p). Fix a rational density $\mu(t) = P(t)/Q(t)$. We can again assume that μ and $-\frac{d\ln\mu(\cdot)}{dt}$ is decreasing on the positive semiaxis. We can also obtain (9) and (10) for the density μ. In particular, from (9), it follows that $\rho^T_{\text{shift}}(0) \to 0$ as $T \to \infty$, i.e., $\frac{1}{r}\mu(q[\mu](1-r)) \to 0$ as $r \downarrow 0$. Since the mappings $q[\mu](r)$ and $\mu(q[\mu](1-r))/r$ are continuous in r, we find that, for every $\lambda \le \mu(0)$, there exists a minimal number r_λ such that $\mu(q[\mu](1-r_\lambda)) = (1-r_\lambda)\lambda$.

Set $\varrho_\lambda \triangleq \mu^{q[\mu](1-r_\lambda)}_{\text{shift}}$ for arbitrary $\lambda \le \mu(0)$; further, $\varrho_\lambda(0) = \lambda$.

Again we reduce the proof to a certain implication from [13], namely, to [13, Theorem 6, $(\exists) \Rightarrow (e)$]. The condition $(\exists)$ means the existence of a density family $(\varrho_\lambda)_{\lambda>0}$ satisfying the following three properties:

1. for sufficiently small λ,
$$\varrho_\lambda(0) = \lambda \ge \varrho_\lambda(t), \qquad \forall t \ge 0; \tag{11}$$
2. for all $\varepsilon > 0$, there exist $\delta_\varepsilon \in (0,1)$ and λ_ε such that, for all positive $\lambda < \lambda_\varepsilon$ and $T \le \delta_\varepsilon/\lambda$, we have
$$\varrho_\lambda(T) \ge \lambda(1-\varepsilon); \tag{12}$$
3. for a certain number $r_0 \in (0,1)$, $V[\cdot] \equiv \mathbb{V}^{i,\sharp}[\mathtt{c}^i[\cdot]]$ and $V[\cdot] \equiv \mathbb{V}^{i,\flat}[\mathtt{c}^i[\cdot]]$ satisfy
$$\lim_{\lambda\downarrow 0}\sup_{\omega\in\Omega}\Big|V[\varrho_\lambda](\omega) - U_*(\omega)\Big| = 0 \tag{13}$$
$$\lim_{\lambda\downarrow 0}\sup_{T\in(0,q[\varrho_\lambda](r_0)),\omega\in\Omega}\Big|V[(\varrho_\lambda)^T_{\text{shift}}](\omega) - U_*(\omega)\Big| = 0. \tag{14}$$

We claim that the family $\varrho_\lambda = \mu^{q[\mu](1-r_\lambda)}_{\text{shift}}$ satisfies all these assumptions for ϱ_λ.

Indeed, since $(\varrho_\lambda)^T_{\text{shift}} \equiv \mu^{q[\mu](1-\lambda)+T}_{\text{shift}}$, conditions (13), (14) follows from (p). Next, (11) follows from construction of r_λ and the monotonity of μ in t. So, we only need to verify (12).

Fix a positive ε and $\lambda < \mu(0)$. Consider an arbitrary positive $T \leq \varepsilon/2\lambda$, and set $s \triangleq \int_0^T \varrho_\lambda(t)\,dt \leq T\varrho_\lambda(0) \leq \varepsilon/2$. From the chain

$$sr_\lambda = \int_0^T r_\lambda \varrho_\lambda(t)\,dt = \int_{q[\mu](1-r_\lambda)}^{T+q[\mu](1-r_\lambda)} \mu(t)\,dt = \int_0^{T+q[\mu](1-r_\lambda)} \mu(t)\,dt - 1 + r_\lambda,$$

we obtain $T + q[\mu](1-r_\lambda) = q[\mu](1-r_\lambda+sr_\lambda)$. Then, we have

$$\frac{\varrho_\lambda(T)}{\varrho_\lambda(0)} = \frac{\mu(q[\mu](1-r_\lambda+sr_\lambda))}{\mu(q[\mu](1-r_\lambda)} \overset{(9)}{=} \frac{((1-s)r_\lambda)^{1+1/k}(1+O(r_\lambda^{1/k}))}{r_\lambda^{1+1/k}(1+O(r_\lambda^{1/k}))}$$
$$= (1-s)^{1+1/k}(1+O(r_\lambda^{1/k})) \geq (1-s)^2(1+O(r_\lambda^{1/k}))$$

So, we can choose a positive λ_0 such that $\frac{\varrho_\lambda(T)}{\varrho_\lambda(0)} \geq 1-2s \geq 1-\varepsilon$ holds for all positive $\lambda \leq \lambda_0$, $T \leq \varepsilon/\lambda$, i.e., the assumption (12) has been verified. Then, we have verified [13, ($\exists$)] for the family ϱ_λ. Applying [13, Theorem 6, ($\exists$) $\Rightarrow$ (e)] for $V[\cdot] \equiv \mathbb{V}^{i,\sharp}\big[\mathtt{c}^i[\cdot]\big], V[\cdot] \equiv \mathbb{V}^{i,\flat}\big[\mathtt{c}^i[\cdot]\big]$, we obtain condition ($e$).

The proof is completed.

References

1. Averboukh, Y.V.: Universal Nash equilibrium strategies for differential games. J. Dyn. Contrl Syst. **21**(3), 329–350 (2015). https://doi.org/10.1007/s10883-014-9224-9
2. Bardi, M.: On differential games with long-time-average cost. In: Pourtallier, O., Gaitsgory, V., Bernhard, P. (eds.) Advances in Dynamic Games and Their Applications. Annals of the International Society of Dynamic Games, pp. 3–18. Birkhäuser, Boston (2009). https://doi.org/10.1007/978-0-8176-4834-3_1
3. Bewley, T., Kohlberg, E.: The asymptotic theory of stochastic games. Math. Oper. Res. **1**, 197–208 (1976). https://doi.org/10.1287/moor.1.3.197
4. Cannarsa, P., Quincampoix, M.: Vanishing discount limit and nonexpansive optimal control and differential games. SIAM J. Control Optim. **53**(4), 1789–1814 (2015). https://doi.org/10.1137/130945429
5. Gaitsgory, V.: Application of the averaging method for constructing suboptimal solutions of singularly perturbed problems of optimal control. Automat. Rem. Contr+ **46**, 1081–1088 (1985)
6. Gaitsgory, V., Parkinson, A., Shvartsman, I.: Linear programming based optimality conditions and approximate solution of a deterministic infinite horizon discounted optimal control problem in discrete time. Discrete Continuous Dyn. Syst. Ser. B **29**(4), 1743–1767 (2019). https://doi.org/10.3934/dcdsb.2018235
7. Jaśkiewicz, A., Nowak, A.S.: Non-zero-sum stochastic games. In: Başar, T., Zaccour, G. (eds.) Handbook of Dynamic Game Theory, pp. 1–64. Springer, Cham (2016). https://doi.org/10.1007/978-3-319-27335-8_33-1
8. Khlopin, D.V.: On asymptotic value for dynamic games with saddle point. In: Bonnet, C., Pasik-Duncan, B., Ozbay, H., Zhang, Q. (eds.) Proceedings of the Conference on Control and Its Applications, pp. 282–289. SIAM (2015). https://doi.org/10.1137/1.9781611974072.39

9. Khlopin, D.V.: Uniform Tauberian theorem for differential games. Automat. Rem. Contr+ **77**(4), 734–750 (2016). https://doi.org/10.1134/S0005117916040172
10. Khlopin, D.V.: On uniform Tauberian theorems for dynamic games. Mat. Sb. **209**(1), 127–150 (2018). https://doi.org/10.1070/SM8785
11. Khlopin, D.V.: Tauberian theorem for value functions. Dyn. Games Appl. **8**(2), 401–422 (2018). https://doi.org/10.1007/s13235-017-0227-5
12. Khlopin, D.V.: On Tauberian theorem for stationary Nash equilibria. Optim. Lett. (published online 22 October 2018). https://doi.org/10.1007/s11590-018-1345-8
13. Khlopin, D.V.: Value asymptotics in dynamic games on large horizons. Algebra Anal. **31**(1), 211–245 (2019). (In Russian)
14. Lehrer, E., Sorin, S.: A uniform Tauberian theorem in dynamic programming. Math. Oper. Res. **17**(2), 303–307 (1992). https://doi.org/10.1287/moor.17.2.303
15. Li, X., Quincampoix, M., Renault, J.: Limit value for optimal control with general means. Discrete Continuous Dyn. Syst. Ser. A **36**, 2113–2132 (2016). https://doi.org/10.3934/dcds.2016.36.2113
16. Lions, P., Papanicolaou, G., Varadhan, S.R.S.: Homogenization of Hamilton-Jacobi Equations (1986, unpublished work)
17. Maliar, L., Maliar, S.: Ruling out multiplicity of smooth equilibria in dynamic games: a hyperbolic discounting example. Dyn. Games Appl. **6**(2), 243–261 (2016). https://doi.org/10.1007/s13235-015-0177-8
18. Mertens, J.F., Neyman, A.: Stochastic games. Int. J. Game Theory **10**(2), 53–66 (1981). https://doi.org/10.1007/BF01769259
19. Monderer, D., Sorin, S.: Asymptotic properties in dynamic programming. Int. J. Game Theory **22**, 1–11 (1993). https://doi.org/10.1007/BF01245566
20. Oliu-Barton, M., Vigeral, G.: A uniform Tauberian theorem in optimal control. In: Cardaliaguet, P., Cressman, R. (eds.) Advances in Dynamic Games. Annals of the International Society of Dynamic Games, vol. 12, pp. 199–215. Birkhäuser, Boston (2013). https://doi.org/10.1007/978-0-8176-8355-9_10
21. Renault, J.: General limit value in dynamic programming. J. Dyn. Games **1**(3), 471–484 (2013). https://doi.org/10.3934/jdg.2014.1.471
22. Solan, E.: Acceptable strategy profiles in stochastic games. Games Econ. Behav. **108**, 523–540 (2018). https://doi.org/10.1016/j.geb.2017.01.011
23. Sorger, G.: Competitive dynamic advertising: a modification of the case game. J. Econ. Dyn. Control **13**, 55–80 (1989). https://doi.org/10.1016/0165-1889(89)90011-0
24. Ziliotto, B.: A Tauberian theorem for nonexpansive operators and applications to zero-sum stochastic games. Math. Oper. Res. **41**(4), 1522–1534 (2016). https://doi.org/10.1287/moor.2016.0788
25. Ziliotto, B.: Tauberian theorems for general iterations of operators: applications to zero-sum stochastic games. Games Econ. Behav. **108**, 486–503 (2018). https://doi.org/10.1016/j.geb.2018.01.009

Open-Loop Strategies in Nonzero-Sum Differential Game with Multilevel Hierarchy

Ekaterina Kolpakova[1,2(✉)]

[1] Krasovskii Institute of Mathematics and Mechanics, Ekaterinburg, Russia
eakolpakova@gmail.com

[2] Ural Federal University named after the first President of Russia B.N.Yeltsin, Ekaterinburg, Russia

Abstract. The paper is concerned with the construction of open-loop strategies for the n-person nonzero-sum differential game with multilevel hierarchy. The dynamics of the first player (leader) is defined by its own position and control. The player of further levels of hierarchy knows the position and control of the players of the upper hierarchical levels. At the same time the dynamics and payoff functional of the player do not depend on the position and control of lower hierarchical levels.

We solve this problem with the help of consequent solutions of optimal control problems for each player. Using the solution of Hamilton—Jacobi equation and the results of optimal control theory we construct the open-loop controls of the players. The specifics of this problem is the construction of solution for the Hamilton—Jacobi equation with the Hamiltonian discontinuous w.r.t. phase variable. In this case we use the notion of multivalued solution proposed by Subbotin. We show that the open-loop controls provide a Nash equilibrium in the differential game with multilevel hierarchy and the set of payoffs of the players is described by the multivalued solution of the corresponding Hamilton—Jacobi equation.

Keywords: Discontinuous Hamilton—Jacobi equations · Nonzero-sum differential game · Nash equilibrium

1 Introduction

The paper is concerned with a construction of a program equilibrium for n-person nonzero-sum differential game with multilevel hierarchy. Under the hierarchical differential game we consider the following differential game: the k-th player is a leader for the i-th player, when $i = k+1, \dots, n$ and is a follower for i- th player, if $i = 1, \dots, k-1$. Therefore the first player is the leader for all other players. Such games are investigated earlier in the case of linear dynamics and quadratic payoffs [1,2].

We construct the open-loop strategies of the players using the approach based on the solution of the system of Hamilton—Jacobi equations. This approach was

M. Khachay et al. (Eds.): MOTOR 2019, LNCS 11548, pp. 620–634, 2019.
https://doi.org/10.1007/978-3-030-22629-9_44

developed in [3–5], where the solution of the system of Hamilton—Jacobi equations is assumed to be smooth. However, in the general case the solution of the system of Hamilton—Jacobi equations belongs to the class of multivalued maps [6]. The particular cases of nonzero-sum differential games with continuous solution of the corresponding system of Hamilton—Jacobi equations are investigated in [7,8]. In these works the dynamics of the players is described by simple motions of $\mathbb{R}$. We consider the n-person nonzero-sum differential game in the space of dimension n. It is assumed that the differential game is composed by n optimal control problems for every player. The k-th player maximizes own payoff knowing the optimal trajectories and optimal controls of $k-1$ players. Thus, the corresponding system of Hamilton—Jacobi equations has a hierarchical structure.

The paper is organized as follows: Sect. 2 is concerned with the statement of the problem; in Sect. 3 we solve the optimal control problem for the first player (leader) and construct an open-loop optimal control; in Sect. 4 we solve the optimal control problem for other players and construct their open-loop optimal controls; in Sect. 5 we design a Nash equilibrium;in Sect. 6 we construct the values of the players on the base of a solution for a system of the Hamilton—Jacobi equations.

2 Statement

Let us consider an n-person nonzero-sum differential game with the dynamics given by

$$\dot{x}_1 = f_1(t, x_1, u),\ x_1(t_0) = x_1^0,\ u \in U, \tag{1}$$

$$\dot{x}_k = f_k(t, x_1, \ldots, x_k, u, v_2, \ldots, v_k),\ x_k(t_0) = x_k^0,\ v_k \in V_k,\ k = 2, \ldots, n. \tag{2}$$

Here $t \in [0, T]$, $x = (x_1, \ldots, x_n) \in \mathbb{R}^n$, U, V_k, $k = 2, \ldots, n$ are compact subsets of $\mathbb{R}$. Denote the set of admissible controls of the leader by $\tilde{U}$:

$$\tilde{U} = \{u : [t_0, T] \to U :\ u \text{ is a measurable function}\}, \tag{3}$$

and the set of admissible controls of the k-th player by $\tilde{V}_k$:

$$\tilde{V}_k = \{v : [t_0, T] \to V_k :\ v \text{ is a measurable function}\}, \text{k} = 2, \ldots, \text{n}. \tag{4}$$

The objective function of the first player is

$$I_1(t_0, x_1^0; u(\cdot)) = \sigma_1(x_1(T)) + \int_{t_0}^{T} g_1(t, x_1(t), u(t))dt \to \max_{u(\cdot)\in\tilde{U}},$$

whereas the k-th player aims to maximize

$$I_k(t_0, x_1^0, \ldots, x_k^0; u(\cdot), v_2(\cdot), \ldots, v_k(\cdot)) = \sigma_k(x_1(T), \ldots, x_k(T))$$
$$+ \int_{t_0}^{T} g_k(t, x_1(t), \ldots, x_k(t), u(t), v_2(t), \ldots, v_k(t))dt \to \max_{v_k(\cdot)\in\tilde{V}_k}.$$

Let us stress that the dynamics and payoff functional of the leader do not depend of the actions of the followers. Hence, the leader solves the optimal control problem. The dynamics and payoff functional of the k-th player depends on the controls of the leader and $k-1$ players of previous levels of hierarchy. Therefore, we solve the optimal control problem for each player consequently, substituting the known control $u(\cdot)$ of the leader and $v_i(\cdot)$, $i=2,\ldots,k-1$ of $(k-1)$ players to dynamics and payoff functional of the k-th player.

We denote by symbol y_k the vector $(x_1,\ldots,x_k)\in \mathbb{R}^k$ and by symbol y_k^0 the vector $(x_1^0,\ldots,x_k^0)\in \mathbb{R}^k$, $k=1,\ldots,n$.

We assume that

A1 the function $f_1:[0,T]\times\mathbb{R}\times U\to\mathbb{R}$ has continuous partial derivatives $\frac{\partial f_1}{\partial t}$, $\frac{\partial f_1}{\partial x_1}$ satisfying the sublinear condition w.r.t. x_1.

A2 the function $\sigma_1:\mathbb{R}\to\mathbb{R}$ is differentiable, the function $g_1:[0,T]\times\mathbb{R}\times U\to\mathbb{R}$ has continuous partial derivatives $\frac{\partial g_1}{\partial t}$, $\frac{\partial g_1}{\partial x_1}$ satisfying the sublinear condition w.r.t. x_1 and the set $(f_1(t,x_1,U),g_1(t,x_1,U))$ is convex for any (t,x_1).

A3 the function $f_k:[0,T]\times\mathbb{R}^k\times U\times V_2\times\ldots\times V_k\to\mathbb{R}$ has continuous partial derivatives $\frac{\partial f_k}{\partial t}$, $\frac{\partial f_k}{\partial x_i}$, $i=1,\ldots,k$ satisfying the sublinear condition w.r.t. y_k, $k=2,\ldots,n$.

A4 the function $\sigma_k:\mathbb{R}^k\to\mathbb{R}$ is differentiable, function g_k has continuous partial derivatives $\frac{\partial g_k}{\partial t}$, $\frac{\partial g_k}{\partial x_i}$, $i=1,\ldots,k$ satisfying the sublinear condition w.r.t. y_k and for $k=2,\ldots,n$, for any (t,y_k) the set $(f_k(t,y_k,U,V_2,\ldots,V_k),g_k(t,y_k,U,V_2,\ldots,V_k))$ is strictly convex.

From assumptions *A*2, *A*4 (the convexity of vectogram) and continuity g_1,g_k it follows that we can use max instead of sup for I_1, I_k, $k=2,\ldots,n$ [9].

3 Optimal Control Problem for the Leader

In this section we construct the open-loop control for the first player, based on the solution of the Hamilton—Jacobi equation.

Let us consider a map

$$\varphi_1:[t_0,T]\times\mathbb{R}\to\mathbb{R},$$

defined by

$$\varphi_1(t_0,x_1^0)=\max_{u(\cdot)\in\tilde{U}} I_1(t_0,x_1^0;u(\cdot)),$$

where $\tilde{U}$ is given by (3). The map φ_1 is called the value function of the first player. Notice that dynamics (1) and the payoff functional I_1 do not depend on x_i, $i=2,\ldots,k$. Thus $\frac{\partial\varphi_1}{\partial x_i}=0$, $i=2,\ldots,k$.

We introduce the Hamiltonian for problem (1), (3) with functional I_1:

$$\begin{aligned}H_1(t,x_1,p)&=\max_{u\in U}[f_1(t,x_1,u)p+g_1(t,x_1,u)]\\&=f_1(t,x_1,u^*(t,x_1,p))p+g_1(t,x_1,u^*(t,x_1,p)),\end{aligned}\qquad(5)$$

where $p = \frac{\partial \varphi_1}{\partial x_1}$, $u^* : [0, T] \times \mathbb{R} \times \mathbb{R} \to U$ satisfies the condition

$$u^*(t, x_1, p) \in \arg\max_{u \in U}\{f_1(t, x_1, u)p + g_1(t, x_1, u)\}. \tag{6}$$

Since Measurable Maximum theorem and conditions $A1$, $A2$ hold a function $u^* : [0, T] \times \mathbb{R} \times \mathbb{R} \to U$ is measurable.

It is known [10] that the value function φ_1 is the unique minimax/viscosity solution of Cauchy problem

$$\frac{\partial \varphi_1(t, x_1)}{\partial t} + H_1(t, x_1, D_{x_1}\varphi_1(t, x_1)) = 0, \quad \varphi_1(T, x_1) = \sigma_1(x_1), \tag{7}$$

where H_1 is given by (5).

We construct the open-loop control with the help of Cauchy method of characteristics [11]. We consider the characteristic system for Bellman equation (7) of the form:

$$\dot{\tilde{x}}_1 = \frac{\partial H_1(t, \tilde{x}_1, \tilde{s})}{\partial \tilde{s}}, \ \dot{\tilde{s}} = -\frac{\partial H_1(t, \tilde{x}_1, \tilde{s})}{\partial \tilde{x}_1}, \ \dot{\tilde{z}} = \frac{\partial H_1(t, \tilde{x}_1, \tilde{s})}{\partial \tilde{s}}\tilde{s} - H_1(t, \tilde{x}_1, \tilde{s})$$

with a boundary condition

$$\tilde{x}_1(T, \xi_1) = \xi_1, \ \tilde{s}(T, \xi_1) = D_{x_1}\sigma_1(\xi_1), \ \tilde{z}(T, \xi_1) = \sigma_1(\xi_1), \ \xi_1 \in \mathbb{R}.$$

The solution $(\tilde{x}_1(\cdot), \tilde{s}(\cdot), \tilde{z}(\cdot))$ is the unique and extendable on time interval $[0, T]$. Let us introduce the mapping

$$\begin{aligned}(t_0, x_1^0) \to \xi(t_0, x_1^0) = \{\xi_1 \in \mathbb{R} : \ \tilde{x}_1(t_0, \xi_1) = x_1^0, \tilde{x}_1(T, \xi_1) = \xi_1,\\ \tilde{s}(T, \xi_1) = D_{x_1}\sigma_1(\xi_1), \ \tilde{z}(T, \xi_1) = \sigma_1(\xi_1), \ \tilde{z}(t_0, \xi_1) = \varphi_1(t_0, x_1^0)\}\end{aligned} \tag{8}$$

It follows from [9,11] that for any point $(t_0, x_1^0) \in [0, T] \times \mathbb{R}$ assumption $A1$ guarantees the existence of optimal open-loop control $u^0(\cdot; x_1^0)$ satisfying the relation

$$\max_{u(\cdot) \in \tilde{U}} I_1(t_0, x_1^0; u(\cdot)) = I_1(t_0, x_1^0; u^0(\cdot; x_1^0)) = \varphi_1(t_0, x_1^0).$$

Pontryagin's Maximum principle implies that the optimal open-loop control $u^0(\cdot; t_0, x_1^0)$ of the first player for the initial point $(t_0, x_1^0) \in [0, T] \times \mathbb{R}$ is defined by the rule $\forall t \in [t_0, T]$

$$u^0(t; x_1^0) \in \arg\max_{u \in U}[\tilde{s}(t, \xi_1^0)f_1(t, \tilde{x}_1(t, \xi_1^0), u) + g_1(t, \tilde{x}_1(t, \xi_1^0), u)]. \tag{9}$$

Here $(\tilde{x}_1(\cdot), \tilde{s}(\cdot), \tilde{z}(\cdot))$ is the solution of the characteristic system for Cauchy problem (7) for any $t \in [t_0, T]$, for any $\xi_1^0 \in \xi(t_0, x_1^0)$ defined by (8).

Denote by $U^0(t_0, x_1^0)$ the set of optimal open-loop controls $u(\cdot) \in \tilde{U}$ for problem (1) with initial condition $x_1(t_0) = x_1^0$ and payoff functional I_1. Designate by $X_1(t_0, x_1^0)$ the set of trajectories $x_1(\cdot; x_1^0) : [t_0, T] \to \mathbb{R}$ such that

$$\dot{x}_1(t) = f_1(t, x_1(t), u(t)), \quad x_1(t_0) = x_1^0, \ u(\cdot) \in U^0(t_0, x_1^0).$$

Assumptions $A1$, $A2$ imply, that for any compact $D_0 \subset [0,T] \times \mathbb{R}$ and $(t_0, x_1^0) \in D_0$, we have that, for some constants M_0, $M_1 > 0$,

$$|x_1(t)| \leq M_0, \quad |\dot{x}_1(t)| \leq M_1, \ t \in [t_0, T].$$

Thus, for each (t_0, x_1^0), the set $X_1(t_0, x_1^0)$ is a compact in $C[t_0, T]$.

4 Optimal Control Problem for the k-th Player

In the previous section we describe the optimal trajectories and optimal control of the first player. Now we shall construct the open-loop control for the k-th player. We suppose that we know optimal trajectory $x_i(\cdot; y_i^0)$, $i = 1, \ldots, k-1$ and optimal controls $u^0, v_2^0, \ldots, v_{k-1}^0$ of $k-1$ players. Further in the paper we fix the initial point $y_k(t_0) = y_k^0$ for optimal trajectories and controls of the players. Define

$$\tilde{f}_k(t, x_k, v_k) = f_k(t, y_{k-1}(t), x_k, u^0(t), v_2^0(t), \ldots, v_{k-1}^0(t), v_k),$$
$$\tilde{g}_k(t, x_k, v_k) = g_k(t, y_{k-1}(t), x_k, u^0(t), v_2^0(t), \ldots, v_{k-1}^0(t), v_k), \ k = 2, \ldots, n.$$

Designate by $Y_k(t_0, y_k^0)$ the set of optimal trajectories $y_k(\cdot; y_k^0) : [t_0, T] \to \mathbb{R}^k$ for problems (2), (4) with payoff functionals $I_1, \ldots, I_k$.

Let us consider auxiliary control problem for the k-th player:

$$\dot{x}_k = \tilde{f}_k(t, x_k, v_k), \quad x_k(t_0) = x_k^0. \tag{10}$$

Put

$$h_k(t, x_k, q_{k,k}, v_k; y_{k-1}(\cdot; y_{k-1}^0)) = \tilde{f}_k(t, x_k, v_k) q_{k,k} + \tilde{g}_k(t, x_k, v_k).$$

Here $q_{k,k} \in \mathbb{R}$. The Hamiltonian for control problem with dynamics (10) and functional I_k under the fixed trajectory $y_{k-1}(\cdot; y_{k-1}^0) \in Y_{k-1}(t_0, y_{k-1}^0)$ and known optimal controls u^0, $v_2^0, \ldots, v_{k-1}^0$ has the form

$$H_k(t, x_k, q_{k,k}; y_{k-1}(\cdot; y_{k-1}^0)) = \max_{v_k \in V_k} h_k(t, x_k, q_{k,k}, v_k; y_{k-1}(\cdot; y_{k-1}^0)) = \tag{11}$$
$$\tilde{f}_k(t, x_k, v_k^*(t, x_k, q_{k,k}; y_{k-1}^0)) q_{k,k} + \tilde{g}_k(t, x_k, v_k^*(t, x_k, q_{k,k}; y_{k-1}(\cdot; y_{k-1}^0)),$$

where

$$v_k^*(t, x_k, q_{k,k}; y_{k-1}(\cdot; y_{k-1}^0)) \in \arg\max_{v_k \in V_k} h(t, x_k, q_{k,k}, v_k; y_{k-1}(\cdot; y_{k-1}^0)). \tag{12}$$

Since Measurable Maximum theorem and condition $A3$, $A4$ hold function $(t, x_k, q_{k,k}) \to v_k^*(t, x_k, q_{k,k}; y_{k-1}^0))$ is measurable. From assumptions $A3$, $A4$ we obtain, that for given $y_{k-1}(\cdot; y_{k-1}^0) \in Y_{k-1}(t_0, y_i^0)$ and fixed $u^0(\cdot; x_1^0)$, $v_i^0(\cdot; y_i^0)$, $i = 2, \ldots, k-1$ the Hamiltonian H_k is Lipschitz continuous function w.r.t. $x_k, q_{k,k}$.

Let us consider a map

$$\varphi_k(\cdot, \cdot; y_{k-1}(\cdot; y_{k-1}^0)) : [t_0, T] \times \mathbb{R} \to \mathbb{R},$$

defined by

$$\varphi_k(t_0, x_k^0; y_{k-1}(\cdot; y_{k-1}^0)) = \max_{v_k(\cdot)\in\tilde V_k} I_k(t_0, y_k^0; u^0(\cdot), v_2^0(\cdot), \ldots, v_{k-1}^0(\cdot), v_k(\cdot)),$$

where $\tilde V_k$ is given by (4), $u^0(\cdot; x_1^0) \in U^0(t_0; x_1^0)$, $v_i^0(\cdot; y_i^0)$, $i = 2, \ldots, k-1$ are optimal open-loop controls for previous $k-1$ players. The map φ_k is the value function of the k-th player under the fixed trajectory $y_{k-1}(\cdot; y_{k-1}^0)$ generated by the control $u^0(\cdot; x_1^0)$, $v_i^0(\cdot; y_i^0)$, $i = 2, \ldots, k-1$. It is known [10] that the value function φ_k is the unique minimax/viscosity solution of Cauchy problem

$$\frac{\partial \varphi_k}{\partial t} + H_k(t, x_k, D_{x_k}\varphi_k; y_{k-1}(\cdot; y_{k-1}^0)) = 0, \tag{13}$$
$$\varphi_k(T, x_k; y_{k-1}(\cdot; y_{k-1}^0)) = \sigma_k(y_{k-1}(T; y_{k-1}^0), x_k).$$

We construct the open-loop control in the same way as in the previous section. The characteristic system for Cauchy problem (13) has the form

$$\dot{\tilde x}_k = \frac{\partial H_k(t, \tilde x_k, \tilde s; y_{k-1}(\cdot; y_{k-1}^0))}{\partial \tilde s},\ \dot{\tilde s} = \frac{\partial H_k(t, \tilde x_k, \tilde s; y_{k-1}(\cdot; y_{k-1}^0))}{\partial \tilde x_k},$$
$$\dot{\tilde z} = \langle \frac{\partial H_k(t, \tilde x_k, \tilde s; y_{k-1}(\cdot; y_{k-1}^0))}{\partial \tilde s}, \tilde s\rangle - H_k(t, \tilde x_k, \tilde s; y_{k-1}(\cdot; y_{k-1}^0))$$

with a boundary condition

$$\tilde x_k(T, \xi_k) = \xi_k,\ \tilde s(T, \xi_k) = D_{x_k}\sigma_k(y_{k-1}(T, y_{k-1}^0), \xi_k),$$
$$\tilde z(T, \xi_k) = \sigma_k(y_{k-1}(T, y_{k-1}^0), \xi_k),\ \xi_k \in \mathbb{R}.$$

The solution $(\tilde x_k(\cdot), \tilde s(\cdot), \tilde z(\cdot))$ is the unique and extendable on time interval $[0, T]$.

Let us introduce the mapping

$$\begin{aligned}&(t_0, y_k^0) \to \xi(t_0, x_k^0; y_{k-1}(\cdot; y_{k-1}^0)) = \{\xi_k \in \mathbb{R}:\ \tilde x_k(t_0, \xi_k) = x_k^0, \tilde x_k(T, \xi_k) = \xi_k,\\ &\tilde s(T, \xi_k) = D_{x_k}\sigma_k(y_{k-1}(T, y_{k-1}^0), \xi_k),\\ &\tilde z(T, \xi_k) = \sigma_k(y_{k-1}(T, y_{k-1}^0), \xi_k),\ \tilde z(t_0, \xi_k) = \varphi_k(t_0, x_k^0; y_{k-1}(\cdot; y_{k-1}^0))\}\end{aligned} \tag{14}$$

It follows from [9,11] that for any point $(t_0, y_k^0) \in [0, T] \times \mathbb{R}^k$ assumption $A3$ guarantees the existence of optimal open-loop control $v_k^0(\cdot; y_k^0)$ satisfying the relation

$$\max_{v_k(\cdot)\in\tilde V_k} I_k(t_0, y_k^0; u^0(\cdot), v_2^0(\cdot), \ldots, v_{k-1}^0, v_k)$$
$$= I_k(t_0, y_k^0; u^0(\cdot), v_2^0(\cdot), \ldots, v_k^0(\cdot)) = \varphi_k(t_0, x_k^0; y_{k-1}(\cdot; y_{k-1}^0)).$$

Pontryagin's Maximum principle implies that the optimal open-loop control $v_k^0(\cdot; y_k^0)$ of the player for the initial point $(t_0, y_k^0) \in [0, T] \times \mathbb{R}^k$ is defined by the rule $\forall t \in [t_0, T]$

$$\begin{aligned}&v_k^0(t; y_{k-1}(\cdot; y_{k-1}^0)) \in \arg\max_{v_k\in V_k} h_k(t, \tilde x_k(t, \xi_k), \tilde s(t, \xi_k), v_k; y_{k-1}(\cdot; y_{k-1}^0)),\\ &\forall\ \xi_k \in \xi(t_0, x_k^0; y_{k-1}(\cdot; y_{k-1}^0)).\end{aligned} \tag{15}$$

Here $(\tilde x_k(\cdot), \tilde s(\cdot), \tilde z(\cdot))$ is the solution of the characteristic system for problem (13) for any $t \in [t_0, T]$, $\xi(t_0, x_k^0; y_{k-1}(\cdot; y_{k-1}^0))$ defined by (14).

5 Design of a Nash Equilibrium

Further in the paper we show that $(u^*, v_2^*, \dots, v_n^*)$ defined by formulas (9), (15) provide a Nash equilibria. We consider the function $x_1(\cdot)$ satisfies (1) with admissible control $u(\cdot) \in \tilde{U}$ and initial condition $x_1(t_0) = x_1^0$; function $x_1^*(\cdot)$ satisfies (1) with optimal control $u^*(\cdot) \in U^0(t_0, x_1^0)$ and initial condition $x_1^*(t_0) = x_1^0$; function $\hat{y}_k(\cdot)$ satisfies (2) with optimal controls $u^*(\cdot) \in U^0(t_0, x_1^0)$, $v_2^*(\cdot) \in \tilde{V}_2, \dots, v_{k-1}^*(\cdot) \in \tilde{V}_{k-1}$, and admissible control $v_k(\cdot) \in \tilde{V}_k$, initial condition $\hat{y}_k(t_0) = y_k^0$; function $y_k^*(\cdot)$ satisfies (2) with optimal controls $u^*(\cdot) \in U^0(t_0, x_1^0)$, $v_2^*(\cdot) \in \tilde{V}_2, \dots, v_k^*(\cdot) \in \tilde{V}_k$, $k = 2, \dots, n$ and the initial condition $y_k^*(t_0) = y_k^0$;

Definition 1 [12]. *We say that $(u^*, v_2^*, \dots, v_n^*)$, where $u^* : [0, T] \to U$ and $v_k^* : [0, T] \to V_k$, $k = 2, \dots, n$ is an open-loop Nash equilibrium of (t_0, x^0) if for any $u(\cdot) \in \tilde{U}$ and $v_k(\cdot) \in \tilde{V}_k$, $k = 2, \dots, n$*

$$\sigma_1(x_1(T)) + \int_{t_0}^{T} g_1(t, x_1(t), u(t))dt \le \sigma_1(x_1^*(T)) + \int_{t_0}^{T} g_1(t, x_1^*(t), u^*(t))dt,$$

$$\sigma_k(\hat{y}_k(T)) + \int_{t_0}^{T} g_k(t, \hat{y}_k(t), u^*(t), v_2^*(t), \dots, v_{k-1}^*(t), v_k(t))dt$$

$$\le \sigma_k(y_k^*(T)) + \int_{t_0}^{T} g_k(t, y_k^*(t), u^*(t), v_2^*(t), \dots, v_k^*(t))dt, \ k = 2, \dots, n.$$

Since u^* is an optimal control of problem (1) with functional I_1, the first inequality is valid. For the following inequalities we have

$$\begin{aligned} & I_k(t_0, y_k^0; u^*(\cdot), v_2^*(\cdot), \dots, v_{k-1}^*(\cdot), v_k(\cdot)) \\ & = \sigma_k(\hat{y}_k(T)) + \int_{t_0}^{T} g_k(t, \hat{y}_k(t), u^*(t), v_2^*(t), \dots, v_{k-1}^*(t), v_k(t))dt \\ & \le \max_{v_k(\cdot) \in \tilde{V}_k} I_k(t_0, y_k^0; u^*(\cdot), v_2^*(\cdot), \dots, v_{k-1}^*(\cdot), v_k(\cdot)) \\ & = \sigma_k(y_k^*(T)) + \int_{t_0}^{T} g_k(t, y_k^*(t), u^*(t), v_2^*(t), \dots, v_k^*(t))dt. \end{aligned}$$

6 Values of the Players

In this section we shall construct the rewards of the players and connect them with the system of Hamilton—Jacobi equations.

6.1 System of Hamilton—Jacobi Equations

We shall construct the payoffs of the players. Put matrix $Q_k = \begin{pmatrix} p & 0 & \dots & 0 \\ q_{2,1} & q_{2,2} & \dots & 0 \\ \dots & \dots & \dots & \dots \\ q_{k,1} & q_{k,2} & \dots & q_{k,k} \end{pmatrix}$, $q_{i,j} \in \mathbb{R}$, $k = 2, \dots, n$, and denote

$$\hat{f}_i(t, y_i, Q_i) = f_i(t, y_i, u^*(t, x_1, p), v_2^*(t, y_2, q_{2,2}), \dots, v_i^*(t, y_i, q_{i,i})).$$

Let us introduce the function

$$\begin{aligned} H_k(t, y_k, Q_k) = \max_{v_k \in V_k} [f_1(t, x_1, u^*(t, x_1, p))q_{k,1} \sum_{i=2}^{k-1} \hat{f}_i(t, y_i, Q_i)q_{k,i} \\ + f_k(t, x, u^*(t, x_1, p), v_2^*(t, y_2, q_{2,2}), \dots, v_{k-1}^*(t, y_{k-1}, q_{k-1,k-1}), v_k)q_{k,k} \\ + g_k(t, x, u^*(t, x_1, p), v_2^*(t, y_2, q_{2,2}), \dots, v_{k-1}^*(t, y_{k-1}, q_{k-1,k-1}), v_k)]. \end{aligned}$$

Here u^* is defined by (6), v_i^*, $i = 2, \dots, k$, is defined by (12). From the properties of minimax solution φ_k [11] for any $(t, y_k) \in [0, T] \times \mathbb{R}^k$ the subdifferential $D_{x_1}^- \varphi_1(t, x_1) \neq \emptyset$, $D_{x_k}^- \varphi_k(t, x_k; y_{k-1}(\cdot; y_{k-1}^0)) \neq \emptyset$, , $k = 2, \dots, n$. We substitute any measurable selector $\partial_{x_1}\varphi_1 : [0, T] \times \mathbb{R} \to \mathbb{R}$ of the map $D_{x_1}^- \varphi_1 : [0, T] \times \mathbb{R} \rightrightarrows \mathbb{R}$, $\partial_{x_i}\varphi_i : [0, T] \times \mathbb{R}^i \to \mathbb{R}$ of the map $D_{x_i}^- \varphi_i : [0, T] \times \mathbb{R} \rightrightarrows \mathbb{R}$, $i = 2, \dots, k-1$ to the matrix Q_k instead of p, $q_{i,j}$, $i, j = 2, \dots, k-1$. Denote new matrix Q_k by symbol $\bar{Q}_k$. Further we shall consider the function $\hat{H}_k$ discontinuous w.r.t. y_{k-1}:

$$\hat{H}_k(t, y_k, q_{k,1}, \dots, q_{k,k}) = H_k(t, y_k, \bar{Q}_k) \tag{16}$$

We note that function $\hat{H}_k$ defined by (16) is Lipschitz continuous w.r.t. $q = (q_{k,1}, \dots, q_{k,k}) \in \mathbb{R}^k$ with Lipschitz constant $\lambda(t, y_k)$. The proof of this assertion is given in [10].

Consider the discontinuous functions

$$H_{k*}(t, y_k, q) = \lim_{\varepsilon \to 0} \inf_{(\tau, \xi_k) \in B(t, y_k; \varepsilon)} \hat{H}_k(\tau, \xi_k, q),$$

$$H_k^*(t, y_k, q) = \lim_{\varepsilon \to 0} \sup_{(\tau, \xi_k) \in B(t, y_k; \varepsilon)} \hat{H}_k(\tau, \xi_k, q).$$

Here symbol $B(t, y_k; \varepsilon)$ denotes the ball with the center at a point (t, y_k) and radius ε.

Lemma 1. *Functions H_k^*, H_{k*} are Lipschitz continuous w.r.t. q with Lipschitz constant $\lambda_k(t, x)$.*

The proof of lemma is the same as in the work [10].

We consider the Cauchy problems

$$\frac{\partial \Phi_k(t, y_k)}{\partial t} + \hat{H}_k(t, y_k, D_{y_k}\Phi_k(t, y_k)) = 0, \; k = 2, \dots, n, \tag{17}$$

with the boundary condition $\Phi_k(T, y_k) = \sigma_k(y_k)$. Here $\hat{H}_k$ is defined by (16). For the Hamilton—Jacobi equation with discontinuous Hamiltonian A.I. Subbotin proposed the notion of M-solution [13].

Let us consider the differential inclusion

$$(\dot{y}_k(t), \dot{z}(t)) \in E_k(t, y_k, q) = \{(\psi, g) \in \mathbb{R}^k \times \mathbb{R} : \psi = (\psi_1, \dots, \psi_k), \quad (18)$$

$$||\psi|| \leq \lambda_k(t, x), \sum_{i=1}^{k} \psi_i q_i - g \in [H_{k*}(t, y_k, q), H_k^*(t, y_k, q)]\}\forall\, q \in \mathbb{R}^k.$$

Recall from [13] that the multivalued map $(t, y_k, q) \rightrightarrows E_k(t, y_k, q) \subset \mathbb{R}^k \times \mathbb{R}$ is admissible.

Definition 2 [13]. *Let $W \subset [0, T] \times \mathbb{R}^n \times \mathbb{R}$ be a closed set. The set W is viable w.r.t. differential inclusion (18), if, for any $(t_0, x_0, z_0) \in W$, there exist $\tau > 0$ and a trajectory $(x(\cdot), z(\cdot))$ of differential inclusion (18) such that $(x(t_0), z(t_0)) = (x_0, z_0)$, $(t, x(t), z(t)) \in W$, when $t \in [0, \tau]$.*

Definition 3 [13]. *A multivalued map $W : [0, T] \times \mathbb{R}^k \rightrightarrows \mathbb{R}$ is called an M-solution of Hamilton—Jacobi equation (17), if the graph W is closed maximal set and it is viable w.r.t. differential inclusion (18).*

The M-solution for Cauchy problem is the M-solution for Hamilton—Jacobi equation, satisfying the boundary condition $W(T, x) = \sigma(x)$ for any $x \in \mathbb{R}^n$.

6.2 Solution of the System of Hamilton—Jacobi Equations

A.I. Subbotin proved the following theorem [13].

Theorem 1. *Let W be a closed set in $[0, T] \times \mathbb{R}^n \times \mathbb{R}$. Assume that $W(t, x) = \{z \in \mathbb{R} : (t, x, z) \in W\} \neq \emptyset$ and*

$$W_*(t, x) = \inf_{z \in W(t,x)} z, \quad W^*(t, x) = \sup_{z \in W(t,x)} z.$$

The map W is the M-solution of Eq. (17) iff epi W_* and hypo W^* are the M-solutions of Eq. (17).

Let us introduce the multivalued map

$$\Phi_k(t_0, y_k^0) = \text{cl} \bigcup_{y_{k-1}(\cdot; y_{k-1}^0) \in Y_{k-1}(t_0, y_{k-1}^0)} \{\varphi_k(t_0, x_k^0; y_{k-1}(\cdot; y_{k-1}^0))\}. \quad (19)$$

Here symbol cl A denotes the closure of the set A. We note that the map Φ_k is compact-valued by definition.

Theorem 2. *The map Φ_k, $k = 2, \dots, n$, defined by (19), is the M-solution of Eq. (17).*

Proof. We fix k. Put $\Phi_k^*(t,y_k) = \max\limits_{\rho\in\Phi_k(t,y_k)} \rho$. We shall show that the hypograph Φ_k^* is viable w.r.t. differential inclusion (18). Choose $(t_0, y_k^0, z_0) \in \text{hypo}\ \Phi_k^*$. There exists an optimal trajectory $\xi : [0,T] \to \mathbb{R}^k$ of systems (1), (2) such that $z_0 \le \varphi_k(t_0, x_k^0; \xi_1(\cdot, x_1^0), \ldots, \xi_{k-1}(\cdot, y_{k-1}^0))$, where φ_k is the value function of the k-th player. From the choice of the point z_0 and the dynamic programming principle we get the following inequality

$$z_0 \le \varphi_k(t_0, x_k^0; \xi_1(\cdot, x_1^0), \ldots, \xi_{k-1}(\cdot, y_{k-1}^0))$$

$$= \varphi_k(t, \xi_k(t); \xi_1(\cdot, x_1^0), \ldots, \xi_{k-1}(\cdot, y_{k-1}^0)) + \int_{t_0}^{t} g_k(\tau, \xi(\tau, y_k^0), u^*(\tau), v_2^*(\tau), \ldots, v_k^*(\tau))d\tau.$$

Here optimal open-loop controls $u^*(\cdot), v_i^*(\cdot)$, $i = 2, \ldots, k$ generates optimal trajectories $\xi_i(\cdot, y_i^0)$, $i = 1, \ldots, k$. Further, for any $t \in [t_0, T]$, we have

$$z_0 - \int_{t_0}^{t} g_k(\tau, \xi(\tau, y_k^0), u^*(\tau), v_2^*(\tau), \ldots, v_k^*(\tau))d\tau$$

$$\le \varphi_k(t, \xi_k(t); \xi_1(\cdot, x_1^0), \ldots, \xi_{k-1}(\cdot, y_{k-1}^0)).$$

Hence

$$z(t) = z_0 - \int_{t_0}^{t} g_k(\tau, \xi(\tau, y_k^0), u^*(\tau), v_2^*(\tau), \ldots, v_k^*(\tau))d\tau$$

$$\le \varphi_k(t, \xi_k(t); \xi_1(\cdot, x_1^0), \ldots, \xi_{k-1}(\cdot, y_{k-1}^0)),$$

that is $(t, \xi(t), z(t)) \in \text{hypo}\ \varphi_k$ and $(\xi(\cdot), z(\cdot))$ satisfies differential inclusion (18). Hence, $(t, \xi(t), z(t)) \in \text{hypo}\ \Phi_k^*$. Since Φ_k^* is upper semicontinuous function, we see that hypo Φ_k^* is a closed set. Therefore hypo Φ_k^* is the M-solution of equation (17).

We consider the case when $z_0 = \Phi_k^*(t_0, y_k^0)$. Let us choose the sequence $z_0^j \to z_0$ as $j \to \infty$ and $z_0^j < \Phi_k^*(t_0, y_k^0)$. For each z_0^j the following inequality is valid:

$$z_0^j - \int_{t_0}^{t} g_k(\tau, \xi^j(\cdot, y_k^0), u^{*j}(\tau), v_2^{*j}(\tau), \ldots, v_k^{*j}(\tau))d\tau$$

$$\le \varphi_k(t, \xi_k^j(t); \xi_1^j(\cdot, x_1^0), \ldots, \xi_{k-1}^j(\cdot, y_{k-1}^0)).$$

Here $\varphi_k(\cdot, \cdot; \xi_1^j(\cdot, x_1^0), \ldots, \xi_{k-1}^j(\cdot, y_{k-1}^0))$ is the value function of the k-th player when the previous players choose trajectories $\xi_1^j(\cdot, x_1^0), \ldots, \xi_{k-1}^j(\cdot, y_{k-1}^0)$.

The sets $Y_i(t_0, y_i^0)$, $i = 2, \ldots, k$, are compacts because there exist constants M_2, M_3:

$$|\xi_i(t)| < M_2,\ |\dot{\xi}_i(t)| \le M_3,\ \forall\ t \in [t_0, T],\ \forall\ \ \xi_i(\cdot, y_i^0) \in Y_i(t_0, y_i^0), i = 2, \ldots, k.$$

Since $X_1(t_0,x_1^0)$, $Y_i(t_0,y_i^0)$, $i=2,\ldots,k$, are compacts and Arzela—Ascoli theorem hold, we obtain $\xi^j(t)\to\xi(t)$, as $j\to\infty$. Besides ξ satisfies differential inclusion (18). Hence,

$$\lim_{j\to\infty}[z_0^j-\int\limits_{t_0}^{t} g_k(\tau,\xi^j(\tau,y_k^0),u^{*j}(\tau),v_2^{*j}(\tau),\ldots,v_k^{*j}(\tau))d\tau]$$
$$\leq \lim_{j\to\infty}\varphi_k(t,\xi_k^j(t);\xi_1^j(\cdot,x_1^0),\ldots,\xi_{k-1}^j(\cdot,y_{k-1}^0)),$$

$$z(t)=z_0-\int\limits_{t_0}^{t} g_k(\tau,\xi(\tau,y_k^0),u^*(\tau),v_2^*(\tau),\ldots,v_k^*(\tau))d\tau$$
$$\leq \lim_{j\to\infty}\varphi_k(t,\xi_k^j(t);\xi_1^j(\cdot,x_1^0),\ldots,\xi_{k-1}^j(\cdot,y_{k-1}^0))\leq \Phi_k^*(t_0,y_k^0).$$

The convergence $u^{*j}\to u^*$, $v_i^{*j}\to v_i^*$, $i=2,\ldots,k$ is proved in the same way as in work [14]. Therefore, $(t,\xi(t),z(t))\in \text{hypo}\ \Phi_k^*$.

Put $\Phi_{*k}(t,y_k)=\min\limits_{\rho\in\Phi(t,y_k)}\rho$. Choose a point $(t_0,y_k^0,z_0)\in \text{epi}\ \Phi_{*k}$. Let trajectory $(\xi(\cdot),\zeta(\cdot))$ be a solution of differential inclusion (18) satisfying the initial condition $\xi(t_0)=y_k^0$, $\zeta(t_0)=z_0$. We fix such optimal trajectory $\xi_i(\cdot,y_i^0)$, $i=1,\ldots,k-1$ and optimal controls $u^*(\cdot,x_1^0)$, $v_i^*(\cdot;y_i^0)$, $i=2,\ldots,k-1$, of optimal control problem (1)–(4) with payoff functionals I_i, $i=1,\ldots,k-1$ that $z_0\geq\varphi_k(t,x_k^0;\xi_1(\cdot;x_1^0),\ldots,\xi_{k-1}(\cdot;y_{k-1}^0))$. For solution $\xi_k(\cdot)$ of control problem (2), (4) with payoff functional I_k we have

$$\varphi_k(t,\xi_k(t);\xi_1(\cdot;x_1^0),\ldots,\xi_{k-1}(\cdot;y_{k-1}^0))$$
$$+\int\limits_{t_0}^{t} g_k(\tau,\xi(\tau;y_k^0),u^*(\tau),v_2^*(\tau),\ldots,v_{k-1}^*(\tau),v_k(\tau))d\tau$$
$$\leq\varphi_k(t_0,y_k^0;\xi_1(\cdot;x_1^0),\ldots,\xi_{k-1}(\cdot;y_{k-1}^0))\leq z_0,\ v_k(\cdot;y_k^0)\in\tilde{V}_k.$$

Therefore

$$\varphi_k(t,\xi_k(t);\xi_1(\cdot;x_1^0),\ldots,\xi_{k-1}(\cdot;y_{k-1}^0))$$
$$\leq z_0-\int\limits_{t_0}^{t} g_k(\tau,\xi(\tau;y_k^0),u^*(\tau),v_2^*(\tau),\ldots,v_{k-1}^*(\tau),v_k(\tau)d\tau=\zeta(t).$$

This means that the trajectory $(\xi(\cdot),\zeta(\cdot))$ lies in epigraph φ_k. Hence, $(\xi(\cdot),\zeta(\cdot))$ lies in epigraph Φ_{*k}. Since the function Φ_{*k} is low semicontinuous, we see that epi Φ_{*k} is a closed set and it is viable w.r.t. differential inclusion (18). Hence epi Φ_{*k} is the M-solution of Eq. (17).

Consider the case $z_0 = \Phi_{*k}(t_0, y_k^0)$. Let us choose the sequence $z_0^j \to z_0$ as $j \to \infty$ and $z_0^j > \Phi_{*k}(t_0, y_k^0)$. For each z_0^j the following inequality is valid:

$$\varphi_k(t, \xi_k^j(t); \xi_1^j(\cdot; x_1^0), \dots, \xi_{k-1}^j(\cdot; y_{k-1}^0))$$
$$\leq z_0^j - \int_{t_0}^{t} g_k(\tau, \xi^j(\tau; y_k^0), u^{*j}(\tau), v_2^{*j}(\tau), \dots, v_{k-1}^{*j}(\tau), v_k^j(\tau))d\tau,$$

where $\varphi_k(\cdot, \cdot; \xi_1^j(\cdot; x_1^0), \dots, \xi_{k-1}^j(\cdot; y_{k-1}^0))$ is the value function for the k-th player under the fixed trajectories $\xi_i^j(\cdot; y_i^0)$, $i = 1, \dots, k-1$ and optimal controls $u^{*j}(\cdot), v_2^{*j}(\cdot; y_2^0), \dots, v_{k-1}^{*j}(\cdot; y_{k-1}^0)$. Consider

$$\lim_{j\to\infty} \varphi_k(t, \xi_k^j(t); \xi_1^j(\cdot; x_1^0), \dots, \xi_{k-1}^j(\cdot; y_{k-1}^0))$$
$$\leq \lim_{j\to\infty} z_0^j - \int_{t_0}^{t} g_k(\tau, \xi^j(\tau; y_k^0), u^{*j}(\tau), v_2^{*j}(\tau), \dots, v_{k-1}^{*j}(\tau), v_k^j(\tau))d\tau.$$

Hence,

$$\Phi_{*k}(t, \xi(t)) \leq z_0 - \int_{t_0}^{t} g_2(\tau, \xi(\tau; y_k^0), u^*(\tau), v_2^*(\tau), \dots, v_{k-1}^*(\tau), v_k(\tau))d\tau.$$

Here $\xi^j \to \xi$ as $j \to \infty$. The convergence $u^{*j} \to u^*$, $v_i^{*j} \to v_i^*$, $i = 2, \dots, k$ is proved in the same way as in work [14]. From Theorem 1 we get that Φ_k is the M-solution of Eq. (17).

Note that epi $\Phi_{*k}(T, y_k) \bigcap$ hypo $\Phi_k^*(T, y_k) = \{\sigma_k(y_k)\} = \Phi_k(T, y_k)$, $y_k \in \mathbb{R}^k$, $k = 2, \dots, n$.

Based on these results we introduce the notion of solution for the system of Hamilton—Jacobi equations.

Definition 4. *The map* $(\varphi_1, \Phi_2, \dots, \Phi_n)$, $\varphi_1 : [0, T] \times \mathbb{R} \to \mathbb{R}$, $\Phi_k : [0, T] \times \mathbb{R}^k \rightrightarrows \mathbb{R}$, $k = 2, \dots, n$ *is said to be the generalized solution for the system of Hamilton—Jacobi equations (7), (17), if* φ_1 *is the minimax/viscosity solution of (7) and* Φ_k *is the M-solution of (17).*

6.3 Link of the Solution of the System for Hamilton—Jacobi Equations with Nash Equilibria

Corollary 1. *Let* $(\varphi_1, \Phi_2, \dots, \Phi_n)$ *is the solution of the system of Hamilton—Jacobi equations (7), (17). Then for any point* (t_0, y_k^0), $\beta_k \in \Phi_k(t_0, y_k^0)$, $k = 2, \dots, n$ *there exists a Nash equilibrium strategies* $(u^*, v_2^*, \dots, v_n^*)$, *where* u^* *is defined by (9),* v_k^*, $k = 2, \dots, n$, *is defined by (15). The payoffs of players equal* $(\varphi_1(t_0, x_1^0), \beta_2, \dots, \beta_n)$.

Indeed, the control u^* defined by (9) maximizes the functional I_1. Hence the payoff of the first player equals

$$\max_{u(\cdot)\in\tilde{U}} I_1(t_0, x_1^0; u(\cdot)) = I_1(t_0, x_1^0; u^*(\cdot)) = \varphi_1(t_0, x_1^0).$$

Let $\beta_k \in \Phi_k(t_0, y_k^0)$. Hence

$$\beta_k \in \mathrm{cl} \bigcup_{y_{k-1}(\cdot;y_{k-1}^0)\in Y_{k-1}(t_0,y_{k-1}^0)} \varphi_k(t_0, x_k^0; y_{k-1}(\cdot; y_{k-1}^0)).$$

There exist $x_i(\cdot; y_i^0) \in X_i(t_0, y_i^0)$, $i = 1, \ldots, k-1$ such that

$$\beta_k = \varphi_k(t_0, x_k^0; y_{k-1}(\cdot; y_{k-1}^0)).$$

From definition v_k^* we obtain that

$$\begin{aligned}\beta_k &= \max_{v_k(\cdot)\in\tilde{V}_k} I_k(t_0, y_k^0; u^*, v_2^*, \ldots, v_{k-1}^*, v_k) = I_k(t_0, y_k^0; u^*, v_2^*, \ldots, v_k^*)\\ &= \varphi_k(t_0, x_k^0; y_{k-1}(\cdot; y_{k-1}^0)).\end{aligned}$$

If $\beta_k \in \mathrm{cl} \bigcup_{y_{k-1}(\cdot;y_{k-1}^0)\in Y_{k-1}(t_0,y_{k-1}^0)} \varphi_k(t_0, x_k^0; y_{k-1}(\cdot; y_{k-1}^0))$ then there exists the sequence $\{\beta_k^j\}$ such that $\lim_{j\to\infty} \beta_k^j = \beta_k$ and $\beta_k^j = \varphi_k(t_0, x_k^0; y_{k-1}^j(\cdot; y_{k-1}^0))$. Consider

$$\begin{aligned}&\lim_{j\to\infty} \varphi_k(t_0, x_k^0; y_{k-1}^j(\cdot; y_{k-1}^0))\\ &= \lim_{j\to\infty} I_k(t_0, y_k^0; u^{*j}, v_2^{*j}, \ldots, v_k^{*j}) = \lim_{j\to\infty} \sigma_k(y_k^j(T)) + \int_{t_0}^{T} g_k(t, y_k^j(t), u^{*j}, v_2^{*j}, \ldots, v_k^{*j})dt.\end{aligned}$$

From Arzela—Ascoli theorem we obtain that

$$\lim_{j\to\infty} y_k^j(t) = y_k(t).$$

The convergence $u^{*j} \to u^*$, $v_i^{*j} \to v_i^*$, $i = 2, \ldots, k$ is proved in the same way as in work [14]. Hence,

$$\begin{aligned}\lim_{j\to\infty} \varphi_k(t_0, x_k^0; y_{k-1}^j(\cdot; y_{k-1}^0)) &= \sigma_k(y_k(T)) + \int_{t_0}^{T} g_k(t, y_k(t), u^*(t), v_2^*(t), \ldots, v_k^*(t))dt\\ &= I_k(t_0, y_k^0; u^*, v_2^*, \ldots, v_k^*).\end{aligned}$$

The following inequality is valid:

$$I_k(t_0, y_k^0; u^{*j}, v_2^{*j}, \ldots, v_k^{*j}) \geq I_k(t_0, y_k^0; u^*, v_2^*, \ldots, v_{k-1}^*, v_k)\ \forall v_k(\cdot) \in \tilde{V}_k,$$

then

$$\lim_{j\to\infty} I_k(t_0, y_k^0; u^{*j}, v_2^{*j}, \dots, v_k^{*j}) \geq \lim_{j\to\infty} I_k(t_0, y_k^0; u^*, v_2^*, \dots, v_{k-1}^*, v_k)\ \forall v_k(\cdot) \in \tilde{V}_k.$$

Therefore, we have

$$I_k(t_0, y_k^0; u^*, v_2^*, \dots, v_k^*) \geq I_k(t_0, y_k^0; u^*, v_2^*, \dots, v_{k-1}^*, v_k)\ \forall v_k(\cdot) \in \tilde{V}_k,\ k = 2, \dots, n.$$

7 Conclusion

We constructed the open-loop Nash equilibria and values of the players in nonzero-sum differential game with multilevel hierarchy. We proved that the solution of corresponding system of Hamilton—Jacobi equations describes the values of the players. Moreover, this solution belongs to the class of multivalued maps. Notice that we derive all open-loop Nash equilibria. This result is based on the fact that differential game (1)–(4) is reduced to the sequence of n optimal control problems.

References

1. Pan, L., Yong, J.: A differential game with multi-level of hierarchy. J. Math. Anal. Appl. **161**(2), 522–544 (1991)
2. Xu, H., Mizukami, K.: Linear feedback closed-loop Stackelberg strategies for descriptor systems with multilevel hierarchy. J. Optim. Theory Appl. **88**(1), 209–231 (1996)
3. Case, J.H.: Toward a theory of many player differential games. SIAM J. Control **7**(2), 179–197 (1969)
4. Friedman, A.: Differential Games. Wiley-Interscience, Hoboken (1971)
5. Starr, A.W., Ho, Y.C.: Non-zero Sum differential games. J. Optim. Theory Appl. **3**(3), 184–206 (1969)
6. Kolpakova, E.A.: On the solution of a system of Hamilton-Jacobi equations of special form. Trudy Inst. Mat. i Mech. UrO RAN **23**(1), 158–170 (2017)
7. Bressan, A., Shen, W.: Semi-cooperative strategies for differential games. Int. J. Game Theory **32**, 1–33 (2004)
8. Cardaliaguet, P., Plaskacz, S.: Existence and uniqueness of a Nash equilibrium feedback for a simple nonzeo-sun differential game. Int. J. Game Theory **32**, 33–71 (2003)
9. Warga, J.: Optimal Control of Differential and Functional Equations. Academic Press, New York-London (1972)
10. Subbotin, A.I.: Generalized Solution of the first Order PDEs. The dynamical optimization Perspectives. Birkhauser, Boston-Berlin (1995)
11. Subbotina, N.N.: The method of characteristics for Hamilton-Jacobi equations and applications to dynamical optimization. J. Math. Sci. **135**(3), 2955–3091 (2006)
12. Basar, T., Olsder, G.J.: Dynamic Noncooperative Game Theory. SIAM, Philadelphia (1999)

13. Lakhtin, A.S., Subbotin, A.I.: The minimax and viscosity solutions in discontinuous partial differential equations of the first order. Dokl. Akad. Nauk **359**(4), 452–455 (1998)
14. Kolpakova, E.A.: Solution for a System of Hamilton—Jacobi equations of special type and a link with nash equilibria. In: Frontiers of Dynamic Games, Static and Dynamic Game Theory: Foundations and Applications, pp. 53–66. Birkhauser, Basel (2018)

On Class of Linear Quadratic Non-cooperative Differential Games with Continuous Updating

Ildus Kuchkarov[1] and Ovanes Petrosian[1,2](✉)

[1] St. Petersburg State University, Saint-Petersburg 199034, Russia
kuchkarov_ildus@mail.ru, petrosian.ovanes@yandex.ru
[2] National Research University Higher School of Economics, Saint-Petersburg 194100, Russia
opetrosyan@hse.ru

Abstract. The subject of this paper is a linear quadratic case of a differential game model with continuous updating. This class of differential games is essentially new, there it is assumed that at each time instant, players have or use information about the game structure defined on a closed time interval with a fixed duration. As time goes on, information about the game structure updates. Under the information about the game structure we understand information about motion equations and payoff functions of players. A linear quadratic case for this class of games is particularly important for practical problems arising in the engineering of human-machine interaction. The notion of Nash equilibrium as an optimality principle is defined and the explicit form of Nash equilibrium for the linear quadratic case is presented. Also, the case of dynamic updating for the linear quadratic differential game is studied and uniform convergence of Nash equilibrium strategies and corresponding trajectory for a case of continuous updating and dynamic updating is demonstrated.

Keywords: Differential games with continuous updating · Nash equilibrium · Linear quadratic differential games

1 Introduction

Most of the real-life conflict-driven processes evolve continuously in time, and their participants continuously receive updated information and adapt. Main models considered in the classical differential game theory are associated with problems defined on a fixed time interval (players have all the information on a closed time interval) [6], problems defined on an infinite time interval with discounting (players have all the information specified on an infinite time interval) [1], problems defined on a random time interval (players have information

Research was supported by a grant from the Russian Science Foundation (Project No 18-71-00081).

M. Khachay et al. (Eds.): MOTOR 2019, LNCS 11548, pp. 635–650, 2019.
https://doi.org/10.1007/978-3-030-22629-9_45

on a given time interval, but the duration of this interval is a random variable) [14]. One of the first works in the theory of differential games is devoted to the differential pursuit game (the player's gain depends on the time of capture of the opponent) [12]. In all the above models and approaches it is assumed that at the beginning of the game players know all information about the game dynamics (equations of motion) and about player's preferences (cost functions). However, these approaches do not take into account the fact that in many real conflict-controlled processes, players at the initial time instant do not have all information about the game. Therefore classical approaches for defining in some sense optimal strategies (for example, Nash equilibrium), such as Hamilton-Jacobi-Bellman equation [2] or the Pontryagin maximum principle [13], cannot be directly used to construct a large range of real game-theoretic models.

In this paper, we apply the approach of continuous updating to a special class of dynamic games, where the environment can be modeled by a set of linear differential equations and the objectives can be modeled by the functions containing affine and quadratic terms. The popularity of the so-called linear quadratic differential games [4] on one hand can be explained by practical applications in engineering. To some extent, this kind of differential games is analytically and numerically solvable. On the other hand, this linear quadratic problem setting naturally appears if the agents' objective is to minimize the effect of a small perturbation of their nonlinear optimally controlled environment. By solving a linear quadratic control problem, and using the optimal actions implied by this problem, players can avoid most of the additional cost incurred by this perturbation.

Most of the real conflict-driven processes are continuously evolving over time, and their participants constantly adapt. This paper presents the approach of constructing Nash equilibrium for game models with continuous updating. In the game models with continuous updating, it is assumed that players

- have information about motion equations and payoff functions only on $[t, t + \overline{T}]$, where $\overline{T}$ – information horizon, t – current time instant.
- receive updated information about motion equations and payoff functions as time $t \in [t_0, +\infty)$ evolves.

In the general form, it is supposed that motion equations and payoff functions explicitly depend on the time parameter. Therefore, in the general form of the differential game with continuous updating information about motion equations and payoff functions updates, because its form changes as the current time $t \in [t_0, +\infty)$ evolves. In this paper, we consider a particular class of linear quadratic differential games with continuous updating, where motion equations and payoff functions do not explicitly depend on time parameter t, but the meaning of the updating procedure is not missed, because the main goal of modeling of behavior of players with continuous updating is reached.

Obviously, it is difficult to obtain Nash equilibrium due to the lack of fundamental approaches to control problems with moving information horizon. Classical methods such as dynamic programming and Hamilton-Jacobi-Bellman equation do not allow to directly construct Nash equilibrium in problems with moving information horizon.

In the framework of dynamic updating approach the following papers were published [5,7–11,15]. Their authors laid a foundation for further study of a class of games with dynamic updating. It is assumed that the information about motion equations and payoff functions is updated in discrete time instants and interval on which players know the information is defined by the value of the information horizon. However, the class of games with continuous updating provides with the new theoretical results.

For the linear quadratic game models with continuous updating Nash equilibrium in closed-loop form are constructed and it is proved that Nash equilibrium in the corresponding linear quadratic game with dynamic updating uniformly converges to the constructed controls. This approach allows concluding that the constructed control indeed is optimal in the game model with continuous updating, i.e. in the case when the length of updating interval converges to zero. The similar procedure is performed for the corresponding trajectory.

The paper is structured as follows. In Sect. 2, a description of the initial differential game model and corresponding game model with continuous updating as well as the concept of a strategy for it are presented. In Sect. 3, the concept of Nash equilibrium is adapted for a class of games with continuous updating and the explicit form of it for a class of linear quadratic differential games is presented. In Sect. 4, the description of the game model with dynamic updating and the form of Nash equilibrium with continuous updating is presented. In Sect. 5, the convergence of Nash equilibrium strategies and corresponding trajectories for a case of dynamic and continuous updating is demonstrated. The illustrative model example and corresponding numerical simulation are presented in Sect. 6. Demonstration of convergence result is as well presented in the numerical simulation part. In Sect. 7, the conclusion is drawn.

2 Game Model

In this section description of the initial linear quadratic game model and corresponding game model with continuous updating are presented.

2.1 Initial Linear Quadratic Game Model

Consider n-player ($|N| = n$) linear quadratic differential game $\Gamma(x_0, T - t_0)$ defined on the interval $[t_0, T]$:

Motion equations have the form

$$\begin{aligned} &\dot{x}(t) = Ax(t) + B_1 u_1(t,x) + \ldots + B_n u_n(t,x), \\ &x(t_0) = x_0, \\ &x \in \mathbb{R}^l,\ u = (u_1, \ldots, u_n),\ u_i = u_i(t,x) \in U_i \subset \text{comp}\mathbb{R}^k,\ t \in [t_0, T]. \end{aligned} \tag{1}$$

Payoff function of player $i \in N$ is defined as

$$K_i(x_0, t_0, T; u) = \int\limits_{t_0}^{T} \left(x'(t) Q_i x(t) + \sum_{j=1}^{n} u_j'(t, x) R_{ij} u_j(t, x) \right) dt, \; i \in N, \quad (2)$$

where Q_i, R_{ij} are assumed to be symmetric, R_{ii} is positive defined, $(\,\cdot\,)'$ means transpose here and hereafter.

2.2 Linear Quadratic Game Model with Continuous Updating

Consider n-player differential game $\Gamma(x, t, \overline{T})$, $t \in [t_0, +\infty)$ defined on the interval $[t, t + \overline{T}]$, where $0 < \overline{T} < +\infty$.

Motion equations of $\Gamma(x, t, \overline{T})$ have the form

$$\begin{aligned} &\dot{x}^t(s) = Ax^t(s) + B_1 u_1^t(s, x^t) + \ldots + B_n u_n^t(s, x^t), \\ &x^t(t) = x, \\ &x^t \in \mathbb{R}^l, \; u^t = (u_1^t, \ldots, u_n^t), \; u_i^t = u_i^t(s, x^t) \in U_i \subset \text{comp}\mathbb{R}^k, \; t \in [t_0, +\infty). \end{aligned} \quad (3)$$

Payoff function of player $i \in N$ in the game $\Gamma(x, t, \overline{T})$ is defined as

$$K_i^t(x^t, t, \overline{T}; u^t) = \int\limits_{t}^{t+\overline{T}} \left((x^t(s))' \, Q_i x^t(s) + \sum_{j=1}^{n} \left(u_j^t(s, x^t)\right)' R_{ij} u_j^t(s, x^t) \right) ds, \quad (4)$$

where $x^t(s)$, $u^t(s, x)$ are trajectory and strategies in the game $\Gamma(x, t, \overline{T})$.

Differential game with continuous updating evolves according to the rule:

Time parameter $t \in [t_0, +\infty)$ evolves continuously, as a result players continuously receive updated information about motion equations and payoff functions under $\Gamma(x, t, \overline{T})$.

Strategies $u(t, x)$ in the game model with continuous updating are defined in the following way:

$$u(t, x) = u^t(t, x), \; t \in [t_0, +\infty), \quad (5)$$

where $u^t(s, x)$, $s \in [t, t + \overline{T}]$ are some fixed strategies defined in the subgame $\Gamma(x, t, \overline{T})$.

State $x(t)$ in the model with continuous updating is defined according to

$$\begin{aligned} &\dot{x}(t) = Ax(t) + B_1 u_1(t, x) + \ldots + B_n u_n(t, x), \\ &x(t_0) = x_0, \\ &x \in \mathbb{R}^l \end{aligned} \quad (6)$$

with strategies with continuous updating $u(t, x)$ involved.

Essential difference between the game model with continuous updating and classic differential game $\Gamma(x_0, T - t_0)$ with prescribed duration is that players in the initial game are guided by the payoffs that they will eventually receive on the interval $[t_0, T]$, but in the case of a game with continuous updating, at the time instant t they orient themselves on the expected payoffs (4), which are calculated using information about the game structure defined on the interval $[t, t + \overline{T}]$.

3 Nash Equilibrium with Continuous Updating in LQ Differential Games

3.1 Concept of Nash Equilibrium for Games with Continuous Updating

Within the framework of continuously updated information in this class of differential games it is interesting to understand of how to model the behavior of players. To do this, we use the concept of Nash equilibrium in feedback strategies. However, for the class of differential games with continuous updating, we would like $u^{NE}(t,x) = (u_1^{NE}(t,x), \ldots, u_n^{NE}(t,x))$ for each fixed $t \in [t_0, +\infty)$ to coincide with the feedback Nash equilibrium in the game (6), (4) defined on the interval $[t, t+\overline{T}]$ at the instant t.

Consider two time intervals $[t, t+\overline{T}]$ and $[t+\epsilon, t+\overline{T}+\epsilon]$, $\epsilon << \overline{T}$. According to the problem statement, $u^{NE}(t,x)$ at the instant t should coincide with the Nash equilibrium in the game defined on the interval $[t, t+\overline{T}]$ and $u^{NE}(t+\epsilon, x)$ at instant $t+\epsilon$ should coincide with the Nash equilibrium in the game defined on the interval $[t+\epsilon, t+\epsilon+\overline{T}]$. Therefore direct application of classical approaches for determining Nash equilibrium in feedback strategies is not possible.

In order to construct such a strategy profile, we define the concept of generalized Nash equilibrium in feedback strategies as an optimality principle:

$$\widetilde{u}^{NE}(t,s,x) = (\widetilde{u}_1^{NE}(t,s,x), \ldots, \widetilde{u}_n^{NE}(t,s,x)),\ t \in [t_0, T],\ s \in [t, t+\overline{T}], \tag{7}$$

which we further use to construct desired strategy profile $u^{NE}(t,x)$.

Definition 1. *Strategy profile $\widetilde{u}^{NE}(t,s,x) = (\widetilde{u}_1^{NE}(t,s,x), \ldots, \widetilde{u}_n^{NE}(t,s,x))$, $t \in [t_0, +\infty)$, $s \in [t, t+\overline{T}]$ is a generalized Nash equilibrium in the game with continuous updating, if for any fixed $t \in [t_0, +\infty)$ strategy profile $\widetilde{u}^{NE}(t,s,x)$ is Nash equilibrium in feedback strategies in the game $\Gamma(x,t,\overline{T})$, $0 < \overline{T} < \infty$.*

Using generalized feedback Nash equilibrium it is possible to define solution concept for a game model with continuous updating.

Definition 2. *Strategy profile $u^{NE}(t,x)$ is called the Nash equilibrium with continuous updating, if it is defined in the following way:*

$$u^{NE}(t,x) = \widetilde{u}^{NE}(t,s,x)|_{s=t} = (\widetilde{u}_1^{NE}(t,s,x)|_{s=t}, \ldots, \widetilde{u}_n^{NE}(t,s,x)|_{s=t}), \tag{8}$$

where $t \in [t_0, +\infty)$, $\widetilde{u}^{NE}(t,s,x)$ is the generalized feedback Nash equilibrium defined above.

Strategy profile $u^{NE}(t,x)$ will be used as a solution concept in the game with continuous updating.

3.2 Theorem on Nash Equilibrium with Continuous Updating for LQ Differential Games

Here we present the explicit form of Nash equilibrium with continuous updating for a two-player differential game.

Theorem 1. *The two-player linear quadratic differential game $\Gamma(x_0, t_0, \overline{T})$ with continuous updating has, for every initial state, a linear feedback Nash equilibrium, if and only if the following set of coupled Riccati differential equations has a set of symmetric solutions K_1, K_2 on the interval $[0, 1]$:*

$$\begin{aligned}\dot{K}_i(\tau) &= -(A\overline{T} - S_jK_j(\tau))'K_i(\tau) - K_i(\tau)(A\overline{T} - S_jK_j(\tau))\\ &\quad + K_i(\tau)S_iK_i(\tau) - Q_i - K_j(\tau)S_{ji}K_j(\tau),\\ &\quad K_i(1) = 0,\ i \neq j \in N,\end{aligned} \tag{9}$$

where

$$S_i = \overline{T}^2 B_i R_{ii}^{-1} B_i', \quad S_{ij} = \overline{T}^2 B_i R_{ii}^{-1} R_{ji} R_{ii}^{-1} B_i',\ i \neq j \in N. \tag{10}$$

In this case there is a unique feedback Nash equilibrium with continuous updating, which has the form:

$$u_i^{NE}(t, x) = -R_{ii}^{-1} B_i' K_i(0) \overline{T} x,\ i \in N. \tag{11}$$

Proof. In order to prove the Theorem we introduce the following change of variables

$$\begin{aligned}s &= t + \overline{T}\tau,\\ y(\tau) &= x(t + \overline{T}\tau),\\ v_i(\tau, y) &= u_i(t + \overline{T}\tau, x),\ i \in N.\end{aligned} \tag{12}$$

By substituting (12) to the motion equations (3), payoff function (4) we obtain

$$\dot{y}(\tau) = \overline{T}Ay(s) + \overline{T}B_1v_1(\tau, y) + \overline{T}B_2v_2(\tau, y) \tag{13}$$

and

$$K_i(y, \tau; v) = \int_0^1 y'(s)Q_iy(s) + \sum_{j=1}^{2} (v_j(s, y))'R_{ij}v_j(s, y)ds,\ i \in N. \tag{14}$$

It is known [4] that the criterion for existence of feedback Nash equilibrium is the existence of symmetric solution for the system of differential Eq. (9). According to [4] feedback Nash equilibrium strategies have the form

$$v_i^{NE}(\tau, y) = -R_{ii}^{-1} B_i' K_i(\tau) \overline{T} y. \tag{15}$$

From (12) we have

$$\tau = \frac{s - t}{\overline{T}},$$

returning to original variables we obtain the following strategies

$$u_i^t(s,x) = -R_{ii}^{-1} B_i' K_i \left(\frac{s-t}{\overline{T}} \right) \overline{T} x.$$

These strategies are Nash equilibrium in feedback strategies in the subgame $\Gamma(x, t, \overline{T})$ by construction.

Task (13), (14) and solution (15) have the same form for all values t in original game with continuous updating. Then a generalized Nash equilibrium in the game with continuous updating has the form

$$\widetilde{u}_i^{NE}(t,s,x) = -R_{ii}^{-1} B_i' K_i \left(\frac{s-t}{\overline{T}} \right) \overline{T} x. \quad (16)$$

Apply the procedure (8) to determine Nash equilibrium with continuous updating using generalized Nash equilibrium (16), $s = t$:

$$u_i^{NE}(t,x) = -R_{ii}^{-1} B_i' K_i(0) \overline{T} x, \; t \in [t_0, +\infty), \; i \in N. \quad (17)$$

This proves the theorem.

4 LQ Differential Game with Dynamic Updating

In this section, we define a game model with dynamic updating in order to later demonstrate the convergence of Nash equilibrium strategies and corresponding trajectories for a case of dynamic and continuous updating.

4.1 LQ Game Model with Dynamic Updating

In papers [5,7–11,16] the method for constructing differential game model with dynamic updating is described. There it is assumed that players have information about the game structure only over a truncated interval and, based on this, make decisions. In order to model the behavior of players in the case, when information updates dynamically, consider the case when information is updated every $\Delta t > 0$ and the behavior of players on each segment $[t_0 + j\Delta t, t_0 + (j+1)\Delta t]$, $j = 0, 1, 2, \ldots$ is modeled using the notion of truncated subgame:

Definition 3. *Let $j = 0, 1, 2, \ldots$. Truncated subgame $\bar{\Gamma}_j(x_0^j, t_0 + j\Delta t, t_0 + j\Delta t + \overline{T})$ is the game defined on the interval $[t_0 + j\Delta t, t_0 + j\Delta t + \overline{T}]$ in the following way. On the interval $[t_0 + j\Delta t, t_0 + j\Delta t + \overline{T}]$ payoff function, motion equation in the truncated subgame and initial game model $\Gamma(x_0, T - t_0)$ coincide:*

$$\begin{aligned}
&\dot{x}^j(s) = Ax^j(s) + B_1 u_1^j(s, x^j) + \ldots + B_n u_n^j(s, x^j), \\
&x^j(t_0 + j\Delta t) = x_0^j, \\
&x^j \in \mathbb{R}^n, \; u^j = (u_1^j, \ldots, u_n^j), \; u_i^j = u_i^j(s, x^j) \in U_i \subset comp\mathbb{R}^k, \; t \in [t_0, +\infty).
\end{aligned} \quad (18)$$

$$K_i^j(x^j, t_0 + j\Delta t, t_0 + j\Delta t + \overline{T}; u^j) = \int\limits_{t_0+j\Delta t}^{t_0+j\Delta t+\overline{T}} \left(x^j(s)\right)' Q_i x^j(s) + \sum_{k=1}^{n} \left(u^k(s, x^j)\right)' R_{ik} u^k(s, x^j) ds, \ i \in N, \tag{19}$$

At any instant $t = t_0 + j\Delta t$ information about the game structure updates, and therefore players adapt to it. This class of game models is called differential games with dynamic updating.

As a solution concept in the differential game model with dynamic updating we will use feedback Nash equilibrium. In the same way as in Sect. 3 we will need to define a special form of it. According to the approach described above, at any time instant $t \in [t_0, +\infty)$, players have or use truncated information about the game structure $\Gamma(x_0, T - t_0)$, therefore classical approaches for determining optimal strategies (cooperative and noncooperative) cannot be directly applied. In order to determine the solution for games with dynamic updating, the notion of resulting feedback Nash equilibrium is introduced:

Definition 4. *Resulting feedback Nash equilibrium*

$$\hat{u}^{NE}(t, x) = (\hat{u}_1^{NE}(t, x), \ldots, \hat{u}_n^{NE}(t, x))$$

of players in the game model with dynamic updating have the form:

$$\{\hat{u}^{NE}(t, x)\}_{t=t_0}^{\infty} = \begin{cases} u_0^{NE}(t, x), & t \in [t_0, t_0 + \Delta t], \\ \ldots \\ u_j^{NE}(t, x), & t \in (t_0 + j\Delta t, t_0 + (j+1)\Delta t], \\ \ldots \end{cases} \tag{20}$$

where $u_j^{NE}(t, x) = (u_1^{j,NE}(t, x), \ldots, u_n^{j,NE}(t, x))$ *is some fixed feedback Nash equilibrium in the truncated subgame* $\bar{\Gamma}_j(x_0^{j,NE}, t_0 + j\Delta t, t_0 + j\Delta t + \overline{T})$, $j = 0, 1, 2, \ldots$ *starting along the equilibrium trajectory of the previous truncated subgame:* $x_0^{j,NE} = x^{j-1,NE}(t_0 + j\Delta t)$.

Trajectory obtained by using motion equation (1) and the resulting feedback Nash equilibrium $\hat{u}^{NE}(t, x) = (\hat{u}_1^{NE}(t, x), \ldots, \hat{u}_n^{NE}(t, x))$ we denote by $\hat{x}^{NE}(t)$ and call the resulting equilibrium trajectory.

4.2 Resulting Feedback Nash Equilibrium with Dynamic Updating

Firstly, consider Nash Equilibrium in truncated subgame $\bar{\Gamma}_j(x_0^j, t_0 + j\Delta t, t_0 + j\Delta t + \overline{T})$.

Theorem 2. *The two-player linear quadratic differential game $\bar{\Gamma}_j(x_0^j, t_0 + j\Delta t, t_0 + j\Delta t + \overline{T})$ has, for every initial state, a linear feedback Nash equilibrium if and only if the following set of coupled Riccati differential equations has a set of symmetric solutions K_1, K_2 on the interval $[0,1]$:*

$$\begin{aligned}\dot{K}_i(\tau) = -(A\overline{T} - S_jK_j(\tau))'K_i(\tau) - K_i(\tau)(A\overline{T} - S_jK_j(\tau))\\ + K_i(\tau)S_iK_i(\tau) - Q_i - K_j(\tau)S_{ji}K_j(\tau),\\ K_i(1) = 0,\ i \neq j,\end{aligned} \tag{21}$$

where

$$S_i = \overline{T}^2 B_i R_{ii}^{-1} B_i', \quad S_{ij} = \overline{T}^2 B_i R_{ii}^{-1} R_{ji} R_{ii}^{-1} B_i',\ i \neq j \in N. \tag{22}$$

In that case there is a unique equilibrium. The equilibrium strategies are

$$u_i^{j,NE}(t,x) = -R_{ii}^{-1} B_i' K_i\left(\frac{t - (t_0 + j\Delta t)}{\overline{T}}\right)\overline{T}x. \tag{23}$$

Proof. To prove this theorem we use similar change of variables as in (12) for each truncated subgame:

$$\tau = \frac{t - (t_0 + j\Delta t)}{\overline{T}}. \tag{24}$$

According to (20) Nash equilibrium for the game model with dynamic updating $\hat{u}_i^{NE}(t,x)$ can be constructed using the Nash equilibrium defined in each truncated subgame $u_i^{j,NE}(t,x)$. Corresponding trajectory $\hat{x}^{NE}(t)$ is constructed using $\hat{u}_i^{NE}(t,x)$ and (1).

5 Convergence of Resulting Nash Equilibrium Strategies and Trajectory

Theorem 3. *For $\Delta t \to 0$ and $x \in X$ (X—limited set) resulting feedback Nash equilibrium strategies $\hat{u}_i^{NE}(t,x)$ in the game with dynamic updating uniformly converge to feedback Nash equilibrium with continuous updating $\widetilde{u}_i^{NE}(t,x)$:*

$$\hat{u}_i^{NE}(t,x) \underset{[t_0,+\infty)}{\rightrightarrows} \widetilde{u}_i^{NE}(t,x),\ i \in N. \tag{25}$$

Proof. Introduce the notation: $t_j \overset{\text{def}}{=} t_0 + j\Delta t$ and let $t \in [t_j, t_{j+1}]$ for some j. According to the definition of $\hat{u}^{NE}(t,x)$ (20) we will need to show that $\|\widetilde{u}_i^{NE}(t,x) - u_i^{j,NE}(t,x)\| \to 0$, when $\Delta t \to 0$.

Consider the expressions for $\widetilde{u}_i^{NE}$ and $u_i^{j,NE}$:

$$\begin{aligned}\widetilde{u}_i^{NE}(t,x) &= -R_{ii}^{-1} B_i' K_i(0)\overline{T}x,\\ u_i^{j,NE}(t,x) &= -R_{ii}^{-1} B_i' K_i\left(\frac{t - t_j}{\overline{T}}\right)\overline{T}x.\end{aligned}$$

From Taylor decomposition for $K(t)$ at the point $t = 0$ we obtain:

$$\|\widetilde{u}_i^{NE}(t,x) - u_i^{j,NE}(t,x)\| \leq \|R_{ii}^{-1}B_i'\|\|x\| \left(\left\|\dot{K}(0)\right\| \frac{\Delta t}{\overline{T}} + o(\Delta t)\right). \tag{26}$$

When $\Delta t \to 0$ the right hand side of (26) converges to zero and as a result the left hand side of (26) also converges to zero. This completes the proof.

Theorem 4. *Equilibrium trajectory in the game with dynamic updating $\hat{x}^{NE}(t)$ pointwise converges to the equilibrium trajectory $\widetilde{x}^{NE}(t)$ in the game with continuous updating $\widetilde{x}^{NE}(t)$ for $\Delta t \to 0$:*

$$\hat{x}^{NE}(t) \underset{[t_0,+\infty)}{\rightarrow} \widetilde{x}^{NE}(t). \tag{27}$$

Proof. Let $t \in [t_j, t_{j+1}]$ for some j. According to the definition of $\hat{x}^{NE}(t)$ we will need to show that $\|\widetilde{x}^{NE}(t) - x_j^{NE}(t)\| \to 0$ when $\Delta t \to 0$.

Trajectories $\widetilde{x}^{NE}(t)$ and $x_j^{NE}(t)$ satisfy the differential equations respectively

$$\dot{\widetilde{x}}(t) = (A - B_1R_{11}^{-1}B_1'K_1(0)\overline{T} - B_2R_{22}^{-1}B_2'K_2(0)\overline{T})\,\widetilde{x}(t),$$

$$\dot{x}_j(t) = \left(A - B_1R_{11}^{-1}B_1'K_1\left(\frac{t-t_j}{\overline{T}}\right)\overline{T} - B_2R_{22}^{-1}B_2'K_2\left(\frac{t-t_j}{\overline{T}}\right)\overline{T}\right)x_j(t).$$

Notice that

$$K_i(0)\widetilde{x} - K_i\left(\frac{t-t_j}{\overline{T}}\right)x_j = K_i(0)(\widetilde{x} - x_j) + \left(K_i(0) - K_i\left(\frac{t-t_j}{\overline{T}}\right)\right)x_j.$$

Let $y_j^{NE}(t) = \widetilde{x}^{NE}(t) - x_j^{NE}(t)$, $\widetilde{A} = A - B_1R_{11}^{-1}B_1'K_1(0)\overline{T} - B_2R_{22}^{-1}B_2'K_2(0)\overline{T}$ and

$$f_j(t) = -B_1R_{11}^{-1}B_1'\left[K_1(0) - K_1\left(\frac{t-t_j}{\overline{T}}\right)\right]\overline{T}x_j(t)$$
$$- B_2R_{22}^{-1}B_2'\left[K_2(0) - K_2\left(\frac{t-t_j}{\overline{T}}\right)\right]\overline{T}x_j(t).$$

Then $y_j^{NE}(t)$ satisfies following differential equation

$$\dot{y}_j(t) = \widetilde{A}y_j(t) + f_j(t).$$

Consider

$$y(t) = \begin{cases} y_0(t), & t \in [t_0, t_0 + \Delta t], \\ \dots \\ y_j(t), & t \in (t_0 + j\Delta t, t_0 + (j+1)\Delta t], \\ \dots \end{cases} \tag{28}$$

and

$$f(t) = \begin{cases} f_0(t), & t \in [t_0, t_0 + \Delta t], \\ \dots \\ f_j(t), & t \in (t_0 + j\Delta t, t_0 + (j+1)\Delta t], \\ \dots \end{cases}$$

then (28) satisfies following differential equation

$$\dot{y}(t) = \widetilde{A}y(t) + f(t).$$

with initial state $y(t_0) = 0$, since $\hat{x}^{NE}(t_0) = \widetilde{x}^{NE}(t) = x_0$.

By the Cauchy formula we have for any $t \geq t_0$

$$y(t) = \int_{t_0}^{t} e^{\widetilde{A}(t-s)} f(s)ds.$$

Taking this into account we have for fixed t

$$\lim_{\Delta t \to 0} \|y(t_j)\| \leq \lim_{\Delta t \to 0} \|e^{\widetilde{A}(t-t_0)}\|(t-t_0)\beta\left(\frac{\Delta t}{\overline{T}} + o(\Delta t)\right) = 0, \qquad (29)$$

where

$$\beta = \left(\|B_1R_{11}^{-1}B_1'\| \left\|\dot{K}_1(0)\right\| + \|B_2R_{22}^{-1}B_2'\| \left\|\dot{K}_2(0)\right\|\right)\overline{T}M(t),$$

$$M(t) = \max_{\tau \in [t_0, t]} \|\hat{x}^{NE}(\tau)\|.$$

According to (29) $y(t) \underset{[t_0,+\infty)}{\rightarrow} 0$, when $\Delta t \to 0$. This proves the theorem.

6 Example Model

6.1 Common Description

Consider the model in which there are two individuals investing in a public stock of knowledge (see also Dockner et al. [3]). Let $x(t)$ be the stock of knowledge at time t and $u_i(t)$ – the investment of player i in public knowledge at time t. Assume that the stock of knowledge evolves according to the accumulation equation

$$\dot{x}(t) = -\beta x(t) + u_1(t,x) + u_2(t,x), \quad x(0) = x_0, \qquad (30)$$

where β is the depreciation rate. Assume that each player derives quadratic utility from the consumption of the stock of knowledge and that the cost of investment increases quadratically with the investment effort. That is, the cost function of both players is given by

$$K_i(x_0, t_0, T; u) = \int_0^T \big(-q_i x^2(t) + r_i u_i^2(t,x)\big)dt, \ i = 1, 2. \qquad (31)$$

Consider the initial game (30), (31) in the terms of LQ-games theory [4]. To find a feedback Nash equilibrium, we need to solve the following set of coupled Riccati differential equations:

$$\begin{cases} \dot{k_1}(t) = -2(-\beta - \frac{1}{r_2}k_2(t))k_1(t) + \frac{1}{r_1}k_1^2(t) + q_1, \\ \dot{k_2}(t) = -2(-\beta - \frac{1}{r_1}k_1(t))k_2(t) + \frac{1}{r_2}k_2^2(t) + q_2, \\ k_1(T) = 0, \\ k_2(T) = 0. \end{cases} \qquad (32)$$

As an example consider the symmetric case $r_1 = r_2 = r$, $q_1 = q_2 = q$. Let $k(t) = k_1(t) = k_2(t)$. We obtain the following differential equation:

$$\begin{cases} \dot{k}(t) = 2\beta k(t) + \frac{3k^2(t)}{r} + q, \\ k(T) = 0. \end{cases} \tag{33}$$

The solution of Cauchy problem (33) is

$$k(t) = \frac{\beta r + v}{3}\left(\frac{2v}{(v-\beta r)e^{\frac{2v}{r}(t-T)} + v + \beta r} - 1\right),$$

where $v = \sqrt{\beta^2 r^2 - 3qr}$. According to [4] feedback Nash equilibrium for the initial game model will have the form:

$$u_i^{NE}(t,x) = -\frac{k(t)x}{r}, \; i = 1, 2. \tag{34}$$

By substituting the value for $k(t)$ in (34) we obtain:

$$u_i^{NE}(t,x) = \frac{\beta r + v}{3r}\left(1 - \frac{2v}{(v-\beta r)e^{\frac{2v}{r}(t-T)} + v + \beta r}\right)x(t).$$

6.2 Game Model with Continuous Updating

Now consider the case of continuous updating. Here we suppose that two individuals at each time instant $t \in [t_0, +\infty)$ use information about motion equations and payoff functions on the interval $[t, t+\overline{T}]$. As the current time t evolves the interval, which defines the information shifts as well. Motion equations for the game model with continuous updating have the form

$$\dot{x}^t(s) = -\beta x^t(s) + u_1^t(s,x) + u_2^t(s,x), \; x^t(t) = x, \quad t \in [t_0, +\infty). \tag{35}$$

Payoff function of player $i \in N$ for the game model with continuous updating is defined as

$$K_i^t(x^t, t, \overline{T}; u^t) = \int\limits_t^{t+\overline{T}} \left(-\left(x^t(s)\right)^2 q_i + \left(u_i^t(s,x)\right)^2 r_i\right) ds, \; i = 1, 2. \tag{36}$$

According to the Theorem 2 defining the form of feedback Nash equilibrium with continuous updating on the first step we need to solve the following differential equation:

$$\begin{cases} \dot{k}(\tau) = 2\beta\overline{T}k(\tau) + \frac{3\overline{T}k^2(\tau)}{r} + \overline{T}q, \\ k(1) = 0. \end{cases} \tag{37}$$

The solution of (37) is

$$k(\tau) = \frac{\beta r + v}{3}\left(\frac{2v}{(v-\beta r)e^{\frac{2v\overline{T}}{r}(\tau-1)} + v + \beta r} - 1\right), \tag{38}$$

where $v = \sqrt{\beta^2 r^2 - 3qr}$. According to (23) feedback Nash equilibrium with continuous updating has the form:

$$\widetilde{u}_i^{NE}(t,x) = -\frac{k(0)x\overline{T}}{r}. \tag{39}$$

By substituting (38) in (39) we obtain:

$$\widetilde{u}_i^{NE}(t,x) = \frac{\beta r + v}{3r}\left(1 - \frac{2v}{(v-\beta r)e^{-\frac{2v\overline{T}}{r}} + v + \beta r}\right)\overline{T}x, \tag{40}$$

by substituting (40) in (30) we obtain $\widetilde{x}^{NE}(t)$ as solution of equation

$$\dot{\widetilde{x}}^{NE}(t) = -\beta\widetilde{x}^{NE}(t) + \widetilde{u}_1^{NE}(t,x) + \widetilde{u}_2^{NE}(t,x), \quad \widetilde{x}^{NE}(0) = x_0. \tag{41}$$

6.3 Game Model with Dynamic Updating

Perform similar calculations for the resulting Nash equilibrium for a game with dynamic updating based on the calculations for the original game and the approach described in Sect. 4.1 and obtain

$$\widetilde{u}_i^{NE}(t,x) = -\frac{k\left(\frac{t-t_i}{\overline{T}}\right)x\overline{T}}{r}, \; t \in [t_i, t_{i+1}]. \tag{42}$$

By substituting (38) in (42) we obtain:

$$\hat{u}_i^{NE}(t,x) = \frac{\beta r + v}{3r}\left(1 - \frac{2v}{(v-\beta r)e^{\frac{2v(t-t_i-\overline{T})}{r}} + v + \beta r}\right)\overline{T}x, \quad t \in [t_i, t_{i+1}], \tag{43}$$

by substituting (43) in (30) we obtain $\hat{x}^{NE}(t)$ as solution of equation

$$\dot{\hat{x}}^{NE}(t) = -\beta\hat{x}^{NE}(t) + \hat{u}_1^{NE}(t,x) + \hat{u}_2^{NE}(t,x), \quad \hat{x}^{NE}(0) = x_0. \tag{44}$$

6.4 Game Model on Infinite Interval

Consider classic approach for Nash equilibrium for the game on infinite interval $[0, +\infty)$. Motion equations have the form

$$\dot{x}(t) = -\beta x(t) + u_1(t,x) + u_2(t,x), \quad x(0) = x_0. \tag{45}$$

Payoff function of player $i \in N$ is defined as

$$K_i(x_0; u) = \lim_{T\to\infty}\int_0^T \big(-q_i x^2(t) + r_i u_i^2(t,x)\big)dt, \; i = 1,2. \tag{46}$$

According to [4] feedback Nash equilibrium strategies have the form

$$u^{NE}(t,x) = -\frac{kx}{r} \tag{47}$$

in our symmetric case ($r_1 = r_2 = r$, $q_1 = q_2 = q$), where k is solution of

$$\frac{3k^2}{r} + 2\beta k + q = 0.$$

By substituting (47) in (45) we obtain $x^{NE}(t)$ as solution of equation

$$\dot{x}^{NE}(t) = \left(-\beta - \frac{2k}{r}\right) x^{NE}(t), \quad x^{NE}(0) = x_0. \tag{48}$$

6.5 Numerical Simulation

Consider the results of numerical simulation for the game model presented above on the interval $[0, 8]$, i.e. $t_0 = 0$, $T = 8$. At the initial instant $t_0 = 0$ the stock of knowledge is 100, i.e. $x_0 = 100$. The other parameters of models: $\beta = 0.9$, $r = 6$, $q = 1$. Suppose that for the case of a dynamic updating (blue solid and dotted lines Figs. 1 and 2), the intervals between updating instants are $\Delta t = 2$, therefore $l = 4$. In Fig. 1 the comparison of resulting Nash equilibrium in the game with dynamic updating (blue line) and Nash equilibrium with continuous updating (red lines) is presented. In Fig. 2 similar results are presented for the strategies.

In order to demonstrate the results of Theorems 3 and 4 on convergence of resulting equilibrium strategies and corresponding trajectory to the equilibrium strategies and trajectory with continuous updating, consider the simulation results for a case of frequent updating, namely $l = 20$. Figures 3 and 4 represent the same solutions as in Figs. 1 and 2, but for the case, when $\Delta t = 0.4$. Therefore, convergence results are confirmed by the numerical experiments presented below.

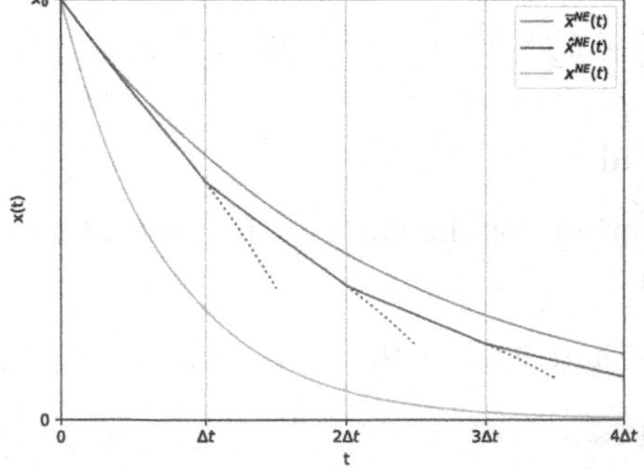

Fig. 1. $\tilde{x}^{NE}(t)$ (41) - red upper line, $\hat{x}^{NE}(t)$ (44) - blue broken line, $x^{NE}(t)$ (48) - green lower line. (Color figure online)

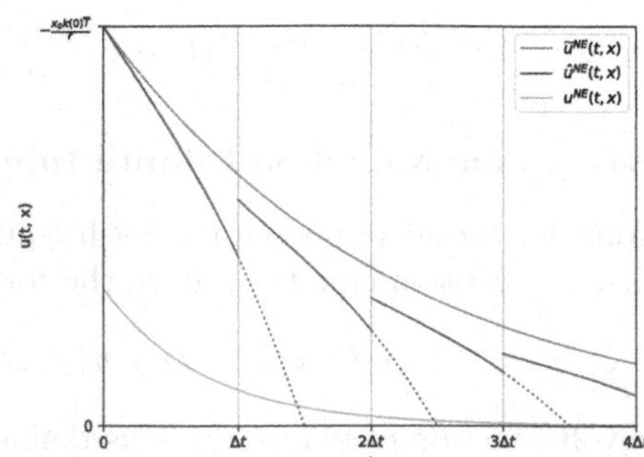

Fig. 2. $\tilde{u}^{NE}(t)$ (40) - red upper line, $\hat{u}^{NE}(t)$ (43) - blue broken line, $u^{NE}(t)$ (47) - green lower line. (Color figure online)

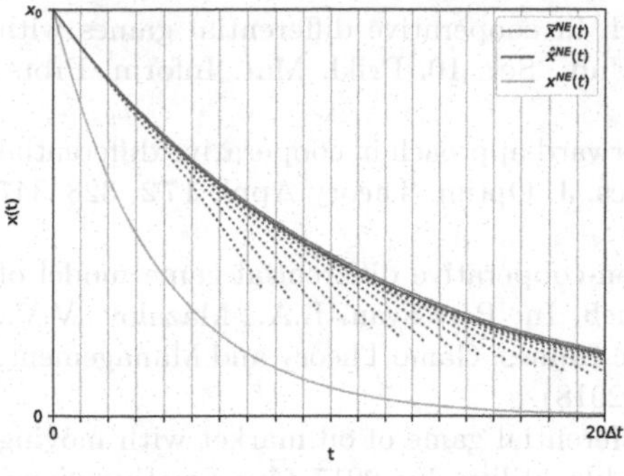

Fig. 3. $\tilde{x}^{NE}(t)$ (41) - red upper line, $\hat{x}^{NE}(t)$ (44) - blue broken line, $x^{NE}(t)$ (48) - green lower line. (Color figure online)

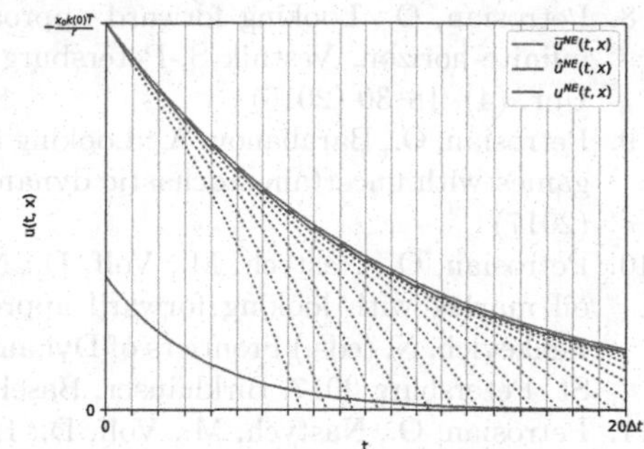

Fig. 4. $\tilde{u}^{NE}(t)$ (40) - red upper line, $\hat{u}^{NE}(t)$ (43) - blue broken line, $u^{NE}(t)$ (47) - green lower line. (Color figure online)

7 Conclusion

The concept of feedback Nash equilibrium for the class of linear quadratic differential games with continuous updating is constructed and the corresponding Theorem is presented. The form of feedback Nash equilibrium for a game model with dynamic updating is also presented and convergence of resulting feedback Nash equilibrium with dynamic updating to the feedback Nash equilibrium with continuous updating as the number of updating instants converges to infinity is proved. The results are demonstrated using the differential game model of knowledge stock. Obtained results are both fundamental and applied in nature since they allow specialists from the applied field to use a new mathematical tool for more realistic modeling of engineering system describing human-machine interaction.

References

1. Basar, T., Olsder, G.: Dynamic Noncooperative Game Theory. Academic Press, London (1995)
2. Bellman, R.: Dynamic Programming. Princeton University Press, Princeton (1957)
3. Dockner, E., Jorgensen, S., Long, N., Sorger, G.: Differential Games in Economics and Management Science. Cambridge University Press, Cambridge (2000)
4. Engwerda, J.: LQ Dynamic Optimization and Differential Games. Willey, New York (2005)
5. Gromova, E., Petrosian, O.: Control of information horizon for cooperative differential game of pollution control. In: 2016 International Conference Stability and Oscillations of Nonlinear Control Systems (Pyatnitskiy's Conference) (2016)
6. Kleimenov, A.: Non-antagonistic Positional Differential Games. Science, Ekaterinburg (1993)
7. Petrosian, O.: Looking forward approach in cooperative differential games. Int. Game Theory Rev. **18**, 1–14 (2016)

8. Petrosian, O.: Looking forward approach in cooperative differential games with infinite-horizon. Vestnik S.-Petersburg Univ. Ser. 10. Prikl. Mat. Inform. Prots. Upr. (4), 18–30 (2016)
9. Petrosian, O., Barabanov, A.: Looking forward approach in cooperative differential games with uncertain-stochastic dynamics. J. Optim. Theory Appl. **172**, 328–347 (2017)
10. Petrosian, O., Nastych, M., Volf, D.: Non-cooperative differential game model of oil market with looking forward approach. In: Petrosyan, L.A., Mazalov, V.V., Zenkevich, N. (eds.) Frontiers of Dynamic Games, Game Theory and Management, St. Petersburg 2017. Birkhäuser, Basel (2018)
11. Petrosian, O., Nastych, M., Volf, D.: Differential game of oil market with moving informational horizon and non-transferable utility. In: 2017 Constructive Nonsmooth Analysis and Related Topics (dedicated to the memory of V.F. Demyanov) (2017)
12. Petrosyan, L., Murzov, N.: Game-theoretic problems in mechanics. Lith. Math. Collect. **3**, 423–433 (1966)
13. Pontryagin, L.: On theory of differential games. Successes Math. Sci. 26, 4(130), 219–274 (1966)
14. Shevkoplyas, E.: Optimal solutions in differential games with random duration. J. Math. Sci. **199**(6), 715–722 (2014)
15. Yeung, D., Petrosian, O.: Cooperative stochastic differential games with information adaptation. In: International Conference on Communication and Electronic Information Engineering (2017)
16. Yeung, D., Petrosian, O.: Infinite horizon dynamic games: a new approach via information updating. Int. Game Theory Rev. **19**, 1–23 (2017)

Spatial Equilibrium in a Multidimensional Space: An Immigration-Consistent Division into Countries Centered at Barycenter

Valeriy Marakulin[1,2](✉)

[1] Sobolev Institute of Mathematics, Russian Academy of Sciences, 4 Acad. Koptyug avenue, Novosibirsk 630090, Russia
marakulv@gmail.com

[2] Novosibirsk State University, 2 Pirogova street, Novosibirsk 630090, Russia
http://www.math.nsc.ru/mathecon/marakENG.html

Abstract. It studies the problem of immigration proof partition for communities (countries) in a multidimensional space. This is an existence problem of Tiebout type equilibrium, where migration stability suggests that every inhabitant has no incentives to change current jurisdiction. In particular, an inhabitant at every frontier point has equal costs for all available jurisdictions. It is required that the inter-country border is represented by a continuous curve.

The paper presents the solution for the case of the costs described as the sum of the two values: the ratio of total costs on the total weight of the population plus transportation costs to the center presented as a barycenter of the state. In the literature, this setting is considered as a case of especial theoretical interest and difficulty. The existence of equilibrium division is stated via an approximation reducing the problem to the earlier studied case, in which centers of the states never can coincide: to do this an earlier proved a generalization of conic Krasnosel'skii fixed point theorem is applied.

Keywords: Migration stable partitions · Barycenter · Tiebout equilibrium · Generalized fixed point theorems

1 Introduction

In a seminal paper [1] a basic model of country formation was offered. In this model, the cost of the population is described as the sum of the two values—the ratio of total costs on the total weight of the population plus transportation costs to the center of the state. This model has been studied in a number of subsequent studies, but the majority of them considered the case of a one-dimensional region and the interval-form "countries" (country formation on an interval). The first progress in the resolution of the problem of existence was obtained in [4,10],

M. Khachay et al. (Eds.): MOTOR 2019, LNCS 11548, pp. 651–672, 2019.
https://doi.org/10.1007/978-3-030-22629-9_46

where well-known Gale–Nikaido–Debreu lemma was applied to state the existence of *nontrivial* immigration proof partition for interval countries, i.e. such that no one has an incentive to change their country of residence. However, in the proof of [4] rather strong assumptions were made. Mathematical part of this approach was significantly strengthened in [6,7], where the modeling was extended to the population distribution, described as a Radon measure (probability measure defined on the Borel σ-algebra); this approach incorporates cities into the model. In [11] (initial version of 2016) a new significant advancement appears, it disseminates the result (existence theorem) to the case of 2 or more dimensional region. The proof of [11] is rather elegant and is based on the application of KKM-lemma (Knaster–Kuratowski–Mazurkiewicz), but the result is essentially limited by the presence of fixed location of the capitals.[1]

Papers [8,9] further develop this approach letting the "capitals" (or other relevant parameters) and individual costs to be changed continuously in space depending on some specific parameters, which is important for example in the context of party formation. The presented proof is based on an original generalization of Brouwer and Kakutani fixed point theorems for the case when a mapping can act outside the domain. However the results of [8,9] still do not cover the model in which the costs are described in a "classical style", as the sum of the two values: the ratio of total costs on the total weight of the population plus transportation costs to the center presented as a *barycenter* (center of mass) of the state. In the literature, this setting is considered a case of especial theoretical interest and difficulty. The present paper fills this gap. The existence of equilibrium division is stated via an approximation, reducing the problem to the earlier studied case, in which centers of the states can never coincide. After that, the limit is carried out, and the desired division is achieved.

The paper is organized as follows. In the second section, I present a theoretical problem and formulate some preliminary results that are the basis for the subsequent considerations. In the third one, I present the main result: a new existence theorem the proof of which involves several auxiliary lemmas and so on. The forth section presents a survey of generalized fixed point theorems applied for earlier studied cases; it requires the existence theorem of Sect. 2 to be proved.

2 The Model of Spatial Equilibrium: Preliminary Analysis and Results

One of the central problems of general spatial equilibrium theory is the existence of an immigration proof partition into n communities, the number of which is initially presented, not only on the plane but in a multidimensional space. This is not just a possible generalization of the one-dimensional case, but also an opportunity to consider in this context more general problems, for example, the

[1] As far as I know, the last version of the paper is more general and admits flexible centers, but still not general enough.

division of society by party affiliation. This is an existence problem for Tiebout type equilibrium [12],[2] that is quite important in the context of various economic theories, e.g. the theory of local provision of public goods or a problem of a society partition according to party affiliation in political sciences and so on. The principle of migration stability of "countries" suggests that the inhabitants have no incentives to change jurisdiction and, in particular, at every frontier point inhabitant has equal (and minimal) costs for all available jurisdictions.

So we need to divide a compact area $\mathcal{A} \subset \mathbb{R}^l$ into n counties, $N = \{1, \ldots, n\}$. In addition, it is also required that the inter-country border was represented by a continuous curve (surface). Here migration stability may be also treated as there is no nonzero mass of the population such that its members benefit from a *continuous* (gradual transformation, homotopy) change of the current inter-country boundary.

The division into the countries can significantly depend on how the population is distributed in a given area of space. A correct mathematical model for this is its description by means of some nonnegative countably additive measure μ, defined on a Borel σ-algebra. Assume that for each country $i \in N$ cost function $c_i(\cdot)$ is specified; this function may depend on the mass $\delta_i \in [0, 1]$ of country i and masses of other countries, individual location $x \in \mathcal{A}$ of an inhabitant and also on additional parameters $y \in Y$ that can be changed according to a partition configuration. In particular, y can be used to specify a center of the country as well as other parameters important for country formation.

An initial basic model for these cost functions is

$$c_i(x, y, \delta_i, r_c(S_i)) = \frac{g_i}{\delta_i} + \rho(x, r_c(S_i)), \quad g_i > 0, \quad i \in N. \tag{1}$$

Here $S_i \subset \mathcal{A}$ is i's jurisdiction which has a weight of population $\delta_i = \mu(S_i)$, $r_c(S_i)$ is a location of its center (fixed or variable), $\rho(\cdot, \cdot)$ is a metric—to determine the distance to the center and the individual location specified by coordinates $x \in \mathcal{A}$. The scalar value $g_i > 0$ presents an aggregated payment (taxes?) for country $i \in N$ formation (price for the government), which has to be paid by country citizens in equal shares. So individual costs are divided into two kinds: equal payment for every inhabitant of the country and individualized costs specified as the distance from inhabitant location to the capital. Notice that assumption (**C**) below is fulfilled now for $l \geq 2$ and if $l = 2$: then for Euclidean metric possible frontier between countries is a branch of hyperbola intersected with $\mathcal{A}$. Now let us consider the concept of spatial equilibrium that is applied in the paper.

[2] The general idea of this equilibrium is that individuals can "vote with their feet" by leaving situations they do not like or going to situations they believe to be more beneficial, i.e. the inhabitants of countries or municipalities are able to move and to choose the place of residence which is more suitable for them. And perhaps it is not necessary to physically move—for example, it is so in the formation of football fans clubs—sometimes it is enough to register and then "consume" the benefits and disadvantages of this membership.

Definition 1. *Let $\mathcal{A} \subset \mathbb{R}^l$ be an area in a finite-dimensional space. A collection of closed subsets $S_i \subset \mathcal{A}$, $i \in N$ is called migration-consistent equilibrium, if it satisfies the requirements:*

(i) $\cup_N S_i = \mathcal{A}$, $\delta_i = \mu(S_i) > 0$, $i \in N$, $\sum_N \delta_i = 1$ *and* $\exists\, y \in Y$ *such that*
(ii) $\forall i \neq j$, $c_i(x, y, \delta_1, \ldots, \delta_n) = c_j(x, y, \delta_1, \ldots, \delta_n)\ \forall x \in S_i \cap S_j$,
(iii) $\forall i \in N\ S_i \supseteq \{x \in \mathcal{A} \mid c_i(x, y, \delta_1, \ldots, \delta_n) < c_j(x, y, \delta_1, \ldots, \delta_n)\, \forall j \in N, j \neq i\}$.

The requirements presented in this definition mean that: (*i*)—subsets $S_i \subset \mathcal{A}$ form a nontrivial division of the area $\mathcal{A}$ into jurisdictions such that each individual is assigned to some jurisdiction and the number of those that have the nationality of several countries is negligible; (*ii*)—border residents of several countries have the same costs; (*iii*)—each jurisdiction includes all those residents who are most profitable to be its members.

In general setting we certainly have to be assumed that the cost functions depend continuously on $\delta \in \Delta^{(n-1)}$ (standard simplex in $\mathbb{R}^n$) and $y \in Y$; moreover Y—the range of y—is convex and compact subset of $\mathbb{R}^m$ for a natural $m > 0$. More specifically, assume that

(P) *The distribution of population on $\mathcal{A}$ is described by an* ***absolutely continuous*** *probability measure μ.*

(**C**) *For each $i \in N$ costs $c_i(\cdot)$ are defined and continuous on*

$$\mathcal{A} \times Y \times (\Delta^{(n-1)} \setminus F_i), \text{ where } F_i = \{\delta \in \Delta^{(n-1)} \mid \delta_i = 0\},$$

and obey

(i) $c_i(x, y, \delta_1, \ldots, \delta_n) \to +\infty\ \forall (x, y, \delta_i, \delta_{-i}) \to (\bar{x}, \bar{y}, 0, \bar{\delta}_{-i})$, *i.e.* $\bar{\delta}_i = 0$[3];
(ii) the set of indifferent agents

$$A_{ij}(y, \delta) = \{x \in \mathcal{A} \mid c_i(x, y, \delta) = c_j(x, y, \delta)\}$$

has zero Lebesgue measure $\forall j \neq i$, and for all fixed $(y, \delta) \in Y \times \Delta^{(n-1)}$.

The assumptions (**C**) is now presented in a strongest form which was applied in previous papers [11], [5] where in addition there was assumed $\mathrm{supp}(\mu) = \mathcal{A}$.[4] Later analysis presented in [9] avoided (**C**)(*ii*) and $\mathrm{supp}(\mu) = \mathcal{A}$. For the case we are interested in $\mathrm{supp}(\mu) = \mathcal{A}$ does not matter, and assumption (**C**)(*ii*) is invalid but for an intermediate approximating model it is true by construction.

In the seminal works devoted to the analysis of Tiebout equilibrium, special attention is paid to the case of an inter-country division under the individual costs $c_i(\cdot)$ defined in (1) in which metric is Euclidean and centers $r_c(S_i)$ are specified as the centers of mass of countries $S_i \subset \mathcal{A}$, $i \in N$. For a multi-dimensional figure S_i the center of mass (barycenter of i-th country) is defined as

$$r_c(S_i) = \frac{1}{\mu(S_i)} \int_{S_i} x d\mu(x), \quad S_i \subset \mathcal{A} \subset \mathbb{R}^l. \tag{2}$$

[3] Here $\delta_{-i} = (\delta_j)_{j \in N \setminus \{i\}}$.
[4] This means that $\mu(B) > 0 \iff \int_B dxdy > 0$ for every measurable $B \subseteq \mathcal{A}$.

It will be so if the distribution of the population is put on the basis of the concept. However, if we want the center of the territory to be understood without taking into account the population, then we arrive at the concept defined by formulas

$$r_c(S_i) = \frac{1}{m(S_i)} \int_{S_i} x dx, \quad m(S_i) = \int_{S_i} dx.$$

Here $m(S_i)$ is Lebesgue measure of S_i, $i \in N$. Clearly, this is a particular case relative to the previous one. Of course, we must take care that considered point-to-set mappings are continuous in some sense in order to ensure the continuity of $r_c(S_i(\delta, y))$. Moreover, it is necessary to carefully consider the case of a potentially possible match of capitals at the domain. This is not trivial, because the assumption that sets $A_{ij}(y, \delta)$ are negligible is violated now. Apparently, this problem can be sorted out by passing to the limit over the family of the individual costs functions that approximate the original ones and obey all necessary properties.

A general idea of the proof is similar to presented in [5,9,11]: for a collection $(\delta_1, \ldots, \delta_n, y)$ of *nominal* parameters one can put into correspondence similar collection of *real* parameters, calculated for an immigration-stable partition defined by nominal ones. In so doing a mapping is defined the fixed point of which obeys all requirements of a country partition that we are looking for. Now we consider this construction in more details.

Recall that

$$\Delta^{(n-1)} = \{\delta \in \mathbb{R}^n \mid \sum \delta_i = 1, \ \delta_i \geq 0 \ \forall i \in N\}$$

is called a standard simplex. Let us specify the mappings

$$S_i : (\delta, y) \to S_i(\delta, y) \subset \mathcal{A}, \quad (\delta, y) \in \Delta^{(n-1)} \times Y, \quad i \in N$$

and mapping $\mathcal{F} : (S_i)_{i \in N} \to (\mu_i)_{\in N}$, defined by formulas:

$$S_i(\delta, y) = \{x \in \mathcal{A} \mid c_i(x, \delta, y) = \min_{j \in N} c_j(x, \delta, y)\}, \quad \mu_i(\delta, y) = \mu(S_i(\delta, y)), \quad i \in N.$$

The following Lemma 1 has been proved in different contexts in [5,9,11] and it specifies a crucial property of the constructed map.

Lemma 1. *Let* (**P**) *and* (**C**) *be fulfilled. Then for some* $0 < \varepsilon < \frac{1}{n}$*:*

(i) $\mathcal{F}(\cdot)$ *defined on* $\Delta_\varepsilon^{(n-1)} \times Y$ *is a Kakutani map*[5] *and*
(ii) $\forall y \in Y$ $\mathcal{F}(\cdot, y)$ *maps the* ε*-sub-simplex*

$$\Delta_\varepsilon^{(n-1)} = \{\delta \in \mathbb{R}^n \mid \sum \delta_i = 1, \ \delta_i \geq \varepsilon \ \forall i \in N\}$$

so that the facets of $\Delta_\varepsilon^{(n-1)}$ *pass into the facets of initial simplex, i.e.*

$$[\delta = (\delta_i, \delta_{-i}) \in \Delta_\varepsilon^{(n-1)} \ \& \ \delta_i = \varepsilon] \Rightarrow$$

$$\mu_i(\delta, y) = 0, \quad \mathcal{F}(\delta, y) = (\mu_1(\delta, y), \ldots, \mu_n(\delta, y)).$$

[5] This is a point-to-set mapping having closed graph and nonempty, convex values, see Sect. 4 and Definition 3 below.

There is also presented a continuous $\mathcal{M} : \Delta^{(n-1)} \times Y \to Y$ and now the resulting map is

$$[\mathcal{F} \times \mathcal{M}](\delta, y) = \mathcal{F}(\delta, y) \times \mathcal{M}(\delta, y), \quad (\delta, y) \in \Delta^{(n-1)} \times Y.$$

Clearly, it suffices to find a *nontrivial* fixed point $\bar{\delta} = (\bar{\delta}_1, \ldots, \bar{\delta}_n) \in \Delta^{(n-1)}$, $\bar{y} \in Y$ of this map, i.e.

$$\bar{y} = \mathcal{M}(\bar{\delta}, \bar{y}), \quad \mu_i(\bar{\delta}, \bar{y}) = \bar{\delta}_i, \forall i \in N \text{ such that } \bar{\delta} = (\bar{\delta}_1, \ldots, \bar{\delta}_n) \gg 0.$$

In [5,9] this fact is established applying generalized Kakutani fixed point theorem. This is Theorem 7 from Sect. 4 below and its corollary, Theorem 9.

Now we formulate the theorem from [8,9] that is general enough and can be applied to establish the main result of the current paper.

Theorem 1 (Marakulin, 2017). *Let $\mathcal{A}$ be a compact subset of a finite-dimensional linear space and μ be a measure on $\mathcal{A}$. If assumptions* (**P**), (**C**) *are fulfilled then the area $\mathcal{A}$ can be nontrivially partitioned into any number of immigration proof communities of non-zero volume. This partition can also obey any consistent continuous requirements.*

Now we state a new theorem in which countries are centered at the center of population mass; it is proved in the next section. First, we consider a particular case of 2-communities, which technical proof can be extended to a general one.

Theorem 2. *In a finite-dimensional space, any compact convex domain equipped with the measure of population absolute continuous relative to Lebesgue measure, can be divided into two immigration proof communities with costs (1) and centered at the center of their mass.*

Now the main result is formulated as follows.

Theorem 3. *In a finite-dimensional space, any compact convex region, with a measure of a population, that is absolutely continuous with respect to Lebesgue measure, can be nontrivially divided into any number of immigration proof communities with costs (1) and centered at the center of their mass. Moreover, these centers are located in the limit of the community.*

3 Partition into Communities Centered in a Barycenter

According to the logic of the previous constructions, in order to include the requirement that the capitals should be at the barycenter, one needs to put

$$\mathcal{M}(\delta_1, \ldots, \delta_n, y_1, \ldots, y_n) = (r_c(S_1(\delta, y)), \ldots, r_c(S_n(\delta, y))) \in \mathcal{A}^n = Y.$$

However, a major difficulty appears here: there is a potential possibility of coincidence of capitals, which leads to a violation of the basic assumption (**C**). Precisely, if a point $(y_i)_{i \in N}$ in $\mathcal{A}^n$ is such that some of its components are equal one to another—there are i, j such that $y_i = y_j$—then it is not clear how one can determine the countries i, j, what are their masses and where the centers of mass are located.

To simplify the problem, let us consider the case of *two countries* and a convex compact region $\mathcal{A}$. Indeed, if centers are equal each other, i.e. $r_1 = r_c(S_1(\delta, y)) = r_c(S_2(\delta, y)) = r_2$, then the equation defining the inter-country border in the form

$$||x - r_1||_2 - ||x - r_2||_2 = \frac{g_2}{\delta_2} - \frac{g_1}{\delta_1}$$

does not depend on x and, therefore, $\mathcal{A}$ coincides with one country (the other one is empty set) or, if $\frac{g_2}{\delta_2} - \frac{g_1}{\delta_1} = 0$, then the whole $\mathcal{A}$ is the "inter-country border". In the latter case, there are infinitely many variants of inter-country division, which does not allow to specify the centers of countries, etc. To resolve the collision, consider the following approximation.

As a new area to be divided into two countries, consider the convex hull of three sets from $\mathbb{R}^{l+2}$: choose arbitrary *noncollinear* $a, b \in \mathbb{R}^l$, $a, b \neq 0$, a real $\varepsilon > 0$ and specify

$$\mathcal{B}_0 = \mathcal{A} \times (0,0), \ \ \mathcal{B}_1 = \mathcal{B}_0 + (a,1,0), \ \ \mathcal{B}_2 = \mathcal{B}_0 + (b,0,1).$$

Next, we consider the convex hull of the union of these sets, whence, because of their convexity, one has

$$\mathcal{B} = \text{co}\{\mathcal{B}_0, \mathcal{B}_1, \mathcal{B}_2\} = \bigcup_{\lambda \geq 0:\, \lambda_0+\lambda_1+\lambda_2=1} (\lambda_0\mathcal{B}_0 + \lambda_1\mathcal{B}_1 + \lambda_2\mathcal{B}_2) \Rightarrow$$

$$\mathcal{B} = \{\mathcal{B}_0 + [\alpha(a,1,0) + \beta(b,0,1)] \mid \alpha, \beta \in \mathbb{R}_+ : \ \alpha + \beta \leq 1\}.$$

Now we specify

$$\triangle = \{(s,t) \in \mathbb{R}^2_+ \mid s + t \leq 1\}$$

and will identify compactum $\mathcal{B}$ with the set $\mathcal{A} \times \triangle \subset \mathbb{R}^{l+2}$. We determine the population distribution using the density $H^\varepsilon : \mathcal{B} \to \mathbb{R}_+$, specified as

$$H^\varepsilon(x,s,t) = w(x) \cdot \gamma^\varepsilon(s,t),$$

where $w : \mathcal{A} \to \mathbb{R}_+$ is the density of "initial" distribution μ, and a simple function $\gamma^\varepsilon : \triangle \to \mathbb{R}$ for some $\varepsilon > 0$, $d > 0$ is defined in the following way. Specify

$$\Box^\varepsilon = \{(s,t) \in \triangle \mid s, t \leq \varepsilon\}$$

and define $\gamma^\varepsilon(\cdot)$ by formula

$$\gamma^\varepsilon(s,t) = \begin{cases} \varepsilon^2, \ (s,t) \in \triangle \setminus \Box^\varepsilon, \\ d, \ \ (s,t) \in \Box^\varepsilon, \end{cases} \tag{3}$$

and satisfying a condition

$$\int_\triangle \gamma^\varepsilon(s,t)dsdt = 1.$$

Now let ν^ε be a measure on $\mathcal{B}$, specified by density $H^\varepsilon(\cdot)$. Due to construction and Fubini theorem we have $\nu^\varepsilon(\mathcal{B}) = \mu(\mathcal{A}) = 1$.

Now consider two projection mappings from $\mathcal{B}$ on its faces:

$$Pr_1(x, s, t) = (x + (2\varepsilon - s)a - tb, 2\varepsilon, 0), \quad Pr_2(x, s, t) = (x + (2\varepsilon - t)b - sa, 0, 2\varepsilon),$$

where $(x, s, t) \in \mathcal{B}$, and vectors $a, b \in \mathbb{R}^l$ are chosen as above. Now we specify a convex compactum

$$X = \Delta^1 \times Y_1 \times Y_2, \; Y_1 = Pr_1(\mathcal{B}) = \mathcal{B}_0 + 2\varepsilon(a, 1, 0), \; Y_2 = Pr_2(\mathcal{B}) = \mathcal{B}_0 + 2\varepsilon(b, 0, 1),$$

where we will look for a fixed point at the first stage of the proof.

For $\mathcal{B}$ we introduce a metric

$$\rho'(\kappa, \kappa') = \sqrt{\langle x - x', x - x' \rangle} + |s - s'| + |t - t'|, \quad \kappa = (x, s, t), \; \kappa' = (x', s', t') \in \mathcal{B}.$$

Now for a given $y_1 \in Y_1$, $y_2 \in Y_2$ and nominal $(\delta_1, \delta_2) \in \Delta^1$ for $\mathcal{B}$ we specify cost functions

$$c_i(\kappa, \delta_i, y_i) = \frac{g_i}{\delta_i} + \rho'(\kappa, y_i), \quad g_i > 0 \;, \delta_i > 0 \quad i = 1, 2, \quad \kappa \in \mathcal{B},$$

which define in $\mathcal{B}$ two "countries"

$$S_i^\varepsilon(\delta_1, \delta_2, y_1, y_2) = \{\kappa \in \mathcal{B} \mid c_i(\kappa, \delta_i, y_i) \le c_j(\kappa, \delta_j, y_j)\}, \quad i \ne j, \quad i, j = 1, 2. \quad (4)$$

Notice that by construction $y_1 \ne y_2$ is always true and the set $S_1^\varepsilon(\delta, y) \cap S_2^\varepsilon(\delta, y)$, i.e. inter-country border is negligible by Lebesgue measure and condition $(\mathbf{C})(ii)$ is satisfied. As usual further we find country masses $\nu^\varepsilon(S_i^\varepsilon) = \nu_i(\delta, y)$ and their centers of mass $r_c(S_i^\varepsilon) \in \mathcal{B}$, where for $\nu^\varepsilon(S_i) = 0$ we put $r_c(S_i^\varepsilon) = \mathcal{B}$.

Lemma 2. *The map* $(\delta, y) \to \nu_i(\delta, y) = \nu^\varepsilon(S_i^\varepsilon(\delta, y))$, $i = 1, 2$ *is continuous on* $\Delta^1 \times Y$ *and for* $\nu_i(\delta, y) = \nu^\varepsilon(S_i^\varepsilon) > 0$ *a value* $r_c(S_i^\varepsilon) \in \mathcal{B}$ *is correctly defined and the function* $(\delta, y) \to r_c(S_i^\varepsilon(\delta, y))$ *is continuous at this point.*

Proof. Standardly. Notice that the point-to-set mapping $(\delta, y) \Rightarrow S_i^\varepsilon(\delta, y)$ is continuous at the point (δ', y') if $\nu_i(\delta', y') = \nu^\varepsilon(S_i^\varepsilon) > 0$ because it implies $\operatorname{int} S_i^\varepsilon(\delta', y') \ne \emptyset$ and, therefore, Slater condition $[\exists x \in \mathcal{A} \mid \; c_i(x, \delta, y) < c_j(x, \delta, y)]$ holds. ∎

Further let us consider a map $\mathcal{R} : \Delta^1 \times Y_1 \times Y_2 \to Y_1 \times Y_2$, specifying new centers of countries: if $\nu_i(\delta, y) = \nu^\varepsilon(S_i^\varepsilon(\delta, y)) > 0$ then

$$\mathcal{R}_i(\delta_1, \delta_2, y_1, y_2) = Pr_i(r_c(S_i^\varepsilon)),$$

and $\mathcal{R}_i(\delta, y) = Y_i$ if $\nu_i(\delta, y) = \nu^\varepsilon(S_i^\varepsilon(\delta, y)) = 0$, $i = 1, 2$.

Specify also $\mathcal{F} : \Delta \times Y \to \Delta$ by formula

$$\mathcal{F}(\delta_1, \delta_2, y_1, y_2) = (\nu_1(\delta, y), \nu_2(\delta, y)).$$

The crucial properties of the mapping $\mathcal{F} \times \mathcal{R}$ are described in the following Lemma 3, which is similar to Lemma 1 presented above, but now it has already been established in the specific context of this section.

Lemma 3. *For any real* $0 < \varepsilon < 1$ *the mapping* $\mathcal{F} \times \mathcal{R} : X \Rightarrow X$ *is a Kakutani map and, moreover, there is* $\varrho > 0$ *such that* $\forall y \in Y$ $\mathcal{F}(\cdot, y)$ *maps faces of subsimplex*

$$\Delta^1_\varrho = \{\delta \in \mathbb{R}^2 \mid \delta_1 + \delta_2 = 1,\ \delta_i \geq \varrho,\ i = 1, 2\}$$

to appropriate faces of initial Δ^1, *i.e. for* $i = 1, 2$

$$[\delta = (\delta_1, \delta_2) \in \Delta^1_\varrho \ \&\ \delta_i = \varrho] \quad \Rightarrow \quad \nu_i(\delta, y) = 0, \quad \mathcal{F}(\delta, y) = (\nu_1(\delta, y), \nu_2(\delta, y)).$$

Proof. Checking of the Kakutani map properties is easy. Let us consider the second part of the lemma statement. Let $D = \max_{\kappa, \kappa' \in \mathcal{B}} ||\kappa - \kappa'||$ be a diameter of $\mathcal{B}$. Then supposition

$$\frac{g_2}{\delta_2} - \frac{g_1}{\delta_1} > 2D > ||\kappa - y_1|| - ||\kappa - y_2||, \quad \forall \kappa \in \mathcal{B},$$

implies $c_2(\kappa, \delta_2, y_2) > c_1(\kappa, \delta_1, y_1)$ $\forall \kappa \in \mathcal{B}$. Therefore one has $\nu^\varepsilon(S^\varepsilon_2) = 0$. Hence if $\delta_2 \leq \frac{1}{2}$ (i.e. $\delta_1 \geq \frac{1}{2}$) for $\nu^\varepsilon(S^\varepsilon_2) > 0$ to be executed one needs

$$\frac{g_2}{\delta_2} \leq 2D + \frac{g_1}{\delta_1} \leq 2(D + g_1) \Rightarrow \ \delta_2 \geq \frac{g_2}{2(D + g_1)} > 0.$$

A similar assessment is easy to get also for the first country. Ultimately, one can put

$$\varrho = \min \left\{ \frac{g_1}{2(D + g_2)}, \frac{g_2}{2(D + g_1)} \right\} > 0.$$

By construction, $\varrho > 0$ does not depend on the choice $0 < \varepsilon < 1$. ■

Combining the statement of the Lemma 3 with the Theorem 9 and via described construction we arrive at

Corollary 1. *For any* $0 < \varepsilon < 1$ *the mapping* $\mathcal{F} \times \mathcal{R} : X \Rightarrow X$ *for some real* $\varrho > 0$ *has a fixed point* $((\delta^\varepsilon_1, \delta^\varepsilon_2), (y^\varepsilon_1, y^\varepsilon_2)) \in X = \Delta^1 \times Y_1 \times Y_2$ *such that*

(i) for some real $\varrho > 0$ *independently on* $\varepsilon > 0$ *one has* $\delta^\varepsilon_i > \varrho$, $i = 1, 2$,
(ii) for $\varepsilon \to +0$ *the centers of mass* $\kappa^\varepsilon_i = (x^\varepsilon_i, s^\varepsilon_i, t^\varepsilon_i) \in \mathcal{B}$ *of states* S^ε_i *obey*

$$(s^\varepsilon_i, t^\varepsilon_i) \leq (\varepsilon + o(\varepsilon), \varepsilon + o(\varepsilon)), \quad i = 1, 2.$$

Proof. (*i*) Constructed above mapping $\mathcal{F}$ satisfies assumption (**C**) in a strong form, i.e. for all $(\delta, y) \in \Delta^1 \times Y$ the set

$$\{\kappa \in \mathcal{B} \mid c_1(\kappa, \delta_1, y_1) = c_2(\kappa, \delta_2, y_2)\}$$

is negligible by Lebesgue measure. Therefore due to Theorem 9 and Lemma 3 the map $\mathcal{F} \times \mathcal{R}$ has a fixed point $((\delta_1, \delta_2), (y_1, y_2)) \in \mathcal{F} \times \mathcal{R}((\delta_1, \delta_2), (y_1, y_2))$ such that $(\delta_1, \delta_2) \gg (\varrho, \varrho)$, for some real $\varrho > 0$ which does not depend on the choice of ε.

(*ii*) Let $i = 1$. Separate S_1^ε into subsets:

$$S_1^0 = \{(x,s,t) \in S_1^\varepsilon \mid (s,t) \in \triangle \setminus \square^\varepsilon\}, \quad S_1^1 = \{(x,s,t) \in S_1^\varepsilon \mid (s,t) \in \square^\varepsilon\}.$$

Now for a given $\varepsilon > 0$ the center of mass can be calculated as

$$r_c(S_1^\varepsilon) = \frac{1}{\nu^\varepsilon(S_1^\varepsilon)} \left[\int_{S_1^0} (x,s,t) d\nu^\varepsilon + \int_{S_1^1} (x,s,t) d\nu^\varepsilon \right].$$

Due to (3), if $D = \max ||\mathcal{B}||$ then the first addend can be evaluated as

$$\left\| \int_{S_1^0} (x,s,t) d\nu^\varepsilon \right\| \le D \cdot \nu(\mathcal{A} \times [\triangle \setminus \square^\varepsilon]) < D \cdot \mu(\mathcal{A}) \int_\triangle \varepsilon^2 dsdt = D \cdot \mu(\mathcal{A}) \frac{\varepsilon^2}{2}.$$

Thus the center of mass of S_1 differs from the center of mass for the state S_1^1 by the value of degree ε^2, i.e.

$$||r_c(S_1^\varepsilon) - r_c(S_1^1)|| = o(\varepsilon).$$

In turn, the country S_1^1 is defined on $\mathcal{B}$ under the additional constraint $(0,0) \le (s,t) \le (\varepsilon, \varepsilon)$, and hence the corresponding components of the center of mass S_1^1 will also be within these bounds, but they differ from the original one by a value of degree not more than $o(\varepsilon)$. ■

The first part of the next lemma can be proven via Theorem 1, but below we especially need in the second one, which is very specific and allows us to state the result.

Lemma 4. *For any $0 < \varepsilon < 1$ there is an equilibrium partition of the area $\mathcal{B}$ on two communities, according to the costs*

$$c_1(\kappa, y_1^\varepsilon) = \frac{g_1}{\delta_1} + \rho'(\kappa, y_1^\varepsilon), \quad y_1^\varepsilon = (r_c^\varepsilon(S_1), 2\varepsilon, 0),$$

$$c_2(\kappa, y_2^\varepsilon) = \frac{g_2}{\delta_2} + \rho'(\kappa, y_2^\varepsilon), \quad y_2^\varepsilon = (r_c^\varepsilon(S_2), 0, 2\varepsilon).$$

Centers $y_i^\varepsilon \in S_i^\varepsilon$ obey $y_i^\varepsilon = Pr_i(\kappa_i^\varepsilon)$, where κ_i^ε are the centers of mass by the measure ν^ε of $S_i^\varepsilon \subset \mathcal{B}$. Moreover $r_c^\varepsilon(S_1) \ne r_c^\varepsilon(S_2)$ and $\nu^\varepsilon(S_i^\varepsilon) = \delta_i > \varrho > 0$, $i = 1, 2$.

Proof. We apply Corollary 1 and fix $\varepsilon > 0$. One can find a fixed point

$$((\delta_1^\varepsilon, \delta_2^\varepsilon), (y_1^\varepsilon, y_2^\varepsilon)) \in X = \Delta^1 \times Y_1 \times Y_2$$

of the map $\mathcal{F} \times \mathcal{R} : X \rightrightarrows X$ such that $(\delta_1^\varepsilon, \delta_2^\varepsilon) \ge (\varrho, \varrho) \gg (0,0)$. Due to construction we have $y_1^\varepsilon = Pr_1(\kappa_1^\varepsilon)$, $y_2^\varepsilon = Pr_2(\kappa_2^\varepsilon)$, where κ_i^ε are the centers of mass of $S_i^\varepsilon \subset \mathcal{B}$ by the measure ν^ε, and therefore

$$\kappa_1^\varepsilon = (x_1^\varepsilon, s_1^\varepsilon, t_1^\varepsilon) \Rightarrow y_1^\varepsilon = (x_1^\varepsilon + (2\varepsilon - s_1^\varepsilon)a - t_1^\varepsilon b, 2\varepsilon, 0),$$

$$\kappa_2^\varepsilon = (x_2^\varepsilon, s_2^\varepsilon, t_2^\varepsilon) \quad \Rightarrow \quad y_2^\varepsilon = (x_2^\varepsilon + (2\varepsilon - t_2^\varepsilon)b - s_2^\varepsilon a, 0, 2\varepsilon),$$

for some $x_i^\varepsilon \in \mathcal{A}$, $s_i^\varepsilon, t_i^\varepsilon \in (0,1)$, $i = 1,2$.

Let us show that $r_1^\varepsilon = x_1^\varepsilon + (2\varepsilon - s_1^\varepsilon)a - t_1^\varepsilon b \neq x_2^\varepsilon + (2\varepsilon - t_2^\varepsilon)b - s_2^\varepsilon a = r_2^\varepsilon$. Indeed, assuming the opposite we have $r^\varepsilon = r_1^\varepsilon = r_2^\varepsilon$ and

$$||(x,s,t) - (r^\varepsilon, 2\varepsilon, 0)|| = ||x - r^\varepsilon||_2 + |2\varepsilon - s| + t,$$
$$||(x,s,t) - (r^\varepsilon, 0, 2\varepsilon)|| = ||x - r^\varepsilon||_2 + s + |2\varepsilon - t|.$$

Consequently, the inequality determining the first (second) country is

$$c_1(x,s,t) - c_2(x,s,t) = t - s + |2\varepsilon - s| - |2\varepsilon - t| + \frac{g_1}{\delta_1} - \frac{g_2}{\delta_2} \leq 0 \Rightarrow$$

$$S_1^\varepsilon = \left\{ (x,s,t) \in \mathcal{B} \mid t - s + |2\varepsilon - s| - |2\varepsilon - t| \leq \frac{g_2}{\delta_2} - \frac{g_1}{\delta_1} \right\}.$$

Hence, by virtue of (2) and due to construction (3), one can conclude that the center of mass of country S_1^ε has the form

$$\kappa_1^\varepsilon = (r_c(\mathcal{A}), m_{st}^\triangle), \quad m_{st}^\triangle = \int_{\triangle'} (s,t)\gamma^\varepsilon(s,t)dsdt,$$

where $r_c(\mathcal{A})$ is a center of mass $\mathcal{A}$ by the measure μ and

$$\triangle' = \left\{ (s,t) \in \mathbb{R}_+^2 \mid t - s + |2\varepsilon - s| - |2\varepsilon - t| \leq \frac{g_2}{\delta_2} - \frac{g_1}{\delta_1} \right\}.$$

For the second country, the center of mass can be found similarly, except the fact that now the second component is obtained by integrating over a set

$$\triangle'' = \left\{ (s,t) \in \mathbb{R}_+^2 \mid t - s + |2\varepsilon - s| - |2\varepsilon - t| \geq \frac{g_2}{\delta_2} - \frac{g_1}{\delta_1} \right\}.$$

Hence, we conclude that the components $x_1^\varepsilon, x_2^\varepsilon \in \mathcal{A}$ of centers of gravity of $S_i^\varepsilon \subset \mathcal{B}$ are equal to each other: $x_1^\varepsilon = x_2^\varepsilon$. Therefore, from the assumed equality $r_c^\varepsilon(S_1) = r_c^\varepsilon(S_2)$ one concludes $(2\varepsilon - s_1^\varepsilon)a - t_1^\varepsilon b = (2\varepsilon - t_2^\varepsilon)b - s_2^\varepsilon a$, that implies

$$(2\varepsilon - s_1^\varepsilon + s_2^\varepsilon)a = (2\varepsilon - t_2^\varepsilon + t_1^\varepsilon)b.$$

By choice, vectors a, b are non-collinear, that means the latter equality is possible only if the coefficients vanish, i.e. if $s_1^\varepsilon = 2\varepsilon + s_2^\varepsilon > 2\varepsilon$ and $t_2^\varepsilon = 2\varepsilon + t_1^\varepsilon > 2\varepsilon$, but this is impossible since via (3) for $\varepsilon > 0$ small enough one has $0 \leq s_i^\varepsilon \lessapprox \varepsilon$ and $0 \leq t_i^\varepsilon \lessapprox \varepsilon$, $i = 1,2$. So the original supposition leads to a contradiction. ∎

Proof of Theorem 2. Without loss of generality, one can think that $\mathcal{A}$ is a solid set, i.e. $\operatorname{int} \mathcal{A} \neq \emptyset$. Now we apply Corollary 1, Lemma 4 and for real $\varepsilon > 0$ we consider a fixed point $((\delta_1^\varepsilon, \delta_2^\varepsilon), (y_1^\varepsilon, y_2^\varepsilon)) \in X = \Delta^1 \times Y_1 \times Y_2$ for any real $\varepsilon > 0$. One has

$(\delta_1^\varepsilon, \delta_2^\varepsilon) \geq (\varrho, \varrho) \gg (0,0)$ and, by virtue of the compactness of X, without loss of generality, one can think that

$$((\delta_1^\varepsilon, \delta_2^\varepsilon), (y_1^\varepsilon, y_2^\varepsilon)) \to ((\delta_1, \delta_2), (y_1, y_2)) = (\delta, y),$$

for $\varepsilon \to +0$ and $(\delta_1, \delta_2) \geq (\varrho, \varrho) \gg (0,0)$. In addition, we recall that

$$y_1^\varepsilon = (x_1^\varepsilon + (2\varepsilon - s_1^\varepsilon)a - t_1^\varepsilon b, 2\varepsilon, 0) \underset{\varepsilon \downarrow 0}{\to} y_1 = (r_1, 0, 0), \quad r_1 \in \mathcal{A},$$

$$y_2^\varepsilon = (x_2^\varepsilon + (2\varepsilon - t_2^\varepsilon)b - s_2^\varepsilon a, 0, 2\varepsilon) \underset{\varepsilon \downarrow 0}{\to} y_2 = (r_2, 0, 0), \quad r_2 \in \mathcal{A}.$$

Let us further study a type of inter-country borders that can be formed between the countries defined by the limiting values of the determining parameters. We want to show that $r_1 \neq r_2$ and, therefore, the boundary is a hyperbolic one. To this end, we assume $r_1 = r_2$, i.e. $0 \neq \varDelta r^\varepsilon = r_1^\varepsilon - r_2^\varepsilon \to 0$ for $\varepsilon \to +0$. Here we have

$$r_1^\varepsilon = x_1^\varepsilon + (2\varepsilon - s_1^\varepsilon)a - t_1^\varepsilon b, \quad r_2^\varepsilon = x_2^\varepsilon + (2\varepsilon - t_2^\varepsilon)b - s_2^\varepsilon a.$$

It obviously follows from the construction (since the center of mass of $\mathcal{B}$ is located on a linear segment $[y_1^\varepsilon, y_2^\varepsilon]$), that

$$r_1 = \lim_{\varepsilon \to +0} r_1^\varepsilon = \lim_{\varepsilon \to +0} x_1^\varepsilon = x_1 = x_2 = r_2 = r_c(\mathcal{A}),$$

where $r_c(\mathcal{A})$ is a center of mass of $\mathcal{A}$ by measure μ. Recall that due to center of mass definition (2) for any measurable $T_1, T_2 \subset \mathcal{A}$, $T_1 \cap T_2 = \emptyset$, $T_1 \cup T_2 = \mathcal{A}$ one has

$$r_c(\mathcal{A}) = \frac{\mu(T_1)}{\mu(\mathcal{A})} r_c(T_1) + \frac{\mu(T_2)}{\mu(\mathcal{A})} r_c(T_2). \tag{5}$$

Define $\sigma^\varepsilon = \frac{g_1}{\delta_1} - \frac{g_2}{\delta_2}$, $h_r^\varepsilon = \varDelta r^\varepsilon / ||\varDelta r^\varepsilon||$, $f(s,t) = (s-t) - |2\varepsilon - s| + |2\varepsilon - t|$, and transform states $S_1^\varepsilon, S_2^\varepsilon \subset \mathcal{B}$ specification:

$$c_1(x,s,t) - c_2(x,s,t) = ||x - r_1^\varepsilon|| - ||x - r_2^\varepsilon|| - f(s,t) + \sigma^\varepsilon \leq 0 \quad \Rightarrow$$

$$(x,s,t) \in S_1^\varepsilon \iff ||x - r_1^\varepsilon|| - ||x - r_2^\varepsilon|| \leq f(s,t) - \sigma^\varepsilon = g^\varepsilon(s,t).$$

Now we divide S_1^ε into subsets:

$$S_1^- = \{(x,s,t) \in S_1^\varepsilon \mid (s,t) \in \triangle : g^\varepsilon(s,t) \leq 0\},$$

$$S_1^+ = \{(x,s,t) \in S_1^\varepsilon \mid (s,t) \in \triangle : g^\varepsilon(s,t) \geq 0\}.$$

Without loss of generality one thinks that $\nu^\varepsilon(S_i^-) \to \delta_i^- > 0$ and $\nu^\varepsilon(S_i^+) \to \delta_i^+ > 0$ for $\varepsilon \to +0$, $i = 1, 2$. Now for a given real $\varepsilon > 0$ the center of mass of the country can be calculated in the following way:

$$r_c(S_1^\varepsilon) = \frac{1}{\nu^\varepsilon(S_1^\varepsilon)} \left[\int_{S_1^-} (x,s,t) d\nu^\varepsilon + \int_{S_1^+} (x,s,t) d\nu^\varepsilon \right],$$

here the first term on the right hand side can be found as

$$\int_{S_1^-}(x,s,t)d\nu^\varepsilon = \int_\triangle \left[\int_{S_1^-(st)}(x,s,t)d\mu(x)\right]\gamma^\varepsilon(s,t)dsdt,$$

where for *fixed* $(s,t)\in\triangle$ we define

$$S_1^-(st) = \{x\in\mathcal{A}\mid ||x-r_1^\varepsilon|| - ||x-r_2^\varepsilon|| \le g^\varepsilon(s,t)\}.$$

Now if $g^\varepsilon(s,t)\le 0$, the set $S_1^-(st)$ is *convex*, and if its mass is nonzero then this set center of gravity is placed in its *interior*; the functional h_r^ε separates the set from the center $\frac{r_1^\varepsilon+r_2^\varepsilon}{2}$ of the hyperboloid:

$$\langle S_1^-(st), h_r^\varepsilon\rangle < \left\langle \frac{r_1^\varepsilon+r_2^\varepsilon}{2}, h_r^\varepsilon\right\rangle.$$

Consequently, integrating the inequality, we obtain

$$\left\langle \int_{S_1^-(st)} xd\mu(x), h_r^\varepsilon\right\rangle < \mu(S_1^-(st))\left\langle \frac{r_1^\varepsilon+r_2^\varepsilon}{2}, h_r^\varepsilon\right\rangle$$

$$\Rightarrow \langle r_c(S_1^-(st)), h_r^\varepsilon\rangle < \left\langle \frac{r_1^\varepsilon+r_2^\varepsilon}{2}, h_r^\varepsilon\right\rangle.$$

Once again integrating, but now by measure with the density $\gamma^\varepsilon(s,t)$, applying standard argumentation one can prove the existence of a real $\tau>0$ such that

$$\left\langle \int_\triangle \left[\int_{S_1^-(st)} xd\mu(x)\right]\gamma^\varepsilon(s,t)dsdt, h_r^\varepsilon\right\rangle + \tau$$
$$< \left\langle \frac{r_1^\varepsilon+r_2^\varepsilon}{2}, h_r^\varepsilon\right\rangle \int_\triangle \mu(S_1^-(st))\gamma^\varepsilon(s,t)dsdt.$$

Here real value $\tau>0$ does not depend on ε for all $\varepsilon>0$ small enough. As a result we have got an estimation

$$\left\langle \int_{S_1^-} xd\nu^\varepsilon, h_r^\varepsilon\right\rangle + \tau < \nu^\varepsilon(S_1^-)\left\langle \frac{r_1^\varepsilon+r_2^\varepsilon}{2}, h_r^\varepsilon\right\rangle. \tag{6}$$

Similar reasoning, but now with regard to the second country and if $g^\varepsilon(s,t)\ge 0$ and

$$S_2^-(st) = \{x\in\mathcal{A}\mid ||x-r_1^\varepsilon|| - ||x-r_2^\varepsilon|| > g^\varepsilon(s,t)\} = \mathcal{A}\setminus S_1^+(st).$$

we obtain

$$\left\langle \int_{S_2^-(st)} xd\mu(x), h_r^\varepsilon\right\rangle > \mu(S_2^-(st))\left\langle \frac{r_1^\varepsilon+r_2^\varepsilon}{2}, h_r^\varepsilon\right\rangle.$$

Now applying (5)[6] and due to $\mu(S_2^-(st))r_c(S_2^-(st)) = \int_{S_2^-(st)} x d\mu(x)$ we obtain

$$\left\langle \left[r_c(\mathcal{A}) - \int_{S_1^+(st)} x d\mu(x) \right], h_r^\varepsilon \right\rangle$$
$$> (1 - \mu(S_1^+(st))) \left[\langle r_c(\mathcal{A}), h_r^\varepsilon \rangle + \left\langle \frac{r_1^\varepsilon + r_2^\varepsilon}{2} - r_c(\mathcal{A}), h_r^\varepsilon \right\rangle \right],$$

that implies

$$\left\langle \int_{S_1^+(st)} x d\mu(x), h_r^\varepsilon \right\rangle < \left\langle r_c(\mathcal{A}) - \frac{r_1^\varepsilon + r_2^\varepsilon}{2}, h_r^\varepsilon \right\rangle + \mu(S_1^+(st)) \left\langle \frac{r_1^\varepsilon + r_2^\varepsilon}{2}, h_r^\varepsilon \right\rangle.$$

Next, we again integrate the inequality, at the same time evaluating the first term on the right by the value $||r_c(\mathcal{A}) - \frac{r_1^\varepsilon + r_2^\varepsilon}{2}|| \to 0$, we come to an estimate

$$\left\langle \int_{S_1^+} x d\nu^\varepsilon, h_r^\varepsilon \right\rangle + \tau' < \nu^\varepsilon(S_1^+) \left\langle \frac{r_1^\varepsilon + r_2^\varepsilon}{2}, h_r^\varepsilon \right\rangle + \left\| r_c(\mathcal{A}) - \frac{r_1^\varepsilon + r_2^\varepsilon}{2} \right\|.$$

It holds for some real $\tau' > 0$, which is independent of the choice of $\varepsilon > 0$ sufficiently small. Summing this inequality with (6) we obtain: for sufficiently small $\varepsilon > 0$ and some $\tau'' > 0$, that does not dependent of $\varepsilon > 0$

$$\left\langle \int_{S_1^-} x d\nu^\varepsilon, h_r^\varepsilon \right\rangle + \left\langle \int_{S_1^+} x d\nu^\varepsilon, h_r^\varepsilon \right\rangle + \tau'' < (\nu^\varepsilon(S_1^-) + \nu^\varepsilon(S_1^+)) \left\langle \frac{r_1^\varepsilon + r_2^\varepsilon}{2}, h_r^\varepsilon \right\rangle.$$

Finally if one divides this inequality onto $\nu^\varepsilon(S_1) = \nu^\varepsilon(S_1^-) + \nu^\varepsilon(S_1^+))$, then due to previous constructions and for some $\tau''' > 0$ we obtain

$$\langle x_1^\varepsilon, h_r^\varepsilon \rangle + \tau''' < \left\langle \frac{r_1^\varepsilon + r_2^\varepsilon}{2}, h_r^\varepsilon \right\rangle.$$

Conducting similar reasoning for the second country, we come to inequality

$$\langle x_2^\varepsilon, h_r^\varepsilon \rangle > \left\langle \frac{r_1^\varepsilon + r_2^\varepsilon}{2}, h_r^\varepsilon \right\rangle + \tau^{iv},$$

that being subtracted the latter one produces

$$\langle x_2^\varepsilon - x_1^\varepsilon, h_r^\varepsilon \rangle > d = \tau''' + \tau^{iv} > 0.$$

However this contradicts to the supposition $r_1^\varepsilon - r_2^\varepsilon = \Delta r^\varepsilon \to 0$, since by Corollary 1(*ii*) (via (3)) we have $0 \le s_i^\varepsilon \lessapprox \varepsilon$, $0 \le t_i^\varepsilon \lessapprox \varepsilon$, $i = 1, 2$ and therefore for $\varepsilon \to 0$

$$x_2^\varepsilon - x_1^\varepsilon = (s_1^\varepsilon - s_2^\varepsilon - 2\varepsilon)a + (2\varepsilon + t_1^\varepsilon - t_2^\varepsilon)b - \Delta r^\varepsilon \to 0.$$

[6] Recall that we assumed $\mu(\mathcal{A})$=1.

So, the points r_1, r_2 obey $r_1 \neq r_2$ and by construction they are the centers of the limiting countries and, at the same time, represent their centers of mass. Wherein

$$S_1 = \left\{ x \in \mathcal{A} \mid ||x - r_1||_2 + \frac{g_1}{\delta_1} \leq ||x - r_2||_2 + \frac{g_2}{\delta_2} \right\}, \quad \delta_1 = \mu(S_1), \quad S_2 = \mathcal{A} \setminus S_1.$$

Theorem 2 is proved. ∎

Proof of Theorem 3. We consider a generalization of the approach outlined in the proof of Theorem 2. Let it be necessary to divide the region $\mathcal{A}$ on n countries. Let us consider a family of nonzero pairwise noncollinear vectors $a_i \in \mathbb{R}^l$, real $\varepsilon > 0$ and a family of convex compacts from $\mathbb{R}^{l+n}$:

$$\mathcal{B}_0 = \mathcal{A} \times (0, 0, \ldots, 0), \quad \mathcal{B}_i = \mathcal{B}_0 + (a_i, e_i), \quad i = 1, \ldots, n,$$

where e_i is a unit vector in $\mathbb{R}^n$. Next, we consider the convex hull of the union of these sets

$$\mathcal{B} = \left\{ \mathcal{B}_0 + \sum_{i=1}^{n} \alpha_i (a_i, e_i) \mid \alpha_i \geq 0, \ i = 1, \ldots, n, \ \sum_{i=1}^{n} \alpha_i \leq 1 \right\}.$$

We identify $\mathcal{B}$ with the set $\mathcal{A} \times \triangle$,

$$\triangle = \left\{ (s_1, \ldots, s_n) \in \mathbb{R}_+^n \mid \sum_{i=1}^{n} s_i \leq 1 \right\}.$$

The population distribution is determined via the density $H : \mathcal{B} \to \mathbb{R}_+$, defined by

$$H(x, s_1, \ldots, s_n) = w(x) \cdot \gamma(s),$$

where $w : \mathcal{A} \to \mathbb{R}_+$ is the density of the measure μ and a simple function $\gamma : \triangle \to \mathbb{R}_+$ is specified for $\varepsilon > 0$ and $d > 0$ by formula

$$\gamma(t) = \begin{cases} \varepsilon^n, & s \in \triangle \setminus \square^\varepsilon, \\ d, & s \in \square^\varepsilon, \end{cases} \quad \square^\varepsilon = \prod_{i=1}^{n} [0, \varepsilon]$$

and satisfies

$$\int_{\triangle} \gamma(s) \bigotimes_{i=1}^{n} ds_i = 1.$$

Now let ν^ε be the measure on $\mathcal{B}$, defined by density $H(\cdot)$. We have $\nu^\varepsilon(\mathcal{B}) = \mu(\mathcal{A}) = 1$.

Consider further the projecting mappings $Pr_i : \mathcal{B} \to \mathcal{B}_i$, $i \in N$ acting as:

$$Pr_i(x, s) = (x + n\varepsilon a_i - \sum_{j=1}^{n} s_j a_j, n\varepsilon e_i),$$

where $(x, s) \in \mathcal{B}$, $s = (s_1, s_2, \ldots, s_n) \in \mathbb{R}^n$, and vectors $a_i \in \mathbb{R}^l$ were chosen above. Next we define a convex compactum

$$X = \Delta^{n-1} \times \prod_{i=1}^{n} Y_i, \quad Y_i = Pr_i(\mathcal{B}) = \mathcal{B}_0 + n\varepsilon(a_i, e_i), \quad i = 1, 2, \ldots, n,$$

in which we are looking for a fixed point at an initial stage of the proof.

For the set $\mathcal{B}$ let us consider a metric: for $\kappa = (x, s)$, $\kappa' = (x', s',) \in \mathcal{B}$ one puts

$$\rho'(\kappa, \kappa') = ||x - x'||_2 + ||s - s'||_1 = \sqrt{\langle x - x', x - x' \rangle} + \sum_{i=1}^{n} |s_i - s_i'|.$$

Now for any $y \in \prod_{i=1}^{n} Y_i$ and nominal $\delta \in \Delta^{n-1}$ for the set $\mathcal{B}$ one can define costs

$$c_i(\kappa, \delta_i, y_i) = \frac{g_i}{\delta_i} + \rho'(\kappa, y_i), \quad g_i > 0\ , \delta_i > 0 \quad i = 1, 2, \ldots, n, \quad \kappa \in \mathcal{B},$$

which specify n "countries" in $\mathcal{B}$:

$$S_i^\varepsilon(\delta, y) = \{\kappa \in \mathcal{B} \mid c_i(\kappa, \delta_i, y_i) \leq \min_{j \neq i, j \in N} c_j(\kappa, \delta_j, y_j)\}, \quad i \in N. \tag{7}$$

Note that for $i \neq j$ it is always true $y_i \neq y_j$ and the set $S_i^\varepsilon(\delta, y) \cap S_j^\varepsilon(\delta, y)$ is negligible by Lebesgue measure. So, in (7) countries are defined correctly, although they can be represented by empty set. Moreover, the statement of Lemma 2 is extended to a current case: the mapping

$$\mathcal{F}_i : (\delta, y) \to \nu_i(\delta, y) = \nu^\varepsilon(S_i^\varepsilon(\delta, y)), \quad i \in N$$

is continuous on $\Delta \times Y$ and if $\nu_i(\delta, y) = \nu^\varepsilon(S_i^\varepsilon) > 0$ then the centers of mass $r_c(S_i^\varepsilon) \in \mathcal{B}$ are defined correctly and, moreover, the function $\mathcal{F}_i$ is continuous at this point.

Let us consider also a mapping $\mathcal{R} : \Delta^{n-1} \times Y \to Y$, specifying new centers of countries: for $\nu_i(\delta, y) = \nu^\varepsilon(S_i^\varepsilon(\delta, y)) > 0$ define

$$\mathcal{R}_i(\delta, y) = Pr_i(r_c(S_i^\varepsilon)), \quad r_c(S_i^\varepsilon) = \frac{1}{\nu^\varepsilon(S_i^\varepsilon)} \int_{S_i^\varepsilon} (x, s) d\nu^\varepsilon(x), \quad S_i^\varepsilon \subset \mathcal{B} \subset \mathbb{R}^{l+n}$$

and put $\mathcal{R}_i(\delta, y) = Y_i$ if $\nu_i(\delta, y) = \nu^\varepsilon(S_i^\varepsilon(\delta, y)) = 0$, $i \in N$. We define also $\mathcal{F} : \Delta \times Y \to \Delta$ by formula

$$\mathcal{F}(\delta, y) = (\mathcal{F}_i(\delta, y))_{i \in N} = (\nu_i(\delta, y))_{i \in N}.$$

The key properties of the map $\mathcal{F} \times \mathcal{R}$ are the same as in two countries case and stated above Lemmas 3, 4, Corollary 1 can now be easily extended to the general case of n countries. Moreover, the proof of Theorem 2 is also workable: assuming that the centers of gravity of any two countries coincide, we come to a contradiction with the fact that the center of gravity of a convex solid set must be located in its interior.

In conclusion, a small remark about the centers of the countries. Like in the two countries case, in general, the capitals, represented as centers of the masses, are located on the territory of their own country—despite the fact that they form non-convex figures! To make sure of this, we will arrange the countries in descending order of individual contribution. Let us assume, without loss of generality, that

$$\frac{g_1}{\mu(S_1)} \geq \frac{g_2}{\mu(S_2)} \geq \cdots \geq \frac{g_n}{\mu(S_n)}.$$

Now one can conclude that the border between the countries $i < j$ is given by the hyperbola branch, which defines a convex fragment in $\mathcal{A}$ that contains the center of the country i and does not contain the center of j, this is so due to the equation

$$||x - r_i||_2 - ||x - r_j||_2 = \frac{g_j}{\mu(S_j)} - \frac{g_i}{\mu(S_i)} \leq 0.$$

Thus, the territory of the 1st country is the intersection of convex regions from $\mathcal{A}$, each of them contains the center $r_1 \in \mathcal{A}$ and their intersections cannot include the centers of other countries. Similarly for the 2nd country $r_2 \in \mathcal{A} \setminus S_1$ and $r_i \notin S_2\ \forall i \geq 3$ etc. Theorem 3 is proved. ∎

4 Appendix: Generalized Fixed Point Theorems

In this section I summarize some results obtained in [5,8,9], which are important for our analysis; all proofs are omitted (the comprehensive source is [9]). I am presenting a series of theorems generalizing both classical Brouwer and Kakutani fixed point theorems and Krasnosel'skii results [2,3].

Below we shall assume that $X \subset \mathbb{R}^n$ is a *convex compact* set and $f : X \to \mathbb{R}^n$. For a first view one can assume, without loss of generality, that $\operatorname{int} X \neq \emptyset$.

We now consider two conditions on the map $f : X \to \mathbb{R}^n$: for every boundary $x \in \partial X$ and every linear functional $h \neq 0$, which supports X at the point x, i.e. if $\langle h, X\rangle \leq \langle h, x\rangle$, the following takes place

$$\langle h, f(x)\rangle < \langle h, x\rangle \tag{8}$$

or

$$\langle h, f(x)\rangle > \langle h, x\rangle. \tag{9}$$

The first requirement I called a "strongly compressive" property, and the second one as a "strongly expanding" one, see Fig. 1.

Theorem 4. *Let $X \subset \mathbb{R}^n$ be a convex compactum and $\operatorname{int} X \neq \emptyset$. Then every continuous map $f : X \to \mathbb{R}^n$ having a strongly compressing or expanding property has a fixed point in X.*

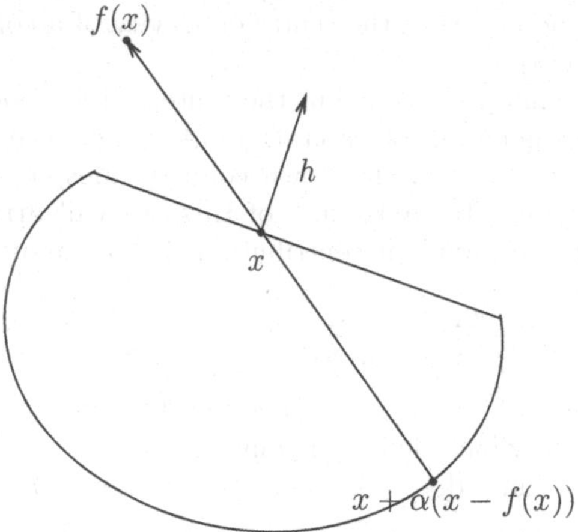

Fig. 1. Strongly expanding property (9).

Based on this result there was obtained the following generalization of Brouwer's theorem. Recall that in the linear space L the *affine* hull $\text{aff}(X)$ of the set $X \subset L$ is specified as

$$\text{aff}(X) = \{\sum_{\Xi} \lambda_\xi x_\xi \mid x_\xi \in X, \lambda_\xi \in \mathbb{R}\ \forall \xi \in \Xi,\ |\Xi| < \infty, \sum_{\Xi} \lambda_\xi = 1\}.$$

The affine hull can also be described as $\text{aff}(X) = x + \mathcal{L}(X - x)$, where $\mathcal{L}(X - x)$ is the *linear* hull of $(X - x)$ for an arbitrarily chosen $x \in X$. Below ∂X is the boundary of the set X in its affine hull, i.e. $\partial X = X \setminus \text{ri}X$, where $\text{ri}X$ is the relative interior of the closed convex set $X \subset L$.

Theorem 5. *Let $X \subset \mathbb{R}^n$ be a convex compactum and $f : X \to \text{aff}(X)$ be a continuous map with values in the affine hull of X. Suppose that f satisfies one of the following conditions:*

(i) Compression

$$\forall x \in \partial X, \forall h \in \mathbb{R}^n \quad [\langle h, x\rangle \geq \langle h, X\rangle \ \Rightarrow\ \langle h, f(x)\rangle \leq \langle h, x\rangle]. \tag{10}$$

(ii) Expansion

$$\forall x \in \partial X, \forall h \in \mathbb{R}^n \quad [\langle h, x\rangle \geq \langle h, X\rangle \ \Rightarrow\ \langle h, f(x)\rangle \geq \langle h, x\rangle]. \tag{11}$$

Then f has a fixed point in X.

Remark 1. The statement of Theorem 5 can be generalized to the case of the Cartesian product of mappings, the first of which satisfies the hypothesis of Theorem 5, and the second—to the conditions of Brouwer theorem or it is reducible to it: for example, the conditions (i) or (ii) are satisfied.

The content of Theorem 5 can be reformulated in a little bit different form. We recall that in a convex analysis with a convex closed subset X of a linear space L and any $x^* \in X$ two kinds of cones are customary associated. First, it is *normal* cone

$$\mathfrak{N}_X(x^*) = \{h \in L^* \mid \langle h, x^* - X\rangle \geq 0\},$$

where L^* is dual space for L. Second, *tangent* cone:

$$\mathfrak{T}_X(x^*) = \{y \in L \mid \langle h, x^* - y\rangle \geq 0 \; \forall h \in \mathfrak{N}_X(x^*)\} = \mathrm{cl}\left(\bigcup_{\lambda > 0} \lambda(X - x^*)\right).$$

Now applying tangent cone, condition (10) can be rewritten in the following way: $\forall x^* \in \partial X \; (f(x^*) - x^*) \in \mathfrak{T}_X(x^*)$. If one take into account that $\mathrm{aff} X = \mathfrak{T}_X(x^*)$ for $x \in \mathrm{ri} X$ and $\mathfrak{T}_X(x^*) \subset \mathrm{aff} X$ for $x^* \in \partial X$, then one can conclude that (10) is equivalent to

$$\forall x^* \in X \quad (f(x^*) - x^*) \in \mathfrak{T}_X(x^*).$$

Similar conclusions can be done for condition (11). As a result we are going to the following reformulation of Theorem 5:

Theorem 6. *Let $X \subset \mathbb{R}^n$ be a convex compactum and $f : X \to \mathbb{R}^n$ be a continuous map. Assume that f obeys one of the following conditions:*

(i) Compression: $(f(x^*) - x^*) \in \mathfrak{T}_X(x^*) \;\; \forall x^* \in X$.
(ii) Expansion: $(x^* - f(x^*)) \in \mathfrak{T}_X(x^*) \;\; \forall x^* \in X$.

Then f has a fixed point in X.

Below I show that classical Kakutani's theorem about multivalued mappings also can be generalized applying more or less similar method.

Kakutani theorem generalizes Brouwer's theorem to multivalued (point-to-set) mappings. Here it is allowed that the values of the mapping are sets, but they are necessarily non-empty convex compact sets, a kind of "generalized point". The requirement of continuity of a mapping is replaced by the closeness of the graph or by its equivalent, upper semicontinuity. However, similarly the Brouwer theorem, the classical version of Kakutani's theorem assumes that the values of the mapping must be subsets of its (convex and compact) domain. In this section, it will be shown that similarly to Brouwer's theorem, Kakutani's theorem on the existence of a fixed point for a point-to-set mapping can be generalized to the case when the last requirement is violated, i.e., the mapping may act beyond the scope of the domain.

As before, we assume that $X \subset \mathbb{R}^n$ is a *convex compact* set and, without loss of generality, $\mathrm{int}\, X \neq \emptyset$. However, now we consider *point-to-set* mapping (correspondence) $F : X \rightrightarrows \mathbb{R}^n$. We recall the following classical definition of upper semicontinuity.

Definition 2. *A point-to-set mapping $F : X \Rightarrow Y \subset \mathbb{R}^n$ is called upper semicontinuous* ***at the point*** *$x \in X$ if $F(x) \neq \emptyset$ and for each open $U \supset F(x)$ there is a neighborhood V_x of the point x such that $F(z) \subset U \; \forall z \in V_x$.*

A mapping $F : X \Rightarrow Y \subset \mathbb{R}^n$ is called ***upper semicontinuous*** *if it is upper semicontinuous at every point of X.*

In the literature one can find also the concept of "Kakutani map".

Definition 3. *A point-to-set mapping $F : X \Rightarrow Y \subset \mathbb{R}^n$ is called a* ***Kakutani map*** *if its graph $GrF = \{(x, y) \in X \times Y \mid y \in F(x)\}$ is closed and for every $x \in X$ the value $F(x) \subseteq Y$ is non-empty and convex.*

It is known that if $X \times Y$ is a compact set, the map $F : X \Rightarrow Y \subset \mathbb{R}^n$ has a closed graph and $F(x) \neq \emptyset \; \forall x \in X$, then this map is upper semicontinuous. Thus, if $X \times Y$ is a compact set, then the Kakutani map is always *upper semicontinuous.*

We now consider additional boundary conditions for $F(\cdot)$. If $x \in \partial X$, one can need compressive condition:

$$\exists y \in F(x), \forall h \in \mathbb{R}^n \quad [\langle h, x\rangle \geq \langle h, X\rangle \;\Rightarrow\; \langle h, y\rangle \leq \langle h, x\rangle], \tag{12}$$

or expansive one:

$$\exists y \in F(x), \forall h \in \mathbb{R}^n \quad [\langle h, x\rangle \geq \langle h, X\rangle \;\Rightarrow\; \langle h, y\rangle \geq \langle h, x\rangle]. \tag{13}$$

The Kakutani mapping for which requirement (12) is fulfilled at *every boundary* point of X is called *compressing.* Similarly, if at every boundary point the mapping satisfies (13), then we call it *expanding.*

Theorem 7. *Let $X, Y \subset \mathbb{R}^n$ be convex compact sets, $Y \subset \mathrm{aff}(X)$ and $F : X \Rightarrow Y$ be a point-to-set Kakutani mapping. Now if the mapping $F(\cdot)$ is* ***compressing*** *or alternatively, is* ***expanding****, then F has a fixed point in X.*

Similarly to the case of one-to-one map the last theorem can be reformulated for multivalued case in terms of tangent cones.

Theorem 8. *Let $X \subset \mathbb{R}^n$ be a nonempty convex compact set and $F : X \Rightarrow Y \subset \mathbb{R}^n$ be a point-to-set Kakutani mapping. Assume that F obeys one of the following conditions:*

(i) Compression: $(F(x^*) - x^*) \cap \mathfrak{T}_X(x^*) \neq \emptyset \;\; \forall x^* \in X$.
(ii) Expansion: $(x^* - F(x^*)) \cap \mathfrak{T}_X(x^*) \neq \emptyset \;\; \forall x^* \in X$.

Then F has a fixed point in X.

The following theorem has applications in the theory of spatial equilibrium and is a direct consequence of Theorem 7.

Let $M \subset \mathbb{R}^n$ be a *convex bounded* polyhedron (polytope) and $\mathrm{aff}(M)$ be its *affine hull.* Let $d \in \mathrm{ri}M$ be some point in the *relative interior* of a polyhedron M, and F_t, $t = 1, \ldots, k$ be its non-trivial faces of a maximum dimension (one less than the dimension of M). With every facet a cone $K_t \subset \operatorname{aff}(M)$ with a vertex at d is associated:

$$K_t = \{d + \lambda(\kappa - d) \mid \kappa \in F_t,\ \lambda \geq 0\} \quad \Rightarrow \quad \operatorname{aff}(M) = \bigcup_{t=1,\ldots,m} K_t.$$

Theorem 9. *Let $\mathcal{F} : M \rightrightarrows \operatorname{aff}(M)$ be a Kakutani mapping defined on a polyhedron M and $d \in \operatorname{ri}M$, $\operatorname{aff}(M)$, F_t, K_t be defined as described above. Let one of the conditions hold:*

(i) Compressive form: $\mathcal{F}(F_t) \subset M, \quad \forall t = 1, \ldots, m.$
(ii) Expansive form: $\mathcal{F}(F_t) \subset K_t \setminus \operatorname{ri}M, \quad \forall t = 1, \ldots, m.$

Then $\mathcal{F}(\cdot)$ has a fixed point in M.

The result of Theorem 9 in its expansive form is illustrated in Fig. 2. Here f continuously maps a smaller simplex $\Delta_\varepsilon^{(n-1)}$ onto its extension $\Delta^{(n-1)}$ and obeys condition (ii). The proof of Theorem 9 follows from Theorem 7 and characterization of boundary points of the polyhedron in terms of supporting hyperplanes.

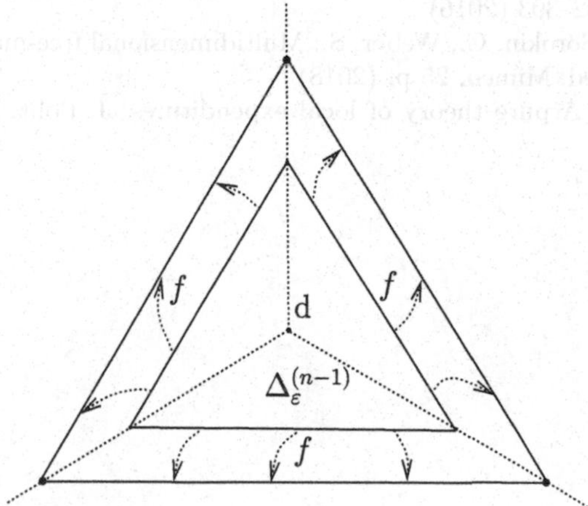

Fig. 2. Initial and embedded sub-simplex $\Delta_\varepsilon^{(n-1)}$ and the mapping $f(\cdot)$.

References

1. Alesina, A., Spolaore, E.: On the number and size of nations. Q. J. Econ. **113**, 1027–56 (1997)
2. Krasnosel'skii, M.A.: Fixed points of cone-compressing or cone-extending operators. Proc. USSR Acad. Sci. **135**(3), 527–530 (1960). (in Russian)
3. Kwong, M.K.: On Krasnoselskii's cone fixed point theorem. J. Fixed Point Theory Appl. **2008**, 18 (2008). https://doi.org/10.1155/2008/164537

4. Le Breton, M., Musatov, M., Savvateev, A., Weber, S.: Rethinking Alesina and Spolaore's "uni-dimensional world": existence of migration proof country structures for arbitrary distributed populations. In: Proceedings of XI International Academic Conference on Economic and Social Development 2010, Moscow, 6–8 April 2010: University-Higher School of Economics (2010)
5. Marakulin, V.M.: On the existence of immigration proof partition into countries in multidimensional space. In: Kochetov, Y., Khachay, M., Beresnev, V., Nurminski, E., Pardalos, P. (eds.) DOOR 2016. LNCS, vol. 9869, pp. 494–508. Springer, Cham (2016). https://doi.org/10.1007/978-3-319-44914-2_39
6. Marakulin, V., M.: Spatial equilibrium on the plane and an arbitrary population distribution. In: Evtushenko, Yu. G. et al. (eds.) Proceedings of the OPTIMA-2017 Conference, Petrovac, Montenegro, 02–06 October 2017, CEUR Workshop Proceedings, vol 1987, pp. 378–385 (1987)
7. Marakulin, V.M.: Spatial equilibrium: the existence of immigration proof partition into countries for one-dimensional space. Siberian J. Pure Appl. Math. **17**(4), 64–78 (2017). (in Russian)
8. Marakulin, V.M.: General spatial equilibrium and generalized Brouwer fixed point theorem. Mimeo, 20 p. (2017)
9. Marakulin, V.M.: On the existence of spatial equilibrium and generalized fixed point theorems. J. New Economic Assoc. (42) (2019, forthcoming). (in Russian)
10. Musatov, M., Savvateev, A., Weber, S.: Gale-Nikaido-Debreu and Milgrom-Shannon: communal interactions with endogenous community structures. J. Econ. Theory **166**, 282–303 (2016)
11. Savvateev, A., Sorokin, C., Weber, S.: Multidimensional free-mobility equilibrium: Tiebout revisited. Mimeo, 25 p. (2018)
12. Tiebout, C.M.: A pure theory of local expenditures. J. Polit. Econ. **64**, 416–424 (1956)

Game of Competition for Opinion with Two Centers of Influence

Vladimir Mazalov[1] and Elena Parilina[2](✉)

[1] Institute of Applied Mathematical Research, Karelian Research Center, Russian Academy of Sciences, 11, Pushkinskaya street, Petrozavodsk 185910, Russia
vmazalov@krc.karelia.ru

[2] Saint Petersburg State University, 7/9 Universitetskaya nab., Saint Petersburg 199034, Russia
e.parilina@spbu.ru

Abstract. The paper considers the model of opinion dynamics in the network having a star structure. An opinion about an event is distributed among network agents restricted by the network structure. The agent in the center of the star is influenced by all other agents with equal intensity. The agents located in non-center nodes are influenced only by the agent located in the center of the star. Additionally, it is assumed that there are two players who are not located in the considered network but they influence the agents' opinions with some intensities which are strategies of the players. The goal of any player is to make opinions of the network agents be closer to the initially given value as much as possible in a finite time interval. The game of competition for opinion is linear-quadratic and is solved using the Euler-equation approach. The Nash equilibrium in open-loop strategies is found. A numerical simulation demonstrates theoretical results.

Keywords: Opinion dynamics · Consensus · Game of competition for opinion

1 Introduction

In modern society informational technologies allow to influence the society opinion on key events. Different centers would like to reach different opinions on the same event and the influence process may be competitive. We introduce a simple model of competition for the society opinion based on DeGroot's model of information diffusion [8], which is represented as a dynamic process where an agent of the society influence each other opinion with the same intensity at any time period. In the paper, the conditions of reaching a consensus are found.

In our paper, we consider a society which is represented by a set of agents who are the nodes in a given network which structure is known. It is assumed the

The work is supported by Russian Science Foundation, project no. 17-11-01079.

M. Khachay et al. (Eds.): MOTOR 2019, LNCS 11548, pp. 673–684, 2019.
https://doi.org/10.1007/978-3-030-22629-9_47

agents exchange information via network and their opinions are influenced by the other agents who have direct connection with them according to the network structure. The intensity of influence does not change over time and is given by a matrix. We propose to consider a star graph as a network. Therefore, there is an agent who is located at the center and any other agent has a unique link with him. All non-central agents are symmetric which means their influence on the central agent and reverse are the equivalent among non-central agents. The star graph of communication may represent the relations in small working societies with a unique leader [17]. Game-theoretical models with given network structures are introduced in [15]. The problem of network partitioning with different approaches is examined in [1,16].

Moreover, we suppose that there exist two players who are not represented in the given network. The players aim society to have an opinion the closest to the given ones. The players may have different target opinions and the problem becomes competitive. The players are also different in costs of influence on the agents and the set of agents they may communicate with. The first player may directly affect only the central agent opinion and the second player may directly affect any non-central agent. We suppose that agents form their opinions with both influence of the neighbours according to communication structure and influence of player 1 (2) if the agent is central or not. The state of the game in discrete time t is represented by a vector of agents' opinions in this period. The state dynamics is linear with respect to the previous period state and players' strategies. As the costs of the players are linear quadratic functions of state and strategies, the game is classified as a linear-quadratic dynamic game. For algorithms of finding solutions of such a class of games see [9–11]. The Euler equation approach of solving liner-quadratic games is described in [12].

The models of reaching a consensus become actual when the variety of social networks like Facebook, Vkontakte, LinkedIn appears and has a popularity in the Internet. The models of informational influence on population and information control models are introduced and examined in the papers [2,14] and in the books [7,13]. A model of opinion dynamics with two principals is presented in [6], in which dynamic process is examined and the limit opinions are obtained for a given matrix of influences. There is a series of papers in which the problem of reaching a consensus is considered as a repeated game [3], a mean field game [4] and an evolutionary game [20]. In the latter paper, several consensus models in which agents have different levels of susceptibility to the inputs received from their neighbours are considered, and the equilibrium points are found. Together with game theoretical models, there exist imitation models of opinion dynamics [5].

In the paper we propose a model of two-player dynamic game of competition for opinion and find the Nash equilibrium in case the players have different target opinions on some event. The competitive models of opinion dynamics are also considered in [18,19]. In the paper [18], the cooperative version of the game is considered and the optimal strategies with open- and closed-loop information structure are found.

In Sect. 2, we describe the model of a game of competition on opinion. In Sect. 3, we formulate the main result of the Nash equilibrium existence. We provide the system of equation to find the players' equilibrium strategies. In Sect. 4, we define a steady state. A numerical simulation is presented in Sect. 5. We briefly conclude in Sect. 6.

2 Game of Competition for Opinion

Let there be a network consisting of an $n+1$ agent with a star structure (see Fig. 1). This network represents how the communication between the agents takes place. In the network, agent 1 is connected to all other agents, and agents $2, \ldots, n+1$ are not directly connected to each other. Assume the agents $i = 2, \ldots, n+1$ are symmetric.

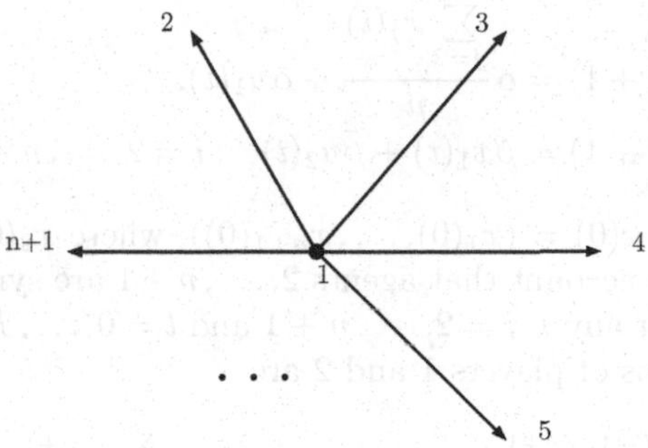

Fig. 1. A communication graph.

We assume that all agents have an opinion on a certain topic. Opinion varies in time, which is assumed to be discrete and finite, $t = 0, 1, \ldots, T$. Opinion of agent i at time t is $x_i(t) \in \mathbb{R}$, $i = 2, \ldots, n+1$.

Suppose there are two centers of opinion control (players 1 and 2) whose goal is to get an opinion as close as possible to $\hat{x} \in \mathbb{R}$ and $\hat{y} \in \mathbb{R}$ for all agents, respectively. The set of agents' opinions $(x_i(t) : i = 1, \ldots, n+1)$ determines the state $x(t) \in \mathbb{R}^n$ at time t. Define the dynamics of agents' opinions. Agents in the network affect each other's opinions at any time with constant intensity, i.e. the matrix the intensity of the impact of player j on player's opinion i, which is $A = \{a_{i,j}\}_{i,j=1,\ldots,n+1}$, where $a_{i,j} \in [0, 1]$, is given. We assume matrix A is of the form:

$$A = \begin{pmatrix} 0 & \frac{1}{n} & \ldots & \frac{1}{n} \\ 1 & 0 & \ldots & 0 \\ \vdots & \vdots & \ddots & \vdots \\ 1 & 0 & \ldots & 0 \end{pmatrix}$$

We notice that agent $2, \ldots, n+1$ affects the opinion of agent 1 with equal intensity which follows from matrix A form. The opinion of agents $2, \ldots, n+1$ are influenced only by the opinion of agent 1.

We define the dynamics of the agents' opinions, taking into account the impact of players 1 and 2 on the network agents. The opinion of agent 1 at time $t+1$ is influenced with probability α by the opinion of other network agents according to matrix A and with probability $\bar{\alpha} = 1 - \alpha$ by the opinion of player 1 who influences on agent 1 with intensity $u_1(t) \in \mathbb{R}$ at time t. Thus, player 1 directly affects only agent 1, but its goal is to make the opinions of all network agents to $\hat{x}$ as closer as possible. The opinion of any agent $i = 2, \ldots, n+1$ is formed by the influence of the opinion of agent 1 with probability β according to matrix A and with probability $\bar{\beta} = 1 - \beta$ by the opinion of the player 2, which affects the agent i's opinion with the intensity $u_2(t) \in \mathbb{R}$ at time t. The state dynamics is

$$x_1(t+1) = \alpha \frac{\sum_{j=2}^{n+1} x_j(t)}{n} + \bar{\alpha} u_1(t), \tag{1}$$

$$x_i(t+1) = \beta x_1(t) + \bar{\beta} u_2(t), \quad i = 2, \ldots, n+1. \tag{2}$$

The initial state $x(0) = (x_1(0), \ldots, x_{n+1}(0))$, where $x_2(0) = \ldots = x_{n+1}(0)$, is given. Taking into account that agents $2, \ldots, n+1$ are symmetric we suppose that $x_i(t) = x_j(t)$ for any $i, j = 2, \ldots, n+1$ and $t = 0, \ldots, T$.

The cost functions of players 1 and 2 are

$$J_1(u_1, u_2) = \sum_{t=0}^{T-1} \left(\sum_{i=1}^{n+1} (x_i(t) - \hat{x})^2 + c_1 u_1^2(t) \right) + \sum_{i=1}^{n+1} (x_i(T) - \hat{x})^2, \tag{3}$$

$$J_2(u_1, u_2) = \sum_{t=0}^{T-1} \left(\sum_{i=1}^{n+1} (x_i(t) - \hat{y})^2 + n c_2 u_2^2(t) \right) + \sum_{i=1}^{n+1} (x_i(T) - \hat{y})^2, \tag{4}$$

where c_i is costs of player i per unit of influence intense.

We define a *game of competition for opinion* as a normal form game of two players with the set of players' strategies U_1, U_2, where $U_j = (u_j(t) \in \mathbb{R} : t = 0, \ldots, T-1)$, $j = 1, 2$, players' cost functions J_1, J_2, defined by formulas (3), (4) s.t. state Eqs. (1), (2) with initial state $x(0) = (x_1(0), \ldots, x_{n+1}(0))$. The game of competition for opinion belongs to the class of linear-quadratic games because the right hand side of Eqs. (1), (2) is linear with players' strategies and cost functions are quadratic with strategies and states.

3 Nash Equilibrium in a Game of Competition for Opinion

We consider the Nash equilibrium as a solution of the game which is the strategy profile (u_1^*, u_2^*) such that the inequalities

$$J_1(u_1^*, u_2^*) \leqslant J_1(u_1, u_2^*),$$
$$J_2(u_1^*, u_2^*) \leqslant J_2(u_1^*, u_2)$$

hold for any $u_1 \in U_1$ and $u_2 \in U_2$.

The following theorem provide conditions to find the Nash equilibrium in a game of competition for opinion which is defined by two players, the set of strategies U_1 and U_2 and cost functions (3) and (4) to be minimized.

Theorem 1. *Let (u_1^*, u_2^*) be the Nash equilibrium and $\{x^*(t) : t = 0, \ldots, T\}$ be the corresponding state trajectory in a game of competition for opinion, then they satisfy the systems:*

$$\begin{cases} \mathcal{A}_1 x_1(t) + \mathcal{B}_1 u_2(t) - \mathcal{D}_1 \beta x_1(t+2) - \mathcal{D}_1 x_2(0) - (1+n\beta)\hat{x} = 0, \quad t = 1, \\ \mathcal{A}_1 x_1(t) + \mathcal{B}_1 u_2(t) - \mathcal{D}_1 \beta x_1(t+2) - \mathcal{D}_1 \beta x_1(t-2) - \mathcal{D}_1 \bar{\beta} u_2(t-2) \\ -(1+n\beta)\hat{x} = 0, \quad t = 2, \ldots, T-2, \\ \mathcal{C}_1 x_1(t) + n\beta\bar{\beta} u_2(t) - \mathcal{D}_1 \beta x_1(t-2) - \mathcal{D}_1 \bar{\beta} u_2(t-2) - (1+n\beta)\hat{x} = 0, \\ t = T-1, \\ \left(\mathcal{C}_1 - n\beta^2\right) x_1(t) - \mathcal{D}_1 \beta x_1(t-2) - \mathcal{D}_1 \bar{\beta} u_2(t-2) - \hat{x} = 0, \quad t = T, \end{cases} \tag{5}$$

$$\begin{cases} \mathcal{A}_2 x_2(t) + \mathcal{B}_2 u_1(t) - \mathcal{D}_2 \alpha x_2(t+2) - \mathcal{D}_2 x_1(0) - (n+\alpha)\hat{y} = 0, \quad t = 1, \\ \mathcal{A}_2 x_2(t) + \mathcal{B}_2 u_1(t) - \mathcal{D}_2 \alpha x_2(t+2) - \mathcal{D}_2 \alpha x_2(t-2) - \mathcal{D}_2 \bar{\alpha} u_1(t-2) \\ -(n+\alpha)\hat{y} = 0, \quad t = 2, \ldots, T-2, \\ \mathcal{C}_2 x_2(t) + \alpha\bar{\alpha} u_1(t) - \mathcal{D}_2 \alpha x_2(t-2) - \mathcal{D}_2 \bar{\alpha} u_1(t-2) - (n+\alpha)\hat{y} = 0, \\ t = T-1, \\ \left(\mathcal{C}_2 - \alpha^2\right) x_2(t) - \mathcal{D}_2 \alpha x_2(t-2) - \mathcal{D}_2 \bar{\alpha} u_1(t-2) - n\hat{y} = 0, \quad t = T \end{cases} \tag{6}$$

where
$\mathcal{A}_1 = 1 + n\beta^2 + \frac{c_1(1+\alpha^2\beta^2)}{\bar{\alpha}^2}$, $\mathcal{A}_2 = n + \alpha^2 + \frac{nc_2(1+\alpha^2\beta^2)}{\bar{\beta}^2}$,
$\mathcal{B}_1 = \beta\bar{\beta}\left(n + \frac{c_1\alpha^2}{\bar{\alpha}^2}\right)$, $\mathcal{B}_2 = \alpha\bar{\alpha}\left(1 + \frac{nc_2\beta^2}{\bar{\beta}^2}\right)$,
$\mathcal{C}_1 = 1 + n\beta^2 + \frac{c_1}{\bar{\alpha}^2}$, $\mathcal{C}_2 = n + \alpha^2 + \frac{nc_2}{\bar{\beta}^2}$
$\mathcal{D}_1 = \frac{c_1\alpha}{\bar{\alpha}^2}$, $\mathcal{D}_2 = \frac{nc_2\beta}{\bar{\beta}^2}$,
taking into account the state Eqs. (1) *and* (2) *and initial state* $x(0) = (x_1(0), \ldots, x_{n+1}(0))$ *with* $x_2(0) = \ldots = x_{n+1}(0)$.

Proof. The game of competition for opinion is a dynamic game with finite horizon and state dynamics Eqs. (1) and (2) whose right-hand sides are linear with states and strategies. The cost functions are linear-quadratic functions with states and quadratic functions with strategies. To find the Nash equilibrium in the game of competition for opinion we use the Euler method and write down the conditions defining the strategy of player 1 as a function of player 2's strategy:

$$u_1(0) = \frac{1}{\bar{\alpha}} \left(x_1(1) - \alpha x_2(0)\right), \tag{7}$$

$$u_1(t) = \frac{1}{\bar{\alpha}} \left(x_1(t+1) - \alpha\beta x_1(t-1) - \alpha\bar{\beta} u_2(t-1)\right), \tag{8}$$

$$t = 1, \ldots, T-1,$$

and the strategy of player 2 as a function of player 1's strategy:

$$u_2(0) = \frac{1}{\bar{\beta}} \left(x_i(1) - \beta x_1(0)\right), \tag{9}$$

$$u_2(t) = \frac{1}{\bar{\beta}} \left(x_i(t+1) - \alpha\beta x_i(t-1) - \bar{\alpha}\beta u_1(t-1)\right), \tag{10}$$

$$t = 1, \ldots, T-1.$$

In expressions (9) and (10), the states $x_i(t)$ are used. As the initial state is such that $x(0) = (x_1(0), \ldots, x_{n+1}(0))$, where $x_2(0) = \ldots = x_{n+1}(0)$ and the agents $2, \ldots, n+1$ are symmetric, we use notation $x_2(t)$ for any $i, j = 2, \ldots, n+1$, because $x_i(t) = x_j(t)$ for any $i \neq j$, $i, j = 2, \ldots, n+1$ and any $t = 0, \ldots, T$.

First, substitute expressions (7) and (8) into cost function $J_1(u_1, u_2)$ defined by (3) and consider it as a function of $x_1(t)$, $t = 1, \ldots, T$, and $u_2(t)$, $t = 0, \ldots, T-1$:

$$\begin{aligned}
J_1(x_1, u_2) &= \left(x_1(0) - \hat{x}\right)^2 + n\left(x_2(0) - \hat{x}\right)^2 + \frac{c_1}{\bar{\alpha}^2}\left(x_1(1) - \alpha x_2(0)\right)^2 \\
&+ \sum_{t=1}^{T-1} \left(x_1(t) - \hat{x}\right)^2 + n \sum_{t=1}^{T-1} \left(\beta x_1(t-1) + \bar{\beta} u_2(t-1) - \hat{x}\right)^2 \\
&+ \frac{c_1}{\bar{\alpha}^2} \sum_{t=1}^{T-1} \left(x_1(t+1) - \alpha\beta x_1(t-1) - \alpha\bar{\beta} u_2(t-1)\right)^2 \\
&+ \left(x_1(T) - \hat{x}\right)^2 + n\left(\beta x_1(T-1) + \bar{\beta} u_2(T-1) - \hat{x}\right)^2.
\end{aligned}$$

Taking the derivative of $J_1(x_1, u_2)$ over $x_1(1)$ and equate it to zero, we obtain the first equation of system (5). Taking the derivative of $J_1(x_1, u_2)$ over $x_1(t)$, $t = 2, \ldots, T-2$, and equate it to zero, we obtain the second group of equations of system (5). The third and the fourth equations of system (5) are obtained by equity of a derivative of $J_1(x_1, u_2)$ over $x_1(T-1)$ and $x_1(T)$, correspondingly, to zero.

Second, substitute expressions (9) and (10) into cost function $J_2(u_1, u_2)$ defined by (4) and consider it as a function of $x_2(t)$, $t = 1, \ldots, T$, and $u_1(t)$, $t = 0, \ldots, T-1$:

$$J_2(x_2, u_1) = (x_1(0) - \hat{y})^2 + n\,(x_2(0) - \hat{y})^2 + \frac{nc_2}{\bar{\beta}^2}\,(x_2(1) - \beta x_1(0))^2$$
$$+ n \sum_{t=1}^{T-1} (x_2(t) - \hat{y})^2 + \sum_{t=1}^{T-1} (\alpha x_2(t-1) + \bar{\alpha} u_1(t-1) - \hat{y})^2$$
$$+ \frac{nc_2}{\bar{\beta}^2} \sum_{t=1}^{T-1} (x_2(t+1) - \alpha\beta x_2(t-1) - \bar{\alpha}\beta u_1(t-1))^2$$
$$+ n\,(x_2(T) - \hat{y})^2 + (\alpha x_2(T-1) + \bar{\alpha} u_1(T-1) - \hat{y})^2 .$$

Taking the derivative of $J_2(x_2, u_1)$ over $x_2(1)$ and equate it to zero, we obtain the first equation of system (6). Taking the derivative of $J_2(x_2, u_1)$ over $x_2(t)$, $t = 2, \ldots, T-2$, and equate it to zero, we obtain the second group of equations in system (6). The third and forth equations of system (6) are obtained by equity of a derivative of $J_2(x_2, u_1)$ over $x_2(T-1)$ and $x_2(T)$, correspondingly, to zero.

The solution of the systems (5) and (6) taking into account the state dynamics Eqs. (1) and (2) with a given initial state $x(0) = (x_1(0), \ldots, x_{n+1}(0))$, $x_2(0) = \ldots = x_{n+1}(0)$ provides the Nash equilibrium in the game of competition for opinion. We can easily prove that the systems (5) and (6) and state dynamics equations form a linear system with respect to $x_1(t)$, $x_2(t)$, $t = 1, \ldots, T$ and $u_1(t)$, $u_2(t)$, $t = 0, \ldots, T-1$, which has a unique solution.

4 Steady State

Suppose there exists a steady state (x_1, x_2) for dynamics (5), (6) when $t \to \infty$. From (8) and (10) we find the limit values for optimal controls:

$$u_1 = \frac{1}{\bar{\alpha}}\,(x_1 - \alpha\beta x_1 - \alpha\bar{\beta} u_2)\,,$$
$$u_2 = \frac{1}{\bar{\beta}}\,(x_i - \alpha\beta x_i - \bar{\alpha}\beta u_1)\,,$$

which yields

$$u_1 = \frac{x_1 - \alpha x_2}{\bar{\alpha}}, \tag{11}$$
$$u_2 = \frac{x_2 - \beta x_1}{\bar{\beta}}, \tag{12}$$

where (x_1, x_2) satisfy the equations:

$$x_1\left(1+n\beta^2+\frac{c_1(1+\alpha^2\beta^2)}{\bar{\alpha}^2}\right)-x_1\frac{c_1\alpha\beta}{\bar{\alpha}^2}+u_2\beta\bar{\beta}\left(n+\frac{c_1\alpha^2}{\bar{\alpha}^2}\right)$$
$$-x_1\frac{c_1\alpha\beta}{\bar{\alpha}^2}-u_2\frac{c_1\alpha\bar{\beta}}{\bar{\alpha}^2}=\hat{x}(1+n\beta),$$
$$x_2\left(n+\alpha^2+\frac{nc_2(1+\alpha^2\beta^2)}{\bar{\beta}^2}\right)-x_2\frac{nc_2\alpha\beta}{\bar{\beta}^2}+u_1\alpha\bar{\alpha}\left(1+\frac{nc_2\beta^2}{\bar{\beta}^2}\right)$$
$$-x_2\frac{nc_2\alpha\beta}{\bar{\beta}^2}-u_1\frac{nc_2\bar{\alpha}\beta}{\bar{\beta}^2}=\hat{y}(n+\alpha).$$

Simplifying we obtain

$$x_1\left(1+n\beta^2+\frac{c_1(1-\alpha\beta)^2}{\bar{\alpha}^2}\right)+u_2\bar{\beta}\left(n\beta-\frac{c_1\alpha(1-\alpha\beta)}{\bar{\alpha}^2}\right)=\hat{x}(1+n\beta),$$
$$x_2\left(n+\alpha^2+\frac{nc_2(1-\alpha\beta)^2}{\bar{\beta}^2}\right)+u_1\bar{\alpha}\left(\alpha-\frac{nc_2\beta(1-\alpha\beta)}{\bar{\beta}^2}\right)=\hat{y}(n+\alpha).$$

Substituting u_1, u_2 from (11) and (12), we obtain the system of equations:

$$x_1\left(1+\frac{c_1(1-\alpha\beta)}{\bar{\alpha}^2}\right)+x_2\left(n\beta-\frac{c_1\alpha(1-\alpha\beta)}{\bar{\alpha}^2}\right)=\hat{x}(1+n\beta),$$
$$x_2\left(n+n\frac{c_2(1-\alpha\beta)}{\bar{\beta}^2}\right)+x_1\left(\alpha-\frac{nc_2\beta(1-\alpha\beta)}{\bar{\beta}^2}\right)=\hat{y}(n+\alpha).$$

The solution of the latter system is a steady state given by

$$x_1=\frac{1}{Q}\left(\hat{x}n(1+n\beta)\left(1+\frac{c_2(1-\alpha\beta)}{\bar{\beta}^2}\right)-\hat{y}(n+\alpha)\left(n\beta-\frac{c_1\alpha(1-\alpha\beta)}{\bar{\alpha}^2}\right)\right), \tag{13}$$

$$x_2=\frac{1}{Q}\left(-\hat{x}(1+n\beta)\left(\alpha-\frac{nc_2\beta(1-\alpha\beta)}{\bar{\beta}^2}\right)+\hat{y}(n+\alpha)\left(1+\frac{c_1(1-\alpha\beta)}{\bar{\alpha}^2}\right)\right), \tag{14}$$

where

$$Q=n\left(1+\frac{c_1(1-\alpha\beta)}{\bar{\alpha}^2}\right)\left(1+\frac{c_2(1-\alpha\beta)}{\bar{\beta}^2}\right)$$
$$-\left(n\beta-\frac{c_1\alpha(1-\alpha\beta)}{\bar{\alpha}^2}\right)\left(\alpha-\frac{nc_2\beta(1-\alpha\beta)}{\bar{\beta}^2}\right).$$

We notice that the steady state does not depend on initial state $x(0)$.

Remark 1. If we consider the case when $\alpha=\beta=0$, i.e. the agents are not influenced by the other agents and influenced only by the players. Then the steady state is $(x_1, x_2, \ldots, x_{n+1})$ coincide with $(\frac{\hat{x}}{1+c_1}, \frac{\hat{y}}{1+c_2}, \ldots, \frac{\hat{y}}{1+c_2})$. All agents will have the same opinion only if the players' target opinions satisfy the condition $\frac{\hat{x}}{\hat{y}}=\frac{1+c_1}{1+c_2}$.

Remark 2. We may consider the case when $\alpha = \beta = 1$, i.e. the agents are not influenced by the players and influenced only by the agents in the network according matrix A. The model represents DeGroot model of opinion dynamics [8]. And we notice that the consensus in this network is not reached because $\lim_{t\to\infty} A^t$ does not exist. In particular, matrix A^t takes one of the forms:

$$A^t = \begin{pmatrix} 0 & \frac{1}{n} & \dots & \frac{1}{n} \\ 1 & 0 & \dots & 0 \\ \vdots & \vdots & \ddots & \vdots \\ 1 & 0 & \dots & 0 \end{pmatrix}, \quad \text{if } t = 2k+1, k = 0, 1, \dots$$

$$A^t = \begin{pmatrix} 1 & \frac{1}{n} & \dots & \frac{1}{n} \\ 0 & \frac{1}{n} & \dots & \frac{1}{n} \\ \vdots & \vdots & \ddots & \vdots \\ 0 & \frac{1}{n} & \dots & \frac{1}{n} \end{pmatrix}, \quad \text{if } t = 2k, k = 1, 2, \dots$$

Considering the networks, in which the consensus is not reached the role of the players is significant.

In the next section we consider a numerical example for which we find equilibrium strategies, corresponding state trajectories and players' payoffs when they use equilibrium strategies.

5 Numerical Simulation

We consider a game with 41 time periods starting from 0. In a star network represented in Fig. 1 there are 7 non-central nodes. Therefore the number of agents $n + 1$ is eight. The initial opinion state is $x(0) = (0.3, 0.7, \dots, 0.7)$, i.e. $x_1(0) = 0.3$ and $x_i(0) = 0.7$ for any $i = 2, \dots, 8$. The probabilities that agent 1 and agent i are influenced by other agents according to matrix A are $\alpha = 0.3$ and $\beta = 0.4$, respectively.

The unit costs for influence intense are $c_1 = 0.1$ for player 1 and $c_2 = 0.05$ for player 2. Their target opinions are $\hat{x} = 0.7$ and $\hat{y} = 0.9$ respectively.

Using Theorem 1 we find the equilibrium strategies represented in Fig. 2. The second player's intense (blue dots) is larger than the first player's one and may be explained by the higher target opinion 0.9 of player 2 contrary to 0.7 of player 1 and smaller unit costs 0.05 contrary to the unit costs 0.1 of player 1. The state trajectories corresponding to the Nash equilibrium are presented in Fig. 3. The steady state is $(0.304164, 0.837986, \dots, 0.837986)$ and is depicted by red (for x_1) and blue (for any x_i, $i = 2, \dots, 8$) lines. As one can notice, the equilibrium state trajectory (red dots — for $x_1(t)$, and blue dots — for $x_i(t)$, $i = 2, \dots, 8$) almost everywhere coincide with the steady state. There are 7 non-central agents, and for the players it is more important to make opinion of these agents closer to the target one rather than to make the opinion of a central agent closer to the target one. The reason is in the form of the total cost functions (3) and (4).

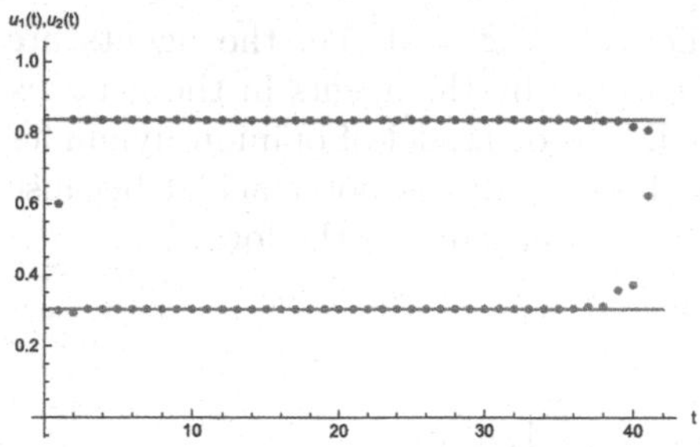

Fig. 2. Strategy trajectories (red — $u_1(t)$, blue — $u_2(t)$). (Color figure online)

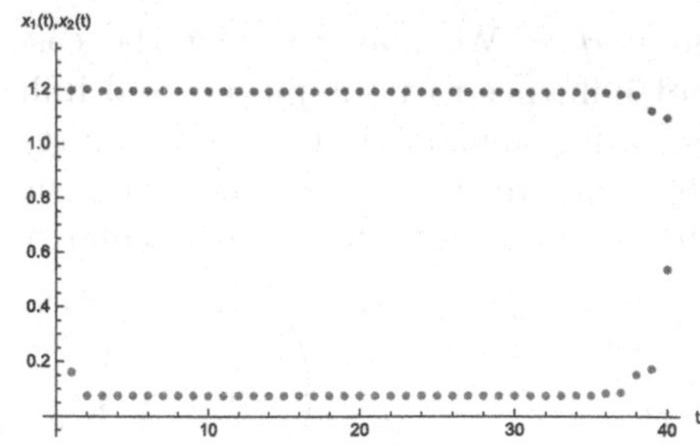

Fig. 3. State trajectories (red — $x_1(t)$, blue — $x_i(t)$, $i = 2, \ldots, n+1$). Red and blue lines are steady states x_1 and x_2 correspondingly. (Color figure online)

In these functions the differences between agent's and target opinions are taken into account with the same weight 1. One may consider cost functions in which these weights are different.

Next, we calculate the total costs of the players defined by formulas (3) and (4) when players use their equilibrium strategies. In Fig. 4 we introduce the total costs $J_1(u_1^*, u_2^*)$ (in red color) and $J_2(u_1^*, u_2^*)$ (in blue color) as functions of the number of agent n. We can easily notice that the costs of the players are increasing functions of the number of agents in the network but they are not linear. Although player 1 has larger unit costs than player 2, his total costs are smaller than the ones of player 2. It may be explained by the closeness of his target opinion $\hat{x}$ to the opinions of all non-central agents.

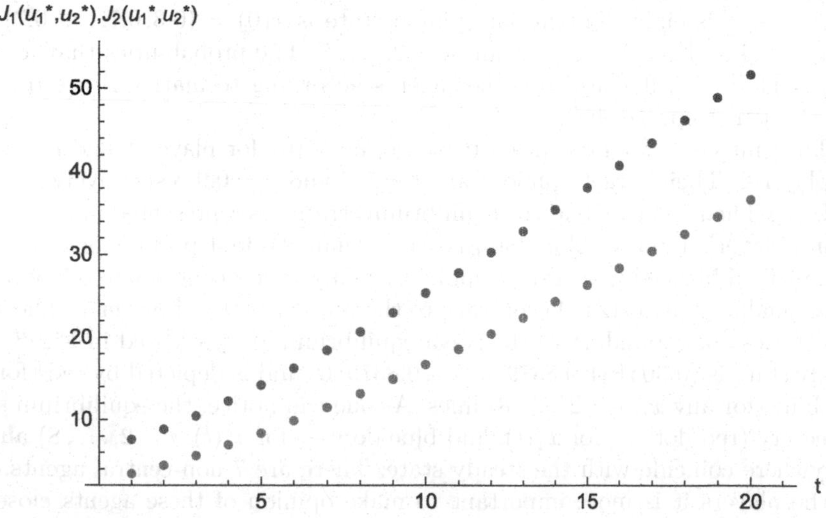

Fig. 4. Players' total costs as functions of the number of agents in the network (red — $J_1(u_1^*, u_2^*)$, blue — $J_2(u_1^*, u_2^*)$). (Color figure online)

6 Conclusions

In the paper we propose a model of a competition for opinion in which two players choose an intense of influence on the agents' opinions at each time period to make their opinions closer to the target opinions of the players. Players may have different target opinions and this fact is a basis of a competition. The players may directly influence different agents in the network. The state of the system is the profile of agents' opinions. We obtain the necessary conditions of the Nash equilibrium, find the steady state for a given state dynamics. Numerical simulations demonstrate the theoretical results. Any other opinion dynamics may be considered to define the game of competition for opinion and it is left for a future research.

References

1. Avrachenkov, K.E., Kondratev, A.Y., Mazalov, V.V.: Cooperative game theory approaches for network partitioning. In: Cao, Y., Chen, J. (eds.) COCOON 2017. LNCS, vol. 10392, pp. 591–602. Springer, Cham (2017). https://doi.org/10.1007/978-3-319-62389-4_49
2. Barabanov, I.N., Korgin, N.A., Novikov, D.A., Chkhartishvili, A.G.: Dynamic models of informational control in social networks. Autom. Remote Control. **71**(11), 2417–2426 (2010)
3. Bauso, D., Cannon, M.: Consensus in opinion dynamics as a repeated game. Automatica **90**, 204–211 (2018)
4. Bauso, D., Tembine, H., Basar, T.: Opinion dynamics in social networks through mean field games. SIAM J. Control Optim. **54**(6), 3225–3257 (2016)
5. Bure, V.M., Ekimov, A.V., Svirkin, M.V.: A simulation model of forming profile opinions within the collective. Vestn. Saint Petersburg Univ. Appl. Math. Comput. Sci. Control Process. **3**, 93–98 (2014)
6. Bure, V.M., Parilina, E.M., Sedakov, A.A.: Consensus in a social network with two principals. Autom. Remote Control **78**(8), 1489–1499 (2017)
7. Chkhartishvili, A.G., Gubanov, D.A., Novikov, D.A.: Social Networks: Models of Information Influence, Control and Confrontation. Springer, Switzerland (2019). https://doi.org/10.1007/978-3-030-05429-8
8. DeGroot, M.H.: Reaching a consensus. J. Am. Stat. Assoc. **69**(345), 118–121 (1974)
9. Engwerda, J.: LQ Dynamic Optimization and Differential Games. Wiley, New York (2005)
10. Engwerda, J.: Algorithms for computing Nash equilibria in deterministic LQ games. Comput. Manage. Sci. **4**(2), 113–140 (2007)
11. Haurie, A., Krawczyk, J.B., Zaccour, G.: Games and Dynamic Games. World Scientific Publishing, Singapore (2012)
12. González-Sánchez, D., Hernandez-Lerma, O.: Discrete-Time Stochastic Control and Dynamic Potential Games. The Euler-Equation Approach. Springer International Publishing, Heidelberg (2013). https://doi.org/10.1007/978-3-319-01059-5
13. Gubanov, D.A., Novikov, D.A., Chkhartishvili, A.G.: Sotsial'nye seti: modeli informatsionnogo vliyaniya, upravleniya i protivoborstva (Social Networks: Models of Informational Influence, Control and Confrontation). Fizmatlit, Moscow (2010)

14. Gubanov, D.A., Novikov, D.A., Chkhartishvili, A.G.: Informational influence and informational control models in social networks. Autom. Remote Control **72**(7), 1557–1567 (2011)
15. Mazalov, V.V.: Mathematical Game Theory and Applications. Wiley, New York (2014). Wiley Desktop Editions
16. Mazalov, V.V.: Comparing game-theoretic and maximum likelihood approaches for network partitioning. In: Nguyen, N.T., Kowalczyk, R., Mercik, J., Motylska-Kuźma, A. (eds.) Transactions on Computational Collective Intelligence XXXI. LNCS, vol. 11290, pp. 37–46. Springer, Heidelberg (2018). https://doi.org/10.1007/978-3-662-58464-4_4
17. Parilina, E., Sedakov, A.: Stable cooperation in a game with a major player. Int. Game Theory Rev. **18**(2), 1640005 (2016)
18. Rogov, M.A., Sedakov, A.A.: Coordinated influence on the beliefs of social network members. Math. Game Theory Appl. **10**(4), 30–58 (2018)
19. Sedakov, A., Zhen, M.: Opinion dynamics game in a social network with two influence nodes. Vestn. Saint Petersburg Univ. Appl. Math. Comput. Sci. Control Process. **15**(1), 118–125 (2019)
20. Smyrnakis, M., Bauso, D., Hamidou, T.: An evolutionary game perspective on quantised consensus in opinion dynamics. PLoS ONE **14**(1), e0209212 (2019)

Equilibrium and Cooperation in Repeated Hierarchical Games

Leon Petrosyan(✉) and Yaroslavna Pankratova

Saint Petersburg State University, Saint Petersburg, Russia
{l.petrosyan,y.pankratova}@spbu.ru
http://www.apmath.spbu.ru/en/staff/petrosjan/index.html

Abstract. In the paper a two-level infinitely repeated hierarchical game with one player (center) C_0 on the first level and $S_1, \ldots, S_n$ subordinate players on the second is considered. On each stage of the game player C_0 selects vector $x = (x_1, \ldots, x_n)$ from a given set X, in which each component represents vector of resources delivered by C_0 to one of subordinate players, i.e. $x_i = (x_{i1}, \ldots, x_{im})$. At the second level, S_i, $i = 1, 2, \ldots, n$, choose the controls $y_i \in Y_i(x_i)$, where $Y_i(x_i)$ depends upon the choice of player C_0.

In this game, a set of different Nash equilibrium also based on threat and punishment strategies is obtained.

In one case, the center enforces special behavior of subordinate firms (vector of manufactured goods), threatening to deprive them of resources on the next steps if the subordinate firms refuse to implement the prescribed behavior.

In another case, the subordinate firms can force the center to use a certain resource allocation threatening to stop production.

Using different combinations of such behaviors on different stages of the game, we obtain a wide class of Nash equilibrium in the game under consideration.

The cooperative version of the game is also considered. The conditions are derived under which the cooperative behavior can be supported by Nash Equilibrium or Strong Nash Equilibrium (Nash Equilibrium stable against deviations of coalitions).

Keywords: Repeated hierarchical game · Nash equilibrium · Cooperation

1 Introduction

There exists an important subclass of multistage nonzero-sum games, referred to as hierarchical games. Hierarchical games model conflict controlled systems with a hierarchical structure. This structure is determined by a sequence of control

Supported by Russian Science Foundation the grant Optimal Behavior in Conflict-Controlled Systems (17-11-01079).

M. Khachay et al. (Eds.): MOTOR 2019, LNCS 11548, pp. 685–696, 2019.
https://doi.org/10.1007/978-3-030-22629-9_48

levels ranking in particular priority order. Mathematically, it is convenient to classify hierarchical games according to the number of levels and the nature of vertical relations. In this paper, we investigate repeated hierarchical games. Using the Folk Theorem's approach [3,5] the different classes of Nash equilibrium are proposed with punishment and threat strategies, which often occurs in real life situation.

The cooperative version of the game is also considered and conditions are formulated under which the cooperative behavior can be supported by Strong Nash equilibrium.

The basic difficulty in constructing the corresponding Strong Nash equilibrium consists in the fact that it is not easy to apply "punishment" for the coalitions deviating from cooperation. In the simple case when only one shot deviations and only of one coalition are allowed the result is published in [3,8]. But the approach which was presented in [3,8] cannot be extended for the general case when different members of deviating coalitions deviate in different time instants, because immediately after the first deviation the no deviating players cannot identify the deviating coalition and thus cannot realize the "punishment" against the coalition which decides to end up with cooperation. To solve this problem we introduce the new "punishment" strategies for players: if they (for instance player i) see the first time (on the next step after deviation) the deviation of one or more players the non-deviating players start to use their optimal strategies in a zero sum game, playing against the coalition of all players (except himself). Which in practice means that each non-deviating player have to start the play against all others immediately after the first deviation by someone take place. We derive conditions under which this way of behavior guarantees that each coalition deviating from cooperation will lose.

2 Two-Level Hierarchical Game

In the beginning, we consider a two-level hierarchical game with one center C_0 and production firms $S_1, \dots, S_n$.

Define a two-level hierarchical game:

- At the first level, the (coordinating) center C_0 selects the control vector $x = (x_1, \dots, x_n)$ from a given set of controls X, where x_i is a control influence of center on its subordinate divisions S_i, $i = 1, \dots, n$ standing at the second level of the hierarchy. Under x_i we can understand a set of resources of m kinds, i.e. $x_i = (x_{i1}, \dots, x_{im})$. In other words the components of the control vector x_i characterize the impact of the center on its subordinate units.
- At the second level, S_i, $i = 1, 2, \dots, n$, choose the controls $y_i \in Y_i(x_i)$, where $Y_i(x_i)$ is the set of controls of firm S_i.

Thus, the center has the right of the first move and can limit the possibility of subordinate firms, directing their actions on track.

We can formalize this problem as a noncooperative $(n+1)$-person game Γ (a coordinating center C_0 and production division $S_1, \dots, S_n$) in normal form [4].

The set of strategies of the player C_0 is

$$X = \{(x_1, \ldots, x_n) : \; x_i \geq 0, \; x_i \in R^m, \; i = 1, \ldots, n, \sum_{i=1}^{n} x_i \leq q, \; q \geq 0\},$$

where $q \in R^m$ can be interpreted as vector of available resources.

The vector x_i, $i = 1, \ldots, n$ is interpreted as vector of resources of m items allocated by center C_0 to the ith production division.

We consider the case of complete information, i.e. each firm S_i knows the selection of C_0 and in accordance with this knowledge selects the vector y_i from some set $Y_i(x_i)$ that has the following form

$$Y_i(x_i) = \{y_i \in R^k, \; y_i P_i \leq x_i, \; y_i \geq 0\},$$

where $P_i = p^i_{lj}$ is interpreted as a technological matrix of the (division) company, $p^i_{lj} \geq 0, l = 1, \ldots, k, j = 1, \ldots, m$. Under y_i we understand a production program of ith production division for different types of products.

Under the strategy of the company S_i we understand the function $y_i(x_i)$, which corresponds to each element x_i vector from the set $Y_i(x_i)$. The set of all such functions is denoted by $\bar{Y}_i$, $i = 1, 2, \ldots, n$.

Define the players' payoff functions in the game Γ.

The payoff function of player C_0 equals to the sum of inner products

$$H_0(x, y_1(x_1), \ldots, y_n(x_n)) = \sum_{i=1}^{n} (\alpha_i, y_i(x_i)),$$

where $\alpha_i \geq 0$, $\alpha_i \in R^k$, $i = 1, \ldots, n$ is a fixed vector.

The payoff function of player S_i is a inner product of vectors β_i and $y_i(x_i)$, i.e.

$$H_i(x, y_1(x_1), \ldots, y_n(x_n)) = (\beta_i, y_i(x_i)),$$

where $\beta_i \geq 0, \beta_i \in R^k$ is a fixed vector.

The goal of center C_0 is to maximize the functional $H_0(x, y_1, \ldots, y_n)$ choosing x, and the goal of the subordinate companies S_i is maximization of $H_i(x, y_i)$, by choosing y_i, $i = 1, 2, \ldots, n$.

In general case, we suppose $\alpha_i \neq \beta_i$, $i = 1, \ldots, n$.

Thus, we construct a game Γ in normal form

$$\Gamma = (C_0, S_1, \ldots, S_n; X, \bar{Y}_1, \ldots, \bar{Y}_n, ; H_0, H_1, \ldots, H_n).$$

In this game, a Nash equilibrium can be constructed [4,8].

2.1 Nash Equilibrium in a Two-Level Hierarchical Game

In game Γ the strategy profile $(x^*, y_1^*(\cdot), \ldots, y_n^*(\cdot))$ is a Nash equilibrium [6,7] where $y_i^*(x_i) \in Y_i(x_i)$ is the solution to a following linear parametric programming problem (where vector x_i as a parameter)

$$\max_{y_i \in Y(x_i)} (\beta_i, y_i) = (\beta_i, y_i^*(x_i)), i = 1, \ldots, n \tag{1}$$

and $x^* \in X$ is a solution to following maximization problem

$$\max_{x \in X} H_0(x, y_1^*(x), \dots, y_n^*(x)). \tag{2}$$

For simplicity assume that the maximum in (1) and (2) are achieved. Note that (2) is a nonlinear programming problem with an essentially discontinuous objective function (maximization is taken over x, and $y_i^*(x_i)$ are generally discontinuous functions of the parameter x_i). Show that the point $(x^*, y_1^*(\cdot), \dots, y_n^*(\cdot))$ is an equilibrium in the game Γ. Indeed,

$$H_0(x^*, y_1^*(x_1), \dots, y_n^*(x_n)) \geq H_0(x, y_1^*(x_1), \dots, y_n^*(x_n)),\ x \in X.$$

Further, for all $i = 1, \dots, n$ the inequality

$$\begin{aligned} &H_i(x^*, y_1^*(x_1), \dots, y_n^*(x_n)) = (\beta_i, y_i^*(x_i^*)) \geq (\beta_i, y_i(x_i)) \\ &= H_i(x^*, y_1^*(x_1), \dots, y_{i-1}^*(x_{i-1}), y_i(x_i), y_{i+1}^*(x_{i+1}), \dots, y_n^*(x_n)) \end{aligned}$$

holds for any $y_i(x_i) \in Y_i(x_i)$. Thus, it is not advantageous for every player C_0, S_1, ..., S_n to deviate individually from the profile $(x^*, y_1^*(x_1), \dots, y_n^*(x_n))$, i.e. it is an equilibrium. Note that this profile is also stable against deviations from it of any coalition $S \subset \{S_1, \dots, S_n\}$, since the payoff H_i to the ith player does not depend on strategies $y_j(x_j)$, $j \in \{1, \dots, n\}$, $j \neq i$.

3 Repeated Game

Consider now an infinitely repeated game G with finite game Γ played on each stage [2].

If on stage l $(1 \leq l < \infty)$ the $n+1$ tuple of strategies

$$(x^l, y_1^l(x_1^l), \dots, y_n^l(x_n^l))$$

is used the payoff of player C is defined as

$$H_0^\infty = \sum_{l=1}^{\infty} \delta^{l-1} \left(\sum_{i=1}^{n} (\alpha_i, y_i^l(x_i^l)) \right)$$

and the payoff of player S_i $(i = 1, \dots, n)$, as

$$H_i^\infty = \sum_{l=1}^{\infty} \delta^{l-1} (\beta_i, y_i^l(x_i^l)),\ \delta \in (0, 1).$$

Denote by $\bar{x} = (x^1, \dots, x^l, \dots)$, where $x^l = (x_1^l, \dots, x_i^l, \dots, x_n^l)$ and by $\bar{y}_i(\bar{x}) = (y_i^1(x^1), \dots, y_i^l(x^l), \dots)$ the strategies of players C_0 and S_i, $i = 1, \dots, n$ in repeated game G.

One can write

$$H_0^\infty = H_0^\infty(\bar{x}, \bar{y}_1(\bar{x}_1), \dots, \bar{y}_n(\bar{x}_n)) = \sum_{l=1}^{\infty} \delta^{l-1} \left(\sum_{i=1}^{n} (\alpha_i, y_i^l(x_i^l)) \right)$$

and

$$H_i^\infty = H_i^\infty(\bar{x}, \bar{y}_1(\bar{x}_1), \ldots, \bar{y}_n(\bar{x}_n)) = \sum_{l=1}^{\infty} \delta^{l-1}(\beta_i, y_i^l(x_i^l))$$

As in the case of finitely repeated game in the game G there is a reach variety of Nash equilibrium [11]. One of them is repetition of the equilibrium $(x^*, y^*(x))$ in each stage game. Denote this equilibrium by $E1$.

As well as the equilibrium based on the strategies of threats or punishment.

- In one case, the center imposes his behavior on subordinate firms (vector of manufactured goods), threatening to deprive them of resources on the next step if the subordinate firms deviate from the prescribed behavior ($E2$).
- In another case, the subordinate firms can impose on center a certain allocation of resources, threatening to stop production at the same step in case of failure to comply with their requirements ($E3$).
- Different combinations of these equilibriums.

Consider now the case $E2$.

In infinitely repeated game G consider in stage game Γ^l the following strategy profile for players C_0, $S_1, \ldots, S_n$

$$(\tilde{x}^l, \tilde{y}_1^l(\tilde{x}^l), \ldots, \tilde{y}_m^l(\tilde{x}^l))$$

Player C_0 solves the following maximisation problem

$$\max H_0(x^l, y_1^l(x^l), \ldots, y_n^l(x^l)) = \max \sum_{i=1}^{n} (\alpha_i, y_i^l(x_i^l)) \tag{3}$$

under conditions

$$\sum_{i=1}^{n} x_i^l \leq q$$
$$y_i^l P_i \leq x_i^l$$
$$x_i^l \geq 0,\ y_i^l \geq 0,\ i = 1, \ldots, n$$

Suppose that maximum in (3) is attained in the point $(\tilde{x}^l, \tilde{y}_1^l(\tilde{x}^l), \ldots, \tilde{y}_m^l(\tilde{x}^l))$. One can see that this profile can be taken the same for all $l = 1, \ldots, \infty$ and we can write it as $(\tilde{x}, \tilde{y}_1(\tilde{x}), \ldots, \tilde{y}_m(\tilde{x}))$.

Derive now the conditions under which C_0 can prescribe other players S_1, ..., S_n, the behavior $\tilde{y}_1^l, \ldots, \tilde{y}_m^l$, which together with $\tilde{x}^l$ is maximizing his payoff. Can this way of behavior form a Nash Equilibrium? It is clear that after getting the resource $\tilde{x}_i^l$, $i \in N$ player S_i can instead of following the instructions made by C_0 (choose $\tilde{y}_i^l(\tilde{x}^l)$) improve his payoff by choosing $\tilde{\tilde{y}}_i^l(\tilde{x}^l)$ such that

$$\begin{aligned}\hat{H}_i = &\max_{y_i^l \in Y_i(\tilde{x})} H_i(\tilde{x}^l, \tilde{y}_1^l(\tilde{x}^l), \ldots, \tilde{y}_{i-1}^l(\tilde{x}^l), y_i^l, \tilde{y}_{i+1}^l(\tilde{x}^l), \ldots, \tilde{y}_n^l(\tilde{x}^l)) =\\ &H_i(\tilde{x}^l, \tilde{y}_1^l(\tilde{x}^l), \ldots, \tilde{y}_{i-1}^l(\tilde{x}^l), \tilde{\tilde{y}}_i^l(\tilde{x}^l), \tilde{y}_{i+1}^l(\tilde{x}^l), \ldots, \tilde{y}_n^l(\tilde{x}^l)) \geq\\ &H_i(\tilde{x}^l, \tilde{y}_1^l(\tilde{x}^l), \ldots, \tilde{y}_{i-1}^l(\tilde{x}^l), \tilde{y}_i^l(\tilde{x}^l), \tilde{y}_{i+1}^l(\tilde{x}^l), \ldots, \tilde{y}_n^l(\tilde{x}^l)) = H_i\end{aligned}$$

But on the next stage and always after (in the infinite game) C_0 can punish S_i by sending him the resource $\tilde{\tilde{x}} = 0$.

Thus we can now define the Nash equilibrium E2.

Definition 1. *In each stage game, players chose strategy profile* $(\tilde{x}, \tilde{y}_1(\tilde{x}), \ldots, \tilde{y}_m(\tilde{x}))$ *if on the previous stage the same profile was chosen. If on the previous stage one of the players* S_i *chooses* y_i^l, $y_i^l \neq \tilde{y}_i^l(\tilde{x})$ *then* $\tilde{x}_k^l = 0$, *for all* $k \geq l$.

Prove that there always exist such $\bar{\delta}$, that for $\delta \in (\bar{\delta}, 1)$ player S_i deviating from the prescribed behavior $\tilde{y}_i^l(\bar{x})$ will loose in the infinitely repeated game. The proof is very similar to the standard proof of Folk theorem. We present it here for better understanding of what follows.

Deviating on stage l player S_i can get at most $\hat{H}_i$, but in all other stages he will be punished by C_0 by sending him the resource $x_i = 0$.

This can be seen from the following inequalities. If player S_i does not deviate he gets

$$H_i^\infty = \sum_{m=1}^{\infty} \delta^{m-1}(\beta_i, \tilde{y}_i(\tilde{x}_i)) = (\beta_i, \tilde{y}_i(\tilde{x}_i))\frac{1}{1-\delta}.$$

If S_i deviates of stage l, he will get at most

$$\sum_{t=1}^{l-1} \delta^{t-1}(\beta_i, \tilde{y}_i(\tilde{x}_i)) + \delta^l \hat{H}_i = \hat{H}_i^\infty.$$

We have

$$H_i^\infty = \sum_{t=1}^{l-1} \delta^{t-1}(\beta_i, \tilde{y}_i(\tilde{x}_i)) + \sum_{t=l}^{\infty} \delta^{t-1}(\beta_i, \tilde{y}_i(\tilde{x}_i))$$

the payoff of S_i if he follows the behavior prescribed by C_0. To prove that $H_i^\infty \geq \hat{H}_i^\infty$ we have to show that

$$\delta^l \hat{H}_i \leq \delta^{l-1} \sum_{l=1}^{\infty} \delta^{l-1}(\beta_i, \tilde{y}_i(\tilde{x}_i)) = \delta^{l-1} H_i^\infty$$

$$\delta \hat{H}_i \leq H_i^\infty = (\beta_i, \tilde{y}_i(\tilde{x}_i))\frac{1}{1-\delta}$$

The last inequality proves the theorem, since $\delta \in (0,1)$ and the existence of $\bar{\delta}$ follows from previous inequality.

Thus the following theorem holds.

Theorem 1. *In two-level infinite repeated game* Γ^∞ *there exists such* $\bar{\delta} \in (0,1)$, *that for all* $\delta \in (\bar{\delta}, 1)$, *the strategy profile* $(\tilde{x}(\cdot), \tilde{y}_1(\cdot), \ldots, \tilde{y}_n(\cdot))$ *will be Nash equilibrium.*

It is possible to modify this Nash equilibrium when the player C_0 dictates his conditions for the first k stages, and the rest of the stages use the Nash equilibrium $E1$ strategies in stage game we considered earlier.

Another type of Nash equilibrium $E3$ consists of using “‘threat” strategies by players $S_1,\dots,S_i,\dots,S_n$. They have the following form in each stage game fix a resource vector $\bar{\bar{x}} = (\bar{\bar{x}}_1,\dots,\bar{\bar{x}}_i,\dots,\bar{\bar{x}}_n)$, and define

$$\bar{y}_i(x_i) = \begin{cases} y_i^*(\bar{\bar{x}}_i), \, x_i = \bar{\bar{x}}_i \\ \quad 0, \qquad x \neq \bar{\bar{x}} \end{cases}$$

Here y_i^* is solution of (1). Players S_i, $i = 1,\dots,n$ can declare in the beginning of stage game, that they will use strategies $\bar{y}_i(x_i)$, $i = 1,\dots,n$. Then feeling the threat C_0 will be forced to use $\bar{\bar{x}} = (\bar{\bar{x}}_1,\dots,\bar{\bar{x}}_i,\dots,\bar{\bar{x}}_n)$, in either case he can get zero payoff. In this Nash equilibrium C_0 is forced to deliver resources $\bar{\bar{x}} = (\bar{\bar{x}}_1,\dots,\bar{\bar{x}}_i,\dots,\bar{\bar{x}}_n)$ to players $S_1,\dots,S_i,\dots,S_n$, since in either case his payoff will be 0.

Theorem 2. *The strategy profile $(\bar{\bar{x}}, \bar{y}_1(x),\dots,\bar{y}_i(x),\dots,\bar{y}_n(x))$ is Nash equilibrium in Γ and the repetition of it in each stage in infinitely repeated game G is Nash equilibrium in G.*

Proof. Since $\bar{y}_i(\bar{\bar{x}}_i) = y_i^*(\bar{\bar{x}}_i)$ is strategy of player S_i which maximaze his payoff under the condition that he gets the resource $\bar{\bar{x}}_i$ (see (1)), the individual deviation of S_i from $\bar{y}_i(\bar{\bar{x}}_i)$ cannot increase his payoff. If C_0 deviates from $\bar{x}$ (if C_0 will not follow the prescribed by $S_1,\dots,S_i,\dots,S_n$ behavior $\bar{x}$) he will get zero payoff since in this case player S_i will stop production by putting $\bar{y}_i(x) = 0$, $x \neq \bar{\bar{x}}$.

4 Cooperation in Infinitely Repeated Game

Consider now the cooperation in infinitely repeated stage game G. Suppose that players decide to cooperate and maximize the sum of their payoffs in G, this is equivalent to the maximization of joint payoff in each stage game. Denote the corresponding strategy profile by

$$(\tilde{\tilde{x}}, \tilde{\bar{y}}_1(x),\dots,\tilde{\bar{y}}_i(x),\dots,\tilde{\bar{y}}_n(x)).$$

We have

$$\max_{x,y_1(x),\dots,y_n(x)} \left[\sum_{i=1}^n (\alpha_i, y_i(x_i)) + \sum_{i=1}^n (\beta_i, y_i(x_i)) \right] = \sum_{i=1}^n (\alpha_i, \tilde{\bar{y}}_i(\tilde{\tilde{x}}_i)) + \sum_{i=1}^n (\beta_i, \tilde{\bar{y}}_i(\tilde{\tilde{x}}_i))$$

under conditions

$$\sum_{i=1}^n x_i \leq q$$
$$y_i P_i \leq x_i$$
$$x_i \geq 0, \; y_i \geq 0, \; i = 1,\dots,n.$$

The cooperative payoff in infinitely repeated game G will be

$$\sum_{l=1}^{\infty}\sum_{i=1}^{n}\delta^{l-1}((\alpha_i,\tilde{\bar{y}}_i(\tilde{\bar{x}}_i))+(\beta_i,\tilde{\bar{y}}_i(\tilde{\bar{x}}_i)))=\sum_{l=1}^{\infty}\sum_{i=1}^{n}\delta^{l-1}((\alpha_i+\beta_i),\tilde{\bar{y}}_i(\tilde{\bar{x}}_i))$$
$$=V(C_0,S_1,\ldots,S_n)$$

Denote by $\bar{N}=\{C_0,S_1,\ldots,S_n\}$ the set of players in game G. Introduce the following notations. Let $S\subset\bar{N}$, $C_0\notin S$ denote by $\bar{y}_S=\{\bar{y}_i,\ i\in S\}$ and similarly $\bar{y}_{N\setminus S}=\{\bar{y}_i,\ i\in N\setminus S\}$. The strategy profile $(\bar{x},\bar{y}(\bar{x}))$ can be written as $(\bar{x},\bar{y}_S,\bar{y}_{N\setminus S})$.

Now we can give the definition of Strong Nash equilibrium (see [1,13]).

Definition 2. *The strategy profile $(\bar{x}^*,\bar{y}_1^*(\bar{x}),\ldots,\bar{y}_n^*(\bar{x}))$ is called strong Nash equilibrium if for all $S\subset\bar{N}$, $\bar{y}_S$, $\bar{x}$ the following inequalities hold*

$$\sum_{i\in S}H_i^{\infty}(\bar{x}^*,\bar{y}^*(\bar{x}))\geq\sum_{i\in S}H_i^{\infty}(\bar{x}^*,\bar{y}_S(\bar{x}),\bar{y}^*_{\bar{N}\setminus S}(\bar{x})),\ if\ C_0\notin S,$$

and

$$H_0^{\infty}(\bar{x}^*,\bar{y}^*(\bar{x}))+\sum_{i\in S\setminus C_0}H_i^{\infty}(\bar{x}^*,\bar{y}^*(\bar{x}))\geq H_0^{\infty}(\bar{x},\bar{y}_{S\setminus C_0}(\bar{x}),\bar{y}^*_{\bar{N}\setminus S}(\bar{x}))$$
$$+\sum_{i\in S\setminus C_0}H_i^{\infty}(\bar{x},\bar{y}_{S\setminus C_0}(\bar{x}),\bar{y}^*_{\bar{N}\setminus S}(\bar{x})),\ if\ C_0\in S.$$

Can this cooperative payoff be attained as a payoff in some specially constructed Nash equilibrium or Strong Nash Equilibrium [9]?

The construction of such Nash equilibrium or Strong Nash equilibrium is based on the so-called "punishment" strategies which will "punish" the deviating player for the deviation from cooperation. The case when the deviating player on stage l became known on next stage is easier to investigate, and classical Folk theorem approaches can be used to construct such type of Nash Equilibrium (see the proof of Theorem 1). But if the deviating player or the deviating coalition is not known, the construction of punishment strategies is more complicated (see [9,10]).

Consider for $C_0\in\bar{N}$, and $S_i\in\bar{N}$, $i\in 1,\ldots,m$ a family of zero-sum stage games $\Gamma_{\bar{N}\setminus C_0,C_0}$ ($\Gamma_{\bar{N}\setminus S_i,S_i}$) based on stage game Γ between coalition $\bar{N}\setminus C_0$ ($\bar{N}\setminus S_i$) as a first player and coalition consisting from a single player $\{C_0\}$ ($\{S_i\}$) as a second. The payoff of coalition $\bar{N}\setminus C_0$ ($\bar{N}\setminus S_i$) is equal to the sum of payoffs of players from this coalition. Denote by $\hat{x}(\hat{y}_i)$ minmax strategy of player C_0 (S_i) in this game. Consider the strategy profile

$$(\hat{x},\hat{y}_1,\ldots,\hat{y}_i,\ldots,\hat{y}_n)$$

and define for each $S\subset\bar{N}$

$$\bar{W}(S)=\max_{y_{S\setminus C_0},x}\sum_{i\in S}H_i(x,\hat{y}_{\bar{N}\setminus S},y_S),\ C_0\in S,$$

where $\hat{y}_{\bar{N}\backslash S} = \{\hat{y}_i,\ i \in \bar{N}\backslash S\}$, $y_S = \{y_i,\ i \in S\}$, and

$$\bar{W}(S) = \max_{y_S} \sum_{i=0}^{n} H_i(\hat{x}, \hat{y}_{\bar{N}\backslash S}, y_S),\ C_0 \notin S,$$

where $\hat{y}_{\bar{N}\backslash S} = \{\hat{y}_i,\ i \in \bar{N}\backslash S\}$, $y_S = \{y_i,\ i \in S\backslash C_0\}$.

Now we shall use the results from [10].

Suppose that there exist solution of the following inequalities

$$\begin{aligned} &\sum_{i\in S} \alpha_i > \bar{W}(S),\ S \subset \bar{N},\ S \neq \bar{N}, \\ &\sum_{i\in \bar{N}} \alpha_i = \bar{W}(N) = V(C_0, S_1, \ldots, S_n). \end{aligned} \tag{4}$$

Consider the modification G^α of game G. The difference between these two games consists only in stage game Γ which is realized on each stage of the game. If the cooperative strategies

$$\tilde{\bar{x}}, \tilde{\bar{y}}_1(x_1), \ldots, \tilde{\bar{y}}_i(x_i), \ldots, \tilde{\bar{y}}_n(x_n)$$

are used the payoffs in stage game of game G^α are equal to $\alpha = (\alpha_1, \ldots, \alpha_n)$, where α satisfies (4). For other strategy profiles in stage games in G and G^α the payoffs coincide [9].

Theorem 3. *Suppose the condition (4) holds and the deviation of a player from cooperation became known to non-deviating players on next stage (but the deviating player may not be identified), then there exists $\bar{\delta} \in (0, 1)$ such that for all $\delta \in (\bar{\delta}, 1)$ in the game G^α there exist a strong Nash equilibrium with payoffs $\alpha_i \frac{1}{1-\delta}$. These payoffs coincide with payoffs in the game G^α when corresponding strategies are played.*

For the proof of the theorem, the following "punishment" strategies for players $C_0, S_1, \ldots, S_n$ are proposed. They include the "punishment" of all players (except themselves) in case the deviation from cooperation by someone became clear, and continue this behavior in all remaining stage games. For example if C_0 on stage l is informed about the deviation of someone he plays $x_i \equiv 0$, $i \in N$, in all stage games in stages $l+1, \ldots, l+k, \ldots$. If S_i is informed about the deviation of someone he plays $y_i(x_i) \equiv 0$ in all stage games on stages $l+1, \ldots, l+k, \ldots$.

This way of behavior generates a Nash equilibrium and also strong Nash equilibrium in the infinitely repeated game G^α and strategically supports the cooperation because if this strategy profile is played the players not deviating from cooperation will get $\alpha_i \frac{1}{1-\delta}$, which is the allocation of cooperative payoff $V(C_0, S_1, \ldots, S_n)$. Denote this strong Nash equilibrium by $E4$ (which is, of course, Nash equilibrium).

Consider now the cooperative version of stage game Γ, $\bar{\Gamma}$. Denote by $V(S)$, $S \subset \bar{N}$ characteristic function in cooperative stage game $\bar{\Gamma}$ defined as value of zero-sum game $\Gamma_{S,\bar{N}\backslash S}$ played between coalition S as first player and $\bar{N}\backslash S$ as second with payoff of coalition S equal to the sum of payoffs of players from S.

If $S \subset \{S_1, \ldots, S_n\}$, then $V(S) = 0$, since player $C_0 \in \bar{N} \backslash S$ can always send zero resources to players from S making their payoff equal to zero. Suppose now that $C_0 \in S$. In this case, the optimal strategy of coalition $\bar{N} \backslash S$ playing in $\Gamma_{S, \bar{N} \backslash S}$ against S will be to put the production vector $y_i(x) \equiv 0$, $i \in \bar{N} \backslash S$. In this case $V(S)$ will be equal to

$$\max_{x, y_i(x), i \in S} \sum_{i \in S} ((\alpha_i + \beta_i), y_i(x_i)) \tag{5}$$

under conditions

$$\begin{aligned} &\sum_{i \in S} x_i \leq q \\ &y_i P_i \leq x_i \\ &x_i \geq 0,\ y_i \geq 0,\ i \in S. \end{aligned} \tag{6}$$

It is clear that the minmax strategy of player $S_i \in \bar{N} \backslash S$ in the game $\Gamma_{S, \bar{N} \backslash S}$ $y_i(x) \equiv 0$, coincides with the minmax strategy $\hat{y}_i(x) = 0$ in the game $\Gamma_{\bar{N} \backslash S_i, S_i}$ when S_i plays against coalition $\bar{N} \backslash S_i$. Then the value of $\bar{W}(S)$, if $C_0 \in S$, as maximal guaranteed payoff of coalition S against players from $\bar{N} \backslash S$ playing $\hat{y}_i(x) \equiv 0$, $i \in \bar{N} \backslash S$, will coincide with $V(S)$ defined by (5) and (6). Also in the case when $C_0 \notin S$, $\hat{x}_0 = 0$, which gives us $\bar{W}(S) = 0$, since in this case C_0 can deliver zero resources to all players S_i, $i \in \bar{N} \backslash C_0$ (playing the game $\Gamma_{\bar{N} \backslash C_0, C_0}$). Thus also in this case $\bar{W}(S) = V(S)$. This gives us the following theorem.

Theorem 4. *In the cooperative game Γ, the function $\bar{W}(S)$, $S \in \bar{N}$ coincides with characteristic function $V(S)$.*

If vector $\alpha = (\alpha_1, \ldots, \alpha_n)$ satisfies (4), then since $\bar{W}(S) = V(S)$ $(S \subset \bar{N})$ this means that α belongs to the core of cooperative version of the game Γ, $\bar{\Gamma}$. Then Theorem 3 can be rewritten in the form.

Theorem 5. *Suppose in game Γ the core is not empty and contains an inner point α, then there exist $\delta \in (0,1)$ such that in game G^α there exist strong Nash equilibrium with payoffs $\alpha_i \frac{1}{1-\delta}$. These coincide with payoffs in game G^α when cooperative strategies are played.*

5 Conclusion

In the paper, we defined different types of Nash Equilibrium in the infinitely repeated hierarchical game ($E1$, $E2$, $E3$, $E4$). It is interesting to investigate Subgame Perfectness of defined equilibriums. Remind that the Equilibrium is called Subgame Perfect [12] if its truncation on the subgame is equilibrium in this subgame.

If the behavior of players in the equilibrium is changing when the game passed from one stage to another the subgame perfectness property may not hold.

It is trivial that the equilibrium $E1$ is subgame perfect, but $E2$ is not since in case of deviation of one of players S_i, player C_0 will use on the next stages (in next stage games) "punishment" strategy against the deviator and this combination of "punishment" and deviation is not an equilibrium in the subgame starting from the next stage game. The same is true also for the $E4$ which strategically supports the cooperation. But it is interesting to mention that the $E3$ in which "threat" strategies are used by players S_i $(i = 1, \ldots, n)$, is subgame perfect. Indeed in any next stage game independent of the behavior of players on the previous stage (was the threat realized or not, was the deviation or not) the $E3$ suggests the same behavior of players in all following stage games. This implies subgame perfectness of $E3$.

If players decide to use $E1$ on some stages of the game and $E3$ in all other cases (in other stages) one can easily verify that this kind of behavior also will be subgame perfect Nash equilibrium in the infinitely repeated hierarchical game. It seems that it is not possible to construct a Strong Nash Equilibrium (or even Nash Equilibrium) which strategically supports the cooperation in the infinitely repeated hierarchical game and in the same time is subgame perfect.

The results of this paper can be extended to the wide class of multistage hierarchical games when the next stage game is not necessary the repetition of the previous one, but the hierarchical game in which the parameters depend on strategies chosen in previous stage game. It seems that the results may be similar.

The considered games are simple game-theoretic models of social behavior and interactions between the government and the population, between the administration and the employers, the chief company and subordinate firms. We can see the realization of $E1$, $E2$, $E3$ and $E4$ in many real life situations around us.

References

1. Aumann, R.J.: The core of a cooperative game without side payments. Trans. Am. Math. Soc. **98**(3), 539–552 (1961)
2. Aumann, R.J., Maschler, M.: Repeated Games with Incomplete Information. MIT Press, Cambridge (1995)
3. Fudenberg, D., Maskin, E.: The folk theorem in repeated games with discounting or with incomplete information. Econometrica **54**(3), 533–554 (1986)
4. Germeyer, Y.B.: Non-Zero Sum Games. Nauka, Moskva (1976). (in Russian)
5. Maschler, M., Solan, E., Zamir, S.: Game Theory. Cambridge University Press, Cambridge (2013)
6. Myerson, R.-B.: Multistage games with communication. Econometrica **54**(2), 323–358 (1986)
7. Nash, J.: Non-cooperative games. Ann. Math. **54**(2), 286–295 (1951)
8. Petrosyan, L., Zenkevich, N., Gromova, E.: Game Theory, 2nd edn. BXV-Petersburg, St Petersburg (2012)
9. Petrosyan, L., Chistyakov, S., Pankratova, Ya.: Existence of Strong Equilibrium in Repeated and Multistage Games, Constructive Nonsmooth Analysis and Related Topics (Dedicated to the Memory of V.F. Demyanov), CNSA 2017, Saint-Petersburg, pp. 255–257 (2017). https://doi.org/10.1109/CNSA.2017.7974003

10. Petrosjan, L.A., Pankratova, Y.B.: Construction of Strong Equilibria in a class of infinite nonzero-sum games. Trudy Inst. Mat. Mekh. UrO RAN **24**(1), 165–174 (2018)
11. Rubinstein, A.: Equilibrium in supergames. In: Megiddo, N. (ed.) Essays in Game Theory, pp. 17–27. Springer, New York (1994). https://doi.org/10.1007/978-1-4612-2648-2_2
12. Selten, R.: Multistage game models and delay supergames. Theory Decis. **44**(1), 1–36 (1998)
13. Vasin, A.A.: Sil'nye situatsii ravnovesiya v nekotorykh sverkhigrakh. Vestnik Moskovskogo Universiteta, ser. Matem. I mekhanika, Vyp. **1**, 30–39 (1978). (in Russian)

Coalition Stability in Dynamic Multicriteria Games

Anna Rettieva[1,2](✉)

[1] Institute of Applied Mathematical Research Karelian Research Center of RAS, Pushkinskaya str. 11, 185910 Petrozavodsk, Russia
annaret@krc.karelia.ru

[2] Saint-Petersburg State University, Universitetskaya nab. 7–9, 199034 Saint-Petersburg, Russia

Abstract. We consider a dynamic, discrete-time, game model where the players use a common resource and have different criteria to optimize. The coalition formation process in dynamic multicriteria games is considered. The characteristic function is constructed in two unusual forms under the assumption of informed players: all players decide simultaneously or members of coalitions are assumed to be the leaders and players decide sequentially. Internal and external stability concepts are adopted for dynamic multicriteria games to obtain new stability conditions. To illustrate the presented approaches a multicriteria bioresource management problem with a finite horizon is investigated.

Keywords: Dynamic games · Multicriteria games · Nash bargaining solution · Internal and external stability · Coalition stability

1 Introduction

Mathematical models involving more than one objective seem more adherent to real problems. Often players have more than one goal which are often not comparable. These situations are typical for game-theoretic models in economics and ecology. For example, in bioresource management problems the players wish to maximize their exploitation rates and to minimize the harm to the environment. Hence, a multicriteria game approach [11] helps to make decisions in multi-objective problems.

In this paper, we consider a dynamic, discrete-time, game model where the players use a common resource and have different criteria to optimize. First, we construct a multicriteria Nash equilibrium applying the bargaining concept (via Nash products) [6]. Then, we obtain multicriteria cooperative behavior as a solution of a Nash bargaining scheme with the multicriteria Nash equilibrium payoffs playing the role of status quo points [7].

This work was supported by the Russian Science Foundation, project no. 17-11-01079.

M. Khachay et al. (Eds.): MOTOR 2019, LNCS 11548, pp. 697–714, 2019.
https://doi.org/10.1007/978-3-030-22629-9_49

The coalition formation process in multicriteria dynamic games is considered. Two ways to construct the players' strategies are presented: all players decide simultaneously (Nash-Cournot strategies) or members of coalitions are assumed to be the leaders and players decide sequentially (Stackelberg strategies). Furthermore, the characteristic function is constructed in an unusual form: the players outside the coalition S determine new Nash strategies in the game with $N\backslash S$ players. This case corresponds to the situation when players know that coalition S was formed.

We extend the internal and external stability concepts [3] to multicriteria dynamic games. The conditions for coalition stability are presented.

To illustrate the presented approaches a multicriteria bioresource management problem with the finite horizon is investigated. In harvesting problems cooperation (partial cooperation) is very important for minimizing the load on the stock. As it was shown (for example, see [8,9]) cooperative behavior is profitable for the players and improves the ecological situation. Hence, from the social point of view, the coalitions that are internally stable are more preferable.

Further exposition has the following structure. Section 2 describes the noncooperative and cooperative solution concepts for a finite horizon multicriteria dynamic game with many players in discrete time. The coalition formation process and coalition stability conditions for a multicriteria dynamic game are presented in Sect. 3. A bicriteria discrete-time game-theoretic bioresource management model (harvesting problem) with a finite planning horizon is treated in Sect. 4. Finally, Sect. 5 provides the basic results and their discussion.

2 Dynamic Multicriteria Game with the Finite Horizon

Consider a multicriteria dynamic game with the finite horizon in discrete time. Let $N = \{1, \ldots, n\}$ players exploit a common resource and each of them wishes to optimize k different criteria. The state dynamics is in the form

$$x_{t+1} = f(x_t, u_{1t}, \ldots, u_{nt}), \;\; x_0 = x, \tag{1}$$

where $x_t \geq 0$ is the resource size at time $t \geq 0$, $f(x_t, u_{1t}, \ldots, u_{nt})$ denotes the natural growth function, and $u_{it} \geq 0$ gives the exploitation rate of player i at time t, $i \in N$.

Denote $u_t = (u_{1t}, \ldots, u_{nt})$. Each player has k goals to optimize. The players' payoffs on finite planning horizon $[0, m]$ are defined as

$$J_i = \begin{pmatrix} J_i^1 = \sum\limits_{t=0}^{m} \delta^t g_i^1(u_t) \\ \ldots \\ J_i^k = \sum\limits_{t=0}^{m} \delta^t g_i^k(u_t) \end{pmatrix}, \, i \in N, \tag{2}$$

where $g_i^j(u_t) \geq 0$ gives the instantaneous utility, $j = 1, \ldots, k$, $i \in N$, $\delta \in (0, 1)$ denotes the discount factor.

2.1 Multicriteria Nash Equilibrium

We design the noncooperative behavior in dynamic multicriteria game applying the Nash bargaining products [6]. Therefore, we begin with the construction of guaranteed payoffs which play the role of status quo points.

The possible concepts to determine the guaranteed payoffs for the game with two players were presented in [6]. It was shown that the variant where the guaranteed payoffs are determined as Nash equilibrium is beneficial for both players and, moreover, improves the ecological situation. Therefore, for the multicriteria game with n players, we adopt this concept of guaranteed payoff points construction. Namely,

$G_1^1, \ldots, G_n^1$ are the Nash equilibrium payoffs in the dynamic game $\langle x, N, \{U_i\}_{i=1}^n, \{J_i^1\}_{i=1}^n \rangle$,

...

$G_1^k, \ldots, G_n^k$ are the Nash equilibrium payoffs in the dynamic game $\langle x, N, \{U_i\}_{i=1}^n, \{J_i^k\}_{i=1}^n \rangle$,

where the state dynamics is in the form (1).

To construct multicriteria payoff functions, we adopt the Nash products. The role of the status quo points belongs to the guaranteed payoffs of the players:

$$H_1(u_{1t}, \ldots, u_{nt}) = (J_1^1(u_{1t}, \ldots, u_{nt}) - G_1^1) \cdot \ldots \cdot (J_1^k(u_{1t}, \ldots, u_{nt}) - G_1^k),$$

$$\ldots$$

$$H_n(u_{1t}, \ldots, u_{nt}) = (J_n^1(u_{1t}, \ldots, u_{nt}) - G_n^1) \cdot \ldots \cdot (J_n^k(u_{1t}, \ldots, u_{nt}) - G_n^k).$$

Definition 1. *A strategy profile $u_t^N = (u_{1t}^N, \ldots, u_{nt}^N)$ is called a multicriteria Nash equilibrium [6] of the problem (1), (2) if*

$$H_i(u_t^N) \geq H_i(u_{1t}^N, \ldots, u_{i-1\,t}^N, u_{it}, u_{i+1\,t}^N, \ldots, u_{nt}^N) \ \forall u_{it} \in U_i, \ i \in N. \quad (3)$$

2.2 Multicriteria Cooperative Behavior

The multicriteria cooperative strategies are obtained as a solution of a Nash bargaining scheme with the multicriteria Nash equilibrium payoffs playing the role of status quo points [7].

First, we have to determine noncooperative payoffs as players' gains when they apply multicriteria Nash equilibrium strategies u_t^N:

$$J_1^N = \begin{pmatrix} J_1^{1N} = \sum\limits_{t=0}^{m} \delta^t g_1^1(u_t^N) \\ \ldots \\ J_1^{kN} = \sum\limits_{t=0}^{m} \delta^t g_1^k(u_t^N) \end{pmatrix}, \ldots, J_n^N = \begin{pmatrix} J_n^{1N} = \sum\limits_{t=0}^{m} \delta^t g_n^1(u_t^N) \\ \ldots \\ J_n^{kN} = \sum\limits_{t=0}^{m} \delta^t g_n^k(u_t^N) \end{pmatrix}.$$

Then, we construct a Nash product where the sum of players' noncooperative payoffs plays a role of the status quo point. To design the cooperative behavior we adopt a Nash bargaining solution, hence it is required to solve the next problem:

$$(\sum_{i=1}^{n} J_i^{1c}(u_t^c) - \sum_{i=1}^{n} J_i^{1N}(u_t^N)) \cdot \ldots \cdot (\sum_{i=1}^{n} J_i^{kc}(u_t^c) - \sum_{i=1}^{n} J_i^{kN}(u_t^N))$$
$$= (\sum_{t=0}^{m} \delta^t \sum_{i=1}^{n} g_i^1(u_t^c) - \sum_{i=1}^{n} J_i^{1N}(u_t^N)) \cdot \ldots$$
$$\cdot (\sum_{t=0}^{m} \delta^t \sum_{i=1}^{n} g_i^k(u_t^c) - \sum_{i=1}^{n} J_i^{kN}(u_t^N)) \to \max_{u_t^c}. \quad (4)$$

Definition 2. *A strategy profile $u_t^c = (u_{1t}^c, \ldots, u_{nt}^c)$ is called a multicriteria cooperative equilibrium [7] of the problem (1), (2) if it solves the problem (4).*

3 Coalition Formation Process

We consider a coalition formation process in dynamic multicriteria games. Let assume that a coalition S is formed. Two ways to construct the players' strategies are considered: all players decide simultaneously (Nash-Cournot strategies) or members of coalitions are assumed to be the leaders and players decide sequentially (Stackelberg strategies). Moreover, the characteristic function is constructed in an unusual form: players outside the coalition S determine new Nash strategies in the game with $N \backslash S$ players. This case corresponds to the situation when players know that coalition S was formed. The sizes of stable coalitions are the subjects of investigation.

3.1 Nash-Cournot Strategies

Under the first approach, players decide simultaneously. Hence, to determine the cooperative behavior of coalition S and the singletons' strategies u_{it}^N, $i \in N \backslash S$, it is required to solve the next problems:

$$(\sum_{i \in S} J_i^{1S}(\tilde{u}_t) - \sum_{i \in S} J_i^{1N}(\tilde{u}_t)) \cdot \ldots \cdot (\sum_{i \in S} J_i^{kc}(\tilde{u}_t) - \sum_{i \in S} J_i^{kN}(\tilde{u}_t))$$
$$= (\sum_{t=0}^{m} \delta^t \sum_{i \in S} g_i^1(\tilde{u}_t) - \sum_{i \in S} J_i^{1N}(\tilde{u}_t)) \cdot \ldots$$
$$\cdot (\sum_{t=0}^{m} \delta^t \sum_{i \in S} g_i^k(\tilde{u}_t) - \sum_{i \in S} J_i^{kN}(\tilde{u}_t)) \to \max_{u_{it},\, i \in S}, \quad (5)$$

$$(J_i^1(\tilde{u}_t) - G_i^1) \cdot \ldots \cdot (J_i^k(\tilde{u}_t) - G_i^k) \longrightarrow \max_{u_{it}^N,\, i \in N \backslash S}, i \in N \backslash S \quad (6)$$

under the dynamics:

$$x_{t+1} = f(\tilde{u}_t), x_0 = x,$$

where

$$\tilde{u}_t = \begin{cases} u_{it}, & i \in S, \\ u_{it}^N, & i \in N \backslash S. \end{cases}$$

Denote the cooperative strategies of the coalition S's members as $u_t^S = (u_{it}^S)_{i \in S}$ and the strategies of singletons as $u_t^{NS} = (u_{it}^N)_{i \in N \backslash S}$.

3.2 Stackelberg Strategies

We assume that members of coalitions are the leaders and players decide sequentially. So, at first, singletons determine the Nash equilibrium strategies under the assumption that cooperative strategies are known. After that, the coalition members determine their behavior.

(I) Coalition members' strategies u_{it}^s, $i \in S$ are fixed. Singletons solve the next problems:

$$(J_i^1(\tilde{u}_t) - G_i^1) \cdot \ldots \cdot (J_i^k(\tilde{u}_t) - G_i^k) \longrightarrow \max_{u_{it},\, i \in N \backslash S}, i \in N \backslash S, \tag{7}$$

where

$$\tilde{u}_t = \begin{cases} u_{it}^s, & i \in S, \\ u_{it}, & i \in N \backslash S, \end{cases}$$

under the dynamics:

$$x_{t+1} = f(\tilde{u}_t), x_0 = x.$$

Denote the obtained strategies as $\tilde{u}_{it}^N$, $i \in N \backslash S$.

(II) To determine the cooperative behavior of coalition S, it is required to solve the next problem:

$$\begin{aligned} &(\sum_{i \in S} J_i^{1S}(\tilde{u}_t) - \sum_{i \in S} J_i^{1N}(\tilde{u}_t)) \cdot \ldots \cdot (\sum_{i \in S} J_i^{kS}(\tilde{u}_t) - \sum_{i \in S} J_i^{kN}(\tilde{u}_t)) \\ &= (\sum_{t=0}^{m} \delta^t \sum_{i \in S} g_i^1(\tilde{u}_t) - \sum_{i \in S} J_i^{1N}(\tilde{u}_t)) \cdot \ldots \\ &\cdot (\sum_{t=0}^{m} \delta^t \sum_{i \in S} g_i^k(\tilde{u}_t) - \sum_{i \in S} J_i^{kN}(\tilde{u}_t)) \rightarrow \max_{u_{it},\, i \in S}, \end{aligned} \tag{8}$$

where

$$\tilde{u}_t = \begin{cases} u_{it}, & i \in S, \\ \tilde{u}_{it}^N, & i \in N \backslash S. \end{cases}$$

Denote the cooperative strategies of the coalition S's members as $u_t^S = (\tilde{u}_{it}^S)_{i \in S}$ and the strategies of singletons as $u_t^{NS} = (\tilde{u}_{it}^N)_{i \in N \backslash S}$.

Note that under presented concepts there is no need to distribute cooperative payoff of coalition S among its members as the vector payoff $J_i^S(\cdot) = (J_i^{1S}(\cdot), \ldots, J_i^{kS}(\cdot))$ of coalition member $i \in S$ is directly obtained from the schemes of characteristic function construction.

3.3 Coalition Stability

The stability concept (internal and external stability) was presented in [3]. Here we adopt these concepts for multicriteria dynamic games to define stable coalitions.

Definition 3. *Coalition S is internally stable if $\neg\exists i \in S$:*

$$J_i^S(u_t^S, u_t^{NS}) < J_i^N(u_t^{S\setminus\{i\}}, u_t^{NS\setminus\{i\}}). \tag{9}$$

Coalition S is externally stable if $\neg\exists i \in N\setminus S$:

$$J_i^N(u_t^S, u_t^{NS}) < J_i^{S\cup\{i\}}(u_t^{S\cup\{i\}}, u_t^{NS\cup\{i\}}). \tag{10}$$

Here $a < b \Leftrightarrow a_j < b_j$, $\forall j = 1, \ldots, k$.

Internal stability means that no coalition member wishes to leave the coalition and become a singleton. External stability means that no singleton wishes to join the coalition.

Definition 4. *Coalition S is stable if conditions (9), (10) are fulfilled.*

Next, we consider a dynamic multicriteria model related with the bioresource management problem (harvesting) to show how the suggested concepts work.

4 Dynamic Multicriteria Bioresource Management Model

Consider a bicriteria discrete-time dynamic bioresource management model with n players and fixed harvesting times. Suppose that the players (countries or fishing firms) harvest a fish stock during finite time horizon $[0, m]$. The fish population evolves according to the equation

$$x_{t+1} = \varepsilon x_t - u_{1t} - \ldots - u_{nt}, \; x_0 = x, \tag{11}$$

where $x_t \geq 0$ is the population size at time $t \geq 0$, $\varepsilon \geq 1$ denotes the natural birth rate, and $u_{it} \geq 0$ gives the catch of player i at time $t \geq 0$, $i \in N = \{1, \ldots, n\}$.

Each player has two goals to optimize: they wish to maximize their profit from selling fish and minimize the catching cost. Suppose that the market price of the resource differs for the players, but their costs are identical and depend on players' catches. Specifically, the payoff functions of the players over the finite time horizon are defined by

$$J_1 = \begin{pmatrix} J_1^1 = \sum\limits_{t=0}^{m} \delta^t p_1 u_{1t} \\ J_1^2 = -\sum\limits_{t=0}^{m} \delta^t c u_{1t}^2 \end{pmatrix}, \ldots, J_n = \begin{pmatrix} J_n^1 = \sum\limits_{t=0}^{m} \delta^t p_n u_{nt} \\ J_n^2 = -\sum\limits_{t=0}^{m} \delta^t c u_{nt}^2 \end{pmatrix}, \tag{12}$$

where, for $i \in N$, $p_i \geq 0$ is the market price of the resource for player i, $c \geq 0$ indicates the catching cost, and $\delta \in (0, 1)$ denotes the discount factor.

4.1 Multicriteria Nash Equilibrium

We begin with the construction of guaranteed payoffs applying one of the variants of their determination [6]. Under this approach, the guaranteed payoff points $G_1^1, \dots, G_n^1$ are defined as the Nash equilibrium in the dynamic game $\langle x, N, \{U_i\}_{i=1}^n, \{J_i^1\}_{i=1}^n \rangle$. As this game is linear the equilibrium exists and is unique. Applying the Bellman principle and assuming the value functions and the strategies have the linear forms, we get the solution

$$u_{1t} = \dots = u_{nt} = \frac{\varepsilon - 1}{n-1} x_t,$$

and the dynamics becomes

$$x_t = \left(\frac{n-\varepsilon}{n-1}\right)^t x_0.$$

Hence, the guaranteed payoff points take the forms

$$G_1^1 = p_1 A x_0, \dots, G_n^1 = p_n A x_0, \tag{13}$$

where

$$A = \frac{\varepsilon - 1}{n-1} \frac{(\delta(n-\varepsilon))^{m+1} + (n-1)^{m+1}}{(n-1)^m (\delta(n-\varepsilon) - n + 1)}.$$

By analogy, determining the Nash equilibrium in the dynamic (linear-quadratic) game with the second criteria of all players $\langle x, N, \{U_i\}_{i=1}^n, \{J_i^2\}_{i=1}^n \rangle$ (linear-quadratic, hence the equilibrium exists and is unique), we get n more guaranteed payoff points

$$G_1^2 = \dots = G_n^2 = G x_0^2, \tag{14}$$

where

$$G = -c\Big(\frac{2n - \varepsilon^2 + \varepsilon\sqrt{4n^2 + \varepsilon^2 - 4n}}{n(-\varepsilon + \sqrt{4n^2 + \varepsilon^2 - 4n})}\Big)^2 \cdot \frac{(2\delta n)^{m+1} - (\varepsilon - \sqrt{4n^2 + \varepsilon^2 - 4n})^{m+1}}{(\varepsilon - \sqrt{4n^2 + \varepsilon^2 - 4n})^m (2\delta n - \varepsilon + \sqrt{4n^2 + \varepsilon^2 - 4n})}.$$

According to Definition 1, in order to determine the multicriteria Nash equilibrium of problem (11), (12) it is required to solve the next problem:

$$p_1(\sum_{t=0}^m \delta^t u_{1t} - Ax)(-c\sum_{t=0}^m \delta^t u_{1t}^2 - Gx^2) \to \max_{u_{1t}},$$

$$\dots$$

$$p_n(\sum_{t=0}^m \delta^t u_{nt} - Ax)(-c\sum_{t=0}^m \delta^t u_{nt}^2 - Gx^2) \to \max_{u_{nt}}.$$

Considering the process starting from one-step till m-step game and seeking the strategies in linear form, we get the multicriteria Nash equilibrium.

Proposition 1. *The multicriteria Nash equilibrium strategies in the problem (11), (12) have the forms* $u_{it}^N = \gamma_{it}^N x_t$, $i \in N$,

$$\gamma_{1t}^N = \ldots = \gamma_{nt}^N = \gamma_t^N = \frac{\varepsilon^{t-1}\gamma_1}{1 + n\gamma_1 \sum_{j=0}^{t-2} \varepsilon^j}. \tag{15}$$

The players' strategy on the last step γ_1^N *is determined from the next equation*

$$-2c\gamma_1 \prod_{i=2}^{m}(\varepsilon - n\gamma_i)\Big[\sum_{i=0}^{m-1} \delta^i \gamma_{m-i} \prod_{j=m+1-i}^{m} (\varepsilon - n\gamma_j) - A\Big]$$
$$+\Big[-c \sum_{i=0}^{m-1} \delta^i \gamma_{m-i}^2 \prod_{j=m+1-i}^{m} (\varepsilon - n\gamma_j)^2 - G\Big] = 0.$$

4.2 Cooperative Behavior

To construct the cooperative payoffs and strategies the Nash bargaining solution is applied [7]. First, we have to determine noncooperative payoffs as the players' gains when they apply multicriteria Nash strategies. Then, we construct a Nash product where the sum of players' noncooperative payoffs plays a role of the status quo points.

According to Proposition 1, the noncooperative payoffs have the forms

$$J_i^{1N}(x) = \sum_{t=0}^{m} \delta^t p_i \gamma_t^N x_0, \; i \in N,$$

$$J_1^{2N}(x) = \ldots = J_2^{2N}(x) = -c \sum_{t=0}^{m} \delta^t \gamma_t^N x_0^2.$$

According to Definition 2, in order to construct the cooperative strategies it is required to solve the problem (4). Hence,

$$(\sum_{t=0}^{m} \delta^t (p_1 u_{1t}^c + \ldots + p_n u_{nt}^c) - Px)(-c \sum_{t=0}^{m} \delta^t ((u_{1t}^c)^2 + \ldots + (u_{nt}^c))^2 - Kx^2) \to \max_{u_{1t}^c, \ldots, u_{nt}^c},$$

where $P = (p_1 + \ldots + p_n) \sum_{t=0}^{m} \delta^t \gamma_t^N$, $K = -nc \sum_{t=0}^{m} \delta^t (\gamma_t^N)^2$.

Considering the process starting from one-step till m-step game and seeking the strategies in linear form, we construct cooperative behavior.

Proposition 2. *The multicriteria cooperative strategies in the problem (11), (12) take the forms* $u_{it}^c = \gamma_{it}^c x_t$, $i \in N$,

$$\gamma_{1t}^c = \frac{p_1 \varepsilon^{t-1} \gamma_{11}^c}{p_1 + \gamma_{11}^c \sum_{j=0}^{t-2} \varepsilon^j \sum_{i=1}^{p} p_i}, \; t = 2, \ldots, m,$$

$$\gamma_{jt}^c = \frac{p_j}{p_1} \gamma_{1t}^c, \; j = 2, \ldots, n, \; t = 1, \ldots, m, \tag{16}$$

and the first player's strategy on the last step γ_{11}^c is determined from the equation

$$p_1\Big[-c\sum_{i=2}^{m}\delta^{m-i}\sum_{l=1}^{n}(\gamma_{li}^c)^2\prod_{j=i+1}^{m}(\varepsilon-\sum_{l=1}^{n}\gamma_{lj}^c)^2-K\Big]$$

$$-2c\gamma_{11}^c\prod_{j=2}^{m}(\varepsilon-\sum_{l=1}^{n}\gamma_{lj}^c)\Big[\sum_{i=1}^{m}\delta^{m-i}\sum_{l=1}^{n}p_l\gamma_{li}^c\prod_{j=i+1}^{m}(\varepsilon-\sum_{l=1}^{n}\gamma_{lj}^c)-P\Big]=0.$$

Let's consider the asymptotic values of the players' strategies and the size of the resource under noncooperative and cooperative behavior.

Applying Nash equilibrium strategies as t tends to ∞ we get

$$\gamma_t^N\to\frac{\varepsilon-1}{n},$$

and for cooperative behavior –

$$\gamma_{it}^c\to\frac{p_i}{\sum\limits_{i=1}^{n}p_i}(\varepsilon-1),\ i\in N.$$

Hence, in both cases, $x_t\to x_0$ and the difference is only a distribution of the total exploitation rate among the players.

The players extract exactly the natural growth increase of the resource $(\varepsilon-1)x_t$, but in Nash equilibrium it is distributed uniformly and under cooperative behavior – proportionally to players' market prices for the resource.

4.3 Coalition Formation

Nash-Cournot Strategies

Under the first approach, players decide simultaneously. Hence, to determine the cooperative behavior of coalition S it is required to solve the problem (5):

$$(\sum_{t=0}^{m}\delta^t\sum_{i\in S}p_iu_{it}^s-P^Sx)(-c\sum_{t=0}^{m}\delta^t\sum_{i\in S}(u_{it}^s)^2-K^Sx^2)\to\max_{u_{it}^s,\,i\in S},$$

where $P^S=\sum\limits_{i\in S}p_i\sum\limits_{t=0}^{m}\delta^t\gamma_t^N$, $K^S=-|S|c\sum\limits_{t=0}^{m}\delta^t(\gamma_t^N)^2$, and the singletons' strategies u_{it}, $i\in N\backslash S$, are defined from the maximization problems (6):

$$(\sum_{t=0}^{m}\delta^tp_iu_{it}-G_i^1)\cdot\ldots\cdot(-c\sum_{t=0}^{m}\delta^t(u_{it})^2-G_i^k)\longrightarrow\max_{u_{it},\,i\in N\backslash S},\,i\in N\backslash S$$

under the next dynamics

$$x_{t+1}=\varepsilon x_t-\sum_{i\in S}u_{it}^s-\sum_{i\in N\backslash S}u_{it},\ x_0=x.$$

Similarly to full cooperative case, continuing the process from one-step till m-step game and seeking the strategies in linear forms, we obtain

Proposition 3. *The strategies of coalition S's members in the problem (11), (12) take the forms* $u_{it}^S = \gamma_{it}^S x_t$, $i \in S$,

$$\gamma_{st}^S = \frac{p_s \varepsilon^{t-1} \gamma_{s1}^S}{p_s + \sum_{j=0}^{t-2} \varepsilon^j (\gamma_{s1}^S \sum_{i \in S} p_i + p_s(n-|S|)\gamma_1^N)}, \; t = 2, \ldots, m,$$

$$\gamma_{jt}^s = \frac{p_j}{p_s} \gamma_{st}^c, \; j \in S, \; j \neq s, \; t = 1, \ldots, m, \qquad (17)$$

and the strategy of player $s \in S$ *on the last step* γ_{s1}^s *is determined from the equation*

$$p_s\Big[-c\sum_{j=1}^m \delta^{m-j} \sum_{i \in S} (\gamma_{ij}^S)^2 \prod_{l=j+1}^m (\varepsilon - \sum_{i \in S} \gamma_{il}^S - (n-|S|)\gamma_l^N)^2 - K^S\Big]$$
$$-2c\gamma_{s1}^S \prod_{j=2}^m (\varepsilon - \sum_{i \in S} \gamma_{ij}^S - (n-|S|)\gamma_j^N)$$
$$\cdot\Big[\sum_{j=1}^m \delta^{m-j} \sum_{i \in S} p_i \gamma_{ij}^S \prod_{l=j+1}^m (\varepsilon - \sum_{i \in S} \gamma_{il}^s - (n-|S|)\gamma_l^N) - P^S\Big] = 0. \qquad (18)$$

The singletons' strategies in the problem (11), (12) coincide and take the forms $u_{it}^N = \gamma_t^N x_t$, $i \in N \backslash S$,

$$\gamma_t^N = \frac{p_s \varepsilon^{t-1} \gamma_1^N}{p_s + \sum_{j=0}^{t-2} \varepsilon^j (\gamma_{s1}^S \sum_{i \in S} p_i + p_s(n-|S|)\gamma_1^N)}, \; t = 2, \ldots, m, \qquad (19)$$

and the strategy on the last step γ_1^N *is determined from the equation*

$$\Big[-c\sum_{j=1}^m \delta^{m-j} (\gamma_j^N)^2 \prod_{l=j+1}^m (\varepsilon - \sum_{i \in S} \gamma_{il}^S - (n-|S|)\gamma_l^N)^2 - G\Big]$$
$$-2c\gamma_1^N \prod_{j=2}^m (\varepsilon - \sum_{i \in S} \gamma_{ij}^s - (n-|S|)\gamma_j^N)$$
$$\cdot\Big[\sum_{j=1}^m \delta^{m-j} \gamma_j^N \prod_{l=j+1}^m (\varepsilon - \sum_{i \in S} \gamma_{il}^s - (n-|S|)\gamma_l^N) - A\Big] = 0. \qquad (20)$$

Stackelberg Strategies

We assume that members of coalitions are the leaders and players decide sequentially. As before, we seek for linear strategies $u_{it} = \gamma_{it} x_t$, $i \in N \setminus S$, $u_{it} = \gamma_{it}^s x_t$, $i \in S$.

(I) Coalition members' strategies u_{it}^s, $i \in S$ are fixed. Singletons solve the problems (7):

$$p_i(\sum_{t=0}^m \delta^t u_{it} - Ax)(-c\sum_{t=0}^m \delta^t u_{it}^2 - Gx^2) \to \max_{u_{it}}, \; i \in N \backslash S$$

under the dynamics

$$x_{t+1} = \varepsilon x_t - \sum_{i \in S} u_{it}^s - \sum_{i \in N \setminus S} u_{it},\ x_0 = x.$$

As in previous case, we obtain that singletons' strategies coincide

$$\gamma_{it} = \gamma_t^N = \frac{\gamma_1^N \prod\limits_{j=2}^{t} (\varepsilon - \sum\limits_{l \in S} \gamma_{lj}^s)}{1 + (n - |S|)\gamma_1^N (1 + \sum\limits_{j=2}^{t-1} (\varepsilon - \sum\limits_{l \in S} \gamma_{lj}^s))},\ t = 2, \ldots, m,$$

and the strategy on the last step γ_1^N is determined from one of the first order conditions.

Denote the obtained strategies as $\tilde{u}_{it}^N = \tilde{\gamma}_t^N x_t$, $i \in N \setminus S$.

(II) To determine the cooperative behavior of coalition S it is required to solve the problem (8):

$$(\sum_{t=0}^{m} \delta^t \sum_{i \in S} p_i u_{it}^s - P^S x)(-c \sum_{t=0}^{m} \delta^t \sum_{i \in S} (u_{it}^s)^2 - K^S x^2) \to \max_{u_{it}^s,\, i \in S},$$

where $P^S = \sum\limits_{i \in S} p_i \sum\limits_{t=0}^{m} \delta^t \gamma_t^N$, $K^S = -|S| c \sum\limits_{t=0}^{m} \delta^t (\gamma_t^N)^2$, under the dynamics

$$x_{t+1} = \varepsilon x_t - \sum_{i \in S} u_{it}^s - \sum_{i \in N \setminus S} \tilde{u}_{it}^N,\ x_0 = x.$$

Continuing the process from one-step till m-step game and seeking the strategies in linear forms, we obtain

Proposition 4. *The strategies of coalition S's members in the problem (11), (12) take the forms $\tilde{u}_{it}^S = \gamma_{it}^S x_t$, $i \in S$,*

$$\gamma_{st}^S = \frac{p_s \gamma_{s1}^S \varepsilon^{t-1} (1 + (n - |S|)\tilde{\gamma}_1^N)}{p_s + \sum\limits_{j=0}^{t-2} \varepsilon^j (\gamma_{s1}^S \sum\limits_{i \in S} p_i + p_s (n - |S|)\tilde{\gamma}_1^N)},\ t = 2, \ldots, m,$$

$$\gamma_{jt}^S = \frac{p_j}{p_s} \gamma_{st}^S,\ j \in S,\ j \neq s,\ t = 1, \ldots, m, \tag{21}$$

and the strategy of player $s \in S$ on the last step γ_{s1}^S is determined from the Eq. (18) with $\gamma_i^N = \tilde{\gamma}_i^N$, $i \in N \setminus S$.

The singletons' strategies in the problem (11), (12) coincide and take the forms $\tilde{u}_{it}^N = \tilde{\gamma}_t^N x_t$, $i \in N \setminus S$,

$$\tilde{\gamma}_t^N = \frac{p_s \tilde{\gamma}_1^N \varepsilon^{t-1}}{p_s + \sum\limits_{j=0}^{t-2} \varepsilon^j (\gamma_{s1}^S \sum\limits_{i \in S} p_i + p_s (n - |S|)\tilde{\gamma}_1^N)},\ t = 2, \ldots, m, \tag{22}$$

and the strategy on the last step $\tilde{\gamma}_1^N$ is determined from the Eq. (20) with $\gamma_i^N = \tilde{\gamma}_i^N$, $i \in N \setminus S$.

Let's consider the asymptotic values of the players' strategies and the size of the resource under both types of coalition formation.

If the players decide simultaneously as t tends to ∞ we get

$$\gamma_t^N \to \frac{\varepsilon - 1}{n - |S|},\ \gamma_{it}^s \to 0,\ i \in S. \tag{23}$$

Hence, in the asymptotic case under Nash-Cournot strategies, cooperative players don't extract the resource. It means that it is not profitable to form a coalition and the players prefer noncooperative behavior.

If the players decide sequentially as t tends to ∞ we get

$$\tilde{\gamma}_t^N \to 0,\ \gamma_{it}^s \to \frac{p_i}{\sum_{i=1}^{n} p_i}(\varepsilon - 1),\ i \in S. \tag{24}$$

Hence, in the asymptotic case under Stackelberg strategies, the opposite result is valid: only cooperative players extract the resource. It means that this type of coalition formation stimulates cooperative behavior that is very important in ecological problems.

As before, in both cases, $x_t \to x_0$ and the difference is only a distribution of the total exploitation rate among the players. The players extract the natural growth increase of the resource $(\varepsilon - 1)x_t$, but under Nash-Cournot strategies it is distributed uniformly among singletons and under Stackelberg strategies – proportionally to the market prices for the resource among cooperative players.

4.4 Coalition Stability

Denote the cooperative strategies of the coalition S's members as $u_t^S = (u_{it}^S)_{i \in S}$ and the strategies of singletons as $u_t^{NS} = (u_{it}^N)_{i \in N \setminus S}$ (Nash-Cournot strategies) or $u_t^S = (\tilde{u}_{it}^S)_{i \in S}$ and $u_t^{NS} = (\tilde{u}_{it}^N)_{i \in N \setminus S}$ (Stackelberg strategies).

The internal stability conditions take the forms:

$$J_i^{1S}(u_t^S, u_t^{NS}) \geq J_i^{1N}(u_t^{S \setminus \{i\}}, u_t^{NS \setminus \{i\}}),$$
$$J_i^{2S}(u_t^S, u_t^{NS}) \geq J_i^{2N}(u_t^{S \setminus \{i\}}, u_t^{NS \setminus \{i\}}),\ \forall i \in S.$$

The external stability conditions take the forms:

$$J_i^{1N}(u_t^S, u_t^{NS}) \geq J_i^{1S \cup \{i\}}(u_t^{S \cup \{i\}}, u_t^{NS \cup \{i\}}),$$
$$J_i^{2N}(u_t^S, u_t^{NS}) \geq J_i^{2S \cup \{i\}}(u_t^{S \cup \{i\}}, u_t^{NS \cup \{i\}}),\ \forall i \in N \setminus S.$$

Nash-Cournot Strategies
The internal stability conditions take the forms:

$$(\gamma_{s1}^S - \gamma_1^N)$$
$$\cdot(p_s + \sum_{j=0}^{t-2} \varepsilon^j (\gamma_{s1}^S \sum_{i \in S} p_i + p_s(n - |S|)\gamma_1^N)) + \sum_{j=0}^{t-2} \varepsilon^j \gamma_{s1}^S (p_s \gamma_1^N - p_i \gamma_{s1}^S) \geq 0,$$

$$((\gamma_{s1}^S)^2 - (\gamma_1^N)^2)(p_s + \sum_{j=0}^{t-2} \varepsilon^j (\gamma_{s1}^S \sum_{i \in S} p_i + p_s(n - |S|)\gamma_1^N))^2 + (\gamma_{s1}^S)^2 \sum_{j=0}^{t-2} \varepsilon^j$$
$$\cdot(p_s \gamma_1^N - p_i \gamma_{s1}^S)(2p_s + \sum_{j=0}^{t-2} \varepsilon^j ((2(n - |S|) + 1)p_s \gamma_1^N + (2\sum_{i \in S} p_i - p_i)\gamma_{s1}^S)) \leq 0.$$

The external stability conditions take the forms:

$$(\gamma_{s1}^S - \gamma_1^N)$$
$$\cdot(p_s + \sum_{j=0}^{t-2} \varepsilon^j (\gamma_{s1}^S \sum_{i \in S} p_i + p_s(n - |S|)\gamma_1^N)) + \sum_{j=0}^{t-2} \varepsilon^j \gamma_1^N (p_s \gamma_1^N - p_i \gamma_{s1}^S) \leq 0,$$

$$((\gamma_{s1}^S)^2 - (\gamma_1^N)^2)(p_s + \sum_{j=0}^{t-2} \varepsilon^j (\gamma_{s1}^S \sum_{i \in S} p_i + p_s(n - |S|)\gamma_1^N))^2 + (\gamma_1^N)^2 \sum_{j=0}^{t-2} \varepsilon^j$$
$$\cdot(p_s \gamma_1^N - p_i \gamma_{s1}^S)(2p_s + \sum_{j=0}^{t-2} \varepsilon^j ((2(n - |S|) - 1)p_s \gamma_1^N + (2\sum_{i \in S} p_i + p_i)\gamma_{s1}^S)) \geq 0.$$

Consider the symmetric case $p_1 = \ldots = p_n$. The internal stability conditions take the forms

$$(\gamma_{s1}^S - \gamma_1^N)(1 + \sum_{j=0}^{t-2} \varepsilon^j ((|S| - 1)\gamma_{s1}^S + (n - |S|)\gamma_1^N)) \geq 0,$$

$$(\gamma_{s1}^S - \gamma_1^N)\Big((\gamma_{s1}^S + \gamma_1^N)(1 + \sum_{j=0}^{t-2} \varepsilon^j ((|S| - 1)\gamma_{s1}^S + (n - |S|)\gamma_1^N))^2$$
$$- \sum_{j=0}^{t-2} \varepsilon^j (\gamma_{s1}^S)^2 (2 + \sum_{j=0}^{t-2} \varepsilon^j ((2|S| - 1)\gamma_{s1}^S + (2n - 2|S| + 1)\gamma_1^N))\Big) \leq 0, \quad (25)$$

and external –

$$(\gamma_{s1}^S - \gamma_1^N)(1 + \sum_{j=0}^{t-2} \varepsilon^j ((|S|)\gamma_{s1}^S + (n - |S| + 1)\gamma_1^N)) \leq 0,$$

$$(\gamma_{s1}^S - \gamma_1^N)\Big((\gamma_{s1}^S + \gamma_1^N)(1 + \sum_{j=0}^{t-2} \varepsilon^j ((|S| - 1)\gamma_{s1}^S + (n - |S|)\gamma_1^N))^2$$
$$- \sum_{j=0}^{t-2} \varepsilon^j (\gamma_1^N)^2 (2 + \sum_{j=0}^{t-2} \varepsilon^j ((2|S| - 1)\gamma_{s1}^S + (2n - 2|S| + 1)\gamma_1^N))\Big) \leq 0. \quad (26)$$

As $\gamma_{s1}^S \le \gamma_1^N$ in symmetric case, only the coalition of size 1 is internally stable and external stability conditions are fulfilled for all coalition sizes. This fact stresses that under Nash-Cournot coalition formation process it is not profitable for players to join coalition. Hence, this concept doesn't stimulate cooperative behavior.

Stackelbers Strategies
We give the results for the symmetric case. The internal stability conditions take the forms

$$(\gamma_{s1}^S - \tilde{\gamma}_1^N)(1+\sum_{j=0}^{t-2}\varepsilon^j((|S|-1)\gamma_{s1}^S + (n-|S|)\tilde{\gamma}_1^N(1-\gamma_{s1}^S))$$

$$+(n-|S|)\gamma_{s1}^S\tilde{\gamma}_1^N(1+\sum_{j=0}^{t-2}\varepsilon^j(|S|\gamma_{s1}^S + (n-|S|)\tilde{\gamma}_1^N)) \ge 0,$$

$$(\gamma_{s1}^S - \tilde{\gamma}_1^N)\Big((\gamma_{s1}^S + \tilde{\gamma}_1^N)(1+\sum_{j=0}^{t-2}\varepsilon^j((|S|-1)\gamma_{s1}^S + (n-|S|)\tilde{\gamma}_1^N))^2$$

$$-\sum_{j=0}^{t-2}\varepsilon^j(\gamma_{s1}^S)^2(2+\sum_{j=0}^{t-2}\varepsilon^j((2|S|-1)\gamma_{s1}^S + (2n-2|S|+1)\tilde{\gamma}_1^N))\Big) - (n-|S|)$$

$$\cdot(\gamma_{s1}^S)^2\tilde{\gamma}_1^N(2+(n-|S|)\tilde{\gamma}_1^N)(1+\sum_{j=0}^{t-2}\varepsilon^j((|S|-1)\gamma_{s1}^S + (n-|S|+1)\tilde{\gamma}_1^N)^2 \le 0,$$

and external stability conditions have the forms

$$-(\gamma_{s1}^S - \tilde{\gamma}_1^N)(1+\sum_{j=0}^{t-2}\varepsilon^j((|S|)\gamma_{s1}^S + (n-|S|+1)\tilde{\gamma}_1^N))$$

$$-(n-|S|)\gamma_{s1}^S\tilde{\gamma}_1^N(1+\sum_{j=0}^{t-2}\varepsilon^j(|S|\gamma_{s1}^S + (n-|S|)\tilde{\gamma}_1^N)) \ge 0,$$

$$(\gamma_{s1}^S - \tilde{\gamma}_1^N)\Big((\gamma_{s1}^S + \tilde{\gamma}_1^N)(1+\sum_{j=0}^{t-2}\varepsilon^j((|S|-1)\gamma_{s1}^S + (n-|S|)\tilde{\gamma}_1^N))^2$$

$$+\sum_{j=0}^{t-2}\varepsilon^j(\gamma_{s1}^S)^2(2+\sum_{j=0}^{t-2}\varepsilon^j((2|S|-1)\gamma_{s1}^S + (2n-2|S|+1)\tilde{\gamma}_1^N))\Big) + (n-|S|)$$

$$\cdot(\gamma_{s1}^S)^2\tilde{\gamma}_1^N(2+(n-|S|)\tilde{\gamma}_1^N)(1+\sum_{j=0}^{t-2}\varepsilon^j((|S|-1)\gamma_{s1}^S + (n-|S|+1)\tilde{\gamma}_1^N)^2 \le 0.$$

As $\gamma_{s1}^S \ge \tilde{\gamma}_1^N$ in symmetric case, the internal stability conditions are always valid and the external stability conditions are not fulfilled for any coalition. This fact stresses that Stackelberg coalition formation process stimulates cooperative behavior and the players have an incentive to join the coalition of large size.

Let's consider the stability conditions in the asymptotic case.

If the players decide simultaneously as t tends to ∞ applying (23) we conclude that internal stability conditions are not fulfilled, but external stability conditions are valid for all the parameters.

Hence, in the asymptotic case under Nash-Cournot strategies, there are no internally stable coalitions at all. It means that it is not profitable to form a coalition and the players prefer noncooperative behavior.

If the players decide sequentially as t tends to ∞ applying (24) we conclude that external stability conditions are not fulfilled, but internal stability conditions are valid for all the parameters.

Hence, in the asymptotic case under Stackelberg strategies, the opposite result is valid: there are no externally stable coalitions. But as the coalitions are internally stable it is not profitable for the players to leave the formed coalition. It means that this type of coalition formation stimulates cooperative behavior. The absence of external stability is not important in the case of formed coalitions. Therefore, from a social point of view, this scheme is more preferable.

4.5 Modelling

We have performed numerical simulation for symmetric case with the following parameters:

$$m = 20,\ n = 10,\ \varepsilon = 1.3,\ p_1 = \ldots = p_{10} = 100,\ c = 50,\ \delta = 0.8,$$

and the size of the formed coalition is 5.

Figure 1 shows the dynamics of the population size for noncooperative, full cooperative (grand coalition formation) and partial cooperative (coalition S is formed under Stackelberg concept) cases. As one can notice even partial cooperation improves the ecological situation as it limits bioresource exploitation.

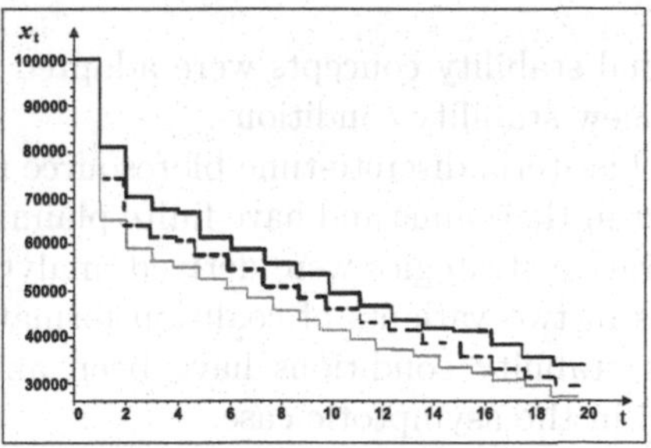

Fig. 1. Population size: dark – full cooperation, dotted – coalition S, light – Nash equilibrium

Figures 2 and 3 show the difference in the players' strategies for two variants of coalition formation. As one can notice, the coalition member's exploitation

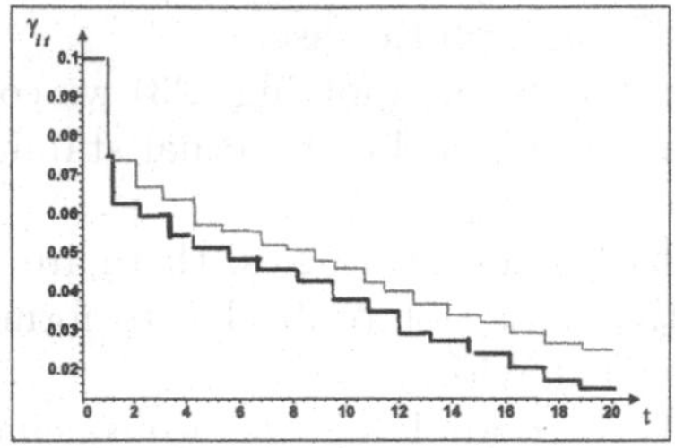

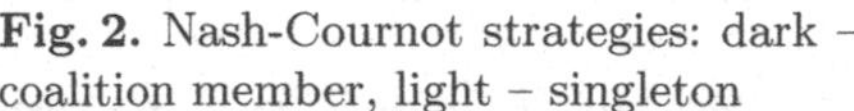

Fig. 2. Nash-Cournot strategies: dark – coalition member, light – singleton

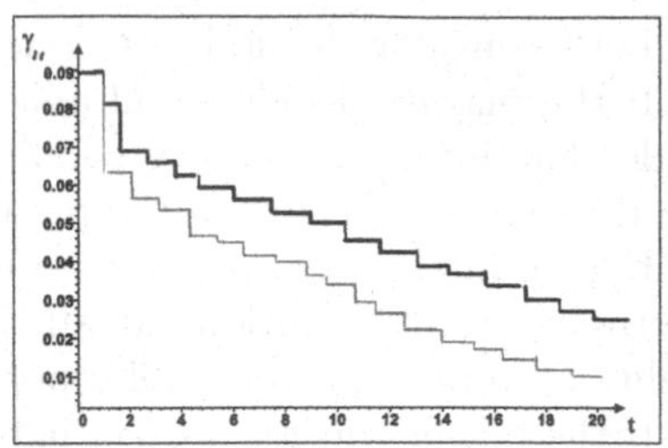

Fig. 3. Stackelberg strategies: dark – coalition member, light – singleton

rate is lower than the singleton's one under Nash-Cournot strategies. For the Stackelberg coalition formation process the opposite result is valid.

Numerical calculations of coalition stability conditions confirm the analytical results: a coalition of size 5 is not internally, but externally stable for Nash-Cournot strategies. For Stackelberg strategies the opposite result is valid. It again stresses that the second variant of coalition formation stimulates cooperation and hence more preferable.

5 Conclusions

The approaches to design cooperative behavior in multicriteria dynamic games with finite horizon are presented. First, we have evaluated the multicriteria Nash equilibrium strategies. Second, we have constructed the multicriteria cooperative strategies and payoffs via the bargaining scheme.

Then we have studied the coalition formation processes in multicriteria dynamic games. Two ways to construct the players' strategies were considered: all players decide simultaneously (Nash-Cournot strategies) or members of coalitions are assumed to be the leaders and players decide sequentially (Stackelberg strategies).

Internal and external stability concepts were adopted for dynamic multicriteria games to obtain new stability conditions.

We have studied a bicriteria discrete-time bioresource management problem, where the players differ in their aims and have finite planning horizons. Multicriteria Nash and cooperative strategies were derived analytically in linear forms. The players' strategies in two variants of coalition formation were also derived analytically. Coalition stability conditions have been analyzed in the case of symmetric players and in the asymptotic case.

It was shown that under Nash-Cournot strategies only coalitions of size 1 are internally stable and external stability conditions are fulfilled. This is the classical result in the literature on IEAs (for example, see [1,2]) that the internal stability concept is valid only for small sized coalitions.

For Stackelberg strategies the opposite result is valid: coalitions are internally, but not externally stable.

In harvesting problems, the cooperative behavior improves the ecological situation. From a social point of view, internally stable coalitions are more preferable. Hence, the Stackelberg scheme of coalition formation is more applicable in bioresource management problems.

The results of numerical modelling showed that the presented approaches stimulate cooperation. Moreover, that is important for ecological systems, even partial cooperative behavior improves the ecological situation.

To minimize the load on the stock the coalition should consist of large number of players and be stable. Here are some advices for ecological managers to improve populations' growth. For Nash-Cournot strategies we cannot guarantee internal stability, but for Stackelberg strategies coalitions are internally stable, but not externally.

That is why the manager first should determine the coalition formation process and then:

- if it is Nash-Cournot, then the manager (referee) should use some mechanisms to internally stabilize the coalitions: it can be fines for breaking off the cooperative agreement, punishment schemas like incentive equilibrium [4] or transfers schemas;
- if it is Stackelberg, the manager should not worry about the external stability because the more players decide to enter coalitions the larger population size will be.

According to aforesaid, there is a need for other stability concepts that enable the formation of coalitions of larger sizes. Hence, future research will consider intercoalition and coalition stability (see [2,5,10]) in dynamic multicriteria games.

References

1. Barrett, S.: Self-enforcing international environmental agreements. Oxf. Econ. Pap. **46**, 78–94 (1994)
2. Carraro, C.: The structure of international environmental agreements. In: Carraro, C. (ed.) International Environmental Agreements on Climate Change. Fondazione Eni Enrico Mattei (Feem) Series on Economics, Energy and Environment, vol. 13, pp. 9–25. Springer, Dordrecht (1999). https://doi.org/10.1007/978-94-015-9169-0_2
3. D'Aspremont, C., et al.: On the stability of collusive price leadership. Can. J. Econ. **16**(1), 17–25 (1983)
4. Mazalov, V.V., Rettieva, A.N.: Incentive conditions for rational behavior in discrete-time bioresource management problem. Dokl. Math. **81**(3), 399–402 (2010)
5. Pieri, G., Pusillo, L.: Multicriteria partial cooperative games. Appl. Math. **6**(12), 2125–2131 (2015)
6. Rettieva, A.N.: Equilibria in dynamic multicriteria games. Int. Game Theory Rev. **19**(1), 1750002 (2017)
7. Rettieva, A.N.: Dynamic multicriteria games with finite horizon. Mathematics **6**(9), 156 (2018)

8. Rettieva, A.N.: Cooperation in dynamic multicriteria games with random horizons. J. Glob. Optim. (2018). https://doi.org/10.1007/s10898-018-0658-6
9. Rettieva, A.N.: A bioresource management problem with different planning horizons. Autom. Remote. Control. **76**(5), 919–934 (2015)
10. Rettieva, A.N.: Stable coalition structure in bioresource management problem. Ecol. Model. **235–236**, 102–118 (2012)
11. Shapley, L.S.: Equilibrium points in games with vector payoffs. Nav. Res. Logist. Q. **6**, 57–61 (1959)

Author Index

Printed in the United States
By Bookmasters